*A Companion to
the History of
American Science*

WILEY BLACKWELL COMPANIONS TO AMERICAN HISTORY

This series provides essential and authoritative overviews of the scholarship that has shaped our present understanding of the American past. Edited by eminent historians, each volume tackles one of the major periods or themes of American history, with individual topics authored by key scholars who have spent considerable time in research on the questions and controversies that have sparked debate in their field of interest. The volumes are accessible for the non-specialist, while also engaging scholars seeking a reference to the historiography or future concerns.

PUBLISHED

PRESIDENTIAL COMPANIONS PUBLISHED

A Companion to the History of American Science

Edited by

Georgina M. Montgomery and Mark A. Largent

WILEY Blackwell

This paperback edition first published 2020
© 2016 John Wiley & Sons Ltd

Edition history: John Wiley & Sons Ltd (hardback, 2016)

The right of Georgina M. Montgomery and Mark A. Largent to be identified as the authors of the editorial material in this work has been asserted in accordance with law.

Registered Offices
John Wiley & Sons, Inc., 111 River Street, Hoboken, NJ 07030, USA
John Wiley & Sons Ltd, The Atrium, Southern Gate, Chichester, West Sussex, PO19 8SQ, UK

Editorial Office
111 River Street, Hoboken, NJ 07030, USA

For details of our global editorial offices, customer services, and more information about Wiley products visit us at www.wiley.com.

Wiley also publishes its books in a variety of electronic formats and by print-on-demand. Some content that appears in standard print versions of this book may not be available in other formats.

Library of Congress Cataloging-in-Publication Data

A companion to the history of American science / edited by Georgina M. Montgomery and Mark A. Largent.
     pages cm – (Wiley Blackwell companions to American history)
   Includes bibliographical references and index.
   ISBN 978-1-4051-5625-7 (cloth) ISBN 978-1-1191-3070-3 (pbk.) 1. Science–United States–History. I. Montgomery, Georgina M., editor.   II. Largent, Mark A., editor.   III. Series: Wiley-Blackwell companions to American history.
   Q127.U6C635 2015
   509.73–dc23 2015019362

A catalogue record for this book is available from the British Library.

Cover design: Wiley
Cover image: Smithsonian Institution Archives. Image 2002–12131

Set in 10/12pt Galliard by SPi Global, Pondicherry, India

Printed and bound by CPI Group (UK) Ltd, Croydon, CR0 4YY

10 9 8 7 6 5 4 3 2 1

*Dedicated to Our Advisor and One of the Founders of the
Study of the History of American Science,
Sally Gregory Kohlstedt*

Thank you.

Contents

Notes on Contributors

James Bergman is Visiting Assistant Professor in the History, Philosophy, and Sociology of Science at Lyman Briggs College, Michigan State University. He received his PhD in the History of Science from Harvard University in 2014 and is currently writing a book on the problem of stability for climatologists and government planners in the twentieth-century United States.

Robert Bernasconi is Edwin Erle Sparks Professor of Philosophy and African American Studies at Penn State University. He is the author of two books on Heidegger and one on Sartre as well as numerous essays about the history of racism and of the concept of race. He is co-editor of the journal *Critical Philosophy of Race*.

Paul D. Brinkman is the Assistant Director of the Paleontology and Geology Research Laboratory at the North Carolina Museum of Natural Sciences. He specializes in history of nineteenth-century natural sciences, especially in museums. His latest book is *The Second Jurassic Dinosaur Rush: Museums & Paleontology in America at the Turn of the*

Twentieth Century. He is writing a new book-length history of the Captain Marshall Field Paleontological Expedition to Argentina and Bolivia, 1922–1927.

Constance Areson Clark is Associate Professor of History at Worcester Polytechnic Institute, where she teaches courses on the history of science and technology. She is the author of *God— or Gorilla: Images of Evolution in the Jazz Age* (2008).

Erik M. Conway is the historian at NASA's Jet Propulsion Laboratory, operated by the California Institute of Technology. He writes on the history of earth, planetary, and space sciences in the twentieth century. His most recent work is *Exploration and Engineering: The Jet Propulsion Laboratory and the Quest for Mars* (2015).

Nathan Crowe is an Assistant Professor at the University of North Carolina Wilmington where he teaches classes in the history of science, technology, and medicine. His work is focused on the history of twentieth-century biological sciences, particularly developmental biology and related biotechnologies.

Stephanie Dick is a Junior Fellow with the Harvard Society of Fellows. She received her PhD from the Department of History of Science at Harvard University in March 2015. She studies the history of mathematics and computing in the postwar United States. Her research is currently supported by a Turing Centenary Fellowship funded by the John Templeton Foundation. She is grateful to Thomas Haigh and Daniel Volmar for commenting on earlier versions of this chapter.

Kevin C. Elliott is an Associate Professor in Lyman Briggs College, the Department of Fisheries and Wildlife, and the Department of Philosophy at Michigan State University. He is the author of *Is a Little Pollution Good for You? Incorporating Societal Values in Environmental Research* (2011), as well as more than 50 articles and book chapters on issues related to environmental ethics and policy, philosophy of science, and research ethics.

Ross B. Emmett is Professor of Political Economy and Political Theory & Constitutional Democracy at James Madison College, Michigan State University. A historian of economic thought, he is currently writing a biography of the Chicago economist Frank H. Knight, co-authoring a book on T. Robert Malthus' contributions to constitutional political economy, and planning a book on the history of the relation of economics and Christian theology since the 1830s.

Sebastián Gil-Riaño is a Postdoctoral Research Fellow with the University of Sydney where he is part of the Race and Ethnicity in the Global South research project sponsored by the Australian Research Council Laureate fellowships program. Sebastián holds a PhD in history and philosophy of science and technology from the University of Toronto and is currently working on a book provisionally titled *Antiracist Science and Internationalism During the Global Cold War*, which tracks how social science experts from Latin America, North America, Europe and Australasia re-categorized "race" from a biological to a cultural, social, and psychological phenomenon in the post-World War II era.

Abraham H. Gibson is a Postdoctoral Fellow at the Philadelphia Area Center for the History of Science. He is the author of several articles on the history of science, and is currently writing a book on the history of the biological sciences between World Wars I and II. He recently earned his PhD in History from Florida State University, where his dissertation examined the evolutionary history of domestic animals in southeastern North America.

Adam M. Goldstein received his PhD degree in philosophy from the Johns Hopkins University. His scholarly interests include bibliography, computational methods in history and philosophy of science, and the role of chance in explanation in evolutionary biology. He is a founding member of the editorial board of *Evolution: Education and Outreach*, where he is senior handling editor and reviews editor.

Matthew R. Goodrum is a Professor of History of Science in the Department of Science and Technology in Society at Virginia Tech. His research explores the history of paleoanthropology, prehistoric archaeology, and theories of human origins from the seventeenth to the twentieth centuries.

Melinda Gormley is Assistant Director for Research with the John J. Reilly Center for Science, Technology, and Values at the University of Notre Dame, which examines the ethical, societal, and policy implications of science and technology. Her historical research covers the life sciences in the United States during the mid-twentieth century, primarily examining scientists engaging with the public and policymakers. In recent years her attention has focused on research integrity, science policy, and communication.

Pamela M. Henson is the Director of the Institutional History Division, Smithsonian Institution Archives, where she is responsible for research and documentation of the history of the Institution. She received her PhD in the history and philosophy of science from the University of Maryland in 1990. Her research focuses on the history of natural history, evolutionary theory, museums, and women in science.

Andrew J. Hogan is an Assistant Professor in the Department of History at Creighton University, Nebraska. His current book project, *Life Histories of Genetic Disease: Seeing and Standardizing Patterns of Abnormality in Postwar Biomedicine*, examines the development of a technological system, conceptual infrastructure, and marketplace for the prenatal diagnosis and prevention of genetic disease since 1970.

Lijing Jiang is a Postdoctoral Fellow at Nanyang Technological University in Singapore. She is currently working on a book manuscript on the history of cell death and aging research in the twentieth century with the working title "Degeneration in Miniature."

Daniel Kennefick is an Assistant Professor of Physics at the University of Arkansas, as well as a contributing editor to the *Collected Papers of Albert Einstein*. His research areas include gravitational waves, supermassive black holes and galactic morphology, and the history of physics and astronomy.

Sally Gregory Kohlstedt teaches at the University of Minnesota in the Program in the History of Science, Technology, and Medicine. Her interest in the intersection of science and society leads her to investigate how the enthusiasm for science and technology became an integral aspect of American culture. Scientific organizations, public schooling, and museums that provided public accessibility played significant roles even as they often also created boundaries and patterns of exclusion. The dynamics continue to inspire her work.

Amy Kohout is a Visiting Assistant Professor of Environmental Humanities at Davidson College, where she teaches courses in environmental and cultural history, including a seminar on natural history, collecting, and display. She received her PhD in history from Cornell University in 2015, and is an editor at *The Appendix*. Her current project, based on her dissertation, explores the intersection of ideas about nature and empire by focusing on the experiences of American soldiers in the US West and the Philippines in the late nineteenth and early twentieth centuries.

Mark A. Largent is a Professor and Associate Dean at Michigan State University and a historian of science and medicine. He earned his PhD from the University of Minnesota's Program in the History of Science and Technology in 2000 and has taught history of science, science policy, and American history classes at Michigan State, Oregon State, and the University of Puget Sound. His first book, *Breeding Contempt* (2008), explores the history of compulsory sterilization in the United States. His *Vaccine: The Modern American Debate* (2012) traces the emergence of parents' current concerns about the modern vaccine schedule. His most recent book, *Keep Out of Reach of Children* (2015) examines the history of Reye's syndrome.

Joseph D. Martin teaches history of science courses at Lyman Briggs College, Michigan State University, and studies the evolution of the ideologies behind scientific communities and institutions, with a focus on the physical sciences in twentieth-century America. He is currently working on a book entitled *Confederated Physics: Discipline, Ideology, and the Solid State Insurrection,* which examines how American solid state and condensed matter physicists negotiated the currents of prestige, political influence, and research funding that defined Cold War physics.

Georgina M. Montgomery is an Associate Professor with a joint appointment in Lyman Briggs College and the Department of History at Michigan State University. Georgina's research focuses on the history of field science and issues of inclusion and science. In her book, *Primates in the Real World: Escaping Primate Folklore and Creating Primate Science* (2015), she argues that the history of field primatology is best understood as a series of attempts by individuals within and outside the traditional scientific community to escape primate folklore and create primate science. She has also published articles in *Endeavour* and the *Journal of the History of Biology.*

Jessica Mudry is an Assistant Professor in Professional Communication at Ryerson University. She is the author of *Measured Meals: Nutrition in America* (2009) that examines role of scientific and quantitative language in crafting the idea of "nutrition" and in American federal food guides. She has published articles in *Food, Culture & Society, Social Epistemology, Environmental Communication,* and *Material Culture Review.* Her research interests are in science, language, and food.

Samantha Muka received her PhD in 2014 from the Department of the History and Sociology of Science at the University of Pennsylvania. She is currently a lecturer in the Critical Writing Program at the University of Pennsylvania and is working on a book on marine biology America between 1850 and 1950.

Christine Neejer is a doctoral candidate in history at Michigan State University. Christine's fields of study are women's history, American history, and the history of science and medicine. Her area of research is American women's activism in the long nineteenth century. Her dissertation explores women's bicycling practices in this era.

Samantha Noll is a doctoral candidate in philosophy studying philosophy of science and agriculture philosophy at Michigan State University. Her research explores the normative dimensions of philosophical and scientific work on agriculture and non-human animals. Specifically, it focuses on better understanding of how values influence agriculture science, farming practices, and theories of animal cognition. She won the Vonne Lund Junior Researcher Prize for her work in global food systems and has published several articles and book chapters.

Donald L. Opitz is Associate Professor in the School for New Learning at DePaul University. His research concerns the role of science in Anglo-American Victorian culture, with an emphasis on gender, class, and sexuality. He is coeditor with Annette Lykknes and Brigitte Van Tiggelen of *For Better or For Worse? Collaborative Couples in the Sciences* (2012) and principal editor, with Staffan Bergwik and Brigitte Van Tiggelen, of *Domesticity in the Making of Modern Science* (2015).

Jim Wynter Porter is a doctoral candidate studying twentieth-century cultural history and history of science with an emphasis on the histories of biology, psychology, and education. He is interested, in particular, in changing definitions of "intelligence": what it was purported to be, where it allegedly came from. Of primary importance to his research is how ideas about – and related measures of – "intelligence" functioned to regulate opportunity in a society organized increasingly around meritocratic ideals.

Stephen Rachman is Associate Professor of English, Director of the American Studies Program, and Co-Director of the Digital Humanities Literary Cognition Laboratory at Michigan State University. He is the editor of *The Hasheesh Eater* by Fitz-Hugh Ludlow (2011), a co-author of the award-winning *Cholera, Chloroform, and the Science of Medicine: A Life of John Snow* (2003), and the co-editor of *The American Face of Edgar Allan Poe* (1995).

Miriam G. Reumann received her PhD in American Studies from Brown University in 1998 and teaches modern United States history at the University of Rhode Island. Her first book, *American Sexual Character: Sex, Gender, and National Identity in the Kinsey Reports* (2005), examines the place of sex in postwar debates about family and politics, while her current research investigates early twentieth-century American sexology, focusing on the work of Adolf Meyer.

Ann E. Robinson is a doctoral candidate (ABD) at the history department, University of Massachusetts Amherst. A former science librarian, Ann is broadly interested in organization, classification, nomenclature, and standardization within the modern physical sciences. She is in the process of completing her dissertation on the creation of a standard periodic table. In 2012–13, she was a Dissertation Writing Fellow at the Philadelphia Area Center for the History of Science and a Herdegen

Fellow at the Chemical Heritage Foundation.

Gina Rumore is a historian of twenty and twentieth-first century American ecology. Her doctoral dissertation, "A Natural Laboratory, a National Monument: Carving Out a Place for Science in Glacier Bay, Alaska, 1879–1959," won the 2010 Rachel Carson Prize for the Best Dissertation in Environmental History from the American Society for Environmental History. She is currently working on several historical projects tied to the National Science Foundation's Long-Term Ecological Research Network.

Michael Ruse is the Director of the Program in the History and Philosophy of Science at Florida State University. He is the author or editor of over 50 books, most recently as author *The Gaia Hypothesis: Life on a Pagan Planet* and as editor *The Cambridge Encyclopedia of Darwin and Evolutionary Thought*. He is now writing a book looking at the history of evolutionary theory through fiction and poetry.

Sara J. Schechner is the David P. Wheatland Curator of the Collection of Historical Scientific Instruments, Harvard University. Her latest books are *Tangible Things: Making History through Objects* (2015) and *Time and Time Again* (2014). Current research focuses on the workshop practices of American telescope maker, Alvan Clark, and pocket sundials as evidence of social change.

Adam R. Shapiro is a Lecturer in Intellectual and Cultural History at Birkbeck, University of London. He is the author of *Trying Biology: The Scopes Trial, Textbooks, and the Antievolution Movement in American Schools* (2013).

Matthew Shindell is Curator of solar system exploration at the Smithsonian

Institution's National Air and Space Museum and has held pre- and postdoctoral fellowships at Harvard University, the University of Southern California, the Huntington Library, the University of California, San Diego, and the Chemical Heritage Foundation. He is a Ph.D. historian of science (UC San Diego, 2011) and has published articles on the histories of planetary geology, isotope geochemistry, Cold War science, and climate change. His current projects include a biography of the American chemist Harold C. Urey and an ongoing collaborative study of expert assessments at the U.S. National Academy of Sciences-National Research Council.

Jacob Steere-Williams is an Assistant Professor of History at the College of Charleston. He received his PhD from the University of Minnesota in 2011. His research focuses on the history of public health, and recent articles have appeared in the *Journal of the History of Medicine and Allied Sciences, Social History of Medicine,* and *Ambix.* He is currently completing a book manuscript on the history of typhoid fever in nineteenth-century Britain

Banu Subramaniam is Associate Professor of Women, Gender, and Sexuality Studies at the University of Massachusetts, Amherst. She is author of *Ghost Stories for Darwin: The Science of Variation and the Politics of Diversity* (2014), and co-editor of *Feminist Science Studies: A New Generation* (2001) and *Making Threats: Biofears and Environmental Anxieties* (2005). Spanning the humanities, social sciences, and the biological sciences, her research is located at the intersections of biology, women's studies, ethnic studies, and postcolonial studies.

Peter J. Susalla is an independent scholar living in Indianapolis. He is

currently working on histories of cosmology in twentieth-century America and the astronomer George Cary Comstock, one of the founders of the American Astronomical Society.

Jeremy Vetter is Assistant Professor of History at the University of Arizona. Working at the intersection of environmental history and history of science, his books in progress include *Field Life: Science in the American West During the Railroad Era* and *Capitalist Nature: Environment and the Sciences of Development in the American West.* He has published numerous articles and is editor of *Knowing Global Environments: New Historical Perspectives on the Field Sciences* (2010).

Mark A. Waddell is a historian of science who has published extensively on the relationship between religion and science, particularly among Jesuits in the sixteenth and seventeenth centuries. He received his PhD in the History of Science, Medicine, and Technology from the Johns Hopkins University in 2006, and his first book is *Jesuit Science and the End of Nature's Secrets* (2015).

Arthur Ward earned his PhD in philosophy from Bowling Green State University. His areas of research are in bioethics and philosophy of science. He teaches science studies classes at Lyman Briggs College at Michigan State University.

Adrian Young is a doctoral candidate in the history of science at Princeton University, interested in the histories of Britain, empire, and the human sciences. His dissertation, "Mutiny's Bounty: Anthropology, Pitcairn Islanders, and the Making of a Natural Laboratory on the Edge of Britain's South Seas Empire," tracks the use of two field sites across the nineteenth and twentieth centuries.

Christian C. Young teaches biology, history of science, and science education at Alverno College in Milwaukee, Wisconsin. He earned his PhD in History of Science and Technology at the University of Minnesota in 1997. Publications include *In the Absence of Predators* (2002) and *The Environment and Science* (2005), articles in *American Biology Teacher* and *Science as Culture*, as well as numerous book reviews in *Journal of the History of Biology*, *Isis*, and *Environmental History*.

Daniel Zizzamia is an instructor at Montana State University. Daniel's historical scholarship addresses the impact of science and technology on environmental perceptions and policies. His dissertation is an analysis of the settlement of the American West as it was effected by the fossils and fossil fuels unearthed and used by scientists, settlers, capitalists, and Native Americans.

Introduction

The History of American Science

Georgina M. Montgomery and Mark A. Largent

Science, by definition, implies universality. As the systematic study of nature, we typically assume that science is as universal as the subjects it examines. Gravity is no different from one nation to another or from one culture to another, so it is reasonable to adopt the notion that the study of it – and the results of that study – is similarly universal. This assumption forms the basis for notions about the capacity for science to transcend boundaries of space and time, to help overcome or even equalize social, cultural, and political differences between individuals and groups, and for the ongoing creation and use of a shared set of tools, terms, and tactics to explore and exploit nature.

Science, however, is not synonymous with nature. Ideally, scientists strive to generate descriptions of natural objects and natural phenomena that are perfectly accurate representations of them. But their explanations of nature are separate from nature itself. Science consists of a complex social system that is invented and undertaken by well-defined groups of human beings in an effort to understand the world around them in ways that are mutually agreed upon by the people engaged in the enterprise. Even this is an idealized vision, with the humans conducting science often falling outside – or between – any clearly defined category like amateur or expert. The methods they use may sometimes be agreed upon while at other times be hotly contested. Nevertheless, those involved in creating science share a common set of assumptions about its capacity to understand and accurately depict nature.

By recognizing the distinction between science on the one hand and nature on the other, we can begin to see how different cultural contexts generate fundamentally different forms of science. The departments, laboratories, journals, and societies formed during the process of professionalization, and the modes of popularization used to disseminate science to the public, vary across disciplines, regions, and indeed countries. However obvious this may seem in the twentieth century, it has been appreciated and carefully explored only since the 1960s.

A Companion to the History of American Science, First Edition.
Edited by Georgina M. Montgomery and Mark A. Largent.
© 2016 John Wiley & Sons, Ltd. Published 2020 by John Wiley & Sons, Ltd.

The history of American science first began to be written in the 1960s and developed to a subfield of the study of history in the early 1980s. Looking back at these formative years, we have identified three books as particularly influential and emblematic of the emergence of the subdiscipline of the history of American science. These three volumes have served as models for us as we have assembled the *Companion to the History of American Science*. All three were edited collections, which we see as a testament to the collaborative nature of the historical study of American science. It is appropriate, therefore, that our generation's summation of its interpretation of the history of science in the United States follows the same format that has proven so important to our predecessors.

Nathan Reingold's 1979 *The Sciences in the American Context: New Perspectives* consisted of 15 essays written by a variety of scholars, many of whom we today regard as among the founders of the history of American science. In addition to Reingold, authors in the volume include Deborah Jean Warner, Stanley Guralnick, Charles Rosenberg, Garland Allen, Spencer Weart, and Carroll Pursell. The volume originated from a group of papers that had been presented at the 1976 meeting of the American Association for the Advancement of Science in a session supported by the National Endowment for the Humanities, the History of Science Society, and the National Science Foundation. The latter two of these organizations would become important incubators for the emerging history of American science.

In his preface to the volume, Reingold explained that the original conference presentations and the current volume reflected a "notable upsurge in research in the history of the sciences in the United States" that had demonstrated "the persistent belief that the pure and applied sciences are particularly important in understanding the course of American history." That is, Reingold argued, that the sciences themselves shaped the nation. As we shall see, Reingold and later authors closed this circle by similarly arguing that there was a particular American-ness to the sciences that were pursued in the United States.

The second seminal work in the history of American science that guided us was the inaugural volume of the second series of *Osiris*, titled "Historical Writing on American Science" (1985). The volume's short foreword was written by Arnold Thackray, the editor of *Isis*, the most prominent journal in the history of science. He described how, a hundred years earlier, the United States "was a junior partner to Europe in both the intellectual power and the social organization of science. However, the twentieth century has seen the rise to dominance of American science, and an associated development of the history of science itself as a profession and a scholarly pursuit". It was therefore fitting, he explained, that the first volume of the new series be devoted to historical writing on American science.

Hidden within Thackray's introduction to the *Osiris* volume was a generational divide between the more senior historians of science who still operated with universalist notions about science and the history of science and the new generation for whom this volume represented a clear articulation of their efforts to convey Americans' style of science. Thackray asserted that the volume represented more than just "the vitality that characterizes the study of their national science by historians in the United States." These essays, he wrote, "stand as testimony to the level of analysis that American historians of science bring to the wiser world of scholarship, in studies of Aristotle, Galileo, Newton, Darwin, and the 'great tradition.'" For Thackray, the study of

American science itself was perhaps not sufficient to warrant the publication, but needed to be augmented by an American interpretation of the "great tradition." In contrast, Sally Gregory Kohlstedt and Margaret Rossiter – the volume's editors – emphasized in their introduction that the volume was a study of the history of American science, and they were mute on the subject of any perceived importance of studying the American interpretation of the history of science.

The third book on the history of American science that helped form the field and has served as a guide for this volume was *The American Development of Biology*, which was edited by Ronald Rainger, Keith Rodney Benson, and Jane Maienschein, and published in 1988. In their introduction, the editors bluntly stated what had only been intimated in earlier volumes: "biology in the United States, due in part to its late nineteenth-century setting and its comparative isolation from Europe, developed a distinctive American character" (Rainger, Benson, and Maienschein 1988: 2). A few years later Reingold's collection of his own essays reflected this in its title, *Science, American Style* (Reingold 1991).

By the early 1990s, the history of American science (and historians of American science themselves) had acquired both legitimacy and authority. No longer was there a need, as Thackray had a decade earlier, to validate their work on the basis of its contributions to more traditional European studies of science. Their capacity to demonstrate both the American character of American science and the influence of science on the nation allowed them to explore the reciprocal relation between science and American culture, politics, economics, and intellectual thought. While throughout the 1990s more traditional historians of science fought the so-called science wars, exchanging blows with postmodernists and arguing about the necessity for a thorough understanding of the technical content of science, historians of American science steadfastly supported the importance of exploring the content of science as well as the assumptions on which it was founded and the contexts within which it operated. They likewise acquired increasing authority in the History of Science Society; beginning with Sally Gregory Kohlstedt in 1992, nearly half of the presidents of the Society were historians of American science.

This volume primarily represents a second and third generation of scholars on the history of American science, many of whom were trained by the first generation or their students. Perhaps not surprisingly, many of the topics featured in earlier edited collections remain as themes in this volume. The process of professionalization, the evolution of science education, museums, the popularization of science, science policy, and science and the military remain as pillars of the new generations' analyses of the history of American science.

Studies of science in the United States also continue to emphasize its applied nature. Although analyzing very disparate disciplines, Samantha Noll's chapter on agricultural sciences, Ann E. Robinson's chemistry chapter, and Stephanie Dick's piece on computer science all center the sciences' applied character. In the biological sciences, the history of genetics, eugenics, and biotechnology all developed in large part in response to national interests and concerns. Similarly, the development of the social sciences in the United States has been contingent on social, cultural, and political forces. As Jim Wynter Porter points out in his analysis of the history of experimental psychology, "It perhaps can be easier to see how applied science is shaped by certain cultural assumptions, but laboratory science, 'pure science,' seems ... more objective," but in fact sciences that are more abstract or theoretical are no less bound by the confines of social and cultural context.

A focus on case studies is another source of continuity between seminal works from the 1970s and 1980s and today, as historians continue to interrogate the local while piecing together the puzzle of national trends. Mark A. Waddell's chapter on science and religion and Adam R. Shapiro's chapter on science education are both great examples of using specific cases to highlight changes taking place across the nation. Other chapters bring the historical gaze to specific local places, including Christian C. Young's chapter on zoos and aquariums and Jeremy Vetter's chapter examining field and laboratory sites. This balance of detailed storytelling and large-scale applicability creates an effective introduction to the history of a scientific topic or discipline in the United States, while the biographical essays that follow each chapter provide a gateway to additional sources.

Certain moments in US history have come into the foreground since the 1970s, emerging as significant factors in the development of a wide range of sciences. Several chapters in this volume discuss eugenics, for example, demonstrating the role of the eugenics movement in the development of disciplines like psychology and anthropology. World War II, the Cold War, and the women's movement also emerge as pivotal events in the development of science and science policy in the United States. The Progressive Era likewise makes appearances throughout the volume, demonstrating its importance in instigating the rapid emergence of the United States as the global leader in science and technology in the twentieth century.

Collaborative groups are also particularly highlighted in this volume, representing the field's continued shift away from narratives focused on individual (commonly white and male) heroic figures and toward the study of scientific teams. Daniel Kennefick's chapter on relativity, Adrian Young's essay on anthropology, and Melinda Gormley's analysis of the history of genetics are just a few of the chapters in this volume that examine collaborations and cohorts when examining the history of American science.

Other categories like gender, race, sexuality, and postcolonialism have built on the foundation began by pioneers like Sally Gregory Kohlstedt and Margaret Rossiter. The volume reflects this growth with wonderful chapters written by Donald L. Opitz (gender), Robert Bernasconi (race), Miriam G. Ruemann (sex), and Banu Subramaniam (postcolonialism). Reflecting the field's shift toward making these categories mainstream or standard areas of analysis, several other chapters in this volume include consideration of these factors to various degrees. Gender, for example, is discussed in chapters ranging from anthropology and astronomy to psychology and molecular biology.

Perhaps one of the characteristics of this volume that is most evident is the complex ways scientific disciplines formed, professionalized, specialized, and cut across into other fields. The "Topics" section of this volume (Part II) embodies this reality with chapters like Nathan Crowe's "Biotechnology" discussing agriculture, warfare, and medicine. Many scientific fields have pushed the limits of their disciplinary identities at various points in time and this trend continues to gain momentum today as sciences move increasingly toward an interdisciplinary mode.

Scientists interested in how contemporary issues in methodology, collaboration, and applications connect with their field's past will find much value in this volume's chapters. Gina Rumore's ecology chapter and Constance Areson Clark's analysis of how science is communicated in popular culture, for example, provide wonderful historical accounts while also shedding light on current trends in science. Equally, historians of science and students alike will find the chapters both accessible and sophisticated, providing ideal summaries of the histories of various disciplines and important topics in the history of science in the United States. Indeed, many of the chapters represent the best introductions to their fields published to date.

We wish to thank the authors and our colleagues for their generous contributions, and we hope the readers enjoy reading the volume as much as we enjoyed editing it.

Part I

DISCIPLINES

Chapter One

AGRICULTURAL SCIENCES

Samantha Noll

Agricultural science may have begun with the Hatch Act of 1887 and the birth of the US land-grant universities in 1862 (Hillison 1996). Actually, it is one of the oldest applications of empirical inquiry, as our current methods of agriculture are the result of thousands of years of trial and error and experimentation in the field. Farming methods slowly improved as humans developed better ways of obtaining reliable knowledge that they then applied. With that being said, however, agriculture science as we know it today really began to take shape between the seventeenth and nineteenth centuries, as the scientific methods born out of Enlightenment thinking were directly applied to farming practices. This form of applied science became institutionalized in the United States in the late nineteenth century with the birth of the land-grant universities in 1862, the establishment of federally funded experiment stations, and extension services meant to communicate scientific breakthroughs to the local farming community (Hillison 1996). This three-way partnership forms the backbone of current agricultural research in the United States (Rosenberg 1997). In point of fact, one could argue that it also forms the backbone of American industry, as technology and farming methods developed by these sciences both increased food supplies and lowered the numbers of workers needed to grow this food, thus providing the workers necessary for industrial development (Thompson and Noll 2014).

This chapter will begin with a brief definition of "agriculture." Before outlining the many different types of agricultural sciences, it is important to recognize the scope of farming practices and thus the varied nature of scientific disciplines that focus on improving these practices. The next section provides a general overview of agricultural science by describing how agricultural science is not one science, but a multidisciplinary field that encompasses work from a multiplicity of scientific disciplines. The third section of the chapter describes the historical movements beginning during the seventeenth century that made the rise and dominance of American agricultural science possible. It

A Companion to the History of American Science, First Edition.
Edited by Georgina M. Montgomery and Mark A. Largent.
© 2016 John Wiley & Sons, Ltd. Published 2020 by John Wiley & Sons, Ltd.

then outlines the history of the development of agriculture science in the United States and how its structure of scientific inquiry differs in this context from that of Europe's. These latter two sections are particularly important, as many of the social consequences of agricultural research can be found in the early history of these sciences. In the final section, current critiques of these sciences are outlined before the chapter ends with a brief overview of agricultural science today.

Agriculture Defined

While agriculture is currently understood as the cultivation of food crops (such as corn, wheat, and soy), the practice also includes raising animals, plants, and other organisms for production and pharmaceutical purposes. The term covers a considerable amount of human activity, including animal husbandry, wine production, biofuel, dairy, hydroponics, and fiber crops, and activities associated with harvesting, distribution, and food processing (Thompson and Noll 2014). The history of agriculture dates back thousands of years and was largely a place-based practice, bound to specific areas, as the development of various methods of production, processing, and storage were influenced by vastly different climates, technological advances, cultural views, and values surrounding the cultivation of food. However, while farmers in disparate areas practiced diverse techniques and methods, most agricultural practices relied on basic activities of land and animal management that still underlie local differences, such as the practice of irrigation, maintaining the fertility of the soil, and general methods of farming, such as intercropping, grassland grazing, and terrace cultivation. Historically a large percentage of the population worked in agricultural production, but current technological developments have greatly reduced the numbers of people working in the field, especially in the United States (Lyson 2004). One such advancement was the development of large-scale monoculture farming, the most common form of field crop cultivation today. However, other forms are still being practiced, such as both large-scale and small-scale organic agriculture, livestock integrated systems, intensive small-scale operations, and traditional farming practices, such as the cultivation of *milpa* originally used throughout Mesoamerica.

Agriculture Science Defined

While agriculture refers to a set of methods or activities used to transform the environment for the production of the above products, the agricultural sciences are grounded in "the application of scientific methods of inquiry to improve the practice of agriculture" (Thompson and Noll 2014: 1021). Very roughly then, one can understand agriculture science as the use of scientific methods and methodologies to improve agriculture practices. Just as agricultural practices are varied, agricultural science can be understood as a multidisciplinary field of biology that encompasses research in the natural and social sciences (Olmstead and Rhode 2008). Traditionally this work was carried out on a multiplicity of topics, such as production techniques, pest control, minimizing the effects of drought, food distribution, selective breeding of plants and animals, the design and implementation of sustainable production methods, and various social and economic topics surrounding food production, storage, and transportation. In the context of the United States, most agricultural research is made possible by, what David MacKenzie (1991) calls, a triangular partnership between farmers, government agencies that fund and sometimes conduct research, and commercial and non-profit public sector research institutions. This arrangement has lasted for over 100 years and has proven highly successful in supporting cutting-edge agricultural research.

Several fields fall under the umbrella of "agriculture science" including agricultural chemistry, economics, geography, philosophy, marketing, agrophysics, animal science, agronomy, aquaculture, biotechnology, microbiology, environmental science, entomology, food science, soil science, waste management, and ecology. Many of these fields focus on a single aspect of agriculture. In addition to being multidisciplinary, or a field that draws upon many distinct disciplines, it is important to note here that much of agriculture science is also interdisciplinary (Jacobs and Frickel 2009), or a field that integrates knowledge originally developed within distinct fields. In truth, the practice of agriculture by its very nature relies upon and integrates varied sources of knowledge as solving problems in agriculture or developing new methods of production, harvesting, and storage often require such integration. For example, an entomologist working on pest control may have expertise on insects but will often have to draw upon and incorporate knowledge from other fields, such as agronomy, ecology, or soil science to properly address the pest problem. Thus the term *agriculture science* is an umbrella term that encompasses work carried out by various disciplines and often across disciplines. For this reason, one can understand agricultural science to signify the entirety of the agricultural sciences that make up this branch inquiry. This chapter will use both terms interchangeably.

In addition, it should be noted here that agricultural sciences are largely applied sciences, in contrast to pure sciences, though not all research is applied in this field. As illustrated above, agricultural sciences are housed in many different departments, as they each draw upon scientific methods and methodologies developed within these fields. When applied to agricultural practices, such methods provide unique and novel insights. In contrast, pure sciences make deductions from mathematics, logic, and previously accepted facts in search of universally applied laws or fundamental principles (Rosenberg 1997). While agricultural sciences are applied sciences, most scientists working in these fields nevertheless accept that pure science is necessary for applied sciences to flourish, as findings in the more abstract sciences open up new avenues for research on the ground. For example, pure research in chemistry opened up the possibility for new fertilizers, and biological research in genetics paved the way for the creation of genetically modified organisms now used in agriculture.

Historical Roots of Agricultural Science

While the agricultural sciences began in earnest in the United States during the late nineteenth century with backing at both the state and national levels (Rosenberg 1997), improving various areas of agricultural practice through the application of components of the scientific method has a long history. Indeed, it is difficult if not impossible to separate the practice of agriculture from technological development and empirical inquiry. For example, Xenophon (*c.* 430–350 BCE) and Aristotle (384–322 BCE), whose texts are foundational in the development of the sciences, both wrote extensively on agriculture. In addition, Roman texts, such as Columella's (4–70 CE) 12 volumes on agriculture, give detailed descriptions of animal husbandry techniques, selective breeding programs for plants and animals, field crop cultivation methods, orchard management regimens, and descriptions of experiments conducted in these areas. The Romans, and before them the Greeks, used highly developed methods and specialized crops, such as those used for fodder, that were lost after the collapse of the Roman Empire and only rediscovered during the Renaissance (Kingsbury 2009). In truth, the rediscovery of these techniques coupled with the scientific, industrial, and agriculture revolutions of the eighteenth century formed the foundation for the current agricultural sciences that we practice today.

Scientific methodologies were applied to agricultural practices throughout the Enlightenment when the various revolutions listed above shifted people's reliance from tradition to the application of scientific methods and cultivated an insistence on change and progress. These factors powerfully influenced subsequent developments in agriculture and the structure of agricultural systems as a whole (Brantz 2011). According to Kingsbury (2009), the Scientific Revolution and industrial enlightenment combined to spur on agricultural developments that then further supported the other revolutions and spurred further research in agriculture. First, the Scientific Revolution was built upon the idea that the natural world is orderly (not controlled by capricious deities or inherently disorderly) and is thus knowable. Through scientific inquiry, it is possible to both obtain reliable knowledge about the world around us and to manage nature for the benefit of humans. The "industrial enlightenment" signifies the technological advances occurring in conjunction with the Scientific Revolution, including the codification of experiments and observations on agricultural techniques that were then made readily available to the intellectual community through translation and printed materials (Kingsbury 2009).

Beginning in the later half of the eighteenth century, the Agricultural Revolution was a culmination of the advancement in farming techniques that greatly increased the crop yields of the day (Kingsbury 2009). During this period, Europe reaped the benefits of the implementation of new agricultural practices, such as the enclosure of pastures, the introduction of hardier plant types, a new four-course rotation schedule (Kingsbury 2009), and the use of composted manures from city centers (Atkins 2012). It is estimated that the production of wheat went up from 19 bushels per acre during the early seventeenth century to over 30 bushels by 1840 (Snell 1985). Subsequent developments pushed these yields even higher, as the four-course rotation method produced on average 80% more food (Kingsbury 2009). Further developments in mechanization and plant breeding increased these yields even more. The subsequent availability of food supported further industrial development, as it freed people from the necessity of working on the land, and thus provided the population necessary for the Industrial Revolution.

In addition, the coupling of agricultural research with commercial interests during this time helped shift the reputation of agriculture, from a practice largely performed by the lower class, to the pursuit of landowners and the educated classes. This shift is directly reflected in the scholarly work of the time, as various philosophers, such as John Locke (1632–1704), Jeremy Bentham (1784–1832) and John Stuart Mill (1806–1873) and economists Adam Smith (1723–1790) and Thomas Malthus (1766–1834) wrote extensively on agricultural practices, technology, and the social factors that influence or are influenced by crop production. In fact, foundational work in most if not all of the disciplines that make up the current agricultural sciences can be found during this time period. For example, the chemist Justus von Liebig (1803–1873), often considered the founder of agriculture science, wrote extensively on using controlled experiments to identify practices useful for improving soil fertility and crop yields; the agronomist Jethro Tull (1674–1741) published on tillage in 1733; and Thomas Jefferson discussed the role that agriculture should play in American higher education (Thompson and Noll 2014).

History of Agricultural Sciences in the United States

Like Europe, the United States began to reap the benefits of the implementation of agricultural research during this time. In this context, however, agricultural science became institutionalized in the late nineteenth century with the birth of land-grant universities in 1862 and the establishment of federally funded experiment stations (Hillison 1996). The first agricultural experiment station was established in Connecticut in 1875 (Rosenberg 1997). A little over a decade later, the Hatch Act provided each state with $15,000 a year to support local experiment stations. The act was passed due to increasing political pressure by the farming lobby and, as Rosenberg (1997) argues, the role of the experiment stations was clear from the beginning: "It was to perform the experiments which the individual farmer, lacking time and opportunity, could not" (p. 154). The average farmer did not have the time or the money to perform experiments in a systematic manner, as the loss of one season's crops could mean the loss of the farm itself. These two developments helped connect two parts of the triangular partnership (farmers, government agencies, and public sector research institutions) that, as MacKenzie (1991) argues, supported cutting-edge agricultural research in the United States for the last 100 years.

Agricultural experiment stations were placed directly under the control of states' land-grant universities that were originally established in 1862 under the Morrill Act (Thompson and Noll 2014). Land-grant universities provided education in agricultural practice, such as animal husbandry and field crop cultivation. Early in their development, the institutions embraced the agricultural sciences, as the application of the scientific method promised to both raise the status of the American farmer and improve the economic viability of farms (Rosenberg 1997). They conducted and continue to conduct research on a multiplicity of agricultural topics, such as soil fertility, cover crops, and farming methods, both at the university proper and at experiment stations. Similar to MacKenzie's (1991) triangular partnership, Rosenberg (1997) argues that the United States' early success in agricultural research was a result of a three-way partnership between universities, experiment stations, and extension services. The last of these was established to disseminate research results, such as new crop types, machinery, and cultivation methods, to the larger farming community (Thompson and Noll 2014). These extension services are state-operated and focus on providing information on advances important within regional contexts, providing training for farmers in all areas of practical farm management, and on recommending efficient fertilizer levels. Today, while the farming landscape has changed dramatically, the three-way partnership forms the backbone of agricultural research and serves as a model for publicly managed and funded approaches to the development and dissemination of research. However, it should be noted that agricultural sciences are also currently being pursued in various commercial and academic contexts outside of this federally funded structure, such as in the large agribusiness corporations.

Ethical and Social Influences

At least two of the current and most serious critiques of American agricultural research and the social consequences of this research can be found in the early history of these disciplines: specifically, the reduction of family farms and the effects of myopically focusing on increasing productivity (MacKenzie 1991; Rosenberg 1997). Indeed, Rosenberg argues that, "we are in retrospect well aware [that] circumstances dictated that the small agricultural producer would not be the most prominent beneficiary of experiment station research and development" (p. 186). Scientists and administrators working in the land-grant system found it more beneficial for their research to work with larger and better funded producers of agricultural products. However, according to Rosenberg (1997), they never considered the consequences of their research as, when successful, it ultimately helped the larger producers gain an advantage over smaller scale producers. Thus the historical structure of American agricultural research could be understood as contributing to the reduction of small farms.

Similarly, according to Thompson and Noll (2014), "the influence of publicly organized research conducted at experiment stations and the organized attempt to extend those results throughout the world provide the basis for viewing agricultural science ... as an applied science with explicit value commitments" (p. 1022). One of the basic commitments of early agricultural scientists included the fundamental belief that increasing productivity was an unproblematic goal of agricultural research (Rosenberg 1997), as most if not all humans are vulnerable to food-borne risk and food and fiber are of paramount importance when meeting basic human needs (Thompson and Noll 2014). The technologies developed from this research placed farmers on what is commonly called the "technology treadmill," as producers who refused or could not afford to implement technological advances commonly found themselves out of business (Hightower 1975; MacKenzie 1991). Again, research focused on improving productivity ultimately favored larger operations with access to more capital and thus helped to increase the numbers of large-scale producers at the cost of family farms. In point of fact, according to Thompson, Ellis and Stout (1991), while current critiques of the agricultural sciences include a wide spectrum of concerns, a group of vocal critics claim that small-scale stakeholder groups were historically ill-served by the three-way partnership above. This last group of critics argues that land grant institutions' predominant goal of increasing yields was the wellspring for many of the current problems faced by American agriculture.

While these are only two of many ethical and social issues currently being addressed in the agricultural sciences, they form the foundation of the first critiques that sparked current debates in this field. In fact, the work of Wendell Berry drew both academic and popular attention to these ethical issues and others surrounding modern agricultural science, industrial agriculture, and the land-grant institution. His best known work, *The*

Unsettling of America (1977), provided a jarring critique of these features of American agricultural research, pointed out the negative social impacts brought about when focusing solely on productivity, and included a lengthy defense of the family farm, thus bringing the plight of small farmers into the public eye (Thompson and Noll 2014).

Over the last 20 years, the agricultural sciences in general and the research priorities of these fields in particular have been subjected to growing criticism (Dahlberg 1988; Johnson 1984; Thompson, Ellis and Stout 1991). These critiques include issues such as the use of genetically modified crops (Doyle 1985), environmental impacts (Jackson and Jackson 2002; Shiva 1992), global hunger (Pimbert 2008), the plight of workers in the United States and abroad (Mizelle 2011), and of various social goals of this research (Hightower 1975; Jackson 1980). However, it should be noted here that the dominant theme in this field is that the agricultural sciences' service to large-scale producers produced benefits that largely outweigh any harms, such as improving the availability of food and fiber products to American citizens and lowering the costs of these staples (Thompson, Ellis, and Stout 1991).

The Green Revolution and Genetic Modification

Both criticisms and beneficial claims concerning agricultural science are clearly found in two common case studies used to outline potential social and ethical impacts of agricultural research: the Green Revolution and genetic modification. The Green Revolution was one of the most controversial applications of the agricultural sciences during the twentieth century (Thompson and Noll 2014). First conceptualized as a strategy for undermining increasing Soviet influence after Word War II, the Green Revolution was primarily a research initiative, funded by the Rockefeller Foundation during the 1950s and 1960s, with the goal of increasing the worldwide production of crops (Anderson and Morrison 1982). While this initiative primarily focused on developing and making available high-yielding crop varieties to farmers in developing countries, it also involved expanding irrigation infrastructure, modernizing farming and management techniques, and making petrochemical inputs available, such as fertilizers and pesticides. If the only factor taken into account is improving access to food and fiber crops, then this initiative can be viewed as a tremendous success, as Green Revolution wheat and rice varieties helped bring about a decade of food surplus in India and other developing countries (Hillison 1996; Thompson and Noll 2014).

However, the initiative is often viewed as a mixed success, as these surpluses came at a price. The Green Revolution crop varieties did improve productivity and thus led to a greater availability of food at lower prices, but small-scale farmers could not stay in business due to lower profit margins. Second, the high-yielding crop varieties were made to be used in conjunction with expensive fertilizers and/or pesticides. Again, small-scale farmers could not afford these inputs and were thus forced out of farming. In fact, according to Pimbert (2008), the decline in agriculture commodity prices coupled with the increase in price for production inputs has led to rising bankruptcies and poverty within rural farming communities. It is estimated that "200,000 farms disappeared between 1966 and 1995" alone (Pimbert 2008: 22). Finally, the increased mechanization of farming led to the large-scale unemployment of landless laborers, as they were no longer needed to perform these tasks. All three of these factors have led to critiques of the Green Revolution centering on the claim that the industrialization of worldwide agriculture caused the disenfranchisement and disempowerment of already marginalized groups. Vandana Shiva (1992) has been particularly vocal in her criticism of the Green Revolution on the above social grounds, as well as the environmental damage caused by increased production. This case study clearly illustrates how the goal of increasing productivity can lead to unintended social consequences and, as will be discussed below, it this tension that influences current agricultural sciences, as factors previously understood as externalities, such as the environmental and social impact of research, are now areas of research themselves in the field.

Second, there are numerous social and ethical debates occurring in the public sphere on various impacts of genetically modified organisms. While most of our production increases within the last 100 years are due to selective breeding and not to genetic modification (Boyd 2001; Greger 2011; Noll 2013), within the last 20 years, new methods of modifying agricultural plants and animals developed directly out of current work in genetics: specifically, genetic engineering. The development of these methods, produced from the direct application of research in biology and genetics, greatly improved crop yields, the drought resistance of plants, and reduced crop predation by pests, thus improving agricultural production as a whole (Chassy 2007). These innovations occurred relatively recently, with the first genetically engineered plant being produced in the early 1980s (James and Krattiger 1996). By 1996, the Environmental Protection Agency had granted 35 approvals to commercially grow 8 genetically engineered crops and, by 2000, scientists developed "golden rice," which was the first genetic engineering project aimed at increasing the nutrient value of food.

Especially prominent debates on genetic modification include disputes over the ownership and patenting of genetic resources, potential health risks of these products, whether or not such products should be labeled, which types of transformations are or are not acceptable, animal welfare issues, and potential ecosystem effects (Lynas and Tudge 2014; Thompson and Noll 2014). While the modification of agricultural plants and animals has a long history, the current technological advances are providing novel new ethical and social issues that will increasingly impact both the application of this technology and work in various agricultural sciences, such as that in genetics, the societal impacts (if any) of such products, and the identification of any potential environmental costs.

Agricultural Sciences Today

At least three influences impacted the shape of current agricultural sciences: (1) new breakthroughs in the pure sciences; (2) social and ethical critiques of agricultural sciences; and (3) the social movements surrounding food that influenced policy and economic factors. The first two influences were touched upon above, with new breakthroughs in genetics paving the way for novel applications in agricultural sciences and social critiques sparking a renaissance in agricultural research focusing on various societal impacts of agricultural practices. The third influence can be understood as another response to the social and ethical critiques outline above. However, this response took the form of social movements that greatly impacted both the political (as some agricultural research is publicly funded) and economic landscapes; landscapes that deeply influence both farming practices and, as will be discussed below, the goals and aims of current agricultural sciences. While this development may at first appear surprising, in fact, it is part and parcel of the historical development of the agricultural sciences for the following reasons: (1) most agricultural sciences are applied sciences and thus cannot be separated from current social developments and (2) as mentioned above, a primary goal of the land-grant institutions and research stations is to perform research useful for local farmers who are themselves effected by their consumers' desires. The final section of this chapter will briefly describe each of the three influences before going on to give a description of the current state of the field.

While ethical critiques began in earnest around 1975 when Glenn L. Johnson (1918–2003) published a series of articles calling for attention to normative issues surrounding agricultural sciences, scholars from a multiplicity of fields now reflect on the social and ethical aspects of agricultural research, such as the methods used, the efficacy of studies, and the social impacts of such research. For example, three sociologists working in agriculture, Lawrence Busch, Frederick Buttel, and William Lacy, have long histories of publishing on various political and economic aspects of agriculture science (Thompson and Noll 2014). In addition, there is a plethora of work that examines normative issues surrounding American agriculture, such as that which focuses on feminist critiques of agriculture (Sachs 1996; Shiva 1992), local food initiatives (Delind 2011; Sbicca 2012; Werkheiser and Noll 2014), selective breeding and genetic modification (Boyd 2001; Greger 2011; Noll 2013; Shiva 2000), and alternative farming systems (Fairlie 2010). In addition, philosophers, such as Paul Thompson, have written extensively on agriculture practices and systems. Paul Thompson was the first ethicist appointed to an agricultural research university, Texas A&M, and is the current W.K. Kellogg Chair in Agricultural Food and Community Ethics at Michigan State University. Indeed, this appointment was the first of many, as today several land-grant institutions have appointed experts working on social and ethical aspects of agricultural science. This trend at institutions and the above scholarly work illustrates how many research administrators and scientists within the land grant institution have accepted the critics' call for agriculture research to address a larger spectrum of social goals (Thompson, Ellis, and Stout 1991).

In addition, social movements occurring over the last 20 years have greatly impacted research currently being performed in the agricultural sciences. While academics largely ignored Wendell Berry's (1977) critiques of agricultural science and land grant institutions (Thompson and Noll 2014), his popular works helped focus public attention on these issues and sparked a citizen lead movement to rebuild and support local food production structures. This work and current bestselling books, such as those by Pollan (2009; Pollan and Kalman 2011) and Kingsolver, Kingsolver, and Hopp (2008) have largely succeeded in mobilizing the public on a wide range of issues concerning agriculture. The result has been the creation of local food movements around the country, with the express aims of rebuilding and supporting small-scale farming operations through community supported agriculture projects, such as community gardens and local food sourcing. These movements are a response to a wide range of critiques of the agricultural system, such as the economic consolation of agriculture, food health scares (see the recent *E. coli* contamination of spinach in 2006), the desire for non-genetically modified foods, the claim that smaller operations are more environmentally "sustainable," a rejection of current industrialized practices and/or selective breeding programs, and a strategy for the revitalization of rural areas (Feenstra 2002; Lyson 2004). Thus these social movements can be understood as at least a partial reaction to some of the technological developments made possible by the agricultural sciences.

While these three impacts are generally seen as a separate issue from the agricultural sciences, in actuality, they have had a large impact on the design and goals of current research projects. For example, the local food movements have greatly increased the economic demand for "organic," "sustainable," and environmentally responsible agricultural products and led to the creation of government certifications to ensure that farmers are using these methods of production. These changes in the economic and social contexts impacted current agricultural science at both land-grant institutions and in the public sector, as they spurred an increased amount of research on organic methods of production, the reintegration of livestock into farming systems, mitigating the environmental impacts of farming, improving the long-term sustainability of diverse types of agriculture, and the various social and economic impacts of small-scale and large-scale operations (Delind 2011; Jackson and Jackson 2002). Certainly, as the needs of the local farming populations shift, land grant institutions are increasingly responding to these economic and social changes. This development is not surprising as, as mentioned above, a primary goal of the land grant institutions and research stations is to perform research useful for local farmers. Thus the types of research performed are meant to reflect the current needs of the larger community.

In reality, current agricultural science is doing just this, with American agricultural scientists conducting cutting-edge research on a multiplicity of topics now aimed at social goals beyond simply increasing production; social goals such as increasing the sustainability of farming methods, reversing or stemming climate change by trapping greenhouse gasses in soil, fighting desertification, and developing area-specific crops and farming methods to help fight hunger in areas with poor growing conditions, be those urban or rural areas. While the land-grant institution was seen in the past, at least by critics, as the wellspring for many of the current problems faced by American agriculture. Today the three-way partnership between universities, experiment stations, and extension service forms the backbone of current research that has far-reaching implications well beyond the United States' borders and is our best hope for addressing many "wicked" or overly complicated issues that we face today.

Bibliographic Essay

Agricultural science has a long history, beginning alongside the birth of empirical methods of knowledge production. In fact, it is one of the oldest applied sciences and for this reason literature on this topic is vast. However, when looking at the history of agricultural science both within the United States and beyond, there are clearly identified developments that served as historical turning points for the field. Thus, a student interested in gaining greater knowledge of the historical development of agricultural science can turn to the following literatures: (1) the history of agricultural development prior to the Enlightenment; (2) agricultural sciences during the Enlightenment; (3) the history of agricultural science in the United States; and (4) critiques and current literature in the field. Studying literature in these areas should provide readers with a detailed overview of the field and how it developed over time.

For example, when reading important texts in the history of agricultural development prior to the Enlightenment, readers should pay particular attention to the first attempts at using rudimentary empirical methods to improve agricultural practices and make note of the agricultural methods used during this time. In this time period, Xenophon and Aristotle both wrote extensively on agriculture. Aristotle's *Politics* includes sections on the soil and Xenophon's *Oeconomicus* discusses daily life on the farm (Strauss 1970). In addition, Roman texts, such as Columella's 12 volumes on agriculture, discussed various agricultural practices and gave descriptions of experiments conducted in these areas (Columella and Ash 1941). See Victor Hanson's (1999) *The Other Greeks: The Family Farm and the Agrarian Roots of Western Civilization* and Signe Isager and Jens Skydsgaard's (1995) *Ancient Greek Agriculture: An Introduction* for a detailed overview of agricultural developments in Greece, and Mark Tauger's (2010) *Agriculture in World History* for an overview of developments during the Classical period and beyond.

The second area of literature about agricultural science during the Enlightenment is also pivotal for understanding the historical development of the field, as this period can be understood as the birth of the modern agricultural sciences (Kingsbury 2009). Here scientific methods born out of the Scientific Revolution both increased agricultural production and shaped the disciplines that we know as agricultural sciences today. Important works during this period include those by various philosophers who wrote on the applications of technology and changing economic and social factors that influenced agriculture, such as John Locke, Jeremy Bentham, and John Stuart Mill and economists Adam Smith and Thomas Malthus. A third literature that is important during this period includes those by the chemist Justus von Liebig, the agronomist Jethro Tull, and the agrarian and politician Thomas Jefferson (Thompson and Noll 2014). Again, see Tauger (2010) for a general overview and Kingsbury (2009) for a superb history of the development of agriculture science during this time.

These first two areas of literature provide the background for a better understanding of the development of agricultural sciences in the context of the United States, as early work in agricultural science was influenced by European thinkers. Unlike Europe, however, agricultural science became institutionalized in the late nineteenth century with the birth of land-grant universities and the establishment of federally funded experiment stations (Hillison 1996). These developments helped form the triangular partnership (farmers, government agencies, and public sector research institutions) that MacKenzie (1991) argues supported cutting-edge agricultural research in the United States for the last 100 years, and continues to support it today. MacKenzie's (1991) essay "Agroethics and Agricultural Research" and Rosenberg's (1997) book *No Other Gods: On Science and American Social Thought* each provide historical overviews of this important period. In addition, Alan Olmstead and Paul Rhode's (2008) *Creating Abundance: Biological Innovation and American Agricultural Development* outlines the historical development of crop specific farming practices (such as tobacco, cotton, and dairy), and technological advances in the United States.

The fourth and final area of literature is the largest by far, as it encompasses key texts in both the numerous social and ethical debates and current developments in the field. Here it is important to note that it is simply not possible to list all of the current literature in agricultural science proper, as the subject is vast, with experts working in many different highly technical fields. However, it is possible to gain an understanding of the larger structure of the field and the social and ethical influences that helped shape current research, again, the literature is vast in this area and continues to grow. For this reason the following review will be cursory at best but is adequate for gaining a basic understanding of the field today.

Over the last 20 years, the agricultural sciences in general and the research priorities of these fields in particular have been subjected to growing criticism (Dahlberg 1988; Johnson 1984; Thompson, Ellis and Stout 1991). The result of these critiques was, in addition to work on increasing productivity and improving farming practices, the expansion of the scope of current agricultural research to include work in the neglected areas identified by the critiques. The work of Shiva (1992) on the Green Revolution, Pimbert's (2008) critique of global food systems, Mizelle's (2011) history of animal husbandry methods, the development of slaughter houses, and worker conditions, and Lyson's (2004) work on the connections between farms and community each provide interesting insights into these critiques. Also, Thompson, Ellis, and Stout's (1991) essay on "Values in the Agricultural Laboratory" and Thompson and Noll's article "Agricultural Ethics" (2014) provide excellent overviews of these critiques and outline how US agricultural scientists are rising to the challenge of addressing these social issues.

Other important works include those on specific subtopics in this field such as Kingsbury's (2009) *Hybrid: The History and Science of Plant Breeding,* which provides an exhaustive history of plant breeding that includes discussions on genetic modification technology, organic agriculture, and other important topics in current agricultural science; Tauger's (2010) *Agriculture in World History* that includes a chapter on current developments in American agriculture; and Fairlie's (2010) *Meat: A Benign Extravagance* that discusses various different farming systems used for both field crop and meat production, the accepted methods of measuring efficiency, and the political and normative influences that are currently shaping farming practices.

Chapter Two

ANTHROPOLOGY

Adrian Young

By the second half of the twentieth century, American anthropology was in a position of tremendous international prominence. Its practitioners fanned out to field sites around the globe, its university departments hosted the world's leading scholars, and its seminars trained hundreds of young anthropologists every year. But for all its successes, by another accounting the discipline was in a perpetual state of crisis. Its subdisciplines drifted inexorably apart. Linguists and archaeologists increasingly worked in separate departments, while physical anthropologists, now rechristened biological anthropologists, took on research problems that were often disconnected from sociocultural anthropology. Even within cultural anthropology, theorists struggled to find common ground – all the more so when waves of poststructuralist theory and postmodern critique broke over the field, prompting anthropologists to rekindle debates about whether or not their discipline was indeed a science at all. Historians of American anthropology must therefore account for twin currents: first, how did American anthropology professionalize into such a prominent academic discipline; and second, why, despite having achieved that status, did it seem at risk of fracture? This chapter will survey the history of American anthropology from its colonial antecedents to its late-twentieth-century forms, paying particular attention to the emergence of that fundamental paradox.

Because this chapter sweeps across generations and is embedded in a complicated national and intellectual milieu, some closer consideration of terms is in order. To begin with, what is "American" about American anthropology? The "science of man" was born in Europe but took root in the United States by growing within an ever-shifting national context. Anthropology's intellectual forbearers, working on the colonial periphery of the Enlightenment world, constructed a science that celebrated the American continent's uniqueness. But the field's institutionalization during the last two centuries was implicitly tied to quintessential American problems, not least American

A Companion to the History of American Science, First Edition.
Edited by Georgina M. Montgomery and Mark A. Largent.
© 2016 John Wiley & Sons, Ltd. Published 2020 by John Wiley & Sons, Ltd.

society's treatment of indigenous people, its history of slavery, its conception of race, its colonization of the interior, and its projection of international power.

Second, what, historically, constitutes "anthropology?" Though most historians avoid the use of anachronistic categories *avant la lettre*, writers have identified anthropological antecedents as distant as Herodotus and note that writing about other peoples has existed under the term since the sixteenth century (Harris 1968; Layton 1997; Liebersohn 2008). A recognizably modern anthropology, however, was a quintessentially Enlightenment science, elaborated as a philosophical and descriptive catalogue of humanity. Many point to figures such as Kant, Herder, and Montesquieu as intellectual founders (Cipolloni 2007; Denby 2005; Zammito 2002). Nevertheless those figures whom we often acknowledge as incipient anthropologists referred to themselves not only as anthropologists but also as ethnologists, ethnographers, geographers, explorers, philosophers – and sometimes avoided academic appellations altogether. It was only as "anthropology" professionalized that these diverse fields coalesced under the auspices of a single department, though even into the twentieth century their unity remained as much an ideal as it did disciplinary reality. Accordingly, this chapter will treat anthropology historically, sensitive to its contingent forms and contexts.

Philology and Native Americans in the Early Republic

Writing about the new world's indigenous peoples began almost from the initial moment of contact. An early generation of historians of anthropology recognized in these colonial texts the origins of a later anthropology (Hodgen 2011; Rowe 1965). However, the search for disciplinary ancestors too often and too easily obscured the historical reality of early modern America; missionaries and colonial writers recorded Native American bodies, languages, artifacts, and rituals for their own reasons. Later histories have accordingly revised our understanding of early modern writing about travel and the colonial encounter by analyzing proto-anthropological texts on their own terms (Campbell 1999; Carey 2003).

Writings about indigenous peoples during colonial and early republican years centered on the question of the American continent's fundamental distinctiveness and vitality (Ellingson 2001; Gerbi 1973). European theorists from Rousseau to Buffon derided the new world and its indigenous people as younger, more savage, and less vibrant in comparison with the old world. Montesquieu's theory of environmental determinism suggested that even transplanted Europeans, and by extension their republican experiment, were doomed to eventual failure in a land that was naturally inimical to civilization. American writers took it as a national mission to refute that notion. Perhaps the best known rejoinder was Thomas Jefferson's *Notes on the State of Virginia*, a wide-ranging treatise on the resources, peoples, and government of his home state (Jefferson 1787). Though reliant on the trope of noble savagery and confident of non-Europeans' innate inferiority, it nonetheless offered a descriptive catalogue of Native American cultures that marked them as objects for study. Jefferson also promoted the scientific exploration of the continent and its people; the Lewis and Clarke expedition, which he instructed to study "the moral and physical circumstances which distinguish the Indians encountered from the ones we know," was only one of many scientific voyages sponsored by the nascent republic (Chamberlain 1907: 509).

Much of the early republic's intellectual interest in Native American life centered on language. American linguistic anthropologists have found ancestors in figures such as Harvard classicist John Pickering (1777–1846), Philadelphia naturalist Benjamin Smith Barton (1766–1815), president of the American Philosophical Society Peter Stephen du Ponceau (1760–1844), and politician and founder of New York University Albert Gallatin (1761–1849). They produced some of the first treatises on Native American languages, though like most of the period's "armchair" scholars, they relied not only on their own experience but also on go-betweens; a network of Moravian missionaries including David Zeisberger (1721–1808) and John Heckewelder (1743–1823), for instance, corresponded with members of the American Philosophical Society about the language of the Delaware people next to whom they lived (Koerner 2004). Though it is tempting to conceive of their work as a proto-linguistics, these writers instead understood their work as philology, especially after the reception of Sir William Jones's (1746–1794) comparative method. American philologists identified Native American languages not only as ready objects for comparison across continents and cultures, but also as evidence of other cultures' philosophical and moral qualities (Gray 1999). Du Ponceau, Gallatin, and their contemporaries described American Indian languages as complex and capable of supporting advanced thought – and thus took a political stance by demonstrating a capacity for "civilizing."

Questions about Native Americans' languages and their aptitude for acculturation were particularly salient ones in the early republic. As the nascent government asserted a violent Indian policy that wavered between coerced assimilation and forced removal, another cohort of "Indian experts" elaborated less universalist and more racialist theories (Harvey 2000). Following brutal frontier wars, the Office of Indian Affairs was created in 1824 under Thomas L. McKenney (1785–1859) as a branch of John C. Calhoun's War Department. It and its successor institutions employed or collaborated with many of the nineteenth-century's prominent experts on Native peoples. The office worked with Lewis Cass (1782–1866), a Michigan politician who led expeditions to the west to search out the source of the Mississippi and wrote about the territories' Native peoples. McKenney also promoted ethnological investigations by Indian agents, especially Henry Rowe Schoolcraft (1793–1864), who with Cass laid the intellectual foundations for the notion that there was a "fixed Indian mind" ultimately unsuited to civilizational progress. It was the beginning of a long and complex relationship between intellectual inquiry, government institutions, and American expansion.

The American School of Ethnology

Albert Gallatin co-founded the American Ethnological Society in 1842, alongside fellow philologist John Russell Bartlett (1805–1886). Ethnologists were interested in language, as they were in all aspects of a peoples' character. However, it was the racialized body that became American ethnology's foremost object of study during the century's middle decades, especially as the science became increasingly embedded in debates about slavery. John Collins Warren (1778–1856), the first dean of Harvard Medical School, wrote on comparative anatomy and Egyptian mummies; his collections became the Warren Anatomical Museum. The most successful of the period's skull collectors, however, was Samuel George Morton (1799–1851), professor of anatomy at the University of Pennsylvania. Morton built up a collection of human crania donated from around the world – many skulls came from Seminole people killed by American soldiers (Fabian 2010; Strang 2014).

Morton catalogued and described his specimens in *Crania Americana* (1839), which became a classic text in nineteenth-century ethnology. In it he argued that humanity was divided into a typological hierarchy of races. The question of how the "races" were related was already an old one in western science; theorists typically fell into one of two broad camps. Monogenists contended that all of humanity was descended from a single set of ancestors in accordance with biblical accounts. Polygenists argued instead that the "races" constituted fundamentally distinct species. Morton was a vehement polygenist, holding not only that "races" were created separately but that on the basis of their morphology they were discernibly inferior.

Morton's polygenesis found widespread support in the antebellum United States, as well as some sympathy abroad. His most notable supporter was Harvard geologist Louis Agassiz (1807–1873), who maintained his adherence to Mortonian polygenesis well after the Darwinian revolution. In the south, his chief adherents were Josiah Nott (1804–1873), an Alabama surgeon and slave owner, and George Gliddon (1809–1857), a diplomat and Egyptologist. Together they wrote *Types of Mankind* (1854), which widely publicized Morton's ideas and brought scientific support to pro-slavery arguments. Morton, Nott, Gliddon, and Agassiz have been treated glossingly, ignored, or justifiably condemned in histories of anthropology, most famously by Stephen Jay Gould (1981). Nevertheless, understood historically, the American School was among the nineteenth-century United States' most prestigious scientific products and played a crucial part in the formation of American anthropology (Dewbury 2007). Indeed, Aleš Hrdlička, himself a founder of modern physical anthropology, wrote in his 1914 history of the field that Morton "may be termed justly and with pride the father of American anthropology" (Hrdlička 1914).

Nineteenth-Century Professionalization

For all their repute or infamy, Morton and Gallatin were not professional anthropologists. It was not until the latter half of the nineteenth century that scholars secured paid positions to study humanity under that appellation. American anthropology professionalized with the support of federal institutions and in the context of continuing westward expansion; it was only after an initial echelon of government anthropologists built the foundations of the discipline that it found a home in universities at the end of the century.

The most crucial early institution for anthropology was the Smithsonian's National Museum, founded in 1846. The Smithsonian became a clearinghouse for research and grew to become one of the nation's foremost anthropological institutions in the years after the US Civil War (Bieder 2003; Hinsley 1981). As western settlement continued, explorers, soldiers, and settlers sent a steady stream of ethnological material to the Smithsonian, ranging from word lists to human remains. Among its most prolific collectors was army major John Wesley Powell (1834–1902), who led a government survey of the Colorado River and wrote about the Indian nations of the southwest (Powell

1875). Powell took over the US Geological Survey and eventually founded the Bureau of Ethnology (later the Bureau of American Ethnology, BAE) in 1879.

Under the imprimatur of the BAE, ethnologists fanned out across the country to collect and catalogue Indian languages and cultures (Powell 1877). It was not an unpractical project in the government's eyes; the Bureau's classifications of Native people rendered them into known, administrable units. Though Powell and his contemporaries were often sympathetic to those they studied, their work nonetheless corroborated the ideology of manifest destiny. Following British theorists Edward Burnett Tylor (1832–1917) and Herbert Spencer (1820–1903), Powell and contemporaries like Lewis Henry Morgan (1818–1881) subscribed to theory of sociocultural evolution that ranked races on a civilizational hierarchy (Hume 2011; Powell 1888). It was an intellectual stance that both validated and was validated by the steady encroachment of American settlement on Native land; races which were understood to occupy lower rungs of the evolutionary ladder were inevitably "uplifted" by superior races or annihilated by them (Brantlinger 2003).

Professional anthropology moved into American universities through museums. Building on the work of the Smithsonian and BAE, universities began to set up their own ethnological collections (Stocking 1988). Their museum anthropology was supported by largesse of gilded age philanthropists. Harvard's Peabody Museum, for example, was founded in 1866 through the patronage of banker George Peabody. It became a prominent research center under the direction of Frederick Ward Putnam (1839–1915), who expanded its collections and began publishing the *Papers of the Peabody Museum*. Following his success in Massachusetts, Putnam went on to establish a number of anthropology museums and departments (Browman 2002). In Chicago, he built up the Field Museum's collections in association with the 1893 World's Fair. Anthropology was often involved in such expositions, which incorporated displays of indigenous peoples and artifacts into its celebrations of manifest destiny (Parezo and Fowler 2007). When disagreements over the fair's organization drove him away, Putnam left to set up the department of Archaeology and Ethnology at the American Museum of Natural History, which together with Columbia made New York a center of American anthropological research. In 1903, Putnam again moved on to found Berkeley's anthropology department and museum with funding from the family of William Randolph Hearst. It was these university–museum pairs, in addition to the University of Pennsylvania, which founded its own museum in 1887, which fostered the field's first cohorts of students and employed many of its first PhDs.

Boas and the Boasians

The first professor of anthropology in the United States was Franz Boas (1858–1942). His status in the history of modern American anthropology cannot be overstated, though the historiography certainly has tried. As the discipline forged its professional and intellectual identity across the twentieth century, it looked to Boas as its intellectual and spiritual founder – his anti-racialist politics and repudiation of sociocultural evolutionism made him conducive to hagiography. But Boas's legacy looms large for more pragmatic reasons, too. Crucially, he was a prolific advisor and his students came to occupy key positions in the growing American university system (Darnell 1998). *Totems and Teachers*, an edited volume offering biographies of foundational anthropologists, includes only two figures who were not students of Boas, or students of his students (Silverman 1981). Moreover, Boas promulgated a broad, interdisciplinary, "four-field" science of humanity that was meant to encompass not only sociocultural anthropology, but also archaeology, physical anthropology, and linguistics (Boas 1904). It was a configuration that granted American anthropology an ambitious ambit, even if only as an ideal.

Anthropology's memory of Boas and his disciples has long been colored by presentist concerns. Because of his status as founding father, later anthropologists pointed to Boas to critique contemporary issues within the discipline (Lewis 2008; Pinkoski 2011; Verdon 2007; White 1968). The question, framed in terms of Thomas Kuhn's philosophy of science, of whether or not anthropology had a "paradigm" and thus produced "normal science," was an active one (Kuhn 1962). Depending on one's commitments, Boas offered either pre-paradigmatic, atheoretical descriptivism or a fully paradigmatic, scientific research program (Darnell 1998, 2001; Stocking 1966). So central is Boas in the anthropological tradition that historians have warned that our telling of anthropology's past might now be distorted, and indeed more critical historical work has sought to reinsert Boas into his intellectual, political, and social context (Castañeda 2003; Stocking 1996). Though Boas was a pivotal figure, it is dangerous to hold Boas, Americanist anthropology, and American anthropology itself as synonymous. That metonymy gives way easily to elision, leaving other traditions in the discipline's past underwritten.

Boas himself was born to a Jewish family in Germany and received a PhD in physics from the University of Kiel in 1881. However, he maintained an interest in geography and, after graduating, conducted an expedition to Baffin Island from which he produced a habilitation thesis at Berlin's Ethnologisches Museum. He emigrated to the United States, taking up a professorship at Clark University, an editorship at the journal *Science*, and fieldwork in the Pacific Northwest. Intellectually, Boas articulated a vision of geography, ethnology, and anthropology as historical sciences, operating at a remove from the epistemological commitments he had absorbed as a physicist while embracing the romanticist tradition of Herder and Humboldt (Stocking 1996). He rejected the period's dominant evolutionist theories of culture in favor of a more nuanced cultural relativism and historical particularism. Societies did not develop along a ranked axis; neither were they solely determined by fixed environmental factors. Rather, he regarded cultures historically and psychologically according to their own particular circumstances (Boas 1911).

In 1899, Boas became a professor of anthropology at Columbia University. There, he trained a profoundly influential cohort of anthropologists. They went on to helm university departments of their own and, collectively, they marked sociocultural anthropology, folklore, and linguistics with a Boasian stamp. His first student was Alfred Kroeber (1876–1960), who went on to head the recently created department and museum at the University of California, Berkeley (Jacknis 2002). Kroeber's fieldwork in the American southwest was wide ranging, in the Boasian mold, and included archaeology and linguistics. He was especially noted for his relationship with Ishi, the last known surviving member of California's Yahi people (Kroeber 2004; Sackman 2010). Theoretically, Kroeber built on Boas's historical particularism to espouse the centrality of "culture" as the quintessential unit of anthropological analysis. "Culture," for Kroeber, was "superorganic," an entity that existed beyond biology or nature (Kroeber 1917).

Other Boasians went on to similar acclaim and success. Edward Sapir (1884–1939), like Boas himself a Jewish immigrant, became a foundational figure in anthropological linguistics. He conducted much of his early fieldwork in California, some of it with Kroeber's informant Ishi, and in Utah, where he developed a close relationship with his own Paiute informant Tony Tillohash (Darnell 1990). On the basis of that work, Sapir argued that, psychologically, language and expression functioned differently across cultures, an approach he and his student Benjamin Lee Whorf expanded into a theory of linguistic cultural relativity (Koerner 1992; Whorf 1956). Many of Boas's other students also expanded on their mentor's relativistic approach with similarly fruitful results. Building on the psychological orientation of Boas's *The Mind of Primitive Man* (1911) Paul Radin celebrated the intellectual capacity of indigenous thought in his broadly comparative *Primitive Man as Philosopher* (1927). Their work reflected not only Boas's sensitivities, but also registered the impact of Freudian psychoanalysis on segments of the American academy. Ruth Benedict gave cultural relativism a near-perfect metaphor in her popular *Patterns of Culture*, which toppled the image of the evolutionary ladder on its side into a great "arc of culture" encompassing every possible expression of human interest or activity, no one more "civilized" than the next (1934: 24). Benedict herself became a mentor to one of Boas's last students, Margaret Meade. Her doctoral fieldwork and resulting monograph, *Coming of Age in Samoa*, suggested that adolescence was culturally contingent and captured a massive public readership, making Mead into perhaps the most well-known anthropologist in the world (Mead 1928).

Collectively, it is worth noting that many of the Boasians were immigrants, women, Jewish, or some combination of the three (Bernstein 2002). It was a background which, coupled with their radicalism and anti-racialist politics, granted them a status as revolutionary outsiders. Indeed, within the context of the early twentieth century, their work held political and cultural implications. In the face of anti-immigrant sentiment, Boas himself took on anthropometric studies of recent arrivals to the country and their children in order to discount racialist theories (Boas 1912). Combating the rise of anti-Semitism, Boas and his students pressured American anthropology to denounce officially typological racial theories as unscientific (Barkan 1992). Melville Herskovits (1895–1963) conducted fieldwork in East Africa and Haiti before founding the anthropology department at Northwestern and, arguably, founding African and African American Studies as an academic field (Gershenhorn 2004; Herskovits 1926, 1941). And, of course, Benedict and Mead became feminist icons (Lutkehaus 2008).

Hooton and the Hootonians

The profound attention afforded to Boas and the Boasians has left other corners of anthropology's history relatively less examined. Among them is physical and biological anthropology, particularly during the twentieth century. Boas's vision was expansive enough to encompass four fields. But because he produced so many talented students of culture, his influence on physical anthropology was lest manifest, even if Boas himself did anthropometric work (Little and Kennedy 2009). Nonetheless, physical anthropology has a complex twentieth-century history, one in which it struggled to reinvent itself in the face of insights from genetics and in response to shifting attitudes toward race in American science and society.

Modern American physical anthropology's first professional leader was Aleš Hrdlička (1869–1943). An immigrant from Bohemia, Hrdlička pursued a medical education in New York before taking up anthropological research. He became a curator at the Smithsonian, where he exerted a profound influence on the shape of the discipline in the United States (Spencer 1979). Though he had no direct students, he created many of the subfield's professional institutions and networks. He established the *American Journal of Physical Anthropology*, which he envisioned as the country's premier journal for the "study of racial anatomy, physiology, and pathology" (Hrdlička 1918: 4) He also methodically expanded the Smithsonian's collection of human remains, both through expansive professional networks and during his own expeditions exhuming bodies without permission from the graves of Native people (Fabian 2010). By the end of his tenure, the Smithsonian's holdings in physical anthropology were unmatched anywhere in the world. Those collections were in service of a perennial project in American anthropology; with them Hrdlička and his collaborators worked to identify connections between American Indian nations and establish the patterns of their ancestors' migration into the Americas.

As anthropology found an expanding home in the universities, Earnest A. Hooton (1887–1954) emerged as a prominent academic leader in physical anthropology. Hooton earned a doctorate in classics at the University of Wisconsin before taking up a professorship at Harvard, where he became an authority on human evolution. His legacy, however, remains burdened by his views on the biological basis of race and his vociferous support for the eugenic movement. While sociocultural anthropology increasingly adopted relativistic stances and abandoned race as a category of analysis, physical anthropology remained wedded to race as its preeminent object of study; the questions of how races originated and were defined constituted its fundamental inquiries. Hooton became the country's most authoritative voice on questions of race, publishing a series of widely read books (Hooton 1937, 1942). Though he joined Boas in offering a moderated denunciation of German race theories, it nonetheless drove his popular and academic work (Hooton 1936).

Hooton's most significant contribution to the field was his mentorship of graduate students. If Boas cemented his hold over sociocultural anthropology through the influence of his academic progeny, then Hooton was his mirror in physical anthropology. His graduate students founded or influenced programs across the country. His first graduate student, Harry Shapiro (1902–1990) became curator of anthropology at the American Museum of Natural History, where he began a long career that included prominent studies of race mixture in the Pacific (Shapiro 1929). Another prominent student, Joseph Birdsell (1908–1994), spent decades conducting work among Australia's aboriginal peoples (Birdsell 1953). But perhaps most crucially, it was Hooton's disciples who articulated new research programs and a new disciplinary identity when scientific theories of race became increasingly disreputable.

After World War II, anthropologists pushed heavily against the typological and hierarchical race theories on which physical anthropology was based. Their efforts led to the publication of the UNESCO statement on race in 1950 (Brattain 2007). At the same time, new theories of population genetics offered by biologists such as Theodosius Dobzhansky (1900–1975) made typological theories of race less tenable. It was a moment of crisis for physical anthropology; in the face of both scientific work and social pressure, their longstanding paradigm was dissolving before their eyes. Hooton's students responded by modifying the race concept. Harry Shapiro authored a second UNESCO statement on race in 1951 that moderated the first statement's claims (Shapiro 1952). Another Hooton student, Sherwood Washburn, called for a "new physical anthropology" that would remake itself to be more relevant to the rest of the anthropological community (Mikels-Carrasco 2012; Washburn 1951a). Other students drew from genetics to articulate a new conception of "races" as variable sets of populations. William Howells (1908–2005) dispensed with typological, polygenic race theory after conducting comparative craniometrical work on populations around the world (Howells 1973). Stanley Garn (1922–2007) published *Human Races* in 1961 to bring physical anthropological understandings of "race" into conversation with insights from biology.

The reconfiguration of physical anthropology by Hooton's students was, as an intellectual project, not divorced from the context of the civil rights movement. That context was especially visible in the case of Hooton disciples who did not sufficiently divest themselves of their mentor's racialism. Carleton Coon (1904–1981) incited a storm of controversy when his 1962 book argued that human races constituted separate strands of human evolution, and that those strands had crossed the threshold of hominization hundreds of thousands of years apart (Coon 1962; Jackson 2001). Coon died bitter and alienated from his field; the model of the discipline he had inherited from his mentor was simply untenable in the postwar world.

New Currents and New Fractures

The decades following World War II witnessed profound growth and change in American anthropology. The federal government had utilized the services of anthropologists during the war itself, producing a sometimes troubled relationship that continued well after 1945 (Price 2008). Ruth Benedict was commissioned to produce an accessible study of Japanese culture, *The Chrysanthemum and the Sword* (1946) and many social scientists conducted fieldwork on Pacific Islands under American occupation. But perhaps the war's most meaningful impact was indirect. The postwar expansion of American tertiary education allowed a marked increase both in the number of anthropology departments and the number of doctoral students they produced. In 1950, American universities produced 22 PhDs in anthropology. In 1974, they produced 409 (Givens, Evans, and Jablonski 1997).

While many of those expanded departments were set up along Boasian, four-field lines, the research within them was becoming increasingly fractured. Not only did the subfields become more separated as each specialization developed its own traditions, but even within the subfields massive rifts began to form. Sol Tax, surveying the state of the field in 1955, asked: "can such heterogeneity be maintained in a single discipline? What do techniques in linguistics have to do with primates, or primates with style or cultural values?" (p. 313).

Many postwar anthropologists made bids for synthesis; Washburn's "New Physical Anthropology" was only one of many attempts at renewed cohesion. Some sociocultural anthropologists turned once more toward materialism, accommodating perspectives from biology, economics, and ecology. Julian Steward (1902–1972) pioneered the study of cultural ecology, developing a systematic, cross-cultural theory of human adaptation built on studies of subsistence regimes (Steward 1955). Leslie White (1900–1975) took that move several steps further. Consciously echoing nineteenth-century social evolutionists like Tylor and Morgan, he defined "technology" and "culture" as humankind's ability to harness the available energy within its environment (White 1959). During the 1960s and 1970s, ecologically grounded analyses became especially prominent and aimed to bring together the work from across the subdisciplines. *Man the Hunter* (1968), for instance, was a massive conference and influential edited volume by Richard Lee and Irven Devore which focused attention on hunting as a "single, crucial stage of human development" and brought together perspectives from sociocultural anthropology, ecology, archaeology, physical anthropology, and even primatology.

Other schools countered the primacy of culture and instead made the social their chief object. The anthropology department at the University of Chicago had long operated outside the Boasian orbit, despite the influence of figures like Sapir. In addition to non-western, indigenous cultures, Chicago's anthropologists studied developing societies in urban, immigrant, and peasant communities. Robert Redfield (1897–1958) made Latin American peasants a career-long preoccupation. Informed by British functionalism, he examined the village not as an isolated cultural unit but instead as a micro-society within a larger system and subject to external forces (Redfield 1930, 1956; Wilcox 2006). Those who followed him adopted more materialist, even Marxist approaches. Julian Steward supervised a number of PhD students at Columbia and Michigan, including Eric Wolf (1923–1999) and Sidney Mintz (1922–), who produced magisterial, Marxist anthropologies that understood local societies within global economies (Mintz 1985; Wolf 1982). That some scholars turned to social anthropology and especially to peasant studies was not unrelated to postwar geopolitics. As American power and American capital expanded its influence in Latin America and the world, Marxist critiques became especially salient.

In still other corners of the discipline, it was neither Marxism nor British functionalism but rather French theory that found ready admirers. Claude Lévi-Strauss (1908–2009) produced a series of broadly comparative studies that revealed the root structures and relationships underlying much of human society, culture, and thought (Lévi-Strauss 1949, 1963). His "structural" anthropology demanded that cultural phenomena be understood not according to their social function, as in British social anthropology, or by their particular history, as in the original Boasian program, but rather by codifying their interrelations and meanings within larger systems. Lévi-Strauss was followed by a wave of French writers, including Michel Foucault (1926–1984) and Pierre Bourdieu (1930–2002), who offered poststructuralist explications of human society, expression, and history. As French semiotics found ready adherents in the United States, David Schneider (1918–1995) at Chicago and Clifford Geertz (1926–2006) at the Institute for Advanced Studies in Princeton developed a "symbolic anthropology," which redefined culture to understand it as a "web of signification" that gave sense to human action (Geertz 1973). The anthropologist's role was to serve as an interpreter of these symbols and meanings.

Of course a complete cataloging of the state of recent anthropology's many subfields and schools would be impossible here. By the end of the twentieth century, anthropologists contributed to gender and performance studies, practiced medical anthropology, and revived applied anthropology. A critical feminist anthropology emerged to critique the discipline's social and intellectual foundations. A renewed interest in "material culture" connected the thinking of archaeologists and cultural anthropologists. Genomics revolutionized biological anthropology and paleoanthropology, and the Chomskyan revolution forever changed the orientation of much of American linguistics. Nonetheless, in many universities these myriad approaches to the study of humanity remained the purview of a single department and, however illusory their cohesion might be in practice, their diverse ambit was the legacy of American anthropology's wide-ranging origins and ambitions. As historian of anthropology James Clifford wrote in 1986, "few today can seriously claim that these fields share a unified approach or object, though the dream persists" (Clifford and Marcus 1986: 4).

Bibliographic Essay

Anthropologists, perhaps more than the practitioners of most other fields, are deeply invested in their own history. That investment inflects scholarship on the history of their discipline; many of the first histories of anthropology were written by professional anthropologists who interpreted the past in light of their present-day concerns. Boas (1904), Hrdlička (1918), and Kroeber (1959), for instance, each penned histories of their discipline in which they laid out their own visions not only of its past, but also of its future. Anthropologists have long been fascinated by ancestors and lineages, none more so than their own. Much of anthropology's history exists as what Henrika Kuklick often called "folk tradition" and the "oral tradition of the seminar room" (Kuklick 2006: 538, 2014: 76). Some traditions were transferred textually too, in the form of books assigned to students in history of anthropology seminars. Marvin Harris's *Rise of Anthropological Theory*, affectionately known as RAT, has been a mainstay for decades (Harris 1968).

Dedicated scholarship on the history of anthropology by anthropologists emerged during the 1960s. The Social Science Research Council sponsored two symposia on the discipline's history in 1962, leading to the publication of Irving Hallowell's famous "The History of Anthropology as an Anthropological Problem" (1965). Hallowell's student, George Stocking, produced a lifetime of scholarship on the history of anthropology. His essay "On the Limits of 'presentism' and 'historicism' in the Historiography of the Behavioral Sciences" (1965) laid out a careful critique of the presentist position, and his *Race, Culture, and Evolution* (1968) remains a classic. Stocking became the dean of history of anthropology in the United States, founding the *History of Anthropology* Newsletter and organizing one edited volume after another (Stocking 1984, 1987, 1988, 1990, 1992, 1996, 2010). Another key figure in anthropology is Regna Darnell, who began studying its professionalization in the United States and Canada soon after Stocking (Darnell 1971, 1990, 1998, 2001).

That is not to suggest that intradisciplinary, presentist perspectives could not produce vital historical scholarship. In the last decades of the twentieth century, anthropology underwent a "reflexive turn" in which anthropologists reconsidered the nature of their method and scrutinized the discipline's politics and history (Trencher 2000). Trenchant criticisms of anthropology's situatedness in the colonial project appeared, beginning with Talal Asad's famous edited volume, *Anthropology and the Colonial Encounter*, in 1973. Johannes Fabian offered a meta-critique of anthropology in his *Time and the Other* (1983), while James Clifford and George Marcus considered anthropological research as a literary form in *Writing Culture* (1986). Clifford himself spent a career producing a theoretically engaged history of anthropology; *The Predicament of Culture* (1988) is a standard point of departure. Feminist critiques also appeared; Clifford's colleague at the History of Consciousness program at UC Santa Cruz, Donna Haraway, produced a masterful, feminist history of primatology in *Primate Visions* (1989). Ruth Behar and Deborah Gordon noted the absence of women in *Writing Culture* in their feminist *Women Writing Culture* (1995).

Contemporary scholarship has looked beyond the history of theory to consider the diverse contexts and politics in which the study of humanity has occurred. Thomas Patterson's *A Social History of Anthropology in the United States* (2001) provides a useful survey of American anthropology's social and political situatedness. Richard Handler, picking up Stocking's torch, has published edited volumes (2004, 2006) that reflect the diversity of interest in the history of anthropology. *Critical Studies in the History of Anthropology* and *Histories of Anthropology Annual*, both published by the University of Nebraska Press, also reflect the range of possible approaches. New scholarship informed by the history of the body continues to unpack the history of scientific racism in anthropology; see Tracy Teslow's *Constructing Race* (2014) and John Wood Sweet's *Bodies Politic* (2007). Andrew and Harriet Lyons wrote a unique book on the *History of Anthropology and Sexuality* (2004). Biographies of anthropologists remain perennially plentiful; Nancy Lutkehaus's treatment of Margaret Mead and her legacy is a particularly good example (2008).

So varied is the recent scholarship in its interests that Henrika Kuklick could label it a "New History of Anthropology" in her edited volume of the same name (2008). It is hoped that future work will continue to look beyond the usual Boasian suspects, producing historically grounded scholarship that reveals the true breadth and scope of the discipline's multi-century past. Kuklick, herself a specialist on Anglo-American anthropology, called for historians to look beyond their national contexts. As the historiography absorbs more postcolonial perspectives, the place of indigenous subjects, go-betweens, and informants like Ishi and Tillohash, should see still more attention. And, regardless of anthropology's internal debates over its status as science, perspectives from the history of science and science studies are too infrequently applied and could reveal a great deal about the development of the discipline's practices, truth-claims, and epistemologies.

Chapter Three

ASTRONOMY AND ASTROPHYSICS

Peter J. Susalla

Astronomy is one of the oldest scientific disciplines and the field with perhaps the longest history in North America. Interest in the sky and celestial objects was prevalent among the numerous indigenous American cultures, and Western astronomical practice – in the form of navigational techniques – arrived with the first European explorers and colonists in the fifteenth and sixteenth centuries. After the founding of the United States, astronomy was one of the first fields in which Americans gained international acclaim. Even though it lacked permanent astronomical facilities in the 1820s, by the end of the nineteenth century American observatories equaled and then exceeded those in Europe and American astronomers took the lead in developing the new field of astrophysics into the twentieth century.

Current historiography allows us to categorize the institutional, conceptual, and practical development of Euro-American astronomy into three eras. In the first three centuries of European colonization (i.e., *c.* 1500–1800) historians have shown American astronomy to have been a combination of imperial endeavor – a tool for navigating, surveying, and conquering the continent – and deeper undercurrent of interest in the workings of the natural world. Although much scholarship has focused on the work of key individual astronomers, the growth of domestic publications, such as almanacs, has revealed a broader cultural interest in astronomy. The most noteworthy American astronomers immediately before and after the Revolution worked primarily in "practical" astronomy, which, in the nineteenth-century meaning of the term, meant observational work with applications ranging from surveying to celestial mechanics (Warner 1979). The popularity of almanacs and widespread cultural fascination with celestial phenomena, such as the appearance of prominent comets, show that astronomy attracted more than just the interest of the elite. However, astronomical research did not extend much beyond elite circles or colonial surveying programs, nor was there

A Companion to the History of American Science, First Edition.
Edited by Georgina M. Montgomery and Mark A. Largent.
© 2016 John Wiley & Sons, Ltd. Published 2020 by John Wiley & Sons, Ltd.

much institutional support for such research, whether in the form of observatories, astronomical associations, or government patronage.

The first substantial period of growth in American astronomical research began in the first half of the nineteenth century, represented by a well-studied observatory-building movement that began in the three decades prior to the American Civil War. Most of these observatories were funded by groups of astronomically interested citizens and were frequently associated with the colleges and universities that proliferated as the young nation expanded westward. As the American economy grew prodigiously and industrialized following the Civil War, so too were astronomers able to convince the captains of industry to invest their money to build ever-larger telescopes, first near the burgeoning urban areas in the East and then in the clearer mountain air of the American West. An additional aspect of the prominence of American astronomy at the turn of the twentieth century was not only large telescopes but also American expertise in the new instrumental and conceptual techniques of astrophysics, a field that had centered on analyzing the light of stars and other celestial objects with a trio of new devices: the spectroscope, the photometer, and the photographic camera. These institutional and research-related developments coincided with an expansion in the size of the American astronomical community and opened new avenues for professional organization and communication.

The observatory movement of the nineteenth century was principally driven by private monies, although one major, publically funded astronomical institution, the US Naval Observatory, was established during an era when Congress was hesitant to devote government funds to scientific institutions that had no explicit practical, economic, or military function. By way of contrast, all aspects of American astronomy in the twentieth century were strongly shaped by federal funding, especially after World War II. The National Science Foundation (NSF) supported both traditional optical observatories and new radio telescopes; the military and America's civilian space program, the National Aeronautics and Space Administration (NASA), helped astronomers to explore bands of the electromagnetic spectrum invisible from the earth's surface (Smith 1997). As observing programs extended along the electromagnetic spectrum, so too did the horizons of astronomical research expand dramatically. As older fields such as planetary astronomy and stellar astrophysics continued to flourish, the discovery of galaxies in the early decades of the century and the cosmic background radiation in the 1960s formed the conceptual core of new cosmological research programs that represented a significant fraction of astronomical research in the year 2000.

The Indigenous Period: Native American Astronomy

The focus of this chapter is the history of astronomy within the portion of the North American continent that became the United States, yet it is important to recognize that practices we might label "astronomy" were important to many of the cultures indigenous to the territory. These practices have recently attracted significant scholarly interest. Although it is impossible, given the diversity of native cultures, to speak of a single "Native American" astronomy, scholars have identified three primary sources of knowledge we can use to make a very brief foray into North American indigenous astronomy. The first and most widespread of sources are the oral traditions that continue to play important roles within Native cultures, often taking the form of accounts of celestial creation and the ordering of the world (that is, cosmologies). Second, and perhaps most controversial, are the physical structures that scholars suggest were built to align with certain celestial phenomena, for instance, the "woodhenges" of the city of Cahokia on the Mississippi River that purportedly mark the position of the sun on important days (Krupp 1994). Finally, we can find astronomical imagery in indigenous material culture, such as the striking star maps of the Skidi Pawnee of the western part of the continent (Chamberlain 1982).

The study of indigenous astronomy has recently acquired the label "cultural astronomy," a sometimes uneasy meeting point between "archaeoastronomers," scholars interested in structural alignments and drawing from a Western tradition of mathematical astronomy, and "ethnoastronomers," cultural anthropologists interested in the cosmologies of contemporary indigenous cultures. Anthropologists have criticized the archaeoastronomical quest for alignments for geometrizing and abstracting structures from their cultural contexts. The existence of objects such as animal-skin sky charts indicates that the pattern of stars held significance to certain cultures, and there were undoubtedly physical sites, like the petroglyphs at Fajada Butte in New Mexico, that marked or represented a link between earth and sky (Carlson 2008). However, as Keith Kintigh has argued, Native American interest in the sky and cosmos "will often have precious little to do with the Sun and stars and lots to do with things like causality and gender differences," that is, a gulf exists between Euro-American and indigenous understandings of nature (Kintigh 1992). Recent controversies over the building of observatories on mountains held sacred by cultures such as the Apache show that distance remains between these worldviews (Swanner 2015).

Colonial American Astronomy

Astronomical practice among the first European explorers to reach North America was largely a practical concern, a set of navigational techniques for exploring the unknown territory. Following the establishment of permanent European settlements along the eastern North American coast, the availability of print matter provides insight into how colonists thought about the universe and shows that the practical value of astronomy was significant but not exclusively so. Almanacs were among the earliest and most popular forms of printed material in the colonies, first appearing in the mid-seventeenth century. A combination calendar and ephemeris, frequently accompanied by prose and poetry, almanacs are one of the most significant sources of astronomical and cosmological information in the American colonies. In addition to their practical functions as timekeepers and guides to the stars, almanacs also revealed perspectives on God's role in the natural world (for instance, the providential purpose of comets) and changing cosmological ideas in the gradual shift from Ptolemaic to Copernican and Newtonian conceptions of the cosmos. Yet since astrology and Ptolemaic cosmology remained important features of colonial almanacs into the eighteenth century (Gronim 1999; Hall 1990; Leventhal 1976; Ruffin 1997), future scholarship will continue to examine how the wider American reading public understood the transition from the old cosmos to the new.

Among the colonial elite, astronomical discourse centered on colleges and universities, most notably Harvard. Harvard affiliates kept abreast of European astronomical and cosmological ideas and joined the discourse of European scientific societies, such as England's Royal Society. Notable early figures include Harvard treasurer Thomas Brattle (1657–1713), whose observations of the Comet of 1680 were mentioned by Newton in the *Principia*, and John Winthrop (1714–1779), who received funding from the colony of Massachusetts to travel to Newfoundland to observe the 1761 Transit of Venus, one the earliest scientific research projects financed by an American government. By the American Revolution, elite colonial astronomers also had a local source for such discourse through the American Philosophical Society, which coordinated colonial observations for the 1769 Transit of Venus (Greene 1954). As John C. Greene (1954) points out, not all the American astronomical elite of the colonial period were academics. David Rittenhouse (1732–1796), one of the most prominent astronomers of the colonial period, was a self-educated craftsman, among the earliest makers of American telescopes, and builder of one of the first American astronomical observatories, a structure to house instruments for the 1769 Transit of Venus. After the Revolution, two of America's leading practical astronomers Nathaniel Bowditch (1773–1838), a professional actuary who later became a Harvard trustee, and Benjamin Banneker (1731–1806), the son of a former slave, were also self-taught. By the Revolution, Americans from almanac readers to surveyors and Harvard faculty had shown a deepening interest in astronomy, and the elite had the beginnings of a local source of scholarly discourse through the American Philosophical Society. However, compared to the large and expensive national observatories in Europe, America lagged behind in permanent astronomical institutions and patronage to support astronomical research.

The Early Republic and Antebellum Years

Federal funding for astronomy began in the early years of the republic in the form of subsidized surveys, such as the Survey of the Federal Territory (i.e., Washington, DC). Although no less a figure than Thomas Jefferson (1743–1826), who made astronomical observations with his own telescopes and sketched plans for an observatory (Bedini 1990; Thomson 2012), worked avidly to promote science in the nascent United States, Congress questioned its ability to support "pure" scientific research with the lack of a clear mandate from the Constitution. Congress found it easier to support science (and astronomy) that had overtly practical, military, or expansionistic aims, such as the Lewis and Clark Expedition (1804–1806) and the US Coast Survey, founded in 1807 (Dupree 1986).

The tension over the role of federal government in scientific research came to a head in the late 1820s and early 1830s, when figures such as John Quincy Adams led the call to establish a national observatory (Portolano 2000). Although Congress forbade the construction of such an observatory, the US Navy established a "Depot of Charts and Instruments" to carry out astronomical work. This institution began with the important but modest task of rating ships' chronometers but soon became a site of astronomical observations complementing those of the Wilkes Expedition, a military and scientific survey of the Pacific. In 1842, Congress voted a large appropriation to make permanent the Depot and its astronomical observatory, which was later named the "US Naval

Observatory" (USNO) while at the same time rejecting the request of Adams, then in the House of Representatives, to use the bequest of the British aristocrat James Smithson to found an explicitly "national" observatory (Dick 2003).

As the Navy established the USNO, a wave of observatory building began across the country, with 20 such institutions completed by 1856 (Dick 1991). In a few cases these observatories were established with public funds, such as the teaching observatory at West Point and the one at Philadelphia Central High School. However, many of the largest were founded by collections of interested citizens who pooled contributions: the Cincinnati Observatory (1843), the Harvard College Observatory (completed with its 15-inch refractor in 1847), the Dudley Observatory (1852) in Albany, New York, and the Detroit Observatory in Ann Arbor (1857) were all funded in this way. The earliest of these American observatories employed European-made telescopes, yet astronomers soon found domestic suppliers. Whereas Harvard College imported its great refractor from Germany in 1846, a decade later the University of Michigan installed a 12-inch made by Henry Fitz (1808–1863) of New York. In the late 1850s and early 1860s, Alvan Clark (1804–1887) and his sons began producing their first refracting telescopes in Cambridgeport, Massachusetts. The Clarks came to dominate the construction of large objective lenses by the end of the nineteenth century (Warner 1968).

American astronomical research during the antebellum period began to attract increased international attention through a series of noteworthy discoveries and inventions. To cite just a few examples, in 1847 the Nantucket astronomer Maria Mitchell (1818–1889) discovered a comet by telescope, for which she won a medal by the King of Denmark and earned significant publicity (Bergland 2008). Sears Cook Walker (1805–1853), the first director of the observatory at Philadelphia Central High School and later an employee of the USNO, calculated the first orbit of the planet Neptune in 1847 and entered an international debate over its recent discovery (Hubbell and Smith 1992). Seven years earlier, the New York physician and chemist John William Draper (1811–1882) took the first astronomical photograph, a daguerreotype of the moon, an event that marked the beginning of one of the central techniques of astrophysics (Lankford 1984).

In addition to such discoveries and the foundation of a large number of permanent observatories, the 1840s were a particularly important decade for the foundation of an American astronomical community. That decade saw the publication of the first American astronomical journals, the establishment of the US Nautical Almanac Office, and the revitalization of the US Coast Survey under Alexander Dallas Bache (1806-1867), who convinced Congress to support a large network of scientific observers, including a number of astronomers. The foundation of the American Association for the Advancement of Science at the end of the decade also provided a new forum for astronomical discourse, for instance, in discussing and promoting the discovery by Pennsylvania astronomer Daniel Kirkwood (1814–1895) of a relationship between the rotation rates, distances, and masses of the planets in the context of Laplace's nebular hypothesis – a discovery that led to Kirkwood being hailed an "American Kepler" (Numbers 1977).

American Astronomy's Gilded Era, 1865–1940

Telescope and observatory building slowed during the Civil War but accelerated in the decades of economic recovery that followed. By 1898, the United States had over 30 telescopes with aperture 12 inches or larger housed in 25 large observatories, with numerous other smaller instruments at these observatories and at institutions elsewhere (Anonymous 1898). Coupled with American leadership in astrophysics, by 1900 the United States had joined the international astronomical community as an equal, owing in large part to the country's growing economic power. The rapid expansion, urbanization, and industrialization of the United States had created enormous amounts of capital concentrated in the hands of the nation's wealthiest citizens, and astronomers did everything in their power to wrest some of this capital for astronomical research. To a large degree, they succeeded. The scale of American astronomy at the turn of the twentieth century, in terms of the size of its instrumentation and research facilities, the production of astronomical and astrophysical data, and the amount of funding astronomers were able to control – roughly one-third of all money from private scientific endowments at the turn of the century (DeVorkin 2000b) – make it an early representation of what historians have labeled "Big Science" (Lankford and Slavings 1996).

Federal interest in astronomy throughout this period continued to be vested in the USNO, which installed a 26-inch Clark refractor, then the largest in the world, in 1871, just six years after the conclusion of the Civil War. Otherwise, the bulk of observatory building and astronomical research took place at observatories established at the nation's private and state-funded universities. In the majority of cases the initial capital investment for observatory buildings and instruments, if not for day-to-day operating costs, came from private benefactors. For these benefactors, financing the construction of an observatory was not simply to advance scientific research but also for reasons of philanthropy, civic duty, or personal memorial (Miller 1970). The sources of such beneficence reveal the range of American economic growth: agriculture (the Washburn Observatory (1878) of the University of Wisconsin and the McCormick Observatory (1884) of the University of Virginia), real estate and westward expansion (the Lick Observatory (1888) of the University of California), and the urban mass transportation boom (the Yerkes Observatory (1897) of the University of Chicago). The Lick and Yerkes Observatories each boasted, in turn, the world's largest refracting telescopes of 36 and 40 inches in aperture, respectively. In rare cases the wealthy financed their own research programs, such as Percival Lowell (1855–1916), the scion of a wealthy Boston family who established a large observatory near Flagstaff, Arizona. Most American observatories were considerably smaller than the Lick or Yerkes Observatories, with their huge telescopes and generous endowments, but played important roles in the growth of American scientific infrastructure as research outposts for America's westward expansion (Bartky 2000). And these observatories did more than make astronomical observations. For instance, Wisconsin's Washburn Observatory ran the clocks at the university and state capitol, sold time signals to railroad companies and local businesses for much-needed income, gathered meteorological data for the US Army Signal Corps, and served as a seismological station (Susalla and Lattis 2009).

George Ellery Hale (1868–1938) was the driving force behind the Yerkes Observatory and many of the large American telescope projects in the early twentieth century.

He was also the leading proponent of the new field of astrophysics. Hale had founded the Yerkes in southeastern Wisconsin with astrophysical problems foremost in mind, but he quickly looked westward to establish a new observatory in clearer skies and with telescopes of even greater light-gathering capability for astrophysical research. Hale secured funding from the Carnegie Institution of Washington for the Mount Wilson Observatory near Pasadena, California. Mount Wilson's 100-inch Hooker reflecting telescope (named for another private benefactor) was the largest in the world until eclipsed after World War II by the Rockefeller Foundation-funded Palomar Observatory near San Diego. Palomar and its 200-inch reflector were also Hale's initiatives.

American industry was not only a source of funding for telescopes but also for expertise in building large and precise astronomical instruments and for directing large-scale research programs. For example, the founders of Cleveland's Warner and Swasey Company, which made the telescope tubes and mountings for the Lick 36-inch and the Yerkes 40-inch, among other large telescopes, began their careers as machine tool makers. The machine-shop-like quality of the astronomical observatory, filled with perfectly engineered and elaborately balanced instruments, served for American physicist Henry Rowland (1848–1901) as the ideal kind of scientific institution on which to model a physics laboratory (Staley 2010). Astrophysical research, an area of seemingly "pure" science divorced from the practical and nationalistic needs that drove institutions like the USNO, also felt an economic influence. Organizational models from the world of business shaped astrophysical research (Nisbett 2007), such as the massive spectral classification program led by Harvard College Observatory director Edward Pickering (1846–1919). Pickering's program, which employed large numbers of unskilled women workers under male supervisors, reflected the gendered division of labor in America's industrial economy (Lankford and Slavings 1996).

Astrophysical research flourished in the decades around the turn of the twentieth century, and Americans led in several areas ranging from instrumentation to theoretical applications. Hale himself made a significant contribution with his invention of the spectroheliograph, and Joel Stebbins (1878–1966), first at the University of Illinois and then at the University of Wisconsin, transformed the photometer into a precise astrophysical tool through the use of sensitive electric and electronic photocells. Mount Wilson and Lick became the focal points of American efforts to test Albert Einstein's general theory of relativity (Crelinsten 2006), and observers at the Lowell Observatory and Mount Wilson contributed to the significant discovery that spiral nebulae appear to be receding from the observer; the discovery of nebular redshifts occurred at a time when the scale of the galaxy and universe were undergoing a radical revision. In addition to these instrumental and observational efforts, new theoretical ideas from quantum and nuclear physics began to take a more prominent role in astrophysical research, such as the work Henry Norris Russell (1877–1957) and Cecilia Payne (1900–1979) in examining chemical abundances in stars (DeVorkin 2000a; Haramundanis 1996).

Hale was able not only to marshal large amounts of industrial money for observatories; he also proved an adept organizer of researchers and sought to bridge gaps between disciplines and subfields. Hale founded the *Astrophysical Journal* in 1895, which quickly became the nation's premier astronomy research publication and featured both physicists and astronomers on its editorial board (Osterbrock 1995), organized the International Union for Cooperation in Solar Research (a progenitor of the International Astronomical Union), and was one of the prime movers behind the Astronomical and

Astrophysical Society of America, later renamed the American Astronomical Society (AAS). The AAS met a need for professional organization and communication in a quickly growing field: Lankford and Slavings have estimated that there were over four times as many American astronomers active in the years 1860–1899 than in the entire period prior to 1860, and the population of the field more than doubled again in the first four decades of the twentieth century (Lankford and Slavings 1997: 288).

Big Science for a Big Universe, 1940 to the Present

Federal involvement and funding changed dramatically for many scientific fields after World War II, and astronomy was no exception. The US government became directly involved in funding explicitly "national" observatories, the kind of institution it was reluctant to build during the antebellum era and instead financing the USNO ostensibly for its practical value to navigation and timekeeping. During the 1950s and with the financial support of the NSF, the federal government helped to found the National Radio Astronomy Observatory (NRAO) in West Virginia and the Kitt Peak National Observatory in Arizona; Kitt Peak later became part of the National Optical Astronomy Observatory (NOAO), which coordinates observing sites in the northern and southern hemispheres. Postwar American astronomers were initially reluctant to accept federal funding for a field that had been dominated by private patronage for the previous century, particularly funding from military sources (DeVorkin 2000b), and private funding continues to play a significant role, such as the millions of dollars from the W.M. Keck Foundation for a pair of 10-meter reflecting telescopes in Hawaii. However, by the end of the twentieth century federal funding accounted for a significant fraction of all American astronomical research (Committee on Astronomy and Astrophysics et al. 2000). Consistent federal funding helped to maintain a field that grew steadily during the space age and avoid the boom-and-bust cycles that plagued postwar physics (Kaiser 2012; Susalla 2013).

Among the most notable conceptual developments arising from federal support for astronomy derived from the United States' military and civilian space programs, which allowed astronomers to see beyond the visible and radio bands of the electromagnetic spectrum accessible from the earth's surface into the portions of the electromagnetic spectrum (infrared, ultraviolet, x-ray, and gamma ray) visible primarily from space. Many space astronomy projects relied on the military not just for funding, but also for launch vehicles and new kinds of instruments, like the now-ubiquitous charge-coupled device (CCD). With the founding of NASA the government had an ostensibly non-military program through which to focus its efforts in space science. However, as Robert W. Smith has argued in his study of the Hubble Space Telescope, "Although NASA was created as a civilian agency, the new technologies that allowed men and machines to move around in space always had national security implications and application" (Smith 1989: 36). Similarly, Joseph Tatarewicz has argued that NASA viewed the field of planetary astronomy as "a potential tool for state ends" in the new era of solar system exploration (Tatarewicz 1990: xii).

America's interest in space and space exploration reformed existing research fields while leading to the creation of entirely new disciplines, such as astrobiology (Dick 1996). Planetary and solar system astronomy were traditional fields that belonged to a small interdisciplinary group of researchers in the decades prior to the space age and were entirely reshaped in the decades after *Sputnik* (Brush 1996b; Doel 1996). Similarly, American stellar astrophysics had made major advances before the space race and received new impetus from the earliest space astronomy programs in the 1960s, for instance, through observations of hot stars in the ultraviolet (Bleeker, Geiss, and Huber 2001). During the 1930s and again in the 1950s, a loose group of Americans and European émigrés and visitors developed compelling theories of stellar energy production and of the birth, evolution, and death of stars (Cenadelli 2010).

One of the most dramatic developments of twentieth-century astronomy in which Americans played a key role was the discovery that most of the enigmatic "spiral nebulae" were in fact systems of stars comparable in size to the Milky Way, that is, galaxies. During the period from 1910 to 1935 a combination of developments, including the revision of the cosmic distance scale, the discovery of nebular (or galactic) redshifts, and the introduction of Einstein's general theory of relativity, which allowed for the creation of mathematical models of the universe as a whole, helped lead to an entirely new field of cosmological research based on the idea of a dynamic, expanding universe populated by billions of galaxies. Following the detection of a pervading microwaveband background radiation in 1965, American physicists and astronomers promoted this radiation as the leftover energy of the earliest phase of an expanding universe: the big bang. By the end of the twentieth century, investigations into the beginning of the cosmos had brought about a close merger between the study of the very large – the expanding universe of galaxies, quasars, and dark matter – and the very small – particle physics and the exotic, high-temperature era of the early universe. Galactic and cosmological research became one of the major branches of American astronomy, accounting for approximately one-fifth of all astronomical research by the year 2000 (Susalla 2013).

Bibliographical Essay

For overviews of the historiography of American astronomy published prior to the early 1980s, the most useful surveys are DeVorkin (1982) and Rothenberg (1985). Hockey (2007) includes numerous entries on individual American astronomers. Although not restricted to the United States, North (2008) is perhaps the best general conceptual survey of the entire history of astronomy. As participant histories, Clerke (1902) and Longair (2006) provide good conceptual overviews for the nineteenth and twentieth centuries, respectively.

Aveni (2008) offers an entry into the methods and debates in North American archaeoastronomy and cultural astronomy, and the contributions to Selin (2000) put North America into a global context of non-Western astronomy. Ruggles and Saunders (1993) give insight into the relabeling of the field as "cultural astronomy" in the early 1990s.

Although historians have mapped some of the basic features of astronomy in colonial America, this era has attracted relatively little scholarly research compared with the more detailed studies of nineteenth- and twentieth-century astronomy. Greene (1954) and Yeomans (1977) offer brief but useful overviews, and Stearns (1970) gives a thorough treatment of astronomy placed in the larger context of colonial science in the seventeenth and eighteenth centuries. Fleming (1964), Wolf (1982), and Ruffin (1997) have all examined Copernicanism in early colonial publications, and Leventhal (1976) and Hall (1990) have shown the persistence of older astrological traditions in many of these same publications. Noteworthy exceptions to the relative quiescence of colonial astronomical historiography have been the more recent work of Schechner (1992, 1997) and Gronim (1999, 2007), who place eighteenth-century American astronomy and cosmology in the changing social and religious contexts of Massachusetts and New York, respectively. Thode (2013) has also examined astronomical practice in American expansion west of the Appalachians.

The observatory movement of the mid-nineteenth century has attracted significant historical attention. Useful listings of observatories are provided in Rufus (1924) and Musto (1967), while Warner (1979), Brush (1979), and Hetherington (1976, 1983) place the movement in general overviews of nineteenth-century American astronomy. Monograph treatments are available for most of the major observatories, including Harvard (Jones and Boyd 1971), Dudley (James 1987), and Cincinnati (Shoemaker 1991). Dick's (2003) treatment of the US Naval Observatory is one of the best institutional histories of American astronomy. Slotten (1994) examines federal support for science in the US Coast Survey, and Stanton (1975) and Philbrick (2003) do the same with the US Exploring Expedition. Plotkin (1980) and Whitesell (1998) also describe pedagogical changes among American astronomers through the importing of new, rigorous methods from Germany in the mid-nineteenth century.

Lankford and Slavings (1997) offer what is perhaps the most sweeping treatment of any era of American astronomy, using statistics to track changes in research interests, patronage, training, and gender among astronomers for the period 1859–1940. However, their analysis does not dive very deeply into the actual practices of the key disciplinary change in this era: the rise of astrophysics. The contributors in Gingerich (1984) outline some of these practices, while Hearnshaw (1996, 2014) describes instrumental developments in spectroscopy and photometry in very fine detail. Osterbrock (1984) uses the career of James Keeler to examine American astrophysics at the turn of the twentieth century, as do his studies of the Lick Observatory (Osterbrock, Gustafson, and Unruh 1988) and Yerkes Observatory (Osterbrock 1997). The Harvard stellar classification program under Edward Pickering has been the subject of a number of studies, including DeVorkin (1982), Plotkin (1978a, 1978b), Rossiter (1982), Lankford and Slavings (1996), and Nisbett (2007). The standard biography of George Ellery Hale is still Wright (1966), although scholars have more recently pursued the different lines of Hale's work in solar astrophysics (Hufbauer 1991) and observatory building (Osterbrock 1993; Sandage 2004). DeVorkin (1999) describes the foundation and development of the AAS, while Williams and Saladyga (2011) detail the history of another important organization, the American Association of Variable Star Observers (AAVSO). Smith (1982) remains a key work on the changing scale of the galaxy and the cosmos during the 1920s and 1930s and the beginnings of modern cosmology, and he has revisited some of these issues more recently (Smith 2006, 2008).

Although historians have tended to focus on the novel astrophysics of the late nineteenth and early twentieth centuries, the history of planetary astronomy has also been the subject of excellent scholarship. Baum and Sheehan (1997) describe planet hunting and celestial mechanics in the late nineteenth century, while Lane (2010) puts Percival Lowell's Mars exploits into a context of American imperialism and interest in geography. Brush's (1996b) three-volume series on the history of planetary physics details the role of Americans in reforming and revitalizing cosmogonical theories during the twentieth century, and see also Doel (1996) for a history of planetary astronomy in the first half of the twentieth century. The history of the positional astronomy during this period has largely been neglected, but see Dick (2003) for astrometry and celestial mechanics at the USNO.

The two best studies of the role of the federal government in shaping and fostering post-World War II space science and space astronomy are Smith (1989) and DeVorkin (1992), but see also Hirsch (1983), Tatarewicz (1990), and Westwick (2007). Malphrus (1996), Sullivan (2009) and Munns (2003, 2013) describe the creation of the National Radio Astronomy Observatory, while Edmondson (1997) examines the institutional structure of the Association of Universities for Research in Astronomy, which manages America's national optical observatories, including the Kitt Peak National Observatory. Edmonson has little to say about the kind of research these observatories conducted, but McCray (2004) has examined the balance between patronage and instrumentation in the optical astronomy community in the second half of the twentieth century. For the conceptual history of twentieth-century cosmology Kragh (1996, 2009) and Smeenk (2003) are good starting points, while Susalla (2013) focuses specifically on the American cosmological community. The historiography of twentieth-century American astronomy has principally treated conceptual and institutional developments, and so cultural themes remain comparatively unexplored territory. Dick (1996), Marché (2005), McCray (2008), Prosser (2009), and Swanner (2015) all prove examples of the various directions pursued in the cultural history of recent astronomy.

Chapter Four

CHEMISTRY

Ann E. Robinson

Chemistry has played an important role in the United States since the first European explorers landed in the new world in the sixteenth century. Spanish explorers sent reports home describing not only the flora and fauna they encountered but also the minerals, metals and other elemental bodies. Early settlers relied on chemical knowledge to heal wounds and cure diseases. Later, chemistry became important in trade as well as an area of interest for experimenters. Despite this long history, the vein of the history of chemistry in the United States is one that has remained largely untapped by historians of science. There has, however, been a recent upsurge of interest in the history of the periodic law and its graphic representation, the periodic table. The chemical elements are often referred to as the building blocks of nature. The periodic law provides a fundamental understanding of how these building blocks function while the periodic table is a visual organizational tool, providing an easy reference to the chemical and physical characteristics of the elements and to their relationships. Given the fundamental importance of the elements and the periodic law to chemistry, exploring their history in the United States can also provide some insight into the history of chemistry in America.

Classifying the Elements

During the eighteenth century, chemists attempted to classify substances in different ways. Some classified materials according to their chemical composition, while others classified according to various qualities. In the early nineteenth century, chemists began to notice mathematical relationships between the atomic weights of the elements. As atomic weight was considered to be the defining characteristic of an element, relationships between atomic weights also revealed relationships between other chemical and

A Companion to the History of American Science, First Edition.
Edited by Georgina M. Montgomery and Mark A. Largent.
© 2016 John Wiley & Sons, Ltd. Published 2020 by John Wiley & Sons, Ltd.

physical characteristics of elements. The German chemist Johann Döbereiner (1780–1849) realized that small groups of elements, usually groups of three, exhibited trends in properties. Most of the classification schemes developed in the decades following this discovery involved groups of these triads. However, it was difficult to incorporate all known elements into one system. If there was a classification scheme that could accommodate all of the elements, then it was also likely that there was an underlying natural law which governed the elements and chemists were eager to ground their science in such laws. Classification schemes were important tools for research, but their development was often tied to education. Systems were developed by students as part of the research required for their degrees and by chemists who were searching for better ways to instruct their students, by means of textbooks, lectures, and tools such as wall charts, chalk boards, and printed cards.

The first classification system developed by an American was that of Oliver Wolcott Gibbs (1822–1908). The system formed a part of his 1845 MD thesis for the College of Physicians and Surgeons in New York. It was not very original, showing great similarity to the work of European chemists (Kauffman 1969), however, among classification schemes for the elements it is generally considered to be the first circular chart (Zapffe 1969). Josiah Parsons Cooke, Jr. (1827–1894) developed another important American classification system. First made publically known in 1854, Cooke's system was one of the most complete at the time, incorporating all 55 known elements. Interestingly, Cooke developed the system "to facilitate the acquisition of chemistry" by students (Kauffman 1969). Cooke's system was known within the United States, having been published in the premier American scientific journal, the *American Journal of Science*, after its presentation to the American Academy of Arts and Sciences. Neither of these classification schemes was well known outside of the United States or made much of an impact in Europe, reflecting the status not only of chemistry in America but of chemical education as well.

Gibbs had received his degree in medicine, and it was not unusual at the time for chemists to receive training as doctors because chemistry was tightly bound to medicine, particularly in colleges and universities. Therefore, medical programs were often the best place to get an education in chemistry. Cooke, on the other hand, received his degree from Harvard College in 1848. Chemical education in the early nineteenth-century United States has been described as having "a largely provincial and traditional character" (Rezneck 1970). College courses usually consisted of lectures, with the occasional demonstration given by the professor at the front of the lecture hall, and recitations by students of material from textbooks. More advanced training was available in Europe, where students were required to work in the chemical laboratory, performing their own experiments under the guidance of the professor. Many Americans traveled to Europe with the intention of pursuing the PhD degree, which at the time was not available from American colleges.

Eben Horsford (1818–1893) was one of those Americans who made the journey to study in Germany and had the honor of being the second American student of the German chemist Justus Liebig (1803–1873). Just before he left Germany in 1846, Horsford wrote of the Americans studying chemistry in Europe, "when all of these have returned home, I see no reason why we should doubt America's instrumentality in the advancement of his department of science" (Rezneck 1970). This statement was optimistic for the time, but Horsford did his part to make it come true. He took up the Rumford professorship at Harvard in 1847 and began implementing the European style of teaching. Both Gibbs and Cooke traveled to Europe, to attend lectures and meet other chemists, although neither obtained the PhD. Upon their return to the United States, they advanced Horsford's efforts to include laboratory work as a part of the courses they taught at Harvard College and at Harvard's affiliated Lawrence Scientific School. Gibbs followed Horsford in 1863 as the Rumford chair at Harvard. Cooke was instrumental in the building of a new chemical laboratory for the use of both students and faculty at Harvard. These three, along with the early faculty at Yale's Sheffield Scientific School, have been credited with creating "pioneer chemistry departments," which influenced departments and programs at other institutions across the country (Beardsley 1964).

Cooke and Gibbs may not be among those often recognized as a precursor to, if not a co-discoverer of, the periodic law, but there is one American on that list. Gustavus Detlef Hinrichs (1836–1923) emigrated to the United States in 1861 and settled in Iowa where he obtained faculty positions at the State University in Iowa City. By 1870, under Hinrichs' direction, the physical laboratory had become one of the four finest in the country. He was also an innovative teacher, using a wall chart of his classification scheme in his lecture hall and providing interactive experiments and assignments for his students, as well as publishing numerous textbooks (Palmer 2007). Hinrichs preferred to publish his own research in European scientific journals rather than American ones, though he did present frequently at the meetings of the American Association for the Advancement of Science (AAAS). As early as 1855, Hinrichs had begun to form a classification system for the elements based on a primal matter he called pantogen. His system, which he called the Atom Mechanics, was reviewed in journals in several countries, however, it was generally not accepted, in part because of the opaque language used and its reliance on Hinrichs' peculiar interpretation of the philosophical notion of the unity of matter.

The periodic law is generally considered to have been discovered in 1869. The Russian chemist Dmitri Mendeleev (1834–1907) was in the midst of writing a chemistry textbook for use in Russia when he realized that when the elements were placed in order by increasing atomic weight, their properties occurred in a periodic fashion. Mendeleev published the first of his many periodic tables in 1869, leaving spaces for several as-yet-undiscovered elements. That same year, the German chemist Julius Lothar Meyer (1830–1895) published a similar table, although he did not leave blank spaces. During the late nineteenth century, Meyer and Mendeleev were often recognized as co-discoverers of the periodic law; they were jointly awarded the Davy Medal by the Royal Society of London in 1882 for the discovery of the periodic relations of atomic weights. However, Mendeleev's successful predictions of new elements has over the past century led to Meyer's role being forgotten.

American chemists played an important role in the acceptance of the periodic law. Foremost among these was Francis Preston Venable (1856–1934). Venable studied chemistry at the University of Virginia under the Irish-born émigré John Mallet (1832–1912), who was highly regarded both in the United States and in Europe, before going to Germany where he earned the PhD. He became professor of chemistry at the University of North Carolina, turning the chemistry department into one of the finest in the south, on par with northern schools, such as Harvard University and Johns Hopkins University (Bursey 1989). Venable had a keen interest in the history of chemical concepts and he wrote the first history of the periodic law, published in 1896. He gave many speeches and wrote numerous articles advocating the use of the periodic system not only in research but in education as well. He believed the teaching of chemistry should be organized around the principles of the periodic system. Venable devised his own modified form of the periodic table for use in instruction. With James Lewis Howe (1859–1955), a chemistry professor at Washington and Lee University who he had met while studying in Germany, Venable co-wrote in 1898 the first American college textbook to be based on the periodic system.

Atomic Weight

By the turn of the twentieth century, the periodic system was firmly in place within chemistry. The periodic law itself was based on atomic weight, which was the defining characteristic of an element. Most chemists in fact regarded atomic weights as constants of nature, much as today we consider such things as the speed of light and Avogadro's number to be constants. Atomic weights, however, were not based on the actual weight of individual atoms of an element because atoms were too small to have been weighed by any means known, but also because many chemists in the nineteenth century believed atoms to be useful theoretical constructs, not physical entities. Rather, atomic weight was an average of the weight of many atoms of an element, determined in relation to the atomic weight of another element. For many elements, this led to sometimes widely varying weights.

Accurate atomic weights were important. Determining how much of what elements were in molecules and compounds was necessary for almost every trade and industry, not to mention in scientific research. The accurate determination of atomic weight was fundamental to the basic understanding of chemistry, although it was not necessarily a path to groundbreaking new research. It was difficult work and the task was a tedious and painstaking one. Despite the fact that chemists considered atomic weights to be constants of nature, there was no one standard upon which atomic weight was based. Of the many standards used, the two most common were based on the atomic weights of oxygen ($O = 16$) and hydrogen ($H = 1$). It was in these areas, the accurate determination of atomic weight and the consensus upon a standard, that American chemists made important contributions.

Two Americans did much to codify the atomic weights and to highlight which elements were the most problematic. The first, Theodore W. Richards (1868–1928), did his graduate work in chemistry at Harvard College under Cooke. After receiving his PhD in 1888, he spent a year in Europe, visiting laboratories and universities, and meeting other chemists; he spent a further year in Germany in 1895 to gain a deeper understanding of the new field of physical chemistry. While working with Cooke, Richards had become interested in basic questions about the elements, particularly in the determination of atomic weights, and he made these questions his life's work. Richards became known, both in the United States and in Europe, for his painstaking and accurate determinations of atomic weight. His research was published in Europe as well as in the United States and was highly regarded. Richards became the first American chemist to win the Nobel Prize, awarded to him in 1914 in recognition of his work on atomic weight. In his Nobel lecture, given in 1919, Richards (1999) said of atomic weights,

"They are the hieroglyphics which tell in a language of their own the story of the birth or evolution of all matter."

The other American codifier was Frank Wigglesworth Clarke (1847–1931). He obtained his BS degree from the Lawrence Scientific School in 1867, where he did research under the supervision of Gibbs. After several years teaching at various high schools and universities, Clarke obtained a position at the University of Cincinnati. Clarke was appointed the official chemist of the US Geological Survey (USGS) in 1883. At both Cincinnati and the USGS, he continued his research into the basic constants of chemistry. The first edition of his *Recalculation of the Atomic Weights* was published in 1882 as a volume in the Smithsonian Institution's *Constants of Nature* series. This work was an attempt to solve, or at least take into consideration, the main problem with atomic weights: "The atomic weight of each element involves the probable errors of all the other elements to which it is directly or indirectly referred" (Clarke 1882). The publication was highly regarded by his American colleagues and soon came to the attention of European chemists, as well, being serialized in such European journals as *The Chemical News*. Ten years later, in 1892, Clarke became the sole member of the atomic weights committee of the American Chemical Society (ACS), charged with publishing an annual table of atomic weights, work which he continued until 1913. He was also an original member of the International Atomic Weights Committee, likewise commissioned with the annual publication of tables of atomic weights (Holden 2004b).

This international committee was originally formed for the purpose of creating consensus on the issue of which of the two most common standards should be used as the basis of atomic weight, $H = 1$ or $O = 16$. Hydrogen was favored by many because of its philosophical implications. In 1815, the English chemist and physician William Prout (1785–1850) first hypothesized that the atomic weights of all of the elements were whole-number multiples of that of hydrogen, making hydrogen the primary matter of the universe. During the succeeding decades, as atomic weights were more accurately determined, it was suggested that all atomic weights were fractional multiples of that of hydrogen, generally one-half or one-quarter. Despite the seeming simplicity of the atomic weight of hydrogen being one and those of the other elements being multiples of some fraction of one, it was not so clear cut. In 1900, the atomic weight of hydrogen was described as "still ... doubtful and may yet be subjected to repeated modifications" (Vogeler, 1900). The atomic weight of oxygen, on the other hand, had been considered to be fairly accurately known for several years.

This accurate atomic weight was the work of Edward W. Morley (1838–1923). A chemistry professor at Case Western Reserve University, he was the first to introduce students to the laboratory as part of their course work. He is most known today for his work with the physicist Albert A. Michelson (1852–1931); the Michelson–Morley experiment is frequently cited as providing early evidence against the ether theory. Among chemists, however, Morley was known for his work on gases and the measurements of various natural constants. He spent 11 years refining experiments which ultimately, in 1895, led to the most accurate determination of the atomic weight of oxygen in relation to the atomic weight of hydrogen (Clarke 1927). This value was widely accepted by chemists everywhere. The work garnered Morley many accolades; in 1907, he became the first American to be awarded the Royal Society's Davy Medal. His work further smoothed the path towards a consensus on a standard upon which atomic weight was based.

A consensus was needed as confusion abounded. Multiple tables of atomic weights were published, each based on one of the many different standards. In their publications, chemists often used atomic weights based on more than one and sometimes did not say which standard they were using. Venable (1889) was much disgusted by this and wrote, "Th[is] evil was and is a crying one and demands the best energies of the wisest chemists to rectify." To this end, in 1899 a group of German chemists called for representatives from 15 countries to vote on several questions, including that of a standard, by mail ballot. The overwhelming majority of chemists, 42 out of 49, voted for the adoption of $O = 16$ as the standard upon which atomic weights should be based. Of the five American delegates, which included Gibbs and Richards, only Mallet voted for $H = 1$. In a report on this vote, Howe (1900) wrote "it is interesting to note that the work of this committee is the final outcome of an agitation which was begun in this country" by Venable's 1889 paper. While Venable was far from the only chemist advocating that a standard be adopted, he was very vocal and his opinions were well known among chemists in both Europe and the United States.

In due course, a small committee was established for the purpose of publishing annual tables of atomic weights, the first of which appeared in 1902. Tables based upon the standard $O = 16$ were published annually, even during World War I. After the war, the International Union of Pure and Applied Chemistry (IUPAC) was established as the organization responsible for setting standards, nomenclature rules, and other conventions for the chemical community. The International Atomic Weights Committee eventually became a commission within the IUPAC. The committee, now titled the Commission on Isotopic Abundances and Atomic Weights (CIAAW), continues to publish regularly updated tables of atomic weights and even a periodic table of the isotopes. Following in the footsteps of Clarke, scientists employed by the USGS have continued to be active members of the commission and its work is currently carried out with the support of the USGS National Research Program.

The Discovery of Elements

One of the signs that a discipline or profession has become established is the creation of associations and organizations, along with the attendant journals, standards, and guidelines. The founding of the American Chemical Society (ACS), the major society for American chemists, was related to the discovery of oxygen. There is still much debate over who discovered oxygen and when, but one of the claimants is Joseph Priestley (1733–1804), an English chemist who came to Pennsylvania in 1794. American chemists in the late nineteenth century claimed Priestley as their own and held a meeting in 1874 to celebrate the centennial of his discovery of oxygen. One outcome of the meeting was the expressed desire to form a national chemical society. The AAAS, the only national scientific society at the time, had a chemical section, however, many chemists felt the need for their own independent society. The ACS was founded two years later, in 1876. The society did not begin in promising fashion and it was a small group of chemists, including Clarke, who turned it into a truly national association with a journal which was held in high regard (Skolnik 1976).

The discovery of new elements by Americans was to come following the discovery of radioactivity in 1896 by the French physicist Henri Becquerel (1852–1908).

Radioactivity opened up entirely new avenues of research that had lasting impact on the sciences of chemistry and physics, as well as on the periodic table. One of the first American chemists to do any significant work in this new field was Charles Baskerville (1870–1922). He earned the PhD at the University of North Carolina, where he was a student of Venable. Baskerville jumped into the study of thorium, despite having little background in the rare earth elements or radioactivity, and fairly quickly determined that he had discovered two new elements among its decay products. He named these elements berzelium, after the Swedish chemist Jöns Jacob Berzelius (1779–1848), and carolinium, after North Carolina. The announcement made him the first American to claim to have discovered an element and as a result he became something of a sensation in the American scientific community (Bursey 1989). Baskerville was subsequently offered a position in New York, at the time a center of chemical research, however, he did little further work in the area of radioactivity.

The major American researcher in the new field of radiochemistry was Bertram Boltwood (1870–1927), a chemist at Yale University. Boltwood did significant work on the decay chains of the radioactive elements uranium, thorium, and actinium. Radioactive elements have half-lives, periods of time during which they decay into other elements, which in turn decay, and so on until the end of the decay chain is reached. Boltwood's research led to the discovery of many radioactive elements, including one he named ionium. His research was instrumental in understanding how radioactivity worked and thrust American science into the forefront of the new field. Boltwood was also among the first to use radiometric dating to determine the age of the earth (Badash 1979). He maintained a close correspondence with Ernest Rutherford (1871–1937), the New Zealand-born physicist who was a pioneer in nuclear physics, as well as with numerous American and European chemists and physicists.

Research in radioactivity led to the discovery of what were thought to be dozens of new elements, including Baskerville's berzelium and carolinium and Boltwood's ionium. These radioactive elements did not fit within the periodic table; often there were more than one with the same or close to the same atomic weight as another element and sometimes the characteristics of these elements did not fit into the table as expected by increasing atomic weight. Chemists began to have grave doubts about the fundamental nature of the periodic law. With the discovery of isotopy by the British chemist Frederick Soddy (1877–1956) in 1913, it became clear that what had been discovered were not new elements but rather isotopes of elements. Baskerville's new elements were in fact isotopes of thorium. Boltwood's ionium was also an isotope of thorium, however, it is an isotope that has been important in research and is still frequently referred to as ionium. An isotope of an element has the same number of protons and electrons but a different number of neutrons. Neutrons have a greater mass than either electrons or protons, thus the different isotopes of an element have different atomic masses. Another discovery in 1913 also calmed fears about the usefulness of the periodic law. The British physicist Henry Moseley (1887–1915) discovered a systematic relationship between the x-ray spectra of the elements and their atomic number, that is the number of electrons within the nucleus of one atom of an element. With these discoveries, atomic number replaced atomic weight as the organizing principle of the periodic table and the periodic law was reworded to reflect the new understanding of the structure of atoms.

One more change to the periodic table was yet to come. The group of elements known as the lanthanides, which incorporate many of the rare earth elements, are not easily accommodated within the periodic table. This is why they are usually found in a separate group below the main body of the table. Some chemists had thought there was another group of elements analogous to the lanthanides, that is sharing the same characteristics, but it was not an idea that was developed. During World War II, Glenn T. Seaborg (1912–1999) was working with the Manhattan Project at the Met Lab in Chicago. He was in charge of the section which researched the chemistry of plutonium, an element which he had recently helped discover. Some of the results obtained by experiments were puzzling, not conforming to the results expected from looking at the placement of elements in the periodic table. Seaborg was a great believer in the use of the periodic table for both research and teaching. He began to explore the possibility that there was indeed an analogous group to the lanthanides, a group he referred to as the actinides. Seaborg explored this idea in seminars held by the chemists at the Met Lab and soon after the end of the war, he published an article explaining his actinides concept and providing a periodic table with the now familiar set of lanthanides and actinides at the bottom (Seaborg 1945).

Seaborg is important to the history of the elements, as well. He was co-winner of the Nobel Prize for Chemistry in 1951, with Edwin McMillan (1907–1991) one of his colleagues at the University of California at Berkeley, for discoveries in the chemistry of the transuranium elements. Seaborg also had the honor of having an element named for him, element 106, seaborgium. The vast majority of the elements that have been discovered since 1940 have been discovered by American chemists and physicists. All of these elements are radioactive, most of them are heavy elements, and all have been discovered with the use of cyclotrons. Beginning with element 101, new elements were produced one atom at a time. As Seaborg (2001) noted, "The age when one could actually do any chemistry with the new elements seemed to be drawing to a close." Although the discovery of new elements has become more about physics than chemistry, IUPAC is still the body which officially names new elements.

During the 1940s and early 1950s, chemists at UC Berkeley discovered nine elements. Research was then shifted up the hill to the Lawrence Berkeley National Laboratory (LBNL), one of several labs which constituted a new system of national laboratories founded in the wake of World War II. The heavy elements group at LBNL was the center of the discovery of new elements in the United States. Through the 1970s, they claimed to have discovered several more elements. However, scientists at the Joint Institute for Nuclear Research (JINR), in Dubna, Russia, countered these claims. The discovery and naming of new elements became a part of the rhetoric of the Cold War. The controversies between LBNL and JINR continued into the mid-1990s when the IUPAC developed a process for adjudicating discovery claims. While the disputes were continuing in the 1980s and 1990s, preeminence in the field of heavy element research shifted to Germany. In the early twenty-first century, the collaboration between the Lawrence Livermore National Laboratory in California and the JINR in Russia has produced several new elements.

American dominance in the field of heavy element discovery broke down in part because of funding issues. The discovery of heavy elements requires millions of dollars to build bigger accelerators and to operate them. There is little practical return for the discovery of new elements; currently the heaviest element with any practical application is americium, element 95, an isotope of which is used in smoke detectors. Federal government funding for such research in the United States began to decline in the

1960s. Although Seaborg was in Washington in the 1960s and lobbied in person for
the funding to build bigger cyclotrons at LBNL, he was unsuccessful and funding was
obtained only to upgrade existing equipment. Funding continued to decline in succeeding decades. In the early 1990s, Congress cut all funding for the Superconducting
Super Collider, although it was already under construction in Texas. The elimination
of funding to maintain and upgrade the Tevatron at Fermilab in 2011 signaled the federal government's unwillingness to continue to provide resources for the discovery of
fundamental scientific knowledge through the use of accelerators. However, American
scientists collaborate with laboratories in other countries, forming large international
research groups which continue to discover more about the fundamental nature of the
universe.

Initially Americans were not at the forefront of chemical science. During the nineteenth century, increasing numbers of American chemists sought advanced training
in Europe, where they made personal connections and built networks. They returned
home, importing new ideas in education and implementing them, and they established
new schools and programs to improve the education available to American students
at home. They sought to recreate the scientific culture they had known in Europe,
establishing local, regional, and national organizations. The development of journals
and other publications helped disseminate their research, bringing it to the attention of
their European colleagues. By the turn of the twentieth century, the results of American
scientific research were increasingly well known and accepted and by the middle of the
century it was often at the forefront. Although textbooks usually portray the story of
the periodic law as a European one, American chemists made important contributions
to its development and acceptance, making the story a truly international one.

Bibliographical Essay

There are few volumes written about the history of chemistry as a whole in the United
States. Beardsley (1964) provides a good overview of chemistry as a profession. The
work provides a brief history of chemical education in America, the rise of professional
organizations and journals, and the venues in which chemists found work during the
second half of the nineteenth century. The footnotes and note on sources provide good
starting points for anyone who is interested in doing research on the topics covered.
For those who find statistics useful, Thackray et al. (1985) will be of great interest.
He and his colleagues collected a wide array of statistics regarding almost every aspect
of chemistry in America. However, they often did not draw more than the most basic
conclusions from this data and left its interpretation open to the reader. Regardless, it
is a treasure trove of statistical data and a useful starting point for research. For those
interested in learning more about chemistry in American academic institutions between
1876 and 1976, Carroll's PhD dissertation (1982) is a good source.

While volumes covering chemistry as a whole are few, there are a significant number
which address specific topics; only a small sampling is included here. Those interested in
the close relationship between medicine and chemistry and the evolution of biochemistry should see Kohler (1982). Servos (1990) provides an examination of the development of the new field of physical chemistry and its import into the United States,
discussing the ways in which disciplines are formed and developed, as well as touching

on the relationship between academic chemists and industry. Badash (1979) likewise examines the new field of radioactivity and its development in America; the conclusion dissects the particularities of science in the United States at the turn of the twentieth century. The rise of the American and European chemical and pharmaceutical industries is the subject of one of the last works of the great business historian Chandler (2005).

Many nineteenth- and early twentieth-century American scientists were not included in the original *Dictionary of Scientific Biography*, although the volumes added in 2008 have filled in some of the gaps. Biographical information is more often found in the *National Cyclopaedia of American Biography*, published between 1898 and 1984. Brief information can be found in *American Men of Science*, which began publication in 1906; women were included in the title and the volumes beginning in 1971. Another source is the *Biographical Memoirs* of the National Academy of Sciences, which provides biographical sketches and discussions of scientific contributions of the deceased members of the NAS. There are many biographies of individual chemists or groups of chemists. Of particular interest for chemistry in America and the periodic system is the short biography of Venable by Bursey (1989). It provides a good account of how Venable turned the chemistry department at the University of North Carolina into a highly regarded research program, from the establishment of a local scientific society and journal to the creation of a science library to the expectations for both students and faculty of research and publication.

The history of the periodic table has recently become the subject of renewed interest. The newest volume covering the topic is by Scerri (2007); there is also a briefer version published as part of Oxford University Press' *Very Short Introduction* series (Scerri 2011). The first history of the periodic law was written by Venable (1896). It provides a particularly good snapshot of the state of chemistry just before the turn of the twentieth century, showing the questions which were of paramount importance to the basic understanding of chemistry at that time. Other early histories of the periodic law include Garrett (1909) and Rudorf (1900).

The response of American chemists to the periodic law is discussed in several articles. Kauffman wrote a pair of relevant articles, one (1970) discussed the relationship between American chemists and Mendeleev, and the other (1969) offered a discussion of American attempts to classify the elements before Mendeleev. Brush (1996b) examined journals and textbooks to determine why the periodic law was accepted in the United States and Britain, concluding it was because the law successfully predicted the discovery of several new elements rather than because it easily accommodated already known facts. The debate over accommodation versus prediction is a lively one within the philosophy of science; articles are easily found in such journals as *Foundations of Chemistry* and *Studies in History and Philosophy of Science*.

Chemical classification is an interesting area of study, one which is related to studies on nomenclature and language. Klein and Lefèvre (2007) examine the different ways in which chemists classified substances during the eighteenth century and how, if at all, this changed after the so-called Chemical Revolution of Lavoisier. Cohen (2004) looks at the various roles played by chemical tables in the century and a half before the discovery of the periodic law. Although his focus is not the periodic table, much of his discussion can be applied to it. See Marchese (2013) for an information visualization look at the periodic table. Those interested in the many different forms of the periodic table should see the work of Van Spronsen (1969) and Mazurs (1974).

There is no lack of articles and books about specific elements. The classic work is by Weeks (1968), although the last edition was published almost 50 years ago. The story of the transuranic elements and the role of American scientists can be found in a volume written by Seaborg and two of his colleagues (Hoffman, Ghiorso, and Seaborg 2000); it is an interesting albeit somewhat biased account. Seaborg was a prolific writer not only of scientific works but also ones of a more biographical nature; several of his books, such as *A Chemist in the White House* (1998) and *Adventures in the Atomic Age* (2001), will be of interest to those who are interested in the role of American scientists in the federal government after World War II.

Atomic weight has a long and controversial history, and a good overview is offered by Holden. A brief version of his article was published in the print edition of the IUPAC's magazine *Chemistry International* (2004a) and a more comprehensive version is available on the IUPAC web site (2004b). The longer version provides citations for those interested in the history of the International Atomic Weights Committee and its current incarnation, as well as the debate over which standard to use as the basis for determining atomic weight, $H = 1$ or $O = 16$. Theories regarding the unity of matter stretch back to the ancient Greeks, but Prout's hypothesis had specific influence on chemistry during the nineteenth and early twentieth centuries; for more, see Brock (1985).

The history of radioactivity in America is often tied to histories of the Manhattan Project and atomic weapons, both of which tend to focus more heavily on physics rather than chemistry. Badash (1979) provides a history of the science of radioactivity during the first two decades of the twentieth century as well as touching on some of the public craze and the response of government and industry which followed the discovery of the element radium. The cultural history of radioactivity has lately become an area of interest for historians. Lavine (2013) looks at the intense fascination Americans had for radioactivity and x-rays in the era before World War II and ties it to the postwar nuclear culture. The role of radioactivity in medicine is also a rich field. Slaughter (2013) examines the transformation of radium therapy from a new wonder drug to a standard treatment for cancer during the first decades of the twentieth century.

The American Chemical Society is far from the only chemical association in the United States, but it is the largest and the oldest still in existence. The ACS is keenly aware of its history and of the history of chemistry and has marked its major anniversaries with the publication of historical volumes. The fiftieth anniversary volume (Browne 1926) included several chapters on the origins and development of the ACS as well as chapters reviewing the progress of various branches of chemistry in America during the previous 50 years. The centennial volume (Skolnik 1976) focused more on the ACS itself and its role in the promotion of chemistry within the United States at various levels, from popular to the federal government, and from education to industry; Reese (2002) follows up on the ACS in the 25 years since the centennial. The Division of the History of Chemistry publishes a quarterly journal, *Bulletin for the History of Chemistry*, which contains articles and reviews, although its content is not limited to the history of chemistry in America. The *Journal of Chemical Education*, the main publication of the Division of Chemical Education, has traditionally included historical articles in its pages and has been a hotbed of debate over the periodic table; in commemoration of the bicentennial, the July 1976 issue included several articles on the history of chemistry in the United States from the colonial era to the twentieth century.

Chapter Five

COMPUTER SCIENCE

Stephanie Dick

"Computer science" is an academic discipline focused on the study of computation, its theory, principles, and problem-solving possibilities. Today, computer science is a vital area of American academia and is institutionalized in some form at nearly every university in the country. However, as late as the 1960s – some two decades after the advent of modern digital computers themselves – "computer science" was still a contested designation.

The discipline was hard won during the 1950s and 1960s by certain communities of computer practitioners, primarily people who came to computing from academic backgrounds in mathematics, physics, and engineering. They sought to create an academic discipline of computer science for professional, intellectual, and practical reasons. They believed that computation ought to be the subject of scientific inquiry for its own sake, rather than a tool used only in service of other domains. They believed that computation raised significant theoretical questions that could not be adequately addressed within existing disciplines. They wanted to forge an academic discipline and with it, respected professional positions, institutional spaces, and independent resources for the study of computation. Eventually they were largely successful. However, their efforts were challenged by people who saw computers differently and by practitioners of other disciplines who had something to lose – students, funding, space, intellectual territory – by the creation of an independent "computer science." The establishment of computer science is best understood in the context of early American computing more generally.

The Computer

The advent of modern digital computing took place primarily in Britain, the United States, and, to a lesser extent Germany, during the 1940s (Rojas and Hashagen 2000). Digital computers of the kind that are today found in nearly every corner of American

A Companion to the History of American Science, First Edition.
Edited by Georgina M. Montgomery and Mark A. Largent.
© 2016 John Wiley & Sons, Ltd. Published 2020 by John Wiley & Sons, Ltd.

culture can be programmed and reprogrammed to perform countless tasks by controlling and manipulating binary storage systems, each unit of which can have a value of "1" or "0" usually made manifest using electric current, hence their designation as "digital" (Ceruzzi 2003: 1–12). Historians of computing are hesitant to identify the first such computer. That designation requires the selection of a set of adjectives – digital, electronic, universal, stored-program, etc. – that include certain machines and exclude others (Haigh, Priestly, and Rope 2013: 8–10). Moreover, identifying the "first" computer obscures the fact that, contrary to abounding narratives of "revolution" that accompanied their early development, computers did not represent a decisive break from the machines and practices that preceded them. Computers rather participate in longer mathematical, technological, and social histories.

Many of computer science's core theoretical tenets were drawn from mathematics, logic, theories of language, engineering, information theory, cybernetics, even theories of mind that predated actual computers by a decade or more (Cordeschi 2002; Davis 2000; Mahoney 2011: 119–96; Mindell 2004). This raises a significant question for the history of the discipline: how is *computer science* related to the *computer*?

Historians of science often question the historical relationships that obtain between theory and experiment, science and technology, idea and application – usually discovering that they are more complex than any standard "top-down" or "bottom-up" models have allowed (e.g. Galison 1987). "Computing" – a word I mean to subtend both the development and use of computers and the academic discipline of computer science – resists both characterizations. (A broader use of the term "computing" would also include automated data-processing infrastructure and information technology of other kinds, particularly in military and business, that I will not emphasize here, e.g., Haigh 2003; Volmar 2015.)

Some historians have suggested that the earliest computers in the United States and elsewhere were born out of attempts to implement, actualize, or apply existing mathematical theories of computation – especially the so-called "Turing Machine," a notional mechanism imagined by British mathematician Alan Turing in the mid-1930s (Copeland 2014; Davis 2000; Dyson 2012; Turing 1937). Turing Machines have become a central tool for theoretical computer science and some early computer developers knew about them in the 1940s, but, as historian Thomas Haigh and others have convincingly shown, "the relevance of [Turing's] work to actual computers was not widely understood in the 1940s" (Haigh 2014: 37; Priestly 2011: 123–56). Early computers were not simply applications of what would later be understood as "computer scientific" theory. They were rather engineering feats undertaken by particular institutional communities in response to their perceived calculation needs (Haigh 2014). Neither was computer science an abstraction incumbent upon the practical developments of computer technology, many of its questions and tools having existed before them.

And yet, computer science would have been an unlikely candidate for academic disciplinary status were it not for the proliferation of actual computing machinery through the academic world during the mid-twentieth century. It would also have been an unlikely development if many academically trained people had not found their way into computer development and research centers during the 1940s and 1950s. Moreover, computer science is linked to the computer because advocates of the former advanced a particular vision of the latter in support of their efforts, one that ran contrary to other contemporaneous visions. Computer scientists were not just advocating for a body of theory but also for a way of thinking about and using those machines.

Early computers were primarily seen as *tools*. Computer science advocates wanted computers to be treated as objects of scientific inquiry in their own right. More, early American computers were used primarily as numerical calculators – they were faster and more powerful than the calculating machines that preceded them, but calculators nonetheless. Advocates of computer science proposed that computers might be used to encode *nonnumeric* information as well – they wanted to program computers to play games, translate languages, and have conversations (e.g. Feigenbaum and Feldman 1963). They wanted to design and experiment with different generalized methods for solving classes of problems by computers. They wanted to explore how much or how little computers and their software could be made to resemble human brains or human minds. They wanted to study the more general question of what could and could not be computationally automated, both in theory and in practice.

The early history of computing is neither a case of "top-down" application of theoretical principles, nor a case of "bottom-up" abstraction of principles from physical systems. Neither "computer science" nor "the computer" should be considered primary and the other secondary either in terms of significance or in terms of historical development (Haigh 2014). Nor did computer science develop in isolation from the development of computers and other computing communities (Ensmenger 2010a). Academic computer science was just one of many communities seeking to harness computing technology for their professional and intellectual ends in the mid-twentieth century.

Early Computing during World War II

In the United States, computing was at first a labor-intensive activity in which human workers – most commonly women – performed calculations by hand and operated mechanized unit record equipment, such as tabulators, keypunches, and card sorters (Abbate 2012: 1–112; Ceruzzi 1991; Grier 1996; Light 1999). As late as World War II, the word *computer* actually referred to the laborer, and not to the machinery. Together, groups of human computers would execute complex calculations that project directors had broken down into sequences of simpler arithmetic steps.

Human computers were often tasked with producing volumes of calculation tables from which scientists, engineers, military personnel, and other technical practitioners could look up solutions to numerical problems. Like the machine computers that came after them, human computers were usually employed in service of government, scientific, or military projects. For example, the federal government funded a large-scale human computer project during the Great Depression, in an effort to create jobs (Fritz 1996: 14). Human computers also created bombing and ballistics tables during World War II and some women computers worked in Los Alamos performing calculations related to the development of atomic weapons (Howes and Herzenberg 1999: 93–114).

America's first computers were also preceded by less complex machines developed to address bookkeeping and administration problems in finance, insurance, accounting, and stock control (e.g., Agar 2003; Haigh 2001; Hughes and Hughes 2000).

Punched Card Accounting Machines, of the sort that contributed to International Business Machine (IBM)'s mid-century successes, were central artifacts in the landscape of 1940s calculation. Along with human computers, these machines and their operators constituted the social and material landscape that preceded the modern computer.

Many of the earliest American computers were developed to take over specifically for those people and machines. In fact, some of the first programmers were women computers recruited to work with the machines that displaced them (Abbate 2012). For example, the ENIAC was developed at the Moore School of Engineering at the University of Pennsylvania to perform ballistics trajectory calculations that previously occupied more than 100 women computers at the US Army Ballistics Research Laboratory during the early war years (Polachek 1997). Six of the first ENIAC programmers were women who had worked on those calculations before (Abbate 2012; Fritz 1996: 13. For a history of the verb "to program" in the context of ENIAC see Grier 1996.) These human computers were members of one of the earliest computing communities, along with ENIAC's more famous architects, Hungarian mathematician John von Neumann and Americans Herman Goldstine, J. Presper Eckert, and John Mauchly.

Like ENIAC, most early American computers were government-funded and were first tasked with calculations for the war effort. Even machines like the IBM ASCC/Harvard Mark I computer – another candidate for the American "first" – that were originally imagined "for application to scientific investigations" were also, even primarily, directed towards military calculations (Aiken 1999: 11; Cohen 1999: 159–67). According to historian Paul Edwards, as late as "the early 1960s, the armed forces of the United States were the single most important driver of digital computer development" (Edwards 1996: 43). This was not a uniquely American phenomenon. In Britain as well, the earliest computers were developed in support of efforts to decrypt machine-encrypted German communications (Copeland 2006). However, the development of computing technology in the United States was of a different scale, especially given the massive mobilization of resources for scientific research and development that followed the war.

During the 1940s, computers in the United States were largely mobilized in service of weapons development and deployment. Rear Admiral Grace Hopper, a programmer of the IBM ACSS/Harvard Mark I described that time as follows:

> Most people have forgotten why we were so greatly in need of computation during the war. World War II saw an almost complete change in our weapons systems. New weapons required numerical techniques for their design and deployment, techniques that depended upon computation. [...] The pressure for that computation was very great; everything was "hurry up, do it yesterday." So pressure to keep the only computer we had running was very, very strong. Mark I ran 24 hours a day, seven days a week – a rough task for a small crew (Hopper 1999: 186–7).

New missiles, naval mines, and depth charges required complex trajectory calculations. The development of atomic weapons during World War II and the more powerful hydrogen bombs of the Cold War required the execution of complex implosion simulations and critical-mass calculations (Bashe et al. 1989: 130–6; Galison 1997: 709–27; Haigh, Priestly, and Rope 2014; Metropolis and Nelson 1982).

Computers enabled the development of weapons and warfare techniques that would not have been previously possible. Computers were also a central component of new defense strategies meant to protect the United States from attack in the contexts of conflict that these new weapons and techniques afforded. Computers came to define the so-called "command and control" infrastructure that would have enacted the American nuclear response envisioned by strategic planners (Bracken 2012; Slayton 2013). The United States became a global military power near the end of World War II and emerged a superpower in its wake. Computing technology was a crucial condition of possibility for the politics, the strategies, the systems, and the weapons that dominated twentieth-century warfare.

Computer Centers and Early "Computer Science"

Early computing was part of the so-called "military–academic–industrial complex" of the postwar period (Edwards 1996: 43–73; Leslie 1993). American developments in computing benefited from the common Cold War belief that American physics and engineering were largely responsible for the Allied victory in World War II and that continued dominance in science and technology would stave off the perceived Soviet threat (Kaiser 2010; Wolfe 2012).

By way of military research and development bodies like that of the US Air Force and through funding organizations like the Defense Advanced Research Projects Agency (DARPA) and the National Science Foundation (NSF), government resources were funneled to important academic computer centers (Aspray 2000a). Early computing practitioners moved with great facility and frequency between universities and key industrial research institutions like IBM Research Labs, the Burroughs Corporation, and Bell Labs. Government and military-funded institutions like the RAND Corporation and the Argonne National Laboratory, often working in close collaboration with the private sector, were also key sites in the early development of computing. Research in these places was not always explicitly or intentionally oriented towards military ends, but all were influenced by the influx of funding and motivation for scientific research that developed in the postwar period.

In some cases, computers were introduced to existing service departments like the Numerical Analysis Department at the RAND Corporation and the Applied Mathematics Division at the Argonne National Laboratory that provided data-processing services to their host institutions. Elsewhere, including at many universities, new centers, laboratories, or bureaus were created for this purpose, usually headed up by faculty from electrical engineering, physics or mathematics departments. Famous examples include Howard Aiken at Harvard University, Mauchly and Eckert at the Moore School, and John von Neumann who immigrated to the United States from Hungary in 1930 to take up a position at Princeton University (Aspray 1990b, 2000b; Ceruzzi 2000; Cohen 1999, 2000; Van der Spiegel et al. 2002).

In these centers, computing was an interdisciplinary activity. Early computers were nodes around which practitioners from all manner of backgrounds assembled, including university faculty, machine operators, businessmen, military personnel, and others (Ensmenger 2010a: 119–24). This computing had in common with other twentieth-century

American research fields that involved the study and development of information technology – cybernetics, operations research, information theory, and systems engineering central among them (Cohen-Cole 2007; Galison 1994; Hughes and Hughes 2000; Kline 2011; Mindell 2004).

These centers were also the sites at which the first research projects in areas that would later be subsumed under "computer science" were undertaken. For example, one of the earliest efforts to develop a computer program that would simulate human thinking – a computer science subfield that would be named "artificial intelligence" in 1956 – was undertaken at the US Air Force-funded RAND Corporation in 1955 (Cordeschi 2002: 176–86; Crowther-Heyck 2005: 215–32). The project reflected the heterogeneous demography of 1950s computing. It was a collaborative effort of Herbert Simon, an economics professor at Carnegie Technical Institute (soon to become Carnegie Mellon University) who began consulting at RAND in 1952, Allen Newell who left a PhD program in mathematics at Princeton University in 1950 for a research position at RAND, and John Clifford Shaw, a member of RAND's Numerical Analysis Department. Newell would eventually also become faculty at Carnegie Mellon University where both he and Simon were instrumental in creating one of the first two "standalone" computer science departments in the country in 1965.

However, research at computing centers was certainly not guided homogeneously or even primarily by the interests of academics, even at universities. These centers remained primarily oriented towards the provision of calculation services for other departments and projects. Moreover, not all members of a computing center or even an individual project agreed about the character and prospects of computers or the meaning of research done with them. For example, Simon was most interested in their early AI program as a step toward modeling human reasoning whereas Shaw was primarily concerned with finding ways to navigate the technological limitations of early computers (McCorduck 1975; Dick 2015). Nathan Ensmenger has aptly identified the 1940s and 1950s centers as an example of historian of science Peter Galison's "trading zones:"

> [T]he electronic computer created around it an interdisciplinary space in which researchers from a variety of backgrounds could productively interact. In such trading zones, researchers did not have to agree on the universal meaning or significance of the instrument but only on local protocols and practices (Ensmenger 2010a; 122; see also Galison 1997).

Computing practitioners who assembled at computer centers coordinated their efforts to keep computers running and develop different uses and applications for them. But, coming from different backgrounds, they recognized quite different professional, intellectual, and practical stakes in computing.

From the heterogeneous assemblies of people that converged at early computer centers, several distinct computing communities emerged, each held together by a particular vision of how computers should be used, how computing professionals should be trained, and what relationship computing should have to existing professions, disciplines, and industries. Computer engineering, software engineering, hackers, IT professionals, management consultants, and of course computer scientists were just some of the computing communities that evolved through the second half of the twentieth century (Ensmenger 2010a; Haigh 2003, 2010a; Hughes and Hughes 2002). The next section suggests that in order to create their communities, practitioners promoted a particular image of the computer and a corresponding identify for themselves, often disassociating the new technology from certain of its inaugural associations.

Changing Associations

Because of their originating contexts, early computers were associated with war, with women, and with clerical and administrative labor; they were primarily identified as *tools* designed to perform calculations in service of military and scientific development. However, almost from the first, some practitioners sought to wrest computers from those early associations. New visions of computing were correlated with the professional ambitions of those who crafted them.

For example, because human computers were primarily women, early computing was associated with "pink collar" work (Davies 1982; Yates 1993; Zunz 1990: 116–21). Programming, in particular, was associated with feminine "*soft*ware" while "*hard*ware" was associated with more traditionally masculine characteristics (Edwards 2003). Further, as mentioned earlier, some early programmers were women. However, throughout the 1950s and 1960s, computer specializations, including programming, became increasingly and eventually overwhelmingly male-dominated (Ensmenger 2010b; Haigh 2001, 2010a). This transpired first "with the transformation of the feminized clerical work of 'coding' into the highly masculine, seat-of-the-pants 'black art' of programming in the 1950s" and later with the "embodiment of certain masculine values into the hiring procedures of the industry" (Ensmenger 2010b: 20). The masculinization of computer specializations disassociated computing from its original context as women's work.

Certain communities – including some feminists and so-called "hackers," largely based in Cambridge, MA and parts of California – believed that computers were not inherently instruments of armed conflict and worked to disentangle digital computers from their original military associations (Levy 2010; Turner 2006). Many saw the computer instead as an instrument of democratization and liberation: "by the 1990s, a metaphor born at the heart of the military research establishment had become an emblem of the sort of personal integrity, individualism, and collective sociability that so many had claimed the very same establishment was working to destroy" (Turner 2006: 16). Especially in light of the emergence of personal computing and the internet through the 1980s, computing was refashioned as an implement of liberation and equalization.

The computer, it turned out, was no single thing after all and it did not force any single experience or transformation on its users. Its significance and its possibilities depended on the intellectual, political, and professional commitments of different user communities. Recognizing this, Michael Mahoney called for a historiography that takes computing communities as the focal point for the history of computing (Mahoney 2005). Computers *are* what different communities have fashioned them to be.

The creation of an academic discipline of computer science was tied to yet another set of efforts to disassociate the computer from its early contexts. Early computers were seen as *tools*, glorified *calculators*, that worked in service of scientific, military, and administrative projects like firing table production to numerical weather prediction (Cohen 2000; Edwards 2010: 111–38). Advocates of computer science wanted to explore what else computers could do. Many worked on nonnumeric computation research projects like game-playing, language translation, and theorem-proving projects, not directly undertaken "in service" of some other discipline or project (Dick 2011; Ensmenger 2012b; Gordin 2014; MacKenzie 1995). Studies of nonnumeric, symbolic, and algebraic computation proliferated during the 1960s and were consolidated by way of conferences and journals like the Symposium on Symbolic and Algebraic Manipulation (SYMSAM), first held in 1966.

Computer science advocates also wanted to study what computers could not do – what the limitations of computation and automation were. Those associated with artificial intelligence in the mid-1950s thought the computer had less in common with ancestral tabulating machines and more in common with human minds or brains – they could do what people could do (e.g., Edwards 1996: 239–75; Heyck 2008; Kline 2011; McCorduck 2004). Others thought that computers would be more like "slave," capable only of performing base mechanical operations that did not capture human creativity or intuition (on the complex and often problematic anthropomorphic language used to describe computers, see Eglash 2007). All wanted a mathematical formalization of what could and could not be automated. They wanted to develop a theory of computing that could illuminate and standardize the ever-expanding prospects of the new technology.

However, the desire for theory was bound up with other professional concerns. Computer science advocates also wanted a stabilized and respected position for themselves in the academic landscape. One influential computer science pioneer, Edward Dijkstra, who emigrated to the United States from his native Netherlands in the early 1980s, described his situation as follows:

> I had to make up my mind, either to stop programming and become a real, respectable theoretical physicist, or to carry my study of physics to a formal completion, only, with a minimum of effort, and to become … what? A programmer? But was that a respectable profession? After all what was programming? Where was the sound body of knowledge that could support it as an intellectually respectable discipline? (Dijkstra 1972, cited in Ensmenger 2010a: 112).

People like Dijkstra wanted to create a "respectable profession" for computing practitioners to recover what autonomy, resources, and respect would be lost in transitioning from an existing discipline to the then service-oriented world of computing.

Professionalization and Institutionalization

Although, as Ensmenger has observed, theory was a "necessary precondition to professional development" in the 1950s and 1960s, it was not sufficient (Ensmenger 2010a: 114). Computer science advocates promoted theoretical concerns in tandem with their mobilization of many other disciplining technologies like the creation of societies, journals, and awards.

Computer science departments began to emerge in universities in the 1960s. But before that, the academic subset of computer practitioners who would eventually seek disciplinary status converged in other professional institutions. Perhaps most notably was the largest American computer organization – the Association for Computing

Machinery (ACM) (Haigh, Kaplan, and Seib 2007). The ACM was founded in 1947 in conjunction with a Symposium on Large-Scale Digital Machinery, hosted at the Harvard University Computation Laboratory (Ensmenger 2010a: 170–5). The proposal to create an association was presented by Samuel Caldwell, a professor of electrical engineering at MIT. His academic orientation was shared by the association's membership, which grew significantly over the next decades, reaching 1113 members by 1951 and 22,761 members by 1969 (Ensmenger 2010a: 170). The ACM was, and remains today, the largest association of academic computer practitioners in the United States and its members organize multiple conferences, journal publications, and awards that serve to coordinate, circulate, and consolidate computer science research.

Also important for the development of American computer science is the Institute of Electrical and Electronics Engineers Computer Society (IEEE CS). IEEE (the umbrella organization) was created in 1963 with a merger between the American Institute of Electrical Engineers (founded in 1884) and the Institute of Radio Engineers (founded in 1912). Beginning in 1946, subsets of that organization took to the creation of subcommittees focused on the place of computing among electrical engineering and its future. IEEE CS emerged out of those subgroups and was formally institutionalized in 1971 (Wood 1995).

Most advocates of computer science were members of one or both of the ACM and IEEE CS. They published in the journals supported by these organizations and presented results at the conferences organized by them. Moreover, these two organizations served as a platform from which members could engage or oppose other computer organizations that sought to professionalize or standardize computing research and training in their own way. For example, the ACM often had a conflicted relationship with another of America's largest computer professional organizations, the Data Processing Management Association (a descendent from the National Machine Accountants Association and later called the Association of Information Technology Professionals) that was (and still is) a powerful organization of computer practitioners with closer ties to business and industry than to academic institutions (Ensmenger 2010a: 170–91; Haigh 2003). Indeed, computer science was always being developed in opposition to other communities – like the Data Processing Management Association – and to other disciplines – like mathematics and electrical engineering.

The battle for academic computer science began in earnest in the late 1950s. By the 1970s "computer science" departments were common at US universities. However, the character of "the discipline" depended a great deal on the contingencies and specificities that obtained as individual institutions carved out a disciplinary place for computing. This is evidenced in part by the fact that the study of computing for its own sake was called many different things – from "computer science" and "computing science" to "informatics" and "synnoetics," and even more names were considered (Ensmenger 2010a: 118).

Computer science manifested quite differently across institutions. At some universities, computer science never became an independent department but was rather incorporated into an existing one. At MIT, for example, the School of Engineering faculty voted in 1974 to rename the existing Electrical Engineering department "Electrical Engineering and Computer Science," instead of creating a new department. Elsewhere, computer science was integrated with mathematics, applied mathematics, or statistics. The creation of standalone departments – first at Purdue University in 1963 and Stanford University in 1965 – were in fact the exception. And where computer science *was* institutionalized as an independent department, it was constituted of people, spaces, students, and funds that had once belonged to other departments. Computer science was not forged from nothing, it was pulled together from existing institutional resources and uniquely so according to the specificities of each institutional context (Aspray 2000a; Ceruzzi 1999; Ensmenger 2010a: 118–24).

In part because computer science would be created by the appropriation or reallocation of resources, there was resistance. Some questioned whether any scientific discipline should be centered around a technological artifact. Others asked whether computer science raised any questions that were not already addressed by existing disciplines and best left that way. At institutions like Harvard and Princeton that emphasized liberal arts education, there was resistance to computer science on the grounds that training in computer science was inherently "industry-oriented" and "applied" in character (Aspray 2000a). Early computing was woven together out of existing technologies and ideas – a "grab bag of theories and techniques drawn from other disciplines" and the custodians of those domains sought to hold on to them (Ensmenber 2010a: 116).

Academic computing gained ground through the 1960s and 1970s, in part through efforts (often fraught) to coordinate and standardize undergraduate computer science training. In particular, the ACM organized many workshops and subcommittees on the topic of curriculum development throughout the 1960s. These culminated in a 1968 ACM report – dubbed "Curriculum 68" – that had a significant impact on the standardization of academic computer training across the country (Austing, Barnes, and Engel 1977; Gupta 2007).

Curriculum and degree program development was not secondary, following upon the creation of an academic computer science. Instead, the perceived need for standardized training throughout the 1950s and 1960s was a key issue on which the creation of the discipline was promoted. As the computing demands of American universities, businesses, and government exploded, so too did the demand for trained computing personnel. At that time, training in computing was anything but standardized (Aspray 2000a; Aspray and Williams 1994; Ensmenger 2010a: 119–21; Mahoney 1990). Many of the "computer centers" discussed in the previous section offered training programs, but these differed widely in content and methodology. In fact, by the 1960s, a veritable crisis had emerged in which there was a serious shortage of computer practitioners compounded by a lack of standardization in their skill-sets, compounded yet again by a growing fear that computer programs were replete with errors – bugs that had the potential to disrupt the financial, military, governmental, and academic institutions that increasingly depended upon them (Ensmenger 2010a: 195–221; MacKenzie 2001: 23–61; MacKenzie and Pottinger 1997). In opposition to more industry-oriented suggestions from the DCMA and others, advocates of academic computer science proposed that the best training would be based on theoretical grounds and administered in an academic setting.

Although even the name of computer science was still up for debate by the 1960s, the discipline had become a presence at many universities and was bolstered by standardized curriculum, professional organizations, conferences, and their proceedings in which most computer science results were circulated. As discussed in the next section, although proponents of the discipline were certainly motivated by questions of professionalism, resources, and pedagogy, a constellation of theoretical and intellectual interests was also central to their efforts.

Theorization

The first *Encyclopedia of Computer Science* was published in 1976, and its 1500-plus pages were meant to give an overview of the research areas that constituted the discipline (Ralston 2004; Ralston et al. 1976). The editorial board offered 10 headings under which the contents of that volume would be organized – "Software," "Hardware," "Computer Systems," "Basic Terminology," "Theory," "Mathematics for Computer Science," "Applications," "Management, Society, Economic and Legal Aspects," "Professional and Educational Aspects," "History" (Ralston et al. 1976: xiv). Each of those 10 headings was in turn elaborated by the enumeration of several pages of sub-headings. Without question, the discipline was heterogeneous and sprawling, even or perhaps especially in its early years.

Computer science now subtends multiple research areas from artificial intelligence to the theory of programming languages. Computer scientists write programs, craft new programming languages, design algorithms for solving different kinds of problems, implement and experiment with them, prove theorems about those algorithms, develop theories about what is computationally possible and what can practically be done – what can and cannot be computed. Computer science defies the traditional dichotomy often seen between "theory" and "experiment," since computer scientists work with abstract mathematical formalism but they also work with machines, write programs, and run experiments.

Many influential pioneers of theoretical computer science – among them John von Neumann, Michael Rabin, Edward Dijkstra, Donald Knuth, and Stephen Cook – came from backgrounds in mathematics and physics. They sought to formalize and explore how complex certain kinds of problems were and whether or not an algorithmic method existed for solving them. In so doing, they drew from a tradition that preceded the computer all together – a logical tradition that went back at least to the work of George Boole in the eighteenth century. In the section "Changing Associations" above, I proposed that many computing communities emerged by intentionally disassociating the computer from its early contexts and connotations. Theoretical computer science, in a sense, hinged on a disassociation of "computing" from the computer itself.

Ensmenger identified an early step in this process with von Neumann's exploration of the so-called "stored-program concept." Von Neumann imagined that computers could store their instructions in memory instead of having operators manually "hard wire them in" for each given task, as was required for many early computers (Haigh, Priestly, and Rope 2014; Von Neumann 1993). The storage of instructions would make computers "universal," meaning that they could be made to perform any computable task by way of stored instructions while their physical architecture remained constant. Ensmenger proposed that "by abstracting the logical design of the digital computer from any particular physical implementation, von Neumann took a crucial first step in the development of a modern theory of computation" (Ensmenger 2010a: 138). However, the logical exploration of computing also drew from a mathematical tradition that long preceded the advent of actual digital computers in the mid-1940s (Aspray 1990a; Davis 2000).

One key piece of that tradition was an imaginary or notional mechanism called a "Turing Machine," presented by British mathematician Alan Turing in 1936 (Turing 1937). Turing created this machine to solve a famous mathematical question called the "decision problem" or *Entscheidungsproblem* posed in 1928 by German mathematicians David Hilbert and Wilhelm Ackermann (Hilbert and Ackermann 1928). Put very simply, the decision problem asked whether or not one could always come up with an algorithm that, when given a proposition from first-order logic, would always return "yes" or "no" in a finite number of steps, where "yes" indicated that the proposition was universally valid and "no" indicated that it was not (Petzhold 2008: 199–321). Hilbert and Ackermann believed there would always be such an algorithm. In 1936, Turing showed this was not the case.

Turing solved this problem by constructing a simple notional machine that could follow instructions consisting of only three basic operations: it could read, write, and erase 1s and 0s on a tape of unlimited length (Turing 1937: 241–6). He then proved the very surprising fact that this simple mechanism "can be used to compute any computable sequence" (Turing 1937: 241). That is to say, any algorithm, any process that can be executed by way of a finite set of rule-bound steps, can be formalized as a set of instructions for this simple mechanism. Of course, it might take this machine a million years or more to complete complex algorithmic tasks (hence its designation as an "imaginary" or "notional" machine rather than a real one), but it could in principle execute them all nonetheless. That meant that anything a modern electronic digital computer could do, this notional machine could do – it was a formalization of what tasks were computable, in principle, and what tasks were not.

Crucially, Turing showed that some tasks were not computable – the decision problem has a negative answer. There are some logical propositions that cannot be "decided" by any algorithm. Turing showed that computation and algorithmic processes have intrinsic formal limitations. Computers could do a lot of things, but they could not do everything. Turing also showed that computers could be treated as mathematical constructs, and their limitations explored. The Turing Machine was understandably taken up by computer science advocates with an interest in the limits of computation, and it remains perhaps *the* central construct of theoretical science (Rabin and Scott 1959).

However, actual computers were not just limited by these, in principle, restrictions on formal systems. There were some processes that were, in principle, computable that would not be feasible for actual execution on a computer. Another central tenet of early theoretical computer science emerged in relation to this problem. Within the domain of all possible computations, some were harder or more complex than others – orders of magnitude harder. Early theoretical computer scientists set out to formalize the study of problem "difficulty" in terms of the measurably different orders of magnitude of computing resources (usually with an emphasis on how many operations, in the worst case) kinds of problems require (Cook 1971).

The formal study of computational limits and of problem difficulty were early and important grounds upon which academic computer science was built. They represented problems that, although they were preceded by work in mathematics and logic, really emerged after the advent of modern digital computing newly enabled and motivated their formulation and exploration. These intellectual and theoretical concerns were given institutional form in tandem with the creation of academic computer science. Many journals and conferences were created for the consolidation of theoretical results – for example the ACM Annual Symposium on the Theory of Computing founded in 1968, and the Journal *Theoretical Computer Science* founded in 1975. Computer science results were also highlighted and canonized by the "Turing Award" – created in

1966 and awarded by the ACM each year for significant contributions to academic computing, but usually favoring contributions to theory.

These were some of the theoretical pillars that enabled computer science advocates to justify the existence of an academic discipline and to navigate the boundary between more concrete concerns of electrical engineering on the one hand and the more abstract concerns of mathematics on the other.

Modern computing did not represent a decisive break from what came before. In it were drawn together existing questions, concepts, technologies, practices, and people from human computation, early twentieth-century calculating machinery, and other academic disciplines. Computing communities emerged throughout the second half of the twentieth century by disassociating, emphasizing, and de-emphasizing different elements of the computer's heritage. Academic computer science emerged through attempts of the academically oriented ACM and IEEE CS membership to accrue improved professional and institutional standing, to establish a sound theoretical grounding for computing, and to standardize and improve the quality of computer pedagogy.

The emergence of academic computer science was at every turn characterized by tandem intellectual, professional, and institutional concerns. It involved the creation and management of multiple disciplining mechanisms including societies, curriculums, conferences, journals, and awards. At every step, the discipline developed to reflect many contingencies within the postwar American academic landscape and its numerous interfaces with industry and military institutions. So often told as a history of "revolution," the history of American computing is, in fact, a story of the selective "association" and "disassociation" of a technological artifact with different people, problems, and places that grounded individual computing communities.

Bibliographic Essay

The history of American computing literature is vast and varied. (Other bibliographic surveys can be found in Ensmenger 2012a and Haigh 2011. A collection of resources is maintained at http://www.sigcis.org/resources.) Key publications for history of computing include *IEEE Annals of the History of Computing*, *Technology and Culture*, the MIT Press, and Springer's History of Computing series.

In general, early histories of computing were histories of the *computer*, often seen as an instrument of technological revolution. Many other histories have focused on particular individuals – either academic computing pioneers or entrepreneurs – seen to have played a significant role in the invention and distribution of computing machinery.

More recently, however, academic computing historians have worked to de-center both the machine and the individual. Rather than characterizing the computer as a revolutionary technology invented by elite individuals, many historians now situate modern computing within longer trajectories of social, technological, and theoretical history. These histories emphasize the many contingencies that shaped the character of computing in particular contexts (e.g. Aspray 1990a; Ceruzzi 1983; Cordeschi 2002; Davis 2000; Grier 2005; Mindell 2004; Priestly 2011; Yates 1993).

Historian Michael Mahoney advanced a very influential vision of "the computer" as a *protean machine* with no singular history or inherent character of its own (Mahoney 2005). He argued that there should be as many histories of computers as there are development and user communities who mobilized computing in service of their ambitions and perspectives. This historiographic project pointed researchers towards *communities* – those who made computers useful and usable in different corners of American culture, from big business to feminism – as the locus of historical investigation. Many community-oriented histories have emphasized the significance of previously underemphasized practitioners like IT workers and women.

Mahoney also called for increased historical interest in *software*, rather than the hardware of computing machines themselves, given that software is the mechanism by which computers can be harnessed to serve the needs of different communities and the means with which most people interact with them (Ensmenger 2012a: 760–71; Mahoney 2005: 129). Historians have only just begun to scratch the surface of software and community-oriented computer histories, but many are currently working in these areas.

Very little has been written about the history of academic computer science in the United States. Some have argued that the disciplining of computer science is not a national story but rather an international one. They have begun to explore the interactions between American and Soviet computer scientists, as well as international efforts to develop standardized theory and programming languages (e.g. Nofre 2010; Tatarchenko 2013). Some historians have explored the intellectual lineage of computation theory in mathematics and logic and the development of intellectual grounds on which computer science was theoretically justified (Davis 2000; Priestly 2011: 67–98, 123–56; Mahoney 2011: 121–96; Tedre 2014). Others have explored particular or comparative institutional stories in which computer science found a departmental home at different American universities (e.g. Aspray 2000a; Minker 2007; Rice and Rosen 2004). Of all subfields of computer science, artificial intelligence, robotics, and programming language development have perhaps received the most attention (e.g. Adam 1998; Barbrook 2007; Cordeschi 2002; Ensmenger 2012b; Kline 2011; Priestly 2011; Suchman 2009).

Chapter Six

CONSERVATION BIOLOGY

Christian C. Young

The history of conservation biology in the United States includes a broad range of related fields and efforts. In the smallest frame, the science of conservation dates only to the mid-1980s, when a group of professional scientists trained in biology began working under the imprimatur of a new organization on the expanding array of issues that loosely fit under the heading of environmentalism. That organization, the Society for Conservation Biology, has helped guide collaborative research efforts worldwide for nearly 30 years. A longer history of conservation biology, however, must come to grips with older usage of the word "conservation." In some contexts, conservation was made synonymous with preservation of nature and protection of wilderness, while in others it was about sustainability and planned harvest of natural resources. The meanings of conservation represent a changing and challenging history. Focusing on the involvement of the scientific community in conservation efforts offers one means of simplifying that history and providing a coherent narrative. Such a narrative also reflects the integration of multiple scientific specialties into conservation biology after 1980, even as conservation biologists worked to apply their research toward explicit human needs.

Nineteenth-Century Conservation and the Continuum of Concern

Early efforts in the United States that blended the impetus for conservation with the reasoning approach of science were spread among a variety of practitioners. Some well-known individuals, like Henry David Thoreau in the middle part of the nineteenth century, seemed especially keen to protect nature in places that provided tranquility. Thoreau was not typical. Still, his writings touched on a growing awareness of how Americans experienced different degrees of connection to the natural world, and that those connections had begun to change rapidly. Especially in long-settled areas in New England, the forests that housed and surrounded previous generations were a distant

A Companion to the History of American Science, First Edition.
Edited by Georgina M. Montgomery and Mark A. Largent.
© 2016 John Wiley & Sons, Ltd. Published 2020 by John Wiley & Sons, Ltd.

memory. The farms that replaced the forests shrunk as cities grew and transportation to other agricultural areas improved. In some places, the trees came back in park-like settings, reminding a new generation of the forest legacy but changed utility. Systematic reports from naturalists of such changes began to accumulate at mid-century. A more comprehensive account from these reports and their importance followed as professionals within the scientific community developed a coherent literature containing observations of their own.

Massachusetts became one of the earliest states to inventory its forests with a report filed by George Emerson in 1846. Emerson described the processes of growth and decay that indicated a healthy forest, and noted that forests in the Northeast had been subject to quite different patterns over the previous hundred years. Human settlement altered the landscape. He suggested, in a speculative flourish, that the lack of trees might correspond with fewer lightning strikes on barren hills, which could lead to less rainfall. His speculations missed the mark, but illustrate the growing awareness of changes that might have farther reaching effects. Emerson's report broadly indicated a loss of natural beauty in the New England forests, connecting with Thoreau's aesthetic and philosophical sensibility. He also provided calculations at the other end of the continuum of concern, estimating that forest products contributed to industries ranging from construction to railroads, in addition to wood for household fuels. The total value would be in the millions of dollars for industries. Without a regular harvest of wood, those industries would founder, and fuel costs of up to $50 per household would result. The basis of the economy in Massachusetts and elsewhere closely followed the fate of its forests (Emerson 1846). The same result would be established repeatedly in conservation studies that followed.

In 1864, George Perkins Marsh provided a groundbreaking account of the relationship between humans and nature. Marsh had the informal training typical of a naturalist at the time, as well as a legal education and significant political experience as a statesman. His concern for the natural world and changes he observed related directly to the consequences of those changes he saw for human communities. He took a synthetic view of human reliance on natural resources, connecting geological features to the living world, with geographic foundations of expanding cities and shifting rural areas. As a native of Vermont, Marsh compared the familiar landscapes of the Northeast with descriptions of parts of the planet he had never visited, but where human settlement had altered plant and animal relationships in significant and seemingly irreversible ways. Amid the expansiveness of American bounty, he observed the same tendencies that led to destruction and waste on other continents, in places where natural resources had once seemed boundless (Marsh 1965).

In describing the conditions of the natural world, Marsh wrote of stability and balance, ideas that relied on a philosophical tradition but had not yet been tested by science. His was a common-sense worldview, like Thoreau's, where changes that upset the balance could be viewed as a threat to the future of certain resources. Forests and other forms of natural vegetation, when replaced with agricultural crops, would not return on their own. The loss of such landscapes had unknown consequences for generations to come. This seemed an obvious enough conclusion to Marsh, though his was a fairly solitary voice in an era of industrial growth and settlement. Marsh was doing more, however, than extending a vague notion into the modern world. He offered descriptions of the ways that one change could cascade into unintended consequences. His studies of geography and forestry showed that cutting trees would measurably alter the moisture level of soils and provide conditions for completely different plant growth. Forests would not simply grow to replace themselves in logged areas. Besides soil moisture, remaining substrates were subject to greater risk of erosion, and accounts of landslides confirmed the fate of hillsides where loggers cleared trees. Through careful consideration of these and other long-range outcomes, he managed to predict a time when activities of his day would lead to far-reaching consequences. His basis for such predictions involved integration of the rapidly expanding literature of natural history from explorers and observers around the world. His was not simply a local study but a model of the synthetic approach to conservation that has endured since (Marsh 1965; Worster 1994).

In the latter part of the nineteenth century, concern shifted farther along a continuum toward practical aspects of understanding the natural world. In the United States, government-sponsored scientific work provided funding to naturalists, explorers, collectors, and resource managers. The network of scientific studies done under this sponsorship expanded the scientific community and linked a variety of institutions and their individual leaders. The Smithsonian Institution and several new agencies including the US Fish Commission, the US Geological Survey, and the Bureau of Biological Survey focused attention on questions that connected with practical issues. Leaders of these agencies regularly shared ideas about the importance and direction of natural history studies and exploration. They attracted the talents of trained naturalists and amateur collectors alike. As the nation's vast resources stretching across the continent became accessible via railways and improved transportation linkages, the scientific network likewise became more coherent (Reidy, Kroll, and Conway 2007). When John Wesley Powell, for example, explored the prairies and mountains along his route to the Southwestern deserts, he created a new sense of what the future might hold for settlers in these areas. It was, at times, a sobering awareness that the bountiful forests of the Northeast and northern Midwestern states represented an exhaustible resource that would not be readily resupplied by mountainous lands. The rich farmland of eastern valleys and Midwestern flatlands gave way to more arid and less productive grasslands or desert (Worster 2001). Powell's initial surveys did not tell the whole story, however, and increased interest was generated by researchers in those areas working from land-grant universities, also funded by the federal government (Dupree 1957).

As the scientific community grew in universities, academic specialties began to supplant the more general affiliations of government naturalists. Universities built their science faculties around the fields of geology, botany, zoology, and later, ecology. The research practices of scientists continued to involve fieldwork, but biological study was increasingly located in laboratories (Allen 1978a; Kohler 2002). For those who focused their efforts on field science, the conservation of places that would provide enduring representations of the natural world expanded the continuum of concern. Plant and animal ecology both depended on the protection of space where populations of organisms persisted in relation to one another. Even as the trend toward specialization in laboratory science gained prominence, field scientists confronted the need to conserve existing and potential new study sites (Cittadino 1980; Hagen 1992a).

Conservation of Sites for Scientific Study

By the turn of the twentieth century, leading botanists at Midwestern universities had formed the framework for plant ecology. Because plant populations yielded more readily than animals to careful measurement, ecologists began to pioneer field experiments in fragments of prairie from Indiana to Nebraska, especially in places within easy distance of the institutions that also housed their teaching responsibilities. Frederic Clements at the University of Nebraska put plant ecology on the map, coining new terms for a wide range of formations and assemblages of plants. His approach established standards of comparison, and his publications served to educate the scientific community as others worked to understand in more systematic ways the relationships among plant populations (Tobey 1981). Clements focused on plants, a reasonable choice on the vast prairies of Nebraska, but his methods influenced ecologists working on smaller parcels to the east. For the expanding school of botanists and zoologists who adopted ecological techniques at other Midwestern institutions, the challenge of adapting those methods to the study of populations of plants, insects, small mammals, and birds led to refinement and expansion of field ecology. In particular, a view of these populations as communities served an urgent interest in science and conservation.

Amid the conceptual growth in ecology and the addition of new techniques came an awareness that ecology had achieved the specialized status of a profession in its own right. Ecologists established a new organization – the Ecological Society of America (ESA) – to help them communicate and coordinate their studies. Arising from the handful of universities that had established ecology departments, practitioners of botanical and zoological studies who embraced ecological techniques quickly joined. In addition to promoting the research agenda of their new profession, most members of the ESA hoped to advance efforts to preserve natural areas. The mission of the ESA, which had initially included integrating preservation, became a source of tension for ecologists who hoped to establish professional credentials through an organization distanced from broader preservation concerns. The first president of the ESA was Victor Shelford, a Chicago-trained assistant professor at the University of Illinois who built his career around the theoretical and practical study of community ecology.

Shelford began his one-year term as president of the ESA in 1915. During that first year of existence, a 25-member committee decided to assemble a list of "all preserved and preservable areas in North America" that might contain enough of what they called "natural conditions" to merit continued protection (Shelford 1926: v). In particular, they hoped to protect areas that would serve as the focus of continued scientific study for ecologists. Successive presidents after Shelford renewed this commitment to varying degrees, and the organization remained focused on scientific study along with protection of sites for years afterward. The list of places to protect required a survey of potential sites across North America (and ultimately beyond). The ongoing work of the committee also included establishment of criteria for what would justify inclusion of a site on the list. In 1926, the committee presented *The Naturalist's Guide to the Americas* as a starting point. The project, which Shelford continued to coordinate in one form or another into the 1960s, provided a backdrop on which to examine the shifting priorities of ecologists in the first half of the twentieth century. Shelford himself remained relevant throughout this period as a mentor and guide to ecologists whose work illustrated those shifting priorities (Shelford 1963).

The 1926 *Naturalist's Guide* summarized natural places currently preserved and accessible for scientific study. It also listed a wide range of lands set aside for specific uses that might yet be turned to preservation and study. These included national parks and forests, grazing lands, national monuments, military bases (some already abandoned), and game refuges. Although settlement and human activities had taken their toll on these lands, for purposes of ecological study, they still maintained an element of simplicity and ties to earlier, less intensive uses. Shelford and others hoped such places could be explored and in some cases refreshed. This provided an important balance to depictions of the spreading development and rapid destruction of natural places documented at the same time (Shelford 1926).

Beginning in the 1920s, a growing community of ecologists, botanists, and zoologists avidly included practical concerns that were often linked to preservation within their research. Indeed, their work appeared prominently in *The Naturalist's Guide*. Their methods represented the foundation of ecological science, even as the broader questions they sought to answer often angled toward preservation of communities. The criteria for preservation of "important areas" originally conceived by the committee, emerged out of Shelford's research agenda and the work of his collaborators. He developed a set of criteria for components of this agenda, and those criteria provided opportunities for scientists to conduct field investigations that would result in useful information for a wider range of groups. The topics ecologists pursued revealed a clear trend toward the study of communities. Organismal physiology, a narrower approach within biology, remained a feature of the studies, but the context and approach aligned more with Shelford's rhetoric about the growth of the discipline and the emergence of professional ecology.

Scientists like Shelford recognized the need for conceptual rigor in biological studies. They were disinclined to take on projects that would appear overly or merely practical to other life scientists. Shelford articulated the need for adherence to conceptual and professional sophistication in ecological research. First, he placed increasing emphasis on quantitative data, and made sure that his own students and colleagues reflected that emphasis in their research and their writing. Second, he expected ecologists to adopt standard data collection techniques, which they would practice consistently and articulate in reports of their research methods. These methods of collecting and sampling would systematically include every animal in a given community. The methods would also provide opportunity to compare and correlate other factors such as climate, soil types, and vegetation types (Shelford 1929).

The Broader Conservation Message

In the midst of efforts to study different habitat areas and at the same time protect them, certain wildlife species attracted additional attention. Wildlife protection, like forest and rangeland conservation, had been undertaken as part of the impulse to sustain natural resources within the Progressive Era. Sustainability, as articulated by Gifford Pinchot, was a form of conservation often at odds with the preservation goals of many naturalists. By the time the ESA was looking to protect study areas for wildlife and vegetation, battles over sustainable use of resources were being waged on public lands across the country. Those early battles also reveal the different conceptual and ideological context of conservation in the formative years of ecology as a biological science.

Where livestock operations had expanded across the country, large predators like mountain lions and wolves, as well as coyotes, became targeted for destruction. Government hunters joined forces with ranchers in the late nineteenth century to systematize predator control, especially on federal land leased for grazing. Those control efforts, trapping and shooting predators by the hundreds every year, in some ways matched the goal of protecting deer and other wildlife on western lands. Near Yellowstone National Park and on the Grand Canyon National Game Preserve, predator control enabled elk and deer populations to flourish for decades. Conservationists valued these game populations, saving them from outlawed wolves and coyotes (Dunlap 1988; Young 2002). Government naturalists took on multiple roles in those days, pursuing and trapping predators as well as carrying out studies and publishing their results in professional journals. Congress allocated modest sums for experiments and demonstrations, which provided a baseline of the changing populations in some areas (Young 2005).

The apparent paradox of destroying some populations in order to save others persisted in mammalian conservation and zoology for decades. By contrast, hunting for birds underwent a dramatic and more consistent shift from the nineteenth into the twentieth century. Market hunters killed birds by the thousands in some areas. The passenger pigeon became best known for its vast numbers and edible meat, as well as the ease with which individuals could be shot out of flocks that darkened the skies. Even when it seemed the flocks might disappear, hunting continued. In the end, the passenger pigeon became a symbol of the need for conservation in the face of human extravagance in consuming natural resources. Other species, in some cases hunted for their meat, and in others for their feathers, came under the watchful eye of early conservation-minded civic leaders. Even hunters who pursued game for sport adopted the argument that too much of a good thing might lead to the loss of their way of life. Market hunting and taking bird feathers for the fashion industry attracted the ire of these conservationists (Barrow 1998; Hornaday 1913). As such, the conservation messages – and there were many from which to choose – took multiple forms and differed among the groups that came to promote them. Attempts to brand those messages differently meant that there could be little coherence within the efforts of those who promoted their work as "preservation," "sustained yield," and "protection" of individual species.

Among those who would eventually articulate a more unified ethic of conservation, Aldo Leopold underwent a transformation that revealed the changing role of the scientific community in relation to conservation work. Raised in a sport-hunting family in the Midwest, Leopold joined the first generation of professionally trained foresters after graduating from Yale in 1909. He took a position with the US Forest Service in New Mexico, where his work included estimating the number of board-feet on a parcel of land with a crew of seasoned lumbermen. He also conducted solitary survey work, spending weeks at a time away from the nearest town by horseback. He learned to shoot wolves on sight as the mandate to protect grazing livestock remained a primary feature of managing federal lands. By the time Leopold moved back to the Midwest in 1924, his interest in wildlife and the potential for establishing principles of management had outgrown the confines of his government job. He created his own project in 1928, funded largely by ammunition suppliers for sport hunting. A survey of game species in Midwestern states led to a groundbreaking publication in 1933, Leopold's comprehensive account of principles derived from his own experience, the survey, and a smattering of ecological concepts, *Game Management*. The book established him as an expert in the field, and earned him a faculty position at the University of Wisconsin, complete with a new department. He served as the first chair of game management in the country (Meine 1988).

Leopold's work in establishing game management as a distinct field of conservation earned him international acclaim. Over the next 15 years, however, another project eventually attracted even broader attention from him. As a kind of retreat from his busy days in Madison, Leopold took his family on weekends to a patch of degraded land along the Wisconsin River. There they worked together to make improvements to the soil and vegetation. He began to carefully record the changes they observed. In addition to those observations, he also narrated his thoughts about conservation into a series of essays, which were accompanied by thousands of pages of journal notes. In 1948, a grass fire threatened to destroy a section of prairie on the land they had worked to restore, and Leopold suffered a fatal heart attack while fighting the fire. From his collected essays, a book – published the following year simply titled *A Sand County Almanac* – contained an accessible account of what conservation had become to a veteran game manager, ecologist, and philosopher (Flader 1994; Meine 1988).

Far from being a comprehensive scientific text on the subject, *A Sand County Almanac* offered a broader integrative foundation for conservation work. At a time when the scientific community was looking to consolidate ecological research related to the subject into a more basic, academic field, Leopold's words provided an opening to continued efforts that would involve the public, government agencies, scientific societies, and emerging conservation groups. The ESA opted to shift conservation activities to its individual members by discontinuing the Committee on Preservation of Natural Conditions. This helped to formalize the Natural Resources Council of America (NRCA), a consortium of organizations that most ecologists supported but that would remain separate from their professional affiliation with the ESA (ESA 2014).

Establishing the Science of Ecology

Official efforts by ecologists to establish professional identities outside of preservation activities coincided with enormous growth in research during the middle of the twentieth century. A groundbreaking paper by Raymond Lindeman was published in the journal *Ecology* in 1942. There, Lindeman combined ideas of community ecology, developed by Shelford and British ecologist Charles Elton, using the movement of energy and nutrients to understand the relationships among species. This approach introduced the trophic-dynamic approach, as Lindeman called it, and cleared the way for understanding biotic and abiotic factors together as part of ecosystems. The study of ecosystems science provided a more integrated conceptual framework as well as experimental and mathematical methods for examining stasis and change (Hagen 1992b). Perhaps most importantly, ecosystem ecology served as a meaningful development for ecologists, rather than a radical departure. Plant and animal ecologists who focused on communities could adapt this framework while continuing their studies.

New techniques also emerged in ecology, especially after nuclear physics successfully demonstrated the destructive power of the atom bomb in ending World War II. After

1946, peaceful applications for nuclear science became the hope of many scientists who feared their efforts in creating the bomb had unleashed an unstoppable force of fear and escalating weaponry. Ecologists soon recognized a significant peaceful use for radioactive materials developed and refined during the war effort. In particular, radioactive isotopes that were taken up by living tissues could be tracked from one trophic level to another within ecosystems. Federally funded research at government labs became available to collaborating ecologists across the country, and the ability to find trace amounts of isotopes allowed those ecologists to see how rapidly these materials moved within communities (Hines 1962). It became important to understand the movement of certain isotopes in order to consider whether dangerous levels might be accumulating, but it was also possible to see how trace amounts – at levels not at all dangerous to the organisms where they might simply pass through – would cycle through communities over time (Hagen 1992a; Klingle 1998).

As the use of radioactive tracers became a standard technique in ecology during the 1950s, the process itself served as the basis for a set of more integrated studies. These studies revealed local impacts of the radiation itself, as well as the global implications of interconnected ecosystems. Ecologists studying remote islands in the Pacific Ocean, where bomb tests continued throughout the decade, also gained insights about the significance of their work more generally. Eugene Odum established a textbook series that served as the basis for curricula in the field for a generation or more, and he also defined and expanded ecological concepts for his peer researchers (Odum 1953). These developments during a period when ecologists worked to focus the scientific foundation of their profession established important opportunities to connect with a broader scientific community as well as conservation efforts in the years that followed.

Another indication of the way scientists organized their work and began to enlist others in making connections to conservation ideals appeared during a conference in Princeton in 1955, entitled "Man's Role in Changing the Face of the Earth." At the conference and in the published proceedings by the same title, scores of scientists from many fields sought to integrate contributions from their specialized disciplines. Along with geographers, land-use planners, and directors of research institutes, a number of ecologists, botanists, and zoologists shared their insights. Their comments reveal a persistent concern about radioactive fallout, a different concern than the fear of nuclear war that had taken hold since Hiroshima, Nagasaki, and the dramatic above-ground tests of even larger bombs that had finally been halted by international bans. This group of scientists modestly pointed to the way interdisciplinary work of science might also apply to current conservation issues. They hoped that such an approach could address these concerns (if not eliminate the larger fears). As part of the critique that emerged, ecologists were charged with paying closer attention to the role of human activities in reshaping ecosystems (Thomas 1956; Young 2005). This signaled a need to revisit the decision a decade earlier to separate professional ecology from its application to conservation and related concerns.

Ecology and the New Environmental Movement

Scientists with training in ecology were drawn to study the impact of human activities, even if some had made an explicit attempt to separate their academic research from applied concerns. A number of important studies in the early 1960s, however, demonstrated the significance of bringing ecological research to bear on broader problems. The best known of these, Rachel Carson's *Silent Spring* (1962), achieved more than any previous scientific or popular account of human impact on wild populations and natural systems. Carson combined her scientific training as a zoologist with an extensive research review of factors that toxicologists, ecologists, chemists, and pathophysiologists had published in previous decades, revealing a deeply troubling connection between pesticides and declining wildlife. Her indictment of pesticides, and by extension the agrichemical industry and government agencies that had done little to regulate chemical use, shook up the scientific community even as it gave the public something new to fear after years of nuclear anxiety. Carson succeeded in making the pesticide issue simultaneously relevant to lovers of nature and those indifferent to the non-human world. The potential impact of a spring when no birds sing meant more than the loss of chirping warblers. Pesticides and other widely distributed chemicals were affecting human health, and the potential for them to become increasingly dangerous seemed undeniable (Lear 1997).

For Carson's message to achieve its goal, the science she summarized had to be converted into action. It was not enough for scientists to publish their findings in professional journals, as some had been doing for years. It was likewise not enough for the public to see the connections and be afraid, or even merely to demand action. It was not enough for government agencies to review enacted legislation and set rules for industry to follow. It was certainly not Carson's intention to call attention to this problem only to have industry shout down her scientific credentials in the interest of economic growth and protection of the marketplace that demanded the increasing agricultural yields that pesticides promised. Carson and her readers saw that science should serve an enlightened public. It should also empower proactive government agencies to protect the future interests of the nation. Immediately for some, and more gradually for others, the lesson of *Silent Spring* sunk in. The scientific community must act in partnership with the public, the regulatory arms of government, and the interests of the commercial sector to review the safety of products and to prevent – wherever possible – future harm to the environment (Lear 1997).

Use of pesticides played a significant role in the early career of entomologist and ecologist Edward O. Wilson. After establishing a theoretical account of diversity based on historical evidence for remote islands that had slowly become inhabited by multiple species, Wilson and Daniel Simberloff attempted an experimental demonstration of the theory Wilson and Robert MacArthur had dubbed "island biodiversity." The experiment involved counting the number of species found on 10 small islands off the Florida coast, each just 1000 square feet. The islands, only about 40 feet across, were then cloaked in specially constructed tents and gassed with chemicals to kill all insects. Under careful observation by Wilson and his collaborators, insects returned to the islands over time, providing the basis for a mathematical model that complemented their earlier theory and was supported by experimental evidence. This work also formed the basis of Wilson's own conservation-minded writings. As prolific an author as any scientist of his generation, he penned multiple Pulitzer Prize winning works of non-fiction as well as several foundational texts for conservation biologists (Wilson 1993, 2006, 2012).

A Synthesis for Conservation Biology

Along with the many individual contributions of ecologists, wildlife biologists, botanists, and resource managers, an expanding core of individuals sensed that beyond the academic achievements, more collective and publicly oriented actions were needed. The urgency that had led scientists to assemble in 1955 at Princeton persisted, and by the 1970s, various groups of scientists working with environmental organizations began to seek an outlet for their specialized expertise.

Among those scientists, a group of biologists trained and mentored by Paul Ehrlich maintained a focus on conservation concerns. Ehrlich's evolutionary explanations attached to advances in the field of ecology helped scientists to advance their work in broader professional circles. Ecological explanations, in turn, became more sophisticated, drawing on a wider array of principles. The insights of island biogeography offered a structure for describing how plant and animal diversity responded to disruptions on increasingly isolated patches of habitat. The fact that those patches of habitat were isolated by expanding human populations, both in size and intensity of land use, fueled a sense that ecological study continued to clash with development. Ehrlich became an outspoken critic of human population growth, and a lightning rod for criticism from advocates of industry and consumer culture. The fact that ecologists identified more threatened and potentially endangered species every year suggested that an important turning point was at hand. Ehrlich required his students and others in his sphere of influence to read the work of Thomas Kuhn as a way of suggesting that scientists themselves were in the midst of a revolution. Among those who heeded Ehrlich's advice, notably Michael Soulé, many agreed that the proper response was for them to remake their science (Meine, Soulé, and Noss 2006). This revolution would involve following the path of explanations that seemed like anomalies to colleagues in a traditional mode, but would point the way around a bend in establishing new directions for science and conservation efforts.

Conservation-minded biologists trained in evolution and ecology saw broader roles for the scientific community in putting their research to work. Seeing more species threatened with extinction, they predicted a crisis of historic significance, brought on by human actions. Ecologists in particular saw how certain plants and animals at crucial nodes in complex communities served as "keystone species" in the larger ecosystem. If such species went extinct, entire ecosystems would collapse. Predictions, though dire, seemed thoroughly supported by the evidence. In order to do what they could to intervene, scientists found partnerships across research institutions, government agencies, zoos, and aquariums. By the mid-1970s, the conservation message of a few large zoos became a coherent and expansive mission of captive breeding programs around the world (Tudge 1992; see Young, this volume). Partnerships with natural resource managers became another potential arena for advancing conservation work. Forestry, wildlife management, range management, and fisheries projects attracted renewed attention from researchers in previously isolated academic institutions (Meine, Soulé, and Noss 2006).

In practice, there had always been meaningful interaction among researchers, land use managers, government agents, and conservationists. By the late 1970s, however, these explicit efforts to articulate the broad opportunities for collaboration and goals of conservation opened additional avenues for cooperation. In 1978, at what was called the

First International Convention on Conservation Biology, the full spectrum of potential contributors to a newly integrated scientific community met to consider their shared vision of future directions. The proceedings of that convention became a synthetic statement of the potential of such work (Soulé and Wilcox 1980). Natural scientists recognized the advantage of joining with social scientists to explore and explain the implications of understanding human behavior as an element of conservation efforts. Changing behavior, as much as identifying threats to ecosystems, would be a significant challenge for the newly integrated profession.

During the 1980s, a growing cadre of scientists calling themselves conservation biologists worked to put their new professional aspirations on the map. With textbooks and trade books bearing the title of conservation biology appearing in greater numbers every year, the field could potentially include anything and everything to do with nature. The rapidly expanding scope meant that conservation biology might come to represent such a wide range of topics and concerns that it would become professionally meaningless. To maintain coherence and provide leadership, Soulé and a few others established a series of meetings to draw up a new organization. The Society for Conservation Biology (SCB) emerged in 1985 from an optimistic but anxious group of biologists who hoped they were acting in time to avert what Soulé called "the worst biological disaster in the last 65 million years" (Soulé 1987: 4). The SCB aimed to integrate better the scientific efforts, such as biological modeling, fieldwork, and experimentation, as well as expand communication of results and connect those results with work in other fields like anthropology and agriculture.

Indeed, the creation of the SCB coincided with a series of significant conceptual advances. In 1986, Walter Rosen from the National Research Council coined the term biodiversity, which helped to frame a forum on biological diversity that Wilson and others were organizing. The published proceedings of that forum gave other scientists and the public a more coherent image of the focused goals of conservation biology (Wilson and Peter 1988). The SCB also provided a more unified scientific voice in responding to sudden crises like the *Exxon Valdez* oil spill and the growing concern over global climate change. Soulé and many of his colleagues, no doubt inspired by their understanding of how the scientific community experiences change, saw this period as a new synthesis, a revolution in the way science was conducted. Engaging with policy in important ways opened a path to the kind of change that conservationists had long envisioned (Meine, Soulé, and Noss 2006).

In recent years, the field of conservation biology, now established under the auspices of a professional organization and with research published in its own peer-reviewed journal, continues to undergo realignments. Supporters and critics alike offer suggestions of what conservation biologists should do, and how their work should be prioritized. In a recent essay, Peter Kareiva and Michelle Marvier (2012) suggested that conservation biology needs to be conceived more broadly than even the founders of the SCB allowed. That broader conception, which they called "conservation science," would more explicitly include social science and humanistic perspectives. It would approach the even larger domain of environmental science, which included attempts to manage the environment to promote human well-being, but conservation science would continue to focus on biodiversity and avoid environmental management. This new sense of conservation science would embrace the evidence-based approach adopted in medical practice. This approach allows for more sustained effort and avoids a purely reactive and crisis-based view of ongoing challenges. Kareiva and Marvier went so far as to suggest that Soulé's version of conservation biology is failing on a number of fronts, and that new functional and normative postulates are needed. These would offer the promise of greater awareness of the resilience of nature, the global scale of problems as well as solutions, and broader partnerships that could transform economic practices in light of long-term ecological costs and benefits (Kareiva and Marvier 2012).

In a sense, the goals expressed in such forward-thinking postulates reflect the earliest impulses of conservation-minded naturalists. Like Thoreau, Marsh, Leopold, and Carson, twenty-first century conservation biologists continue to envision a community of scientists who are simultaneously informed by their study of the natural world and committed to applying that understanding for the sake of humanity. Over the past 200 years, amateur observers and naturalists contributed alongside professional scientists to recognize the potential of humans to change, consume, sustain, and restore natural systems. They increasingly see the need to engage children in conservation work, both to improve environmental education and to offset the deficit many children face in firsthand encounters with nature (Louv 2005). Most recently, expanding "citizen science" movements reflect the ongoing involvement of amateurs in the history of conservation biology. These citizen science programs often use technology to connect species monitoring with broader databases and professional research programs (Newman et al. 2011). The tradition of wide participation in natural history collection continues to connect information gathering with a focus on addressing concerns in the natural world. The future, in the eyes of a conservation biologist, remains open. No matter how dire the problems may seem, there is within the scientific community a view to solutions. In recognizing the connections within that community, conservation biologists demonstrate the problem-solving genius of collaborative effort toward a shared goal.

Bibliographic Essay

Conservation biologists have contributed significantly to explanations of the recent history of this field. Their involvement in the legacy of the distinctive scientific enterprise they have established is a useful starting point for understanding the forces that have shaped collaboration and integration of scientific ideas since 1980. That involvement can also obscure contributions from influences outside of science. It is also possible to overlook efforts that did not emerge within the particular sphere of the most active recorders of conservation biology in the last three decades. That said, an excellent account of the first 20 years of conservation biology and its most widely recognized origins comes from Curt Meine, Michael Soulé, and Reed Noss (2006).

The history of conservation biology is also examined within the broader field of environmental history. While many environmental historians view science as a kind of black box, leading them to focus instead on social and political movements, elements of the scientific work appear in most accounts. A useful and comprehensive introduction is provided by Donald Worster (1994). Worster also discusses Thoreau at length. Hal Rothman examines the development of environmentalism in the twentieth century and provides easy access to the ideals and politics of Americans in *Saving the Planet* (2000). Among the best recent histories of what later came to be seen as environmental activism in the 1950s is Mark W.T. Harvey's *A Symbol of Wilderness* (2000). In this book, Harvey distinguishes the popular activism based on aesthetics of wilderness of the 1950s from that based more substantially in environmental science of the 1960s and later.

Historians of science, focusing on ecology and natural history, offer additional insights into the development of the scientific community that coalesced around conservation issues. For more on the broader context of the naturalist tradition, Paul L. Farber has developed a particularly readable and comprehensive narrative (2000). Keith R. Benson summarizes the transition from nineteenth-century natural history, especially zoology, to twentieth-century life science (Benson 1988). The transition in botany is described by Eugene Cittadino (1980). The best book on the development of American ornithology and the role of scientists in conservation is Mark V. Barrow's *A Passion for Birds* (1998). Barrow also examines extinction in detail across a wider range of species (2009). Ronald C. Tobey details much of Clements' work and the history of plant ecology (1981). Joel B. Hagen provides an excellent overview of the history of ecology (1992b).

Wildlife protection is addressed broadly in Thomas R. Dunlap's *Saving America's Wildlife* (1988). More recent study, including analysis of the scientific community involved in wildlife management in Yellowstone National Park, appears in James A. Pritchard's *Preserving Yellowstone's Natural Conditions* (1999). Christian C. Young examines the case of the Kaibab deer in the Grand Canyon National Game Preserve in the early twentieth century (2002). Biographies of key contributors to the scientific community focused on conservation issues include several excellent additions to the literature. Curt Meine has written the most complete biography of Aldo Leopold (1988). Additional and more accurate insights into the development of Leopold's thinking are provided by Susan L. Flader (1994). Linda J. Lear has written a thorough biography of Rachel Carson's life (1997). The rhetoric of Carson's *Silent Spring* is the subject of analysis in a volume edited by Craig Waddell (2000). In addition, Edward O. Wilson's memoir, *Naturalist* (1994), provides details of his life and the foundations of his many contributions to conservation biology.

Among scientists who have offered insights of their own regarding the history of conservation biology, several examples stand out as primary texts that contain elements of synthesis of how certain conceptual advances were achieved. Eugene Odum's work includes the pioneering textbook, *Fundamentals of Ecology* (1953), with revised editions appearing in 1959 and 1971. Edward O. Wilson and Frances M. Peter edited the first report of the National Forum on BioDiversity, simply titled *BioDiversity* (1988); the second report, *Biodiversity II* (Reaka-Kudla, Wilson, and Wilson, 1997), followed Wilson's own exploration of the topic, *The Diversity of Life* (1993), which he has updated in more recent works (2006, 2012). The recent critique of Kareiva and Marvier (2012) has prompted additional comparison of island biogeography as a conventional approach in conservation biology to reveal that additional insights from more varied landscapes may be valuable. An initial comparison is offered by Chase Mendenhall, Daniel Karp, Christoph Meyer, Elizabeth Hadly, and Gretchen Daily (2014).

Chapter Seven

ECONOMICS

Ross B. Emmett

The central divide that has created modern economics is not any of the ones you probably expect. It is not the one between those in favor of government interventionists and those in favor of markets. It is not the divide between Keynesians and free marketers. It is not the one between theory and empirical investigation, or the one between theorist and practitioner. As important as those debates are to economists, policymakers, and practitioners, all of these groups are, today, on the same side of the divide. Rather, the central divide that makes modern economics what it is, is the divide between an economics told largely in words, and an economics told largely in models. Until the mid-twentieth century, economists did economics with words. Today, economists do economics with models. That is, they seek to combine elements of theory and evidence through mathematical formalism. Their debates hold the vestiges of economics told in words, but all their debates are told in stories about models. An economic problem is explained by use of a model. The evidence relevant to the problem will be collected through insights gleaned from a model. The story of economic change is told in the relationships among a model's variables. And recommended policies will be evaluated by use of a model. The model used may be Marshallian, Walrasian, Keynesian, Cournot, behavioral, or experimental. But it is not economics today if it is not a model. The way economists work and think since the mid-twentieth century is captured in the title of Mary Morgan's (2012) recent book *The World in the Model*. Such was not always the case. How then did economics in the United States get from words to models?

In the Beginning was the Word *Laissez-Faire*

Adam Smith, an eighteenth-century Scottish moral philosopher, began the modern study of economics by speaking of "the obvious and simple system of natural liberty" (Smith 1976, Bk. IV, Ch. 9); in a word, *laissez-faire*. In nineteenth-century America,

A Companion to the History of American Science, First Edition.
Edited by Georgina M. Montgomery and Mark A. Largent.
© 2016 John Wiley & Sons, Ltd. Published 2020 by John Wiley & Sons, Ltd.

Protestant moral philosophers were the primary teachers of "clerical *laissez-faire*" (May 1949: 14, emphasis in original; see Heyne 2008); that uniquely American reflection on political economy that combined our Puritan heritage with the economic insights of Adam Smith and his successors. While dissenting moral traditions were certainly present in nineteenth-century America (Frey 2009), the argument that individual freedom and responsibility were best reinforced by free markets, Christian morality, and republican government was reinforced by popular texts like Francis Wayland's (1837) *Elements of Political Economy*.

Meanwhile, practical economic problems were rampant in the new republic, and practical men of affairs, often politicians and lawyers, sought their solutions. Thus, it is not surprising that the first two volumes of Joseph Dorfman's (1946–1959) magisterial study of *The Economic Mind in American Civilization* is filled with names like William Penn, Benjamin Franklin, Alexander Hamilton, John Adams, Thomas Jefferson, Thomas Paine, George Tucker, Jacob Nunez Cardoso, and Henry C. Carey. Only Carey, among this group, argued against the *laissez-faire* arguments. Carey built the case for a "national" economic system that would fit the needs of the new nation. To end conflict among the competing interests of American enterprise, industrial America, Carey (1851) argued, needed tariff protection to promote the nascent manufacturing sector, a banking system that would facilitate saving and local investment, and a strong government that would provide transportation and communication improvements beneficial to the agricultural and commercial interests as much as they would be to manufacturing. Carey's national system was well received politically in the industrial Middle Atlantic states; both New England and the southern states, however, continued to promote *laissez-faire*.

Free markets might sound incongruous with the presence of slavery, around which the economies of southern states were oriented. However, despite their personal opposition to slavery, those writing and teaching in the clerical *laissez-faire* tradition argued that it was primarily a legal and moral issue, rather than an economic one. One economist who did not shy away from the topic was Daniel Raymond, whose *Thoughts on Political Economy* (1820) was the first economics treatise written in the United States. Arguing that the nation's economic uniqueness lay in the extent to which labor, rather than capital, was instrumental to the creation of wealth, Raymond found slavery to have a significant negative effect on economic growth. Slaves were less efficient producers than they would be if freely employed in labor markets, and their owners were a less productive group than were owners of other forms of capital. While he largely agreed with others in the *laissez-faire* tradition about economic policies, and concurred with their abolitionist moral sentiments, Raymond also argued that the end of slavery would have a positive benefit for economic progress in the United States (Barber 2003: 232–3, 235; Frey 2002; Petrella 1987).

Thus, between the United States' general *laissez-faire* orientation, the inclination in middle America to promote national manufacturing, and southern slavery, "the broad contours of American economic discourse," in the first half of the nineteenth century were "heavily moulded by sectionalism" (Barber 1988: 9). One might add that the authority dispersed to states through the jurisdictional mandates of the Constitution meant that practical debate over economic issues from tariff policies to money and banking and industrial development was fragmented and fraught with the politics of states' rights (Barber 2003: 232). Clearly, antebellum America was a world of words, and not of models.

Postbellum to *Fin de Siècle*: Advocacy, Objectivity, and Religion

Only after the Civil War did the separate science of economics begin to emerge in the United States. But it was still a world of words. Professorships in political economy became commonplace by the 1870s, although the first independent department of economics was not launched until 1879, at Harvard University. After that, more departments appeared at American colleges practically every year (see Barber 1988, 2003; Coats 1988). While most were still primarily oriented toward undergraduate instruction, the establishment of full-time professors in the field inevitably led to an increase in specialized research. And the problems that economists could study multiplied as the population grew and migrated westward, economic growth was reinvigorated, and technological change introduced entirely new industries. In addition, the clerical *laissez-faire* language of individual responsibility gave way, even before the Progressive Era, to concerns about monopolies, social cohesion, and the need for social control, which, in turn, spawned calls for universities and governments to address "the social question" (Rodgers 1982). The Morrill Act (1862) had made the promotion of practical higher education a priority, and new state universities responded with appointments to address both existing and new economic issues. Hence, the late 1800s saw the rise of independent agricultural economics departments, and the creation of appointments in banking and insurance, transportation economics, and industrial relations.

Still, the emerging science had to compete with other sources of economic argument for attention in the public realm. The best-selling economics book of the nineteenth century was written, not by a professor, but by a journalist, whose own independent study of classical economics led him to a theoretical conclusion with a practical policy application. In *Progress and Poverty*, Henry George (1879) argued that most taxes had dilatory effects on economic productivity, and hence that all government functions should be funded by a tax on the one source of value that came only from demand, not any kind of investment – the value of unimproved land. The simplicity of George's policy prescription certainly helped to make it popular. But more importantly, the single tax argument appealed to the notion that land was a common resource. George's innovation was the idea that society could reap the benefits from land better by taxing away the rents from private property rather than directly socializing the land.

Between George's popular arguments and new academic approaches to the practical problems of postbellum American society, traditional *laissez-faire* theory seemed less and less relevant to the modern complex of institutions, organizations, interests, and responsibilities. Richard T. Ely's *Outlines of Economics* (1893), a standard American economic text from the 1890s until the 1930s, provided the framework for the new economics. The text had a short introduction to theoretical concepts, and then moved on to deal in depth with the economic development of the United States, public finance, transportation, insurance, municipal development, and agriculture, among other topics. Throughout the book, the role government could play in addressing the issues was emphasized.

Adherents to classical *laissez-faire* resisted this movement; several of the "old school" formed the Political Economy Club in the early 1880s to provide a forum in which economists and leading public figures could meet "for discussion, [and] … for encouragement of the study in the proper scientific spirit" (J. Laurence Laughlin, quoted in Coats 1961: 626; see also Furner 1975: 65–7). At about the same time, open conflict emerged between adherents of the "old" school and the "new" school at Johns Hopkins University. The focus on graduate education at Johns Hopkins provided an alternative to the German universities American graduate students had been going to for scientifically oriented graduate education in the social sciences. One of the recent German trained graduates was the aforementioned Ely, who joined Hopkins' economic department in the early 1880s. Along with his criticism of *laissez-faire,* Ely was a proponent of the approach of the German Historical School. Another Hopkins faculty member, Simon Newcomb, professor of mathematics and astronomy, was *a laissez-faire* proponent, and a staunch opponent of government intervention. Newcomb had created one of the first mathematical models of the quantity theory of money; a key theme of both classical political economy and neoclassical economics in the twentieth century. Ely thought little of Newcomb's mathematics, arguing in print that such models were not essential to the future development of economic research (Ely 1884; see also Ely 1886). Newcomb responded (1884, 1886), and the two men's disagreement launched a *Methodenstreit* that carried the debate between *laissez-faire* and new approaches into the twentieth century (Furner 1975: 57–65).

In the context of the debate between old and new schools, the American Economics Association (AEA) was born in 1885. The impetus came from Ely. Although counseled by older economists who were friendly to his criticisms of *laissez-faire* to broaden the Association's range of views and to work within the Political Economy Club, Ely chose to form a new association, united in its advocacy of economics as a science of social reform. The founding document of the Association, then, combined "the powerful strands of modern science, reform activism, and social Christianity" (Furner 1975: 70). For Ely and some of the others who signed on in 1885, the methods they had adopted from the German Historical School fit well with a program of advocating radical reforms of American economic policy, informed by the ethical stance of the Christian Social Gospel movement. Indeed, a significant portion of the early membership was Christian ministers and other social activists, not just economists. The days of clerical *laissez-faire,* were over, but Christian zeal still lay near the heart of the social science (Bateman and Kapstein 1999).

Over the next decade, however, discussions among economists both inside and outside the AEA led gradually to a rejection of radical activism, a willingness to recognize as fellow economists those from both schools, and, perhaps most importantly, secularization of the profession. Histories of the AEA often focus on the resolution of the tension between science and reform which brought members of the old school into the Association before the turn of the century, setting the stage for the struggles over science and social control that characterized American economics in the first half of the twentieth century (Furner 1975 is the classic study; but see also Fourcade 2009). But the demise of the Social Gospel in the first two decades of the new century, presaged by the embarrassed reaction of economists of both schools to Ely's Social Gospel infused language and his zealous subjugation of science to reform in the name of Christianity, led the economists to enter the new century in pursuit of a secular, professional social science (Bateman 1998; Coats 1960).

Pluralism and Social Control: The First Half of the Twentieth Century

The usual story of the history of economics after the 1890s is a story of the dominance of the newly formed blend of classical theory and marginal analysis – a blend first named "neoclassical" by the American theorist Thorstein Veblen (1900: 261) – until the Keynesian revolution of the 1930s (e.g., see Blaug 1997). Of course, neoclassical economics was not just *laissez-faire* redux. While marginalism provided a new way of understanding how market allocations benefit participants, it also provided the basis for understanding how market allocations fail their participants. Market failure, in fact, became a key theme of twentieth-century economics (Medema 2009).

However, whether the story about neoclassical dominance is true in regards to economics in Britain and Europe (Black, Coats, and Goodwin 1973), it is not a useful account of American economics from the 1890s to the end of World War II. Instead, American economics was a conflicted pluralism, with competing understandings of theory and of the role of historical and empirical investigation vying with each other. The period saw the emergence and decline of the American institutionalist movement, which was the secularized inheritor of Ely's efforts to create a historical economics that would inform social reform in the United States (Morgan and Rutherford 1998; Rutherford 2011). The "revolutionary" ideas of J.M. Keynes' *General Theory* (1936) had little impact initially in the United States, because both neoclassical theorists and institutionalists interpreted Keynes through the lens of their own work on the structure of the economy and the causes of cycles (Rutherford 2011: 298–306). American economists had, after all, been stabilizing industries and incomes by shaping economic policy in state and federal agencies since the beginning of the century, and had largely rejected the notion of macro-solutions through short-term fiscal and monetary policy changes (Barber 1985; Rosenof 1997). What Keynesianism in the United States became, then, was more the product of the mid-century synthesis between Keynes and neoclassical theory produced by Alvin Hansen, Paul Samuelson, and their followers.

The Progressive Era was in full swing during the first two decades of the twentieth century. As we saw in the last section, *laissez-faire* was out, the creation of a new society via scientific research in the service of social reform was in. Ely moved from Johns Hopkins to Wisconsin, where his reforming spirit and scientific orientation were welcomed. New voices emerged to join Ely: Veblen, John R. Commons, Wesley Mitchell, Morris Copeland, and others (Rutherford 2011). The annual meetings of the AEA provided opportunities for discussions on a wide range of practical topics: the programs for the meetings in 1911 and 1912, for instance, included sessions on the relation of economics and accounting; workingmen's compensation schemes; the comparative costs and benefits of waterways, canals, and railroads; immigration; diseases among industrial workers; income tax reform; tariff reform; the measurement of the cost of living; and, slipped among these, a few theory-oriented discussions of socialism, the centenary of classical economist David Ricardo, and the relation of price to value.

The variety of these themes, and the manner in which they were approached, reflected the economists' understanding of what scientific study in their discipline meant. Economists immersed themselves in a particular economic problem. They studied its history, the industries affected by it, and the legislation that addressed it. General economic principles seemed largely irrelevant. The motivations of individuals and businesses in the relevant industries were far more complex than simple self-interest. The property rights regime of the nineteenth century, touted by *laissez-faire* advocates as spontaneous and self-defining, was instead created by the state, and could be remade to benefit those disaffected by the distribution of its rights. Demand and supply did not respond to changes in price as theory indicated it would. And, in any case, prices and wages were poor indicators of the value of particular industries or people to national economic welfare. The scientific knowledge gained by the economists' expertise could be used to direct policies that would contribute to real improvements in welfare (Bernstein 2001; Fried 1998; Ross 1991: 172–218, 371–86; Rutherford 2011). Walton Hamilton (1919) finally gave the new approach a name when he called it "institutionalism" at a special session of the AEA's annual meeting in December 1918. Coalescing the various threads of the movement, Hamilton argued that, for the institutionalists, the only "science" of economics worthy of the name was one that would (1) provide a unified scientific approach; "(2) be relevant to the modern problem of control, (3) relate to institutions; (4) be concerned with matters of 'process'; and (5) be based on an acceptable theory of human behavior" (quoted in Rutherford 2011: 18).

The American economists who did adopt marginalism all bore sympathies with their institutionalist colleagues' ethical concerns about social reform. At the end of the nineteenth century, German-educated John Bates Clark, professor at Columbia University, shared with progressive economists a Christian-inspired sympathy for social reform (J.B. Clark 1886). But George's proposal for a land tax, as well as more radical socialist challenges, led him to reconsider the arguments for the distributional consequences of market organization. Clark's marginal productivity theory of the distribution of incomes became an integral component of neoclassical analysis; the first major American contribution to pure economic theory (J.B. Clark 1899). It also provided the basis for resistance to reforms that would increase labor union power.

Clark's contemporary, Irving Fisher, an early contributor to the theory of general equilibrium, formulated an important theory of interest and capital, and was a quantity of money theorist who became widely known among the American public for his willingness to speak on about almost any issue – economic or not – and frequently advocated controversial policy proposals. Fisher was probably the best-known American economist in the decade before the 1929 stock market crash (Dimand and Geanakoplos 2005).

Clark's son, John Maurice, also straddled the boundary between theorists and institutionalists. J.M. Clark, who was a professor at the University of Chicago before succeeding his father at Columbia, was a key early figure in a new wave of interest in theory inspired by the notion that neoclassical economics could be used to examine aspects of market activity that *laissez-faire* economists often downplayed. J.M. Clark focused on oligopolies and monopolies in capital-intensive industries, explaining how they obtained market power despite large fixed costs, and how their regulatory control could be improved (J.M. Clark 1923; J.M. Clark 1926). A contemporary of J.M. Clark's, Allyn A. Young, became interested in another understudied aspect of neoclassical theory: the role of increasing returns in economic progress (Young 1928). Young taught theory at Harvard, but, in 1927, became the first American economist to hold a position in England, teaching for one year at the London School of Economics before he unexpectedly died.

The new willingness to combine neoclassical theory, the concerns of institutionalism, and social reform, reached its most famous American statement in *The Modern Corporation and Private Property*. Adolf Berle and Gardiner Means (1932) combined Clark's theoretical basis for social control of large organizations with John R. Commons' (1924) focus on the role of legal institutions in shaping the economic consequences of market activity. Berle and Means argued that the legal separation of ownership and control in the corporation undermined the positive social consequences of private property, and created the potential for society to replace private ownership with public control that would direct corporations toward socially beneficial outcomes.

When J.M. Clark left the University of Chicago in 1926, institutionalists were a majority of the department's faculty. But if you include as theorists Clark and future US Senator Paul Douglas – creator of the Cobb–Douglas (1928) production function that has played a central role in neoclassical theory ever since – and add Jacob Viner and Lloyd Mints, the department could be said to have had a majority of theorists. Perched then, as it were, on the cusp of the theory/institutional divide, Chicago was poised to lead economics in a new direction. And it did.

Frank H. Knight, whom Chicago hired as a replacement for J.M. Clark, was the author of *Risk, Uncertainty, and Profit* (Knight 1921), which clarified J.B. Clark's theory of profit while managing simultaneously to undermine the contribution of neoclassical theory to a moral defense of market outcomes. Yet Knight also was able to open the door for the integration of economics and organizational theory within the framework of neoclassical theory, not institutionalism. Indeed, he spent the 1920s criticizing every key aspect of the institutional agenda (see Knight 1935). Thus, the addition of Knight, who alternated teaching the required graduate courses in price theory with Viner, became the key piece that led Chicago economics to gradually stand apart from other departments, in the strength of their price theory teaching and, more importantly, in the fact that Chicago economists began to see all economic problems through its lens (Reder 1982); Models and words, but still models.

Doing Economics with Models

To understand the transition from doing economics with words to doing economics with models, we can look at a debate that raged at the University of Chicago between 1948 and 1954. At its center was the question of what made economics a science. On one side of the debate was Tjalling Koopmans, research director of the Cowles Commission for Economic Research; on the other was Milton Friedman, newly appointed professor in the university's department of economics. The Cowles Commission was the most important center of econometric research in the United States in the 1930s and 1940s, and continues even today to be an important research center for policy-oriented econometric research. Koopmans was awarded the Nobel Memorial Prize in Economic Science in 1975. Friedman had started his study of economics at Chicago, but finished his doctorate at Columbia University. His return to Chicago in 1948 ushered in the second generation of the Chicago School of Economics, which challenged Keynesianism with monetarism, rewrote economic history, reformed the study of industrial organization, and created a new approach to the study of law and economics. Friedman also was awarded the Nobel Memorial Prize, one year after Koopmans.

Friedman was ever the provocateur; he loved debate, and stirred the pot in this case by challenging the Cowles' approach to theory and empirical evidence. Koopmans' acceptance of the Walrasian principle – that every market was connected through the price system to every other market – led him to use general equilibrium theory to build models whose parameters could be empirically estimated, and, thereby, made useful for financial market analysis and macroeconomic policymakers. Friedman, on the other hand, focused on Marshallian partial equilibrium theory – the economics of particular markets – and judged the empirical evidence about the consequences of government interventions in those markets against the predictions provided by theory. For instance, before arriving in Chicago, Friedman had completed an empirical study of professional incomes. In the study, licensing and regulation correlated with higher incomes for members of the profession; Friedman argued the empirical results supported the claims economic theory made about artificial constrictions of supply. (Friedman and Kuznets 1954) The Chicago–Cowles debate was centered upon the relation of theory and empirical evidence. Both accepted the primacy of theory, but differed over what theory – Walrasian or Marshallian – was relevant to policy analysis. They also differed over whether econometric analysis focused on the estimation of model parameters (Koopmans) or on the testing of model results (Friedman).

To complicate matters, there was a third side to the debate: Simon Kuznets, professor at the University of Pennsylvania, research associate at the National Bureau for Economic Research (NBER) at Columbia University, and winner of the 1971 Nobel Memorial Prize in Economics. Kuznets was silent in the Chicago–Cowles debate because he was miles always, but his presence always lingered in the background. For one thing, Friedman's study on professional incomes was conducted under the guidance of, and co-authored by, Kuznets. Although Kuznets was never the free market advocate that Friedman became – he was ideologically, perhaps, to the left of the Keynesianism of Cowles Commission members – he was affiliated with the NBER, which Koopmans had criticized a couple of years before. In his review of Arthur Burns and Wesley Mitchell's NBER study, *Measuring Business Cycles* (1946), Koopmans concluded that the authors were engaged in "measurement without theory," and said their program would benefit from making fuller use of "the concepts and hypotheses of economic theory … *as a part of the processes of observation and measurement.*" (Koopmans 1947: 162, emphasis in original) Kuznets allowed his work to counter Koopmans' argument, rather than respond directly, but we can see that the Cowles–NBER debate was also over the relation between theory and empirical investigation. The Cowles position was that theory informed the construction of the economic model that was estimated; for Kuznets the relationship was more complicated. While he was not as theory-neutral as NBER director Wesley Mitchell tried to be, he was cautious in his use of theory to inform, or be informed by, empirical study. His work on economic growth, national income, and income inequality all reflect the balance he strove to achieve between theory and evidence.

The NBER–Chicago relationship completed the triangular relationship: for Friedman (1953), positive economic analysis involved comparing the results of simple theoretical models against the empirical evidence collected by NBER-type research. But much of the research Chicago School economists provided over the next couple of decades undermined the policy arguments of the tradition from which the NBER had come. Another Chicago Nobel laureate, Robert Fogel (co-winner in 1993), signaled Kuznets' importance to the Chicago tradition by devoting his last book to an appreci-ation of Kuznets' role in empirical economics (Fogel et al. 2013). As we can see, the NBER–Chicago–Cowles debates are some of the first to be waged largely in models. All of the participants in these debates are model-builders; indeed, each of them is among the first economists to think entirely in terms of models.

When the Cowles Commission left Chicago, Friedman's approach to the relation between theory and empirical evidence became the standard for Chicago economists. A good applied policy science like economics, members of the Chicago School came to argue, would base predictions about the consequences of changes in market out-comes on the assumptions of economic rationality, stable preferences, and information efficiency (Fama 1970; Friedman 1953; Stigler and Becker 1977). Assuming individu-als were rational and markets reasonably competitive, adjustments would therefore be rapid enough to discount short-term distortions. Also, the institutional and historical differences between industries, individuals, and political units were less important for the economist's task than keen insight into the use of price theory. From trade theory to agriculture and economic development, from monetary theory to industrial organiza-tion, Chicago faculty and students systematically replaced the underlying assumptions of institutionalism with a framework of analysis which marked the Chicago School as a distinct approach to economic theory and policy analysis (Emmett 2010; Van Horn, Mirowski, and Stapleford 2013; Van Overtveldt 2007). The Chicago School also, of course, challenged Keynesian economics. But the story of their challenge requires us to look first at how American economics became Keynesian.

Doing Economics with Keynesian Models

The initial American response to Keynes' *General Theory* was varied, and not as enthu-siastic as it is often portrayed. Institutional economics focused on the underlying struc-ture of the economy, and thought long term; it favored institutional reform rather than short-term fixes. Even before the New Deal a number of institutionalist-inspired public institutions at both the federal and state level had been created, and a vast array were soon to be on the way (Bernstein 2001; Rosenof 1997). Keynes' theory, on the other hand, focused on short-term use of government investment in public works to spur consumption and increase total output via the multiplier effect (Backhouse and Bate-man 2011). Also, in order to keep his book on message, Keynes avoided the use of ideas that theorists close to the institutionalists had recently introduced, such as monopolistic competition (Chamberlin 1933) and the modern enterprise's separation of ownership and control (Berle and Means 1932).

The first major move toward accepting Keynes came in 1938, when Alvin Hansen, a Harvard business cycle theorist with Wisconsin institutionalist roots, used the occa-sion of his presidential address to the AEA to make an argument, drawing partly on Keynes, that the American economy could be facing secular stagnation (Hansen 1938). Hansen went on to create, with John Hicks, the IS-LM model that began the synthesis of neoclassical theory with Keynesian theory (Hansen 1953). Hansen's stu-dent, Paul Samuelson (1948), furthered Hansen's integration of Keynesian logic with neoclassical theory, and also made the Keynesian synthesis intelligible to students, researchers, and public policymakers (Klein 2006; Laidler 1999) via the Keynesian models he built. Samuelson was the first American winner of the Nobel Memorial Prize, in 1970.

By the 1960s, the Keynesian synthesis was the basis for standard macroeconomic models, and concerns about the long-term institutional structure of the economy had been relegated to the periphery of economics. When Milton Friedman remarked in the December 31, 1965 issue of *Time* that, "we are all Keynesians now," he was not far wrong. However, his remark was also ironic, because there continued to be an undercurrent of dissent, especially from Friedman and his Chicago School colleagues.

The debate between Chicago and the Keynesians slowed, coalescing around the question of whether fiscal policy tools or monetary policy tools were more effective in controlling the macro-economy. Keynesians favored fiscal policy – short-term changes in government spending and taxation; Chicago economists favored the use of long-term rules for expansion of the money supply, earning thereby the label "monetarist." While the policy difference was substantial, the theoretical debate led to recognition on both sides that the models the Keynesian synthesis had created could produce either Keynesian or monetarist results. The theoretical difference between the two sides reduced to a disagreement over the stability of different variables, and the size of their effects in the world of the model, even if the ideological divide that informed their differences in public policy advice made them seem worlds apart.

Since the early 1980s, there has been substantial agreement on the basic model underlying both the Federal Reserve's control of the money supply and governmental debate over fiscal policy. In that sense, Friedman was right: since the 1970s, most macro-economists have worked within the context of the same Keynesian models, even if they greatly disagree over key issues regarding their structure and use.

Models to Live By

But model building in economics is not limited to the fields that the public immediately recognizes, nor does it depend upon ideological orientation. Three examples from American economics at the turn of the twenty-first century can help us understand what economics has become. The first involves University of Chicago economist Ronald H. Coase's essay "The Problem of Social Cost" (1960). One of the essays for which Coase was awarded the Nobel Memorial Prize in 1991, "The Problem of Social Cost", uses a thought experiment to set up a discussion of the economic and legal issues related to externalities, usually understood as a common source of market failure. The thought experiment did not set up a model, but George J. Stigler accomplished the task in his price theory textbook when he put what he called "the Coase theorem" into the framework of a Pigouvian social cost partial equilibrium model. Assuming the absence of bargaining costs among market participants, the Coase theorem specifies that the legal assignment of rights and liability will have no effect on the private costs and benefits of the product. In other words, in the absence of transaction costs, social costs will equal private costs, and private bargaining will reach an optimal equilibrium outcome (Stigler 1966: 13-14).

Prior to Coase's work, economic implications of legal decisions had largely been dominated by legal realism, an approach related to institutionalism in economics. But once Coase's insights could be rendered in a model, they provided the means for an economic analysis of legal decisions that furthered academic research and informed judicial decisions.

A second example comes from labor economics. One of the first uses of the partial equilibrium neoclassical model of supply and demand that every student learns is that minimum wage laws lead to increases in unemployment. A voluminous literature on the theoretical possibilities that would either weaken or strengthen that prediction exists, but at the beginning of the 1990s, a large majority of economists agreed with the proposition that minimum wage laws had negative employment effects. A surprising test case, studied by David Card and Alan Krueger (1994), provided empirical evidence that the majority opinion might be wrong. Numerous subsequent studies have extended Card and Krueger's analysis; shown that it could not be replicated using other data sources for the same case; and tried to reconcile the various outcomes. In other words, labor economists have done what modelers do: examine all the options with the model for explaining anomalous outcomes. The political implications of the debate are enormous, of course, and economic pundits weigh in energetically. Yet among labor economists who know the literature, the disagreements come down to different estimations of parametric values, especially regarding the elasticities of demand and supply in the market, and the specification of some of the variables. The consequences are significant, but everyone in the debate is working with the same model.

Finally, the economic analysis of education has several models at its disposal, but a common one is built upon the human capital arguments of T.W. Schultz and Gary Becker, another pair of Chicago economists who have won the Nobel Memorial Prize; Schultz in 1979 and Becker in 1992. Human capital is the marketable traits and skills possessed by labor force participants that can increase their productivity. In his presidential address to the AEA in 1960, Schultz (1961) argued that investments in education and training for workers was as fundamental to economic growth as investments in other capital goods. One reason to emphasize public investments in human capital, Schultz argued, was education's ability to help individuals deal with periods of uncertainty about future employment (or other) prospects. Human capital models soon appeared, estimating the returns to investment from education, on-the-job training, and one's family background. Becker's (1975) use of the human capital model extended it into issues of discrimination and wage inequality. Today, almost all economic discussion of education accepts the human capital model as the framework of analysis. The human capital model is used to defend public funding and provision of education, the payment of rewards to students for educational achievement, the adoption of common standards for all students, and the adjustment of teacher contracts to reduce the market power of seniority. It does not matter which side of the debate one is on, the only scientific analysis acceptable is one built on the human capital model.

The examples provided here simply point to the way economics and model building have become synonymous with each other, both within the discipline and by the public. Since the 1950s, to be an economist is to use models that share certain assumptions to explain the behavior of individuals, and their behavior's consequences, in the world outside the model. In the process, the science of economics in the United States has come to live in its models.

Bibliographic Essay

The starting point for any investigation of the history of economics in the United States remains Joseph Dorfman's (1946–1959) magisterial and comprehensive (up to the mid-1930s) *The Economic Mind in American Civilization*. Because he wrote his five-volume history before models completely took over economics, Dorfman has a broad under-standing of contributions to economics, which assists the reader to see the connec-tions between the development of academic economics and its use in business, finance, and government. Between Dorfman and the 1990s, much of the history of economic thought was rational reconstruction or *Dogmengeschichte*, and hence paid little atten-tion to institutional and cultural context, or the practices and tools of economics. Parts of Dorfman's story are revisited and recast in Dorothy Ross' *The Origins of American Social Science* (1991), and Mary Furner's *Advocacy and Objectivity* (1975).

The first step in updating Dorfman's treatment is a reading of Malcolm Ruther-ford's *The Institutionalist Movement in American Economics, 1918–1947* (2011). While Rutherford only takes us forward about 15 years, he has the benefit of 50 additional years of reflection on the pluralism of American economics during the critical interwar years, and the unique role of institutional economists. That period was also the focus of a *History of Political Economy* conference; the papers from which were published in *From Interwar Pluralism to Postwar Neoclassicism*, edited by Rutherford and Mary Mor-gan (1998). Morgan's book *The World in the Model: How Economists Work and Think* (2012) is a guide to the history of the relation between economists' models and nar-ratives from the eighteenth to mid-twentieth centuries. She has also written one of the best histories of econometrics (Morgan 1991). Roy Weintraub's (1991, 2002) work on how economics became a mathematical science should also be consulted because much of the history he tells occurs in the United States. When we think of the turn to mathematical formalism in the Cold War period, we have to include game theory, created by John Nash (Nasar 1998) and European émigrés John von Neumann and Oscar Morgenstern (Leonard 2012) and integrated into strategic studies at places like the RAND Corporation (Amadae 2003; Mirowski 2002).

William Barber's studies of economics in American academic institutions and in American government provide systematic treatments of the intersection of economists' ideas and their practices. His edited volume on the emergence of economics as a disci-pline in universities in the United States at the end of the nineteenth century is essential reading (Barber 1988), as is his treatment of the role of economists in government lead-ing up to and in the New Deal (Barber 1985, 1996). Other studies of economists in the New Deal include Bernstein (2001), Colander and Landreth (1996), Dimand and Geanakoplos (2005), Phillips (1994), Rosenof (1997), and David Laidler's *Fabricat-ing the Keynesian Revolution* (1999). The latter three items all deal with the extent to which American economists adopted Keynesianism during the New Deal period. *The Battle of Bretton Woods* tells the story of Keynes' role in the debates over the creation of post-World War II international financial systems (Steil 2013).

The standard account of the founding of the American Economic Association lies in the series of articles A.W. Coats published as the official historian of the Associa-tion (Coats 1960, 1961, 1963). Fourcade's (2009) cross-national comparison of social science association histories usefully complements and expands on Coats' work. Brad Bateman has provided a sustained treatment of the relation of the Social Gospel to the emergence of the American Economic Association in a series of articles (Bateman 1998, 2001; Bateman and Kapstein 1999). And Donald Frey (2009) surveys the range of American ethical perspectives on economics.

Chicago economics has become the subject of significant study. Alongside van Overtveldt's (2007) somewhat sensationalized general overview, one should place my own set of essays and edited volume (Emmett 2009, 2010), Dan and Clare Hammond's collection of the correspondence of Milton Friedman and George Stigler (2006), and the *Building Chicago Economics* volume of collected essays (Van Horn, Mirowski, and Stapleford 2013). Stories of Chicago are inevitably linked to the re-emergence of free-market arguments in the postwar era. Burgin's (2012) history is currently the best, connecting changes in economics in the United States to liberal movements in Europe. But other studies that should be consulted include Caldwell (2003), Mirowski and Plehwe (2009), and Jones (2012).

Chapter Eight

EXPERIMENTAL PSYCHOLOGY

Jim Wynter Porter

There is no clear beginning to the history of psychology. Its "subject matter … is as old as reflection" (Robinson 1995: 12). Long before there was any organized body of knowledge by its name, people have wondered about their minds and the minds of others. What is consciousness? Where do thoughts and ideas come from? Different eras have approached such questions in different ways. Was it "inspiration" from the gods? Was it an animating spirit? Was mind a manifestation of a soul or a subtle form of matter produced by the brain of "L'homme machine"? What is our mind's relation to our actions and how then are we to interpret our own and others' behavior? Though the answers may change with time and place, these basic questions are as old as human history.

Another, more recent, less universal, and decidedly epistemological brand of question was addressed to psychology as it took systematic shape in post-Enlightenment Europe: was psychology a *science*, or did it belong to metaphysics? In fact, there was a decisive shift in the nineteenth century to resettle psychology from what appeared to be the fog-bound provinces of philosophy and address its questions in the clear light of science. Yet historically, the "variables" in many if not all of psychology's formulations have been, to say the least, hard to quantify and measure.

Physics, the paragon of modern science, can tell us with stupefying reliability how an apple will behave as an object-in-the-world. Newton's law of universal gravitation

$$F = \frac{Gm_1m_2}{r^2}$$

predicts, time and again, with how much force and at what rate any given apple will fall from any given branch of any tree.

But what is an apple as it is experienced by someone? Take just the taste when you bite into one – a pretty rudimentary issue psychologically speaking. Can you measure

A Companion to the History of American Science, First Edition.
Edited by Georgina M. Montgomery and Mark A. Largent.
© 2016 John Wiley & Sons, Ltd. Published 2020 by John Wiley & Sons, Ltd.

the taste of an apple? Can you even describe it adequately? Some measure of tart and sweet, yes, but so is an orange, and you are really missing something, according to the aphorism, if you confuse the two. At some point, we can say an apple tastes like an apple, and that is about as far as we can get. It is an *experience*, a quality, and this quality is both elusive and yet also definite. In fact, most all experience is like this. We can describe some of its features in some measure, but there is always something that slips description, something that can only be experienced – not rendered – something inherently *subjective*. This may seem like airy sophistry or a philosopher's pseudo-dilemma, but I hope to suggest over the course of this chapter that this problem (or source of wonder, depending on your point of view) has not only serious scientific consequences but also profound political bearing.

Historians of science in fact are committed to exploring intersections between the scientific, the social, and the political. They generally hold that science does not float free above social concerns, objective and impartial. Rather it is saturated with – or perhaps just another aspect of – culture. These historians also point out that science earns its keep by proposing – or prophesying its researches will one day yield – solutions to social problems. Yet, scientists do not simply unlock these solutions, timeless and pristine, from some vault of universal truth. Rather the solutions and problems (and the very terms in which they are understood) are framed for science *a priori* by a historically evolving social order (Shapin and Schaffer 1989: 14–15).

A history of psychology should therefore strive to know what social or political purposes scientific psychology has aligned with. Social harmony and self-understanding? Or state building and social engineering? Has it provided a more complete definition of what it means to be human, or helped us quarantine the "deviant" from the "normal"? Has it, in short, worked in the interest of liberation or control? (Herman 1996: 9–12.) Answers to these questions are hotly contested, rarely "yes or no," but attending to basic assumptions about the relative importance of mind and subjectivity can be revealing.

Psychology is an eclectic and evolving field, however, this chapter will focus on only *experimental* or laboratory psychology, beginning in the late nineteenth-century United States and tracing its development until about 1950. The panoply of twentieth-century schools – psychometrics, psychoanalysis, gestalt, educational, cognitive, comparative, developmental, social, industrial – will not be discussed. There simply is not space to tell these stories here, but the accompanying bibliographic essay sketches the terrain and suggests further reading.

This chapter will limit its coverage in this way because experimental psychology was, in the first half of the twentieth century, considered the most scientific of all the discipline's branches, and, in a burgeoning age of science, it is therefore pivotal in explaining the rapid development of the entire field as a science. The aim here is to explore how and why experimental psychology grew so rapidly and took the shape it did. This chapter will follow its early developments, then its coalescence under the banner of behaviorism, a school-cum-dynasty within experimental psychology during the first half of the twentieth century. There is a general consensus among historians that behaviorism was the most powerful intra-disciplinary influence shaping the growth of the entire and diverse field of psychology in the United States, so it forms a natural locus for this analysis (Mills 2000: 1).

This chapter will trace a branching network of influence through select figures in early scientific psychology: Wilhelm Wundt, Edward Titchener, William James, G.S. Hall, J.B. Watson, and Clark Hull. They are all important in their own right, but they appear here not because they are titanic actors who single-handedly steered the course of the discipline, but rather because they vividly illuminate broader trends, and decisive contrasts and continuities between one another. Only the last two of these – Watson and Hull – are behaviorists in name, but to better understand their assumptions and methods, it is necessary to examine their predecessors.

To deepen this analysis of why experimental psychology grew as it did, this chapter will also explore sympathies and correspondences between the purportedly "objective" space of the laboratory and the broader surrounding matrix of cultural norms and values. It perhaps can be easier to see how *applied* science is shaped by certain cultural assumptions, but laboratory science, "pure science," seems indeed more objective. You measure this phenomenon to test that hypothesis, draw a conclusion: cut and dry. To see these connections between lab and society, we'll need to unearth a set of beliefs so taken for granted – and indeed so useful in their particular way – that they often go unnoticed. I am referring here to the culturally conditioned assumptions that what we can see, touch, count, and measure is more real, significant, or knowable than what we cannot. This set of assumptions is often referred to as matter theory or materialism. This materialist tradition undoubtedly preceded – but was certainly augmented and institutionalized by – the Scientific Revolution, Cartesian mind–matter dualism, and the Industrial Revolution. It was also attended by the rise of modern transatlantic economies that depended increasingly on the ability to control and take calculated risks on the exchange of material goods (Dear 2009: 8–88). Materialism has been moreover buttressed by the allied assumption that when we measure something, we free it in its essence from subjective experience.

Robert Young (1970: 1–2) notes that in the Cartesian cleaving of mind from matter, the natural sciences claimed matter as its rightful domain of interest and designated mind a trap, a problem, a field of bias that needed correcting to do science. Scientific psychology then took up the unenviable task of studying all the things that the natural sciences bracketed out as noetic, ephemeral, not worth – or even inimical to – knowing. Moreover, experimental psychology set out to produce from this study knowledge that had the same degree of predictive power as the laws of biology and physics. This is perhaps an impossible task. Nonetheless, in attempting it, psychology has made fascinating inroads into that unknown country with which everyone is intimately acquainted.

This impetus in psychology to be another among the "hard" sciences – to be as predictive and utilitarian – rides stowaway as a part of the field's larger materialist freight. As such it constitutes another important and oft-overlooked governing cultural norm. And that it was good to be a "hard" science and decidedly less good to be a "soft" one derived in part from historically conditioned assumptions about what were "reason" and "emotion," and what were the roles of men (as "reasoners") and women (as "emoters") in culture and nature. Thus developments in this story are also propelled by even deeper attachments to gender norms and beliefs about "male science" and "female nature" (Merchant 1990).

All these cultural norms and many others are deeply cemented into the bedrock of a supposedly value-neutral science. But to explain just what happened to experimental psychology, we will need to not only keep these broad cultural norms in mind but also examine specific developments. To accurately account for the growth of this new science in the United States in the early twentieth century we must sojourn briefly in Germany in the late nineteenth century.

Wilhelm Wundt and Edward Titchener: The Laboratory Model

The development of American laboratories in the natural sciences and in psychology lagged about a generation behind their German forbears. By the mid-nineteenth century, Germany had become the model for Europe and the United States for how to develop and professionalize the scientific disciplines within the university system. Perhaps the most important figure in mid- to late-nineteenth-century German psychology was Wilhelm Wundt (1832–1920). He was not the first to conduct a psychological experiment, but most credit him with building the first psychology lab. More than this, Wundt established a systematic experimental approach to the study of psychology, produced an enormous volume of publications during his career, and trained a small army of students in his research methods (O'Donnell 1985: 15–31). When Wundt established his laboratory for psychological research in Leipzig in 1879, students from around the globe flocked there, and numerous other practicing psychologists (William James and G.S. Hall among them) visited to study how his lab worked. In total, 180 psychologists earned their PhDs with Wundt. Thirty-three of these were US students (Benjamin et al. 1992).

While Wundt was ultimately interested in more voluntaristic aspects of consciousness, his experimental practice was decidedly atomistic. Like the physicist's search to uncover the indivisible building blocks of matter, Wundt sought to break the complex process of consciousness into what he supposed were its immutable units. How did he do this? By measuring what he could and demonstrating these measurements revealed reliable patterns.

Wundt's methods of investigation were introspection and reaction-time study. Introspection involved presenting stimuli – visual, auditory, tactile – in quick series to observing subjects who would then describe what they experienced. Numerous trials would be made in an effort to standardize and find patterns within the results. Reaction-time measures required a subject-observer to tap a switch in an electrical relay as soon as they experienced a stimulus, or as soon as they could discriminate between two simultaneous stimuli (Hilgard 1987: 39–47). In this way, Wundt hoped to separate and compare the sensory-motor and decision-making components of an action (Benjamin 2006: 42).

Because of his systematic, atomistic approach and his mentoring of so many American psychologists, Wundt dominated the field for generations and profoundly shaped how psychological research would emerge in the United States. He demonstrated a replicable approach to fundamental problems such as how to design a rigorous experiment and develop a laboratory to facilitate such research. He defined what sorts of psychological phenomena one could measure and what could be inferred from such measurements. In short, Wundt offered a "logical," straightforward model for the experimental study of a complex subject. Wundt's philosophy – his atomism applied to mind – would soon pose problems for the next generation of researchers, but his techniques for making psychology a laboratory science blazed a clear trail forward.

Consider Edward Titchener (1867–1927), a student of Wundt. Like Wundt, Titchener held that complex consciousness was composed of stable units of simple perception and sensation. These "mental atoms," Titchener believed, could be sifted from the swarm of thought by introspection conducted in a carefully controlled laboratory environment. In 1892, Titchener set up a laboratory at Cornell that employed isolation rooms and acoustic baffles to limit unwanted background stimuli. He also rigorously trained all his introspecting subjects to respond in highly controlled ways to a range of sensory stimuli. Yet, rather than isolating fewer and fewer fundamental units of consciousness, Titchener's index spooled outward, eventually cataloguing more than 44,000 discrete perceptions distinguishable by such factors as duration, intensity, and quality (Benjamin 2006: 79; Titchener 1910).

The interest in the structure of consciousness and an atomism that nonetheless seemed to yield an infinitude of "fundamental particles" would soon be challenged in different ways by psychologists more interested in the *function* of consciousness – what it did in the world. One group of challengers, represented here by William James, preserved an interest in the primacy of consciousness – however elusive to study it might be – in determining that functional behavior. The other and later group, represented here by J.B. Watson, decided the only way forward, *scientifically,* was to ignore consciousness and study only the exterior: observable behavior.

William James: Darwinian Consciousness

In the decades following Darwin's *On the Origin of Species* (1859), biology acquired a great deal of glamour as a theoretical system, and rapidly developed as a laboratory science that required new levels of funding and achieved a new degree of specialization (Coleman 1971: 160–6). Biology had come of age, while psychology was, comparatively, a fledgling science.

When William James (1842–1910) began work as a psychologist in the mid-1870s the subject had no designated home of its own within American universities and was typically taught as an annex to philosophy. During James' time, however, the decided impulse was to separate psychology from the realm of philosophy and entrench it more firmly in the foundation of science. James himself, though, was something of an exception to this trend. James is less known for his experimental work, but rather his probing philosophical questioning of contemporary problems in the field.

Many have commented that his early education was disorganized and peripatetic but also sumptuous: a sort of moveable (and moving) feast of subjects and instructors. While it was a meandering path, Menand (2002: 94) notes that "it permitted him to approach intellectual problems uninhibited by received academic wisdom." In 1875, James offered the first course in psychology taught in the United States, and began to draw and train students like G. Stanley Hall, who were soon to help establish the new American field.

James was initially enthusiastic about the prospects of new experimental approaches to psychology but he also insisted that *phenomenological* introspection – not the experimental brand practiced by Titchener – could yield crucial empirical (though non-testable) insights (James 1918: 297–302). Put more simply, he wondered: when we look inside, what do we see? What he found was a stream, "a stream of consciousness."

He meant by this that mind, as he saw it, was always on the move. It comprised what appeared to be elements – discrete and observable – flickering into the foreground one moment, and then dissolving into one another the next. This "stream" for James connected the exterior and interior, the private and the social. It was a waterway in which memory, anticipation, physical sensation, thought, and emotion were dissolved, and whose banks were indelibly sculpted by all the cultural norms and taboos to which we were heir. It was a stream, ultimately, channeled by our identity, our sense of self, and it all slipped from moment to moment along a continually unfolding present (James 1918: 224–90, 291–6).

James was keen moreover to connect this phenomenological description of consciousness with harder science. He had read Darwin's *On the Origin of Species* with great interest and found natural selection attractive for its wide-reaching explanatory power and its anti-teleological description of evolution and speciation in nature. Though the Darwinian worldview increasingly proscribed purposive intentionality from the natural world, James felt the study of psychology should maintain a central place for mind and intention. This put him in opposition with other post-Darwinian thinkers like T.H Huxley and W. Clifford who held that our sense of purpose was an illusion, and mind a passive byproduct of a brain-machine running on a dizzying complex of instincts honed over eons of evolution (James 1879: 1–2).

Mind, James insisted, was not the passive shadow of matter, but instead a powerful tool – wrought by evolution no less – that not only functioned as the seat of our selfhood and awareness, but also provided super-added adaptive value. Consciousness, James noted, involved a discriminating and selecting capacity. This conscious selection was for James an extension of Darwinian natural selection that acted on the mental plane. What behaviors or ideas – in any particular context – were useful to a person in bringing about his or her desired ends? One would select from an array of alternatives those ideas or behaviors that got one closer to one's goals. Like natural selection, then, conscious selection was in this sense adaptive. James felt that all aspects of our selves – beliefs, values, reasons, and so on – were enlisted into this act of conscious selection (James 1918: 138–44).

James' interest in complex mental processes put him increasingly at odds with reductive experimental approaches to the study of mind and behavior like Wundt's and Titchener's. How could one come to know mind by looking at isolated sensations or the speed of neuro-physiological impulses taken out of context? How could one break the stream of consciousness into "atomic" units and examine them piece by piece? These questions would in a sense be forgotten by many in the field for the next 50 or 60 years and then asked again, and in a new way, by a new generation of psychologists in the post World War II years. But in James' day the drive to make psychological questions amenable to laboratory study was stronger than the philosophical cautions he raised. James stepped away from psychology later in his career and left the building of a laboratory-based research program to a new generation.

G.S. Hall: The Profession Builder

Perhaps James' most important student, and one of the most important of this new generation of laboratory builders, was G.S. Hall (1846–1924). Like his mentor James,

Hall had a strong impetus to connect the new scientific psychology with biology. Hall was an energetic builder of the profession of psychology and perhaps, more than anyone else, should be credited with institutionalizing the discipline in the United States. Hall founded the first department of – and graduate program in – psychology in the United States at Johns Hopkins University in 1883. He also then set up the first lab for American experimental psychology – patterned on the Wundtian model – in the same year. Hall established the first US journal for psychology, *The American Journal of Psychology* in 1887 and then went on to found the country's first national-level professional organization, the American Psychological Association in 1892. In 1888 he left Hopkins for Clark University. Under Hall's direction, Clark trained over half of all American psychologists who took their degree between 1890 and 1900 (Fancher and Rutherford 2012: 328).

Hall also helped inaugurate the child study movement, which grew later into developmental psychology. The cornerstone of Hall's developmental theory was his own brand of recapitulationist psychology adapted from biologist Ernst Haeckel's theory of embryological development (*vis.* "ontogeny recapitulates phylogeny"). It was a theory that, in Hall's hands, imported and "scientized" the racism, sexism, and gender norms of his day (Bederman 2008: 77–120). Though his child study and developmental theory were criticized by some, including James, Hall's program resonated sufficiently with the cultural currents of his time. He was one of the period's leading psychologists, and a dominant shaper and spokesperson for the American field.

Beginning with Hall's lab at Hopkins, nearly 40 psychology labs opened along with the founding of departments at various colleges and universities in the United States in the 17 years between 1883 and 1900. Even as early as 1893, one close observer estimated that the United States already had twice as many labs as all of Europe (Nichols 1893: 409). Perhaps even more remarkable than this rapid growth is the fact that very close to half of these 40 laboratories were founded by students of either Hall or Wundt (Hilgard 1987: 32–4).

We should not forget the power of the laboratory itself to shape a discipline and perpetuate certain ways of knowing. Labs are assemblages of equipment and trained personnel, coordinated in specific ways to ask very particular questions of the world. They attract funding (to the lab and the university), if the lab produces reliable results; and this selective process in turn shapes the sorts of future questions that can be reliably asked and answered. In this way labs exercise both a catalytic and normalizing function over the field (Kuhn 1996: 23–42). They garner funds and grow the profession but also direct research with increasing inertia in particular directions. They inherit their ways of working from other practitioners past and present (mentors, colleagues, and competitors), and are moreover symbols of the growth and power of the discipline (Capshew 1992: 132). And while laboratories seem on the one hand to be the pinnacle of objective science – highly controlled, hermetically closed spaces, magic-carpet-fact-generators that float above the din and disorder of culture – they in fact import whole worldviews and systems of assumptions that become a tacit part of their everyday functioning. Latour has remarked in this regard that "no one can say where the laboratory is and where the society is" (Latour 1983: 154). For this reason – the porous boundaries between lab and society – we also should not fail to take note of how psychology's rapid emergence as a laboratory science in the United States coincided with a particularly volatile period of industrial growth and urbanization in US history known as the Progressive Era.

Explosive demographic growth of the nineteenth century, coupled with the recent emancipation of African Americans and a surging influx of immigrants from all parts of Europe, was rapidly re-embroidering the fabric of early twentieth-century American cultural life. More and more people, and from a greater diversity of backgrounds than ever before, were moving to US cities. Many US commentators looked at this rapid change and saw problems. Some even heralded a civilization in decline. Cities, increasingly dense and variegated showcases of human life, made visible poverty, overcrowding, and pollution on a scale many had never witnessed before (Rodgers 1998: 48). These same cities were likewise becoming cultural repositories of a new level of ethnic pluralism that was a threat to white hegemony and often a goad to nativist tendencies in white American culture (Paul 1995: 97–114). Progressives, as a social and political movement, responded to these rapid cultural and demographic changes by pushing to reform, rationalize, and make more efficient all aspects of society: manufacture, labor, urban planning, education, and so on. Importantly, Progressives argued that science and scientific expertise should play a central role in imagining and implementing this rational reform of society. Thus Progressives bent their considerable political will to forge networks between various levels of government and private philanthropy. These new private–public corporations directed more money toward scientific research than had ever been available before (Cravens 1993: 60–71).

Thus psychology was emerging as a laboratory science in the United States right when the cultural value of science and the scientific expert were dramatically on the rise. Many psychologists felt they could offer powerful solutions to numerous social problems, but first they had to complete the reformation of psychology itself. Psychology had to be pried once and for all from the hands of philosophy and allowed to stand confidently alongside biology, chemistry, and physics as a natural science in its own right.

Behaviorism: Psychology without a Mind

Such a psychologist-reformer was John Broadus Watson (1878–1958). He took aim at experimental programs like Titchener's that sought to systematically explore consciousness through introspection. How did you know your introspective subjects were really reliable in their reporting of their internal world? How did you know if they were indeed experiencing the same thing? What good was a taxonomy of mental elements that did not reveal any unifying underlying patterns but instead multiplied outward in seemingly endless variation? Yet, unlike James, Watson was loath to theorize about consciousness. In fact, he held, this was precisely the problem: psychology's preoccupation with an essentially unknowable and – in all likelihood – irrelevant mind.

In his 1913 "Psychology as the Behaviorist Views it," Watson, in sweeping iconoclastic terms, lay the groundwork for a new psychology, one that was even more conducive to laboratory investigation and thus more like a natural science. He wrote:

> I believe we can write a psychology, define it as the science of behavior, and never go back on our definition: never use the term consciousness, mental states, mind, content, introspectively verifiable, imagery and the like ... (Watson 1913: 166)

Mind was a minefield, murky and noetic, and psychology, if it were to be a natural science, would do well to skirt the edges of this limbo. If you took mind from psychology, what was left to study? The most knowable, observable part: behavior. This was precisely Watson's proposed object of study: "Psychology as the behaviorist views it is a purely objective experimental branch of natural science. Its theoretical goal is the prediction and control of behavior" (Watson 1913: 158). Buckley and others have argued that this new vision of psychology resonated deeply with a body of practitioners that sought for its discipline the status and respect of a natural science (Buckley 1989: 86).

There is some debate about precisely when behaviorism tipped experimental psychology decisively in its direction, but Benjamin (2006: 144) dates its command of the field from the teens to the 1960s, noting "behaviorism would come to dominate American psychology like no school before or since for a period of nearly fifty years." Its assumptions and research practices would occupy the efforts of vast ranks of experimental psychologists whose work can only be summarized here.

Watson's original aim – the prediction and control of behavior – became the cornerstone of the behaviorist research program. Crucially, behaviorists defined their subject without reference to any internal mental states or processes; rather behavior was the observable material organism in action, responding to measurable environmental stimuli. Each response (R) might elicit a new set of stimuli (S) so that S–R–S chains could form (or be engineered by the experimenter) indefinitely. Stimulus and response were "glued" together in the behaviorist conception by a phenomenon they called conditioning. Different forms of conditioning had been well established experimentally by the research of Edward Thorndike (law of effect) and Ivan Pavlov (classical conditioning).

Conditioning theory held that certain patterns of response would be strengthened if positively reinforced by a subsequent reward–stimulus, or weakened by a subsequent punishment–stimulus. Stimulus–response conditioning theory also held that various stimuli could be arbitrarily fused or substituted for one another by association. In this way responses could be entrained to stimuli that originally had no such effect on the organism (e.g., Pavlov's dogs conditioned to salivate to the sound of a bell).

Conditioning might seem synonymous here with our colloquial notions of learning: an organism gradually learns to respond in different ways to its environment depending on the reactions it gets. In fact, a great volume of behaviorist research was pointedly interested in learning and learning theory. It is important to remember though that learning, for a behaviorist, was not an internal mental event. Because it could not be experimentally defined as a kind of insight or problem-solving process or stored memory, it was not conceptualized in any of these ways. Rather, learning was simply the experimentally observable result of the conditioning interaction between a stimulus-generating environment and responsive organism. Thus, what was practicable in the lab began to set the limits of what was conceptually permissible, and even over time to hem the boundary of what was theoretically imaginable.

The rise of behaviorism was attended by a major shift in experimental psychology toward animal research, and the modest white rat, though certainly not the only animal model used, became king of this laboratory jungle. Animals made possible the lab-based experimental testing of conditioned learning in a way that would not have been possible or ethical with humans. Various kinds of mazes, "multiple-choice" pens and puzzle boxes – outfitted with loops, levers, and trip switches the animals could manipulate – became standard equipment in the behaviorist lab (Boakes 1984: 145–57). These sorts of experiments, over many trials, yielded highly empirical results which could be graphed as curves of data points, often showing diminishing times to successful completion of various tasks, like running through a maze.

Though many later (in the late 1950s and early 1960s) began to question the usefulness of its theories and data, the behaviorist program generated highly reliable results and gained considerable impetus as a research program. The field grew correspondingly. From 1920 to 1939, the heyday of early behaviorism, 1700 people graduated with PhDs in psychology, more than quadruple the number between 1900 and 1919 (Harper 1949). Notably this period of rapid growth spanned the economic contraction of the Great Depression. Total money directed toward laboratory research in experimental psychology research mushroomed exponentially as well: $30,000 in 1893 to well over $1,000,000 in 1925 (Capshew 1992; Ruckmich 1912, 1926).

Though behaviorists chiefly experimented with animals, the hope was always to extrapolate laws of behavior leading to the better understanding – or prediction and control – of human behavior. Indeed, Watson (1913: 158) expressed a desire for "a psychology that concerns itself with human life" and made the sweeping and famous solicitation: give him any infant and he would train it for any profession (Watson 1930: 82). Psychologists were increasingly confident they had achieved for their science a rigorous experimental foundation that would allow them to produce knowledge readily applicable to human development and the engineering of society. Degler (1992: 154) notes that a "wide spectrum" of psychology united behind a belief that "through behaviorism a better social order would be feasible."

While the movement from animal experiment to engineering human behavior depended on an affirmation of the evolutionary continuity between humans and animals, behaviorism took a decisive step away from the heavily biologized psychology of their forbears. Indeed many in the field vigorously denied a primary role for instinct in behavior. Instinct was just another hypothetical feature of an unknowable, perhaps illusory interior (Degler 1992: 152–4). Behaviorists instead staked the power and broad applicability of their approach on the assumption that the organism – animal or human – was largely plastic and its behavior could be shaped in any way the experimenter desired by stringing together various concatenations of stimuli and responses. This brand of environmentalism buttressed behaviorism as a science in its own right, more a peer of biology, and less a tag-along second sibling (Degler 1992: 153). It was a radical environmentalism, moreover, that sought to place the power to shape and control in the hands of the expert.

When considering what sort of "better social order" to which behaviorists imagined their work might contribute, it is well worth exploring connections and shared contexts between the laboratory and contemporary American society at large. It was an era marked not only by new social anxieties and a quest for scientific efficiency, but by unheralded economic growth as well. In the years between 1890 and 1929, the US economy was, in addition, utterly transformed by a "revolution in production." The country witnessed a mind-boggling surge in manufacturing output, much of which was accomplished by the new assembly line process and readily available fossil fuel sources (Leach 1994: 15–17). Yet, this new kind of production heralded a need for a new kind of control of the worker as well. As Gramsci (1971: 302) observed in his *Prison Notebooks*, this emergent American-style mass production was "the biggest collective effort to date to create, with unprecedented speed and with a consciousness of purpose unmatched in history, a new type of worker and a new type of man."

Workers on the new "continuous process" line no longer crafted a single product from start to finish. Instead they now performed a much simpler series of repetitive tasks, as they in turn became a fixed point in a larger assembly process that steadily trundled past them. Workers were deskilled and retrained to perform as efficiently as possible for this new mode of labor by a process referred to as "scientific management." Architects of this new management style like Henry Ford and Frederick Taylor deployed a photographic technique, time-motion analysis, to study a worker's task as an incremental series of motions that could then be tweaked and tuned gesture by gesture, making the haptics and kinesthetics of the overall task performance more efficient.

It is likely no coincidence that behaviorism emerged as a science at the same time Tayloristic management practices came to dominate the politics and economics of production. Both Taylorism and behaviorism enforced a bodily micromanagement of their subjects in the interest of molding more efficient behavior. Like photographic time-motion analysis of the worker, behaviorism emphasized study of only the outward, exterior, observable actions of the organism. Under both disciplines the subjects became bodies in motion, entrained to a sequential series of cues: the next unit on the assembly line, the chain of reinforcing stimuli. Both Taylorism and behaviorism were, in short, training regimes that assumed learning and performance to be mindless and that consequently treated their subjects as automata.

Merchant (1990: 292) has directly linked behaviorism with mechanical materialism, and Mills (2000: 23–54, 70) has observed a general overlap between behaviorism, social science theory, and certain attempts at social reform between 1890 and 1920. Likewise, Buckley (1989: 145) has noted that when Watson left experimental psychology for a career in advertising he brought a Tayloristic shop-floor management style to the offices of J. Walter Thompson. But no one has yet to examine in detail what appear to be these intriguing sympathies between Tayloristic scientific management and the behaviorist research program itself.

Indeed, where is the lab and where is the society? This is certainly not to say that either behaviorism or Taylorism directly brought about the other. Rather both seem to have emerged from and acted as mutually supporting parts of the same worldview: one where doctrines of control and efficiency prevailed and where – for the sake of this efficiency – larger social systems could be envisioned as a mechanical coordination of their fundamental component parts. At least in the world of work (laboratory science and manufacture), the material thing – the commodity, the working body-in motion, the behaving experimental subject – was what was worth having, knowing, and controlling. The thoughts, beliefs, feelings, identities, and ideologies pulled into the matrix of these processes were neither here nor there. They were immaterial.

Clark Hull: A Mechanics of Behavior

If, under Watson, behaviorism attempted to establish itself as a peer to, not dependent of, biology, in the following decades under Clark Hull (1884–1852), the prediction and control of behavior, reached even further, aspiring to Newtonian-mechanical certitude. Hull – whose work represents the apotheosis, if not the terminus, of behaviorism

– labored in the 1930s to erect an explicitly quantitative and experimentally testable behavioral theory (Amsel and Rashotte 1984: 54–63). Here is an example of how his program worked. One of his learning experiments – grounded in stimulus–response conditioning theory – required human subjects to attempt to verbalize a random series of nonsense syllables in the proper sequence: e.g., KEM–FAP–ZIT–YEV–JUD–KEX–POF (Hull 1935: 502). In the Wundtian tradition, Hull assumed the difficulty of verbalization at any point in the series to be a direct function of reaction time. Bear in mind, Hull essentially viewed these responses as observable verbalizations entrained to a series of stimuli, not the outcome of stored "memories" or mnemonic "maps" or anything so mentalistic.

Copious data from other researchers running rats in mazes had long made clear that conditioned learning was generally most difficult to achieve in the middle of the maze. Hull reasoned then that there was a general law of behavior at work that could be applied across a wide range of species and learning tasks. He expressed this putative law as a theorem, applied here to humans and syllable learning:

> THEOREM VII: The rate of reaction times of the syllables of a rote series learned by massed practice will be shortest at the end positions and progressively longer the farther the syllable from the ends of the series. (Hull 1935: 507)

Hull then rendered this theorem as a mathematical statement employing derivative calculus for the function

$$\frac{dR_1}{dn} = m[-2n + (N + 1)]$$

to express the rate of reaction times as a difficulty curve at any point in the ordinal series of syllables, with difficulty peaking in the middle. Had Hull mathematically expressed a fundamental law of behavior? Perhaps he had, if the data, after a sufficient series of trials, approximated the curve his calculus projected. The proof then was in the deductive experimental test of this very explicitly defined formula.

The influence of Hull's program cannot be overestimated. It dominated experimental psychology well through the 1950s, and continued to be well cited even into the early 1970s (Amsel and Rashotte 1984: 10). Countless practitioners joined in the Hullian project of testing and making further derivations from his numerous behavioral–mathematical postulates (Benjamin 2006: 149). Between 1941 and 1950 nearly half of the articles published in behaviorist flagships, the *Journal of Comparative and Physiological Psychology* and the *Journal of Experimental Psychology*, explicitly referred to and relied on Hull's research (Amsel and Rashotte 1984: 2).

Benjamin (2006: 149) observes that in the end, however, Hull's postulates were perhaps too precisely testable for their own good. The ballooning Hullian project deflated through the late 1950s as its postulates, produced "oceans of anomalies" and were gradually but more or less completely falsified by experimental trial (Webster and Coleman 1992: 1070). Behavior, not to mention mind, appeared now to be not so lawful as Hull's postulates.

The power of Hull's work, however, and its ability to sway the field toward its methods cannot be ignored. It was heralded by advocates as psychology's arrival as a hard – even Newtonian – science. In fact Hull (1935: 495) explicitly drew comparisons between his and Newton's program: "In a truly scientific system ... the theorems must constitute specific hypotheses capable of concrete confirmation or refutation. This was eminently true of Newton's system." Hull (1950: 221) moreover claimed he sought principles of behavior as certain as those "governing the law of falling bodies." This of course brings us back to the problem of apples: how reliably they fall, howsoever they taste.

Once psychology took hold on US soil, how and why did it grow as rapidly as it did? The answers are both material and ideological. Clearly underlying economic and demographic forces played a role. To some extent psychology mounted the rising shoulders of an industrializing nation-state. But economic factors were not at all overdetermining. Recall the period of rapid growth for the discipline despite overall economic contraction of the Great Depression. Nor do economics ever function in an ideological vacuum.

In some sense the laboratory was a more proximate lever governing the fate of the discipline. The field followed behind the lab, and went where it went. Even as James' philosophical psychology was left behind, his theory of conscious selection perhaps predicted precisely why. The constellation of beliefs, values, and functions that inhered in laboratory work were more useful to the profession – at least in the short run – and were thus assiduously selected by its practitioners. Again the reasons for this are material and ideological. Lab work generated reliable knowledge and garnered resources. Lab work was also predicated on belief-driven choices about what sorts of problems were worth solving and which were not. Evaporating off the subjective aspects of behavior and then arguing they had not mattered in the first place was certainly one such set of choices that hastened the growth of laboratories and the field.

And as Capshew and Latour remind us, precisely because of their particular materiality, labs performed a symbolic function as well. Laboratories broadcast the aura of "hard science" (an aura ringing with gendered overtones), and increasingly came to crowd the landscape of a field aspiring to such a status. In fact this ideal – this quest to be a hard science – impelled US psychology rapidly forward out of the nineteenth century. It had sought its own fundamental particles, its own embryology, its own selection theory, its own mechanics.

And if laboratories have symbolic value, then so do mathematical equations. Above the ordered play of variables, there is another kind of symbolism at work. Mathematical formulae glint with recondite certainty. They gleam like a world stripped down to its scaffolding, its indigestible working parts.

If individual consciousness is indeed like a stream, then experimental psychology has navigated a larger and shared waterway. It has floated down the decades like a barge of materials, techniques, and theories – all modifiable and interchangeable – on a river of deeper tacit beliefs and assumptions: assumptions about what sorts of knowledge are most needed to address changes in the social order; assumptions about how best (and who is most able) to reliably produce this knowledge; and finally assumptions about what is mind and what is matter, and which matters more.

Bibliographic Essay

The secondary literature on the history of psychology is as diverse as psychology itself. There are numerous surveys offering highly accessible broadsheet maps of this fascinating terrain. Ernest Hilgard's *Psychology in America* (1987) is widely used and Ludy

Benjamin's *A Brief History of Modern Psychology* (2006), while not as internalist and technical as Hilgard's, offers more social context and biographical color.

For those readers interested in pre- and early scientific psychology, Daniel Robinson's *An Intellectual History of Psychology* (1995) is a necessary starting place. Robinson follows diffusely psychological bodies of thought from early Greece to the early twentieth century, tracing their philosophical shifts through rationalism, materialism, and empiricism. Robert Young's *Mind, Brain, and Adaptation* (1970) proceeds through the nineteenth century from phrenology to early neurophysiology, demonstrating a mounting pressure to abandon holistic explanations of mental function for more reductive and verifiable studies of reflex. Robert Richards, in *Darwin and the Emergence of Evolutionary Theories of Mind and Behavior* (1987), foregrounds a current of thought running contrary to the larger experimental trend Young documents. Here we find a set of thinkers – including Romanes, Morgan, and James – who sought to preserve a place for mind in the emerging science of behavior. All three of these books explore tensions between the "material" and the "mental" proceeding from Cartesian mechanical materialism.

The secondary literature on various branches of applied psychology is eclectic. Benjamin and Hilgard cover the breadth of the canopy admirably. The history of IQ and psychometrics is undoubtedly the most substantial body of secondary literature on applied psychology. Stephen J. Gould's *The Mismeasure of Man* (1981), Lelia Zenderland's *Measuring Minds* (2001), and John Carson's *The Measure of Merit* (2007) are essential reading here. These works are all richly contextual and suggest the social construction of IQ. While all explore overlapping contexts and causal factors in World War I and interwar years, Gould links the reification of IQ more to racism and eugenics, Carson to the needs of a burgeoning meritocracy, and Zenderland to the Progressive Era quest for scientific efficiency. Nicholas Lemann's *The Big Test* (2000), while more journalistic and less historiographical, is a riveting and thoroughly researched account of the progress of testing in the post World War II years. Hamilton Cravens' *Before Head Start* (1993) intersects in many ways with this history of psychometrics, and also provides an incredibly informative entrée to the history of developmental psychology. A good place to start for the history of industrial psychology would be John Waller's "The Hawthorne Studies" from *Fabulous Science* (2002). Waller offers a concise and rigorous analysis of the science and the social context around the series of oft-cited, and frequently misunderstood, Hawthorne Factory experiments.

There's a wealth of historical literature covering Freudian and neo-Freudian psychoanalysis. The works of Nathan Hale and Frank Sulloway are foundational and offer useful contrast to one another. Hale's two-volume opus, *The Rise and Crisis of Psychoanalysis in the United States* (2000), treats the genesis of psychoanalysis and then the complicated story of its reception and rejection by academic psychology in the United States. With experimental psychology rapidly on the rise, psychoanalysis was deemed untestable and therefore unscientific. On the other hand, Frank Sulloway, in *Freud, Biologist of the Mind* (1992), explores the way Freud's thought was deeply impelled by science, by the biological theory of his day. Freud, Sulloway argues, went to pains to conceal these connections – to remain a "cryptobiologist" – in the interest of establishing his psychoanalysis as a new science of mind, *sui generis*.

The decades before and after World War II saw the emergence of a new stripe of psychologist: the social psychologist. Ugly international spectacles of racism – Nazi atrocities and the US Jim Crow regime – compelled a new attention to the way power relations and ideologies influenced human behavior and the structuring of the social order. Social psychologists like Otto Klineberg, Kurt Lewin, Mamie Clark, and Kenneth Clark, experimentally tested theories seeking to explain how human behavior was bound up in social interactions, and mediated by power relations and identity. Experimental questions about social aspects of mind and subjectivity occupied the field in a way they never had before. There are a number of important books that tell different aspects of this story.

Ellen Herman's *The Romance of American Psychology* (1995) is a seminal history of experimental and social psychology during and after World War II. Psychology's relation to state power and the multifarious political uses to which it was put are central to her analysis. She demonstrates how the same theories deployed in an effort to harmonize race relations at home were also put to use in pro-American Cold War era "Third World development" and counter-insurgency strategies. Additionally, Herman's history covers the second-wave feminist reappropriation of Freudian theory which was, in its original form, in many ways misogynistic. Jessica Grogan's recent *Encountering America* (2012) explores the emergence of humanistic psychology as a part of the larger counterculture movement of the 1960s. Humanistic psychology posed itself as an alternative to behaviorism and psychoanalysis, and as an approach that helped individuals achieve "self-actualization" by embracing the "primacy of subjectivity."

Carl Degler's *In Search of Human Nature* (1992) provides perhaps one of the most large-scale synoptic accounts of twentieth-century psychology. Degler depicts – very persuasively and on the basis of an exhaustive literature review – psychology's off-again, on-again relationship with biology. He holds that while the new social turn in experimental psychology, underway since the early 1930s, came to fruition in the years following World War II, there was a decided move among psychologists in the late 1960s and early 1970s back to biological explanations of human behavior. Degler suggests this was in many ways due to the stunning successes of genetics and molecular biology in the 1950s and 1960s. Thus, though Clark Hull's mechanics did not bear lasting fruit, the quest for laws that would sculpt psychology into a harder science continued into the late twentieth century.

Chapter Nine

GENETICS

Melinda Gormley

Genetics is the scientific study of heredity and variation in living organisms. Genetics traverses several scientific fields, which has contributed to its growth and predominance as a discipline and therefore what follows is an interdisciplinary history. A complete picture of the history of genetics in the United States requires knowledge of intellectual pursuits, disciplinary infrastructure, and practical applications. The application of genetic principles to biology, medicine, agriculture, and engineering has stimulated examinations of the ethical, legal, social, and policy implications of innovative research. Complementary chapters in this volume that inform the history of genetics in the United States include Molecular and Cellular Biology, Medical Genetics, Anthropology, Sociobiology and Evolutionary Psychology, and the American Eugenics Movement.

The originator of genetics is usually identified as Gregor Mendel, however, attempts at understanding heredity and devising theories about hereditary mechanisms predate him and the rediscovery of his work in 1900. Mendel worked in a monastery near Brünn (now Brno) of the then Austro-Hungarian Empire when he performed extensive experiments with pea plants and published his findings in 1866. He established nomenclature and laws that influenced early experiments of transmission genetics. Transmission, or classical, genetics examines how traits pass from one generation to the next. Classical genetics research predates the understanding that nucleic acids hold genetic material and it is based on phenotype (i.e., expression of genes). In contrast, modern genetics marks the shift to research on nucleic acids and genotypes (i.e., one's full genetic material whether expressed or not). Dominant factors, according to Mendel, occurred more often and were transmitted in their entirety; recessive traits stemmed from factors occurring less frequently and were latent. In 1900, three botanists, Hugo de Vries in the Netherlands, Carl Correns in Germany, and Erich von Tschermak-Seysenegg in Austria, independently verified aspects of Mendel's work prompting a number of American researchers to perform investigations that confirmed and extended Mendel's laws

A Companion to the History of American Science, First Edition.
Edited by Georgina M. Montgomery and Mark A. Largent.

(Bowler 1989; Falk 2009; Müller-Wille and Rheinberger 2007; Olby 1985; Paul and Kimmelman 1988; Robinson 1979; Stubbe 1965).

Genetics drew the interest of embryologists, morphologists, agriculturalists, breeders, and other biologists upon the rediscovery of Mendel's laws. Some recognized genetics' potential to provide information vital to evolutionary theory and many realized the benefits of genetic principles for improving society. The inability to explain inheritance, specifically how information passes from parent to progeny, was one inadequacy of the Darwinian theory of evolution. Taking many of its first adherents from the naturalist tradition, genetics and evolution experienced a period of disconnection before synthesis became possible in the mid-twentieth century (Farber 2000; Hagen 1999; Larson 2004). Knowledge of genetics made little headway during the first decade of the twentieth century; nonetheless the discipline began taking shape. In the 10 years following the rediscovery of Mendel's work, genetics and genes were coined and researchers showed that Mendel's laws applied to plants, animals, and humans.

Even before the rediscovery and expansion of Mendel's work, eugenics gained traction, first in Britain and soon thereafter in the United States. Eugenics, a word coined by Francis Galton in 1883, translates to "good in birth" or "noble in heredity." The eugenics movement sought to improve the physical and mental capabilities of human beings through selective breeding. Positive eugenics is the promotion of breeding among physically and mentally superior human beings and negative eugenics prevents procreation among those exhibiting traits considered undesirable (Bowler 1989; Kevles 1985; Paul 1995). Eugenics both helped and hindered the professionalization of genetics in the United States during the first decades of the twentieth century when the two endeavors were closely intertwined. In the early twentieth century, eugenics had two meanings – the study of human genetics and the application of genetic laws to guide human reproduction. The Progressive Era compelled a significant number of social and health reforms, and eugenics was an aspect of this larger movement. American biologists gained social authority and funding by aligning themselves with the eugenics movement (Allen 1986; Largent 2008). By the late 1920s eugenicists were succeeding at legally implementing their ideas, and eugenics had strong associations with social reform. Eugenics developed negative connotations after the civil rights movements in the latter half of the twentieth century because of its relationship to sterilization laws at the state level and immigration laws at the national level. The emerging understanding of the Holocaust in the 1960s and 1970s was increasingly linked to eugenics, and a growing number of authors asserted that the eugenics movement had led to prematurely enacted policies that had been based on either incomplete scientific knowledge or a poor understanding of the science (Kevles 1985; Largent 2008; Ludmerer 1972). Eugenics came to be defined as the practical application of genetic principles to improving the human population, whereas human genetics came to denote the scientific study of heredity and variation in human beings. Scientists' attempts to undermine the American eugenics movement contributed to the professionalization of genetics as well as the discipline's growing reputation during the twentieth century (Gormley 2009). Human genetics witnessed a period of decline beginning in the 1930s regaining a solid foothold in the United States by the 1960s (Ludmerer 1972).

Private foundations and philanthropists supported the eugenics movement in large part because its work fit well with long-held notions about inequalities among human races and it supported the motivations of public health and urban reform movements of the late nineteenth and early twentieth centuries. The Eugenics Record Office at Cold Spring Harbor, for example, was funded by the Carnegie Institute of Washington and a wealthy widow, Mary Harriman. Workers based at the Eugenics Record Office helped shape eugenic research and laws during its 30-year existence (Allen 1986). Eugenicists found a favorable social and political environmental in the United States until the mid-twentieth century and even after eugenics fell into disfavor some researchers and granting agencies supported endeavors with aims similar to eugenics (Allen 1991; Ludmerer 1972; Paul 1991, 1995). The nature–nurture debate that broaches the influence of heredity and environment has been an ongoing conversation related to evolution and development on which several fields including eugenics and psychology have weighed in (Cravens 1978; Gould 2006; Keller 2010).

Professionalization of Genetics

To be academically successful, a discipline needs not only intellectual contributions but also infrastructure in the form of journals, societies, and university programs that serve the collective needs of its members (Appel 1988). Founded in 1903, the American Breeders' Association focused on genetics and eugenics as they applied to agriculture and hosted the *American Breeders' Magazine* from 1910 to 1913 at which point it became the *Journal of Heredity* (Kimmelman 1983, 1987). The *Journal of Heredity* and *Genetics*, which was started in 1916, as well as the Genetics Society of America founded in 1932 and the American Society for Human Genetics established in 1948, contributed to the discipline's growth and centralization. E.W. Sinnott and L.C. Dunn published the highly regarded and frequently used textbook *Principles of Genetics*, keeping the information current by producing five editions between 1925 and 1958. The textbook's historical case study approach aimed to teach genetics information and scientific processes and it would serve as a model that authors of subsequent genetics textbooks adopted (Skopek 2011). Geneticists were a small community during the first third of the twentieth century and most American contributions originated from one of a few academic centers (Sapp 1987). Thomas Hunt Morgan and E.B. Wilson, for example, oversaw genetics education at Columbia University. William E. Castle and E.M. East directed the Bussey Institution at Harvard University. H.S. Jennings and Raymond Pearl were at Johns Hopkins University. From these three institutions emerged the first generation of biologists trained as geneticists and these PhDs in genetics went primarily into agricultural careers at experiment stations and with the government in the years before genetics positions and departments were available (Kimmelman 2006).

At Columbia University after 1910, T.H. Morgan and his students established the fundamentals of so-called "classical" transmission genetics and greatly expanded genetic knowledge by performing experiments with *Drosophila*. Their contributions laid foundations for the modern evolutionary synthesis that merged genetics and Darwinian evolution. In the early years researchers sought abnormal traits that could be tracked over several generations. After breeding many generations of fruit flies, Morgan found one fly with white rather than red eyes. Tracking white-eyed flies over several generations, he found that certain traits occur together routinely. Morgan concluded that genes were physical entities residing on chromosomes like beads on a string and when a chromosome underwent an aberration such as crossover, the genes located closely to one another remained on the same chromosome. Morgan's conclusion led two members of his laboratory, A.H. Sturtevant and C.B. Bridges, to begin mapping the specific locations of more genes on chromosomes. Understanding the importance of abnormalities to genetic studies, H.J. Muller, also Morgan's student, used x-rays to induced mutations in organisms and he demonstrated that the frequency of gene mutations was proportional to the organism's exposure. Morgan and Muller received Nobel Prizes in Physiology and Medicine in 1933 and 1946, respectively. Morgan and his students took a reductionist approach believing that with a better understanding of the basics, genetics would eventually elucidate issues involving evolution and development (Allen 1978a, 1978b; Bowler 1989; Brush 2002; Dunn 1965; Holmes 2006; Kohler 1994; Roll-Hansen 1978). Chromosome theory and the reductionist approach were attacked by William Bateson, Wilhelm Johannsen, and Richard Goldschmidt. Critics such as these typically trained in European countries and Goldschmidt illustrates the German context in which practitioners concentrated on developmental genetics, thinking about the whole organism and the role of genes in the evolution of species and the development of individuals. This approach stemmed naturally from the German educational system that discouraged specialization and produced scholars with a wide breadth of knowledge (Allen 1974; Dietrich 2011; Gilbert 1988; Harwood 1993; Roll-Hansen 1978).

Political events contributed to the growing professionalization of genetics in the United States, and again, Goldschmidt's career is representative. He had been director of the Kaiser Wilhelm Institute for Biology for 15 years before losing his job because of Nazi laws and in 1936 he moved permanently to the United States working at the University of California (Dietrich 1996). Geneticists Curt Stern and Ernst Caspari are two other German émigrés who relocated permanently to the United States in 1932 and 1938, respectively. Beginning in 1944, Stern and Caspari collaborated on experiments examining the genetic effects of radiation on *Drosophila*, research that was paid for by the US Army's Manhattan Engineering District (Gormley 2007). An area ripe for research would trace how scientists who relocated to the United States shaped the intellectual trajectory of genetics. For example, Stern started a temporary position in T.H. Morgan's laboratory in 1932 and did not return to Germany because Adolf Hitler had come to power. Stern's serious entry into human genetics, according to his student and human geneticist James V. Neel, dates to 1939 when Stern first offered a seminar on the topic. Moreover, Neel stated that Stern's textbook, *Principles of Human Genetics* (published in 1949 with revised editions in 1960 and 1973), was influential to the development of human genetics (Neel 1987). Also important was the US government's genetics research on survivors of the atomic bombings at Hiroshima and Nagasaki through the Atomic Bomb Casualty Commission to which Neel was a major contributor (Beatty 1991; Lindee 1994; Neel 1994). Fueling the growth and development of genetics in general was the small size and close ties between members of this international community (Cain 1993, 2002; Kohler 1994; Provine 1986). In addition to a number of foreign geneticists permanently relocating to the United States, many participated in exchange programs spending extended periods in foreign laboratories and attended conferences such as the International Congress on Genetics held about every five years (Krementsov 2005).

Quantitative approaches were pivotal to understanding how genetic changes affected populations. Statistical analysis had been used to study heredity in individuals since the mid-twentieth century through the work of Francis Galton and his student Karl Pearson in England. Englishman R.A. Fisher's 1930 book, *Genetical Theory of Natural Selection*, laid the basis for relating population genetics to evolutionary theory, specifically selection, and the American Sewall Wright built on this foundation. Trained as a geneticist and adept at mathematics, Wright's shifting balance theory showed that new species could emerge from a subgroup isolated from the larger population. A subgroup after much inbreeding might develop unique traits that, when reintroduced to the larger population, give it a competitive edge and therefore supplant the original population (Larson 2004; Mayr and Provine 1980; Palladino 1996; Provine 1971, 1986; Smocovitis 1996).

Theodosius Dobzhansky, one of the architects of the modern synthesis that unified plant, animal, and human evolution, employed field, laboratory, and mathematical approaches, and merged knowledge of evolution on micro- and macro-levels. Dobzhansky had trained in Kiev and brought an appreciation for population genetics with him when he moved permanently to the United States in 1927 that proved pivotal to the contributions he made after relocating. Spending more than 10 years in Morgan's laboratory, Dobzhansky performed field and laboratory studies of *Drosophila* and in his 1937 book, *Genetics and the Origin of Species*, explained how the principles of genetics when applied to populations allowed one to study mechanisms of evolution. Not particularly skillful at mathematics, Dobzhansky collaborated with Wright after meeting him at the 1932 International Genetics Congress in New York. Biologists began applying the principles of genetics to evolution during the 1930s and by the end of World War II a new understanding of evolution had been achieved with genetics at its center. Other notable contributors to the evolutionary synthesis were paleontologist George Gaylord Simpson and systematist Ernst Mayr (Adams 1994; Ayala 1985; Cain 2002; Farber 2011; Laporte 1991; Smocovitis 1996).

The evolutionary synthesis influenced the course of other disciplines, as demonstrated by the "new" physical anthropology devised in the early 1950s. The evolutionary synthesis merged principles of Mendelian genetics and Darwinian evolution and the new physical anthropology called for the incorporation of principles from the evolutionary synthesis into anthropological studies. Anthropologist Sherwood L. Washburn merged information from Dobzhansky and other biologists to devise the new physical anthropology and in the process shifted anthropological research from claims based on morphology and phenotype to ones based on genetics and genotype. The connections that anthropology had with eugenics in the first half of the twentieth century were remedied with the development of a new physical anthropology and anthropological genetics in the 1950s and 1960s (Gormley 2009; Farber 2011; Marks 2012; Smocovitis 2012; Sommer 2012). See the chapter on Anthropology in this volume.

The biological sciences established a collective identity during mid-century. The modern evolutionary synthesis intellectually unified many biological research programs and World War II provided a reason to capitalize on the burgeoning unification. In the United States during the first half of the twentieth century the biological sciences were largely segregated, with many academic centers operating separate botany and zoology departments. Funding for the biological sciences favored research with applications to improve society that ranged from agriculture and livestock to eugenics and medicine. Geneticists had long enjoyed monetary support not only from private foundations and philanthropists but also from federal agencies like the Public Health

Service, National Institutes of Health, and Department of Agriculture. When the American government was deliberating postwar funding for scientific research, biologists testified that they deserved comparable federal support for basic research as was enjoyed by physicists, chemists, and mathematicians. Scientists' Congressional testimonies in 1945–1946 about the establishment and structure of a governmental agency to support scientific research, what is today the National Science Foundation, resulted in making biological research an entity separate from medicine and on par with physics, chemistry, and mathematics (Appel 2000).

Cold Spring Harbor, after closing the Eugenics Record Office in 1939, continued to be a major site for genetic research and a place where scientists made foundational contributions to molecular biology. Phage research undertaken by Yugoslav-born Milislav Demerec and cytogenetics pursued by American-born Barbara McClintock contributed to the Cold Spring Harbor Laboratory's influence in molecular genetics. Demerec studied mutations in genes in the 1930s and after being named the Laboratory's director in 1941 shifted his research to mutations in bacteria. In the 1930s McClintock provided proof of crossover, a concept previously suggested by Morgan, and in the 1950s showed that genes can jump locations. She won a Nobel Prize in 1983 for her work (Comfort 2001; Keller 1983). The Laboratory's annual symposium covering a range of topics in biophysics and genetics incited original research and the dispersion of genetic principles across an array of research areas. Closely associated to Cold Spring Harbor Laboratory and its annual courses were pioneers of bacteriophage genetics, namely German-born Max Delbrück, Italian-born Salvador Luria, and American-born Alfred Hershey. Working independently and together, they enhanced the understanding of the genetic structure and replication mechanism of viruses using bacteriophage, a virus that infects and replicates in bacteria, for which they shared a Nobel Prize in 1969 (Endersby 2007; Holmes 2006).

Investigations of the chemical properties of hereditary substances were another growing area of genetic research beginning in the 1930s that fell under the umbrella of molecular biology. Before researchers turned their attention to deoxyribonucleic acid (DNA), proteins were considered the key to inheritance. Chemist Linus Pauling of the California Institute of Technology (Caltech) seemed likely to discover the structure of DNA because of his intimate knowledge of the structural chemistry of proteins and their building blocks, amino acids. Pauling and three colleagues showed in 1949 that individuals suffering from sickle cell anemia, an inherited disease of the blood, have hemoglobin with an abnormal structure. Using electrophoresis to analyze blood samples, they determined that individuals with the full-blown disease were homozygous recessive and those with sickle cell trait were heterozygous. A few months earlier Neel of the University of Michigan had come to the same conclusion comparing blood samples of children with sickle cell anemia to that of their parents. Tracing traits through a family's lineage dominated studies of human heredity before the mid-twentieth century for lack of other tactics and was a common approach to eugenic investigations. A significant feature of Pauling and his colleagues' accomplishment was that their technique required an individual and not two or more familial generations. Pauling called sickle cell anemia a "molecular disease" to connect form and function: a molecule with an abnormal structure can impair a person's health (Gormley 2005; Kay 1993; Neel 1994). Both serology and medical genetics have historically had close connections to studies of human heredity and eugenics

(Comfort 2012; Marks 2012). Medical Genetics is discussed in another chapter of this volume.

Key Discoveries and Breakthroughs in American Genetics

A major turning point came in 1953 when American James D. Watson and Englishman Francis Crick announced that the structure of DNA was a double helix formed from four nucleotides (adenine, thymine, guanine, and cytosine) and speculated that DNA's structure led to an understanding of its replication (Olby 1974). In 1957 Matthew Meselson and Franklin W. Stahl of Caltech confirmed Watson and Crick's hypothesis that one strand of DNA produces a complementary strand. Like the rediscovery of Mendel's laws, Watson and Crick's findings opened up several rewarding research pathways including techniques that would eventually be used to decipher the genetic codes of many organisms and the human genome. Biochemical research of protein synthesis led investigators to figure out the genetic code during the 1960s. Shortly after the structure of DNA was announced, physicist George Gamow suggested that a string of three nucleotides in DNA determine which one amino acid is produced. Crick responded with what would turn out to be an accurate hypothesis that genes code proteins. Figuring out the steps of this process was the result of investigations conducted in several countries including England and the United States (Judson 1979; Morange 2001).

The German Johann Heinrich Matthaei, while at the National Institutes of Health and with some assistance from Marshall Nirenberg, made an important breakthrough in 1961. Matthaei synthesized a long chain of RNA comprised entirely of uracil, which he called "poly U." Like DNA, RNA has adenine, cytosine, and guanine but instead of thymine it has uracil. Matthaei's work proved that poly U made a protein, confirming that many triplets of nucleotides synthesize a string of amino acids which then fold into configurations to make proteins. In additional experiments Matthaei demonstrated that poly U made only one amino acid, phenylalanine, and not any of the other 19 amino acids. The poly U experiment convinced Crick and others that a biochemical approach led to answers. Crick working with Sydney Brenner soon verified that a triplet of uracil, UUU, codes for phenylalanine. By late 1966 it was known which nucleotide triplet codes for which amino acid and which triplets mark the end of a chain. The genetic code had been deciphered (Judson 1979; Morange 2001).

Genetic engineering developed in the 1970s and 1980s with the advent of recombinant DNA technology. Genetic engineering, which is also referred to as gene manipulation, gene cloning, recombinant DNA technology, and genetic modification, is the application of genetic principles to create products that do not exist naturally. Genetic engineering tends to get more attention than other biotechnology subdisciplines. Paul Berg of Stanford University accomplished the first step in 1971 with a gene-splicing experiment that linked DNA fragments of two different organisms. In 1973 Stanley Cohen of Stanford University and Herbert Boyer of the University of California at San Francisco spliced together DNA fragments and cloned them inside a bacterium host. The Cohen–Boyer experiment helped initiate the biotechnology industry because cloning opened up new possibilities in gene therapies for diseases and disorders. Boyer became a major player in the corporate side founding Genentech in south San Francisco, California in 1976 (Bud 1993; Hughes 2011; Maienschein 2003).

Technological developments and investment were crucial to genetic engineering and genomics, the branch of molecular biology focusing on aspects of the structure, function, and mapping of genomes. In the 1970s Walter Gilbert and Allan Maxam at Harvard University and Fred Sanger at Cambridge University developed procedures for sequencing nucleic acids, and in the early 1980s Marvin Carruthers of the University of Colorado devised a technique for adding new bases one by one and producing a predetermined DNA sequence. Carruthers and Leroy Hood of Caltech invented new technologies that automated DNA sequencing and soon it was possible to conceive of deciphering an organism's entire genome. The Human Genome Project, which lasted from 1990 to 2003, aimed to decode the 20,000–25,000 human genes and the three billion bases in human DNA. It was an international effort to which the United States made a large financial commitment (Davies 2001; Kevles and Hood 1992; Maienschein 2003).

Genetics, Policy and Ethics

The applications of genetic principles and genetic engineering have long raised ethical and policy issues. Before genetics was a rigorous science, eugenics and scientific racism gained followers as discussed above. A recent event in the nature–nurture debate involves sociobiology, as discussed in the chapter on Sociobiology and Evolutionary Psychology. Also genetic screening and counseling have raised concerns that couples will "play God" and design their children, which is a claim that some scholars have countered (Cowan 2008; Green 2007). In addition to controversies involving human enhancement, there are also concerns about genetically engineered plants and animals. Attempts to regulate genetically engineered seeds have stemmed from the potentiality for an environmental disaster if the modified genes that make crops heartier were to be picked up by weeds and concerns that large corporations are overpowering small-scale farming.

A change witnessed over the course of history is the advent of preemptive discussions on the ethical, legal, social, and policy issues generated by new scientific and technological findings in genetics and genetic engineering. After the invention of recombinant DNA technology, scientists worried about potential hazards of genetic engineering and pushed for the development of research guidelines. A moratorium on using recombinant DNA technology was issued until after holding a conference to draft guidelines. Many scientists whose contributions had shaped this field including Paul Berg, Herbert Boyer, and James D. Watson authored the request that appeared in *Science* on July 26, 1974. About 100 molecular biologists met in 1975 at the Asilomar Conference Grounds to deliberate on how research could go forward safely. Two aims were to have scientists (and not governmental officials) devise the guidelines and to focus on the responsible conduct of research (more so than the larger social and ethical issues of recombinant DNA technology). More recently, the Human Genome Project instituted an Ethical, Legal, and Social Implications (ELSI) research program that addressed the complex issues associated with the projected scientific achievement. A goal was to address the implications while simultaneously performing the scientific research so that solutions to controversial issues could be devised before the science and technology became available (Kevles and Hood 1992).

Genetics and genetic engineering have repeatedly proven to be contentious topics in the public sphere and a growing trend among not only geneticists and other biological scientists but also some historians, philosophers, and science, technology and, society (STS) scholars is pursuing research into the ethical, legal, societal, and policy implications alongside the development of scientific and technological innovations (Berry 2007; Cowan 2008; Green 2007; Parthasarathy 2007). The study of genetics has long captivated the public's attention, which is not surprising given the promises touted. Statements about the hopes and hype surrounding genetics were especially prevalent in the decades before and after the turn of the twenty-first century because of the Human Genome Project (Maienschein 2003; Nelkin and Lindee 2004). From the rediscovery of Mendel's laws to sequencing the human genome, the twentieth century witnessed radical transformations in biology, medicine, and industry as a result of genetics and genetic engineering. Scholars have examined the role of genetics in the advancement of various biological and medical sciences as well as the connections of hereditary studies and genetics with social Darwinism, eugenics, and scientific racism. These histories are quite varied examining the discipline's intellectual developments, social contexts, and organizational infrastructures as well as explaining international dimensions and national idiosyncrasies. Historians of science find two main problems with the historiography on genetics. Scientists and science writers tend to wax hagiographic, which of course is not confined to genetics. Genetics dominates historical narratives to the detriment of more inclusive fields. For example what should be titled a history of molecular genetics is presented as a history of molecular biology. The latter point reflects on the universality and prominence of genetics.

Bibliographical Essay

The history of genetics is part of several larger narratives, including the history of evolution and evolutionary theories. Edward J. Larson's *Evolution: The Remarkable History of a Scientific Theory* (2004) presents a broad and approachable history of evolution that weaves in genetics, molecular biology, sociobiology, and other topics. Genetics plays an important role in the modern evolutionary synthesis, which is a fundamental feature of the history of evolutionary biology. For connections between the modern synthesis and natural history see Paul Lawrence Farber's *Finding Order in Nature* (2000). Vassiliki Betty Smocovitis concentrated on the history of the evolutionary synthesis in *Unifying Biology* and in it provides a discussion of the early historiography on the evolutionary synthesis (1996).

Connections between genetics and other disciplines including embryology, agriculture, paleontology, molecular biology, medicine, and anthropology demonstrate the influence that genetics and geneticists have had in the development of biological and medical sciences. Michel Morange discusses the relationship between genetics and molecular biology in *A History of Molecular Biology* (1998). Nathaniel Comfort concentrates his examination on medicine in *The Science of Human Perfection* (2012). Two lengthy works produced in the 1970s draw heavily on interviews with scientists contributing to discoveries in molecular biology. They are Horace Freeland Judson's *The Eighth Day of Creation* (1979) and Robert Olby's *The Path to the Double Helix* (1974).

Additional connections have been explored on a discipline-by-discipline basis (Cain 2002; Keller 1995b; Maienschein 2003; Smocovitis 2012).

Eugenics and related topics have been covered by many scholars. Publications from the late twentieth century continue to stand as fundamental works in the area including those by Garland E. Allen, Daniel J. Kevles, Kenneth M. Ludmerer, and Diane B. Paul. Kevles' *In the Name of Eugenics* (1985) and Paul's *Controlling Human Heredity* (1995) and *The Politics of Heredity* (1998) cover scientific and social aspects in both Britain and the United States since the mid to late 1800s. Ludmerer looks at *Genetics and American Society* from the 1900s to 1930s. Additional studies of human heredity and evolution covering aspects of social Darwinism, eugenics, and scientific racism provide information about the history of human genetics (Barkan 1992; Cravens 1978; Gould 1981; Marks 1995; Reardon 2004; Smedley 1999). Scholars have paid more attention to the history of the study of genetics in human beings than in plants and animals, although recently there has been a surge of studies on plant and animal breeding as they relate to issues such as imperialism and intellectual property (Bugos and Kevles 1992). A fair amount of this recent research, however, concentrates on the European context (Berry 2014; Charnley 2013; Charnley and Radick 2013; Theunissen 2014).

Biographies of geneticists and monographs on model organisms provide an understanding of genetics and its related fields as well as discuss interactions between scientists. William Provine performed interviews and used other sources to tell a history of evolutionary biology through the life of Sewall Wright (1986). Frederic Lawrence Holmes demonstrates how Seymour Benzer helped to make sense of the relations between genetics and molecular biology through research on bacteriophage (2006). Evelyn Fox Keller relied heavily on interviews for her feminist critique of science and biography of Barbara McClintock (1983). Nathaniel Comfort offers another view of McClintock while also putting Keller's work into context (2001). *Drosophila melanogaster* is one of several model organisms commonly used in experimental genetics research. Other model organisms, such as corn, guinea pigs, and mice, have also received attention and by following model organisms scholars elucidate aspects of laboratory practices and professional networks (Endersby 2007; Kohler 1994; Rader 2004; Rheinberger 2010; Smocovitis 2009).

Historians are not the only scholars making contributions to the history of genetics. Scientists have tended to concentrate on the discipline's intellectual developments (Davies 2003; Dunn 1965; Falk 2009; Sturtevant 1965). STS scholars tend to pay greater attention to recent political and social controversies than historians because controversies have been a focus of STS since the discipline began and for many historians such events are too recent (Gaudillière 2006, 2009; Jasanoff 2006, 2013; Parthasarathy 2007).

Many people in society largely misunderstand genetics. Authors whose publications outline a grand narrative of the history of genetics while also providing perspective on public (mis)perceptions are Michel Morange (1998, 2001) and Evelyn Fox Keller (1995b, 2000). Staffan Müller-Wille and Hans-Jörg Rheinberger provide a longer historical survey examining the cultural understanding of heredity within two books that collectively span from the beginning of the sixteenth century to the end of twentieth century (2007, 2012).

Chapter Ten

GEOPHYSICS

Matthew Shindell

It is difficult to define geophysics, and even more challenging to determine how its story should be traced through time. As Gregory A. Good has observed, the geosciences present two characteristic problems to historians: one of scale and one of method. In terms of scale, the geosciences have concerned themselves with matters as small as elements and their isotopes and as large as the formation of the solar system (sometimes on the same day and in the same laboratory); the timescales involved range "from the flash of a lightning strike to the billions of years of geological time" (Good 1998: xxi). On the problem of method, geophysics has combined the universalist methods used in physics and physical chemistry with the historical methods (not to mention varying disciplinary approaches) developed in geology, oceanography, meteorology, and the other earth sciences. One could make Good's problematic duo into a trio and add the observation that geophysicists regularly and unapologetically have violated the porous boundaries between pure and applied, and science and technology; not only have some of the best theorists in geophysics also been among its best instrumentalists and tinkerers (or have employed and worked closely beside them), they have also had to learn how to pilot submarines, interpret satellite data, and forge relationships with the agencies that provide these technologies. These problems make it difficult to devise a holistic approach to the history of geophysics, even in one national context.

The simplest way to define geophysics is as the physics of the Earth. But this is too simple to be historically useful. David Oldroyd adds some specificity to the definition, calling geophysics "the branch of experimental physics concerned with the Earth, atmosphere, and hydrosphere" (Oldroyd 2009: 395). However, while the Earth, atmosphere, and hydrosphere do broadly define geophysics' areas of interest, labeling geophysics a branch of anything implies more disciplinary unification than actually is warranted. Moreover, many of its methods and its most notable practitioners did not originally come from physics. As Stephen G. Brush and C. Stewart Gillmor have

A Companion to the History of American Science, First Edition.
Edited by Georgina M. Montgomery and Mark A. Largent.
© 2016 John Wiley & Sons, Ltd. Published 2020 by John Wiley & Sons, Ltd.

observed, the bigger problem for the historian lies in geophysics' relationship to what they called its "adjacent scientific communities" within the earth sciences (Brush and Gillmor 1995: 1943). One might argue that these earth science "communities" are more than just adjacent to geophysics; even describing them as overlapping understates the extent to which they are intertwined. Many of the greatest accomplishments of geophysics have involved the assemblage of observations from seemingly disparate places and techniques – geological observations of strata, submarine or satellite imagery or remote-sensing datasets, physical or mathematical models of the internal dynamics of the Earth – with the theoretical and technical contributions of scientists working far afield both in distance and tradition.

Geophysics is necessarily a motley affair; so too is its history. It is a "complex constellation of sub-disciplines," but perhaps will never be a unified discipline (Good 1990: 37). This should not be taken as evidence that geophysics is not yet a mature science in its own right. Geophysics may be an example of how a science can be both motley and mature (a more nuanced version of this argument can be found in Good 2000). Unified or not, it is a subject worthy of our attention, as Naomi Oreskes and James R. Fleming have noted, because "nations felt it worthwhile to spend billions of dollars on it, and to worry about who had access to its secrets" (Oreskes and Fleming 2000: 254). In the United States, geophysics has received levels of patronage that place it among the most successful sciences of the nineteenth and twentieth centuries, and it has been key to great discoveries and achievements in all of the earth sciences.

There is an American story to be told in the history of geophysics. On one level it is the story of how American geophysical talent and resources developed over the course of the nineteenth and twentieth centuries, how this growth was related to various national projects and patronage, and the institutions that developed as a result. On another, perhaps more significant, level it is also the story of what resulted socially and intellectually from this growth. This chapter covers the history of American geophysics from the early nineteenth century into the Cold War. It is broken into three periods: geophysics up to the Civil War, geophysics between the Civil War and World War II, and geophysics in the Cold War. This chapter mainly addresses geophysics of the solid Earth, since other chapters in this volume already address oceanography, meteorology, and space science.

Geophysics Before the Civil War

Relative to other sciences in the United States, geophysics had an early start. The first era of development occurred prior to the Civil War, beginning in the early nineteenth century. During this period, the practical concerns of geography – a mixture of natural history (within which an American geological tradition was developing) and geophysics – dominated American science. William Goetzmann termed this period of geographical exploration an "age of classification" (Goetzmann 1991: 309), within which a romantic, Humboldtian approach to the study of nature met the practical needs of a young government. According to Nathan Reingold, those branches of American science that grew most dramatically during this initial period were those that could prove some practical benefit to the geographical project. The "pure" laboratory scientist or theoretician was a rarity, as nearly every notable member of the nascent American scientific community was "to some extent concerned with the physics of the land, the oceans, the atmosphere" (Reingold 1985: 61). Much of the work they did was in the assembly of the sciences we know today as geophysics.

Geophysics during this period also benefited from the nature of American education. Scientific education, such as it was, was primarily geared toward civil engineering and surveying. One of the most advanced scientific educations available in the United States before the Civil War was to be found in the US Military Academy at West Point, founded in 1802. Here science was presented as a cultural enterprise linked to the practical concerns not only of engineering but also of the military, commerce, the economy, and American prosperity; the curriculum was a mixture of French science, mathematics, and engineering, along with a Jeffersonian model of the enlightened "man of science" (Slotten 1994: 8). This curriculum included chemistry, mineralogy, dynamics, and astronomy – all sciences that would serve a young geophysicist well.

The Army Corps of Engineers oversaw West Point, and many of the academy's graduates went on to careers within the Corps. From 1838 to the Civil War, the Corps also established a Corps of Topographical Engineers, charged with mapping the American West, while at the same time collecting a wide range of measurements of various natural phenomena. Much of this work they undertook in the spirit of the romantic tradition mentioned above. They also applied this knowledge to the construction of roads, the improvement of rivers and harbors, locating subsurface aquifers, and other projects of improvement. In the course of their surveying, they in fact accomplished "a mammoth project in geodetic mapping" (Goetzmann 1991: 11) that involved not only the accurate determination of precise locations but also measurements of variations in the Earth's gravity in the presence of mountain ranges and basins. This work was not done in isolation; in their surveys of the Great Lakes, as well as the Pacific Railroad Surveys, the Corps worked in collaboration with the Smithsonian Institution (established in 1846, only seven years before the Railroad Surveys began), as well as some of the nation's foremost scientists (Goetzmann 1991: 305). In this way, Goetzmann suggests, there existed an unofficial government patronage of science in a Jacksonian America that seemed otherwise openly opposed to such patronage. That the money spent on projects of exploration could be tied directly to the fulfillment of "Manifest Destiny" no doubt helped to make such patronage appealing (Reidy, Kroll, and Conway 2007).

But geophysical work was not limited to the exploration of continental North America. During this pre-Civil War period, America also conducted its first overseas scientific venture, the 1838 United States Exploring Expedition, headed by Naval Lieutenant Charles Wilkes. While the sea may not have been manifestly American, it was nonetheless imagined as a "frontier" that Americans chose to explore. Here the United States found a chance to compete with European rivals. Promoters of the Wilkes Expedition – newspaper editor, author, and amateur explorer Jeremiah Reynolds chief among them – combined the romantic pursuit of natural history with the understanding that a successful national expedition would bring economic and political rewards to the young nation. Before Congress, Wilkes argued that such an expedition would raise American geographical prestige to a level rivaling that of any European nation (Robinson 2010: 24). Indeed, the expedition explored and staked economic claims on over 200 of the Pacific Islands on its way to Antarctica. Wilkes' interest in the expedition was also an expression of inter-service rivalry. Wilkes was a naval officer with scientific interests of his own – primarily in the area of geophysics, which was as important to naval navigation as it was to terrestrial geography. Wilkes wanted to use the expedition to put the navy on equal scientific footing with the army and its Corps of Engineers. Rather than simply agreeing to transport the scientists whom Reynolds had chosen for the expedition to Antarctica, Wilkes insisted that the navy use the opportunity to boost its own scientific prestige. While Wilkes succeeded in collecting a wealth of data, the expedition was ultimately deemed a failure because its data was not widely circulated and little was done with it (Reingold 1985: 110).

The United States was building up a base of geophysical talent. The first federal scientific agency founded was its Survey of the Coast (later renamed the Coast Survey, then the Coast and Geodetic Survey, and known today as the National Geodetic Survey). While the Coast Survey received its charter in 1807, it was not until the 1840s, under the direction of Alexander Dallas Bache, that the Survey began proactively expanding its duties to include any and all geophysical phenomena to which it could lay claim. Bache was a graduate of West Point and a well-heeled and well-connected member of the burgeoning scientific circles of Philadelphia (one of Bache's closest allies was Joseph Henry, the first Secretary of the Smithsonian Institute). Well before taking the reins of the Survey, Bache had developed a particularly strong interest in the science of terrestrial magnetism and was inspired by the international effort to collect a global dataset of geophysical observations in the service of understanding the mystery of the Earth's magnetic field. Bache brought to the Survey the commitment that large government-sponsored survey programs, when done well, could also contribute to larger, international projects that could not otherwise be supported. Vice versa, Bache believed that by participating in projects like the global "magnetic crusade," the Survey would also serve the nation's commercial interests by contributing to the practical improvement of navigation. After taking over leadership of the Survey, Bache introduced the regular measurement and recording of magnetic data at all of the Survey's primary stations and at major commercial ports (Good 1985; Slotten 1994: 123). As superintendent of the Coast Survey, Bache was far more successful than his predecessor, the Swiss geographer Ferdinand Rudolph Hassler. Hassler's efforts had given the Survey a good scientific foundation; however, under his leadership the Survey had languished. Bache was aided by political and scientific connections that supported his vision both in Congress and in Philadelphia. In this way, Bache managed to forge a connection between Philadelphia and Washington DC. By the 1850s, Bache had in effect made himself "the superintendent of the largest and strongest segment of the federal scientific establishment" (Dupree 1986: 116). Geophysics was at the heart of this establishment.

In his history of physics in the United States, Daniel Kevles found that the federal government had been something of a reluctant hot house for science, a trend that continued well beyond the Civil War. In addition to the Coast Survey, Kevles singled out the US Geological Survey, the Weather Service (1870), and the Naval Observatory (1830) (Kevles 2001: 49). Within these agencies, physical science methods were brought to bear upon practical economic and military problems; what resulted was a mixture of pure and applied methods, as well as a community of practitioners who used these methods to understand the Earth. While these scientists worked within government and the military, they were not bureaucrats; the Washington circle gathered together in their own societies, held scientific meetings, and attempted, such as they were able, to use their authority for the promotion of American science. In these circles – which included the university physicists, chemists, and mathematicians who approached the problem of the Earth (there were not yet any departments of geophysics), and the scientists employed in federal service – an American geophysical "tradition" developed. It did not develop in isolation from geology. Still, although both geology and geophysics were involved in geographic surveying, the two approaches remained relatively distinct and, at least within universities, evolved along their own paths from this early period up to the Cold War (Oreskes and Doel 2003).

From the Civil War to World War II

In 1879, at the urging of the National Academy of Sciences (which, during the Civil War, Bache had helped to found), a new national survey project came into being – the US Geological Survey (USGS). Under the Academy's plan, the Coast Survey now became the Coast and Geodetic Survey, was moved from the Department of the Treasury to the Department of the Interior, and was tasked with completing geodetic, topographic, and land-parceling surveys. The independent Geological Survey was also established in the Department of the Interior; its focus was the nation's geological structure and economic resources. Like the Coast Survey, the Geological Survey was under civilian management. While the USGS employed a great number of hard rock geologists, it nonetheless had geophysical ambitions as well. The USGS's first and third directors, Clarence King and Charles D. Walcott, both attempted to found a geophysical laboratory within the Survey. Working under King and Walcott, George F. Becker, a field geologist with training in physics, and Arthur L. Day, an American scientist trained in Berlin, organized and conducted geophysical work within the USGS on a small scale (Yoder 1994).

Meanwhile, outside of the laboratory, USGS field geologists also engaged with geophysical methods and questions. Clarence Dutton and G.K. Gilbert both joined the Survey at its founding and became models of the USGS field scientist. Both men began working for the USGS after having worked under John Wesley Powell's 1875 survey of the Rocky Mountains. Powell favored taxonomic mapping over geophysics (when he became the second director of the USGS, geophysical laboratory work all but ceased), but this deterred neither Dutton nor Gilbert from approaching the landforms of the West with a physical mindset. During the course of his mapping of the West, Dutton wrote several classic papers that melded the inductive methods of field geology with the theoretical concerns of geophysics – particularly in the area of volcanism. Dutton sought to uncover and describe the evolutionary laws that governed volcanic processes. According to Stephen J. Pyne, Dutton's search for evolutionary laws in his study of volcanoes displayed a "romantic intelligence," through which "he was able to bring igneous stratigraphy and landform studies into conformity with theories on the geophysical evolution of the earth" (Pyne 2007: 80). In addition to being one of America's first world-class volcano experts, in 1889 Dutton proposed the enduring term "isostasy" to describe the concept of a general balance within the Earth's crust, and the relative buoyancy of continents and mountains on top of the mantle (an idea already articulated by John Henry Pratt and George Biddle Airy in the 1850s). Dutton suggested that the mantle, although composed of solid rock, might behave as a viscous fluid on a geological timescale.

Gilbert, who joined the USGS as its senior geologist and spent time as its acting director, along with Dutton helped to establish the Survey's approach to mapping the West. As Wallace Stegner described him, Gilbert was "Powell's right hand," and "as far removed from a laboratory drudge," but in working on the problems Powell handed him, Gilbert "built a bridge of equations where Powell leaped by intuition" (Stegner 1992: 155). Pyne, describing Gilbert's approach, notes that "His insights were physical rather than chemical, dynamical rather than topographical, typological rather than historical" (Pyne 2007: 92). Gilbert took Powell's observations in the Henry Mountains, for example, and wrote a remarkable monograph, *The Geology of the Henry Mountains* (1877), in which he "described and dissected them so precisely and exactly" (Stegner 1992: 156), that in them he discovered a new kind of mountain structure formed when strata are domed by the upward pressure of lava – structures that he named "laccolites." Gilbert's explanations of the landforms of the West were dynamic; he brought the perspectives of physics and engineering (primarily mechanics) to bear upon geomorphological questions. Dutton and Gilbert both became major figures in reconciling the geological and geophysical theories of global evolution. Their work for the Survey marked these initial years of the USGS as an era in which exploratory science was "dedicated to the solution of specific scientific problems of wide-spread implication," and not merely to the Humboldtian collecting and classifying of observations and measurements (Goetzmann 1991: 309).

In addition to their mapping work, both Dutton and Gilbert became involved in earthquake studies in California. Dutton in particular is credited with bringing modern seismology to the United States, but he did not do this alone. Beginning in the 1870s hundreds of scientists and engineers from the United States and England, at the invitation of the Japanese government, served as instructors and professors in Japanese universities. Many of these scientists became interested in earthquakes after experiencing the Yokohama earthquake of 1880. One American scientist teaching in Japan at the time was Thomas C. Mendenhall, a visiting professor of physics at Tokyo Imperial University. In Tokyo, Mendenhall became one of the founding members of the Seismological Society of Japan, after becoming acquainted with John Milne, James Alfred Ewing, and Thomas Lomar Gray. These three British scientists developed the first modern seismometers. Milne in particular "elevat[ed] 'scismology' from a geological pastime into a modern (instrument-based) science" (Clancey 2006: 63). The Ewing–Gray–Milne seismograph "became the instrumental kernel around which the 'new' science of seismology was subsequently built, yielding as it did a product (a 'seismogram') that was not only highly readable, but … reproducible in scientific reports and papers" (Clancey 2006: 72). The new instrument not only registered an earthquake's occurrence, it produced a record of physical information about the passing seismic wave.

Mendenhall returned to the United States in 1881, founded the Ohio State Meteorological Service, and also became a professor in the US Signal Corps. The Signal Corps, under the direction of astronomer and meteorologist Cleveland Abbe, had in the 1870s been transformed into a national weather service. As head of the service's research department, Abbe instructed weather observers to report on both meteorological phenomena and earthquakes. Mendenhall came to the Signal Corps in 1884 and over the next several years worked to develop a simple seismograph that could be used for the routine recording of earthquakes. He would eventually become superintendent of the Coast and Geodetic Survey in 1889 (*Rose Polytechnic Institute: Memorial Volume* 1909).

American activity in seismology – particularly in California – picked up after the devastating San Francisco earthquake of 1906. At the dawn of the twentieth century there were only a handful of seismographs in use throughout the entire United States – only a few had been installed in California, in the Lick Observatory and on the University of California campus in Berkeley (thanks to the university's president, the astronomer Edward S. Holden). However, there was already a small set of Survey geologists interested in earthquakes and what they might reveal about geomorphology, the Earth's structure, and its processes, including Dutton, Gilbert, and their colleague William M. Davis. There were also university scientists who saw earthquake studies as a key to larger questions in geology, like the structure and origin of continents. One such geologist was Yale University's James Dwight Dana, who as a young man had traveled with the Wilkes Expedition, and like Dutton had spent much of his career studying volcanoes. The Johns Hopkins University geophysicist Harry Fielding Reid, who had begun collecting seismological data for the Geological Survey in the early 1900s, used USGS data in the wake of the San Francisco earthquake to develop his elastic rebound theory, which related earthquakes for the first time to activity along geological faults. By Reid's theory, earthquakes were due to the sudden release of the tension built up by the gradual creeping of the Earth's surface on either side of the fault. Reid's theory was based primarily on the Survey's geodetic observations, which demonstrated that between 1851 and 1907 there had in fact been significant movement and distortion of locations to the west of the San Andreas fault (Geschwind 2003).

In the early twentieth century, private research institutions were critical to America's growing geophysical community. One of the foremost among these was the Carnegie Institution of Washington, founded in 1902 (Good 1994; Yoder 2005). The Carnegie Institution devoted itself early on to geophysical studies; it opened its Geophysical Laboratory in 1905. Although the new Laboratory built upon Walcott, Day, and Becker's experience with geophysical research in the Geological Survey and also solicited the expertise of geophysicists worldwide when developing its research program, the Carnegie Institution in fact mainly supported geochemical work on its Washington campus (Servos 1983). The Institution's support of seismological research in California, however, certainly advanced geophysics. In 1921, under the direction of Harry O. Wood, the Institution supported the establishment of a Seismological Laboratory at the California Institute of Technology (they also contributed to the establishment of Caltech's Division of Geology in 1925). The Seismological Laboratory became home to the German-American seismologist Beno Gutenberg, who in 1912 (while still in Germany) had used seismological data to detect a discontinuity 2900 km below the Earth's surface – later determined to be the boundary between the Earth's mantle and its core. Wood also brought the young theoretical physicist Charles Richter to the Seismological Laboratory; Richter would of course develop one of the Laboratory's most publicly known and used geophysical tools, the Richter scale for measuring earthquake intensity. In its first decade, the Seismological Laboratory was primarily interested in the study of California's earthquake phenomena. Deborah R. Coen argues that even Richter's scale, which would eventually be adopted as a universal measure of earthquake intensity, was at first developed only as a local tool that could be used among scientists as well as in public outreach and education. When Gutenberg replaced Wood as the head of the Laboratory in the 1930s, only then did the Laboratory begin pursuing earthquake studies beyond their local implications, looking more toward the study of the Earth's deep interior (Coen 2013: 263; Goodstein 1984: 227).

Seismological research produced complex models of the Earth's interior and the boundaries between its various layers, as well as arguments about the solid or liquid nature of the core. The Harvard geophysicist Francis Birch suggested in 1940 that the temperature and pressure in the core could allow it to be composed of solid iron. Caltech seismologist Hugo Benioff detected evidence of a solid core in 1952 in the Earth's free oscillations. In addition to his theoretical contributions, Benioff also designed seismographs and other instruments that became standards of the profession around the world.

The Carnegie Institution also funded international work, supporting a worldwide survey of terrestrial magnetism under the direction of Louis Agricola Bauer. Bauer, a physicist who had trained in Germany and had become acquainted with that country's leading magnetic researchers, had already organized and begun the first US magnetic survey under the auspices of the Coast and Geodetic Survey's Division of Terrestrial Magnetism (founded in 1899 with Bauer as its chief). His work with the Carnegie Institution initiated the Institution's Department of Terrestrial Magnetism (DTM) in 1905. Bauer saw the limitations of national surveys, in their inability to provide a truly global dataset or scientific infrastructure that could "bring together the great facts concerning the Earth's magnetism," for the production of "decisive deductions of theory" (Bauer quoted in Good 2013: 31). One of the first scientists to recognize the dynamic nature of the Earth's structure, the German-American physicist Walter M. Elsasser, first proposed his geomagnetic dynamo theory at a 1940 DTM symposium. Other big names in American geomagnetism, such as Teddy Bullard, also worked for DTM during its pre-World War II surveying.

These were not distinctly American achievements. None of these scientists worked in isolation; in addition to their colleagues within the United States, they all had collaborators or counterparts in other countries without whom their work would not have been possible. For example, it was the Danish seismologist Inge Lehmann, using her own seismographic network in Denmark and Greenland, who first suggested that the Earth's core might not be entirely solid or liquid, but rather composed of two layers with different physical properties. Nonetheless, these achievements illustrate the growing base of American geophysical talent, which was becoming world-class – by foreign import, as in the cases of Gutenberg and Elsasser, and through "homegrown" talent such as Benioff. In fact, although geophysics had been somewhat an international pursuit even since Bache's day, the United States was in the twentieth century becoming more involved than ever before in international geophysics projects and forming new organizations to do so. In 1919 the US National Research Council (within which Day was now the home secretary) helped to found the American Geophysical Union (AGU) to represent the nation in the International Research Council's International Union of Geodesy and Geophysics. Heading up the AGU was William Bowie of the Coast and Geodetic Survey. The AGU included sections devoted to geodesy, seismology, meteorology, terrestrial magnetism and electricity, oceanography, volcanology, and geophysical chemistry (today the AGU consists of 23 sections and focus groups).

Geophysical science remained central to American science in these years prior to World War II. However, geophysics did not emerge as a discipline *per se* during this time, even as other scientific disciplines were gaining momentum within America's burgeoning research universities. Geophysics was developing a common language, publications, professional societies, institutions, and a few firm footholds in academia; however, at the beginning of the twentieth century, "geophysics was suspended in the same colloidal state that had prevailed for several generations" (Good 2000: 275). This was in part due to the nature of the geosciences themselves – the wide variety of problems being addressed by geophysicists and the non-uniformity of approaches to these topics made it difficult for a unified discipline to emerge. But it was also due to the sometimes contentious divide between the various scientific communities interested in an organized study of the Earth. For every step taken to bridge this divide, a significant amount of work was spent shoring up boundaries – work that no doubt seemed important in these formative years for American science. Individual practitioners and research programs may have been willing to adopt the methods of other disciplines to make progress in their work, but on the whole not many were willing to cede intellectual territory or disciplinary control to outsiders (Doel 1996; Oreskes and Doel 2003).

University geologists in particular – whose methods had mostly evolved from the observational and inductive traditions of natural history – were often happy enough to work with colleagues and data from the Geological or Geodetic Survey, but nonetheless resisted the interventions of the physicists and astronomers who attempted to answer longstanding geological questions. When Lord Kelvin set a limit for the age of the Earth based on the physics of a cooling body, for example, geologists reacted strongly against him not only because the limit he set was far too young to accommodate their own understanding of deep time and their uniformitarian approach, but also because he seemed to presume that his methods were superior to theirs, not complementary (Oreskes 1999). This might explain why geologists were sometimes reluctant to cede ground to those whom they viewed as interlopers. In 1918, for example, the American astronomer Harlow Shapley argued with great enthusiasm that astronomy should turn its attention to the Earth, proclaiming that "[t]he progress of science frequently demands and utilizes close cooperation of its many branches. We may study the stars, indeed, with the aid of fossils in terrestrial rocks, and acquire knowledge of atomic structure from the climates of Precambrian times" (Shapley 1918: 283), only to admit in the same essay that he found geologists less eager than he to see physical methods applied to their work, mistaking their disciplinary hesitance for a lack of interest.

Any history of geophysics would be remiss if it failed to mention the support of industry in the development of geophysical methods or the employment of American geophysicists. In fact, oil companies were one of the largest employers of geophysicists during the years between World War I and World War II. Oil companies used magnetic, electrical, and gravimetric methods to search for oil deposits. US exploration geophysicists developed the technique of seismic reflection, which used the seismic echoes from exploded dynamite to map underground deposits (Doel 2003a: 401; Lawyer, Bates, and Rice 2001).

Cold War Geophysics

World War II and the ensuing Cold War brought a dramatic increase in funding for geophysical research in America and a corresponding growth in the geophysics communities. This support created what Ronald E. Doel described as a "new intellectual map" for the geosciences, "a new set of challenges, guided by military and national security needs, which elevated the fortunes of certain fields of the physical environmental sciences and decreased opportunities in others" (Doel 2003b: 636). Both Doel and Naomi Oreskes observe that one of the key developments to come from the military push was the increasing domination of deductive physical methods, particularly those of geophysics. Oreskes has described this push as a move from the field to the laboratory, involving an adoption of "the concomitant values of exactitude and control that laboratory work suggests" (Oreskes 1999: 289). Doel and Oreskes further asserted that the rise of physical laboratory methods in the earth sciences during this period had little to do with their historical successes at settling controversies within the geosciences, but rather came as "the result of an abstract epistemological belief in the primacy of physics and chemistry, coupled with strong institutional backing for geophysics premised on its concrete applicability to perceived national-security needs" (Oreskes and Doel 2003: 538). While American geologists before the war had attempted to protect their discipline from the wholesale invasion of physical and deductive methods, the tide of funding for such work in the geosciences caused a seemingly irreversible shift in the discipline's priorities.

This trend is evident in the history of American seismology, the Cold War chapter of which looks dramatically different from that of the early twentieth century. While seismology before World War II had relied primarily on funding from the Surveys and private money from organizations like the Carnegie Institution, after the war seismology came to depend heavily on the military and the newly formed National Science Foundation (NSF). The consequence of this infusion of funds was to transform "the young, small, and underdeveloped academic field of seismology into a large academic-military-industrial endeavor" (Barth 2003: 744). Earthquake research in this period became "tied to that central technological artifact of the Cold War years, the nuclear bomb" (Geschwind 2003: 127). Earthquake experts found that their expertise was valuable in the remote detection and monitoring of nuclear weapons tests, and their work became increasingly entangled in the Cold War national security state.

Emblematic of this shift in funding for seismology is Vela Uniform, a project sponsored by the Department of Defense's Advanced Research Projects Agency (ARPA, later DARPA) and the Atomic Energy Commission (AEC) with the explicit purpose of developing seismological science in the service of nuclear weapons test detection. Vela Uniform led to an increase in federal funding for seismology by a factor of 30 from the years between 1958–1961. Funding remained at this high level from 1960–1971, with a total government expenditure of $250 million for seismology during these years (Barth 2003: 744). The sponsors' priority of remote detection led the field to shift its research and instrumental emphasis further from the detection and analysis of local earthquakes, favoring instead the detection of distant earthquakes. Vela Uniform funding led to an increase in instrumental capabilities and scale – in addition to the productive combination of increasingly powerful computers with large seismic arrays, the Coast and Geodetic Survey installed 120 seismic stations around the world in what became the World-Wide Standard Seismograph Network (Barth 2003: 759).

New institutions also came to the fore as the result of Cold War funding. An example par excellence is Columbia University's Lamont Geological Observatory, established in 1949 under the direction of oceanographer and geophysicist Maurice Ewing. The Office of Naval Research (ONR), a sponsor primarily interested in the logistics of submarine warfare, supported Ewing's research program at Lamont, focused as it was on studying ocean basins using seismology, sonar, photography, and deep sea coring. Vela Uniform also contributed to the success of Ewing's research program, supporting the design and implementation of ocean floor seismographs, and providing data that Lamont seismologists Bryan Isacks, Jack Oliver, and Lynn Sykes could apply toward the emerging theory of plate tectonics.

Seismology was not the only field within geophysics to be transformed in the Cold War. Geodesy became increasingly important to national security; intercontinental ballistic missile guidance systems require precise "knowledge of the distance, direction and gravity field between any given points A and B" (Cloud 2000: 372). The regional scale of geodetic mapping prior to World War II would not do in an era in which a missile fired from one continent was expected to hit accurately its target across an ocean. The Department of Defense and the Intelligence Community became the principal sponsors of the measurement and mapping of gravity variations at sea, the development of theoretical geodesy, and the design of new instrumentation (including, eventually, satellites). Much of the work they sponsored was coordinated by the Mapping and Charting Research Laboratory at Ohio State University, which also established an Institute of Geodesy, Photogrammetry, and Cartography for the training of a new generation of Cold War geodesists.

Another field that developed rapidly in the Cold War was isotope geochemistry, which evolved from the purview of a handful of physicists and physical chemists into a transformative force for university geology departments throughout the United States (Shindell 2015). More often than not, government or military contracts paid for the physical instrumentation used in the new geochemistry – mass spectrometers and cyclotrons – and the salaries of those who used them. The questions asked using these technologies tended to be related in one way or another to the concerns of the contracting agency. Thus geochemistry research programs, at least on paper, often evolved around questions central to such activities as the search for and understanding of nuclear fuel sources, the use of isotopes as tracers for explosions, and, when the navy paid for research, the characteristics of the sea floor and ocean circulation (Hamblin 2005; Oreskes 2015).

One of the earliest institutions to benefit from the increased funding geophysics and geochemistry were receiving after the war was the University of Chicago. Here a collection of physicists and physical chemists who had participated in the Manhattan Project founded the Institute for Nuclear Studies, within which they developed a program in isotopic studies of the physics and chemistry of the Earth. The University of Chicago had been no stranger to geophysics prior to the war. The geologist Thomas C. Chamberlain, who had geophysical inclinations toward questions of the Earth's origin and formation, as well as to climate change, had organized the department of geology in the late nineteenth century and left his mark (Fleming 2000). Before the war, Chicago's geologists already tended to be more lab- than field-oriented. The Chicago geologists considered theirs to be one of the strongest geophysical programs in the country, housing one of the only working high-temperature petrology labs outside of the Carnegie Institution's Geophysical Laboratory, to which it had close ties. As early as 1947 the department was receiving ONR contracts to do geophysical and geochemical research (Goldsmith 1991). Increased funding for the use of isotopes and physical methods in geochemistry led to the development of research programs that produced data on paleoclimate in the labs of Harold C. Urey and Cesare Emiliani (work that required the deep sea core samples retrieved by Ewing's team), and, in Harrison Brown's laboratories in Chicago and later Caltech, the modern determination of the age of the Earth by Clair C. Patterson of 4.55 billion years.

Geophysicists and geologists around the world contributed to a new theory that unified older seismological observations with continental drift and sea-floor spreading – the theory of plate tectonics – a theory which became a foundation for the "modern synthesis" in the earth sciences (Doel 2003a: 391) and remains today the "reigning theory in the earth sciences" (Frankel 2009: 385). One particularly fruitful area of research was the study of paleomagentism – remnant magnetism in rocks – which eventually yielded geophysical evidence for the emerging theory. One key American contribution to this geomagnetic work came from the Geological Survey's Allen Cox, G. Brent Dalrymple, and Richard Doell. By the early 1960s, using potassium-argon dating methods developed by the isotope geochemists, these USGS scientists established a timeline for past geomagnetic reversals, consisting of four major periods. Matching the original location and orientation of the rocks dated gave a rough sketch of how continents might have wandered (the other alternative was that the magnetic poles themselves had wandered).

Oceanographic evidence for drift was also mounting. Since the early twentieth century, echo-sounding techniques had yielded topographic information about the ocean floor. Beginning in the 1950s, ever more research vessels were sent to map the ocean floor, primarily at the behest of the ONR and the NSF. Collectively these ships discovered a global system of sea-floor ridges and trenches. At Lamont, Ewing, along with Bruce Heezen and Marie Tharp, were surprised to find that the sediments they collected from the sea floor were much younger than expected. It was an American geophysicist, Harry Hess, who in 1960 synthesized this new information about the sea floor to construct a model of sea-floor spreading within which material from convection currents in the Earth's mantle rises through the ridges, moving the sea floor away from the ridges toward the trenches, where it descends back into the mantle. American geophysicist Arthur D. Raff confirmed Hess' model in 1955 when he found a distinct pattern of alternating magnetic stripes parallel to a ridge off the coast of the Pacific Northwest. Two British geophysicists, Frederick Vine and Drummond Matthews, investigated the striping phenomenon further and in 1963 proposed that the pattern resulted from the extrusion of mantle material through the ridges. When this material solidified on the ocean floor, it acquired a magnetic polarity matching the Earth's magnetic field. When the Earth's magnetic field reversed, one stripe ended and a new stripe began. The Vine–Matthews theory predicted that each stripe was produced during a certain period in the history of the Earth's magnetic field, and that stripes equidistant from one another were produced during the same period. The US scientific deep-sea drilling vessel, the *Glomar Challenger* (an NSF-funded vessel), provided confirmation of this prediction later in the decade. Dating the materials retrieved by the vessel showed that the material furthest from the Mid-Atlantic Ridge was indeed older than that found nearest to the ridge.

Using the relative ages of the material, geophysicists determined that the Atlantic sea floor was spreading at a rate of about 2 cm per year. By the late 1960s the core concepts of plate tectonics had crystallized and taken hold in the American geological and geophysical communities.

Much of this work would not have been possible without the funding that military interest in geophysics provided. Indeed, geophysicists made close alliances with the state during this period, as evidenced by the career of the seismologist Frank Press, who began his career under Ewing at Lamont and succeeded Gutenberg as head of the Seismological Laboratory at Caltech (where he expanded the budget through funding from Vela Uniform and other contract research), and went on to serve as a science advisor to four US presidents. According to Doel, even the International Geophysical Year (IGY) of 1957–58, which involved several thousand scientists from nearly 70 countries, must be understood in this Cold War context. The IGY was promoted as a collaborative and apolitical example of science's international character, and yet "federal and military planners saw considerable advantage in using the IGY to collect synoptic, global data with potential military applications, and U.S. geopolitical strategy involving Antarctica was furthered by allowing this continent to be 'constituted for science'" (Doel 2003a: 404). Nonetheless, the IGY initiated a new era of global monitoring, which in turn contributed to an increased understanding of the Earth's environment and its vulnerability.

Bibliographic Essay

Brush, Doel, Gillmor, Good, Oldroyd and Oreskes have all produced excellent introductory essays to the history of geophysics, all of which were drawn upon in this essay and are cited in the bibliography. Brush has also published a three-volume *History of Modern Planetary Physics* (1996b). Brush and Landsberg's *The History of Geophysics and Meteorology: An Annotated Bibliography* (1985) is a great resource for materials published prior to 1985. Lee C. Lawyer, Charles Carpenter Bates, and Robert B. Rice, in their volume, *Geophysics in the Affairs of Mankind* (2001), provide a valuable insider's account of the history of exploration geophysics. Gillian M. Turner's *North Pole, South Pole: The Epic Quest to Solve the Great Mystery of Earth's Magnetism* (2011) and Shawna Vogel's *Naked Earth: The New Geophysics* (1995) are both interesting and informative histories of geophysics written at a more popular level.

On the history of the Army Corps of Topographical Engineers, Goetzmann's *Army Exploration in the American West, 1803–1863* (1959) provides a thorough examination of the Corps' activities and interests in geodesy. On the history of Bache and the Coast Survey, Hugh R. Slotten's *Patronage, Practice, and the Culture of American Science* (1994) along with Dupree's *Science in the Federal Government* (1957) are excellent resources. Michael S. Reidy, Gary R. Kroll, and Erik M. Conway's volume, *Exploration and Science: Social Impact and Interaction* (2007), also contains chapters relevant to the history of geophysics and exploration in America. This essay draws upon Reingold's documentary history, *Science in Nineteenth-Century America* (1985); his monograph *Science, American Style* (1991) is also essential reading. Kevles' *The Physicists* (2001) addresses the pre-Civil War history of geophysics as somewhat of a prelude to the establishment of American physics.

For the history of the USGS, this chapter has drawn upon two monographs: Pyne's biography *Grove Karl Gilbert: A Great Engine of Research* (2007) and Stegner's *Beyond the Hundredth Meridian: John Wesley Powell and the Second Opening of the West* (1992). Carl-Henry Geschwind's *California Earthquakes* (2003) provides the story of how earthquake studies and seismology evolved in California.

The five volumes of the AGU's *History of Geophysics* are excellent resources for anyone interested in the history of geophysics. The fifth volume, edited by Good and focused on the theme of the Carnegie Institution's contributions to the development of American geophysics, is the only volume in the series to contain original historical articles.

Doel's *Solar System Astronomy in America* (1996), while not devoted to the history of geophysics per se, nonetheless illustrates well the tension between geophysicists and scientists from other disciplines during the first half of the twentieth century.

The 2003 special issue of the journal *Social Studies of Science* devoted to the earth sciences in the Cold War is required reading for anyone seeking to understand the impact of the Cold War on the development of geophysics. The 2000 special issue of *Studies in the History and Philosophy of Modern Physics* devoted to the history of geophysics is equally valuable. This chapter draws upon articles from both of these issues.

On the history of plate tectonics in America, Oreskes' *The Rejection of Continental Drift* (1999) is the authoritative source. Insiders' accounts of the plate tectonics revolution can be found in Oreskes' edited volume, *Plate Tectonics: An Insider's History of the Modern Theory of the Earth* (2001). A good companion to Oreskes is Cox's *Plate Tectonics and Geomagnetic Reversals* (1973), which collects some of the seminal papers that established the theory.

An excellent, first-person account of the International Geophysical Year can be found in J. Tuzo Wilson's *I.G.Y.: The Year of the New Moons* (1961).

Chapter Eleven

Marine Biology

Samantha Muka

Marine biology is a broad and inclusive interdisciplinary field, encompassing a wide range of subfields. The cohesion of the field, if there is any, comes not from methodological similarities, but the environments in which subjects reside. Marine biology is a field primarily defined by the locations around which it is structured: biologists who work with organisms that live in the ocean environment, regardless of their particular interest in that organism, fall under this disciplinary umbrella. The term marine biology, therefore, brings together under one title a wide range of research techniques and goals, including ecology, evolutionary biology, taxonomy, physiology, and animal behavior, to name but a few.

While interest in marine organisms and their environment is ancient, the definition of marine biology is fairly new, becoming an established discipline in the post World War II scramble for government-based research funding. Early American exploration of ocean resources focused on both deep sea exploration and fisheries concerns. On the open ocean, shipboard scientists combined collection of hydrographical data with collection of marine organisms. Closer to land, aquaculturists and fisheries experts focused on crafting human techniques for effectively producing fishes for market. By the turn of the twentieth century, dwindling fisheries resources and an influx of German academic laboratory techniques brought researchers to newly established marine laboratories along the shorelines of the United States. Researchers at these institutions came from a variety of fields, drawn together in a single location. Until the end of World War II, scientists interested in working on or with marine organisms remained an undefined group, referring to themselves by a variety of titles, including zoologist, physiologist, and only occasionally qualifying this with the word *marine* in the title. However, postwar funding opportunities pushed researchers interested in the marine environment to examine their commonalities and relationships to both each other and their subject. By

A Companion to the History of American Science, First Edition.
Edited by Georgina M. Montgomery and Mark A. Largent.
© 2016 John Wiley & Sons, Ltd. Published 2020 by John Wiley & Sons, Ltd.

the early 1960s, marine biology became loosely defined as a field based on location, regardless of a researcher's methodology.

This chapter will outline the history of American marine biology from the eighteenth century to the present. Until the mid-twentieth century, both fisheries concerns (fisheries biology) and quantitative analysis of marine fauna (biological oceanography) intertwined with general studies of marine organisms (marine biology). However, by the 1930s, these fields began to differentiate themselves as separate fields of study, creating not just new disciplines, but defining marine biology in the process. First, it will highlight the growing interest in marine organisms in America in the mid-nineteenth century. Increased utilization of deep sea spaces and concerns about declining wild fish stocks both led to new examinations of marine organisms and their environments. Next, it will detail the infusion of German biological ideas into the American academic system and the subsequent rise of marine laboratories. These spaces allowed a wide range of researchers interested in marine organisms to congregate and share information. Next, it will detail how oceanography and fisheries biology split from marine biology in the mid-twentieth century and the eventual definition of marine biology post World War II. Finally, it will suggest several historiographical gaps that, if filled, would allow a deeper understanding of the place of marine biology in American scientific and popular culture today. While the histories of both fisheries and biological oceanography have been written separately, it is only through the combination of these narratives that the history of marine biology becomes clear.

Open Ocean Research: Onboard Naturalists and Marine Zoologists

Through out the eighteenth century, increased ocean travel and exploration led to an exploration of the marine environment. The European extension of exploration throughout the world took shipboard naturalists not only to new lands but also over new waterways. Basic surveying techniques, including sounding and trawling, allowed onboard naturalists limited access to pelagic (deep sea) organisms. However, during the early and mid-nineteenth century, increased commercial, cultural, and scientific interest in the deep sea pushed scientists to more closely examine and accurately describe the open ocean environment. These increasingly in-depth descriptions produced large numbers of tomes dedicated to identifying deep sea flora and fauna and also contributed to the rise of marine biology and oceanography in the United States.

The age of exploration brought shipboard naturalists into contact with new marine organisms and environments. Exploratory expeditions included at least one onboard naturalist dedicated to collecting and cataloging the new species of animals and plants encountered. These voyages were not necessarily focused on collecting marine organisms, but they did result in increased understanding of marine organisms and environments (Deacon 1997). For example, naturalists and geologists aboard ships encountered coral island formations in various stages of growth. Charles Darwin, acting as an unofficial onboard naturalist of *The Beagle*, built upon theories about the formation of these islands from earlier expeditions and came up with a working theory of their formation based on the understandings of the natural history of coral polyps. Theorizations of the formation of these islands led to a deeper biological understanding of the lifecycle of the organisms responsible (Sponsel 2009). However, until the nineteenth century, little direct focus was given to thoroughly surveying the marine environment.

The nineteenth century saw an increase in interest in the marine environment from commercial, cultural, and scientific communities. While early landlubbers and sailors alike often viewed the ocean environment as "a great void that was empty and feature-less, the antithesis of civilization" (Rozwadowski 2005: 6), increased travel and labor on the open ocean, including the extension of the American whaling industry further and further ashore, led to the desire to understand the deep sea and how it might provide utilizable resources (Kroll 2008). From previous voyages, scientists developed various theories of the pelagic environment, ranging from a barren landscape devoid of, in fact incapable of sustaining, life to a primordial location from which all life on earth had sprung (Deacon 1997; Rozwadowski 2005). To settle these debates and begin to utilize untapped marine resources and spaces, the American government and private interests sent naturalists and marine zoologists on voyages meant to survey the depths.

Early marine biological investigations occurred in conjunction with both coastal and deep sea surveys. Throughout the nineteenth century, both the US Coastal Survey and private expeditions, such as the Harriman expedition to Alaska, performed extensive marine surveys in American waters. The work of surveying consisted of taking tem-perature and depth readings (soundings) and also dredging and netting in order to accurately describe the flora and fauna at given depths in specific locations. Sound-ing equipment enabled accurate readings of water depth; dredging equipment gave a limited, but essential, view of the construction of the ocean and the organisms that lived in a region. While depth of water and ocean floor construction were impor-tant, identifying organisms that existed in a given location was equally so. Different forms of dredging arose to satisfy curiosity about ocean floor makeup and the organ-isms that lived at the bottom of the ocean. Dredges that scooped up ocean mud brought up worms, crabs, and other bottom dwellers to be examined aboard ship. Researchers combined these with rake and mop-like dredges that picked up other types of species, including clams, oysters, starfishes, sponges, and jellyfish. Deployment of a wide range of dredges and nets helped researchers marry depth and temperature record-ings with the type of organisms found in a given location (Deacon 1997; Rozwadowski 2005).

Scientific examination of organisms taken during dredging surveys was limited to tax-onomical and morphological identification. The onboard naturalists of the eighteenth-century voyages gave way to scientists who identified themselves as marine zoologists. However, they performed similar functions. Organisms brought up during dredging were often dead or dying by the time scientists examined them. Dredging, even in shallow coastal waters, damaged the delicate marine organisms and limited their abil-ity to survive and thrive in captivity. Quick field sketches were performed, the location at which specimens were taken was noted, and it was placed in a jar of alcohol to be preserved. The collection of specimens was then parceled out to specialists; the jellyfish went to one naturalist, the fishes to another for identification. Slowly, the specimens were identified as either existing species, named if they were newly discovered, and large volumes of findings were published with images and descriptions of these organ-isms found during surveys. While surveying led to many discoveries of new organisms, and a recognition that a wide range of organisms could and did thrive in deep sea envi-ronments, a deeper understanding of the nature of these specimens and their natural histories were not generally explored due to limited onboard facilities and the nature of dredging.

Fisheries Research: Culturists and Fisheries Biologists

Native peoples relied on coastal fisheries in the Northeast and Northwest for harvesting salmon, oyster, abalone, and other near shore specimens long before the arrival of Europeans to American shores (Cronon 1983; Taylor 1999). These groups employed an array of fishing technologies, including poisons, nets, and, in the Northwest, salmon runs, "weirs were the ultimate fishing devices." These "salmon dams" were permeable net walls stretched across a river that let water flow downstream but kept salmon from swimming upstream (Taylor 1999: 19). Immigrant fishermen recognized the abundance of fishes found on these coasts and nearby waterways and built robust fishing villages and industries around these traditional fishing grounds. European-born and first[generation American fishermen utilized similar technologies as their Native counterparts, and also began increasing their catches with the use of boats, trawling, and other intensive fishing practices. These increased catches eventually led to a decrease in available fish stocks (Keiner 2010; McEvoy 1990; Wadewitz 2012).

Fish culturing was the keystone in American fisheries research in the early nineteenth century. The practice of culturing involves taking mature fishes from their native environments, milking them of their reproductive materials, and fertilizing these materials to rear large numbers of fishes in captivity. American culturists developed a wide range of craft knowledge to rear fish from fertilized egg through the fry and mature stages of their lifecycles. Knowledge of the availability of mature and ripe males and females was combined with an understanding of feeding practices that would sustain the fishes without wasting important resources on their survival. Of primary importance to their craft was an understanding of the fishes' lifecycle and how to cultivate them through these lifecycles in artificial environments. Utilizing hatching jars they coaxed fertilized eggs into fry form and then released them into ponds to grow to maturity. Culturists had success with species like cod, but struggled with high demand species such as salmon and other marine species (Taylor 1999). Disputes over rapidly decreasing species would turn government and scientific interest to more clearly understanding these wild stocks in their environment.

In the 1860s, American biologists and fisheries experts became aware of the rapidly decreasing fish stocks in Eastern fisheries. Spencer Fullerton Baird, assistant secretary of the Smithsonian, was asked to investigate the claims by fishermen in both Massachusetts and Rhode Island that the use of certain types of nets was decreasing the fish stocks in these areas. Baird was given a limited amount of time and resources to investigate these claims; he spoke with local fishermen in both states to gauge stock depletion based on local knowledge and presented these findings to the states. Each state ruled differently (Rhode Island banning certain nets; Massachusetts seeing no evidence to do so), and the outcomes convinced Baird that more systematic investigation was required to make any conclusions. According to Baird, "this remarkable contradiction in the results of the two commissions showed the necessity of a special scientific investigation on this subject, to be prosecuted in the way of direct experimentation on the fishes themselves, their feeding and their breeding grounds" (Allard 1978; Baird 1873: viii).

The US Commission of Fish and Fisheries was established in 1871 to "investigate, promote, and preserve" the fisheries of the United States. Baird's vision of a Commission that would both propagate known species in captivity and investigate the natural history of those species in the wild resulted in a tri-part research program: the Division of Inquiry (sometimes referred to as the Division of Scientific Inquiry), the Division of Fisheries, and the Division of Fish Culture. The Division of Fisheries oversaw existing fisheries in the United States and its territories. This group arbitrated disputes such as the one Baird worked on between Massachusetts and Rhode Island. In addition, they sought to work internationally to save dwindling resources, such as the fur seal rookery in the Aleutian Islands. The Division of Fish Culture funded work previously performed by private culturists and sought new avenues for rapidly increasing stocks of fishes in American waterways. By experimenting with new ways to rear and ship species, Baird hoped to restock American waterways with hardier fish varieties. In addition to culturing research, Baird recognized the gaps in knowledge about the natural history of species, especially marine species, and funneled money into the Division of Inquiry to fill in some of these gaps (Allard 1978; Muka 2014a).

Baird imagined the Division of Inquiry as the basic science branch of the Fish Commission. Scientists in this Division would observe and experiment upon wild populations in order to compile scientific knowledge that would fill in gaps for the Divisions of Fish Culture and Fisheries to operate successfully. In Baird's vision, the Division of Inquiry was as important as the other two branches of the Commission and he fought for the money to build permanent research stations to serve as epicenters for this research. Baird's push for permanent research stations occurred at a time of flux in the American academic biological community where researchers flocked to the seaside to access new organisms for experimentation. The confluence of fisheries and academic biology concerns resulted in an explosion of permanent research stations and the growth of marine biology (Allard 2000).

Marine Stations: Fisheries and Academic Laboratories

Beginning in the 1860s, governments, universities, and private natural history groups interested in surveying and studying marine resources established permanent marine stations throughout the world. Russia, France, Japan, England, Canada, Germany, Italy, the Netherlands, Sweden, and the United States all established permanent station locations by the end of the nineteenth century. Built by a local scientific society in 1867, the marine laboratory of Arcachon on the Bay of Arcachon, Southwest France, may be the oldest laboratory of this kind, but it was quickly followed by the founding of similar stations throughout the world. Russia's privately funded Sevastopol Station, founded in 1871, was quickly followed by the Stazione Zoologica Anton Dohrn (1872) in Naples, Italy and the Station Biologique de Roscoff (1872) in Brittany, France. Others swiftly followed and new stations opened in Sweden (Kristiniberg, 1877), Japan (Misaki, 1887), Scotland (Gatty, 1896), England (Plymouth, 1888), the United States (Penikese Island, 1877 and USBF Woods Hole, 1888), Canada (New Brunswick, 1899), and the Netherlands (Helder, 1890) throughout the 1880s and 1890s. By the turn of the twentieth century most of these countries had established multiple stations. While traditional shipboard and fish culturist research continued, marine stations allowed a wide range of individuals interested in marine organisms to congregate and interact in the same location (Greze 1971; Muka 2014a; Oppenheimer 1980).

The growth of marine stations can be linked to the growth of American experimental biology during this period. Laboratory-based experimental biology, centered primarily in the German university system, migrated into American universities during this period, where it was common for American biologists to travel to German institutions to take graduate degrees in science. Post-Darwin, German biologists moved to the seashore to examine and experiment upon invertebrates in an attempt to understand the evolution of organisms. As these men returned to the United States to teach, and European scholars immigrated to the United States, the laboratory-based biology practiced in Europe followed (Allen 1978a; Maienschein 1991a; Pauly 1988). Two epicenters of this new biology emerged: the Johns Hopkins University and Harvard University. Both of these universities became, not just epicenters of experimental biology, but Petri dishes for the burgeoning interest in marine research.

At Johns Hopkins, W.K. Brooks, who studied with several European-trained biologists at Harvard, focused his particular combination of morphology and physiology training on researching at marine laboratories (Benson, this volume). At Harvard, Brooks trained with Alexander Agassiz while working at Penikese Island in Massachusetts During his tenure at Hopkins, he took students to the Chesapeake Bay, Woods Hole, Beaufort, North Carolina, and the Tortugas laboratories to conduct research. Brooks trained E.G. Conklin, E.B. Wilson, and T.H. Morgan in the new experimental methods and highlighted the importance of marine research in this endeavor. In turn, these investigators became professors at universities throughout the United States and spread the new experimental method, and strengthened the link between this methodology and marine research (Benson 1979; Maienschein 1991a).

E.L. Mark, who studied in Leipzig and experienced the German biological link with the seashore of the Marine Zoological Laboratory of the Austrian government at Trieste, trained a large group of professional biologists at Harvard from 1877 to 1921. Many of these experimental biologists shared Mark's interest in working with marine organisms. A number of his students would go on to become major figures in marine research throughout the twentieth century. Alfred Goldsborough Mayer was the first director of the Carnegie Institution of Washington's Tortugas Laboratory and William Emerson Ritter was the first director of the San Diego Marine Biological Association's Laboratory at San Diego (later renamed Scripps Institution of Oceanography) (Muka 2014a).

German laboratory ideals also affected fisheries research and led to the establishment of fisheries marine stations. Spencer Baird's vision for the Division of Scientific Inquiry (DSI) was for a branch of the US Fish Commission (USFC) that would perform intensive observation and experimentation on marine species. Long-term work with marine species required permanent stations to facilitate these studies. Following the international example of marine station founding, and especially that of the Naples Zoological Station, Baird established a marine station for the DSI in a house in Woods Hole, Massachusetts; he procured government funding to build a permanent marine station there in 1885 (Allard 1978, 2000). Woods Hole served as the base for the DSI; each summer, the laboratory invited students and professors from Northeastern universities to utilize laboratory space for their research in the hopes that any work on the coast would result in useful data on the marine environment in that region. Subsequent laboratories were founded in Beaufort, North Carolina and Key West, Florida. The DSI opened its marine stations to a wide range of researchers, ranging from self-taught culturists to academically trained experimentalists. Well-known academic zoologists, including Caswell Grave, David Starr Jordan, T.H. Morgan, and William Keith Brooks spent time at DSI stations (Muka 2014a).

These stations were simultaneously important for both the strengthening of American biology and the eventual development of marine biology. Marine stations served as an institutional base for American biology by allowing easy and consistent access to fresh and living organisms for observation and experimentation. Researchers interested in studying the basic physiology of organisms worked with marine organisms because they were simple in structure, extremely abundant, and offered new glimpses into consistencies between invertebrates and vertebrates. These researchers were not necessarily interested in studying the marine environment, but instead utilized marine organisms to examine a host of biological questions. The "engineering ideal" in biology pushed researchers to find organisms that had enough natural plasticity to survive and thrive during experimental procedures (Pauly 1987). This ideal favored specimens that could go on "living, synthesizing proteins, moving, reproducing, and so on despite catastrophic interference in their constitution, environment, or form" in the laboratory (Landecker 2007: 10). Many marine organisms satisfied this form of plasticity.

In addition to the importance to the wider biological community, these stations also served as the crucible of American marine biology. Keith Benson (2001) has pointed out that marine biology as a professional group did not exist during this period and therefore it would be anachronistic to say that marine stations was part of this profession's "institutional identity." While it is true that the term "marine biology" and the profession of "marine biologist" were not in use during this period, the stations facilitated the growth of knowledge about the marine environment and jumpstarted research that would blossom into an established discipline post World War II. Biological surveying of the area surrounding stations was the largest ongoing project at marine stations; each year, stations amassed data on the local flora and fauna available in that area. In addition to surveys, specimen collection and observation of organisms in their natural environment was an important step in many experiments. These observations contributed to knowledge of the marine environment and the construction of major questions about that environment that would form the basis of marine biological investigations throughout the twentieth century. Before the establishment of permanent marine stations, laboratory biologists working with marine organisms did so at private residences and temporary dwellings near the shore. These spaces were isolated and ephemeral. In contrast, the permanent marine station provided space where researchers working on a wide range of scientific problems could interact over extended periods of time, forming research communities and networks of knowledge. The formation of these groups, comprising individuals approaching marine problems from a variety of positions, led to the growth of a marine science community (Muka 2014a).

Marine Biology Defined

By the early twentieth century, there was a wide range of paths to travel to explore the marine environment. Deep sea exploration from ships was joined by fisheries research and biology at marine stations to form somewhat separate but intersecting areas of marine knowledge. However, by the interwar era, this braid of marine science began to unravel. In an effort to more clearly define their research goals, communities that focused on fisheries and deep sea concerns split off from the main body of marine science, officially forming fisheries biology and oceanography.

During World War I, all marine stations turned their attention to producing information that would support the American war effort. The American navy conscripted the boats from each station and researchers worked for the war effort on land. Alfred Goldsborough Mayer sought to contribute by finding the cause of shell shock; the Puget Sound Biological Laboratory in Washington began studying and harvesting sphagnum moss, which the Red Cross used for bandages throughout the war. The change to private and university stations during the war was marked, but there was a larger shift at fisheries stations that lasted long after the war concluded (Muka 2014a).

During the war, the government began to slowly shift the research at their marine stations from the general life sciences to a more distinct focus on experimenting on fish stocks and gathering statistical information on them. In 1917, Hugh Smith, the Commissioner of Fishes, stated in his yearly report that

> In biological work the year has been marked by substantial readjustments. These have arisen partially from enlarged responsibilities and opportunities coming from an increase in personnel, partly from the fact that some of the investigations have progressed to a stage justifying or requiring a rearrangement of plans, and partly from the conditions of national exigency. On the whole, the changes and the new undertakings have the effect of concentrating the efforts of the Bureau upon problems of most immediate practical importance. (Smith 1918: 79)

The DSI station at Woods Hole began to focus on rearing and stocking lobster throughout the Northeast and Beaufort turned its attention more fully to farming black terrapin and understanding wood-boring marine worms in order to protect American ships from destruction. No longer did they send open invitations to universities and researchers, but instead started to train their own researchers in specific fisheries methods. In 1926, the Bureau of Fisheries announced that, "A review of the progress made in fishery investigations during recent years indicates that a distinct branch of scientific study that may be termed 'fishery science' has been developed." The Fish Commissioner described it as a hybrid science, combining zoology, geography, ichthyology, marine ecology, and oceanography with the methods of biometrics and vital statistics (O'Malley 1927: xxxii; Wolfe 2001). The Bureau of Fisheries diverted much of its funding to this new scientific discipline and closed its laboratories to researchers not performing specific work on fisheries concerns, effectively breaking ties with much of the rest of the network (Hubbard 2006, 2014; Smith 2007).

In addition to the establishment of fisheries biology as a coherent discipline, American biological oceanographers emerged during the interwar period. As previously seen, marine zoologists and shipboard naturalists performed deep sea explorations of marine flora and fauna alongside geological surveys starting in the eighteenth century. During World War I, it was seen that knowledge of the larger marine environment, including statistical understandings of marine inhabitants, was vitally important to American military power. These shipboard surveys shifted in focus, morphing from an emphasis on discovery and identification to a numerical analysis of the ocean and its inhabitants. Drawing upon theories from ecology as well as the statistical focus of fisheries biology, biological oceanography emerged as a specific discipline (Mills 1989).

The formal emergence of this discipline can be most easily pinned to the renaming of the San Diego Marine Station in La Jolla, CA to the Scripps Institution of Oceanography. The marine station at La Jolla was founded in 1903. Its director, William Emerson Ritter, a student of E.L. Mark's, relocated onto the West Coast, had a bifurcated goal of general ocean exploration and specific examination of marine issues, such as diminishing fish stocks. Throughout his tenure as director, the station experienced a crisis of mission: how to focus on large questions about the ocean environment while contributing to the useful knowledge about dwindling ocean resources. While Ritter never succeeded in establishing a coherent mission for the station, his successor T. Wayland Vaughn, did. Vaughn, a geologist who frequently worked on coral at the marine laboratory in the Tortugas, succeeded Ritter as the head of the marine station in 1924. In 1925, the name Scripps Institution of Oceanography was adopted to reflect the growing importance of statistical methods and shipboard research as research goals in La Jolla (Benson 2002). Oceanography continued to differentiate itself as a separate discipline throughout World War II by strategically linking itself with the war effort (Hamblin 2005).

Both fisheries biologists and biological oceanographers identified their disciplinary communities during the interwar period through methodological similarities. Fisheries biology and biological oceanographers pointed to their use of large datasets and statistical analysis of data to separate them from the larger body of marine work being done during this period. Whereas fisheries researchers sought to utilize their statistical analysis for applied purposes, namely to perpetuate and manage wild fisheries, biological oceanographers shifted focus to technological and military applications of large-scale understandings of the marine environment. The identification and cohesion of these two fields left a large question in their wake: what was left for marine biology?

Marine biologists asserted their claim to a broad disciplinary identification based on coherence of location in which their research was performed and a shared understanding of the importance of discovery and description in marine research. After World War II, government funding of marine research, much of which was going toward oceanography, forced marine biologists to try to identify the commonalities of their field. During World War I and World War II, oceanography became closely aligned with military applications. After the wars, the increase in government funds for marine research was sought by those who did not identify as oceanographers, but did work within the marine context. Unlike fisheries biology and biological oceanography, marine biologists did not identify themselves based on experimental or quantitative similarities. Marine biologists worked both on shore and on ship, they worked in fields with heavy observation as well as experimentation, and researchers worked in a variety of disciplines, including animal behavior, zoology, embryology, chemistry, and physiology. "Rather than trying to promote a single methodological approach – emphasizing experimental techniques – these marine biologists articulated a broad and inclusive vision of an interdisciplinary field built on discovery and description" (Ellis 2007: 470). By asserting that any researcher in any field could lay claim to the title of marine biology as long as their work was performed in a given location, the definition of marine biology did not define a coherent field as much as give a wide variety of researchers access to post World War II funds.

Marine Biology in the Twentieth and Twenty-First Centuries

Because of the expansive definition of marine biology formulated in the late 1950s and early 1960s, it is difficult to undertake a coherent history of the field in the second half of the twentieth century. However, there are several fields of study that can be undertaken to fill in the gaps in historiographical knowledge we currently have in the history of marine biology. Understanding the impact of the rise of the conservation movement, the importance of the cultural identity of the marine biologist, and the role of public aquaria and theme parks in the public's understanding of marine science are all future avenues that the literature can take.

After World War II, marine biology became inextricably linked with the conservation movement. As fisheries biology and biological oceanography separated from marine science, ecological and environmental sciences became deeply embedded in the public's understanding of marine biology. In particular, the passing of the Marine Mammal Protection Act (MMPA, 1972) and the Endangered Species Act (ESA, 1973) both called attention to the field of marine biology and inextricably linked it with conservation initiatives. The growth in popularity of marine biology might be linked to the recognition of "glamour" species in the environmental movement, including sea turtles, manatees, and blue whales. While some historians have traced the history of marine biology from its ties with whaling and fisheries research to the inaction of the MMPA and the growth of the conservation movement, the linking of marine biology with conservation in the postwar period has somewhat obscured its prewar links with fisheries biology (Burnett 2012). Examinations of the impact of the MMPA and ESA on the work performed by marine biologists are also needed. Understanding the legal ramifications of federal laws on scientific research is imperative for historians interested in the production of scientific knowledge (Benson 2012; Davis 2007). Finally, it is important that historians of science understand how the linking of marine biology with conservation may have changed professional identifications and the construction of the disciplinary boundaries of marine biology. It is possible that a reexamination of the field in the twenty-first century would reveal a cohesive discipline formed around conservation goals.

While the conservation movement may have catapulted marine biology into the scientific limelight, the superstardom of individual marine biologists also contributed to the shaping of the discipline. Famous marine explorers and researchers introduced the American public to the work of marine biology. At the turn of the twentieth century William Beebe rose to fame with his deep sea submersible "bathysphere" (Gould 2004). By mid-century, Eugenie Clark's *Lady with a Sphere* (1953) introduced readers to an exotic globetrotting scientific lifestyle and this was closely followed by the immersion of the public in Jacques Cousteau's *Silent World* (1956). Today, the association of marine biology with biologists such as Sylvia Earle and explorers like Robert Ballard and James Cameron has done much to push marine biology into the public light. While some work has been done on the history of popular images of the marine environment, and the role of the working biologist/celebrity biologist in this understanding, much more work needs to be done (Kroll 2008). In particular, it is important to know how the image of the celebrity biologist has drawn attention to the field, and specific areas of the field of marine biological investigation, while obscuring areas that lack big name scientific celebrities or identifiable human outreach potential. In addition, historians can use the popularity of these individuals to analyze the impact of celebrity scientists on the recruitment of new researchers into scientific disciplines.

Finally, an underexplored historiographical thread is the concurrent growth of the public aquarium and theme park community alongside marine biology throughout the twentieth century. The establishment of public aquaria occurred at roughly the same time that marine stations were founded throughout Europe and the United States. In fact, many marine stations contained public aquariums to increase revenue for their research. The Naples Zoological Station was often referred to by its nickname "The Aquarium." Its saltwater tanks could be viewed by the vacationing public for a small fee; they drew huge crowds (Ghiselin and Groeben 2000). In the United States, the New York, Miami, Boston, Steinhart, and Philadelphia aquaria, all founded in the early twentieth century, employed scientists trained by the US Fish Commission's Division of Scientific Inquiry and the zoological and morphological departments of Harvard and Johns Hopkins (Muka 2014a). Histories of these spaces have categorized them as outside or slightly offset from academic marine biology and fisheries biology (Davis 1997), but recent literature suggests that they can be viewed as an extension of these fields (Muka 2014b).

Bibliographic Essay

Because of the interdisciplinarity inherent in the study of marine biology, literature on the field comes from a wide range of sources. Few works, with the exception of Margaret Deacon's *Scientists and the Sea, 1650–1900: A Study of Marine Science* (1997) and Helen Rozwadowski's *Fathoming the Ocean: The Discovery and Exploration of the Deep Sea* (2005), have tried to trace the history of the field throughout the nineteenth and twentieth centuries. Instead, those interested in the history of marine science should read literature that examines the general growth of American biological sciences, oceanography, and fisheries science. Marine sciences intersections with ecology and environmentalism and cultural understandings of the sea can also provide interesting insights into marine science in the United States throughout the twentieth century.

To understand the history of marine biology in the United States, it is important to understand the general growth of experimental and field biology from the mid-nineteenth century to the present. It is through the lens of the history of biology, and especially physiology and organismal biology, that the history of marine biology has been told. Garland Allen's *Life Science in the Twentieth Century* (1978), Jane Maienschein's *Transforming Traditions in American Biology* (1991a), and the edited volume *The Expansion of American Biology* by Maienschein, Ronald Rainger, and Keith Benson (1991) provide a good analysis of the growth of the field of biology during this period. For a cultural and social history of American biology, Philip Pauly's *Biologists and the Promise of American Life: From Merriwether Lewis to Alfred Kinsey* (2000) provides a glimpse of the depth and breadth of American biology at the turn of the twentieth century. Gerald Geison's edited volume *Physiology in the American Context, 1850–1940* (1987) brings together a particularly rich set of papers on the field.

Historians have debated the relative importance of field versus laboratory investigations in biology throughout the twentieth century; this debate intersects with the history of marine science because of the importance of location in the discipline. For an overview of the lab/field debate, see Robert Kohler's *Landscapes and Labscapes: Exploring the Lab-Field Border in Biology* (2002). Philip Pauly's biography of Jacques Loeb, *Controlling Life: Jacques Loeb & the Engineering Ideal in Biology* (1987), provides an interesting portrait of an experimental biologist using marine organisms during this period.

Studies on marine stations and oceanography paint a picture of the biological interest in the marine environment from the mid-nineteenth century to the present. For information and resources on American marine stations, see Keith Benson's chapter in this volume. Eric Mills' *Biological Oceanography: An Early History, 1870–1960* (1989) provides a fantastic overview of the growth of biological oceanography. Keith Benson and Philip Rehbock's edited volume *Oceanographic History: The Pacific and Beyond* (2002) is an encyclopedic overview of a wide range of topics in the history of oceanography, including several articles on the intersection of marine biology and oceanography.

The history of fisheries biology is intricately intertwined with marine biology in the first half of the twentieth century. This history has been written from two perspectives: the environmental and ecological history of the field and the rise of the specialized science. For an overview of the environmental and ecological history, see Arthur McEvoy's *The Fisherman's Problem: Ecology and Law in California's Fisheries, 1850–1980* (1990) and Joseph Taylor's *Making Salmon: An Environmental History of the Northwest Fisheries Crisis* (1999). For the rise of the discipline of fisheries biology, see Tim Smith's *Scaling Fisheries: The Science of Measuring the Effects of Fishing, 1855–1955* (2007) and Jennifer Hubbard's *A Science on the Scales: The Rise of Canadian Atlantic Fisheries Biology, 1898–1939* (2006).

Postwar marine science intersects heavily with ecology and environmentalism. D. Graham Burnett's *The Sounding of the Whale: Science and Cetaceans in the Twentieth Century* (2012) traces the history of whales in America, from a source of fisheries wealth to a symbol of environmental concern. In addition, Frederick Rowe Davis' *The Man Who Saved Sea Turtles: Archie Carr and the Origins of Conservation Biology* (2007) demonstrates the postwar intertwining of marine concerns with early conservation efforts in the United States.

Public understandings of the ocean impact the science performed in that space and vice versa. A wide array of literature highlights public perceptions of marine life. Gary Kroll's *America's Ocean Wilderness: A Cultural History of Twentieth-Century Exploration* (2008) explores cultural constructions of the ocean from William Beebe to Rachel Carson and Eugenie Clark. Stefan Helmreich's *Alien Ocean: Anthropological Voyages in Microbial Seas* (2009) examines both scientific work and the scientists who perform that labor, showing the impact of conceptions of the ocean on the work performed in the field. Both Susan Davis' *Spectacular Nature: Corporate Culture and the Sea World Experience* (1997) and Judith Hamera's *Parlor Ponds: The Cultural Work of the American Home Aquarium, 1850–1970* (2012) explore the ways that non-scientists understand and construct the marine environment. Finally, biographies and autobiographies of marine researchers offer a glimpse into the process of marine exploration and experimentation. Carol Gould's *The Remarkable Life of William Beebe* (2004) gives interesting insight into the career of a public intellectual and diving pioneer. Lester D. Stephens and Dale R. Calder's *Seafaring Scientist: Alfred Goldsborough Mayer, Pioneer in Marine Biology* (2006) follows the career of Alfred Goldsborough Mayer, a jellyfish biologist and the director of the Carnegie Institution of Washington's Tortugas Marine Laboratory. Kenneth R. Manning's Pulitzer-prize nominated biography of E.E. Just, *Black Apollo of Science: The Life of Earnest Everett Just* (1983), offers an important portrait of the marine biology career of a minority in the twentieth century. In addition to biographies, autobiographies such as Eugenie Clark's *Lady with a Spear* (1953) are important primary sources that add much to our understanding of marine science.

Chapter Twelve

MEDICAL GENETICS

Andrew J. Hogan

The field of medical genetics in the United States has multiple origin stories. One, often repeated by Victor McKusick, "the father of medical genetics," begins with the stabilization of the human chromosome number in 1956 (Collins 2008; McKusick 1982, 2001; Rimoin 2008). This important advance led to the recognition that Down syndrome and other clinical disorders were associated with extra or missing chromosomes. A second origin story, favored by many historians, points to the 1930s and 1940s when physicians and geneticists with strong ties to the American eugenics movement engaged in research aimed at better understanding the hereditary basis of intellect and other traits, and sought to use this knowledge to improve the human genetic stock (Comfort 2012; Kevles 1985; Paul 1998; Stern 2012). The extent to which the medical and scientific work of this earlier generation of eugenicists contributed to the growth of medical genetics from the mid-1950s onwards remains a point of disagreement (Cowan 2008). Some assert that medical genetics began anew during the postwar period under a younger generation of practitioners who were largely untainted by eugenic thinking, while others argued that medical genetics represented the continuation of eugenic thinking, only under a different name.

Attempts to address historical disputes about the origins of modern medical genetics face basic questions. Who, to begin with, counts as a medical geneticist? Is a PhD trained biologist doing medical genetics when studying hereditary factors with the goal of reducing certain traits in the population? Is medical genetics at its core a clinical specialty that is primarily practiced by physicians and oriented toward diagnosing, treating, and counseling individuals with little interest in population level impacts? Along similar lines, one may question whether medical genetics must always involve a clinical encounter, or if basic laboratory research also constitutes its practice? In many ways, the history of medical genetics is coextensive with human genetics. Distinctions between

A Companion to the History of American Science, First Edition.
Edited by Georgina M. Montgomery and Mark A. Largent.
© 2016 John Wiley & Sons, Ltd. Published 2020 by John Wiley & Sons, Ltd.

these two areas are often idiosyncratic, and may have more to do with who is doing the research (a physician or a scientist) than the focus of it.

Despite differing accounts of when the field began and what it is today, histories of medical genetics nearly universally acknowledge turn-of-the-century British physician Archibald Garrod as an influential figure. Between 1890 and 1910, Garrod studied the blood and urine of patients to identify various "inborn errors of metabolism," which he eventually linked to Mendel's newly rediscovered patterns of inheritance. Garrod, with input from the geneticist William Bateson, traced the inheritance pattern of primarily non-pathological biochemical variants in individuals. His approach influenced medical geneticists later in the twentieth century as they examined the inheritance of genetic traits and disease. Indeed, historical accounts of this foundational work have noted that, like Mendel before him, Garrod's research was largely overlooked for many decades after its publication (Weatherall 2012). The eventual recognition among physicians of the importance of Garrod's work on inborn errors of metabolism has often been pointed to as an important marker of when medical genetics truly began (Comfort 2012; Cowan 2008).

The Emergence of Medical Genetics

While American scientists rose to dominance in many fields during the late nineteenth century, they lagged behind their European counterparts in medical genetics until the 1960s. Though American researchers like T.H. Morgan and Barbara McClintock contributed significantly to examining the basic genetics of flies and corn (Comfort 2001; Kohler 1994), during the 1920s and 1930s many geneticists distanced themselves from the scientific and social complexities of studying human heredity. This area had been tainted for many genetics by the American eugenics movement (for more on this, see chapter 27) and in particular the Eugenics Record Office (ERO). Many geneticists believed that the ERO collected and housed hereditary data that was insufficiently rigorous for scientific study.

Another impediment to American medical genetics before World War II was a lack of government funding in the area. In comparison, Britain's Medical Research Council offered significant support for medical genetics from the 1930s (Harper 2008b). At this time, the ERO received funding from the Carnegie Foundation. Another major US-based private funder of research in medical genetics was the Rockefeller Foundation. It, however, was more generous in funding British medical geneticists than their America colleagues (Kevles 1985). As a result, American medical geneticists primarily relied on funding from individual donors whose support of eugenics was rooted in racial and anti-immigrant biases (Comfort 2012; Stern 2012).

America's first institution to adopt the term medical genetics was Raymond Pearl's Division of Medical Genetics, in the Institute for Biological Research at Johns Hopkins University in 1925. Pearl, a biologist, took a constitutional approach to studying common diseases like tuberculosis, and was influenced by Garrod's conception of inherited chemical individuality. The Division of Medical Genetics was a "combination of Eugenics Record Office and medical clinic," which collected data on the biological and environmental attributes of individuals and their families, including racial characteristics (Comfort 2012: 75). Even as Pearl actively spoke out against eugenicists for being insufficiently scientific (Rosenberg, 1997, 96), he conducted surveys similar to those of the ERO, with the goal of improving public health on the population level. While Pearl's Division of Medical Genetics only lasted five years, his data collection methods influenced future practice in the field (Comfort 2012).

During the 1920s and 1930s, eugenicists came from all walks of life, and had a variety of interests and biases. Broadly speaking, eugenicists coming from a background in genetics or medicine were interested in reducing the incidence of disease and "feeblemindedness" through discouraging reproduction among those thought to harbor hereditary susceptibilities. Pearl's pursuit of eugenics as public health was continued by likeminded professionals, including clinician Madge Macklin, who sought to bring eugenics into medicine by expanding the teaching of genetics in medical schools (Ludmerer 1972). Macklin found a compatriot in Laurence Snyder, a Harvard-trained geneticist at Ohio State University who in 1931 began teaching the first genetics course in an American medical school, and soon after also became the nation's first professor of medical genetics. At this time, most physicians had very little interest in genetics, were concerned about links to the eugenics movement, and in some cases mocked the teaching of this seemingly irrelevant topic in medical schools (Comfort 2012; Cowan 2008; Kevles 1985).

As an undergraduate, Snyder had spent a summer working in the ERO, where he learned the basics of human heredity field research. Later in his career, Snyder's research focused on genetic traits known to follow Mendelian inheritance patterns, like blood groups and the ability to taste the chemical phenylthiocarbamide (PTC), (Kevles 1985). During the 1930s, Snyder worked with the physician William Allan to sample populations in Appalachian North Carolina. These two researchers, one a geneticist with a well-honed knowledge of statistical analysis and the other an affable country physician, drew on each other's complementary strengths to improve their approaches to field-based medical genetics. Snyder, Allan, and Macklin were at once eugenicists and medical geneticists, who sought to identify and prevent the passage of certain hereditary traits through research, education, and sterilization, in much the same way that public health officials attempted to slow the spread of germs using quarantine (Comfort 2006, 2012).

The Heredity Clinics

During the 1940s the medical genetics of Allan and Snyder was institutionalized in heredity clinics, which brought together geneticists and clinicians to provide medical information and encourage eugenic decision making aimed at reducing the incidence of hereditary disease and feeblemindedness. The aims of these heredity clinics were in line with the various educational campaigns that American eugenicists had been promoting for decades. In North Carolina, Snyder aided Allan as he worked to create a department of medical genetics in the Bowman Gray School of Medicine at Wake Forest University. Opened in 1941, the department was home to Allan's ongoing field research, clinical counseling, and the genetics training of medical students. In 1942 Claude Herndon, a young physician who was also interested in human genetics research, joined the department. Herndon initially accompanied Allan on his Appalachian research trips. Following Allan's death in 1943, however, Herndon took over and shifted to primarily doing research that began with clinical patients. Herndon also introduced genetics to medical students in the classroom at Wake Forest, teaching future clinicians to take family histories (Comfort 2012; Ludmerer 1972; Paul 1998; Stern 2012). This technique remains foundational in medical genetics to the present day (Harper 2008b; Pyeritz 2011).

Additional heredity clinics were also opened in 1941 at the University of Minnesota and University of Michigan. The biologist Lee Raymond Dice and mathematical geneticist Charles Cotterman, who had done his PhD under Snyder, led the heredity clinic in Michigan. As biologists, Dice and Cotterman were interested in studying all human traits, but worked to develop connections with the University of Michigan School of Medicine in order to see patients and access their medical records (Comfort 2012; Paul 1998; Stern 2012). These links were no doubt strengthened through the hiring of physician/geneticist James Neel in 1946. Neel had earned a PhD in genetics at the University of Rochester where he studied flies, but had taken an interest in studying humans, which was a risky professional move at the time. To pursue this career path, Neel decided to attend medical school. Seeking to distinguish medical genetics from earlier eugenics research, Neel took a conservative approach to studying human genetic traits, sticking to those that showed clear Mendelian inheritance patterns. Neel went on to lead the University of Michigan's heredity clinic, overseeing its move to the medical school and expansion into a major research institution during the late 1950s, when it became the first center in the United States for medical genetics that rivaled Lionel Penrose's Galton Laboratory in London (Kevles 1985: 230).

Elsewhere in the Midwest, the University of Minnesota's heredity clinic was funded by a gift from the physician and eugenicist Charles Freemont Dight, who charged the institute with promoting "biological race betterment" (Paul 1998: 136). The Dight Institute was initially led by fly geneticist Clarence Oliver, and after 1946 was taken over by Sheldon Reed, another geneticist who soon thereafter coined the term genetic counseling. Under both Oliver and Reed, the primary role of this heredity clinic was to provide information to families about genetic risk. While Oliver was active in discouraging at-risk parents from having more children, Reed was not as eugenically directive in his counseling. In this sense, his approach was more in line with present-day genetic counseling. However, Reed retained the expectation that patients would make the "right decision," namely one that would reduce the incidence of hereditary disease. Like Michigan's heredity clinic, the Dight Institute also began outside of a medical school with PhD biologists at the helm, but sought to develop closer ties with medicine (Comfort 2012; Stern 2012: 82).

During the 1940s and 1950s, the hereditary clinics at Michigan, Minnesota, and Wake Forest had varying and complex ties to the American eugenics movement. Allan, Herndon, and Dice were all strong and vocal proponents of improving human heredity and breeding through genetics research and clinical counseling. Each took advantage of available funding from anti-immigration and racially biased eugenicists like Wickliffe Draper and Charles Goethe, though often anonymously, in order to conduct their research. At Minnesota, Reed was somewhat more cautious about where his money came from. He rejected Draper's offers, but still accepted funding from Goethe in small and targeted infusions. Dice and Neel at Michigan also took money for their research from Draper, because of the open-ended projects in human genetics that the funding allowed. Neel, however, sought to avoid ties with eugenics; this included turning down the offer of a leading role in the American Eugenics Society (Comfort 2012; Stern 2012).

Many of the leaders of the heredity clinics during the 1940s became actively involved in building and overseeing the first professional organization in human and medical genetics, the American Society for Human Genetics (ASHG). Discussions about how to name the society and its journal reflected tensions over whether physicians should be accepted as full members. The society's first president Hermann J. Muller was against the full inclusion of MDs in ASHG. His opposition was rooted in a preference for population-level eugenic approaches aimed at combating a deteriorating human gene pool, rather than individualized clinical action. A debate ensued over whether to include medical genetics in the title of the society's journal. Ultimately, the *American Journal of Human Genetics* was chosen, based on the biased view of geneticists that human genetics subsumed all medical genetics research (Comfort 2012; Paul 1998).

In its early years, the ASHG membership consisted primarily of geneticists, though by 1952, 35% of its members were physicians, and in 1964 physicians comprised close to half of the society (McKusick 1975). Over these years, many ASHG presidents had close ties to eugenics and the heredity clinics, including Snyder, Dice, Oliver, Neel, Herndon, Reed, and Macklin. The society's most prestigious award, first given in 1962, is named after Allan (Comfort 2006; Paul 1998). Historians have disputed claims that this older generation of medical geneticists had retired or been discredited by the time that medical genetics had become an institutionalized discipline in the United States. As Diane Paul (1998: 141) has described, many medical geneticists in the 1950s and 1960s continued to see themselves as practicing a "worthy form of eugenics," aimed at benefiting society, while also preventing suffering in individual families.

Human Cytogenetics in the United States

During the 1950s, the study of human genetics had almost no direct applicability to disease diagnosis in the clinic. Besides identifying the potential Mendelian inheritance patterns of disorders using pedigree analysis, heredity clinics had little to offer their patients. An important advance, which helped genetic medicine transform itself from a research backwater in 1955 into an exciting field of study during the early 1960s, involved improvements in the microscopic study of human chromosomes, also known as human cytogenetics (Lindee 2005). American laboratories, however, lagged behind European researchers in human cytogenetics during the late 1950s, when this field suddenly rose to prominence (Harper 2008b).

Human cytogenetics represented a new approach for examining hereditary disease. Rather than making predictions based on family histories, the microscopic visualization of human chromosomes offered physical markers for the diagnosis of certain diseases. Human cytogenetic analysis during the late 1950s led to the association of multiple clinical disorders with visible chromosomal abnormalities. This included the link between Down syndrome and trisomy 21 by the French physician Jerome Lejeune, and the association of Turner syndrome with a missing X chromosome by Charles Ford and colleagues in Great Britain (Cowan 2008; Kevles 1985).

While American researchers were not at the forefront of human cytogenetics in the late 1950s, they did contribute significantly to developing the techniques that made these medical advances possible. In the early 1950s, human chromosomes were both difficult to see under the microscope and required invasive procedures to obtain. T.C. Hsu, a postdoctoral fellow in Galveston, Texas, solved the first problem in 1952, following the accidental realization that a low salt (hypotonic) solution made chromosomes much easier to see under the microscope (Kottler 1974; Martin 2004). This advance led to the stabilization of the human chromosome number by Joe-Hin Tijo and Albert Levan in Sweden. On prominent display at the first International Congress of Human Genetics in 1956, this finding has been pointed to as an important taking-off point for the diagnostic achievements of medical genetics (McKusick 2001).

A second major American contribution to human cytogenetics greatly expanded the accessibility of human cells for chromosomal analysis. Before 1960, tissues from living patients were most often derived from painful spinal taps or testicular biopsies. This meant that few people were likely to volunteer for large-scale chromosome studies. A new technique developed by Paul Moorhead, Peter Nowell, David Hungerford, and others in Philadelphia allowed for skin and blood samples to be used for human chromosomal analysis. With this advance, tissues could be more easily derived from a wide array of clinical patients and the broader population for cytogenetic studies of human chromosomal variation. Soon thereafter, cytogenetics laboratories were set up in Baltimore and across the United States. In Philadelphia, Nowell and Hungerford took advantage of their new technique to identify a chromosomal abnormality in chronic myeloid leukemia patients, which came to be known as the Philadelphia chromosome (Lindee 2005).

Building Genetic Medicine in Baltimore

In Baltimore, Maryland, another important center for medical genetics was under construction. In 1957, Victor McKusick, a cardiologist at Johns Hopkins University who had developed an interest in the genetics of connective tissue disorders, founded the Division of Medical Genetics. The division inhabited a small space where teaching, research, and clinical care were conducted in close quarters. McKusick saw this situation as a virtue, because it encouraged collaboration (McKusick 2006). Having inherited a tradition of bringing in postdoctoral fellows from the United Kingdom to research chronic disease, McKusick began pushing fellows toward medical genetics. He eventually trained 100 students from 26 countries, including many major figures in postwar medical genetics, such as Samuel Boyer, David Weatherall, and Peter Harper (Harper 2008b: 287). In 1959, physician and cytogeneticist Malcolm Ferguson-Smith came to Johns Hopkins University from Glasgow to do a postdoctoral fellowship with McKusick. During his three years at Johns Hopkins University, Ferguson-Smith began the first clinical cytogenetics laboratory in the United States (Malcolm Ferguson-Smith, interview by Peter Harper, 2003). In the 1960s, a PhD granting program was also started up, allowing physicians to gain significant additional training and credentialing in genetics during their time at Johns Hopkins (McKusick 2006).

When it came to educating physicians in genetics, however, McKusick sought to reach a broader population of medical professionals than he could at Johns Hopkins University alone. He accomplished this beginning in 1960 through the annual Short Course in Medical and Mammalian Experimental Genetics, held at the Jackson Laboratory in Bar

Harbor, Maine. The short course attracted over 100 students from around the world every year for two weeks of coursework on genetics taught by researchers from Johns Hopkins University, the Jackson Laboratory, and many other institutions (McKusick 2006). With the short course, McKusick sought to create a uniform vision and identity among medical geneticists. Always a field that drew from a variety of backgrounds, the Bar Harbor courses helped to establish a sense of cohesion (Harper 2008b).

The Bar Harbor Short Course received significant funding from the March of Dimes Foundation. Beginning in the 1960s, the March of Dimes invested in research aiming to prevent birth defects, and in doing so provided extensive support for the organization, expansion, and standardization of American medical genetics. In addition to the Bar Harbor course, the March of Dimes provided funding for conferences on the delineation of birth defects, chromosomal nomenclature, human gene mapping, and genetic counseling (Comfort 2012; Lindee 2005; Stern 2012). Throughout the 1960s and 1970s McKusick maintained a close and productive relationship with the March of Dimes, which provided the funding necessary to bring many of his discipline-building ideas to fruition. The March of Dimes' periodical *Birth Defects: Original Article Series* offers a valuable overview of the various interests and ambitions of medical genetics during this era.

Diagnostic Medical Genetics

In addition to McKusick at Johns Hopkins University, departments of medical genetics and laboratories for clinical cytogenetics sprung up quickly in North America during the late 1950s and early 1960s. Some of the earliest were located at the University of Washington in Seattle, the University of Wisconsin in Madison, McGill University in Montreal, and New York University (Harper 2008b). Even as medical genetics expanded institutionally, its preventative capabilities remained limited. A major advance came in 1966, when Yale University physicians Mark Steele and W. Roy Breg cultured amniotic fluid cells from a prenatal sample and examined their chromosomes for the first time. This advance made it possible to diagnose Down syndrome and other chromosomal disorders prenatally using amniocentesis. These were not traditional hereditary disorders, which were passed down through families, but rather were caused by random errors in chromosome reproduction. Nonetheless, the prenatal diagnosis of chromosomal disorders would become central to the growth of medical genetics (Löwy 2014).

In the late 1960s, the clinical uptake of amniocentesis remained slow, and limited to the pregnancies of women over age 40, who were at high risk for Down syndrome. This was primarily because the safety of amniocentesis, especially when performed without ultrasound guidance, was poorly established. As the availability of ultrasound increased during the 1970s, the uptake of prenatal amniocentesis grew along with it (Cowan 2008). Once again, Johns Hopkins University was at the forefront of this process with the help of a major grant from the March of Dimes. The Johns Hopkins Prenatal Birth Defects Center opened in 1969, and by 1973 had conducted 156 amniocenteses. Many of these procedures were performed to look for Down syndrome, though a few also sought to identify inborn disorders like Tay–Sachs disease, which could be identified based on biochemical abnormalities (Lindee 2005; Stern 2012). Over the course of the 1970s, the number of chromosomal and metabolic disorders that could be identified prenatally increased rapidly, as new tests and improved techniques were developed.

A postnatal preventative option was also created during the 1960s for interrupting the development of phenylketonuria (PKU), an inborn metabolic disorder that caused intellectual disabilities. PKU involves a biochemical defect that leads to the build up of phenylalanine in the body. If PKU is detected early, however, and if the afflicted infant is put on a special diet, intellectual disability can be minimized. In 1961, Buffalo microbiologist Robert Guthrie developed a postnatal blood test for PKU that would eventually allow for the mass screening of newborns and widespread prevention of this disorder's most devastating effects. While the PKU diet was difficult and expensive, and how long it needed to be followed was unclear, Guthrie's biochemical screen did facilitate successful population-level prevention of PKU (Lindee 2005; Paul and Brosco 2013).

PKU and Down syndrome were just a few of the inborn disorders that medical geneticists hoped to one day treat or prevent following diagnosis. During the early 1970s, the promise of identifying, testing for, and perhaps even repairing individual disease causing genes seemed to be on the horizon. Evoking the climate of scientific optimism surrounding the moon landing, Victor McKusick (1970) called for the human chromosomes to be thoroughly mapped. Along with Yale geneticist Frank Ruddle, McKusick helped to organize the large-scale mapping of human genes throughout the 1970s and 1980s. Beginning in 1973, McKusick and Ruddle played a prominent role in organizing regular human gene mapping workshops, which brought together physicians, cytogeneticists, and molecular biologists interested in mapping genes to each human chromosome (Comfort 2012). McKusick regularly published updated versions of the growing human gene map, depicted on standardized representations of chromosomes. In addition, during the early 1980s, McKusick created a similar diagram showing the location of diseases along the chromosomes. He called this the "morbid anatomy of the human genome," referencing the medical tradition of examining the body for markers of disease. McKusick believed that this morbid anatomy best represented the goals of medical genetics (Hogan 2014; McKusick 1986).

McKusick's human gene and disease maps were part of a much larger project: a catalog of all human genetic traits known as *Mendelian Inheritance in Man* (MIM). During the late 1950s, McKusick took advantage of the large pool of postdoctoral fellows he had at the Division of Medical Genetics to begin performing an annual review of the published medical genetics literature. By the mid-1960s, this had evolved into his regularly updated catalog. With MIM, McKusick sought to organize and systematize human genetic knowledge, with a focus on discrete genetic traits and their inheritance patterns. MIM grew quickly in volume alongside the burgeoning field of medical genetics throughout the 1970s and 1980s, eventually becoming too large to print. Having existed in computerized form from its very beginning, McKusick's catalog was made publically available online in 1987 as OMIM, and continues to be a key reference for medical geneticists worldwide (Comfort 2012; Harper 2008b; Lindee 2005).

MIM is one of many examples of textbooks developed during this era that sought to bring order to genetic knowledge and diagnosis. In 1960, American physicians James Stanbury, James Wyngaarden, and Donald Frederickson published the first edition of *The Metabolic Basis of Inherited Disease*, which examined Mendelian traits that caused biochemical errors, resulting in clinical disorders. This text went through seven editions over the next 35 years, with both Frederickson and Wyngaarden rising to become directors of the National Institutes of Health, a testament to the rising importance of the study of inherited disease in medicine during the 1970s and 1980s. Other significant organizing texts during this era included Minneapolis dentist Robert J. Gorlin's *Syndromes of the Head and Neck*, published in 1964, and American physician David W. Smith's *Recognizable Patterns of Human Malformation* (1970). These texts contributed to the development of dysmorphology, a medical area that sought to improve the delineation of clinical disorders, many of which were likely caused by genetic mutations (Harper 2008b). The creation of standardized ways of recognizing and reporting patterns of abnormality in the clinic were very important to the development of medical genetics. While laboratory testing has come to play a central role in the field, scholars have shown that clinical knowledge production remains equally significant in medical genetics research and practice (Featherstone and Atkinson 2012; Keating and Cambroiso 2004; Latimer et al. 2006; Rabeharisoa and Bourret 2009).

During the 1970s and 1980s, reference manuals traced the quickly growing base of knowledge about individual genetic disorders, and facilitated the growth of clinical and prenatal diagnostics. McKusick's vision of all diseases being associated with specific genomic locations helped to attract funding for medical genetics, from organizations including the Howard Hughes Medical Institute, and served as an important justification for the Human Genome Project during the late 1980s (Cook-Deegan 1994; Hogan 2014). Locating disease genes, or as McKusick put it, doing a "morbid anatomy of the human genome," remains an important concept in medical genetics to this day, the impacts of which can be seen in various online genomic databases, such as the University of California Santa Cruz Genome Browser (Hogan 2013b).

Professionalization in Medical Genetics

Between the late 1970s and early 1990s, medical genetics also became increasingly institutionalized as a distinct medical specialty. The *American Journal of Medical Genetics* began in 1977 under the leadership of John M. Opitz, an American physician who had emigrated from Germany as a teenager. Opitz intended for the journal to attract submissions from both clinicians and basic researchers (Opitz 1977). Three years later, clinical genetics became a board certified medical subspecialty. This move had been initially opposed by McKusick, who saw genetics as a body of knowledge that belonged in all medical fields rather than encapsulated in its own subspecialty (McKusick 1975). However, while McKusick always maintained a generalist vision of medical genetics, his opposition to its institutionalization was short lived (McKusick 1984). In 1991, the American College of Medical Genetics (ACMG) was established, and the field of medical genetics was awarded membership in the American Board of Medical Specialties. This placed medical genetics on equal footing with other well-established medical specialties like internal medicine, surgery, and radiology. The Board of Medical Genetics is the only medical specialty that has been added as a member since 1980.

While medical genetics has always been a diverse area, bringing together physicians and scientists with backgrounds in biochemistry, medicine, cytogenetics, and molecular genetics, its relationship with the field of genetic counseling has been fraught at times.

Like medical genetics, the origins of genetic counseling can be traced back to the 1940s era heredity clinics. Sheldon Reed, who led the Dight Institute, coined the term genetic counseling in 1947. Over the next 25 years, genetic counseling was primarily done by MDs or PhDs. In 1969, however, Melissa Richter, a psychologist, began a master's level program for genetic counseling at Sarah Lawrence College outside New York City. The initiation of nine additional master's programs followed over the next decade. Immediate pushback towards master's level education in genetic counseling came from the ASHG, which began a committee on genetic counseling in 1970. The committee went on to question the ability of anyone with fewer years of training than an MD or PhD to provide genetic counseling. Over the course of the next decade, however, genetic counseling programs and their newly minted students persevered, leading to a reversal of ASHG opinion. Nonetheless, tensions between the two groups continued (Stern 2012).

The National Society of Genetic Counselors (NSGC) was founded in 1979 after eight cohorts of master's level genetic counselors had graduated. This new professional organization maintained an at times troubled relationship with the medical genetics community (Harper 2008b). A debate raged about the inclusion of physicians and dentists in NSGC. Physicians were not happy about the possibility of being excluded; after all, MD and PhD trained clinicians had been calling themselves genetic counselors for decades. The new and growing population of master's trained genetic counselors, however, wanted to consolidate their power in a homogeneous professional organization, which would be largely free from the influence of medical geneticists. Ultimately, full membership in the NSGC was limited to individuals whose primary profession was genetic counseling. From 1980, the NSGC was offered two positions on the American Board of Medical Genetics, which continued to oversee the credentialing of genetic counselors. In the early 1990s, as the ABMG was restructured to allow for it to become a fully fledged medical specialty, genetic counseling was dissociated from the board and formed the fully independent American Board of Genetic Counselors (Heimler 1997; Stillwell 2013).

Medical Genetics and the Private Sector

With the introduction of large-scale genetic sequencing into medical practice in recent years, genetic counseling and medical genetics have developed increasingly close ties with the private sector. Before the 1990s, genetic diagnosis was primarily done in small academic laboratories. The increasing use of molecular techniques in genetic medicine, however, has made diagnostics increasingly expensive, meaning that it must be done on a large scale to be affordable. In recent years, many small academic diagnostic laboratories have either closed or become involved in doing specialty testing for large diagnostic corporations like LabCorp and GeneDx. Genetic testing, molecular and chromosomal, is now often a nationalized process, in which samples are sent daily across thousands of miles for processing and testing, with electronic results being returned to providers for diagnosis and patient counseling (Milunsky 1993; Rapp 1999). Both for legal and liability reasons, diagnostic genetic testing corporations have increasingly relied on consulting from medical geneticists and have themselves become major employers of genetic counselors (Biesecker and Marteu 1999; Harris, Kelly, and Wyatt 2013).

Corporate interest in genetic diagnostics has been both driven by, and encouraging of, the increasing demand for genetic testing in recent decades. This includes preconception screening, prenatal diagnosis, postnatal screening, and pediatric genetic testing, as well as pharmacological and cancer genetics. Prenatal genetic diagnosis, once primarily targeted to women over age 35 due to increased Down syndrome risk, has expanded significantly. This has been due to the expansion of the number of disorders and abnormalities that can be tested for, and growing encouragement for broad based genetic testing by the obstetrics community. In 2007, the American College of Obstetricians and Gynecologists (ACOG) recommended that prenatal diagnosis for chromosomal disorders be offered to all pregnant women, regardless of age (ACOG 2007). Although this ACOG statement was specific to chromosomal analysis, once a prenatal sample has been acquired, women may also be offered much more extensive genetic testing for specific markers and mutations associated with clinical abnormalities. Moves like this one have contributed to establishing prenatal diagnosis as a standard procedure in all pregnancies, rather than only those known to be at increased risk for certain genetic disorders.

As genetic testing has become increasingly powerful, affordable, and commonplace, its use has moved beyond targeted genetic diagnosis and toward undirected screening, which seeks to uncover any genomic abnormality that may have clinical relevance (De Jong et al. 2014; McGillivray 2012). This evolution in genetic testing has largely gone unregulated by federal and state governments in the United States. In addition, the use of high resolution genetic testing, such as DNA sequencing and microarray, has become increasingly accepted by influential professional organizations like ACOG and the ACMG (ACOG 2009, 2013; Manning and Hudgins 2007).

To overcome some of the challenges that emerge from its historical links to eugenics, postwar medical geneticists have sought to highlight the scientific accuracy and targeted nature of genetic diagnosis, treatment, and prevention. With the recent growth of untargeted testing, however, the breadth of what counts as genetic abnormality has become increasingly unclear. This has raised concerns that the bar of acceptability for going forward with a pregnancy has become inappropriately high, even as the true clinical implications of many genetic traits labeled as "abnormal" remain uncertain (Shuster 2007; Turrini 2014). Chastened by its eugenic past, medical genetics has been careful to acknowledge the limits of what can be known about the well-being of future generations. Going forward, professional caution and independence from corporate interests will be important to ensuring that medical genetics does not once again overestimate its predictive power.

Bibliographic Essay

In recent years, scholars have shown increasing interest in American human and medical genetics before 1955. Earlier historical accounts were few, but important. Ludmerer (1972) provided a valuable resource on mid-twentieth-century human genetics in America, and its complex relationship with eugenics. Daniel Kevles (1985) also offered an extensive contribution to this literature, but primarily focused on human genetics and its links to eugenic thinking in the United Kingdom during this era. Kevles examined the important role of the Medical Research Council in funding human genetics research in the UK from the 1930s, and the influential role of Lionel Penrose, J.B.S. Haldane, and Julian Huxley. Nathaniel Comfort (2006, 2012), Diane Paul (1998), and Alexandra Minna Stern (2012) have since added significantly to our knowledge of human and medical genetics in pre-1955 America. These scholars have done important work in tracing the earliest American institutions in human and medical genetics, including heredity clinics and the American Society for Human Genetics (ASHG), with a specific focus on their financial and intellectual links to American eugenics. Each of these books also offers significant insights on the development of human and medical genetics since 1955.

Many scholars have examined the quick rise and golden age of human cytogenetics between 1955 and 1965. Malcolm Jay Kottler (1974) provided an important early historical analysis of uncertainty over, and the eventual stabilization of, the human chromosome number. Aryn Martin (2004) has more recently offered an account of this history, with a particular focus on methods of categorization and counting. In a chapter on human cytogenetics during the 1960s, Susan Lindee (2005) examined the socially negotiated processes of standardizing chromosomes across international boundaries. Soraya de Chadarevian (2010: 180) has since considered the important role of chromosomal analysis in providing an early "glimpse" of the human genome. In the same volume, Maria Jesus Santesmases explored the clinical value and complexity of chromosomal diagnosis around 1960. Andrew J. Hogan (2013b) provided a history of the use of chromosomal analysis in the clinical setting from the late 1960s to the present, and has analyzed the role of cytogenetic standardization in establishing the human genome as an object of study in basic genetics and medical genetics (Hogan 2013a, 2014).

In the area of human molecular genetics, Bruno Strasser (1999) has provided an early account of the establishment of sickle cell anemia as a molecular disease during the late 1940s. While the 1953 elucidation of the double helical structure of DNA was an important moment in all genetics fields, de Chadarevian (2006) demonstrated that studies of DNA were absent from human genetics before the late 1970s. Peter Harper (2008b) extensively traced the rise of human molecular genetics after the development of Southern blotting in 1975, through the completion of the Human Genome Project. In an edited volume, Kelves and Leroy Hood (1992) offered a multifaceted analysis of the origins of the Human Genome Project, which is complemented by the contemporary research of Robert Cook-Deegan (1994). McKusick (2001) also traced the history of the Human Genome Project, in particular highlighting his important role in facilitating the integration of cytogenetic and molecular methods in human gene mapping since 1973.

The development of medical genetics as a scientific and clinical discipline from the late 1950s was thoroughly traced by Harper (2008b), who examined the establishment of medical genetics at Johns Hopkins and the University of Washington. Comfort (2012) offered a complementary account, which also highlighted the importance of the Moore Clinic, Bar Harbor Short Course, and significant funding from the March of Dimes. Internal accounts of the growth of medical genetics as an interdisciplinary field, which was home to physicians and geneticists, can be found in societal presidential addresses by McKusick (1975) and Arno Motulsky (1987). Additional insights on the central role of McKusick in this history were provided by Dronamraju and Francomano (2012) in an edited volume and autobiographically by McKusick (2006).

Many scholars have examined the important part that prenatal diagnosis has played in the history of medical genetics (Lowy, 2014). Ruth Schwartz Cowan (2008) traced the development of prenatal diagnosis as a technological system, integrating many technical and social components, such as access to abortion. Stern (2012) highlighted the establishment and early work on the Prenatal Birth Defects Center at Johns Hopkins. Both of these scholars have recounted that the uptake of prenatal diagnosis happened quite slowly during the 1970s. Cowan (1994) has demonstrated that the diffusion of ultrasound and legal action by women played an important role in increasing the prevalence of amniocentesis. Rayna Rapp (1999) offered a groundbreaking ethnographic account of the differing experiences of women undergoing prenatal diagnosis due to varying conceptions of risk and cultural backgrounds. Also focusing on risk perception, Hogan (2013c) examined the role of women, physicians, government regulators, and the media in shaping the perception and uptake of chorionic villus sampling (CVS), as an alternative to amniocentesis.

The development of genetic counseling as a professional field also has an important place in the history of medical genetics. Stern (2012) offered a much-needed historical account of genetic counseling from its origins in World War II era heredity clinics. She argued for greater internal awareness of the field's historical links to American eugenics. In addition, Stern traced early disagreements about training qualifications in genetic counseling. Harper (2008b) also noted historical tensions between American genetic counselors and their medical genetics colleagues. Audrey Heimler (1997) provided an internal perspective on these debates based on her own role in them as a genetic counselor. A dissertation by Devon Stillwell (2013) has also offered significant insight on this history. The genetic counseling profession has continued to grow in recent years, and has increasingly expanded from the clinical setting into the private sector. Biesecker and Marteau (1999) offered valuable insights about the future of the field.

Over the past decade, a number of interviews have been conducted with late twentieth-century practitioners of human and medical genetics. Transcripts are available online for many of these interviews. One series of interviews with primarily British geneticists was conducted by Peter Harper of Cardiff University, and are available at: http://www.genmedhist.info/interviews/. Another collection, the Oral History of the Human Genetics Project, is a joint venture of the University of California, Los Angeles and Johns Hopkins. Transcripts of interviews conducted primarily with American geneticists can be found at: http://ohhgp.pendari.com/Collection.aspx. Together, these interview collections offer valuable opportunities to examine international variations in the history and practice of human genetics, as well as to learn about the frequent interchange of knowledge and personnel between the United States and United Kingdom.

Chapter Thirteen

METEOROLOGY AND ATMOSPHERIC SCIENCE

James Bergman

To write a chapter on the American atmospheric sciences is an odd assignment, in some ways. After all, as Harper (2003: 687) has written, "Almost by definition meteorology is an international science. The atmosphere knows no boundaries; data have long been shared except during times of conflict between nations." Climate scientists, also, have adopted as a main indicator the global average temperature. And finally, when meteorologists in Bergen, Norway, were studying the movement of air masses, they posited the existence of a "polar front," a confrontation between arctic and tropical air that wrapped around the globe (Friedman 1989).

On the other hand, the atmospheric sciences have just as often been regional or even local. The original meaning of the term "meteorology" was the study of any sublunar phenomena. Meteors or storms, for example, were all fair game in describing meteorological phenomena, and so were the moral transgressions that might have invited these phenomena (Jankovic 2000, ch. 3). Climate, also, was a means of explaining regional and cultural differences (Livingstone 1999, 2002).

This brief sketch will examine the history of meteorology, climatology, and the atmospheric sciences in general in the United States. Existing scholarship has found that concepts of "weather" and "climate" often tracked with political and infrastructural priorities at the time, particularly the political desire to forecast the weather for commerce and provide long-term guidance for farmers. By basing meteorology and climatology on observational systems, however, the atmospheric sciences have often developed in tension with the theoretical commitments of its practitioners. At the same time, those observational systems have been crucial in shaping the American public's understanding of local, regional, national, and global weather and climate. For the periodization of meteorology, I take, as my starting point, Fleming's (1990) periodization of Emerging Systems (1800–1870), Government Services (1870–1920), and the Disciplinary/Professional Period (1920–). At the same

A Companion to the History of American Science, First Edition.
Edited by Georgina M. Montgomery and Mark A. Largent.

time, I have added my own periodizations where necessary, particularly the transition from national to global networks in the post-World War II period and the period of long-range forecasting and climatological classification between the years 1900 and 1940.

In many ways, the national level of analysis is appropriate for the history of the atmospheric sciences. The encounter of settlers with the American climate, for instance, forced a revision of the simple correlation between climate and latitude. (The etymology of climate, after all, is from the Greek for "incline.") The weather was a crucial element of the imagination of settlers as they read the land through which they travelled (Valencius 2002, ch. 4). Ideas about the atmosphere expanded as letter exchanges grew into correspondence networks, and telegraph wires were strung between military outposts. Meteorology in the United States often developed in conversation with the development of state and federal institutions. James Fleming (1990: 170), in his seminal work on the development of American meteorology in the nineteenth century, observed that: "Perhaps more than most sciences, natural history excepted, meteorological research was (and is) shaped by national resources, both cultural and geographical. Indeed the expansion of meteorological systems into the frontier mirrored the development of the nation as a whole." Indeed, the atmospheric sciences sometimes did more than "mirror" the development of the nation. Concerted efforts in the atmospheric sciences sometimes led to the consolidation of national institutions. The adoption of standard time, for instance, was originally initiated not by the railroads, but by Cleveland Abbe's lobbying of Congress for standard times by which weather observers could set their clocks (Bartky 1989).

Even when meteorological work crossed national boundaries, that crossing required conscious effort. When a meteorological school was imported from other countries, historians have described the action as just that: an import, an application of methods developed in one place to the atmospheric phenomena of another. Moreover, efforts at international cooperation, and eventually global cooperation, required a substantial amount of political work between countries to support material and computational efforts to "make data global" (Edwards 2010: 251). Attempts to recapture an understanding of weather and climate as local phenomena have required similarly concerted efforts. Hulme (2008: 6) has specifically called for a project "to reveal that discourses about global climate change have to be re-invented as discourses about local weather and about the relationships between weather and local physical objects and cultural practices."

Meteorology to 1870: The Emergence of a National Network

As the term "meteorology" implies, the science originally had as its object any sublunar (and hence imperfect and less predictable) phenomena (Jankovic 2000). "Climate," by contrast, was considered far more regular: A characteristic that was determined primarily by the latitude of a place. *Climata* were the parallel bands into which classical scholars divided the world (Kupperman 1982: 1265), and were taken from the Greek word *Klima,* meaning slope or incline (Fleming 1998: 11–12).

The colonization of the New World and island states challenged that conception of climate. Settlers in Virginia – a state with the same latitude as southern

Spain – were surprised to find a climate that was not in the least bit hospitable to the olive groves and vineyards that grew there (Fleming 1998: 22; Kupperman 1982: 1267). Such discrepancies between expectations and realities had economic ramifications, and the need for better understanding of climate created spaces for some of the first climate "models." For many mercantilist thinkers, to understand climatic diversity was to maintain an advantage in global trade (Jankovic 2010a: 202).

At the same time that trade and colonization were changing conceptions of climate, informal correspondence networks facilitated the identification of patterns in American weather, particularly in the occurrence of the violent storms for which the American colonies were becoming well known. Settlers from Northern Europe were surprised by the frequency and intensity of storms that occurred in the Eastern portion of the United States (Fleming 1998: 22). Benjamin Franklin's collection of weather observations from correspondents facilitated his observation that those storms generally traveled from southwest to northeast (Fleming 1990: 7).

The connection between land use, climate, and weather was also a topic of interest in the United States. Although the effects of clearing the land for agriculture were not as stark on the continent as they were on island colonies such as Mauritius, St. Helena, and in the Caribbean (Grove 1995), observers such as Thomas Jefferson noticed that a decline in the frequency of storms followed the clearing of large swaths of land for agriculture (Fleming 1998: 26). They perceived a correspondence between the "civilization" of an area through cultivation and the creation of a more "temperate" climate.

Informal networks continued to play an important role in weather observation into the nineteenth century. With the westward migration of settlers, weather observation networks expanded as well. As Valencius (2002) has argued, the "airs" of a place were an integral part of the medical geography of a region, a key part of settlers' "reading" of the land. Physicians on the frontier found medical reports from the area to be a way of maintaining their intellectual relevance while building their new homes and practices (Valencius 2002: 189). This was not merely a preoccupation of settlers, but also those in charge of defending the frontier. After the War of 1812, the US Surgeon General Tilton, reflecting on the experience of the Army Medical Office during the war through the lens of Hippocratic theories of disease, instructed all medical officers to keep a journal of the weather. This led to the publication of the Office's first *Meteorological Register*, which earned the respect of the scientist Alexander von Humboldt, who recommended that Russia, with its considerable territorial expanse, should emulate such a register (Fleming 1990: 16).

The need to manage an ever-widening network of colonization and control was certainly a motivating factor in the development of observation networks, but scholars have noted the importance of theory in driving many of these networks. Fleming (1990: xx and 24) has argued that meteorology was not nearly as Baconian as earlier historians – I.B. Cohen, George Daniels, and Greg de Young – had argued. Rather, many observation networks were fueled by competition between theories of storm formation advanced by William Redfield, James Espy, and Robert Hare. Redfield argued that gravity was the main driver of storms, Espy that it was heat, and Hare that it was electricity (Fleming 1990: 23–54). The importance of the controversy lay not in its resolution and consequent advancement of any one theory, but rather the new focus of meteorology on short-term meteorological phenomena and new networks of observers taking simultaneous observations (Fleming 1990: 58). Also crucial was the attention of Joseph Henry, who would later be appointed as the first head of the Smithsonian Institution (Fleming 1990: 76).

The Smithsonian meteorological project recruited participants to conduct simultaneous observations at distinct times of the day. The project provided standardized instruments, uniform procedures, and a shift from private to public weather records (Fleming 1990: 93). It also provided a focal point for American meteorology, even in cases when agencies were uncooperative. When Matthew Fontaine Maury sought to recruit observers for his meteorological network, Smithsonian observers noted that it would conflict with their duties to the Smithsonian (Fleming 1990: 108). It was also a focal point of cooperation for state networks, the US Coast Survey, the agricultural branch of the US Patent Office, and the Army Medical Office, although the latter considered its data proprietary (Fleming 1990: 110–29). At this point, meteorological observation was coalescing into a national system of observation.

The nascent Smithsonian network facilitated the incorporation of a development crucial to the growth of meteorology and the application of weather observations to forecasts: the telegraph. Whereas before, observers would mail their observations, the telegraph enabled observations to be transmitted quickly, and for forecasts and warnings to be transmitted if storms were on the horizon. At the same time, it ushered in a period in which theory in meteorology would be overshadowed by the possibility of weather forecasts and warnings. Fleming notes that "the efficiency of the telegraph in bringing timely and accurate warnings to the public took priority over more theoretical concerns" (1990: 141). This is in keeping with Katharine Anderson's noted tension between forecasting and theoretical meteorology. The former attracted a great deal of funding, but all too often, it came at the cost of the credibility of the latter (Anderson 2005). Frederik Nebeker also draws a line between forecasting, dynamic meteorology, and climatology that would remain in place until the advent of numerical weather prediction in the postwar era (Nebeker 1995: 1).

Correspondence networks, first by mail, then by telegraph, went some way toward an *American* meteorology. And it was not only the Smithsonian or Espy that fueled these interests. Two state agricultural societies lobbied Congress in 1856 to have Maury's observation network brought to land. Although Henry opposed the extension of Maury's network, the interest in weather knowledge flowing from a nationally centered network was clear (Pietruska 2009: 99). By the time of the Civil War, the Smithsonian network looked like the most likely candidate for an agricultural weather service.

However sophisticated Henry's network was on the eve of the Civil War, the war itself disrupted Henry's network beyond repair. Telegraph lines between North and South were cut, and Union lines were appropriated for military uses. The war highlighted the degree to which the integrity of "American" meteorology was tied to the very definition of "America" in general. During the Reconstruction Era, Henry, his network gone, had to rely on the newly formed Department of Agriculture for weather observations, and even that department had difficulty furnishing observations (Fleming 1990: 146–50).

1870–1919: The Signal Service, the Weather Bureau, and the Political Contingency of Data

The Civil War interrupted weather networks for approximately 10 years. Fleming has designated the period from 1870 to 1920 the era of government service in meteorology (1990: xvii). Whitnah, in his institutional history of the Weather Bureau, has identified the administration of Chief Willis Moore, 1895 to 1913, as a period of expansion of public services (1961: 82–101). Mergen (2008) characterizes the US Signal Service and its successor in 1891, the US Weather Bureau, as "marketers" of weather services to a very skeptical public. And Pietruska describes the public during this period as characterized by "popular and scientific preoccupation with prediction" (2009: 1).

The development of meteorology as a "government service" in the gilded age was embedded in a larger discourse about prediction and uncertainty. Pietruska (2009) describes the "crisis of certainty" that historians have identified in the latter half of the nineteenth century. The same technology that enabled weather prediction – the telegraph – also changed the way communities understood the connection between their future and that of the nation (Pietruska 2009: 32). Indeed, Anderson has observed that the synoptic weather maps produced by the Signal Service in the 1870s and 1880s indicated an "official, tightly-coordinated global enterprise, staffed by observers who were subordinated to the national leadership of a chief meteorologist." The former volunteer observers were subordinated to salaried military officers, and the Signal Service was "without parallel in the European systems" (2006: 85).

At the same time, the rigid centralization of the Signal Service, and its tight discipline and control, were resisted by a number of different interests. Agricultural interests, in particular, felt that the needs of farmers, many of whom did not have access to telegraphs, were not being met by the Signal Service, which had been set up to give storm warnings on the Great Lakes. The uncertainty among agricultural organizations about markets conditioned the developments and controversies surrounding the control of the Signal Service. The transfer of the service from the US Army to the Department of Agriculture had a great deal to do with its failure to reach farmers without access to telegraph services (Pietruska 2009: 97–138). According to this interpretation, the development of an observation network at the national level occurred to the detriment of local communities.

Another side to the story, however, is the response to public demand at the Weather Bureau under the leadership of Willis Moore. Pietruska (2011) argues that the Weather Bureau sought to distance itself from traditional methods of long-range forecasting, first by trying to discredit it, and then by rationalizing its uncertainty by transitioning from a "culture of certainty," in which forecasters could be dismissed for too many inaccuracies in their forecasts, to a culture of probability, in which uncertainty was embraced but rationalized (Pietruska 2011: 104). At this point, meteorology at the Weather Bureau was driven primarily by the demands of agriculture, and private meteorological observatories were conducting upper air observations. By the end of World War I, though, observations of the upper atmosphere would begin to take center stage, as air travel became a greater feature of the American transportation infrastructure, both civilian and military.

1900–1939: Climatology, Long-Range Forecasting, and the Persistence of Agricultural Interests

Although meteorology appeared to be abstracting itself from local circumstances, climatology was, in many ways, maintaining its connection to regions, albeit in ways that made meteorologists and geographers uneasy. Nebeker has argued that the proliferation of data gave a "continuing push to the new science of climatology" and ultimately made climatology a statistical science. To some extent, this meant the creation of climatology as a subset of meteorology, as an atmospheric science. In his *Handbook of Climatology* the Austrian meteorologist Julius Hann wrote, "'By climate we understand the totality of meteorological phenomena which characterize the average state of the atmosphere at some place on the earth's surface'" (quoted in Edwards 2010: 63). Edwards (2010: 63) has identified the publication of Hann's *Handbook* in 1883 as the beginning of professional climatology, and Robert DeCourcy Ward's English translation of the *Handbook* in 1903 facilitated its transition to the United States (Miller 1988: 353). But the appeal of climatology lay in its potential to facilitate long-range forecasting, as well as its utility in identifying regions of agricultural productivity.

The possibility of analyzing regions of productivity was greatly facilitated by the climatic classification system of Wladimir Köppen. Because Köppen's classification relied only on numerical data, geographers were able to use Weather Bureau data to map out climatic regions of the United States. Even more importantly, though, Köppen's classification enabled the mapping of the *frequency* with which different areas fell into a particular climatic category. This was absolutely crucial to understanding variation in climate. According to the geographer R.J. Russell,

> Climate, as ordinarily considered, implies long experience. With reference to a particular place, it is usually expressed in the form of statistical averages and departures. These may be fitted into quantitative definitions and thus be used to outline climates in the distributional sense. The climatologist interested in this phase of the subject usually tries to delimit regions on the basis of records extending back through the greatest possible number of years. In doing so he tacitly assumes a fixity of climatic boundaries that is open to question (Russell 1934: 92).

Climatic boundaries were in no way fixed, argued Russell. Statistical analysis of weather data facilitated the identification of "transition zones" – zones that had been deemed unsuitable for agriculture.

Long-range forecasting was an area of considerable ambivalence for meteorologists but of great interest for policymakers and farmers alike. The possibility of identifying cyclical regularities in climatic fluctuation held considerable allure for those interested in stabilizing fluctuations in agricultural markets. Additionally, the tools for such analyses proliferated in the first 30 years of the twentieth century. C.E.P. Brooks wrote in the preface to his 1926 survey *Climate Through the Ages*, it would not have been possible, 20 years previous, to write a history of the previous 5000 years, given the advancements in those 20 years (Brooks 1970: 12). Studies of tree ring widths by A.E. Douglass (1919) of the University of Arizona, and the analysis of pollen in peat bogs by Paul Sears (1935), made it possible to determine variations in rainfall and temperature over a much longer period of time. Nebeker (1995: 94) has argued that meteorologists were abandoning the search for cycles by the 1930s, although cyclical analysis still enjoyed political support in a number of offices in the Department of Agriculture.

Understanding changes in climate over such a long time facilitated the explanation of historical events in terms of climate, and it also held out the hope that climatic cycles might be found that would allow long-term weather forecasting. The Carnegie Institution (1929) is an example of the conferences and publications on the study of cycles that it sponsored, and several of the annual meetings of the National Academy of Sciences included panels on climatic change. Geographers such as Ellsworth Huntington (1922) and Ellen Churchill Semple (1911) relied on the earlier work of Friedrich Ratzel to find the key to differences between civilizations in climatic change and regional differentiation in climate.

1919–1945: The Upper Atmosphere and the Change in Focus of Meteorology

Developments in meteorology in the nineteenth century had been primarily a matter of increasing the network of ground-based observations. The 1900s saw substantial interest in the upper atmosphere. In the United States, Willis Moore, chief of the Weather Bureau, wrote,

> Having reached the highest degree of accuracy possible with our present instrumental readings, it becomes necessary to invade new realms if we desire to improve the character of the forecast and to make it of greater utility (quoted in Monmonier 1999: 69).

The development of the box kite and instruments that could be mounted on the kites facilitated these upper air measurements (Monmonier 1999: 72–4). But it was the rise of air travel, for civilian and military purposes, that necessitated measurements and three-dimensional modeling of the atmosphere.

Interestingly, it was in Norway that many of the techniques for modeling and forecasting for aviation were pioneered. With the loss of weather data from the countries at war during the time, the Norwegian meteorological service had to develop new methods for forecasting the weather. Led by Vilhelm Bjerknes, a former physicist who sought to apply his expertise in hydrodynamics to meteorology, members of the emerging "Bergen School" of meteorology developed a dense network of weather stations in Norway. The observation system developed out of necessity more than anything – World War I had cut the country off from data from the rest of the world, and German interference with shipping channels constricted its food supply. However, with such a dense network, they were able to identify discontinuities in pressure, temperature, and wind direction; with their observations of cloud formations such as the "squall lines" that aviators had already learned to identify as indicators of storms, they were able to identify similar discontinuities in the vertical dimension. The result: a meteorology that modeled "fronts," confrontations of two air masses, one colder, one warmer, that had migrated from different regions of the world (Friedman 1989).

To many reformers of meteorology, implementing a forecasting method that depended on the analysis of air mass movements required the development of a meteorology that was based more solidly on physics and mathematics. The absence of mathematical rigor in weather forecasting had been a matter of insecurity among meteorologists since the nineteenth century (Harper 2006; Nebeker 1995: 40–2). The possibility of making the science "more like physics" was enticing to many, even though there were other meteorologists, particularly in the Austrian empire, who saw more promise in statistical forecasting (Coen 2007: 289–90).

But the possibility of exporting the Bergen School to the United States did not depend merely on the prestige of the school. It depended, also, on the place-based expertise of the practitioners. Turner (2010: 35–71) has noted the connection between the increasing importance of air travel for civilian and military purposes, the consequent strategic importance of the polar region as a faster flight route, and the entry of Bergen-trained meteorologists into American meteorology. The Bergen School had expertise in both. First, the vertical dimension of the atmosphere had been an integral part of its weather forecasting. Second, a number of its members had participated in expeditions toward the North Pole and had lived in conditions that prepared them for the expedition (Turner 2010: 30). The experience of Bergen School meteorologists in working with aviators, the military, and forecasters positioned them well to influence meteorological practice (Turner 2010: 11).

Despite the experience of the Bergen School forecasters in aeronautical and polar endeavors, importing them was not so simple. In part, this was because the term "import" is not entirely metaphorical. The Weather Bureau controlled the main data-gathering network in the United States, and the Bureau was skeptical of Bergen techniques. This was not mere conservatism. The Bureau had invested a great deal in a standardized data network that gathered information for weather on the ground, a network that was well geared to the agricultural interests its parent department served (Turner 2010: 103). It was also underfunded, and Congress was slow to appropriate additional funds for upper air data collecting (Harper 2008a: 31). Finally, as Roger Turner has discussed, the Weather Bureau was rightly skeptical of new methods of forecasting, given the number of "amateur" weather services that claimed to provide better forecasts than the Weather Bureau but nevertheless did not base their predictions on meteorology (Turner 2010: 109).

A number of events began to strengthen the case for importing Bergen techniques for forecasting, or at least reforming the forecasting practices of the Weather Bureau. First was the risk of air travel. Shipwrecks and storms fueled some of the earlier inquiries into meteorology; planes were even more vulnerable to weather hazards. The period saw accidents, such as the crash of the rigid airship USS *Akron* in 1933, that were traced back to poorly forecast weather conditions (Harper 2008a: 28).

Bergen meteorology was introduced through other channels, as well. Francis Reichelderfer, for instance, mounted an effort to improve education in meteorology for naval officers. Reichelderfer lobbied Harvard to create a meteorology program, and when that failed, successfully lobbied MIT. The first professor of meteorology to be hired at MIT, Carl-Gustaf Rossby, was grounded in the Bergen School and would ultimately develop one of the most important innovations in modeling the general circulation of the atmosphere: Rossby waves, tongues of low pressure extending downward from the poles that moved in a period of one to two days, generally the time between substantial changes in the weather (Turner 2010: 112–13).

Additionally, the development of new science advisory bodies to the US government during the New Deal brought academic interests into conjunction with policymakers and government agencies. The Science Advisory Board was set up by executive order in 1933, and the first committee it established was one to reform the Weather Bureau (Harper 2008a: 28–32). The committee, composed of Caltech president Robert Millikan, MIT president Karl Compton, and the geographer Isaiah Bowman, recommended the adoption of air mass analysis techniques in weather forecasting. It had the ear of the president and the secretary of agriculture, and its recommendation to replace the Weather Bureau chief was implemented; it was hamstrung by lack of funding during the Great Depression (Harper 2008a: 32–4).

The major watershed in the organization of American meteorology was World War II. The heavy dependence on aerial warfare necessitated a massive increase in staff to make forecasts for military operations. The famous D-Day forecast enabled the Allies to catch the Germans unprepared; the Allies predicted a small opening in the generally poor weather, while the Germans forecast no such opening in the weather and therefore did not expect an invasion on that day (Fleming 2004: 59; Turner 2010: 141). Academic meteorologists, especially at MIT, were hired to teach the cadets, many of whom had come from mathematics or physics. The training effort infused thousands of meteorologists into a community that had numbered only a few hundred before the war, and these new forecasters looked at forecasting as a matter of technical expertise, rather than long experience (Harper 2003: 670).

World War II left a number of legacies for individuals in the postwar era, particularly a denser upper air network and a large corps of meteorologists. What it did not leave was a clear mission for these institutions (Harper 2003: 689). A possibility, however, was opened up through another legacy of the war: the electronic computer (Harper 2003: 671). Computing power had been a major obstacle to dynamic weather forecasting using differential equations. Even the Bergen School relied primarily on maps and diagrams for its forecasts, rather than algorithms and equations (Edwards 2010: 88; Friedman 1989: 155–69; Nebeker 1995: 57). The one attempt that had been made to forecast the weather using equations – Lewis Fry Richardson – took six months, and it was a forecast for six hours into the future. *And* the forecast lacked accuracy – Richardson ended up with a value of atmospheric pressure of 145 millibars when the actual change was less than one (Lynch 2006: 119; Nebeker 1995: 76).

1945–Present: From National to Global

The advent of the digital computer presented numerous options that meteorologists had not had before, and Nebeker (1995: 2–3) describes the use of the digital computer as the point when climatology, forecasting, and physical meteorology all converged into one science with a common set of practices. Once again, however, the application of computer technology depended on the active participation, and political negotiation, of meteorologists. Harper (2008a) has argued that the key to the development of numerical weather forecasting was the collaboration of forecasting expertise and a theoretical understanding of meteorology. In this case, the tension between forecasting and theoretical meteorology, although still present, proved productive both in providing public services and as an effective tool for dynamic meteorologists.

Harper (2003: 688–9) observes that, in the United States, expertise in forecasting and theoretical meteorology – dynamic meteorology – rarely intersected before World War II Forecasters, many of whom worked for the Weather Bureau, depended more on a "feel" for the weather, based on long experience examining synoptic maps and forecasting on that basis, rather than on calculating it. Theoretical meteorologists, on the other hand, had little experience in forecasting. It was the entry of meteorologists from outside the United States, especially meteorologists from the Scandinavian countries who had been trained in Bergen School techniques, that brought these fields together. These meteorologists prided themselves on a grounding in theoretical models of the dynamics of the atmosphere, as well as on-the-ground forecasting (see also Harper 2008a, chapter 5).

The Meteorology Project used data gathered on the national level, if only because the computers could not handle more data than that (Harper 2003: 687). But numerical weather prediction served as a gateway for models of the general circulation of the atmosphere and ultimately global climate (Edwards 2010: 111). The computer only provided the starting point, though. Modelers had to deal with far greater obstacles before their models could be considered "global," starting with their data. The development of national observation networks allowed for data sharing among countries, but data gathering tended to follow national priorities. The *persistence* of infrastructure necessitated the investment in models that would normalize the data being collected (Edwards 2010: 25). The atmosphere might lack borders, but Weather Bureau stations reported to Washington, not Ottawa or Mexico City.

This "nationalization" of climate had, itself, been a substantial shift in the understanding of climate, which, for most of its conceptual history, was thought of as a way of understanding regional differences. At the same time, the concept of "circulation" served as a way to understand the interconnectedness of different cultures that were part of the same political unit (Coen 2010) and disparate regions of the world (Jankovic 2010a). By the first half of the 20th century, climatologists such as the Austrian immigrant to the United States, Victor Conrad, began including general circulation in their books on climate, even in the 1930s (Edwards 2010: 71). After World War II, that concept of circulation took on even more importance. Before the rising temperature of the globe was a broad concern, the circulation of fallout from nuclear tests provided a clear justification for better modeling of the circulation of the atmosphere over the globe (Edwards 2010: 208). Additionally, Cold War politics, and the concern about Communist encroachment in many corners of the world, meant that global knowledge was paramount (Edwards 2010: 189).

Another factor in global climate modeling, of course, was the Cold War diplomacy of the United States, of which science was a substantial part. Miller (2001: 167) has examined the coproduction of advances in meteorological observation – in the years following World War II, it became possible to view the global atmosphere in its entirety – and the Cold War foreign policy that linked science to the postwar order. The technical assistance provided by the World Meteorological Organization in Libya after decolonization, for instance, was intended to guide its transition to statehood as well as provide it with meteorological assistance (Miller 2001: 193). The Antarctic Treaty System was another example of the role of science as diplomacy by other means, except in this case, the "frozen" territorial rights provided by the treaty to its members turned Antarctica into a laboratory for "elite" nations, while excluding non-members (Howkins 2011: 196–7). Likewise, global observation systems, such as World Weather Watch, depended not only on the development of satellite technology but also on the competition between Cold War superpowers to fund the project (Edwards 2010: 231).

As Edwards (2010) has argued, the creation of *global* weather, and *global* climate, also reinforced, perhaps more than most institutions, international cooperation. Although participation in the system of weather observations was voluntary at first, it became a quasi-obligatory system based on the shared permanent infrastructure that developed. The material culture of modeling, the normalization of data, and a whole host of other practices, made it difficult for countries to *leave* the cooperative network once they were a part of it.

From Global Back to Local?

The atmosphere is a global entity. But meteorology and climatology had to be *made* into that, from the description of local airs, waters, and places in the eighteenth and nineteenth centuries, to national weather observation and forecasting networks in the late nineteenth and early twentieth centuries, and finally to the analysis of movements of large air masses and ultimately the general circulation of the atmosphere. Much of this required the lobbying of stakeholders, including American agricultural and shipping interests, airlines, the military and intelligence agencies, and meteorologists themselves. A large amount of this movement had to do with the importance of data gathering to meteorology, and advances in communication and transportation that enabled faster circulation of that data.

At the same time, a number of counter-narratives to the "globalization" of the atmosphere have emerged. James Fleming and Vladimir Jankovic's *Osiris* offering, "Klima," endeavors to recapture the older conception of climate as a "biospatial frame" rather than the standard definition now, "average weather" (Fleming and Jankovic 2011: 2). Climate change still looms over many of the articles in that *Osiris* volume, but the issue is often refracted through national and local priorities, such as drought (Morgan 2011), arctic science (Sörlin 2011), and regional weather (Endfield 2011). Additionally, Mike Hulme, in a now-famous essay on climate change, has argued that identifying climate change with rising global average temperature is a barrier to a good public understanding of climate change (Hulme 2008).

Even in government atmospheric science, data gathering was less homogeneous than the Weather Bureau network might have shown. The Works Progress Administration did not just fund the punch card tabulation of climatic data; it also sponsored small networks of observers in drought and flood-stricken regions. These networks proved shorter term than originally intended, but the resulting data invited interpretations that would eventually be integrated into the field of microclimatology. The climatologist who ran that project, C.W. Thornthwaite, went on to consult for agribusiness firms, where he set up smaller scale networks to collect data on water availability of farms and the need for irrigation. Access to the firms' proprietary weather measurements was an advantage his consultancy cited in luring military contracts for microclimatology (Bergman 2013, chs. 3 and 5). While this in no way contradicts the narrative of "infrastructural globalism" or permanence, scaling back to local scientific practices opens up new questions about the relationship of the public to the climate, and indeed, redefines the relationship between weather and place. Indeed, the geographical concept of climate and "place" becomes much more complicated in a world of global data and exchange.

Additional examples abound, from the pre- and postwar periods. Although many weather consultants were denounced as outright quacks, others were credited as meteorologists. Henry Helm Clayton, for instance, developed a system of weather forecasting from cyclical patterns that were surprisingly accurate (Pietruska 2010: 89). Likewise, General Electric funded a substantial amount of work on weather modification in the 1950s (Fleming 2010: esp. 137–164). And the Travelers Insurance Companies started a Weather Research Center in the 1950s that precipitated a significant amount of research into the atmospheric science for a variety of clients (American Meteorological Society 1964).

Of particular importance in an age of ever increasing information and data insecurity, however, is the challenge to the model of meteorology and climatology based on data-gathering networks. Randalls (2010), for instance, has examined the commercialization of the British Meteorological Office and the emergence of the weather derivatives community. The emergence of these new models has conditioned new standards of data gathering and dissemination, one in which actors are less interested in whether the data was "accurate," that is, corresponded to the real conditions that emerged in the atmosphere, than whether the data have allowed them to forecast the market reaction to that data. According to Randalls (2010: 716), "[d]ata become facts by virtue of being accepted by prominent actors within the weather derivatives community. This accords with the notion that deciding which data are acceptable is done through social and economic interactions, and not through a purely rational scientific process." For Randalls and his co-editors of a special *Social Studies of Science* volume entitled "STS and Neoliberal Science," these developments illustrate the emergence of "neoliberal science," loosely defined as an organization of science based on market incentives and funding structures (Lave, Mirowski, and Randalls 2010). Weather and climate have thus become more fragmented; not necessarily "local," but not necessarily reflecting an image of the general circulation of the atmosphere, either. One might also examine the work of the Climate Corporation, just recently purchased by Monsanto. The Corporation uses the free data from the National Weather Service to provide forecasts and crop insurance to farmers (Hardy 2013; see also Bergman 2013: 256).

* * *

Existing scholarship has rightly focused on the co-creation of atmospheric, political, conceptual, and infrastructural spaces. New communications technologies, more extensive infrastructure, more powerful methods of computing, all undergirded by hard-fought political relationships, enabled the visualization and analysis of the atmosphere on regional, national, and ultimately global scales. The young field of the history of the atmospheric sciences has only begun, however, to understand how these pictures of the atmosphere mediate the everyday collective experience of weather. More importantly, the examination of the way broad-based networks have mediated perceptions of the weather has paved the way for a complementary question: How do the extant and new fissures and barriers in these networks, whether barriers of secrecy, finance, or infrastructure, shape our collective debate over the future of the climate? If much of the emphasis in the last 25 years of scholarship has traced the increase in scale of these networks, the next few years will zoom in on their local cleavages and conflicts.

Bibliographic Essay

Overviews of the history of the atmospheric sciences, American or otherwise, are scarce. Brant Vogel, "Bibliography of Recent Literature in the History of Meteorology: Twenty Six Years, 1983-2008" (2009), should be the first stop for anyone interested in the history of meteorology in any part of the world. As an overview for chronologies and dates, Aleksandr Khrgian's *Meteorology: A Historical Survey* (1970), is adequate. James Fleming's *Historical Perspectives on Climate Change* (1998), although never intended as an overview of the atmospheric sciences, spans 300 years of climate science and provides useful context for anyone interested in the subject. Finally, the *Sciences of the Earth: An Encyclopedia of Events, People, and Phenomena,* edited by Gregory A. Good (1998), is a useful resource for many topics in meteorology.

A large part of this chapter has dealt with problems of scale in meteorology. The volume dealing with this issue most directly is *Intimate Universality: Local and Global Themes in the History of Weather and Climate,* edited by James Fleming, Vladimir Jankovic, and Deborah R. Coen (2006). Other useful sources on this topic include Mike Hulme, *Why We Disagree About Climate Change: Understanding Controversy, Inaction and Opportunity* (2009); articles by Deborah Coen, including "Climate and Circulation in Imperial Austria" (2010) and "Imperial Climatographies from Tyrol to Turkestan" (2011); and Vladimir Jankovic and Michael Hebbert, "Hidden Climate Change – Urban Meteorology and the Scales of Real Weather" (2012).

For the broader cultural implications of weather and climate, Benjamin Strauss and Sarah Orlove's edited volume, *Weather, Climate, Culture* (2003), contains a variety of treatments of that intersection. Classic works on the intersection of climate and history over the *longue durée* include Clarence Glacken's *Traces on the Rhodian Shore: Nature and Culture in Western Thought from Ancient Times to the End of the Eighteenth Century* (1967); Emmanuel Ladurie's *Times of Feast, Times of Famine: A History of Climate Since the Year 1000,* translated by Barbara Bray (1971); and Hubert Lamb, *Climate, History, and the Modern World* (1982). Cultural histories of early modern weather and climate include Vladimir Jankovic's *Reading the Skies: A Cultural History of English Weather, 1650–1820* (2000) and *Confronting the Climate: British Airs and the Making of Environmental Medicine* (2010b); and Jan Golinski's *British Weather and the Climate of Enlightenment* (2007) and "American Climate and the Civilization of Nature" (2008). Post-1800 cultural histories include William B. Meyer, *Americans and Their Weather* (2000), and Bernhard Mergen, *Weather Matters: An American Cultural History Since 1900* (2008).

As in most histories of science, a great deal of the literature on the history of the atmospheric sciences has dealt with the professionalization of the field. Katharine Anderson's *Predicting the Weather: Victorians and the Science of Meteorology* (2005) deals head-on with a surprising, but almost ubiquitous, tension between the science of meteorology and the service of weather forecasting. Anderson focuses on Victorian Britain. Works that examine the professionalization of meteorology at other places and times include James Fleming's discussion of nineteenth-century American meteorology, *Meteorology in America, 1800–1870* (1990); Robert Marc Friedman's discussion of the rise of the influential Bergen School in *Appropriating the Weather: Vilhelm Bjerknes and the Construction of a Modern Meteorology* (1989); Kris Harper's discussion of the use of computer modeling in meteorology post World War II in *Weather by the Numbers: The*

Genesis of Modern Meteorology (2008a); and Frederik Nebeker's overview of twentieth-century meteorological theory and practice, *Calculating the Weather: Meteorology in the Twentieth Century* (1995). Also useful is a collection of essays written largely by practitioners who experienced the period of professionalization, *Historical Essays on Meteorology, 1919-1995: The Diamond Anniversary History Volume of the American Meteorological Society*, edited by James Fleming (1996).

Those interested in the observation and data networks that were so crucial to the sciences of meteorology and climatology would again do well to examine Fleming (1990) and Friedman (1989). Those interested in early meteorological observation should consult Klaus Wege and Peter Winkler, "The Societas Meteorologica Palatina (1780–1795) and the Very First Beginnings of Hohenpeissenberg Observatory" (2005). For discussions of later observations, particularly satellite imaging, see Mark Monmonier, *Air Apparent: How Meteorologists Learned to Map, Predict, and Dramatize Weather* (1999); Pamela Mack, *Viewing the Earth: The Social Construction of the Landsat Satellite System* (1990); and Eric Conway, *Atmospheric Science at NASA: A History* (2008).

For scholarship on modeling, calculation, and prediction, Nebeker (1995) and Harper (2008a) are again important works, and Paul Edwards' *A Vast Machine: Computer Models, Climate Data, and the Politics of Global Warming* (2010) has set the standard for work on climate modeling. Those interested in the evolution of meteorological modeling in the nineteenth century should consult Gisela Kutzbach's *The Thermal Theory of Cyclones* (1979). Those interested in consulting climate modeling should consult the contributions to the forum "Changing Climate – Modeling Climate," *Historical Studies in the Physical and Biological Sciences* 37, 1 (September 2006). Cultural histories of prediction and probability in the late nineteenth and early twentieth centuries include Jamie Pietruska "Propheteering: A Cultural History of Prediction in the Gilded Age" (2009) and Deborah Coen, *Vienna in the Age of Uncertainty: Science, Liberalism, and Private Life* (2007). Gary Fine, *Authors of the Storm: Meteorologists and the Culture of Prediction* (2007), provides an ethnography of current forecasting practices at the National Weather Service.

Unsurprisingly, much of the literature on current work in meteorology and climatology deals with climate change. For overviews on the discovery of global warming, Spencer Weart's *The Discovery of Global Warming* (2008); Fleming (1998); and Fleming, *The Callendar Effect: The Life and Times of Guy Stewart Callendar (1898–1964)* (2007), all discuss important aspects of the path to discovery. For discussions of the politics of the climate change debate, see, Naomi Oreskes and Eric Conway, *Merchants of Doubt: How a Handful of Scientists Obscured the Truth on Issues, from Tobacco Smoke to Global Warming* (2010) and *Changing the Atmosphere: Expert Knowledge and Environmental Governance,* edited by Clark A. Miller and Paul Edwards (2001). And a now-classic article by Brian Wynne and Simon Shackley, "Representing Uncertainty in Global Climate Change Science and Policy: Boundary-Ordering Devices and Authority" (1996), examines the ways in which uncertainty has been represented in the climate change debate.

Chapter Fourteen

MOLECULAR AND CELLULAR BIOLOGY

Lijing Jiang

Today, academic departments of life sciences around the world often combine the study of the cell and that of biomolecules into single research programs or clusters with the overarching name of "molecular cell biology." The textbook, *Molecular Biology of the Cell* (Alberts 1983), has been among the canonical textbooks for undergraduate students of life sciences since the 1980s. Such naming practice, to some extent, reflects a reality that cell biology and molecular biology are usually intertwined concerns when it comes to research and teaching. The two fields nevertheless emerged through parallel and relatively independent paths in their historical formations. Integration of the two fields did not happen until molecular biology gained significant momentum in the 1970s, especially with the promotion of cancer research and a wider utilization of technologies of radioisotope labeling, immunofluorescence, recombinant DNA, and cell cultures. This chapter reviews the separate emergences and developments of modern cell biology and molecular biology before making tentative suggestions about how technologies and research interests since the 1970s led to more direct contacts and interactions between the two fields that led to the "molecular cell biology" of recent years. As the developments of the two fields, particularly that of molecular biology, had been well studied by historians, each section also attempts to introduce relevant historiographical debates, or the lack thereof. At the end of each section, I will also suggest several characteristics of American developments in these areas and point to potentially fruitful questions for future studies.

The Cell and the Emergence of American Cell Biology

Although morphological structures of what we now recognize as cellular vesicles and cell walls already became apparent under simple microscopes during the seventeenth century, early users of microscopes were usually bewildered by what they observed.

A Companion to the History of American Science, First Edition.
Edited by Georgina M. Montgomery and Mark A. Largent.

Robert Hooke, who famously coined the word *cell* to describe the dead cork structures as recorded in his 1665 *Micrographia*, engaged in extensive descriptions of a range of structures, including a 17-inch drawing of a gnat, in the same book (Hooke 1964). To see cells as basic subunits of life only took place almost two centuries later and had more to do with philosophical changes than technological advances. In the early nineteenth century, botanist Matthias Jakob Schleiden and zoologist Theodor Schwann independently proposed the cell as a universal basic component for plants and animals, respectively. The combination of the two proposals became what is now known as the cell theory.

Although reductions of chromatic and optical aberrations had been achieved in the finer craft of the compound microscope since Hooke, the highly speculative German *Naturphilosophie* that attempted to give general theoretical structure to nature catalyzed the formulation of the cell theory in more important ways. Champions of the cell theory not only described cells as subunits, but also suggested how the formation and activities of the cell follow physiochemical regularities. Schwann, for example, asserted that the cell must be the ultimate seat of metabolic activity. He also speculated that cells were self-generated from "cytoblastema," a hypothetical, formless organic matter. Although Schwann and Schleiden intentionally distanced themselves from *Naturphilosophie*'s vitalism by suggesting that physical and chemical processes, such as organic crystallization, were forces behind the formation of new cells, the very concept of "cytoblastema" nevertheless smacked of the kind of wild speculations characteristic of *Naturphilosophie* (Coleman 1971). Others continued to investigate about cellular structures, and by the end of the twentieth century, chromosomes, cell membranes, and the basic process of cell division had been observed, reported, and discussed (Harris 1997: 128–65). Pathologist Rudolf Virchow's axiom *Omnis cellula e cellula* (All cells come from cells) eventually replaced Schwann's vitalist explanation about cell formation. At the time, however, studies of the cell were mostly concerned with morphological changes of the cell in division, development, and diseases. Since little work was done on the chemical changes that were associated with the morphological changes being observed, historians saw the field of cell study formed at the end of the nineteenth century as part of a traditional cytology.

At the turn of the twentieth century, such cytological work was picked up and incorporated into studies of embryology, genetics, and evolution by American biologists, at a time when professionalization and institutionalization of experimental biology were taking place. In view of the larger change, Garland Allen, in his *Life Sciences in the Twentieth Century* (1978a), famously argued that American biologists of the early twentieth century had revolutionarily turned away from a European morphological tradition and shifted to an experimental focus in their work. Allen's much-debated "revolt from morphology" thesis provided a takeoff point from which other historians discussed continuities of American experimental biology from traditions of morphology and natural history (Maienschein, Rainger, and Benson 1981; Rainger, Benson, and Maienschein 1988). Regarding the study of the cell, Jane Maienschein (1991b) showed that American cytologists had actually borrowed ideas, methods, and questions from European morphological traditions. The series of studies done by German biologist Theodor Boveri entitled "Zellenstudien" had inspired American biologist Edmund Beecher Wilson's *The Cell in Development and Inheritance*, an influential book that documented the cellular and nuclear changes in fertilization and early embryogenesis with meticulous drawings (Wilson 1897). Bringing cytology to embryological questions and taking issue with Ernest Haeckel's recapitulation argument, Wilson, Charles Otis Whitman, and Edwin Grant Conklin investigated detailed cell fates and cell lineages during the development of leeches, earthworms, and a range of marine organisms such as *Nereis* and *Crepidula* during their summer research at the Marine Biological Laboratory, Woods Hole. Ross Harrison at Yale University around 1907 carried out the first tissue culture experiment with frog nerve fibers, which has since been recognized as a "crucial experiment" that definitively demonstrated that the nervous system was composed of cells, showing the superiority of Santiago Ramón y Cajal's neuron hypothesis over Camillo Golgi's reticular hypothesis. These American cytologists sought for a middle ground between morphological and physiological traditions, while attempting to raise hypotheses about specific questions with sound evidence, a kind of "positive knowledge," other than the kind of grand theories that were the hallmark of *Naturphilosophie*. In doing so, they applied a variety of questions from genetics, embryology, and evolution, and sought to answer these questions through the study of the cell.

The American practice of cytology was given sufficient space to grow in the country's academia that provided more positions than its counterpart in Europe and allowed a range of topics to develop at the turn of the twentieth century. A new generation of American cytologists trained in such an environment developed diverse, somewhat fragmented research interests about the cell. In 1925, Wilson's third edition of *The Cell* came out with the bulk of 1232 pages, almost three times the size of its 483-page second edition. By the 1920s, cytology had become so expansive and diverse in its terrain that perhaps no one except Wilson would be adequate to single-handedly author a comprehensive, definitive monograph to review the whole field. The Marine Biological Laboratory at Woods Hole, Massachusetts, an appealing place for a summer resort, provided an important annual meeting spot for American cytologists to exchange ideas and to collaborate, although their interests diverged at the time. In 1924, Edmund Vincent Cowdry, a Canadian cytologist, collected a series of original contributions on topics including the heredity, differentiation, chemistry, physical structure, permeability, subcellular organs, and behaviors of the cell and published them with a multi-authored volume *General Cytology* (Maienschein 1991a). The contributors of the influential volume had met at Woods Hole in the summer of 1922 to discuss their work. Many of them viewed cytology as a kind of morphology which valued precise descriptions of the internal structures of the cell in different species, tissues, physiological conditions and environments, and thus called for contributions from different specialties such as those in anatomy, physiology, and pathology. The authors nevertheless also included biologists with more experimental bent such as geneticist Thomas Hunt Morgan and embryologist Edwin Conklin, who further widened the disciplinary span of the volume. Cowdry himself had envisioned multidisciplinary cooperation a key to the success of the volume and had asked contributors to complete their chapters according to the general outline of the book resulting from previous group discussions. The collaborative, interfiled style of book editing and disciplinary vision also influenced how Cowdry approached the emerging topic of biology of aging when he organized the Woods Hole Conference on Aging in 1937 (Park 2008).

In the 1920s, basic processes of cell division, structures and functions of cell membranes, mitochondria, and the Golgi apparatus were all established based on observations under the light microscope. However, dissentions regarding whether these structures were artifacts under the microscope still abounded. It was not until the combined applications of ultracentrifugation and electron microscopy became more available that the fine structures and functions of these subcellular organelles were further elucidated. William Bechtel (2006) argues that the bona fide modern cell biology only began to emerge once the wider application of ultracentrifuge and electron microscope brought biochemistry into contact with old-fashioned cytology in the 1940s. By then, biochemists had already discovered a host of reactions in cellular respirations such as glycolysis, alcoholic or lactic acid fermentation, and aerobic respiration. They realized that cellular and organelle membranes offer particular locations for certain reactions to take place. By the early 1930s, the centrifuge had been developed sufficiently to be available for use in separating subcellular components. The technique evolved further in the 1950s into multiple centrifugations at increasing speeds aided with sucrose gradient. The ultracentrifuge helped biochemists to locate where different respiration reactions took place within the cell. It was also through such high-speed centrifuge that Albert Claude working at the Rockefeller Institute in the late 1930s obtained cell fractions that contained what he believed as cancer-causing particles, whose sizes and shapes Claude examined further with the electron microscope in the 1940s. These particles eventually turned out to be ribosomes, cellular organelles responsible for protein synthesis that cancerous cells happened to have a great number of. Although Claude left the Rockefeller Institute in 1949, the fame of his work on ribosomes piqued further developments on cell biology at the Institute. The Institute eventually trained a number of scientists who used ultracentrifuge and electron microscope readily to unveil the fine structures of subcellular organelles and detailed localization of where their functional processes took place. Well-known achievements included Keith Porter's work on endoplasmic reticulum in connection to ribosomes and protein synthesis, George E. Palade and James Jamieson's work on Golgi apparatus and special structures of the mitochondrial enzyme systems, and Porter and Palade's work on the exact morphology of mitochondria (Bechtel 2006: 162–257).

Institutional developments in cell biology followed the flurry of research on subcellular organelles. In 1962, the American periodical, *Journal of Biophysical and Biochemical Cytology,* changed to its current title, *Journal of Cell Biology.* The field's society, the American Society for Cell Biology, was established in the same year. Although these institutional developments helped to shape a disciplinary and professional niche for cell biology, in Bechtel's formulation, cell biology had a philosophical underpinning provided by the intellectual product of its research. That is, modern cell biology aims to look for cellular mechanisms, which Bechtel defines as the "structures performing a function in virtue of their component parts, component operations, and their organizations." Combination of ultracentrifuge that could separate cellular fractions (as component parts), biochemical analysis (to elucidate component operation), and microscopic visualization of fine structures proved to be extremely useful for such investigation on mechanisms (Bechtel 2006: 24–44).

What Bechtel left out from his picture of cell biology based on a twentieth-century conception of biological mechanism, however, were a range of phenomena of the cell that did not get decomposed to their components, operations, and organizations. Particularly, the cell culture technologies that were maturing into mass operations in the later part of the twentieth century had played important roles in revealing *in vitro* cellular life, the cell's potentials in growth and duration, its representational power of certain diseases, and the very malleability of life itself. In this regard, Hannah Landecker has shown how cellular display under glass about growth, sex, drinking, and death had triggered scientists' imaginations and generated new understandings, practices, and metaphors about life (Landecker 2001, 2007). The technological spread of the cell culture provides a continuing thread that linked different episodes of cell studies ranging from Ross Harrison's tissue culture to HeLa cell lines, even to animal cloning. Together these examples show the technological settings of cell biology were not only about machines, but also about the handling of biological materials themselves. Another relatively little explored issue is the role of metaphors about cellular life in the making of cellular science. With the exception of a few pioneering works that examine how metaphors ranging from "crystals," "cellular factories," and the more recent biobricks played into articulating ideas and expanding research interests in cell biology, the complex roles played by metaphors in cell biology remains to be fleshed out (Haraway 2004; Reynolds 2007).

The Creation and Early Development of Molecular Biology

The very success of the molecular paradigm of biology has attracted a number of historical works focusing on individual scientists, institutions, materials and methods, as well as controversies of ideas in the history of molecular biology. In view of the large number of related works, it becomes impossible to review the field in one chapter without leaving some important literature out. To effectively make use of space, this section focuses on several historiographical issues about the foundation and development of molecular biology without elaborating on some well-known events. Readers interested in the details of the construction of the double-helix model, the phage group, the molecular work done at the Pasteur Institute, for example, can consult works listed in the bibliographical essay.

Although popular accounts of molecular biology usually held James Watson and Francis Crick's discovery of the double-helix structure of DNA and the construction of the DNA model at Cambridge as a watershed, historians have noted that the significance of the double-helix model in the development of the field was quite ambiguous around 1953 and was only reconstructed as a milestone after the mechanism of protein synthesis became clearer (Olby 2003, also see Gingras 2010). Instead of paying attention to this and other individual discoveries, recent historiographical debates have focused on the roles of patronage, the influences from physicists and biochemists, as well as the particular epistemic roles of research materials of viruses, radioisotopes, and various experimental systems in the making of the disciplinary trajectories (more in the bibliographical essay).

The extent that the Rockefeller Foundation's funding played in the development of molecular biology was a lasting and much-debated issue (Morange 1998: 79–87). Before the 1920s, the Rockefeller Foundation had supplied financial support to general teaching and applications of science. However, the rapidly expanding American academic system soon made it unrealistic for the Foundation to offer wholesale funds to unselected programs. In the 1930s and 1940s, the Foundation started to offer support to individual scientists whose interests fell into selected criteria favored by its officers. One of the Foundation's foremost interests was the kind of biology that would incorporate a scientific rigor similar to physics to reveal the chemical nature of the gene, an approach that Thomas Hunt Morgan had suggested as the future directions for genetics.

Particularly, Warren Weaver, director of the natural sciences division of the Foundation, was responsible for channeling a number of funds to "molecular biology," a term Weaver first coined in 1938. These funds were usually project based, with medium-sized dollar amounts. They were allocated to selected projects in biochemistry, cell physiology, and genetics, often for the use of radioisotopes, ultracentrifuges, x-ray crystallography, and electron microscopes. For example, such grants were given to Niels Bohr for his use of radioactive isotopes in physiology, and to Theodor Svedberg to develop ultracentrifuge for biochemical work. According to Robert Kohler, since the emergence of the "new biology," Weaver was interested in shaping sciences broader than the established approaches of biochemistry, and the Rockefeller Foundation actually funded a number of physical and organic chemists, physiologists, physicists, as well as biologists, among whom traditional biochemists constituted only a small portion (Kohler 1976, 1991: 303–91).

Most historians acknowledged the essential role of the Rockefeller Foundation in shaping an emerging molecular biology. Several have emphasized the importance of the Foundation's support in promoting technologies vital to developments in molecular biology (Bechtel 2006; Rasmussen 1997). Lily Kay argued that the most important aspect of the Rockefeller Foundation's support was nevertheless its facilitation of an *ethos* of scientific cooperation. Projects the Foundation supported cultivated a kind of "cooperative individualists" against the pre-World War 1 laissez-faire state of science and helped to forge the beginning of a big life science based on a hegemonic agenda of cultural dominance by physical science (Kay 1993: 6–11). However, Pnina Abir-Am disagrees with this general assessment and points out that benefactors of the Rockefeller Foundation were usually already established scientists who would have done their projects even without Foundation funding and that the Foundation had actually limited the growing molecular biology field by not funding a Cambridge group of theoretical biologists and embryologists that included Joseph Woodger, Joseph Needham, and Conrad Waddington (Abir-Am 1982).

A second much-debated question was what was the physicists' and biochemists' share in the making of the new molecular life science. For both chemistry and physics, historical actors offered their own narratives about the importance of their own disciplines (Fruton 1992). Interestingly, physicists usually quoted Erwin Schrödinger's *What is Life?*, first published in 1944, as an important inspirational source for their move to tackle scientific questions about life. The book's legacy nevertheless remains hard to analyze. On the one hand, it was a short book that was fun to read and easy to finish. On the other hand, most of the points made in Schrödinger's account proved to be false, even absurd, as molecular biology progressed in later years, such as his analogy that made the stability and mutability of the gene comparable to the dual particle-wave nature of light (Schrödinger 1944: 32–55). A no-less compelling reason for physicists to move toward biology was a growing distaste against being "conscripted" to a science of death that notoriously demonstrated its power in the bombing of Hiroshima and Nagasaki. Nuclear physicist Leo Szilard, for example, had viewed molecular biology as an outlet for pursuing a science of life, and he offered a number of clarifications of ideas in the new biology and helped cross-pollinate these ideas between different research communities (Lanouette and Szilard 1994). Furthermore, physicists offered technical and cognitive expertise, ranging from making and using machines to offering information theory and cybernetics. Physicists and mathematicians Norbert Wiener, Claude Shannon, and John von Neumann in the 1940s, for example, spearheaded the use of cybernetics in thinking about biology. Among others of the RNA Tie Club, a "gentleman's club" formed between physical scientists and biologists who aspired to resolve how RNAs synthesize proteins, George Gamow and Leslie Orgel also used complex calculations to try to "crack the code" (Kay 2000: 73–192). Although physicists did take part in the beginning of molecular biology, historians have debated about whether their contributions were more or less technical, or whether physicists had actually offered important epistemic contributions through scientific reductionism and their information vision (Fuerst 1982; Yoxen 1979). Evelyn Fox Keller, for example, acknowledged their contributions in both areas, but argued that the social authority and authorization that the powerful science of physics seeped into molecular biology were the influences that mattered most (Keller 1990).

Biochemists have been particularly keen to stress the importance of their work for the making of molecular biology, especially their contributions to revealing structures of macromolecules and processes of protein synthesis. Because of the more ambiguous boundaries between biochemistry and molecular biology, biochemists sometimes felt molecular biologists were enjoying public celebrations without giving due credits to biochemists, as epitomized by Erwin Chargaff's remark that molecular biologists were "biochemists practicing without a license" (Abir-Am 1992; De Chadarevian and Gaudillière 1996). At a collective level, there were a number of exchanges of instruments, materials, and experimental systems between self-identified molecular biologists and biochemists in the 1950s and 1960s that led to developments of experimental materials and institutions for molecular biology. For example, Wendell Stanley's vision of a biochemistry laboratory for virus study eventually evolved into part of the department of molecular biology at Berkeley; Paul Zamecnik's rat-liver cell system at the Huntington Memorial Hospital in Boston became the favored experimental system for studying protein synthesis and information transfer even if it might have initially seemed a messy material tailored to biochemists' taste (Creager 1996; Rheinberger 1996).

Although case studies regarding the creation of molecular biology at a couple of laboratories in the United States and England had generated a critical mass that produced meaningful conversations by the late 1990s, the predominant historical focus on limited geographical locations has incurred criticism. Several recent volumes that turned attention to the European story of molecular biology had enriched the field. For example, Bruno Strasser has shown that the development of molecular biology in Europe was much shaped by postwar science policy in reaction to American developments, and that in Switzerland particularly, the creation of a social identity for molecular biologists was concurrent of the country's development of electron microscopy (Strasser 2006). A recently edited volume published a translation of the "Three-Man Paper," a paper written in German that proposed the target theory of the gene, along with a series of essays regarding the paper's larger contexts (Sloan and Fogel 2011). This unveiled important European networks of research that have contributed to the theoretical grounding of molecular biology. The linguistic barrier of historical studies of molecular biology proved more difficult to break regarding the non-Western world. With few exceptions, little was known about how molecular biology took root outside the confine of Euro–American regions (Uchida 1993). The funding from the Rockefeller Foundation, postwar rechanneling of resources to sciences, as well as immigration of scientists from Europe made American molecular biology undergo its strongest development in the postwar era. Therefore, postwar developments of molecular biology outside the United States usually took American science as a reference point, and often had American-trained scientists as major players. Further studies of the history of molecular biology of elsewhere would not only help us appreciate a molecular biology in global contexts, but also to analyze the wider influences of American developments.

Expansion of Molecular Biology and Molecular Cell Biology

Now let us return to the question of why people often combine the two relatively independently originated sciences about the cell and the molecule under the name "molecular cell biology." One simple answer is that the combination is convenient. The increasing compartmentalization of life sciences in the last century has made it cumbersome to carve out each subfield with its own name for individual research programs and general textbooks. Biological inquiries regarding neighboring biological levels – namely, the cell and the molecule – may well be lumped together to ease academic administration and publication of books. In addition, the study of the cell and that of biomolecules often intersected historically. E.B. Wilson, for example, had tried to analyze the chromosomes through observing cells. Self-identified molecular biologists had manipulated cellular nuclei and cytoplasm in order to probe the different roles of the DNA and cytoplasmic proteins in various cellular functions. Since developments of molecular biology brought molecular visions and methods to the study of the cell in ways similar to how molecular biology encroached other areas in life sciences, some confluences of cell biology and molecular biology were revealed in general accounts of the expansion of molecular biology. This section selectively reviews the developments of molecular biology with particular relevance to studies of the cell since the 1960s. Both historians and practitioners have characterized the period as a time of "normal science" in a Kuhnian sense for molecular biology in which scientists focused on resolving puzzles rather than trying to revolutionize the field with new fundamental principles. The time was featured with an exodus of molecular biologists toward problems beyond basics of biological information and into problems of biological function, development, and disease.

For the integration of molecular biology and cell biology, technologies that could label biomolecules without killing the cell became important. Bechtel has showed how wider uses of ultracentrifuge and electron microscopy had integrated biochemistry and cytology in the 1940s. These were technologies shared by molecular biologists as well. Other technologies with reaches to molecular biology and cell biology at the same time included electrophoresis, radioisotope labeling, nuclear magnetic resonance (NMR), immunofluorescence, and polymerase chain reaction (see Biotechnology chapter, this volume). Particularly, radioisotopes and heavy isotopes were heavily funded and distributed by the US Atomic Energy Commission in the 1950s and 1960s. They were used, among other aims, to label molecular transformations in living cells. Using heavy isotopes, the Meselson–Stahl experiment eventually illuminated the semi-conservative manner in which DNA replicates, a process that had occupied cytologists' interest for some time (Creager 2013: 239–59; Rheinberger 2010: 170–202). Since the 1980s, molecular cell biology has developed quickly, with the illumination of a number of molecules related to cell division, signaling, immortalization, aging, and death. Partly, this was due to the development of efficient and simple methods, such as immunofluorescence, that could reveal the architecture of cellular components without the need of breaking the cell. The cytoskeleton was one such structural component of the cell whose architecture was brought to view through the immunofluorescence technique.

Apart from technological upgrades that made molecular studies of the cell possible, in the 1960s and 1970s many molecular biologists migrated from their initial focuses on the phage, viruses, and molecular structures to questions in neurobiology, embryonic development, and human diseases. Although historian of molecular biology Michel Morange has characterized the period as the inception of an age of normal science in molecular biology, for other fields that were being intruded by molecular biology, the sea change was felt far more exciting than simply having scientific details filled out (Morange 1998: 167–83). Some applications of molecular biology to problems in cellular biology elucidated concrete correspondences between particular molecular changes and cellular functions. In cancer research, previous studies on propagations of animal viruses stimulated theoretical proposals on how some viruses would induce cancer, which eventually made studies of virus-induced immortalization of cells a major research direction in Nixon's "War on Cancer." For example, Renato Dulbecco, a former phage geneticist, moved into biomedical studies of tumorigenesis in mammalian cells (Kevles 1993). Paul Berg, as one of those who moved to study tumor viruses in the 1960s, also began to move into studies of gene expression and regulation in eukaryotic cells. For Berg, an analogy between the transformation of bacteria through bacteriophage λ and the transformation of eukaryotic cells by tumor viruses such as SV40 and polyoma virus proved most useful in devising a new experimental system to probe how cells regulate gene expression. As Doogab Yi's fine narrative of Berg's scientific trajectory shows, viruses became vectors to explore and alter the genetic makeups of the cell in Berg's research, which eventually evolved into one of the earliest schemes of recombinant DNA operation (Yi 2008). Compared to his colleagues interested in cancer etiology, Sydney Brenner's move to higher organisms was more open-ended. Having painstakingly monitored the development of neurons and eventually mapped the cell lineage of all individual somatic cells during development of the nematode *Caenorhabditis elegans*, Brenner, John Sulston, and H. Robert Horvitz eventually uncovered a number of gene expressions responsible for various cellular fates. Among them was the famous series of "death genes" responsible for initiating a process of programmed cell death to obliterate cells that the growing worm no longer needed (Jiang 2012).

Expansions of molecular biology to the wider fields of biology and biomedicine generated new networks between molecular biology and embryology, physiology, biophysics, pharmacology, neurobiology, endocrinology, and, certainly, cell biology. Many of these fields used cell cultures as substrates for observations and assays. Some held cellular phenomena as primary objects of inquiry, such as investigations about how cells differentiate, how cell cycles are regulated, and how they modify and regulate protein secretions. Molecularization of these inquiries meant molecularization of cellular activities and structures at the same time. However, it is important to note that these lines of work usually took several decades to mature. Horvitz's work in molecularization of cell death, for example, reached its golden age in the late 1980s and early 1990s when he was teaching at the Massachusetts Institute of Technology, more than 10 years after his postdoctoral work of mapping cell lineages of the worm at Cambridge (Jiang 2012). In the 1960s and 1970s, many lines of research that tried to link molecular changes to cellular activities actually went through trials that proved to be futile. On the other hand, during that time, molecular explorations of topics beyond the cell such as immunology and neurology often simply ignored what was happening inside the cell all together (Morange 1998: 180–3).

Although the post-genomic age we are living in today testifies to the successful expansion of molecular biology, we still know very little about the development of molecular biology and cell biology as well as their integrations after the 1970s. For example, although we have good works on the development of the recombinant DNA technology and the controversies it incurred, we know little about how wider application of the recombinant DNA technology brought about concrete changes in molecular biology of the cell. Historical works that depict how biologists of different stripes in the 1970s and 1980s adopted molecular methods to investigate a diverse range of biological problems would contribute to mend the relative lack of knowledge about the period. More importantly, such work would help us untangle the complex relations between increasingly compartmentalized yet interacting biological research programs of the late twentieth century. In this area, studies of the development of molecular biology of cancer have offered an exception. Both sociological and philosophical studies are available about how the "oncogene paradigm" emerged and spread within the biomedical community (Fujimura 1996; Morange 1993). Further works that illuminate how molecular studies of, to name a few, cell differentiation, cell signaling, cell death and aging emerged and fared in the late twentieth century, would tell us much about the shifting epistemic grounds of both molecular biology and cell biology. The seemingly decline of the disciplinary identity of molecular biology, as shown through the declining use of the term, further underscores the importance of studying these cases that might teach us about the changing epistemic roles of the cell and the molecule (Morange 2008).

Bibliographical Essay

The success of molecular biology as a discipline in the late twentieth century led to a multiplication of historical studies of it. In comparison, histories of cytology and modern cell biology have been understudied and generated only a handful of monographs. For a history of cytology before the early twentieth century, see Henry Harris' *The Birth of the Cell* (1997), William Coleman's "Form: Cell Theory" in his book *Biology in the Nineteenth Century* (1971: 1–34), J. Andrew Mendelsohn's "Lives of the Cell" (2003), and Jane Maienschein's *Transforming Traditions in American Biology, 1880–1915* (1991a). For the development of modern cell biology, see William Bechtel's *Discovering Cell Mechanisms* (2006) and Hannah Landecker's *Culturing Life: How Cells Became Technologies* (2007).

For a foundational history of molecular biology and overviews of the discipline's general history, see Horace Freeland Judson's *The Eighth Day of Creation* (1979), Michel Morange's *A History of Molecular Biology* (1998), and Soraya de Chadarevian's *Designs for Life* (2002). For a commentary of the famous phage group, see Angela Creager's essay review "The Paradox of the Phage Group" (2010). For expansions of molecular methods to fields of biology and medicine, articles in the edited volume by de

Chadarevian and Harmke Kamminga, *Molecularizing Biology and Medicine: New Practices and Alliances, 1910s–1970s* (1998) offer a range of case studies.

There are a number of biographies of individual molecular biologists or works that use extensive biographical materials. Some of them are scholarly in nature while others target a general audience. Such works include Brenda Maddox's *Rosalind Franklin: The Dark Lady of DNA* (2002), Ernst Peter Fischer and Carol Lipson's *Thinking About Science: Max Delbrück and the Origins of Molecular Biology* (1988), Errol Friedberg's *Sydney Brenner: A Biography* (2010), and Catherine Brady's *Elizabeth Blackburn and the Story of Telomeres: Deciphering the Ends of DNA* (2009).

Because the changing concept of the gene and ways of researching it were at the core of the development of molecular biology, certain studies on the history of genetics in the twentieth century dealt with important scenarios of molecular biology. They include Evelyn Fox Keller's *The Century of the Gene* (2000), Lily Kay's *Who Wrote the Book of Life? A History of the Genetic Code* (2000), and Stephen Hall's *Invisible Frontiers: The Race to Synthesize a Human Gene* (1987).

Due to the material and experimental underpinnings of cellular and molecular biology, works on the organisms or technologies of special importance for its development have generated vibrant discussions. Historians have shown that material models operated in biology in ways distinct from how theoretical paradigms worked in conventional physics, a traditional subject of inquiry for history and philosophy of science. Notably, Hans-Jörg Rheinberger (1997) has described how a group of scientists at the Collis P. Huntington Memorial Hospital of Harvard University had turned the study of a cancer-causing particle into an *in vitro* study of ribosomes and protein synthesis. Borrowing scientists' own parlance, Rheinberger used the term "experimental systems" to define evolving units of experimental research that embed epistemic objects, produce investigative questions, and leave traces for future research. Rheinberger's argument that the evolving, generating features of experimental systems constitute the driving force of modern experimental life sciences has since had an influential reach. For example, in *The Life of a Virus,* Angela Creager (2002) depicted how experimental operations about biochemical analysis, etiological investigation, and sequencing of the tobacco mosaic virus (TMV) led to points of departure for new directions of research. More importantly, Creager described how technical (using ultracentrifuge to prepare viruses and vaccines), disciplinary (biochemistry versus molecular biology), and political (funding for infantile paralysis or cancer) concerns played into the representations of the TMV as a model and exemplar as well.

The body of literature on experimental and model organisms also include Robert Kohler's *Lords of the Fly: Drosophila Genetics and the Experimental Life* (1994), Rachel Ankeny's "Wormy Logic: Model Organisms as Case-Based Reasoning" (2007), Karen Rader's *Making Mice: Standardizing Animals for American Biomedical Research, 1900–1955* (2004). In addition to Landecker's *Culturing Life*, books that dealt with the technological foundations of molecular biology include Nicolas Rasmussen's *Picture Control* (1997), Creager's *Life Atomic* (2013) and Hallam Stevens' *Life Out of Sequence* (2013).

As molecular biology gained much public attention and involved fashioning its methods, objects of research, and research communities as distinctive from conventional disciplines such as biochemistry, social and cultural dimensions of molecular biology have generated continuing discussions. Related works include Kohler's *Partners in Science* (1991), Pnina Abir-Am's "The Politics of Macromolecules: Molecular Biologists,

Biochemists, and Rhetoric" (1992), Kay's *The Molecular Vision of Life* (1993), and Dorothy Nelkin and Susan Lindee's *The DNA Mystique* (2004). This chapter does not have space to discuss works that tackled gender issues in molecular biology. Interested readers could consult, to begin with, Bonnie Spanier's *Im/partial Science: Gender Ideology in Molecular Biology* (1995) and Evelyn Fox Keller's *Refiguring Life: Metaphors of Twentieth-Century Biology* (1995b).

Chapter Fifteen

NUCLEAR, HIGH ENERGY, AND SOLID STATE PHYSICS

Joseph D. Martin

When Henry Rowland and 35 of his colleagues established the American Physical Society (APS) in 1899, the fields of nuclear, high energy, and solid state physics did not exist. Ernest Rutherford would not postulate the atomic nucleus until 1911, the cyclotron of Ernest Lawrence and M. Stanley Livingston would not accelerate its first particles until 1931, and no one would earnestly propose grouping physicists according to the phase of matter they studied until 1943. At the Society's centenary in 1999, its membership topped 41,000 and these fields represented its largest and most powerful constituencies.

Understanding the transformation of American physics through the twentieth century requires examining how these specialties grew together and how they grew apart. American physics blossomed as European émigré scientists bolstered an ascendant domestic physics community in the 1930s. The prestige and political influence physicists earned with their contributions to Allied victory in World War II fueled rapid postwar expansion. The specialties discussed here, once established, benefited alike from favorable mid-century conditions, but they charted different, mutually reinforcing pathways to success within American science and Cold War society.

Common historical roots helped nuclear, high energy, and solid state physics excel in the United States, but so did the rivalries that developed between these fields. In becoming an internationally recognized scientific force, American physics also became larger and more diverse, supporting parallel visions of how best to wield the discipline's prestige, funding, and political influence. Although they identified with the same broad scientific discipline, these three fields evolved different notions about the purpose and scope of that discipline and measured success in different ways. The viability of each of these trajectories, and the diversity they represented, combined to make physics the defining scientific endeavor of Cold War America.

A Companion to the History of American Science, First Edition.
Edited by Georgina M. Montgomery and Mark A. Largent.
© 2016 John Wiley & Sons, Ltd. Published 2020 by John Wiley & Sons, Ltd.

American Physics Ascendant

In 1964, John Van Vleck, by then a hoary elder statesman of American physics, recalled the 1920s as a watershed decade: "*The Physical Review* was only so-so, especially in theory, and in 1922 I was greatly pleased that my doctor's thesis was accepted for publication by the *Philosophical Magazine* in England [...]. By 1930 or so, the relative standings of *The Physical Review* and *Philosophical Magazine* were interchanged" (quoted in Duncan and Janssen 2007: 566). The reversal of *Physical Review*'s fortunes owed much to its editor, John Torrence Tate, and his opportunistic reaction to the quantum revolution. Tate ensured that articles on quantum phenomena were published quickly. Under his stewardship, the previously lackluster American journal rode the cresting quantum wave to become a prestigious venue for both American and European physicists, and, in the 1930s, for European émigrés settling in the United States (Nier and Van Vleck 1975).

The reputational gains American physics made in the 1920s are remarkable in light of the community's youth. American scientists had done little to contradict Alexis de Tocqueville's observation that Americans had a knack for practical problems but little patience for abstract theoretical reflection. The few physical theorists the United States produced, such as Josiah Willard Gibbs who made foundational contributions to statistical mechanics and physical chemistry, labored largely in isolation. Americans who cultivated international repute typically did so on the strength of their experimental accomplishments.

These accomplishments included Rowland's custom-made diffraction gratings, which were highly sought after in Europe (Sweetnam 2000). Albert Michelson's meticulous interferometer experiments, designed to detect long-postulated signs of the earth's motion through the luminiferous æther, won him continental accolades including the 1907 Nobel Prize (Goldberg and Stuewer 1988). But experiment and application have traditionally been less esteemed than theory in physics, and until recently in history of science as well (Johnson 2008). The judgment that the United States lagged behind in physics should be understood as an observation that American physicists did not make theoretical contributions that were recognized as valuable by their more established continental counterparts.

It would be fair to say that the rise of American physics in the 1920s was a quantum phenomenon. Quantum mechanics provided Americans multiple routes by which to achieve theoretical respectability. First, as the Rockefeller, Carnegie, and Mellon fortunes were directed toward eleemosynary ends, the sciences benefited. Foundation fellowships allowed many to travel to Europe either for their doctoral degrees or for postdoctoral fellowships (Kohler 1991). The likes of J. Robert Oppenheimer, who would go on to lead nuclear weapons research at Los Alamos, future president of MIT Julius Stratton, and physical chemist Harold Urey learned their quantum physics in Göttingen, Zurich, and Copenhagen, bringing the latest European developments with them when the returned.

Second, the growth of research universities and the expansion of graduate education in the United States produced the first crop of American-educated theorists who could compete internationally (Geiger 1986). Edwin Kemble at Harvard advised students such as Van Vleck and John Slater who, like others of their generation, took great pride in their domestic training (Assmus 1992). The pragmatic outlook common among American-trained physicists meant that they were less bothered than their counterparts across the Atlantic by the counterintuitive philosophical implications of quantum mechanics (Schweber 1986). The reputations Van Vleck, Slater, and their European-trained compatriots established during the 1920s ensured that when a wave of European physicists alit in the United States they put new roots into fertile soil.

American physics had matured in time to mobilize for World War II. As the United States entered the war in December of 1941, European savants were assimilating into American universities. Their expertise buoyed the war effort. Enrico Fermi migrated his nuclear research program from Rome to Columbia University, and then to the University of Chicago where he oversaw the first controlled nuclear chain reaction. Hans Bethe, established for some years at Cornell University, brought the latest in quantum theory to bear on bomb construction at Los Alamos. Felix Bloch, by then a fixture at Stanford University, proved critical for the radar research carried out largely in Cambridge, Massachusetts. In each case, European luminaries worked alongside able and energetic American counterparts.

Wartime research defined the complexion of postwar physics and the identity of postwar physicists. Nuclear, high energy, and solid state physics all trace their roots to groups of researchers and sets of problems brought together by the war. Nuclear weapons were the most visible and psychologically powerful legacy of wartime physics, but radar and operations research also drove the large-scale federal investment in physics through the 1940s and 1950s that supported its rapid growth. The National Science Foundation (NSF), established in 1950, was the official federal organ for supporting basic research, but the service agencies, in particular the Office of Naval Research and the Air Force Office of Scientific Research, supported a wide range of projects while the legislative machinery behind the NSF worked deliberately through the late 1940s (Geiger 1992). Physicists, flush with funding, also found unprecedented influence in the halls of government. Advisory committees within the National Academy of Sciences formed to guide the distribution of federal largess and scientists took on key advisory roles in the legislative and executive branches of government where they shaped funding decisions and defense policy (Kevles [1971] 2001).

The end of World War II heralded a new age for American physicists, who enjoyed a measure of social and economic stability that war-ravaged Europe could not match. After competing as underdogs against their better-established European colleagues for much of the early twentieth century, American physicists emerged from the war with an intact and cohesive community, tremendous momentum, generous federal support, and widespread social approbation. This was the environment in which new specialties with distinct professional interests began to form. If World War II instilled a sense of unity and common purpose in the American physics community, the Cold War presented the opportunity for newly established specialties to go their own ways.

Nuclear Physics

The notion that the atom confined tremendous amounts of energy originated in the early twentieth century. Physicists, though, considered applications remote even after Rutherford demonstrated in 1911 that the bulk of an atom's mass resides in a small nucleus (Weart 2012). Understanding of the nucleus grew by leaps and bounds after

James Chadwick's discovery of the neutron in 1932. Italian physicist Enrico Fermi paced the investigations of nuclear structure, using neutrons to bombard various elements and examine their behavior. His most interesting results came from bombarding the heavy elements thorium and uranium, the results of which he interpreted as being new elements, but which were actually products of fission – the splitting apart of the atomic nucleus (Cooper 1999). In 1938 Otto Hahn and Fritz Strassman, building on Fermi's experiments, published conclusive evidence of nuclear fission. They relied heavily on correspondence with Lise Meitner who by that time had fled Berlin for Sweden where, with her nephew Otto Frisch, she employed the liquid drop model of the nucleus to articulate a theoretical model of fission (Sime 1996; Stuewer 2010).

In the same year that Hahn and Strassman published their findings, the Hungarian-born Leó Szilárd joined the exodus of physicists from Europe to the United States, where he took up a post at Columbia University. For Szilárd, the discovery of fission was both a vindication and an ill omen. He had been promoting the idea of a self-sustaining nuclear chain reaction since 1933, shortly after learning about the neutron. Fission showed that his instincts were correct, but he also recognized that a fission-based weapon could decide the looming war. Szilárd's concern that Germany had a head start prompted him to seek out another recent refugee from fascism; Szilárd and Albert Einstein drafted a now-famous letter to President Franklin Roosevelt urging him to commit the resources of the United States to nuclear research. Thus began the process that would lead to the establishment of the Manhattan Engineer District in 1942 and the eventual detonation of three nuclear weapons, one a test conducted at Alamogordo, New Mexico and two used against urban targets in Japan.

Shortly after its inception, nuclear research was measured by its military applications. It was conducted with an urgency to match. European physicists arrived in the United States carrying the knowledge that nuclear weapons were possible and a grim tenacity born of first-hand experience of the danger Adolf Hitler posed. They provided the impetus for the Manhattan Project, which turned two billion government dollars toward synthesizing the knowledge and talents of European émigrés, the know-how of American scientists and engineers, and industrial infrastructure of previously unimagined scope (Rhodes 1986). This synthesis accelerated theoretical understanding of the nucleus. Knowledge of the number and velocity of the neutrons emitted in nuclear decay, for instance, was necessary to accurately predict the critical mass at which a fissile element would explode. In that sense, war research accelerated physics along a path it was already following, but it is most notable for the new directions it opened.

The Manhattan Project was a benchmark for postwar physics in several ways. Along with radar, it set a blistering standard for blackboard-to-battlefield turnaround time that conditioned expectations for subsequent technological development. It sensitized a generation of researchers to science on a grand scale. It eroded the boundaries physicists had tried to maintain between science and politics. Most critically, it normalized large government expenditures for scientific research. Science spending had not been a major budget item before the war (Dupree 1957; Kleinman 1995). Demobilization from World War II, however, coincided with a different kind of mobilization for a different type of war. By the middle of the 1950s, federal funding for research and development was more than 50 times its prewar levels (Westfall and Krige 1998). Nuclear physicists had demonstrated that seemingly arcane research might quickly result in earth-shaking technological advances. Wary of the Soviet Union's global aspirations, the United States government was determined not to be caught napping on the next big breakthrough.

Nuclear physics was the foremost beneficiary of the federal government's newfound enthusiasm for science, but financial support came at a price. Previously driven by curiosity about the composition and structure of matter and a spirit of open communication, nuclear physics in the Cold War was inextricably bound to US defense interests and became subject to pointed mission directives and intricate secrecy regimes (Wellerstein 2008). A field that began with theoretical questions about atomic structure became a technical enterprise directed at exploiting atomic energy either through bombs or reactors. Los Alamos, which had housed the laboratory charged with theoretical research and bomb assembly, remained a weapons lab. Aware that the basic principles of nuclear weapon design were well known by the end of World War II, physicists estimated that it would be no more than five to ten years before the Soviet Union produced its first bomb. The estimate proved conservative, as the Soviets detonated their first nuclear weapon in 1949.

If the Manhattan Project catapulted nuclear physicists into a position of influence, the arms race ensured they would maintain it and that nuclear research would remain handsomely funded. Following the 1949 Soviet test, President Harry Truman made development of a fusion bomb – the so-called "super" that obsessed Hungarian émigré Edward Teller – a high priority. Teller and Stanislaw Ulam eventually overcame the technical obstacles that had made fusion weapons unfeasible and the first thermonuclear weapon, code-named "Mike," was detonated at Enewetak Atoll in 1952 (Rhodes 1995).

Whereas the physics community had been united behind the bomb work carried out as part of the war effort, it was divided over the wisdom of pursuing such an aggressive weapons program in peacetime. Many Manhattan Project veterans were appalled by the use of nuclear weapons against civilian targets in Japan and resisted efforts to expand the nuclear arsenal. The Red Scare exacerbated tensions between weapons researchers and skeptics in the early 1950s, when any opposition to a more powerful nuclear deterrent could be spun as support for communism.

Both the virulence of American anti-communism and the political clout of the defense establishment were tested in the Atomic Energy Commission (AEC) hearing on Oppenheimer's security clearance. The former head of Los Alamos had opposed Teller's push for fusion weapons. His visibility after the war as a public intellectual and advocate for international control of nuclear weapons made Oppenheimer a target for hawkish politicians and scientific administrators, who moved to suspend his clearance. The hearing saw both Teller and General Leslie Groves, who had recruited Oppenheimer to manage Los Alamos, testify that he was a security risk. His clearance was revoked in May 1954 (Cassidy 2005).

The defeat Oppenheimer and the long list of colleagues who testified on his behalf suffered speaks to the extent to which the defense establishment had captured nuclear physics, a trend that extended to so-called peaceful uses of nuclear power (Hewlett and Holl 1989). After the war, Manhattan Project research sites scattered across the country were converted into National Laboratories, administered by the AEC. Although science pursued in the National Laboratory System, from reactor and cyclotron research to, somewhat later, biological and ecological investigations, was considered non-military, it was nonetheless calibrated to advance the strategic interest of the United States, especially vis-à-vis the Soviet Union (Westwick 2003). Research on nuclear reactors, such as that conducted at Argonne National Laboratory outside of Chicago, led to civilian nuclear power. On the other hand, nuclear supremacy in all arenas, from weapons to electrical power generation, was seen as critical to Cold War economic and defense goals.

This duality defined Cold War nuclear research. A field that seemed remote from daily life, not to mention geopolitics, in the early part of the century became deeply enmeshed in a global ideological struggle by the 1950s. Nuclear physics no longer meant simply the study of the nature and structure of the atomic nucleus, but rather the exploitation of a few heavy elements in the service of national aims. The influence physicists could have over these aims shepherded them from obscurity to the center of American civic and political life. Nuclear physics became an outgrowth of national defense during the Cold War. Some physicists, however, were not content to let postwar public approbation and federal funding be directed entirely toward technological projects. No field exemplifies the new opportunities the postwar boom enabled better than high energy physics.

High Energy Physics

It took some time after World War II for nuclear and high energy physics (HEP) – known as particle physics in its early days – to become differentiated. Well into the 1950s, "nuclear physics" referred both to the weapons and reactor research and to investigations of the particles that composed the atomic nucleus. Two factors cleaved these traditions apart. The first was rooted in laboratory practice. Early cyclotron designs became the template for a family of new machines, which rapidly grew in size and power. Large particle accelerators required dedicated facilities, and those drawn to identifying and classifying the properties of elementary particles had little use for bombs or reactors. The second factor was ideological. Many Manhattan Project veterans were disillusioned with war work. Whereas nuclear physics became intertwined with the Cold War military–industrial complex, high energy physicists explored the most abstruse corners of the physical world, where they would be little troubled by the ethical quandaries weapons work posed (Stevens 2003). In the absence of clear practical justifications for building expensive accelerator facilities, high energy physicists evolved a rhetoric of fundamentality, suggesting that the most profound and important truths about the physical world could be found among the smallest constituents of matter and energy.

Although HEP would come to be known for its theory of elementary particles and their interactions, the Standard Model, the field originated in an experimental innovation, the cyclotron. Today, we associate particle accelerators with vast underground tunnels and enormous detectors, but the first cyclotron fit in the palm of Lawrence's hand. The cyclotron, when it appeared in 1931, offered meaningful advantages over linear accelerators, also new to the scene. By accelerating particles in a spiral, rather than a straight line, the cyclotron limited the space it took up and brought higher energies within reach.

The cyclotrons in Lawrence's Radiation Laboratory, or "Rad Lab" as his facility at Berkeley was known, grew rapidly. Following the first diminutive 4.5″ device, the 1930s saw 11″, 27″, 37″, and 60″ iterations as Lawrence's team pushed for higher energies. Like most of American physics, this rapid progress was channeled into the war effort in the early 1940s. Glenn Seaborg used the 37″ cyclotron to bombard uranium, leading to the discovery of plutonium. As the pressing need to separate isotopes of uranium became clear, Lawrence used his machines to separate U-235 and U-238 electromagnetically (Heilbron and Seidel 1989).

Cyclotron research would blaze a very different trail than that of bombs and reactors after the war. The Rad Lab saw the same expanded funding and increased visibility that nuclear physics did by virtue of its wartime contributions, but as HEP split from nuclear physics it defined its own research agenda. The 184″ cyclotron that went online in 1946 would be dedicated not to problems of immediate defense or economic importance, but rather to exploring a vast and puzzling menagerie of new particles. Larger machines meant that physicists could artificially generate new particles called "mesons" that had previously – and then only recently – been observed in cosmic ray research.

Through the 1950s accelerators became more powerful and high energy physicists developed more sophisticated methods for manipulating the data they produced. The invention of the bubble chamber by Donald Glaser in 1952 supplanted older cloud chamber and photographic emulsion detection by making it easier to produce high-quality photographs of particle collisions. Iconic images of the intricate swirling lines generated as charged particles forced hydrogen bubbles to dissolve from solution underscored both the precision and the beauty high energy physicists sought. These images proved a seemingly inexhaustible source of discovery. New particles such as the Xi-naught, Sigma-naught, and Omega-naught were first detected in bubble chambers and bubble chamber data was instrumental in the detection of processes called weak neutral currents, which allowed theoretical unification of the weak nuclear force and the electromagnetic force (Brown, Dresden, and Hoddeson 1989; Galison 1987).

As bountiful as this data harvest was, the particle zoo troubled physicists who maintained ironclad conviction that the underlying structure of the universe should be simple and elegant. High energy theorists, irked by the expanding family of elementary building blocks and by the fact that experimentalists, with whom they were fiercely competitive, were driving the new physics, looked to make sense of the chaos (Traweek 1988). The result was a new theoretical model proposed by Murray Gell-Mann and George Zweig in 1964, which suggested that most observed particles, including protons and neutrons, were composed of smaller particles. Gell-Mann dubbed these "quarks," borrowing the spelling from a line in James Joyce's *Finnegan's Wake*.

The quark theory was not immediately accepted, but it did become a working model for many theorists, who saw in it a way to cull the particle zoo. When the dust settled, the result was the Standard Model of particle physics. HEP has subsequently focused on elaborating the Standard Model, culminating recently with the discovery of the Higgs boson at CERN in 2012. This elaboration called for larger accelerators, huge research groups, and billions of dollars in funding – the hallmarks of big science.

The emphasis on larger, higher-energy machines led in the late 1960s to the establishment of the National Accelerator Laboratory, better known as Fermilab, outside Chicago. The United States' investment in Fermilab testifies to the power and influence of physics in Cold War America. Not only was the lab hugely expensive, but its first director, Robert Wilson, took care to justify it on anything other than practical grounds. When asked in Congress what the lab would do to benefit national defense, Wilson famously replied that physics at Fermilab, like great art or literature, "has nothing to do directly with defending our country except to make it worth defending" (quoted in Stevens 2003: 174).

The tendency to justify HEP on the basis of the ennobling knowledge it produced represented a break from the practical mindset that consumed nuclear physics. As HEP uncovered ever smaller bits of matter and energy, becoming ever more remote from technical applications, the field's philosophical grounding adapted to fit. High energy physicists became outspoken proponents of reductionism, the view that knowledge about the smallest scale of the universe represented the most, and perhaps even the only truly fundamental truth (Cat 1998). In contrast to both nuclear and solid state physicists, high energy physicists wore their distance from practical concerns as a badge of pride, trading on the argument that modern society had a duty to support knowledge for its own sake.

The reductionist rational worked well during the Cold War. Tension with the Soviets ensured that support for basic research proceeded with the assumption that the next game-changing discovery might, like fission, come from unexpected quarters. The Cold War also meant that high energy physicists could deploy nationalistic rhetoric, as Wilson did, without invoking specific technological or economic outcomes. HEP maintained generous levels of funding by spending the political capital that nuclear physics earned during the Manhattan Project and leveraging the stranglehold it held on the intellectual prestige physics as a whole had earned after World War II. Its practitioners parlayed these circumstances into a license to pursue their research largely unfettered by the deliverables the federal government demanded from other sciences.

The unspoken agreement between the HEP community and the federal government that sustained large expenditures for blue-skies research collapsed with the Berlin Wall and had devastating consequences for the Superconducting Super Collider (SSC). The enormous accelerator slated to be built in Texas was designed to complete the Standard Model, and, many hoped, push beyond it. When Congress pulled its funding in 1993, the machine came with a $11.8 billion price tag. Its demise resulted both from internal management difficulties and from post-Cold War political shifts. Administering a project of such scale posed insurmountable challenges even for seasoned scientific administrators, and with the Soviets no longer lurking in the background, legislators who had been elected with a mandate to cut spending began to demand economic justifications more specific and weighty than any the HEP community could provide (Hoddeson and Kolb 2000; Kevles [1971] 2001; Riordan 2000, 2001).

The erosion of financial and political support for the SSC was a turning point for American HEP and for American physics as a whole. Debates over whether to fund the SSC coincided with the establishment of the Human Genome Project, which would anoint biology the standard bearer for American science (Kevles 1997). Although the Tevatron at Fermilab would remain the world's most powerful accelerator until the Large Hadron Collider overtook it in 2009, high energy physicists understood the SSC's demise as a sign that the United States was no longer prepared to underwrite big physics. As of this writing, the future of HEP is being hotly debated, but it seems clear that whatever form it takes in the future, the combination of ever larger machines running at ever higher energies and an unabashedly reductionist ideology that caused the field to burn so brightly for the second half of the twentieth century was the product of distinctive, now extinguished, Cold War conditions.

Solid State Physics

Solid state physics (SSP) was a professional innovation; it began as a strategy for bringing academic and industrial researchers into closer contact. This made it unlike nuclear physics, which coalesced around questions about the atomic nucleus and applications of nuclear energy, and high energy physics, which centered on a type of experimental investigation and the theories used to codify its results. The intended scope of the field covered the properties of solid matter, encompassing thermodynamics, optics, acoustics, electromagnetism, mechanics, and quantum mechanics as they manifested in solids, a list that includes almost all of the main topical divisions of physics that existed before World War II. This breadth was unusual within a community that traditionally identified its topics with well-defined problems or methods. The reasons SSP was unorthodox, and the manner in which it navigated that unorthodoxy, are critical for understanding how it emerged as a successful area of American physics, and the largest, during the Cold War (Hoddeson et al. 1992).

SSP first became a recognizable professional entity in the late 1940s, when the American Physical Society's Division of Solid State Physics (DSSP) was established. The society's third topical division came into being circuitously. In 1943, Roman Smoluchowski, a research physicist at General Electric (GE), began mustering support for a Division of Metals Physics within the APS. Like many of his colleagues at GE and other industrial laboratories where physicists were employed in growing numbers, Smoluchowski resented the lack of representation industrial researchers had within the APS and worried that physics fields with industrial relevance might branch off into new engineering specialties. The APS was largely an ivory tower institution and had not adapted to the needs of industry. The APS Council shot down several proposals for a Division of Industrial Physics on the grounds that it was not a "subject," as required by the 1931 amendment to the Society's constitution that had made divisions possible. Metals, Smoluchowski reasoned, were close enough to a subject of physics to pass muster with the Council, but also central enough to the day-to-day work of industrial researchers that such a division would attract them (Weart 1992).

The Council was reluctant, in no small part because of vocal opposition by some powerful members to any and all divisions. John Van Vleck decried the "Balkanization" of American physics and favored a community subdivided as little as possible (Weart 1992). Nonetheless, Smoluchowski persisted and his division was approved on the condition that it devote itself to "solid state" instead of "metals" on the grounds that the theoretical and methodological differences between metals and other solids were too slight to merit cordoning off the former. SSP thus represented a compromise between forces defending a traditional pure science ideal and those who favored embracing the applications of physics as part and parcel of the discipline.

The early years of SSP cemented its relationship with industry. In 1948, John Bardeen, Walter Brattain, and William Shockley invented the transistor at Bell Laboratories. The implications of a solid-state device that could amplify and rectify electrical currents was immediately clear in a technological environment that relied heavily on vacuum tubes, which were clunky by comparison and prone to breaking (Riordan, Hoddeson, and Herring 1999). Other breakthroughs came in nuclear magnetic resonance (NMR) and superconductivity, which permitted applications in magnetic resonance imaging, and maser and laser technology, which were also adapted for industrial and commercial purposes (Bromberg 1991). With these contributions to its credit, SSP could make a strong claim to strategic importance in an increasingly technological economy and evolved from a broad and messy constellation of specialties into a field dominated by a smaller set of research programs, notably but not exclusively semiconductor, superconductor, NMR, and laser research. The success of these research programs made the DSSP the largest division of the APS by the early 1960s. By the end of the 1980s it enrolled a full quarter of the Society's total membership.

Segments of this large, fractious community were not content to be typecast as technical specialists. Although technical relevance ensured reliable funding, that funding was often tied to specific outcomes. Solid state physicists resented the comparatively free rein high energy physicists enjoyed. That some high energy physicists derided the intellectual contributions of solid state work did little to ease tensions between the sibling fields. The tendency to regard research on complex systems as inelegant and intellectually inferior traced to the quantum revolution and one of its architects, Wolfgang Pauli, who reportedly referred to the physics of solids as "*Schmutzphysik*," or "dirt physics." The English equivalent of the slight, in which "solid state physics" becomes "squalid state physics," is attributed to Gell-Mann. These insults were often repeated as rallying cries by solid state physicists themselves as they fought for intellectual esteem (Joas 2011).

Foremost among those concerned about the intellectual reputation of the field was Philip Anderson, a Bell Laboratories theorist and PhD student of John Van Vleck. Anderson's paper "More Is Different" (1972) launched a headlong assault on the high energy physicists' strong reductionist picture of the physical world, which excluded other enterprises from access to fundamental knowledge. Anderson acknowledged that physicists might learn a great deal by reducing systems to their constituents, but argued that it was folly to leap to the conclusion that knowledge of those constituents alone would suffice to reconstruct the reduced whole. Behavior at higher levels of complexity, in other words, is novel, unpredictable from lower level data, and just as fundamental as anything happening at the quark scale.

By this time the DSSP, having attracted industrial physicists along with their academic counterparts, was the largest division of the APS by a comfortable and widening margin. "More Is Different" resonated with a large proportion of this plurality, many of whom had previously seen dreary prospects for sharing in the prestige and influence other physical fields wielded. The revived effort to boost SSP's intellectual status and create distance from its industrial roots adopted a new name, "condensed matter physics." The term "solid state physics" had long had its critics. Dwight Gray, editor of the *AIP Handbook of Physics*, summed up the difficulty many had with the moniker, writing: "Adding [SSP] to the conventionally labeled group of mechanics, heat, acoustics, and so forth is, of course, a little like trying to divide people into women, men, girls, boys, and zither players" (Gray 1963: 40). "Condensed matter" acknowledged that the methods physicists used to study solids were often just as suitable to other dense materials, such as liquids, gels, polymers, and plasmas. Defining a field that emphasized this class of methods, proponents of condensed matter hoped, would highlight intellectual contributions over industrial applications (Martin 2015).

The high-profile political scraps that preceded the SSC's demise gave condensed matter physicists a chance to showcase their ideology. Before Congress and in the popular press, Anderson and his colleagues attacked the reductionist rhetoric used to justify the

SSC, maintaining that their field was just as fundamental as HEP, while also being more technologically and economically relevant. This was a delicate dance. Condensed matter physicists did not want to undermine funding for exploratory research. At the same time, they recognized that some claim to short-term relevance was essential to survive in the post-Cold War environment.

The failure of the SSC was in some measure a result of the new priorities of a new context, one to which SSP was better adapted, despite efforts from some segments of the community to shed the field's association with technology. A close and successful collaboration with industry had helped SSP grow into a powerful force, and although efforts to reimagine the field as a basic research enterprise did change the internal character and constitution of the community, they were less successful in branding solid state research for more general consumption. Solid state physicists, who envied the public recognition enjoyed by their nuclear and high energy colleagues, never managed to attain the same level of visibility, in part because they lacked a compelling public spokesperson. John Bardeen, winner of two Nobel Prizes in 16 years, might easily have traded his success for celebrity akin to that of an Oppenheimer or a Richard Feynman had he not been notoriously humble and unremittingly laconic (Hoddeson and Daitch 2002). Nevertheless, the desire solid state physicists retained to partake of the acclaim physics enjoyed kept their field from migrating into engineering. When combined with an impressive track record of technical accomplishment, that meant that SSP permitted physics as a whole, including HEP, to present itself to its patrons and to the public as a field that both probed the inner secrets of the universe and exploited them to practical ends.

*   *   *

Between 1939 and 2014, 86 physicists from the United States received the Nobel Prize in Physics. Before 1939, only four had been similarly recognized. The starkness of these numbers indicates a phase change in American physics around mid-century when the fields described above asserted themselves. The independent paths to success they plotted, when braided together, gave American physics tremendous strength and versatility. For all their squabbling and ideological differences, these were interdependent endeavors.

Nuclear physicists reaped the rewards of physics' centrality to national defense. High energy physicists exploited its newfound intellectual prestige to focus the world's largest machines on the universe's smallest constituents. Solid state physicists forged connections with industry to exploit physics' untapped technological potential. Neither of these trajectories could have been fully realized except against the background of the others. Nuclear and high energy physics gave solid state a level of esteem and political influence to which to aspire and kept it grounded in the search for general physical principles when it might easily have diffused into engineering. High energy physics relied on the practical benefits of nuclear and solid state to conduct research that could not have commanded the huge quantities of funding it required without the political will secured by other areas of physics. Nuclear physics retained its strong policy influence in part because it was not narrowly diverted into weapons and reactor design, but, on the strength of its sibling fields, showed potential for new and unexpected breakthroughs.

American physics established itself in the second half of the twentieth century because of its diversity. Physicists fleeing a fractured Europe injected new perspectives into a fledgling American community as new specialties expanded opportunities for growth. World War II galvanized American physics and sold the public on a field about which it previously knew little, but physics succeeded after the war precisely because it abandoned a common purpose. By developing competing ideologies, nuclear, high energy, and solid state physics gave physics writ large a range of influence it lacked when it existed as a more uniform entity. American physics did not succeed because of its defense applications, its intellectual achievements, or its industrial relevance alone. It succeeded because these elements of the physics community, even when at loggerheads, worked in consonance.

Bibliographic Essay

The core interpretive frameworks in the history of American physics over the past decades have focused on the factors that made American science distinctive (Reingold 1991). Literature on "big science" or "big physics" emphasizes the scale of the machines, research groups, funding, and impact of Cold War physics (Heilbron and Seidel 1989; Hoddeson, Kolb, and Westfall 2008; Kevles 1997; Westfall 2003b). A lively segment of the historiography examines the relationship between basic and applied research (Crease 1999; Johnson 2008; Kevles [1971] 2001; Kleinman 1995; Leslie 1993), a distinction that exerted considerable policy influence in the Cold War. A segment of the historiography that focuses on conceptual developments has devoted considerable attention to the unity and fundamentality of physical knowledge (Cat 1998; Galison and Stump 1996; Kragh 2011; Pais 1988; Stevens 2003).

Aside from these traditions, histories of American physics are broadly split between studies of the community of American physics and attention to its conceptual development. Kevles ([1971] 2001) provides a touchstone for many of the community studies. The influence of the European diaspora of the 1930s has been thoroughly examined, as have the strides the American physics community made before receiving a boost from European émigrés (Assmus 1992; Holton 1981; Schweber 1986). Considerations of the Cold War era have emphasized the influence of the distinctive pressures competition with the Soviet Union placed on American physicists (Forman 1987; Wang 1999; Wolfe 2012).

Conceptual historians have addressed both the theoretical and experimental aspects of American physics. The former have turned to quantum mechanics to craft a large-scale narrative of twentieth-century physics (Kragh 1999, Pais 1988). Recent scholarship has challenged the longstanding assumption that quantum mechanics was developed in terms of fundamental problems and only later applied to more complex systems. Joas (2011) argues that the so-called "applications" of quantum mechanics to molecules, plasmas, and solid state systems was integral to the development and articulation of the theory. Histories of experiment in the early twentieth century have tended to focus on major individuals in American science and their influence, including Albert Michelson (Goldberg and Stuewer 1988), Arthur Holly Compton (Stuewer 1975), Henry Rowland (Sweetnam 2000), and Ernest Lawrence (Heilbron and Seidel 1989). Studies of the post-World War II era shift their attention to apparatus and material culture (Galison 1997; Riordan, Hoddeson and Herring 1999; Westfall 2003a).

Recent historiographical concerns about overspecialization have promised to change the way historians approach American physics. Whereas previous historical scholarship generally focused on well-defined topical areas that mirrored those used by scientists themselves, recent scholarship calls for historians of science to articulate their work in terms of broadly relevant historical themes (Kaiser 2005b; Kohler 2005). Many of the foundational works on American physics addresses themes such as textbooks and pedagogy (Kaiser 2002, 2004, 2005a; Midwinter and Janssen 2013), experimental practice (Crease 1999; Galison 1987, 1997; Westfall 2003a, 2003b), and the influence of patronage on scientific knowledge (Bromberg 2006; Forman 1987; Hounshell 1997; Kevles 1990). This variety of thematic – as opposed to topical – categorization represents the current *modus operandi* for the history of American physics.

Chapter Sixteen

NUTRITION

Jessica Mudry

The development of the American idea of nutrition was influenced greatly by the import of the ideas, practices, and structures of education of European chemistry laboratories. Prior to the establishment of the US Department of Agriculture in 1862, it could be said that formal studies of nutrition did not exist in the United States. In the late 1700s and early 1800s Americans farmed, ate what they farmed, and had formalized ways of understanding what they ate through almanacs, farming community publications, and laboratory notes. The scientific complexity of these publications varied as laboratory science developed and the experiential knowledge of farmers gave way to more formalized agricultural chemistry. Some of the publications were both scientific and political, as were the "agricultural essays" of John Taylor's *Arator*, in which both food production and yield shared space with political comments about the virtues of diligence and hard work. Other publications like *The American Farmer* and *The New England Farmer* tended to be more farmer-focused and included writings about the rural economy alongside the prices of produce (Cohen 2009; Rossiter 1975).

The early model of nutrition came with the importation of the politics, scientific practices, and establishment of American university research mandates around the relationship between food and eating. In 1862, Abraham Lincoln established the US Department of Agriculture (USDA) so that the government could have a "science-producing agency" that would test the agricultural output of farms, and publish the results (Cochrane 1993; Congress, Act 37). The US government's commitment to the scientific research of food production in the fields worked in tandem with the Morrill Land-Grant College Act, also passed in 1862, that gave states the right to sell public land and use the profits to fund colleges of agriculture. These agricultural colleges housed experiment stations that provided a space for scientific research in agriculture. But the space itself was just the physical architecture

A Companion to the History of American Science, First Edition.
Edited by Georgina M. Montgomery and Mark A. Largent.
© 2016 John Wiley & Sons, Ltd. Published 2020 by John Wiley & Sons, Ltd.

that gave rise to the beginning of American nutrition. The Land-Grant College Act and the establishment of the USDA encouraged researchers to identify themselves as scientists, publish their research in journals (put out by the USDA and circulated among the laboratories), wrest agriculture from the hands of farmers who were "hicks, yokels and ignorant bumpkins," and place it firmly in the hands of scientists who could approach food production in objective, quantifiable, and verifiable ways (Rosenberg 1977). This national effort to encourage a more professional institution that governed food production influenced the future of nutrition in a myriad of ways.

At Wesleyan University in Storrs, Connecticut, a young chemist named Wilbur Atwater and his former advisor Samuel Johnson established the first experiment station in the United States. Both men actively promoted agricultural research and, having trained at agricultural chemistry laboratories in experiment stations in Europe, had a decidedly scientific mandate for the stations' development. Both Atwater and Johnson were heavily influenced by Justus Liebig's research in nutritional science. Liebig trained scientists from around the world at his laboratory in Giessen, Germany and specialized in the quantitative analyses of organic elements of food. Johnson and Atwater's research at Wesleyan reflected their training in Germany and in the late 1800s they began using bomb calorimetry to determine the heats of combustion of various foods using a modified Bertholet bomb calorimeter. As such, the Wesleyan calorimeter was a fundamental tool for establishing the scientific and quantitative framework of nutrition research in the United States. Building involved apparatuses, and training personnel to use these technologies was expensive, thus Johnson and Atwater were vocal promoters of federal funding for more institutional research centers. Both men were advisors to Congress from 1885–1887, encouraging the adoption of European-style scientific laboratories in university research stations, and they were instrumental in Grover Cleveland's passage of the Hatch Experiment Station Act of 1887. The Hatch Act allowed for an alliance between the USDA and the experiment stations, allowing the government agency to track, coordinate, standardize, and influence the research and the research scientists. It was the combination of Johnson and Atwater's research experience in nascent European nutrition laboratories, the passage of the Morrill Land-Grant College Act, the adoption of the Hatch Act, and the ethos and institutional values of science taking root in the newly established experiment stations that allowed for American nutrition to develop in a scientific, objective, and rational manner (Mudry 2009).

A central figure in the establishment of nutrition, as an academic field, was Wilbur Olin Atwater. The principal scientist at the Wesleyan experiment station, Atwater was greatly influenced by German chemists and physiologists, who worked to establish the relationships between and among food, metabolism, and health. Atwater worked at the University of Munich with Carl von Voit (a former student of Justus Liebig), who established the metabolic effects of proteins, carbohydrates, and fats on the body. Much of the German research that Atwater imported understood the body as a machine that was fueled by food and outputted work (Carpenter 1994; Cravens 1976). Atwater's appreciation for this model of the body influenced his research at the Wesleyan experiment station, and his articles in the USDA publications reflected that. As well, the body-as-machine model provided an organizing principle for both a politics and economics of nutrition research.

Pioneering Calorimetry: Technologies of American Nutrition

During his time in Voit's laboratory in Munich, Atwater experienced a number of technologies that helped him shape the quantitative model of nutrition and, in turn, influenced the USDA's understanding of what American nutrition ought to be. Atwater's lab colleague, Max Rubner, had begun experimenting with a new laboratory apparatus, the human calorimeter: a technology that married nutritional information about food and the energetic output of the human body. Along with the bomb calorimeter, an apparatus that Atwater established in his Wesleyan laboratory and that furnished scientists with data about food's caloric content, the human calorimeter allowed scientists to establish an objective and numeric relationship between food and human activity.

The rationality of this relationship impressed Atwater, and he saw this relationship between food, eating, and the human body as a boon to American society. Atwater had extensive experience determining nutrient ratios of foods, but the calorie allowed him to establish a direct relationship between doing and eating. Establishing this relationship could allow scientists and policymakers to understand nutrition as a political and economic issue, as well as a health issue. If Americans ate only what they output energetically, this could reduce waste and increase efficiency in the workforce. In a *Century* magazine article of 1887 Atwater encouraged, in the name of efficiency, an education in the science of nutrition:

> Our task is to learn how our food builds up our bodies, repairs their wastes, yields heat and energy, and how we may select and use out food-materials to the best advantage of health and purse. (Atwater 1887: 59–60)

As Atwater advocated publicly for education in nutrition, he continued to push for federal monies to fund nutrition research in his own laboratory. After the passage of the Hatch Act, Atwater published articles in the USDA bulletins and yearbooks suggesting the establishment of an American food laboratory because he claimed: "Of the fundamental laws of nutrition we know as yet too little" (Atwater 1893) and "among the things essential to health and wealth, to right thinking and right living, is our diet" (Atwater, 1895: 357). Atwater's alliance between eating, thinking, and living "right" demonstrates the ways in which nutrition became a social issue that fit neatly into both political and economic arguments for eating right (Biltekoff 2013).

In 1895, Atwater was successful in securing $10,000 from Congress and proceeded to build the first American respiration calorimeter that was based on the design of Voit and Rubner's calorimeters in Germany. Atwater and his collaborator, Wesleyan physicist Edward Rosa, built a small copper-lined room that measured temperature changes when a human subject was inside. The heat produced by the human subject was measured as calories expended, and the scientists carefully monitored the calories of the meals eaten, and the waste produced. These early experiments allowed, with impeccable accuracy, for the establishment of the relationship between expended energy, metabolized energy, and metabolic rates during physical activities like sleeping, reading, cycling, and performing mental tasks. The Atwater–Rosa calorimeter made the calorie the currency of eating and, with future nutrition research, allowed both the calorie and the consumer to be assessed qualitatively. The development of American nutrition relied heavily on a quantitative framework with objective means of research and assessment of both food

and people. Food analysis through calorimetry, or elemental ash analysis, allowed for
that (Atwater and Rosa 1899). Atwater paired his room calorimetry work with his work
in bomb calorimetry where he arrived at the "Atwater units" for carbohydrates, fats,
and proteins at four calories/gram, nine calories/gram, and four calories/gram, respectively. These Atwater units reflect the amount of energy per nutrient that can be metab-
olized by the human body, and their establishment represents a watershed moment in
American nutrition. The Atwater units were established as nutritional benchmarks, and
could be used by subsequent scientists to gauge foods, people, and diets.

To create a system of assessment, the USDA began to establish compendia of food's
chemical composition. In 1896, the *Office of Experiment Stations Bulletin* published
its first compendium of calorimetric analysis and chemical components breakdown of
some 2600 food items. This publication charted fat, carbohydrate, protein and ash content, and caloric value per pound of food (Atwater 1896). All told, USDA publications
between 1887 and 1902 charted over 8000 foodstuffs. These, and early dietary studies
of groups of people in various jobs and stations in life, became the definitive boundaries
for what nutrition ought, and ought not, to look like. As well, these references helped
create a structure for public education campaigns for nutrition that involved self-analysis
of metabolism, monitoring of food intake, and the qualitative rating of one's diet based
on nutrition data. The normalization of the eater, and the rationalization of the quality
of one's diet through comparison and calculation, shaped the mandate of nutrition as
wholly scientific and impartial. The idea of the taste of food, the whims of one's palate,
or the enjoyment of one's job or physical exercise became irrelevant in the framework
of nutrition. Also, an understanding of the science of nutrition, and comprehension of
the nutritional composition of foods, allowed eaters to be economically savvy, justifying
their purchasing habits as frugal if they purchased the greatest number of calories and
nutrients for the least amount of money (Atwater 1902).

As the USDA began publishing compendia of food data and dietary studies of various American populations, a strictly scientific framework of nutrition began circulating
among USDA experiment stations. The dietary studies, conducted by the USDA, were
meant to provide the government with "the most reliable data concerning the food
consumption of people of different nationality, sex and occupation, and under different
financial and hygienic conditions" (Atwater and Bryant 1898: 1). Atwater conducted
many of the studies, under the aegis of the "Food and Nutrition of Man" series published by the USDA. While the dietary studies did not produce or publish data about
food, they did provide data about Americans, and put in place the scientific framework
for assessing people through their diets. The studies published the "correct" consumption of calories and factors of calculating meals. For example, a woman required 0.8
the food of a man, children under two years old 0.3 the food of a man. These factors,
among others, provided a moralizing tone for nutrition that allowed experiment station
workers to assess both the diet and the people eating it. In the case studies of families in
Chicago in 1898, "[T]he need of training in housekeeping and cooking was very apparent … The conditions of the families in the congested districts of Chicago and other
cities can undoubtedly be improved by education. The housekeeper should be taught
how to prepare and serve food. In this way the diet may be made more attractive and
more wholesome" (Atwater and Bryant 1898: 72).

The idea of nutrition grew up through the agricultural experiment stations but just
after the turn of the century, schools of home economics began to recognize the

fundamental role that food played in public hygiene and public health, as well as recognizing the importance of knowing "what things the body needs in its food and how these needs can be filled by the ordinary food materials" (Hunt 1917).

Nutrition in the United States

Various doctors, health reform theorists, and religious types, who saw food as the ultimate way to stay healthy often countered "official" nutrition advice that came from the government through the USDA. Sylvester Graham, of the eponymous "graham" cracker, was a Presbyterian minister who became a spokesperson for the Pennsylvania Temperance Society in 1830. Alongside William Alcott, who preached that humans had a duty to "understand the structure and function of the human body" (Whorton 1982: 60), Graham felt that maintaining a diet full of vegetables, exercising and dressing appropriately, and being abstinent were the cornerstones to being moral and healthy. Meat, spices, alcohol, coffee, and tea overstimulated the body and were the gateways to poor health. The idea of quashing stimulation, long associated with physiological and sexual desire, was easily justified in America and, as such, Graham's bland diet of vegetables, grains, and not much else, gained popularity in the early mid-1800s. While Graham did not use the word "nutrition," the word "hygiene" became an all-encompassing term for correct patterns of eating and being (Gevitz 1988).

Graham's legacy of a "hygienic" diet was carried on in the latter half of the nineteenth century by health reformer John Harvey Kellogg. A Seventh Day Adventist, Kellogg became the resident physician at Michigan's Battle Creek Sanitarium. Like Graham, Kellogg encouraged temperance, abstinence, and vegetarianism. Kellogg introduced granola as a health food, as well as cereals and grains for breakfast. Kellogg understood the idea of nutrition as beginning with food and ending with a satisfying trip to the toilet, and thus promoted regularity through the intake of bran. Another health reformer concerned with food intake and food output was Horace Fletcher who, while less concerned with moral purity, did agree with the Progressive Era's drive to efficiency and waste reduction. Fletcher was known as the "Great Masticator" who preached that the secret to good health was to chew food thoroughly so as to extract every nutrient and optimize digestion. While Kellogg thought that healthy eating should result in frequent daily bowel movements, Fletcher thought that healthy eating should result in no more than one bowel movement every fortnight.

Although these popular dietary reformers differed in their physiological activities, their goals were the same: the achievement of health through eating. While Graham and Kellogg's earlier diets had a moral tone (through temperance), Fletcher's drive to optimize the function of the body through the efficient extraction of nutrients by chewing better represented the Progressive Era in the United States. The themes of efficiency and waste reduction in nutrition were also seen in the launch of the federally published food guides, the first of which was published by the USDA in 1917.

At the turn of the twentieth century the USDA began its public campaign of nutrition education. In 1916 the agency published its first food guide aimed at codifying a healthy diet. Nutrition departments in universities did not yet exist but the scientific structures of nutrition, in the form of food groups, calories, and vitamin analyses, moved into the public sphere in home economics departments. Such departments promoted hygiene and rules for living through their curricula and home economics further developed in publications like the *Journal of Home Economics*, which published articles such as "Nutrition for the People," written by Max Rubner (Rubner 1913). Much of the research in food chemistry, early physiology and human metabolism, and nutrition found their applications popularized in schools of home economics. The domestic sciences were taught in the land-grant colleges and many of the women's colleges, and the semantic weight of "science" in home economics afforded many of the programs both political and financial support from the government. A scientific approach to food and its interaction with the body was, according to home economists, a worthy goal and education programs encouraged the public's fluency in a language of proteins, calories, and energy. The move toward a scientific, as opposed to a moral, ideal home, meant that the justification of food choices was quantitative and able to be externally referenced. Federally published food guides provided such a reference.

Federal Food Guides Codify the Science

Because nutritional research from the laboratories of the experiment stations fell under the jurisdiction of the USDA, the department's public education materials focused on creating a scientific framework for a healthy diet. Knowing groups of foods according to their scientific function and how much of each food to eat according to one's activity level or socioeconomic status became a way in which the USDA could promote nutritional science as the arbiter of food rules. As well, federally published food guides became a way in which the USDA could reframe eating as a scientific process aimed at improving public health, instead of a pleasurable, cultural, or social activity. The published food guides often used the science of nutrition as the justification for adhering to whatever political or social cause reflected the era (Biltekoff 2013; Mudry 2009).

The first official food guide published in 1917 reflected the currents in nutrition research and encouraged "a simple method of selecting and combining food materials to provide an adequate, attractive and economical diet" (Hunt 1917: preface). Standards, order, and economics began to take hold as justifications for particular food choices. Home economists like Ellen Swallow Richards, who ran an experiment station in Boston, as well as Isabel Bevier at the University of Illinois Urbana-Champaign experiment station, championed such justifications. Scientific advice usurped family traditions and lay -knowledge as validations for raising children, cleaning one's home, or preparing meals, in particular ways. As such, domesticity "expanded into an objective body of knowledge that has to be actively pursued" (Shapiro 2001: 7). Food preparation and feeding one's family, then, became the domain of nutrition science. Federally published food guides were a way to disseminate that science.

In the first USDA food guide, published by Carolyn Hunt in 1917, five food groups were introduced: fruits and vegetables, meat and meat substitutes, starches, sugars, and fats. The co-author of the guide was Helen Atwater, Wilbur's daughter, and his influence was evident. The food guide grouped food into either their botanical category or their scientific function in the body. The idea of "fuel value" in the form of calories further grouped the food and thus required both knowledge of basic nutrition and food groups and calculation on the part of the housekeeper to buy and prepare enough "fuel" for the family. Improper calculation became subject to moralized judgment about the quality of the diet and of the citizen.

Vitamins – A New Era in Nutrition

By the 1920s the scientific foundation of nutrition was growing, owing largely to the uptake of scientific techniques that could isolate and determine the content of vitamins and trace elements. This ability of scientists to isolate and identify biologically active agents in food changed the role of food in nutrition. While Atwater's research treated food as fuel, the isolation of vitamins ushered in a new, smaller, and increasingly quantitative understanding of foods. Vitamin research began in America in 1913 with Wisconsin biochemist Elmer McCollum isolating fat-soluble vitamin A. Following the isolation of vitamin B in 1912 by Polish biochemist Casimir Funk and Hungarian chemist Albert Szent-Györgyi's isolation of vitamin C, research in vitamin isolation allowed for the generation of qualitative judgments of food based on their vitamin content. As scientists determined more biological functions of vitamins (vitamin B prevented beriberi, vitamin C prevented scurvy, and vitamin D prevented rickets), the USDA could functionalize foods, group them and make claims about the correct, scientifically prescribed intake of certain foods. Specific foods could now be designated as nutritious, and others not, depending upon their vitamin and mineral content.

In 1928, the American Institute for Nutrition (AIN) and the *Journal of Nutrition* were established, providing both a focus and a forum for research and researchers interested in the relationships between food and the body. Early issues of the journal focused largely on energy intake and basal metabolism, as well as the role of vitamins and minerals in preventing disease. Article titles like "Meat in Nutrition: Preliminary Report on Beef Muscle" (Nelson et al. 1930), "Food Intake in Pregnancy, Lactation and Reproductive Rest in the Human Mother (Shukers et al. 1931) and "A Comparison of Apricots and their Carotene Equivalent as Sources of Vitamin A" (Morgan and Madsen 1933) demonstrate the longstanding influence of Atwater's scientific and mechanical models of food and the body on the field of nutrition. In 1934, the American Institute for Nutrition held its first meeting at Cornell University's Medical School and in 1940, the Institute joined the Federation of American Societies for Experimental Biology. Nutrition then became officially aligned with scientific experimentation, and the research published in the pages of the journal reflected this alliance. The link with biology also meant that nutrition researchers could compete with the other human sciences for funding from the National Science Foundation.

By the 1930s the social landscape of the United States had undergone widespread changes, and there was rampant unemployment, hunger marches, and breadlines. From the late 1920s to the early years of the 1930s, household incomes plummeted and a third of the American workforce was unemployed. Nutrition in the United States was reduced to concerns about stretching calories to feed the population. How to choose those calories wisely and economically became important, and food guides from this era reflected this. The 1933 food guide, written by USDA home economist Hazel Stiebeling, had 12 food groups, but also 4 eating groups. The food groups reflected the nutritional information available at the time, and grouped foods based on minerals, vitamins, calories, proteins, and fat. In addition to this, there were groups of eaters who followed the nutrition guide in one of four ways: restricted, adequate at minimum cost, adequate at moderate cost, or liberal. Whether someone ate at a "restricted" level or a "liberal" level did not matter, the guide made sure that their nutritional needs were satisfied. As such, a restricted eater was allotted 240 pounds of flour a year and 8 dozen eggs, and a liberal eater was allotted 100 pounds of flour and 30 dozen eggs. Restricted eaters were recommended to get almost half of their calories from flour, with only 5% of their calories coming from proteins: meat, fish, and eggs. Liberal eaters had only to eat 15% of their calories from flour and 21% from meat, fish, and eggs (Stiebeling 1933: 3). Prominent nutrients at the time included calcium, phosphorous, iron, and vitamins A, B, and C.

Despite encouraging the impoverished to eat more flour, the flour was not fortified. Early fortification programs in the United States included only iodide in salt to prevent goiter – begun in 1924 – and vitamin D to milk begun in 1933, the same year that the USDA published its food guide (Quick and Murphy 1982). Still, if the poor ate flour, they were guaranteed "excellent returns on calories, proteins, phosphorous, and iron" (Stiebeling 1933: 10). The addition of fats to the cereal made it palatable, according to the guide, and gave it a "staying quality" (Stiebeling 1933: 10). By the late 1930s and early 1940s several deficiency diseases were identified and fortification became commonplace. In 1940, the Committee on Food and Nutrition recommended that flour be fortified with thiamin, niacin, riboflavin, and iron (Foltz, Barbooka, and Ivy 1944; McLester 1939; Williams, Mason, and Wilder 1943). The US government's hopes here were to improve the nutritional status of the nation. This was timely, since World War II had forced the government to generate information about the nutritional health of Americans. Fighting in the war meant the recruits needed to be strong and healthy, but the Depression had taken a physical toll on the population, and the nutrition of the nation was grim, despite government efforts at fortification, rationing, and nutritional education campaigns. In 1941, President Roosevelt called a National Nutrition Conference for Defense in order to gather information about the American population's nutritional health and the pool of soldiers called upon to fight in Europe. The results were so bad that then surgeon-general Thomas Parran warned that the ill-health of the male population was "a danger to military strength" (Mudry 2009).

Good nutrition, again, became a politically charged cause during World War II and the White House sought a technocratic solution. President Roosevelt called on the Committee on Food and Nutrition to conduct studies to determine the exact amounts of each nutrient that Americans ought to be eating to be "healthy." These standards, born out of the exigency of war, became the recommended dietary allowances, or RDAs. Generating the RDA numbers was fraught with dissention, and the "allowances" were, in fact, a set of numbers that exceeded the minimum daily requirements of any nutrient or vitamin by about 30%. As governments established minimum requirements for specific nutrients (recommended dietary allowances, or RDAs, were first established by the United States in 1941), foods became explicitly prophylactic to a degree that they had never been before. Federally mandated minimums for nutrient intake provided the food industry an angle with which to market its products as essential tools for maintaining national health, safety, and security. In an advertisement from the 1940s, Kix cereal advertised that it "consulted leading nutrition authorities before making Corn Kix," and as such, it contained vitamin B, vitamin D, calcium, and phosphorous. Brer Rabbit Molasses claimed that: "Scientific tests have shown Brer Rabbit is second only to liver as a rich food source of iron the body can use" and the ad revealed a stamp of acceptance from the American Medical Association (*Life Magazine* 1943). To boot, the American government endorsed "Vitamin Donuts" that were each fortified with "a minimum of 25 units of Vitamin B1."

The establishment of the RDAs placed quantitative nutrition at the forefront of good health and created a framework for healthy eating. In 1943, the USDA published the *National Wartime Food Guide,* which codified the RDAs, and further grouped foods according to their nutritive content: carbohydrates, proteins, fats, dairy products, and vegetables. American rationing programs and "victory gardens" further inculcated the ideas of a scientifically justified diet among the population. Vitamin-packed vegetables were to be grown and preserved; meats, sugars, and coffee were to be rationed (Bentley 1998). According to the USDA, America's victory in World War II relied on the nutritious eating patterns of the population, and while Americans at home were forfeiting their meat, sugar, and coffee for the war effort, the nutrition of Americans overseas was greatly improved. The servicemen became accustomed to larger portions of food, and a greater variety of it.

After World War II, the field of food chemistry burgeoned. Food preservation and food additive techniques, alongside advances in refrigeration, agricultural mechanization, transportation infrastructure and architecture meant that the American supermarket offered more and different kinds of foods to eat. In much of the United States, making sure that Americans ate enough and received the RDAs seemed less important than celebrating the abundance of what was available. Poor nutrition became racialized and localized mostly in the rural South and Appalachia, and largely ignored until Robert Kennedy visited rural Mississippi in 1967 and saw farm workers who had been displaced by agricultural technologies and who survived on one meal a day. By the 1950s, food fortification reflected currents in nutrition research, with margarines, rice, and cereals fortified with folate, iron, and niacin. But overall, concerns about having enough food or sufficient nutrients in the American diet had subsided. New food guides counseled what constituted an "adequate" diet, which consisted of four food groups: meats, grains, dairy, and fruits and vegetables. These "Basic Four" groups were considered the backbone to good nutrition, but nutritionists at the USDA expected that they be a foundation for a good diet and supplemented with fats, oils, and refined grains to "round out the meals" and "satisfy the appetite" (Page and Phipard 1956: 1).

The landscape of the supermarket began changing in the 1950s and many of the "supplemental" foods that were available were in the form of processed, preserved, or prepared foods. Refrigerators with freezers meant that the frozen dinner became a reality. The availability of cheap sources of saturated fats and widespread oil hydrogenation allowed for greater shelf lives of snacks, dairy and dairy-like products (Critser 2003). In the strict framework of quantitative nutrition, products like Twinkies, Cheez-Whiz, and Cool Whip did not provide much beyond calories, but with techniques that extended "mouthfeel" and gave the eater a shot of sugar or salt, they provided tastes new to the American public.

Nutrition research published in the *Journal of Nutrition* in the 1960s reflected the zeitgeist of the era; there was enough food to go around and, as such, researching deficiencies became less important. Instead, the *Journal of Nutrition* and the *Journal of Clinical Nutrition* published research on the effects of the patterns of intakes of certain nutrients: "Assessment of Nicotinic Acid Status of Population Groups" (De Lange and Joubert 1964), "Chromium, Cadmium and Lead in Rats: Effects on Life Spans, Tumors and Tissue Levels" (Schroeder, Balassa, and Vinton 1965); and the exposure of subjects to abnormal nutritional circumstances: "Tolbutamide-induced hypoglycemia" (Brown and Stone 1964) and "Effect of Progressive Starvation on Rat Liver Enzyme Activities" (Freedland 1967). It was in this decade, however, that nutrition scientists began to publish about obesity, a harbinger of the shift in attitudes toward food and the body that begins in the late 1960s and early 1970s.

In the 1960s, the American Red Cross and the American Heart Foundation published literature aimed at informing the public about the dangers of being overweight. In 1960, though not changing the Basic Four food guide, the USDA published a circular entitled "Food and Your Weight," which, in addition to providing the reader with calorie counts for popular foods, included a section for "Basic Weight-Control Facts" (Page and Fincher 1960: 5) and "Suggestions for Reducers" (Page and Fincher 1960: 15), which cautioned: "Persons who are considerably overweight should probably reduce." The guide cautioned not to eat a diet that was too limited or that was "inadequate in essential nutrients." (Page and Fincher 1960: 15) This circular demonstrated the shift in concerns about nutrition in the United States. Concerns about malnutrition because of a lack of food were replaced with concerns about malnutrition, disease, and ill-health due to overconsumption. The scholarly journals of the American Institute for Nutrition track this shift as well. In 1965, for example, the *Journal of Nutrition* and the *Journal of Clinical Nutrition* published a total of 26 articles on obesity, by 1968, this number had almost doubled to 44, and by 1978 and 1979, the number of articles pertaining to obesity in these journals had risen to 91 and 102, respectively. In 1968, the Senate Select Committee on Nutrition and Human Needs held its first meeting. Still concerned about malnutrition because of a lack of nutrients, the committee was surprised when Dr. Jean Mayer, nutrition professor at Harvard University, dispelled this notion by stating that the problem of nutrition in American was *over*consumption of food, not underconsumption. Mayer pointed to an increase in cardiovascular disease, poor physical activity programs, and diets high in saturated fats as the culprit. By the time the USDA published the *Dietary Goals for the United States* in 1977, hunger was no longer a concern for nutritionists. Rather, Americans needed a good education in nutrition so that they could "begin to take responsibility for maintaining their health and reducing their risk of illness" (USDA 1977: iv).

Codifying this new approach to nutrition, one of eating less instead of eating more, was difficult for both eaters and industry. In the Senate Select Committee Meeting, the assistant secretary of health, Dr. Julius Richmond, stated that "many experts now believe that we have entered a new era in nutrition, when the lack of essential nutrients is no longer the major nutritional problem facing most American people" (USDA 1977: xxxiii). Many of the written goals in the report contained words such as reduce, avoid, and limit. Lobbies and industries like the beef and sugar industry were unhappy with the report, as federal science indicated that meat and sugar were the culprits in poor nutrition habits that led to heart disease, high cholesterol, and diabetes (Nestle 2002).

In 1980, after the USDA published the McGovern report, it issued a stopgap food guide called "Nutrition and Your Health: Dietary Guidelines for Americans." The message for good nutrition herein was to eat a variety of foods, but to avoid too much fat, sodium, cholesterol, and sugar. This guide gave suggestions for the number of servings per day as a limit to consumers. In the years between the stop-gap guide and the "Food Guide Pyramid," published in 1992, research in nutrition shifted even more dramatically from diseases of lack of food to diseases of abundance. The number of overweight and obese Americans was growing at an exponential rate and nutrition science began to focus on what to avoid eating, as opposed to what to eat, to become healthier. The learned "Obesity Society" was established in 1982, to "promote research, education and advocacy to better understand, prevent and treat obesity" and the society began publishing the US-based journal *Obesity: The Journal for Health and Social Behavior* in 1993. Corollary medical foci that developed alongside the shift in new nutrition advice included advances in bariatric surgery (lap-band, stomach-stapling), pharmaceutical intervention for the overweight and obese (Orlistat, Xenical), and type-2 diabetes treatment protocols for the overweight and obese.

In 1990, the US government passed the Nutrition Labeling and Education Act that required all packaged foods to be printed with a nutrition facts label. The bill mandated that all food packaging contain an information panel that indicated the serving size, cholesterol, calories, fats, sugar, protein, sodium, and carbohydrate content of the food in the package. The goal of this label was to "provide information regarding the nutritional value of … food that will assist consumers in maintaining healthy dietary practices." The government's hope was that the nutrition facts label could structure the food industry's use of a product claim to be nutritious, and be of use to the consumer who would want to make healthy food choices. The US Food and Drug Administration monitors the food industry's claims to be "nutritious" and these claims are mostly associated with quantifiable chemical contents of a food and how much of any given nutrient the food contributes to a mandated "Daily Value" of calories, minerals, fats, proteins and carbohydrates. For example, the claim of "excellent source of" and "rich in" can only be applied to foods containing 20% or more of a particular nutrient in each federally mandated "serving size" (Stehlin 1993).

In 1992, the secretary of the USDA, Edward Madigan, released the "Food Guide Pyramid," an illustrated food guide that represented the currents in American nutrition. The overview of American nutritional science was metabolically reductive but easy to understand: fat on the plate must mean fat on the body. As such, the food guide encouraged Americans to eat mostly low-fat breads, cereals, and pasta and fewer fatty meats and dairy products. The guide advised that fats and oils were to be used "sparingly." The low-fat advice seemed to make sense to the public and to food processors who began removing fat from their products and replacing it with sugars which had, of course, no fat. Food chemists began to make synthetic fats that were unmetabolized by the body (Olestra) and extending mouthfeel in products whose fat had been removed. The goal of many food chemists was to try to remain within the nutritional guidelines delineated by the food guide, but to allow consumers to eat what they wanted. What was confounding was that while Americans began to eat less fat, obesity rates in the United States continued to climb. Between 1991 and 2010, rates of overweight and obesity rose from just over 20% to just shy of 40% of the American population (www.cdc.gov). Diet-related diseases like type-2 diabetes, and metabolic syndrome (first referred to as "Syndrome X" by endocrinologist Dr. Gerald Reaven in 1988) which refers to the unhealthy profile of high insulin, high blood sugar, elevated triglycerides, and low levels of HDL cholesterol, began to increase exponentially.

Though popular medical opinion continued to attribute high intake of red meat and saturated fat to heart disease, Reaven blamed sugar and high glycemic indexed carbohydrates, whose consumption was rampant in the 1990s because of the "low-fat" nutritional advice. Between 2004 and 2006 the ubiquity of high-fructose corn syrup (HFCS) in sodas and processed snack foods became fodder for both research scientists and science journalists, who drew links between the widespread consumption of HFCS and the so-called obesity epidemic (Bray Nielsen, and Popkin 2004; Jacobson 2004).

Over the twentieth century, sugars and carbohydrates have been, perhaps, the most polarizing nutrient with respect to health and diet. The drive to reduce and remove sugars from food began in the late 1960s when bench chemists created sugar substitutes, cyclamate, acesulfame K, and aspartame, but until 1982 – when aspartame was approved for wider use in Europe and the Americas – only saccharin was government approved as a sugar replacement. Chemical companies and food and beverage manufacturers promoted calorie-free sweeteners as healthier alternatives to sugar. In 1972 Robert Atkins published *Dr. Atkins' Diet Revolution*, which promised weight loss by eschewing all carbohydrates, sugars in particular. However, the Atkins diet did allow, and even encourage, the use of artificial sweeteners. This villainizing of sugar may have been more cosmetic and cultural than purely health related, but as "sugar" in the form of high-fructose corn syrup, dextrose, maltose, cane syrup, and juice concentrates was added to virtually all industrial food products, doctors, dieticians and nutritionists became concerned with the role of sugar consumption in diabetes, obesity, and metabolic syndrome.

As academics, policymakers, and industry attempt to agree upon what makes a nutritious diet, American legislation attempts to draw distinctions between foods that are nutritious and newer foods that "function" through the use of isolated additives. Functional foods are part of nutrition that professor of consumer behavior and nutritional science Brian Wansink calls "consequence-related" (Wansink 2007). According to Wansink, functional foods will act in the body beyond their simple "attribute-level" of calories, fat, proteins, and carbohydrates. We might consider a food "functional" in its natural, unprocessed state – oats are functional because their soluble fiber content has been shown to reduce cholesterol – or foods may be functional through the addition of various nutrients. Fruit juices with added soluble fiber, granola bars with added probiotics, and peanut butter with omega-3 fatty acids are all cases of nutrient fortification making a food "functional."

The popularity of functional foods demonstrate that the early model of nutrition in the United States remains and is continually reinforced even into the twenty-first century. The basic understanding of nutrition is that food functions biochemically in the body to promote health or cause disease. As nutrition researchers continue to understand the chemical, cellular, and even genetic interactions between food and the body, consumer literacy in the complexities of human metabolism do not necessarily keep pace. With a barrage of messages that are always changing, knowing what constitutes a nutritious diet is difficult. Health claims of a food being "nutritious" have always been part of nutrition in the United States. And, as the science of nutrition continues, yesterday's vitamin donuts and today's kale chips will undoubtedly give way to new superfoods promising to bestow health upon their eaters.

Bibliographic Essay

An analysis of nutrition in the United States, its key figures, and historically relevant moments involves understanding the histories of several components, all of which contribute to making nutrition a formal discipline in American academia, politics, and medicine. While it depends somewhat on an understanding of what nutrition is, the general consensus in the United States is that nutrition is a science, with a formal curriculum of education, standard research procedures and protocols, and formal regulatory bodies that advise governments on policy.

Early texts on American nutrition begin as histories of agriculture, and Benjamin Cohen's *Notes from the Ground: Science, Soil and Society in the American Countryside* (2011), Margaret Rossiter's *The Emergence of Agricultural Science – Justus Liebig and the Americans, 1840–1880* (1975) provide historical arguments for how and why agriculture became scientized in the United States. Early documentary histories of the state of nutrition among Americans, across the years, can be found in Elaine McIntosh's *American Food Habits in Historical Perspective* (1995). A dietitian, McIntosh attempts to provide data for interested researchers. In a similarly analytical approach, Waverly Root and Richard de Rochemont's *Eating in America: A History* (1976) provides readers with information about what America was eating and while this is not a history of nutrition, per se, it might be a useful starting point for researchers who are interested in specific food consumption patterns. More critical examinations of patterns of consumption and dietary practices are found in Harvey Levenstein's *Paradox of Plenty: A Social History of Eating in Modern America* (1993) and *Revolution at the Table: The Transformation of the American Diet* (2003). Levenstein, an historian, provides critical, social, and political reasons for why Americans ate in the manners they did. *Revolution* covers the period from 1880–1930, and *Paradox* covers the period from 1930 to the twenty-first century.

The role of science in American politics, and the establishment of American university programs in nutrition cannot be understated and, as such, researchers interested in this role should go to Charles Rosenberg's article "Rationalization and Reality in the Shaping of American Agricultural Research" in *Social Studies of Science* (1977), as well as his work *No Other Gods: On Science and American Social Thought* (1977). As research-institution-based science became the *lingua franca* of nutrition in American universities less formal "nutritions" could be found in privately owned health and hygiene institutes. Gevitz's *Other Healers: Unorthodox Medicine in America* (1988) points to the importance of other, alternative health movements, and their social influence in the United States, and James Whorton's more documentary approach, *Natural Cures: The History of Alternative Medicine in America* (2002) provides researchers with contexts for alternative approaches and definitions to "nutrition." The text with the finest point on alternative approaches to nutrition comes also from Whorton in *Inner Hygiene: Constipation and the Pursuit of Health in Modern Society* (2000), wherein he points to the influences that religion had on ideas of health and eating in the United States. Daniel Sack's *Whitebread Protestants: Food and Religion in American Culture* (2001) will also provide researchers with a critical material culture approach to religion and American food.

Beyond religion, researchers seeking to understand how the political context and culture affects nutrition policies should look to Charlotte Biltekoff's *Eating Right in*

America: The Cultural Politics of Food and Health (2013) for a historical, American Studies approach to nutrition. Biltekoff examines both identity- and subject-making in and through discourses of food and nutrition in the United States. A more rhetorical approach to the ways in which language shapes nutrition and health through discourses of science can be found in Mudry's book *Measured Meals: Nutrition in America* (2009). Finally, for a nutritionist's critical opinion of modern nutrition politics, and the resulting policies, Marion Nestle's *Food Politics: How the Food Industry Influences Nutrition and Health* (2002) is a popular and thorough account of nutrition in American with special attention paid to the influential role of industry in nutrition policy.

Scholars Jessica and Allison Hayes-Conroy's edited volume on critical nutrition studies *Doing Nutrition Differently: Critical Approaches to Diet and Dietary Intervention* (2013) contains valuable readings on issues of gender, colonialism, and race's roles in nutrition, and while not entirely focused on American nutrition, it is a good place for scholars to find critical approaches to nutrition that certainly apply to an American context. One of the scholars therein Gyorgy Scrinis, who coined the phrase "nutritionism" and who wrote a text of the same name *Nutritionism: The Science and Politics of Dietary Advice* (2013), will provide scholars with a critical, though not exclusively American, approach to the ideology of nutrition and the scientific practices subsumed by such an ideology. A history of such reductionist approaches to food can be found in Rima Apple's *Vitamania: Vitamins in American Culture* (1996). Apple's text, while dated, points clearly to the patterns of reductive science and nutrition research that, through nutrient isolation, created a place for both policies and marketing of "functional foods."

Finally, commentaries on how and why the American obesity crisis arose are plenty and range in opinion as to *why* obesity happened in America. While researchers will be able to engage with primary materials on the subject, political, social, and cultural commentaries on the subject may be found in journalist Greg Critser's *Fat Land: How Americans Became the Fattest People in the World* (2003), scientists Kelly Brownell and Katherine Horgen's *Food Fight: The Inside Story of the Food Industry, America's Obesity Crisis, and What We Can Do About It* (2004), and law professor Paul Campos' *The Obesity Myth: Why America's Obsession with Weight is Hazardous to Your Health* (2004). Each author has a distinct methodological approach that is informed by his/her background. While the data in the text may change, the approaches of the writers will provide scholars with methodological frameworks to perform their own work.

Chapter Seventeen

PALEOANTHROPOLOGY AND HUMAN EVOLUTION

Matthew R. Goodrum

Long-held beliefs about human origins underwent profound change during the nineteenth century as groundbreaking works introduced Americans to a suite of new discoveries and ideas about human evolution and the cave dwellers of the Ice Age. Prehistoric archaeologists identified a succession of Stone, Bronze, and Iron Ages reaching back thousands of years. Crude stone artifacts found with extinct Ice Age animals extended this prehistory even further into the geologic past. Charles Lyell and John Lubbock in England, Gabriel de Mortillet in France, and Carl Vogt in Germany systematically assembled this growing archaeological evidence to argue that culturally primitive humans existed during a vast period in the deep past. Excavators then began to discover the fossilized remains of these Ice Age humans. In Germany Hermann Schaaffhausen, professor of anatomy at the University of Bonn, examined a partial human skeleton unearthed from a cave in the Neander valley in 1856. He concluded that the Neanderthal skeleton belonged to a barbarous race that inhabited Europe at a remote period in the past. Maximin Lohest excavated two Neanderthal skeletons from a cave near Spy in Belgium, which contributed to the growing debate over whether the Neanderthals were merely an extinct Ice Age race or a distinct human species. French anthropologists examining human bones found with extinct animals from a number of French sites concluded that they represented yet another type of Ice Age human they called Cro-Magnon Man.

If all of this was not shocking enough, Charles Darwin's theory of evolution transformed the scientific study of human origins. Darwin's theory that natural selection caused the evolution of new species raised the disturbing prospect that humans were not created by God but had evolved from an animal ancestor. European zoologists had already identified many similarities between ape and human anatomy. Darwin used this evidence to argue that humans had evolved from an ape ancestor. Supporters of evolutionary theory, such as Thomas Huxley in England and Ernst Haeckel in Germany,

A Companion to the History of American Science, First Edition.
Edited by Georgina M. Montgomery and Mark A. Largent.
© 2016 John Wiley & Sons, Ltd. Published 2020 by John Wiley & Sons, Ltd.

conducted comparative anatomy studies of apes and humans to strengthen the evidence that humans evolved from an ancient form of anthropoid ape.

Human Evolution Comes to the United States

These dramatic events occurring in Europe were quickly communicated to the United States. American editions of Charles Lyell's *The Geological Evidences of the Antiquity of Man* (1863), Thomas Huxley's *Evidence as to Man's Place in Nature* (1863), and Charles Darwin's *The Descent of Man* (1871) all appeared in the same year as the English editions, and John Lubbock's *Prehistoric Times* (1865) was printed in New York in 1872. Soon, American scientists were composing books on these subjects for the general public. Charles Rau, curator of antiquities at the Smithsonian Institution, reviewed recent discoveries of Paleolithic artifacts, art objects, and human fossils in *Early Man in Europe* (1876). Thomas Wilson, curator of prehistoric anthropology at the Smithsonian, frequently communicated the latest European discoveries to American readers. He attended the International Congress of Prehistoric Anthropology and Archaeology held in Paris in 1889 and, while summarizing the events of the meeting, was the first American to use the term *paleoanthropology* in print (Wilson 1891). This term was introduced in the 1860s in France to refer to the study of extinct types of humans such as Neanderthal and Cro-Magnon. It is important to note that at that time one could accept extinct species of humans without accepting the idea of human evolution. Both were controversial and many American intellectuals rejected the notion of primitive Ice Age peoples (see, for example, Southall 1875). Some tried to find ways of reconciling these ideas with Christianity (Livingstone 1992). Alexander Winchell, professor of geology at Vanderbilt University, resorted to the old but controversial idea of pre-Adamites, people created by God before the creation of Adam. Winchell (1880) proposed that the original pre-Adamic peoples were dark skinned but over time, by the law of progressive development, one group of pre-Adamites gave rise to a light-skinned Adamic race in central Asia, which then migrated to many parts of the world.

Race was a significant scientific and social problem in Europe and the United States, and the emerging discipline of anthropology focused much of its attention to gathering data about the existing human races and to explaining the origin of racial diversity. One of the major questions dividing anthropologists was the question of whether the human races form a single species with a single origin (monogenism) or are distinct species each having separate origins (polygenism). Besides the moral and political consequences of this debate, it also had implications for scientific theories about human origins. By mid-century anthropologists had developed instruments and techniques to take careful measurements, particularly of the size and shape of the skull (called craniometry), in order to establish the distinctive characteristics of different human races. These same techniques were used to examine the skulls and skeletons of fossil humans to establish their relationship with existing human races. Meanwhile, supporters of human evolution compared the anatomy of living apes with modern humans and the Neanderthal and Cro-Magnon fossils to construct scenarios of human evolution.

Edward Drinker Cope was among the most influential Americans to address the question of human evolution in the late nineteenth century. Cope studied comparative anatomy at the Academy of Sciences of Philadelphia and served as a pale-ontologist with the US Geological Survey. He became a prominent paleontologist and promoted a form of evolution called neo-Lamarckism, which downplayed natural selec-tion and argued that organisms could acquire new traits directly in response to their environment and pass those traits to their offspring. Cope believed that the differ-ent human races evolved at different rates and that the darker skinned "Negro" and "Mongolian" races possessed more ape-like traits than the "Indo-European" race, with Negroes being the most ape-like and least evolved. He thought Neanderthals were a distinct species that was even more ape-like than the "lowest" modern human races, and that humans had evolved from Neanderthals (Cope 1893). Unlike Darwin, Cope argued that humans and the anthropoid apes diverged from lemurs in the Eocene, meaning that humans had not evolved from apes (Cope 1888). Theories and debates about human evolution continued to proliferate through the turn of the century, with most of the major discoveries and theories coming from Europe. But American scien-tists began to play a more prominent role after the turn of the twentieth century.

The most influential physical anthropologist in the United States in the early twen-tieth century was Aleš Hrdlička. Born in Bohemia, Hrdlička immigrated to the United States in 1881. He studied medicine and in 1896 traveled to Paris to study physi-cal anthropology with Léonce-Pierre Manouvrier. Hrdlička was offered a position in the Division of Physical Anthropology at the Smithsonian Institution in 1903 and was appointed its curator in 1910. He assembled an extensive collection of human skeletons from around the world and made the Smithsonian a center for physical anthropology research. Hrdlička was largely responsible for establishing the early institutions dedi-cated to physical anthropology in the United States. He founded the *American Journal of Physical Anthropology* in 1918, serving as its editor from 1918 to 1942. The journal had an international outlook and became one of the primary venues for announcing new hominid fossils and discussing human evolution. Hrdlička founded the American Association of Physical Anthropologists in 1930, which helped to professionalize and organize the discipline in the United States (Spencer 1981).

Besides his research on the native peoples of the Americas, Hrdlička closely followed the discoveries of new hominid specimens. He was among the few Americans to travel throughout Europe to examine hominid fossils and excavation sites, and there were many new fossils to consider. In 1894 the Dutch anatomist Eugène Dubois announced a new species he called *Pithecanthropus erectus,* based upon fossils he discovered on the Indonesian island of Java. A wealth of new Neanderthal fossils were unearthed in Europe and Africa, and in 1907 quarry workers in Germany excavated a massive jaw that was attributed to a new hominid named *Homo heidelbergensis.* In South Africa, Raymond Dart announced in 1925 that a small skull belonging to a new hominid he named *Australopithecus africanus* represented the evolutionary link between the apes and humans. And then there were the notorious Piltdown fossils found in England in the 1910s. Drawing upon his direct observation of many of these specimens, Hrdlička published a thorough summary of the existing hominid fossil record titled *The Skeletal Remains of Early Man* (1930). Hrdlička supported a linear version of human evolution from an anthropoid ape ancestor. In the Huxley Memorial Lecture, Hrdlička outlined his Neanderthal phase hypothesis. He defined Neanderthals as the makers of Mouste-rian artifacts and unlike many European anthropologists he argued that modern *Homo sapiens* evolved directly from Neanderthals (Hrdlička 1927; Spencer and Smith 1981).

Hrdlička exerted a powerful influence in American anthropology, but he had little training in biology or evolutionary theory. By the 1920s other American researchers were entering the discussion about human evolution. Henry Fairfield Osborn, curator of vertebrate paleontology at the American Museum of Natural History in New York, rose to prominence studying the evolution of mammals. In biology, he supported a form of evolution called orthogenesis and proposed the idea of "adaptive radiation," where early generalized forms of an organism would migrate into an area and if conditions were right spread out and evolve into a series of more specialized forms. Osborn was drawn into paleoanthropology when Nebraska state geologist Edwin Barbour announced the discovery near Omaha in 1906 of a partial human cranium that appeared to be geologically old. They called the find Nebraska Man and thought it might be the earliest type of human known in North America. Osborn examined the specimen, concluding that it indicated the early appearance of humans in the New World. It fit well with his conviction that a land bridge existed in the Pleistocene connecting North America to Siberia allowing mammals to migrate across. Embarrassingly for Osborn the evidence for Nebraska Man eventually dissolved, but he devoted ever more time to the question of human evolution (Regal 2002:. 88–96).

Although Osborn initially appeared sympathetic to the idea of an anthropoid ape ancestry for humans, he gradually rejected the idea that the human lineage had passed through an ape-like stage. He suggested that apes and humans had diverged early in the evolution of primates and evolved separately. Osborn also believed that humans had evolved in Central Asia. Drawing upon William Diller Matthew's theory, presented in *Climate and Evolution* (1914), about the changing climate in Asia from the Pliocene to the Pleistocene, Osborn argued that proto-humans in Central Asia responded to the new harsher climate by evolving more human-like traits while apes stayed in the forests where they remained apes. In *Men of the Old Stone Age* (1925), Osborn argued that early human species, such as Neanderthals and Cro-Magnons, evolved successively and independently in Central Asia over long periods. Like other mammals, each of these species radiated outward to various parts of the world, displacing preceding groups. Osborn also helped to shape American public opinion about human evolution through the popular Hall of the Age of Man exhibit that opened at the American Museum of Natural History in 1924. Using artifacts and replicas of hominid fossils, with dioramas painted by the artist Charles Knight, Osborn outlined human evolution, but the prevalence of racial prejudices was also evident in the way the exhibit depicted the long separation of the human races (Regal 2002).

While Osborn was advocating his particular theory of human evolution, William King Gregory arrived at a very different conclusion. Gregory studied zoology under Osborn at Columbia University and later came to work with Osborn at the American Museum of Natural History. He became an expert on fossil primate teeth and published a book on the *Origin and Evolution of the Human Dentition* (1922). In 1924, Gregory examined a fossil ape specimen recently found in India called *Dryopithecus*, concluding that it represented a common ancestor of humans and gorillas. Raymond Dart sent Gregory casts of the *Australopithecus africanus* specimen in 1929 and after studying the teeth Gregory was struck by how similar they were to human teeth. At a time when most researchers did not consider *Australopithecus* a hominid, Gregory was convinced that it represented a human ancestor. This was a period when many different scenarios of human evolution were competing with one another. Gregory strongly believed that humans had evolved from an anthropoid ape ancestor. In his influential book *Man's Place among the Anthropoids* (1934) he argued that the gorilla is the best living example of what our ape ancestor looked like. He did not think *Pithecanthropus* was a direct ancestor although it probably represented what our more immediate ancestor looked like. Gregory believed that the great apes and humans ultimately originated from a gibbon-like ancestor and that brachiation was a pre-adaptation to bipedalism. Unlike many of his peers, he thought bipedalism evolved before our large brain did.

Americans first became involved in a significant way in paleoanthropological excavations in the 1920s. George Grant MacCurdy, curator of archaeology and anthropology at the Peabody Museum at Yale, and Charles Peabody, curator of European archaeology at the Peabody Museum at Harvard, founded the American School of Prehistoric Research in 1921. The school undertook joint excavations (from 1929–1934) with the British School of Archaeology in Jerusalem of caves at Mount Carmel in Palestine, which were led by English archaeologist Dorothy Garrod. These uncovered a sequence of Stone Age levels spanning about 600,000 years. MacCurdy recruited many American students into prehistoric research through the school, among them Theodore McCown. From 1931 to 1937 McCown participated in the excavations at Mount Carmel. In Skhūl cave he was part of the team that unearthed 10 Neanderthal skeletons that possessed features more like modern humans than classic European Neanderthals. These fossils and Neanderthal fossils from the nearby Tabūn cave were sent to the Royal College of Surgeons in London where McCown and English anatomist Arthur Keith collaborated in examining this remarkable collection of fossils. They considered the Skhūl and Tabūn Neanderthals to be about the same age and to belong to a single but highly variable population that was in the process of an evolutionary transition, where the Skhūl specimens possibly represented a sort of Neanderthaloid form of Cro-Magnon that evolved in the Middle East (McCown and Keith 1939).

Meanwhile, excavations at Zhoukoudian in China, conducted by Canadian anatomist Davidson Black and Chinese paleontologist Wenzhong Pei between 1927 and 1934 produced a wealth of hominid fossils that Black named *Sinanthropus pekinensis*. Anthropologists quickly accepted "Peking man" as a human ancestor, since anatomically and geographically it fit expectations. When Black unexpectedly died in 1934, the Rockefeller Foundation, which funded the research, chose Franz Weidenreich to replace him. Weidenreich, a German-born Jew who fled Nazi Germany to the United States, arrived in Beijing in 1935 where he continued to uncover fossils and artifacts until the advancing Japanese army forced excavations to cease and Weidenreich to flee China in 1941. He arrived in New York and spent the remainder of his career at the American Museum of Natural History. Besides publishing detailed studies of *Sinanthropus*, noting their similarity with the *Pithecanthropus* fossils from Indonesia, Weidenreich also proposed a theory that a single hominid species had spread throughout the Old World and populations inhabiting particular areas tended to remain there, producing regional types or races that remained fairly constant over time, while limited interbreeding between these regional populations allowed them to remain a single species. In this scheme *Homo erectus* (with regional variants including *Sinanthropus* and *Pithecanthropus*) evolved into Neanderthals who in turn evolved into modern humans throughout the Old World, but with populations in different geographical regions maintaining distinct racial variations throughout this entire process. While Weidenreich purported to see evidence of racial continuity in the fossil records of different regions, his theory faced scientific skepticism as well as opposition due to changing postwar attitudes about the biological validity of "race" (Caspari and Wolpoff 1996).

Race remained a fundamental but problematic concept in American anthropology. Physical anthropology was taught at few universities and had few trained specialists until after World War II. Most were trained as anatomists and they treated human races as ideal types. These ideas profoundly influenced the study and interpretation of hominid fossils. Earnest Hooton, as professor of physical anthropology at Harvard University, exemplified this tradition. He trained many of the United States' anthropologists and his *Up from the Ape* (1937) was a widely read introduction to human evolution and human racial diversity. Hooton's *Man's Poor Relations* (1942) was among the first comprehensive treatises on primate taxonomy and behavior, reflecting their importance for understanding human evolution. It is also worth mentioning that Hooton drew attention to geographical variation in Neanderthals and distinguished between what he called the "classic Neanderthals" of Western Europe and more "modern" looking Neanderthals from Mount Carmel and elsewhere.

Modern Biology and Paleoanthropology Meet

The theoretical framework within which paleoanthropology operated began to change in the 1950s as a result of the rise of the Modern Evolutionary Synthesis (Tattersall 2009: chapter 7). During the early twentieth century several different mechanisms were proposed to explain how evolution operated (Bowler 1986). With the integration of Mendelian genetics and population genetics into evolutionary theory in the 1930s a new generation of biologists applied mathematical techniques to investigate how changes in the frequency of genes in populations combined with natural selection could produce species change. This demonstrated that Darwinian natural selection was the primary mechanism for evolution and that other models of evolution, such as neo-Lamarckism and orthogenesis, were invalid. The Modern Synthesis emphasized that all species are composed of populations that possess significant genetic variation. This variation, operated upon over time by natural selection within the context of the natural environment, produce new species through adaptation. This new way of thinking changed biology and paleontology in the 1940s, but it initially had little affect on paleoanthropology. The "Genetics, Paleontology and Evolution" conference held at Princeton University in 1947, instrumental in introducing the Modern Synthesis into American paleontology, laid the foundation for the Cold Spring Harbor symposium on the "Origin and Evolution of Man" held in 1950. This meeting brought together anthropologists and paleontologists studying human evolution with leading advocates of the Modern Synthesis who made the case for why paleoanthropology needed to integrate the new thinking about evolution into their interpretations of the hominid fossil record.

Theodosius Dobzhansky, a geneticist at Columbia University and an architect of the Modern Synthesis, was one of the first to apply this new thinking to human evolution. Beginning with the premise that any species possesses considerable variability, Dobzhansky urged fundamental changes to the way hominid fossils were interpreted taxonomically. He suggested that *Pithecanthropus* and *Sinanthropus* should not be considered distinct species but racial variations of a single species. He believed Neanderthals also possessed racial variation (Dobzhansky 1944). Ernst Mayr, a zoologist at the American Museum of Natural History who made important contributions to the Synthesis, wanted to go even further. He criticized former anthropologists who frequently created new species or even genera for each new hominid fossil. This had produced a plethora of hominid species often distinguished by very minor differences. Mayr applied the principles of the Synthesis to simplify hominid taxonomy, placing all known hominids into only three species, *Homo transvaalensis* (australopithecines), *Homo erectus*, and *Homo sapiens* (Mayr 1950). Few paleoanthropologists were willing to adopt Mayr's extreme taxonomic reform, but they were induced to reduce the number of hominid species. Francis Clark Howell (1960) more successfully proposed instead to group all hominids into one of two genera (*Australopithecus* and *Homo*).

Physical anthropology adopted new methods and a new perspective in the way it interpreted hominid fossils through the influence of the Modern Synthesis. William Howells, who succeeded Hooton at Harvard in 1954, played a major role in transforming physical anthropology into a population-based biological science and helped free anthropology from its preoccupation with typological classifications of human races. He introduced into physical anthropology the multivariate analysis and quantitative methods used by evolutionary biologists and was among the first to use complex statistical methods to understand hominid skulls. After comparing the crania of modern human and extinct hominids, Howells rejected the idea that Neanderthals evolved into humans and he dismissed Weidenreich's regional continuity hypothesis. He proposed the "Noah's Ark" hypothesis that imagined several migrations of hominids out of Africa, the last migration consisting of modern humans who replaced the previous populations of hominids that had migrated out of Africa.

At the University of Chicago, Sherwood Washburn also worked to reform physical anthropology by incorporating the principles of the Modern Synthesis. In an influential paper Washburn (1951a) outlined the principles of "The New Physical Anthropology." It rejected the focus on racial classification and craniometry and instead stressed the need to study living primates and humans in order to explain their anatomical features through their function and adaptation to specific environments. He encouraged the study of primates in their natural environment in order to reconstruct the behavior of early hominids, as exemplified in the "The Social Life of Early Man" conference he organized in 1959. While Howells and Washburn downplayed the biological significance of racial variation, Carleton Coon, professor of anthropology at the University of Pennsylvania, used the Modern Synthesis' understanding of adaptation, especially in response to climate, to explain human racial variation. In *The Origin of Races* (1962) Coon used the hominid fossil record to argue that five geographical subspecies or races of *Homo erectus* existed and that each of these races evolved independently and at different rates into the five primary races of *Homo sapiens* living today. The differing rates of evolution also explained why existing races had attained such different levels of cultural development. Although Coon presented his theory as a strictly biological and anthropological examination of the evolution of the human races, the book quickly created a furor since many anthropologists saw its claims as racist and scientifically unwarranted. Coon denied these claims and continued to explore the biological mechanisms that could explain human geographical variation, but this episode highlights how sensitive the study of human races had become.

The principles of the Modern Synthesis and the New Physical Anthropology influenced a new generation of students who began to reexamine the hominid fossil record. Francis Clark Howell, who studied with Washburn at the University of Chicago, analyzed Neanderthal specimens from this new perspective. He considered Neanderthal evolution and their geographical variation within the context of what was known about the Pleistocene climate (Trinkaus and Shipman 1993: 283–90). Howell came to view the "classic Neanderthals" as a distinct population inhabiting Western Europe during the early portion of the last glacial period. He explained their distinctive features as adaptations to the particularly cold climate of that time. Neanderthal specimens dating from the preceding warmer Riss-Würm interglacial period were more geographically widespread and anatomically were less specialized than the classic Neanderthals. Howell argued that geographic isolation and adaptation to extreme climate conditions caused the evolution of the distinctive "classic Neanderthal" characters (Howell 1952). Howell was the first anthropologist since Hrdlička to visit museums and institutions around the world containing Neanderthal fossils so he could examine them first hand. This allowed him to see the Neanderthals as a population rather than as isolated specimens and this dramatically changed his perception of them.

Howell (1958) proposed the idea that early generalized Neanderthals in Western Asia, like those from Tabūn (what he called pre-Neanderthals) evolved into the more sapiens-like Skhūl population, which he called "Proto-Cro-Magnons." When the climate later warmed, the impediment preventing this population from migrating was removed and they entered Western Europe, thus explaining the rapid appearance of Cro-Magnons in the European fossil record. Meanwhile, in Western Europe, the pre-Neanderthals evolved into classic Neanderthals. While Howell was formulating this new understanding of Neanderthal evolution, Ralph Solecki excavated Shanidar cave in Iraq between 1953 and 1960. This produced Mousterian artifacts and in 1953 the skeleton of an infant. After Solecki unearthed three intentionally buried adult Neanderthal skeletons in 1957 he invited T. Dale Stewart, a physical anthropologist at the Smithsonian Institution, to examine the skeletons. Then, during August 1960, the team excavated four additional Neanderthal skeletons. Soil samples collected from the site contained large quantities of pollen from one burial, leading Solecki to suggest this individual was buried with flowers. He interpreted this as evidence that these Neanderthals were more human-like than was previously thought (Solecki 1971).

New Techniques and New Ideas

While paleoanthropologists were discovering new hominid fossils and debating the details of human evolution, the way paleoanthropology was practiced began to change after 1950. Paleoanthropology received important institutional support with the establishment of the Wenner-Gren Foundation for Anthropological Research. Founded in 1941 with its headquarters in New York, the Foundation financially supported anthropological research and from its inception human evolution was a significant focus. Under the directorship of Paul Fejos (from 1941–63), the Foundation built laboratories, funded projects around the world, and sponsored conferences. From 1946 to 1953 the Viking Summer Seminars in Physical Anthropology, organized by Sherwood Washburn, fostered the introduction of genetics and the Modern Evolutionary

Synthesis into paleoanthropology. Beginning in 1959 the Foundation also sponsored influential symposia on major issues in human evolution research. One of the early projects the Foundation supported was radiocarbon dating.

Archaeologists and paleontologists had relied on stratigraphy to provide a relative chronology for the objects they discovered, but few means existed for determining absolute dates for most of prehistory. When American physicist Willard Libby developed the radiocarbon dating method in the late 1940s researchers finally had a means for dating organic material up to 50,000 years old. This was useful for late prehistory, but for older objects other methods were needed. A solution came in the mid-1950s when Garniss Curtis, Jack Evernden, and John Reynolds at the University of California at Berkeley created the potassium–argon dating method. This method could only date volcanic rock but it had a range of billions of years. First used in 1959 to date the *Zinjanthropus* hominid Louis and Mary Leakey discovered at Olduvai Gorge, this method has since proved invaluable. In 1953 Farrington Daniels at the University of Wisconsin suggested that the phenomenon of thermoluminescence might be used to date certain prehistoric ceramic and stone artifacts. George Kennedy and Leon Knopff in the United States and Norbert Grögler and Friedrich Houtermans in Europe independently formulated the thermoluminescence dating method in the 1950s. Other dating methods followed and the consequences were profound since for the first time paleoanthropologists had a time frame for human evolution and could confidently arrange hominid fossils in their chronological order.

The rise of molecular anthropology in the 1960s transformed paleoanthropology in other ways (Marks 1994; Morgan 1998; Sommer 2008). British biologist George Nuttall first suggested in 1902 that the chemistry of blood proteins might be used to determine the phylogenetic relationships between humans and apes. In the early 1960s American biochemist Morris Goodman compared blood proteins of humans and apes and the results showed the close relationship of chimpanzees and gorillas to humans (Hagen 2010). Following Goodman's lead, Allan Wilson and Vincent Sarich at the University of California at Berkeley used a new technique, micro-complement fixation, to compare the albumin molecule in humans and apes. They found that over time genetic mutations changed blood proteins, but more significantly they found that these mutations accumulate at a constant rate. This so-called "molecular clock" allowed them to calculate how much time had passed since two species shared a common ancestor. They showed that gibbons and orangutans share a distant ancestry with humans but that gorillas and chimpanzees share a more recent common ancestor with humans (Wilson and Sarich 1967). The advent of molecular anthropology offered new ways to approach the evolutionary history of hominids and apes but at the beginning paleoanthropologists were skeptical of the claims made by these "outsiders" who were employing unfamiliar techniques and proposing notions contrary to long-held views of hominid phylogeny.

The most objectionable claim was that humans and chimpanzees shared a common ancestor about five million years ago. Many paleoanthropologists believed apes and hominids had diverged much earlier. Few ape fossils existed, but Kenyan anthropologist Louis Leakey pointed to *Proconsul* and *Kenyapithecus*, Miocene apes from East Africa, as important clues to the evolution of apes and hominids. At Yale University, Elwyn Simons and David Pilbeam collaborated during the 1960s in studying specimens of *Ramapithecus*, a fossil ape first discovered in 1932. Certain features of these very fragmentary remains led them to argue that *Ramapithecus* was probably a direct ancestor of hominids and they believed that apes and hominids diverged 15 million years ago or more. But when paleontologists found more complete specimens in the 1970s it became clear that *Ramapithecus* resembled orangutans and could not be part of the hominid lineage (Lewin 1987: 85–127).

While paleoanthropologists searched for the origins of hominids in the fossil record primatologists were studying apes in the wild, not only to learn about their behavior and social organization but also to draw inferences about the likely behavior and social structure of early hominids. Louis Leakey encouraged Jane Goodall and Dian Fossey's field studies of chimpanzees and gorillas in the 1960s that led to breakthroughs such as the discovery of chimpanzee tool use. Others used studies of primates to speculate about the factors that led our ape ancestors to evolve into hominids. Sherwood Washburn was especially influential in this regard. Working with his student Irven DeVore, who studied baboon behavior, Washburn formulated the "Man the Hunter" hypothesis in 1968. It argued that the shift from foraging for fruits to hunting and the eating of meat was a crucial factor in transforming our ape ancestors into hominids. Hunting large game animals, they argued, fostered bipedalism, tool use, language, and intelligence. This hypothesis tended to emphasize the centrality of male hunting in human evolution, while females were relegated to child rearing and minor food foraging. But this was a period when growing numbers of women were obtaining degrees in the sciences and the feminist movement was influencing American thought. The Man the Hunter hypothesis was countered by a group of women anthropologists who saw in it the influence of male dominance in science. Sally Slocum drew upon her studies of Rhesus monkeys to propose the "Woman the Gatherer" hypothesis. It suggested that female foraging was not only an important source of food but could also have spurred the manufacture of tools for gathering and carrying food. Nancy Tanner and Adriennne Zilhman expanded and refined this hypothesis and anthropologists continue to debate the relative roles of men and women in hominid societies and human evolution (Dahlberg 1981).

By the 1970s, new ideas originating in biology once again influenced paleoanthropology. The Modern Evolutionary Synthesis' depiction of gradually evolving lineages confronted paleontological evidence for the stability of some species over long periods followed by the rapid appearance of new species. This prompted Niles Eldredge and Stephen Gould to propose the theory of punctuated equilibrium in 1972. While grounded in the Darwinian view of evolution, this theory argued that species might remain relatively unchanged for long periods, but under intense selective pressure new species can emerge quite quickly. Eldredge and paleontologist Ian Tattersall were the first to apply the theory to the hominid fossil record (Eldredge and Tattersall 1975). They argued that hominids did not slowly evolve incrementally from one species into another. Rather, hominid species could persist relatively unchanged for long periods only to be quickly replaced by new species. This would produce discontinuities in the fossil record that would make it difficult to reconstruct hominid phylogenies. Significantly, Eldredge and Tattersall also conducted one of the first cladistic analyses of hominid phylogeny. German entomologist Willi Hennig introduced cladistics, a new approach for determining the evolutionary relationship between species in 1950, but his ideas attracted little attention until his book was translated into English in 1966.

The Focus on Africa

Meanwhile, partially as a result of Louis and Mary Leakey's discovery of *Zinjanthropus boisei* (1959) and *Homo habilis* (1960) at Olduvai Gorge, paleoanthropologists increasingly turned their attention to Africa as the cradle of hominid evolution. Promising geological deposits along the Omo River in Ethiopia prompted the International Omo Research Expedition, which operated between 1967 and 1974. Clark Howell led the American contingent and from the start the expedition reflected the changes that were occurring in paleoanthropology. This was a large multidisciplinary project that included geologists, paleontologists, archaeologists, paleoanthropologists, and an array of other scientists that collected hominid fossils and data about the environment they lived in. This project served as a model for the large paleoanthropological projects that followed. One of the most successful of these is the Koobi Fora Research Project.

Kenyan paleoanthropologist Richard Leakey, son of Louis and Mary Leakey, established the East Rudolf Research Project (its original name) in 1969 to search for hominids along the shores of Lake Turkana in Kenya. From a base camp at Koobi Fora, Leakey assembled a multidisciplinary group of scientists that included American geologist Kay Behrensmeyer and English born anatomist Alan Walker. Walker joined the project in 1970 and for most of the time of his involvement he was successively a professor at Harvard University (1973–1978), Johns Hopkins University (1978–1995), and Pennsylvania State University (since 1995). Walker was part of the team that reconstructed the contentious hominid cranium (KNM-ER 1470) discovered in 1972 that has alternatively been considered *Homo habilis* or the type specimen for *Homo rudolfensis*. Leakey and Walker considered another cranium (KNM-ER 3733 discovered in 1975) to represent an African *Homo erectus*, although now it is often designated *Homo ergaster*. Walker collaborated with Leakey again in the analysis of a nearly complete 1.6 million year old hominid skeleton (KNM-WT 15000) found at Nariokotome along the western shore of Lake Turkana in 1984 (Walker and Leakey 1993). In 1985 Walker discovered a 2.5 million year old cranium (KNM-WT 17000), nicknamed the Black Skull (Tattersall 2009: chapter 10).

Not long after work at Koobi Fora began French and American researchers explored deposits at Hadar in Ethiopia. Among them was Donald Johanson, a graduate student at the University of Chicago. In 1973 they found parts of a hominid knee and the following year they recovered portions of a 3 million year old hominid skeleton nicknamed Lucy. Johanson unearthed the remains of approximately 13 additional individuals in 1975 but political turmoil in Ethiopia prevented further work. Unsure whether these fossils should be considered *Homo* or *Australopithecus*, Johanson and Tim White (a graduate student at the University of Michigan) daringly decided to establish a new species *Australopithecus afarensis*, which they considered the common ancestor of *Homo habilis* and *Australopithecus africanus* (Gibbons 2006: chapter 5).

Paleoanthropologists at the University of California at Berkeley established the Middle Awash Research Project in 1981 in collaboration with Ethiopian researchers. Tim White, who had just joined the faculty, attained prominence through a series of important discoveries. In the early 1990s, White's team unearthed a 4.5 million year old hominid they named *Ardipithecus ramidus*, followed at the end of the decade by a

5.5 million year old related species they named *Ardipithecus kadabba*. Other members of the team discovered remains of a 2.5 million year old hominid that White named *Australopithecus garhi* in 1999. These new hominids from the Middle Awash represented some of the oldest hominids known and expanded the growing human family tree. Equally remarkable, White's team excavated *Homo sapiens* remains in 1997 near the village of Herto that were 155,000 years old. When Alan Walker and Meave Leakey announced a 4.1 million year old hominid from Kenya (*Australopithecus anamensis*) in 1998 they added one more piece to the remarkable record of hominid evolution in Africa. The growing number of new hominid species raised contentious questions about hominid phylogeny and taxonomy though (Gibbons 2006).

During the 1980s many paleoanthropologists adopted the Out of Africa Hypothesis. This argued that *Homo sapiens* had evolved from *Homo ergaster* in Africa about 200,000 years ago and these humans subsequently migrated out of Africa into Europe and Asia where they replaced the hominid species living there. Supporters of this idea rejected the so-called Multiregional Hypothesis promoted by Milford Wolpoff, a paleoanthropologist at the University of Michigan. Wolpoff, with Australian paleoanthropologist Alan Thorne, cited the morphological continuities in *Homo erectus* and *Homo sapiens* fossils in different parts of Asia to argue that after *Homo erectus* left Africa about a million years ago they spread throughout Asia and Europe. As populations settled in regions with differing environments they adapted to local conditions, thus producing regional racial variations. Wolpoff argued that *Homo erectus* throughout the Old World evolved into *Homo sapiens*, with each region retaining local racial features that are visible in the fossil record. Wolpoff acknowledged his debt to the earlier hypothesis of Weidenreich and criticized the Out of Africa Hypothesis (Wolpoff and Caspari 1997). Research conducted in molecular anthropology seemed to support the Out of Africa Hypothesis. Rebecca Cann, a graduate student at the University of California at Berkeley, conducted research with Mark Stoneking and Allan Wilson to examine the mitochondrial DNA taken from women from different ethnic groups in order to produce a phylogeny of different human populations. Their influential paper argued that all modern humans derived from a population that lived in Africa about 200,000 years ago (Cann, Stoneking, and Wilson 1987).

* * *

The history of paleoanthropology in the United States is the history of the scientific and public responses to the idea of human evolution and to the discovery of hominid fossils. The United States' place in that history changed from being largely an observer and commentator in the late nineteenth century on ideas and events occurring in Europe to becoming instigators of excavations and contributors to theories in the twentieth century. An important part of that history consists of the creation of institutions, journals, and sources of funding for paleoanthropological research. These helped to professionalize the discipline and to promote research. The United States assumed greater leadership in paleoanthropology after 1950 and in recent decades American universities have trained large numbers of paleoanthropologists, many of whom are citizens of African and Asian nations where this research is conducted. Yet paleoanthropology and human evolution remain contentious subjects for many Americans and creationism continues to be a powerful cultural force. The history of paleoanthropology illuminates the wide range of scientific, institutional, cultural, and social interactions of this discipline with other sciences and with the public in general.

Bibliographic Essay

Paleoanthropologists, not historians, have largely written the history of paleoanthropology. Consequently, these histories have tended to focus on the discovery of hominid fossils, changing notions of hominid phylogeny, and evolutionary theory. Recently, however, historians of science have begun to explore a wider range of disciplinary, institutional, and social aspects of the history of paleoanthropology. America's contributions figure prominently in many histories, yet few works are devoted solely to the history of American paleoanthropology. Tattersall (2000), Henke (2007), and Foley (2001) offer brief summaries of the major developments in paleoanthropology in the twentieth century. Goodrum (2013) does the same but from the perspective of a historian of science. While Reader (2011) recounts the events surrounding major hominid fossil discoveries, Tattersall (2009) provides a detailed and comprehensive history of paleoanthropology that includes the influence of biological theory. Bowler (1986) examines the relationship between competing conceptions of evolution and early theories of human evolution. Gundling (2005) traces the discovery and changing interpretations of the australopithecines. Boaz (1981, 1982) focuses specifically on American paleoanthropologists' contributions to the debate about the australopithecines and other early hominids. Gibbons (2006) explores recent paleoanthropological research in Africa and highlights the social aspects of scientific practice. Spencer (1984) and Trinkaus and Shipman (1993) investigate the changing interpretations of the Neanderthals. Spencer and Smith (1981) discuss early American contributions to Neanderthal research. Trinkaus (1982) summarizes American studies of *Homo erectus*.

Delisle (2007) and Fleagle and Jungers (2000) explore the factors that have shaped thinking about hominid phylogeny and its centrality to paleoanthropology. Fleagle and Jungers (1982) discuss this specifically within the American context. Histories of American physical anthropology, such as Spencer (1981, 1982) and Little and Kennedy (2010), examine disciplinary, methodological, and institutional relationships between physical anthropology and paleoanthropology in America. Race remains a fundamental issue in both disciplines and while many works examine the history of race and anthropology in America, Caspari (2003) and Wolpoff and Caspari (1997) examine this issue within the context of paleoanthropology. As developments in biology changed evolutionary theory these changes influenced ideas about human evolution. Delisle (1995) and (2001) respectively examine the influence of the Modern Evolutionary Synthesis and cladistics on paleoanthropology. Hagen (2009, 2010), Marks (1994, 1996), Morgan (1998), and Sommer (2008) discuss the rise of molecular anthropology and the important role played by American scientists in this new discipline.

Increasingly historians have begun to examine the social history of paleoanthropology. Livingstone (1992) investigates the relationship between human evolution and religion within the context of ideas about pre-Adamism, while Astore (1996) describes American Catholic responses to early scientific theories of human origins. Livingstone

(2008) explores the complex relationships between theories of human origins and race, religion, and politics in America. Clark (2008) explores the social, political, and religious aspects of the debate over human evolution in early twentieth-century America. Haraway (1988b) and Proctor (2003) represent two superb examples of social histories of specific episodes in paleoanthropology. The role of women in paleoanthropology and in human evolution is a significant new subject of discussion. Hager (1997) offers a general overview of the subject. Fedigan (1986) investigates the role of women in models of human evolution, while Wiber (1997) explores the way gender is depicted in visual reconstructions of hominids.

Chapter Eighteen

PALEONTOLOGY

Paul D. Brinkman

The methods of scientific paleontology were first established in Western Europe at the end of the eighteenth century, but they were influenced by the input of fossils from the European colonial hinterland, including British North America. American fossils, distinct from their Old World counterparts and often spectacularly well-preserved, had a profound impact on the development of paleontology in France and Great Britain. European savants, who were keen to incorporate them into their various theories of the earth, eagerly sought American fossils. New World fossil riches, however, soon fostered the development of American science. Naturalists working in British North America, who had comparatively easy access to fossil resources, were quick to adapt European methods, to compare their specimens to Old World analogs, and to correlate American strata in relation to the geological system then being developed in Britain and France. In short order, American naturalists were making increasingly valuable contributions to science. By the middle of the nineteenth century they had graduated from mere collectors and providers of the raw materials of science to respected colleagues on par with their European peers. By the turn of the twentieth century, thanks in part to the embarrassment of fossil riches recovered from the arid and thinly settled areas of the trans-Mississippi West, American paleontologists – as they were now called – were leaders in their young field.

The First American Fossils

American fossils were intensely curious objects to European naturalists. Fossil shark teeth from the Carolinas were among the first vertebrate fossils from North America to get noticed in the scientific literature when, in 1705, a brief mention of them appeared in a short, descriptive paper by James Petiver in the *Philosophical Transactions of the Royal Society*. The Frenchman, Jean-Etienne Guettard, published the first illustration of

A Companion to the History of American Science, First Edition.
Edited by Georgina M. Montgomery and Mark A. Largent.
© 2016 John Wiley & Sons, Ltd. Published 2020 by John Wiley & Sons, Ltd.

a North American fossil mammal in 1756. The specimen, not identified by Guettard but now recognizable as the molar of a mastodon, had been collected along the Ohio River in 1739 by Baron Charles de Longueuil, a French Canadian army officer. Longueuil shipped his specimens to Paris and placed them in the royal collection at the Jardin des Plantes. In 1767, an Irish adventurer named George Croghan sent an assortment of American fossils, including portions of elephant-like tusks, to London. (Croghan's locality, now known as Big Bone Lick, Kentucky, is a sulfurous spring with a salt lick that attracted Pleistocene animals and then preserved their fossil remains in huge numbers. It has been billed as the cradle of American paleontology.) Benjamin Franklin, then living in London, acquired some of these fossils and identified them as the remains of a carnivorous elephant in a letter to Croghan. Later, he decided that the animal must have been too bulky for stalking prey, and that its knobby molars would have been useful instead for grinding twigs and branches. Franklin, who never published his ideas about the mysterious fossils, also speculated that elephant remains indicated a warmer climate regime for ancient North America. (Buffetaut 1987: 26–7, 34–8; Hedeen 2008) British anatomist William Hunter described fossils from Croghan's locality in 1769. Hunter interpreted the fossil remains, which he dubbed the American incognitum, as gigantic, elephant-like, meat-eating monsters, and argued that they were likely extinct: "And if this animal was indeed carnivorous," he wrote, "which I believe cannot be doubted, though we may as philosophers regret it, as men we cannot but thank Heaven that its whole generation is probably extinct" (quoted in Buffetaut 1987: 39). Despite Hunter's evidence, many natural philosophers were reluctant to accept the idea of extinction, which, by breaking a link in the Great Chain of Being, cast doubt on the perfection of God's creation.

The interpretation of American fossils took on political significance after the United States won its independence. In his only book, *Notes on the State of Virginia* (1785), Thomas Jefferson famously defended the denizens of North America against the outrageous claims of French naturalist Georges-Louis Leclerc, Compte de Buffon, who argued that the cold and humidity of America stunted growth and produced smaller, inferior beings relative to the Old World. Buffon offered the runty native wildlife of North America as his proof. Jefferson, then governor of Virginia, responded with an impassioned refutation. Central to his argument was the "mammoth," an elephant-like creature many times larger than its Old World counterpart (Bedini 1990). He made his most important foray into vertebrate paleontology in 1799, while serving as third president of the United States. In that year, he published a description of some fossils found in a cave in western Virginia. Enormous unguals inspired the name "Great-Claw or Megalonyx," and conjured up visions of a terrible lion that preyed on the mammoth. A lifelong skeptic of extinction, Jefferson was certain that enormous, unknown creatures still wandered the sparsely settled regions of North America. He sent Meriwether Lewis and William Clark to explore this wilderness in 1803, hoping they would return with more evidence of these American monsters (Bedini 1990; Semonin 2000: 304–5, 343–4). "Our entire ignorance of the immense country to the West and North-West, and of its contents, does not authorise us to say what it does not contain," he reasoned (quoted in Semonin 2000: 304). Although they did not find additional evidence of mastodon or megalonyx, they did find fossils. Lewis and Clark discovered several fossil localities and brought back the first fossil specimens from the American West. Later, Jefferson hired Clark to collect fossils for him at Big Bone Lick (Semonin 2000: 350).

Naturally, a serving president practicing science brought the prestige of that office to American paleontology, but it also left Jefferson vulnerable to ridicule from his political opponents, who mocked his obsession with natural history (Semonin 2000: 352–3).

Caspar Wistar, a Quaker doctor and anatomist from Philadelphia and a friend of Jefferson, was the preeminent American authority on fossils of the eighteenth century. In 1787, at a meeting of the American Philosophical Society (presided by Franklin), Wistar presented a paper describing "a large thigh bone" found in a marl bed near Woodbury Creek, New Jersey. Though this specimen was never published, it seems likely from accounts of Wistar's talk that it was the first known American example of a group that would later be called dinosaurs. Unfortunately, the specimen is now considered lost (Thomson 2008: 41–2). By the turn of the nineteenth century, Wistar was adjunct professor of anatomy, midwifery, and surgery at the University of Pennsylvania and curator of the American Philosophical Society. In 1799, he wrote the first technical study of Jefferson's megalonyx, in which he skillfully reconstructed the functional anatomy of the animal's foot. Paleontologist George Gaylord Simpson characterized this achievement as "almost incredible," given the lack of available comparative materials in the United States. Published together with Jefferson's less sophisticated contribution, Wistar's paper reinterpreted the fossil as a new genus of giant ground sloth, similar in many details to French paleontologist Georges Cuvier's well-known *Megatherium* (Buffetaut 1987: 123; Semonin 2000: 310–12; Simpson 1942: 153). Cuvier had established the methods of scientific vertebrate paleontology on the basis of his description of *Megatherium* – Cuvier's first – in 1796 (Buffetaut, 1987 : 53). Thus, Wistar should be considered a true paleontological pioneer.

The first American to realize the popular appeal of fossil vertebrates was the artist and museum innovator Charles Willson Peale. Peale supervised the exhumation of first one, then another, spectacularly complete fossil skeletons from swampy farm land in the Hudson River valley of New York. Late in 1801, he exhibited his fully reconstructed "mammoth" – now known to have been a mastodon – with great fanfare at his private museum in Philadelphia's Philosophical Hall. The world's second mounted fossil vertebrate skeleton, Peale's exhibit created a sensation. Mammoth fever swept the young nation. The second skeleton toured England with Peale's son Rembrandt, to mixed reviews and meager door receipts. The Peales interpreted the bones as "symbols of the immense power of the nation's newly discovered natural antiquity" (Semonin 2000: 339). Yet, to emphasize the fearsome symbolic strength of their animal, Peale and his son – in the face of mounting evidence to the contrary – continued to portray it as "exclusively carnivorous," even mounting the long, curving tusks backward, so that, seemingly, it could skewer smaller animals (Semonin 2000). Rembrandt Peale wrote a lengthy disquisition on the mammoth in 1803, but this work was derided by European naturalists (Rainger 1992: 4).

Beginning in the 1820s, a number of American naturalists had begun to apply Cuvier's methodology to their investigations of local fossils. Cuvier's procedure involved the reconstruction of whole animals from fragmentary fossils through an extensive knowledge of comparative anatomy. Soon, American naturalists were making important contributions to vertebrate paleontology – contributions that their European peers consulted and took seriously. Richard Harlan – another Philadelphia physician – and his *Basilosaurus* featured prominently in European debates about the antiquity of fossil mammals, for example. In Paris, Henri de Blainville cited Harlan's description of *Basilosaurus*, noting its double-rooted teeth, and argued that the so-called Stones-field fossils, which came from the Middle Jurassic of England and also featured double-rooted teeth, were likely reptiles and not the earliest known fossil mammals. To help resolve the controversy, Harlan brought specimens to Great Britain in 1839. Work-ing together with London comparative anatomist Richard Owen, they determined that *Basilosaurus* was a fossil whale, and not a marine reptile as Harlan had originally con-cluded. The reinterpretation of *Basilosaurus* uprooted de Blainville's argument alto-gether. This example shows that Europeans had greater respect both for American fos-sil descriptions and their interpretations after 1830 (Gerstner 1970). In 1835, Harlan became the first American to describe a fossil from England, *Ichthyosaurus coniformis*. This incident is also indicative of the changing status of American science (Simpson 1942: 162).

Americans also made important contributions in invertebrate paleontology, again by adapting established European methods to New World fossils. Thomas Say, a skilled entomologist and conchologist and one of the original founders of Philadelphia's Academy of Natural Sciences (established in 1812), was the first American to advocate the efficacy of Englishman William Smith's pioneering stratigraphic methods in an arti-cle in the inaugural issue of Benjamin Silliman's *American Journal of Science* in 1819. Among the first American geologists to apply these methods was Lardner Vanuxem. Trained at the Ecole des Mines in Paris, Vanuxem taught chemistry and mineralogy in Columbia, South Carolina. He retired from teaching in 1826 and devoted him-self to an intensive investigation of the geology and stratigraphy of the Eastern states. Samuel George Morton, a Philadelphia physician better known for his controversial work in craniometry, then edited Vanuxem's notes on American fossil invertebrates and published a landmark study that distinguished the various strata of the Atlantic Coast region and correlated those strata with their European counterparts. Morton's paper established that the green sands and marls of New Jersey were equivalent to the Sec-ondary strata of England, for example. Morton was largely responsible for shaping the Academy of Natural Sciences into the preeminent center for the study of vertebrate paleontology in the United States by the mid-nineteenth century (Simpson 1942: 166; Stroud 1992; Thomson 2008: 98–9).

Another important institution in the development of American paleontology was the state geological survey, several of which were established in the early nineteenth century to systematize the exploitation of natural resources. Geologist and paleontologist James Hall and his colleagues – Vanuxem included – of the State Geological Survey of New York, made many important contributions to American stratigraphy and invertebrate paleontology beginning in the 1830s and 1840s. Hall later worked for other state sur-veys and amassed enormous fossil collections. His encyclopedic *Palaeontology of New York* (1847–1894) was largely responsible for elucidating the stratigraphy of the East-ern United States and for establishing standards of nomenclature for American geology (Merrill 1924; Thomson 2008: 98–9).

Fossil ichnology – or the study of fossil footprints – was perhaps the one subfield of paleontology in which nineteenth-century American naturalists were true innova-tors, and Edward Hitchcock was the first pioneer. Hitchcock studied theology at Yale, where, having fallen under the influence of Silliman, he became a devoted student of geology. Hitchcock established the great paleontological tradition of Amherst College, in Massachusetts, where he taught for many years. While serving as director of the state's geological survey, Hitchcock developed an interest in the Triassic rocks of the Connecticut River Valley. Fossil footprints, which were preserved in great profusion in these sediments, had attracted tremendous popular interest. Some locals suggested that these three-toed prints were the sacred marks of Noah's raven. Hitchcock made a collection of fossil footprints for Amherst College and produced the first scientific study of them in 1836, showing that some, at least, of these footprints had been made by giant reptiles. So little did the fields of geology and theology come into conflict in Hitchcock's worldview that he wrote a book on the religion of geology and served for a time as professor of geology and natural theology (Simpson 1942: 166–7).

Paleontology and Western Exploration

The federal government funded numerous surveys of the Western states and territories both before and after the American Civil War. Organized primarily for more practical purposes, such as the search for natural resources or railroad routes to the Pacific Coast, these surveys also yielded an abundance of fossils. Explorer and geologist Ferdinand Vandeveer Hayden, one of the most successful of the survey scientists, had a tremendous impact on the development of American paleontology. Hayden had studied theology at Oberlin College, and medicine in Cleveland, Ohio and Albany, New York. In Albany, he apprenticed himself to Hall, who sent him west in the summer of 1853 with Fielding Bradford Meek, an invertebrate paleontologist. Together they explored, drew sections, and made fossil collections near the Black Hills and along the White River in Nebraska Territory. But Hayden bristled under Hall, and soon he attached himself to other patrons, including Spencer Fullerton Baird of the Smithsonian Institution. In 1867, he was appointed leader of the Geological Survey of Nebraska. Hayden was a remarkably successful explorer, promoter, and collector, but he described few fossils. Instead, he maintained a long-term partnership with Meek, who penned most of the invertebrate fossil descriptions with Hayden listed as junior author. Over a period of two decades, Hayden and Meek established the basic structure of Cretaceous and Tertiary formations in the West, and their results were read in Europe with great interest (Foster 1994; Thomson 2008: 136–43). Federal paleontology was ultimately organized around the US Geological Survey (established in 1879 by consolidating Hayden's survey with other federally funded exploration), which sponsored field-oriented research, and the Smithsonian, where fossils were curated. Meek initially took charge of these collections.

The majority of Hayden's vertebrate fossils were described by Joseph Leidy, a Philadelphia physician and natural scientist of wide interests and talents who is often considered the founder of American vertebrate paleontology. In 1856, Leidy described the first American dinosaurs on the basis of an assortment of fossil teeth collected by Hayden in the Judith River basin of present-day Montana. Two years later he described and named another dinosaur, *Hadrosaurus foulkii*, from a more complete specimen found in the Upper Cretaceous green sands near Haddonfield, New Jersey. According to Leidy, *Hadrosaurus* walked with an erect, kangaroo-like posture as a consequence of its greatly reduced forelimbs. This was in contrast to the prevailing European view of dinosaurs as lumbering, elephant-like behemoths. The relative completeness of *Hadrosaurus*, as well as the abundance and excellent preservation of other fossils collected subsequently in the US West, conveyed a competitive advantage to American paleontologists over their European rivals, whose fossils were often fragmentary and fewer in number. Leidy was among the first Americans to exploit this advantage. In 1868, he created a sensation at Philadelphia's Academy of Natural Sciences when, together with British sculptor Benjamin Waterhouse Hawkins, he mounted *Hadrosaurus* in a dramatic, life-like pose and exhibited it to the public. The world's first mounted dinosaur was so popular that the Academy had to implement an admission charge in an effort to keep the crowds down. In 1869, Leidy published a lavishly illustrated monograph on the extinct fauna of Dakota and Nebraska, which added more than 70 new genera and numerous species to America's ever-increasing fossil fauna. Leidy did little fieldwork. Many of his fossils were collected by Hayden, including the earliest dinosaurs and most of the fossil mammals from Dakota and Nebraska. A devoted empiricist, Leidy had relatively little interest in writing about the more theoretical aspects of paleontology. Nevertheless, his work attracted the admiration of many prominent European scientists, including Charles Darwin (Rainger 1992; Warren 1998).

Othniel Charles Marsh and Edward Drinker Cope usurped Leidy's role as the dominant figures in American vertebrate paleontology in the late 1860s. Marsh's maternal uncle, George Peabody, was one of the wealthiest Americans and perhaps the greatest philanthropist of the nineteenth century. Peabody gave Yale a sizable donation to construct a natural science building. Yale then appointed Marsh as America's first professor of paleontology and curator of the new Peabody Museum of Natural History in 1866. Marsh led the first of four Yale College Scientific Expeditions – manned by Yale undergraduates and escorted through the West by the US Army – beginning in 1870. In 1882, he was appointed vertebrate paleontologist of the US Geological Survey, which provided an appropriation for fieldwork and an outlet for publications and relieved him of much of the financial burden of his research. Marsh published numerous paper reconstructions of the extinct animals he named and described, including many bizarre-looking American dinosaurs. Some of the most spectacular of these reconstructions were reproduced in popular books and newspapers – these served as important visual aids in a time when mounted fossil vertebrate exhibits were almost unknown in American museums. One of Marsh's most important scientific contributions was his 1880 monograph *Odontornithes*. Darwin lauded this book on toothed birds as the best empirical support for evolution to appear in two decades. However, certain government officials, including Congressman Hilary Herbert of the Allison Commission, were less favorably impressed. In 1892, "birds with teeth!" became a rallying cry for opponents of wasteful federal spending on science. Another important contribution was Marsh's *The Dinosaurs of North America* (1896), the first comprehensive account of America's abundant dinosaur fauna. Marsh's fossils formed the nucleus of the vertebrate paleontology collections at the Yale Peabody Museum and the National Museum of Natural History (Schuchert and Levene 1940; Wallace 1999).

Edward Drinker Cope was Marsh's contemporary and arch-rival. Born into a family of prosperous Philadelphia Quakers, Cope showed an early interest in science, publishing his first paper at the age of 19. Though largely self-taught, Cope benefited from Leidy's patronage and instruction at the Academy of Natural Sciences. Membership in the Academy and access to its collections provided Cope with his earliest opportunities for research. Cope made many fossil collecting trips to the American West, often with the aid of Hayden and other federally funded surveys. He also employed a stable of talented fossil collectors, including Charles H. Sternberg, Jacob Wortman, and others. Brilliant, combative, and extremely productive, Cope published more than 1400 articles and named and described more than 1000 vertebrate species, both fossil and extant. He bought a controlling interest in a struggling scientific journal, the *American Naturalist*, in order to have an outlet for his papers. In the 1880s, disastrous mining investments drained his finances, forcing him to accept a teaching position and to sell his fossil collections. He exhibited an enormous, life-sized paper reconstruction of his sauropod dinosaur *Camarasaurus supremus* at a meeting of the American Philosophical Society in 1877. He showed it again in Europe in 1878, where it excited tremendous interest. Cope's best-known work in paleontology was *Vertebrata of the Tertiary Formations of the West* (1883). With more than 1000 pages and 80 plates, this hefty book was often referred to as "Cope's Bible." Perhaps his most important contribution to American paleontology was the discovery and description of vertebrate fossils from what he called the Puerco Formation, in northern New Mexico, which filled a long time interval between the oldest Eocene and youngest Cretaceous deposits of North America. Cope also made important fossil discoveries in the Texas Permian (Brinkman 2010; Osborn 1931; Wallace 1999).

Cope and Marsh feuded bitterly in the second half of the nineteenth century. No one knows exactly when or how their feud began. They met in Europe in the 1860s. They exchanged friendly letters, collected fossils together, and even named new species after one another. But, by the early 1870s, they were fighting over access to fossil localities and priority of publication. Marsh accused Cope of stealing fossils. Cope countered that Marsh was guilty of "misrepresentations." In 1877, enormous, well-preserved Jurassic dinosaur fossils were discovered simultaneously in several localities, including Morrison and Garden Park, Colorado and Como Bluff, Wyoming. Cope and Marsh raced to name these new forms first, authoring hasty and poorly illustrated descriptions of *Camarasaurus, Stegosaurus, Allosaurus* and many other iconic American dinosaurs, often on the basis of incomplete remains. Their slipshod work left a knot of synonymous forms for the next generation of American paleontologists to untangle. In 1890, both of their reputations suffered when they traded accusations of plagiarism, incompetence, and dishonesty in the *New York Herald*. When Marsh lost his lucrative federal subsidy in 1892, he was forced to scale back his research. Cope's research, meanwhile, had already been greatly reduced. Both died before the turn of the century, leaving a legacy of heated competition and taxonomic confusion (Brinkman 2010; Wallace 1999).

Cope and other American paleontologists made important theoretical contributions to evolutionary biology. Alpheus Hyatt, for example, was an important invertebrate paleontologist and, along with Cope, America's most significant neo-Lamarkian evolutionary theorist. Hyatt, zoologist and paleontologist at the Massachusetts Institute of Technology, worked primarily on fossil cephalopods, particularly ammonites. In the 1860s and 1870s, he found apparent parallels between the evolutionary history of some of his fossil organisms and the ontogeny of later individuals. In other words, the same patterns of progress and degeneration over geological time seemed to be repeated in the shells of individual animals – which grow by accretion over an individual's life – found in later geological periods. He argued an exact parallel between the development of an individual and the evolutionary history of the group, including the idea of racial senility followed by extinction (Bowler 1996: 82–3, 347).

Paleontology Moves into Museums

Beginning in the 1890s, the institutional setting for American vertebrate paleontology shifted from private collections into large, urban museums funded by Gilded Age philanthropists like Andrew Carnegie and Marshall Field. Henry Fairfield Osborn, a protégé of Cope's, began his career at Princeton, but moved to New York's American Museum of Natural History in 1890. Osborn's Department of Vertebrate Paleontology, which benefited from lucrative funding from J. Pierpont Morgan and his peers – Morgan was Osborn's uncle – was the most successful of the museum-based paleontology programs. Success inspired imitation, however, and a competition to collect exhibit-quality fossils and to build collections erupted between rival museums in New York, Pittsburgh, and Chicago. This competition – less acrimonious but no less intense than the contest between Cope and Marsh – led to a number of important innovations. For example, an intensive search for new field sites turned up a number of previously unknown fossil localities in the US West. Improved field techniques like quarry mapping and plaster field jackets yielded exhibit-quality fossils for museum paleontologists. Osborn's staff developed better techniques for displaying mounted fossils in life-like poses, and for reconstructing prehistoric life with models and colorful murals by artists like Charles R. Knight. Museum paleontologists developed efficient, labor-saving techniques for cleaning fossils, many of which are still used today. Adam Hermann, for example, experimented with acids and sand-blasting equipment. Elmer S. Riggs, at Chicago's Field Columbian Museum, adapted pneumatic tools to fossil preparation, and then shared details of his new technique with preparators at rival museums. Crowd-pleasing dinosaur exhibits, especially *Apatosaurus, Diplodocus* and other gigantic sauropods, as well as Osborn's *Tyrannosaurus rex*, brought record numbers of visitors into museums. Newspapers, movies, and popular books helped make dinosaurs and other extinct animals into pop-culture icons. At the same time, replicas of the Carnegie Museum's *Diplodocus* donated to museums in Europe and Latin America, acted as vectors for the international spread of *dinomania*. Carnegie Quarry, near Vernal, Utah, scene of the Carnegie Museum's most active Jurassic dinosaur collecting, was later designated Dinosaur National Monument, a unique, *in situ* display of dinosaur fossils (Brinkman 2010; Rainger 1991).

American museums organized ambitious expeditions to search for fossils in other parts of the globe beginning at the turn of the century and continuing through the end of the 1920s. Osborn's program was first to send its collectors abroad. In 1907, Osborn himself led an expedition with Walter Granger to collect early Tertiary mammals in Egypt. Later, in the 1920s, Granger collected dinosaur eggs and several new kinds of fossil mammals and Cretaceous dinosaurs during a high-profile and hugely expensive expedition to China and Mongolia. Barnum Brown was the best travelled of Osborn's associates, having collected fossils in Argentina, Canada, Cuba, Abyssinia, India, Burma, Greece, and elsewhere. Osborn's ambition had been to dominate the field of American vertebrate paleontology – these expeditions show that his program had global reach (Dingus and Norell 2010; Rainger 1991). Chicago's Field Museum sent a party to collect Cretaceous dinosaurs in Canada in the summer of 1922. This was a trial run for a more ambitious expedition to make a representative collection of all known fossil mammal faunas of southern South America, which lasted from 1922–1927 (Brinkman 2013). In 1909, invertebrate paleontologist Charles Doolittle Walcott, secretary of the Smithsonian Institution, discovered a unique fossil fauna of well-preserved, soft-bodied invertebrates from the Cambrian Period high in the Canadian Rockies of British Columbia. The importance of this discovery became clearer after a reassessment of the bizarre fossils was done in the 1960s (Gould 1989). By bringing together collections of comparative fossil materials from around the world, these high-profile expeditions were a boon for American paleontology.

From Modern Evolutionary Synthesis to Paleobiological Revolution

George Gaylord Simpson, a vertebrate paleontologist specializing in fossil mammals at the American Museum of Natural History, was arguably the most influential paleontologist of the twentieth century. An architect of the Modern Evolutionary Synthesis of the 1940s and 1950s, Simpson introduced rigorous quantitative methods into paleontology and stressed the importance of causal explanations, thereby bringing his field into closer contact with biology. With Anne Roe, a psychologist, he wrote an influential textbook on statistical methods for biologists and paleontologists titled *Quantitative Zoology* (1933). His *Tempo and Mode in Evolution* (1944) is considered a classic in the field. Simpson's most important insight was that paleontologists could bring the additional dimension of time – via the fossil record – to population biology, and thus they could derive the mechanisms that drive evolution. He emphasized that paleontologists could track the distribution and transformation of fossil organisms through time and space. In fact, following *Tempo and Mode*, temporal biogeography became the focus of paleontological evolutionary theory (Sepkoski 2012: 37–44).

Despite a decided turn toward biology, some American paleontologists – particularly those who were members of university or museum geology departments – remained somewhat committed to geology, or found creative ways to weave geological and biological research threads together. Beginning in the mid-1940s, Everett C. Olson, a vertebrate paleontologist in the geology department at the University of Chicago, incorporated the methods and tools of sedimentology and paleoecology to address new questions regarding paleobiological interrelationships and ancient animal communities, especially in the Clear Fork Permian beds of north-central Texas. Olson did not abandon his interest in more traditional questions of morphology and systematics. Yet his shift of emphasis reflected the growing American interest in paleoecology in the 1930s and 1940s. His research developed into an interest in taphonomy, which was then being pursued in Europe. Olson was largely responsible for introducing his American colleagues to this subfield in the 1960s. By the early 1950s, Olson's research had attracted a coterie of talented graduate students. He became the head of an exciting, interdisciplinary research program that drew on the strengths of both the biology and geology departments at the university. He had also cultivated cooperative working relationships with colleagues at the Field Museum, including paleontologist Paul O. McGrew, which invigorated his teaching and research (Rainger 1997).

In invertebrate paleontology, meanwhile, a similar movement toward greater quantification and a more biological emphasis took place in the mid-twentieth century. Simpson's American Museum colleague, Norman Newell, for example, was an invertebrate paleontologist who worked to promote the biological aspects of paleontology in the 1950s and 1960s. In the hands of Hall, Meek, and others, invertebrate paleontology in the United States had developed initially in the context of stratigraphy and (later) petroleum geology. Newell, by contrast, applied statistical methods to study broad patterns in the fossil record in order to address biological problems, including the significance and possible causes of mass extinctions. For his work on the latter, he raised the hitherto taboo topic of catastrophism in the fossil record. He was also a pioneer in the subfield of paleoecology. Newell's appreciation of the quality of the fossil record stands in stark contrast to Darwin's preoccupation with its deficiencies. Because Newell trained so many of the next generation of paleobiologists, his impact on the discipline was profound and long-lasting (Sepkoski 2012: 54–67).

Quantification in invertebrate paleontology was aided by the growth of vast fossil collections and the compilation of taxonomic data into readily usable formats. For example, G. Arthur Cooper was an invertebrate paleontologist at the National Museum of Natural History who specialized in the taxonomy and stratigraphy of Paleozoic brachiopods. By conducting fieldwork virtually every year, and by developing an acid-etching laboratory to prepare silicified fossils, especially the Permian brachiopods of Texas, Cooper greatly enlarged the size and range of the national collections. In the 1950s, he contributed to the monumental, multi-volume *Treatise in Invertebrate Paleontology*, which evolved into a vast encyclopedia of taxonomic information on invertebrate fossils and was a precursor of paleontological databases. The University of Kansas was the institutional locus for work on the *Treatise* (Sepkoski, personal communication).

By the second half of the twentieth century, evolutionary paleoecology had become the main research interest for the majority of American invertebrate paleontologists. James W. Valentine, for example, was a pioneer of evolutionary paleoecology at the University of California in the 1960s and 1970s who studied diversity patterns among marine fossil invertebrate faunas. By the mid- to late 1960s, an influential paleoecology research group emerged at Yale University, which was committed to integrating biology and ecology into paleontology. They helped to change the focus of paleoecology from a geological interest in sedimentation to a more biological emphasis on extinct organisms and their environments (Sepkoski 2012: 123–7).

An American dinosaur renaissance began in the 1960s that radically changed the way paleontologists interpret these charismatic animals. In 1964, John Ostrom of Yale University discovered a small theropod dinosaur in the Early Cretaceous sediments of south-central Montana, which he dubbed *Deinonychus*. Based on fossil evidence of a stiff tail, long, powerful hind limbs, and a sickle-like, retractable claw, Ostrom concluded that *Deinonychus* "must have been a fleet-footed, highly predaceous, extremely agile and very active animal, sensitive to many stimuli and quick in its responses. These in turn indicate an unusual level of activity for a reptile and suggest an unusually high metabolic rate" (quoted in Desmond 1975: 84). Ostrom was suggesting that some dinosaurs, at least, must have been endothermic or warm-blooded. Ostrom's student, Robert T. Bakker, then authored a series of papers in the early 1970s that further explored the implications of dinosaur endothermy (Desmond 1975). When Bakker finally compiled his ideas in a book in the mid-1980s, he gave it the provocative title *The Dinosaur Heresies* (1986), even though these ideas were no longer considered particularly heretical by the majority of dinosaur paleontologists.

The hypothesized agility, speed, and pack-hunting abilities of small theropods like Ostrom's *Deinonychus* seemed inconsistent with prevailing ideas about dinosaur dimwittedness. At the same time, new ideas about small theropod behavior demanded greater nervous coordination and a larger, more complex brain than most dinosaurs appeared to possess. In 1969, the American Museum's Edwin H. Colbert and a Canadian paleontologist named Dale A. Russell wrote a detailed description of the skull and braincase of *Dromaeosaurus*, another small theropod. (Brown had collected this specimen for the American Museum in the Late Cretaceous of western Canada, in 1914.) Their conclusion was that dromaeosaurs and other small theropods boasted brains that were proportionally much larger than other types of dinosaurs, which bolstered the hypothesis of complex and agile hunting behaviors (Desmond 1975).

Although morphological and taxonomic work continued to be done, new studies with an emphasis on behavior became the mainstays of dinosaur paleontology by the 1980s. In 1978, John R. Horner and Bob Makela discovered and described the remains of what they interpreted as a vast dinosaur nesting site in west-central Montana. From this site, they recovered eggs and eggshell fragments and an abundance of fossil bones, large and small and representing a growth series of dinosaurs from hatchling through adults. A new dinosaur, *Maiasaura* – the good mother lizard – was established on the basis of these remains. Most provocatively, they used a profusion of unusual trace evidence to draw conclusions about dinosaur nesting, parenting, and social/herding behaviors. These novel conclusions, which appeared to lend additional support to the idea of dinosaur endothermy, attracted a storm of media attention and popular interest (Horner and Dobb 1997).

The period from 1970 through 1985 has recently been styled a paleobiological revolution. A landmark paper published in 1972, "Punctuated Equilibria: An Alternative to Phyletic Gradualism," by invertebrate paleontologists Niles Eldredge and Stephen Jay Gould, set much of the agenda for this movement. Important contributions by David Raup, J. John Sepkoski, Jr., Steven Stanley, and others on species diversity, taxonomic survivorship, and rates of evolution and extinction drew heavily on quantitative techniques adapted from population biology. Technological advances in computing, new methods of quantitative modeling and analysis, and powerful, new statistical tools have made quantitative, computer-aided research a hallmark of paleobiology. Meanwhile, vast numbers of new fossils – overwhelmingly fossil invertebrates – have provided the raw materials for paleobiologists to investigate large-scale patterns in the fossil record. One of the most remarkable contributions of the paleobiological revolution was to stretch the fossil record back some two billion years to the earliest microbial stages of life on earth. In 1975, a new journal, *Paleobiology*, provided an outlet for theoretical, quantitative paleontology. The paleobiological revolution was driven by American paleontologists, but has, by the present day, evolved into an international endeavor (Sepkoski 2012; Sepkoski and Ruse 2009).

When the methods of scientific paleontology were first being developed in France and Great Britain in the late eighteenth century, America was merely a source for fossils, many of them unique and well preserved and thus unusually informative. Indeed, American fossils were crucial to the development of the discipline. By adapting the methods first established in Europe and applying them to an abundance of local fossils, American naturalists were soon able to contribute meaningfully to the field, although London and Paris continued to be the centers of the paleontological world. By the 1830s, American fossil interpretations were as eagerly sought as the fossils themselves. By the end of the nineteenth century, due to its copious fossil resources and a favorable institutional environment for the development of science, the balance had shifted – American paleontologists had become acknowledged leaders in their field. This trend continued through the twentieth century, when American paleontologists were the innovators and instigators of the dinosaur renaissance and the paleobiological revolution.

Bibliographical Essay

There is a rich and ever-growing literature on the history of American paleontology. Certain classic texts, including a number of outstanding primary sources, remain in print today, while new articles and books are published every year. Naturally, there are biases in this literature. For example, vertebrate paleontology, and especially dinosaur paleontology, has been covered very thoroughly, while paleobotany and invertebrate paleontology have received considerably less attention. The nineteenth and early twentieth centuries have been the subject of innumerable books and articles, while the second half of the twentieth century, by contrast, has only recently begun to attract the interest of historians of science. The best, most useful and most accessible of these sources are reviewed here.

The best general survey of the history of vertebrate paleontology is arguably Eric Buffetaut's concise and very readable *A Short History of Vertebrate Palaeontology* (1987). Unfortunately, this book, which is no longer in print, is surprisingly scarce. Peter Bowler's *Life's Splendid Drama: Evolutionary Biology and the Reconstruction of Life's Ancestry, 1860–1940* (1996) provides a comprehensive account of the history of evolutionary biology with a special emphasis on scientists who specialized in morphology and paleontology of fossil and living animals over an 80-year period following the publication of Charles Darwin's *On the Origin of Species* (1859). Although both of these books are global in scope, there is much in each of them that will be of interest to students of American paleontology.

Perhaps the best account of the origins and early development of vertebrate paleontology in America is George Gaylord Simpson's thoroughly researched and well-illustrated "The Beginnings of Vertebrate Paleontology in North America" (1942), which covers the period from the very earliest fossil discoveries in North America through 1842, when the Western states and territories were just beginning to be explored for fossils. Simpson's article has been an important secondary source for many subsequent histories of American paleontology. Patsy Gerstner's short article "Vertebrate Paleontology, an Early Nineteenth Century Transatlantic Science" (1970) is an important resource on Richard Harlan's contributions to American paleontology. Gerstner's thesis is that European naturalists learned to respect the work of their American peers by the 1830s, after Americans had adopted Georges Cuvier's pioneering comparative approach to fossil anatomy. Paul Semonin's *American Monster: How the Nation's First Prehistoric Creature Became a Symbol of National Identity* (2000) relates the eighteenth- and early nineteenth-century story of the discovery and interpretation of the fossils of enormous, elephant-like mastodons and mammoths in the United States. Semonin argues that the mysterious American monsters became emblems of the power and potential of the young nation in which their fossil remains were found. Stanley Hedeen's *Big Bone Lick: The Cradle of American Paleontology* (2008) is a brief and very readable account of the history of Big Bone Lick, Kentucky, an abundant source of many of the earliest fossil mammals found in America. Finally, a relatively recent and very engaging book by Keith Thomson, *The Legacy of the Mastodon: The Golden Age of Fossils in America* (2008), tells an adventurous story of American paleontology from the late eighteenth through the end of the nineteenth centuries, beginning with Thomas Jefferson and ending with the infamous feud between Edward Drinker Cope and Othniel Charles Marsh.

There is an enormous body of literature on the nineteenth-century history of American paleontology, with much of the focus directed at the sordid story of the Cope–Marsh feud. Of all the books about Cope–Marsh, one of the best (and most recent) is surely David R. Wallace's *The Bonehunters' Revenge: Dinosaurs, Greed and the Greatest Scientific Feud of the Gilded Age* (1999). There are many other books on Cope–Marsh, some good, some bad. Url Lanham's classic *The Bone Hunters: The Heroic Age of Paleontology in the American West* (1973) provides a colorful account of American bone diggers. There are many excellent published primary sources from this period of American paleontology. Charles H. Sternberg's classic autobiography, *The Life of a Fossil Hunter* (1909), was republished in 1990 and remains one of the liveliest and most interesting accounts of late nineteenth-century paleontology fieldwork. Michael F. Kohl and John S. McIntosh edited and published the field journals of Marsh's fossil hunter Arthur Lakes, who discovered Jurassic dinosaurs at Morrison, Colorado and later worked the classic Como Bluff, Wyoming locality with William Harlow Reed. *Discovering Dinosaurs in the Old West: The Field Journals of Arthur Lakes* (1997) includes vivid descriptions of the places where Lakes collected and the people he worked with. The book also includes black-and-white reproductions of many of Lakes' watercolors and drawings of the West. Lastly, there are three worthy biographies, Leonard Warren's *Joseph Leidy: The Last Man Who Knew Everything* (1998), Henry Fairfield Osborn's *Cope: Master Naturalist* (1931), and Charles Schuchert and Clara M. LeVene's *O.C. Marsh: Pioneer in Paleontology* (1940), about the three most important American paleontologists of the nineteenth century.

The literature on the history of American paleontology in the early twentieth century is likewise very good. Ronald Rainger's *An Agenda for Antiquity: Henry Fairfield Osborn and Vertebrate Paleontology at the American Museum of Natural History, 1890–1935* (1991) is an excellent source on Osborn's vertebrate paleontology program at the American Museum in New York. Paul D. Brinkman's recent book, *The Second Jurassic Dinosaur Rush: Museums & Paleontology in America at the Turn of the Twentieth Century* (2010), which draws heavily on archival sources, tells the lesser-known story of American paleontology in the post Cope–Marsh era. It is an excellent resource on fieldwork at the turn of the twentieth century, and on the development of museum paleontology in the United States. Lowell Dingus and Mark A. Norell have written a very readable biography called *Barnum Brown: The Man Who Discovered Tyrannosaurus Rex* (2010), which documents the adventurous life of one of the most important dinosaur hunters of the twentieth century.

The late twentieth century was a period of revolutionary change in American paleontology – its history deserves more attention. Stephen Jay Gould's *Wonderful Life: The Burgess Shale and the Nature of History* (1989) is a very readable account of the discovery, interpretation, and later re-interpretation of well-preserved, soft-bodied fossils from the Cambrian of British Columbia. Gould also writes provocatively about the role of chance in the unfolding history of life on earth. Adrian Desmond's book, *The Hot-Blooded Dinosaurs* (1975), is a carefully researched and well-written account of the dinosaur renaissance of the 1960s and 1970s, with some attention also devoted to the early history of dinosaur paleontology in Europe. Two recent contributions have begun to turn the tide. David Sepkoski and Michael Ruse's *The Paleobiological Revolution: Essays on the Growth of Modern Paleontology* (2009) is an invaluable edited collection of essays, many written by the very scientists who fomented the paleobiological revolution. Sepkoski's excellent follow-up volume, *Rereading the Fossil Record: The Growth of Paleobiology as an Evolutionary Discipline* (2012), is the first book-length examination of the rise of paleobiology as a discipline in the 1970s.

Chapter Nineteen

ECOLOGY

Gina Rumore

In 1926 plant ecologist William Skinner Cooper used the metaphor of a "braided stream" to describe the patterns of vegetation of the earth. This stream, according to Cooper, was "one of enormous complexity, with its origin in the far distant past," whose "more or less separate and definite elements branch, interweave, anastomose, disappear"(Cooper 1926: 397). His dynamic viewpoint provides an apt metaphor for the history of ecology in the United States. Since its official adoption by the Botanical Society of America in 1893 as a branch of botany, through its professionalization in the early twentieth century, to its popularization in the decades after World War II, American ecology has always been a dynamic science, combining many different specialties that have at times converged, diverged, died off, and even reappeared. The current mainstream of American ecology arose from many branches of biological science, brought together by a common interest in the interactions between organisms and their environment, both biotic and abiotic.

While the origins of an ecological approach to studying nature date back at least to seventeenth- and eighteenth-century Europe (Worster 1994), ecology as an organized field of scientific research began to take form in the relatively uninhabited spaces of the American landscape and American science in the late nineteenth century. In 1893 the Botanical Society of America officially adopted the term ecology, coined by the German Darwinist Ernst Haeckel in 1866, to differentiate the study of "the general phenomena of plant development, the plant's relations to its environment, and its relations to other organisms" from the then dominant field of laboratory physiology (quoted in Cittadino 1980: 175). Both the term and the research questions it suggested rapidly spread among American biologists studying the interactions among flora, fauna, and their abiotic environment, giving rise to a diverse community of self-identified ecologists.

A Companion to the History of American Science, First Edition.
Edited by Georgina M. Montgomery and Mark A. Largent.
© 2016 John Wiley & Sons, Ltd. Published 2020 by John Wiley & Sons, Ltd.

Limnologist Stephen Alfred Forbes' "The Lake as a Microcosm" (1887), one of the foundational works of ecology in the United States, anticipated many of the major themes and concepts that would come to define American ecology throughout the twentieth century: continual biotic and abiotic change and equilibrium, the study of food webs and energy flows, competition and niches, community formation and population dynamics, model systems, and, ultimately, ecosystems. Forbes's research also foreshadowed the (often uncomfortable) ties between environmentalism, politics, and the science of ecology in the United States (Schneider 2000). While none of the branches of American ecology can truly be isolated from one another, to make sense of the current state of ecology in the United States, it is helpful to first consider the history of these branches as somewhat isolated streams and then to return to the common themes that have drawn them together into mainstream ecology as reflected by shared professional societies, scientific journals, and university departments.

Physiological and Physiographical Ecology

Haeckel coined the term "ökologie" in his 1866 *Generelle Morphologie der Organismen* (286–9) to define a new branch of physiology: the study of the physical and chemical processes of an organism and the relationship between organisms and their environments, both biotic and abiotic, and between mutually dependent organisms. While Haeckel never pursued this branch of physiological research himself, several self-identified European "plant biologists" did. In the late nineteenth century, the German plant biologists Oscar Drude and Andreas Schimper and their Danish colleague Eugenius Warming laid the foundation for a highly influential branch of American ecology, applying their laboratory training in physiology to studies of biogeography and plant adaptations.

Warming's *Plantesamfund,* published in 1895, and translated into English as *The Oecology of Plants* in 1909, analyzed how plants compete for and divide up the limited resources they need to survive, carefully differentiating among levels of interdependence between different species of plants and animals. Warming also addressed biogeographical questions, analyzing the impact of light, heat, humidity, soil, terrain, and animals on species abundance and distribution, and he developed the concept of succession to describe change in community composition over time. The concept of succession became the cornerstone of the dynamic ecology that took root in the United States and Britain in the late nineteenth century.

The dynamic field studies of plant physiology conducted by the European plant biologists found a receptive and eager audience with two young American botanists who were conducting their dissertation research in the 1890s: Henry Chandler Cowles at the University of Chicago and Frederic Clements at the University of Nebraska. For both men, ecology was a study of the dynamic processes of ecological succession, or vegetational change over time. Yet, whereas Cowles derived his research program and theories from the physiographic study of land formations, Clements derived his research program and theories from the physiological study of organisms. The ecological science of both Cowles and Clements strongly reflects the contexts in which they conducted their doctoral studies and the broader social and environmental changes taking place in the United States in the late nineteenth and early twentieth centuries.

The physiographic ecology program developed by Cowles grew out of his education in the geology department at the newly founded University of Chicago. Cowles began his doctoral research at the University of Chicago in 1893 in geology; he switched his focus to botany after taking several courses from the prominent botanist John Merle Coulter who joined the Chicago faculty in 1895. Coulter promoted the ecological viewpoint as adapted from the German plant biologists in his classes, and Cowles found that the ecology of Warming meshed nicely with the physiographic approach espoused in the geology department at the University of Chicago. Physiography is the study of geological processes that give rise to current landforms. At Chicago, the geologist Thomas Chamberlin and geographer Rollin Salisbury worked together to create a physiography program within the geology department that highlighted the role of the physical environment in shaping organic life on earth.

For his dissertation, Cowles conducted a study of ecological succession of vegetation on the sand dunes of Lake Michigan. Cowles, relying predominantly on inferential field methods for his research, claimed that as he walked away from the shores of Lake Michigan and over the dunes that he could see over physical space the same patterns of succession he would expect to see in one place over time. Scientists today refer to this type of study as a chronosequence (Cowles 1899). Three major ecological concepts stand out from Cowles's early publications: the centrality of constant change to the study of ecological succession, the parallels between ecological succession and physiographic changes in topography, and the importance of field studies for testing and revising theories (Cowles 1901). These concepts strongly influenced the botany and zoology, including ecology, taught at Chicago in the early twentieth century (Mitman 1992). And through Cowles, this physiographic perspective influenced a large part of the succeeding generation of American ecologists: Cowles taught ecology at Chicago for over 30 years and was renowned as a great teacher. In turn, many of his students went on to become influential teachers in ecology programs across the United States (Rumore 2009).

The University of Chicago ecology program evolved contemporaneous to, but in a quite different environment from, the physiological ecology program of Clements, who began his career at the University of Nebraska and then briefly headed the botany department at the University of Minnesota before spending the remainder of his career as a researcher at the Carnegie Institute of Washington's Desert Laboratories in Tucson, Arizona. The ecology developed by Clements grew out of the "new botany" outlined by his advisor, Charles Bessey (Tobey 1981). For Clements, this link to physiology suggested two main ecological theories: (1) that plant formations should be studied as a "superorganism" and (2) that the superorganism has a lifecycle analogous to a real organism ending in a climax stage determined by the climate (Clements 1916). Through Clements's extensive publishing of research articles and textbooks, his theories promoted and often plagued ecological research in the United States through the first half of the twentieth century (Barber 1995; Hagen 1992a).

Animal Ecology

Animal ecology in the United States had its headwaters with Forbes's studies of Darwinian evolution in lake systems, which through the influence of the mainstream of plant ecology at the University of Chicago grew into its own strong branch of ecology in the 1910s and 1920s. Despite the success of Forbes's ecological approach to the study of lake fauna, however, animal ecology took longer than plant ecology to set root in the United States. Whereas plant ecologists could easily count plants and follow the change in plant types and ratios in a relatively small area over time – per the methods developed by Clements and Cowles – animal ecologists had to develop techniques to capture change over time to study objects that frequently moved and often had fairly large ranges. Two early American animal ecologists, Charles A. Adams and Victor Ernest Shelford, both worked closely with Cowles at Chicago, and both founded their ecological studies on the process of succession within communities.

The major breakthrough that gave shape to zoological research in American ecology came from overseas, finding an open field in the United States in which to set roots, adapt, and flourish. In 1927 British ecologist Charles Elton published *Animal Ecology*, popularizing the niche concept, a term coined by the American naturalist Joseph Grinnell in the 1910s, and focusing it on food relationships among organisms in a community. The concepts of the niche – an organism's defined role in a community – and the food chain have remained fundamental to ecology ever since. These concepts allowed animal ecologists to study the functional roles played by animals in the larger community. "When an ecologist says, 'there goes a badger',," Elton wrote, "he should include in his thoughts some definite idea of the animal's place in the community to which it belongs, just as if he had said, 'there goes the vicar'" (Elton 1927: 63–4). While much of Elton's work touched on many of the same points Forbes had argued 40 years before, regarding the roles organisms play in maintaining balance in the lakes he studied, Elton's work functionalized this discussion. In emphasizing the significance of the food chain, Elton defined energy as the major commodity of exchange in biological communities. And energy flow, combined with another British-born concept, the ecosystem, would come to dominate ecology in the American setting in the second half of the twentieth century.

Ecosystem Ecology

Until the 1940s, the community approach to ecological studies espoused by Cowles and Clements defined the mainstream of American ecology. But a 1942 paper by Yale postdoctoral researcher Raymond Lindeman served as a harbinger of a major shift in ecological studies that came to define American ecology in the second half of the twentieth century. Lindeman's famous paper, "The Trophic Dynamic Aspect of Ecology" (1942), appeared in print several months after his untimely death at the age of 27. Lindeman, a student of Cooper (who was a student of Cowles) argued in this paper that ecological succession could be studied as a process of change in energy flow through an ecosystem over time.

The British ecologist Arthur Tansley coined the term ecosystem in a 1935 essay in the journal *Ecology* (not insignificantly, in an issue celebrating the career and influence of Cowles). Tansley, arguing against the dominance and abuse of the organism view of plant communities, wrote, "But the more fundamental conception is, as it seems to me, the whole *system* (in the sense of physics), including not only the organism-complex, but also the whole complex of physical factors forming what we call the environment or the biome ... It is the systems so formed which, from the point of view of the ecologist, are the basic units of nature on the face of the earth" (400). Lindeman's paper grabbed hold of Tansley's concept and gave it meaning through the application of Elton's ideas about energy flow through communities.

Yet Lindeman's paper almost never made it into print. Significantly, Lindeman submitted his paper to *Ecology*, the flagship journal of the Ecological Society of America (ESA), which, since 1920, has published research from across the diverse branches of the science. The debate surrounding whether or not Lindeman's paper should be published clearly reveals the tensions forming among the several branches of American ecology in the 1940s. Lindeman conducted his PhD work at the University of Minnesota in limnology. As his paper was based on his study of Cedar Bog Lake, the editor of *Ecology* sent the paper to two prominent limnologists – believed to be Paul Welch and Chancey Juday – for review. Both reviews came back overwhelmingly negative. The reviewers felt that Lindeman's work was representative of a dangerous trend in American limnology – applying mathematics to develop theory – led by the Yale limnologist G. Evelyn Hutchinson, Lindeman's postdoctoral advisor at Yale (Cook 1977).

The limnology program at Yale led by Hutchinson, an Englishman, greatly challenged standard limnological and ecological thought in the 1940s, advancing ecosystem ecology through its focus on biogeochemistry. As Juday wrote to his colleague Robert Pennak in 1942:

> The Yale school of mathematical-limnologists is having a high time displaying their mathematical abilities. The interesting part about it is that they are applying mathematical formulae used in sub-atomic physics where all of the forces are presumably uniform to limnological problems where there are all sorts of un-uniform factors involved ... Apparently they do not have brains enough to see the point in the two very different situations ... In a short time I shall expect them to tell all about a lake thermally and chemically just by sticking one, perhaps two, fingers into the water, then go into a mathematical trance and figure out all of its biological characteristics. As the next stage in their evolution they will probably be able to give a lake an "absent treatment" similar to a spiritualist, so it will not be necessary to visit a lake at all in order to get its complete chemical, physical, and biological history (quoted in Beckel 1987).

While Hutchinson's theorizing did not prove popular with many of his contemporaries, it had a strong impact on a number of his students, including Lindeman, who showed through his paper how the ecosystem concept could work, and Howard Thomas (Tom) Odum, who, with his brother Eugene, would go on to make the ecosystem the dominant concept in ecology in the second half of the twentieth century.

Ecology in the "Atomic Age"

The building and dropping of the atomic bombs on Hiroshima and Nagasaki had an enormous impact on the mainstream of American ecology. In the post-World War II fallout from nuclear physics, American ecologists discovered new questions, new tools, and new sources of funding. The Odum brothers took the lead in shaping this "new ecology," so proclaimed by the older brother, Eugene Odum. The new ecology, as defined by the Odums, focused on the flow of energy and nutrients through ecosystems. Eugene Odum codified this approach to ecological study through his 1953 textbook, *The Fundamentals of Ecology*, which remained the standard textbook for the field for over two decades.

The Odums gained renown among ecologists for their 1954 study of the Enewetak Atoll, a project for which they received the ESA's Mercer Award in 1956. Enewetak, located among the Marshall Islands in the South Pacific, had been a site of nuclear bomb testing by the United States in the early 1950s, and the US government was interested in the environmental impacts of nuclear bombs. Funded by the American Energy Commission (AEC), the Odum brothers measured energy flow through this closed coral reef system in much the same way biologists would study the metabolism of a single organism. In many respects the Odums – Eugene was trained in physiology at the University of Illinois – approached their ecosystem studies as had Clements half a century prior, by viewing the ecosystem as an organism with energy inputs and output. At the same time, however, they incorporated the ideas of Tansley, Lindeman, and Hutchinson, taking a broader view of the ecosystem, measuring inputs and outputs (Kingsland 2005).

Concurrent with funding the Enewetak research, the AEC also began funding ecological research at several of the national laboratories involved in the Manhattan Project, including Oak Ridge National Laboratory, which had hosted a plutonium-producing reactor and a uranium isotope separation plant during the war. After the war, research at Oak Ridge shifted to focus on peaceful uses for atomic energy as well as the environmental and health effects of radioactive pollution. Led by Stanley Auerbach, ecologists at Oak Ridge developed a program of radioecology, using radioactive isotopes of essential nutrients, like phosphorus, to follow, using radiation detectors, how nutrients cycle through ecosystems. This approach allowed Oak Ridge ecologists to measure both rates and efficiency of nutrient transfer among organisms and between the abiotic environment and organisms. In addition to radioecology, ecologists at Oak Ridge took advantage of the newly developed computers to model ecosystems electronically, developing programs that allowed them to make theoretical alterations to ecosystems and then predict the outcomes (Bocking 1995).

The ecological research projects begun with AEC and Office of Naval Research funding in the decades after the war revealed a tension seemingly inherent in the science of ecology since its inception in the 1890s: ecology as a science and ecology as a means of critiquing technological and human impact on the natural world. Just as ecologists in the 1950s and 1960s received funding and tools – radionucleotides and computers, for example – to advance their studies, they were also faced with the negative impacts these funding sources and many of the technological advances were having on the systems they studied. This tension would lead ecologist Paul Sears, in 1964, to famously dub ecology a "Subversive Subject" (Sears 1964).

Ecology and Environmentalism

In the wake of Rachel Carson's 1962 *Silent Spring*, ecology became a household term no longer limited to a field of scientific research but suddenly synonymous with environmentalism. In her book, Carson, a US Fish and Wildlife Service biologist, presented scientifically sound and damning evidence against the horrific environmental impacts of many of the pesticides widely used in the United States at the time. This public alignment between ecology and environmentalism took to a much higher level a tension that had existed between often complicated commitments of ecologists in the United States – scientific research and preserving the environments and organisms they study.

As noted in the introduction to this article, the roots of this tension were visible in the research of Forbes in the late nineteenth century as the Illinois lakes he studied were threatened by development and conflicts between stakeholders – in this case scientists, fishermen, and developers. The conflicted relationship was formalized in 1917 when the fledgling Ecological Society of America (founded in 1915) formally recognized a Committee for the Preservation of Natural Conditions of Ecological Study (later renamed the Committee for Preservation of Natural Conditions in the United States). Hoping to avoid the conflict (perceived or real) between scientific research and lobbying for environmental protection, the ESA disbanded its Preservation Committee in 1945. Several of the ecologists who participated in the disbanded ESA Preservation Committee would form a group called the Ecologists' Union, which, in 1950, renamed itself the Nature Conservancy (Rumore 2012).

As the formation of the Nature Conservancy proved, removing the Preservation Committee from the ESA did not remove the environmentalist from many ecologists. In 1968 Stanford ecologist Paul Ehrlich published *The Population Bomb*, raising concerns over the rapid growth of the human population and imminent depletion of the natural resources needed to support life on earth. The same year ecologist Garret Hardin published "The Tragedy of the Commons," which argued that technological solutions were not sufficient in dealing with the population crisis and for the need to impose "mutual coercion mutually agreed upon" in curbing population growth and the resulting destruction of the environment (Hardin 1968). Through their books and lobbying actions, ecologists found an influential political voice in the 1960s. In 1969 the US government passed the National Environmental Protection Act and created the Environmental Protection Agency (EPS). In 1972 the EPA banned the use of DDT – one of the deadly chemicals discussed by Carson in *Silent Spring* – in the United States.

Population Ecology

As the success of the Odums' "new ecology" and Eugene Odum's popular textbook made clear, ecosystems ecology had come to dominate the science. Yet many ecologists saw populations, not ecosystems, as the proper level of ecological study. Whereas ecosystems ecologists focused on functional explanations – the role of organisms in the ecosystem – population ecologists studied how these organisms evolved to fill the roles. To population ecologists, the explanations of ecosystem ecologists seemed teleological and to completely ignore evolution. While population ecology really began to take off in the United States in the 1960s, it grew from the headwaters of mathematical ecology spurting up in the United States and Europe beginning in the 1920s.

In the 1920s statistician Raymond Pearl discovered the work of Belgian mathematician Pierre-François Verhulst, who had first described what he called the "logistic curve" and what Pearl called the population growth curve. Pearl solved the equation that describes this curve, calling it the Verhulst–Pearl equation. This equation describes the change in the rate of population growth over time. In the 1930s Italian mathematician Vito Volterra built on the Verhulst–Pearl equation to model competition between two species, resulting in an equation that models predation in a two-species system. Yet, like Pearl, Volterra had been preceded in his discovery. The American demographer Alfred J. Lotka had discovered the two-species predation equation in 1925. This model is known as the Lotka–Volterra equations. Building on these mathematical equations and Elton's description of the competitive exclusion principle – that no two species can fill the same niche – Russian ecologist Georgy Gause discovered a continual competition among similar species for limited resources. This principle greatly shaped the field of evolutionary ecology, which began to grow in the United States in the 1960s.

Robert MacArthur took the lead in the late 1950s in applying mathematical modeling to population ecology. After receiving his PhD from Yale under Hutchinson, MacArthur spent a postdoctoral year in England working with ornithologist David Lack, who studied the role of competitive exclusion in the evolution of finches on the Galapagos Islands. Lack returned to the United States eager to develop mathematical models describing the laws underlying the niche concept and competitive exclusion principal and tested by using data he collected in the field. Together with Harvard entomologist E.O. Wilson, MacArthur developed the island biogeography theory, which they published in 1963 (MacArthur and Wilson 1963). According to this theory, the number of species on an island (its diversity) is a dynamic equilibrium between colonization and extinction. Though this theory remains controversial, it has succeeded in building MacArthur's vision for population ecology: encouraging ecologists to employ an approach that combines theory, mathematics, and experimentation.

Long-Term Ecological Research

The program of ecosystem modeling begun in the 1950s took off in the late 1960s with the International Biological Program (IBP). Following the success of the International Geophysical Year (1957), biologists worldwide began to lobby for their own international research program. Officially starting in 1968, the main American component of the IBP featured five biome studies funded by the National Science Foundation (NSF): coniferous forests, deciduous forests, deserts, grasslands, and tundra. The biome projects aimed to exhaustively model these five representative biomes over the course of six years. Yet the complexities of both the systems under study and of coordinating such a large-scale research effort proved far greater than anticipated. Ecologists at the time and historians of ecology have generally regarded the success of the IBP with great skepticism (Boffey 1976; Hagen 1992b: chapter 9; Kingsland 2005: 221–3). Despite failing to meet its major goals, the IBP in the American context did create several opportunities for ecologists. First, the program involved over 1800 ecologists, training an entire generation of ecologists in ecosystem ecology. Second, the program established a precedent for large-scale federal funding for ecological research. Finally, through their failure to meet their system modeling goals, ecologists realized the value of long-term ecological research projects for advancing their science.

During the 1960s and early 1970s a smaller-scale research project provided another vision for how ecosystem research might move forward in the wake of IBP. At the Hubbard Brook Experimental Forest in New Hampshire, ecologists Gene Likens and

Herbert Bormann were carrying out their own long-term ecosystem study, testing long-held assumptions about the processes of ecological succession. Using a watershed as their study area, Likens and Bormann conducted experiments including denuding small areas within the forest and then measuring regrowth over succeeding decades. Based on their results, Likens and Bormann hypothesized a "shifting mosaic steady state" in place of the stable climax proffered by ecologists from Clements to Eugene Odum. Where others saw a uniform climax, Likens and Bormann saw an old growth forest where patches existed in different stages of succession. As a whole, the system appeared to be balanced, but within the system, change was the only constant (Bormann and Likens 1979).

Building on the success of the Hubbard Brook study and the precedent of financial investment in large-scale ecological research by the NSF established during the IBP, ecologists successfully lobbied the NSF in the mid- to late 1970s to fund a series of long-term ecological research sites. Whereas a typical NSF grant worked on a two-to-three-year cycle, meaning scientists had to have major results in this short time frame to get more funding, Long Term Ecological Research (LTER) grants would fund research sites on a five-year cycle with the understanding that the funding would continue in perpetuity so long as the sites remained (the funding is now on six-year cycles). Since the first year of funding in 1980, the LTER program has grown from six to 26 sites, spanning the United States, Puerto Rico, Tahiti, and Antarctica (including Hubbard Brook, which joined the Network in 1986). In 1993, LTER went international with the founding of the International LTER. Today the ILTER has member networks in 40 countries spanning the globe.

In many ways the LTER sites represent the current mainstream of ecology in the United States and around the world. These sites represent the myriad ecosystems of the earth and they bring together ecologists representing the diverse branches of ecology that have arisen in the United States and globally: plant, animal, ecosystem, evolutionary, population, landscape. In addition, they draw in scientists who might not otherwise self-identify as ecologists, but whose work addresses ecological problems: limnologists, biogeographers, geologists, genericists, sociologists, economists, foresters, conservation biologists, anthropologists. And in the twenty-first century, the US LTER Network Network has increasingly engaged humanists – historians, philosophers, and artists – as they strive to be relevant to environmental and social problems in the twenty-first century United States.

Ecology in the United States today is a complex, transdisciplinary enterprise defined by a shared perspective: the study of the interaction between organisms (humans included) and their biotic and abiotic environment. Yet the braided stream of American ecology remains dynamic and its course subject to current political, scientific, and environmental influences.

In an era of big data, ecologists once again find themselves having to adapt to a new landscape. In 2004, the NSF began working with ecologists to develop a new continental-scale observatory: the National Ecological Observation Network (NEON). When fully operational in 2017, NEON promises to collect massive amounts of environmental data through remotely monitored electronic sensors, airborne instruments routinely deployed via aircraft, and field sampling. This approach represents a radical shift in ecological research from direct observation, if not rich immersed observation, to data collection in anticipation of analysis. Many questions about an ecology based on big data merit attention: Who will analyze all this data and how will such massive datasets change the science of ecology? Will the ecologists of the twenty-first century give ecosystems the "absent treatment" as Juday predicted and feared? Or will they continue to prefer place-based research as exemplified by the approach taken by Forbes, Cowles, and Clements to the nearly 2000 LTER scientists today? The answers to these questions will shape the "new ecology" of the twenty-first century.

Bibliographic Essay

Historians of ecology have tackled this broad subject from many different perspectives. The most comprehensive treatment of the subject comes from Donald Worster's *Nature's Economy: A History of Ecological Ideas* (1994), which explores the roots of ecology from Gilbert White and Linnaeus in the eighteenth century through conservation biology in the late twentieth century. For a more encyclopedic approach to the history of ecological thought, see Robert P. McIntosh, *The Background of Ecology: Concept and Theory* (1985). Sharon Kingsland provides a broad, narrative treatment of ecology in the United States in *The Evolution of American Ecology, 1890–2000* (2005), beginning with botanical research at the New York Botanical Garden and ending with urban ecology and Baltimore's Long-Term Ecological Study.

Several works on the history of American ecology have focused on one or more of the many overlapping branches discussed above. Joel Hagen's *An Entangled Bank: The Origins of Ecosystem Ecology* (1992b) focuses on Clements and Cowles and their programs of plant ecology as the progenitors of ecosystem ecology. Hagen has also written several article-length treatments on the history and influences of Clements's theory of the "superorganism," including "Organism and Environment: Frederic Clements's Vision of a Unified Physiological Ecology" (1988) and "Clementsian Ecologists: The Internal Dynamics of a Research Group" (1992a). For a recent discussion of Tansley's break with Clements and the ecosystem concept, see Arnold G. Van Der Valk "From Formation to Ecosystem: Tansley's Response to Clements' Climax" (2014).

Other historians have focused more on the influence of the University of Chicago and Henry Chandler Cowles on the development of American ecology, including Gregg Mitman's *The State of Nature: Ecology, Community, and American Social Thought, 1900–1950* (1992), Eugene Cittadino's "A 'Marvelous Cosmopolitan Preserve': The Dunes, Chicago, and the Dynamic Ecology of Henry Chandler Cowles" (1993), and Gina Rumore's "A Natural Laboratory, a National Monument: Carving Out a Place for Science in Glacier Bay, Alaska, 1879–1859" (2009).

For a thorough and thoughtful treatment of population ecology and the mathematics that gave birth to it, see Sharon Kingsland, *Modeling Nature: Episodes in the History of Population Ecology* (1985).

Two large themes in the historiography of American ecology are the history of the tension between preservation/environmentalism and ecology as a professional science and also the struggles to establish field research as a legitimate way to conduct biological research in a country where laboratory research was increasingly coming to be accepted as the gold standard for scientific research. For analysis of the former theme, see Rumore, "Preservation for Science: The Ecological Society of America and the Campaign for Glacier Bay National Monument" (2012); Abby Kinchy, "On the Borders of Post-War Ecology: Struggles over the Ecological Society of America's Preservation Committee, 1917–1946" (2004); Daniel W. Schneider, "Local Knowledge, Environmental Politics, and the Founding of Ecology in the United States: 'Stephen Forbes and the Lake as a Microcosm' (1887)" (2000); Stephan Bocking, *Ecologists and Environmental Politics: A History of Contemporary Ecology* (1997); Sarah Tjossem, "Preservation of Nature and Academic Responsibility: Tensions in the Ecological Society of America, 1915–1979" (1994); and Robert A. Croker, *Pioneer Ecologist: The Life and Work of Victor Ernest Shelford, 1877–1968* (1991).

For analysis of the tensions between laboratory and field research in ecology, see Kingsland, "Fritz Went's Atomic Age Greenhouse: The Changing Labscape on the Lab–Field Border" (2009); Rumore, "A Natural Laboratory, a National Monument" (2009), Robert Kohler, *Landscapes and Labscapes: Exploring the Lab–Field Border in Biology* (2002); Keith R. Benson, "Experimental Ecology on the Pacific Coast: Victor Shelford and His Search for Appropriate Methods" (1992).

Chapter Twenty

SOCIOBIOLOGY AND EVOLUTIONARY PSYCHOLOGY

Abraham H. Gibson and Michael Ruse

American scientists have made a number of important contributions to the field of sociobiology, which examines the evolution of social behavior among all living things, and the closely related field of evolutionary psychology, which focuses more explicitly on the social evolution of humans. To understand American contributions to these two fields, however, one must first understand the intellectual baggage that American scientists inherited from their forebears. For most of Western civilization, sociobiological research was stunted by the Judeo-Christian conviction that other species had been created distinct from humans and that God had not bothered to endow these other species with souls. It was simply taken for granted that nonhuman organisms did not possess psychic lives of their own. René Descartes famously reaffirmed this position in the early seventeenth century, when he wrote that nonhuman organisms were little more than machine-like automatons, and that they had no minds of which to speak (Merchant 2006). His opinion that other species were devoid of mental life carried considerable weight, and his ideas enjoyed paradigmatic status for the next several centuries.

As historians of science are well aware, the previously unbridgeable divide separating humans from other animals began to quickly disappear after Charles Darwin published *On the Origin of Species* in 1859. Darwin, of course, was English rather than American, but his work was quickly published in the United States, where enthusiastic champions like the professor of botany at Harvard, Asa Gray, promoted his ideas and defended him against critics. From our perspective, three points should be stressed in the adoption of a Darwinian worldview by Americans. First, a majority of American scientists converted to an evolutionary perspective rather quickly. Notwithstanding holdouts like the Swiss-born Harvard biologist Louis Agassiz, and the Princeton-based Presbyterian Charles Hodge, most Americans quickly accepted evolution as a biological phenomenon, including evolution extending to humankind (Bowler 1984; Ruse 1996). This was true among scientists, the general public, and even among church

A Companion to the History of American Science, First Edition.
Edited by Georgina M. Montgomery and Mark A. Largent.
© 2016 John Wiley & Sons, Ltd. Published 2020 by John Wiley & Sons, Ltd.

leaders (Roberts 1988). It is true that the more evangelical wing in the American South had, as it continues to have, strong negative sentiments about evolution. It is true also that, primarily because of political difficulties in Italy, Catholic reaction swung from cautious acceptance to a distrust and at times rejection of Darwinism (Artigas, Glick, and Martinez 2006).

Second, there was Darwin's mechanism of natural selection. Because more organisms are born than can survive and reproduce, there is a consequent struggle for existence that produces a winnowing process that is akin to the selection of animal and plant breeders. Since success in the struggle will on average be a function of the superior qualities of the survivors – the fitter organisms – over time, and since offspring tend to resemble their parents in many ways, there will be change and evolution. One should say that this mechanism had a somewhat mixed and low-key reception in the scientific world. No one wanted to deny that it can and does work, including on humans, and there were indeed some important empirical studies using natural selection, such as the work done by Henry Walter Bates (1863) on mimicry among brightly colored insects like butterflies. Nevertheless, even though they generally accepted evolution as a natural phenomenon in the years following the publication of *On the Origin of Species*, American scientists were not overly enthused by natural selection. They tended to support ideas like directed variation and the inheritance acquired characteristics, ideas that more readily allowed for purposive action and direction. Those for whom adaptation was insignificant tended to opt for large variation – what we today would call macromutations – this was the position of Darwin's great supporter, Thomas Henry Huxley.

What is interesting is that *outside* the American scientific community, one senses that natural selection found more acceptance in the United States than it did even in England. In particular, American philosophers who advanced the Pragmatisms showed a keener appreciation of selection than one finds even today among British thinkers (Ruse 2009). William James (1880) particularly made much of selective ideas in his thinking about culture and the nature of knowledge. This was a time when people like the English philosopher Henry Sidgwick (1876) were denying absolutely the significance of evolution itself to philosophical inquiry. In the twentieth century British philosophy conquered the United States and Pragmatism never made the gains that one might have anticipated. To this day, American philosophers lead the charge against the relevance of evolution, Darwinian or otherwise, to problems of epistemology or ethics. Recognizing that in these matters it is always difficult to keep separate the contributions of Darwin and Spencer, it is well known that toward the end of the nineteenth century, Americans – academics, businessmen, and others – showed great enthusiasm for what has been labeled Social Darwinism, basically a variant of laissez-faire economics (Bannister 1979). Life is a struggle for existence. Some will win. Some will lose. And that is how nature intended it, whether it be in the animal and plant world or in the world of commerce. Ideas like these were picked up by the novelists, particularly Jack London, whose *Call of the Wild* remains (at least for its rather thrilling writing) a deserved favorite.

The third thing done by Darwin, of great importance to us here, was to tackle the issue of social behavior. Darwin always recognized that behavior was just as important in the struggle for existence as physical characteristics, and he knew that this extended to social behavior. Well before he put pen to paper, the activities of the social insects like ants and bees had long been a subject of much interest and discussion, often in a natural theological context, as observers tried to show that hives and nests are paradigmatic examples of God's designing concern. The problem with social behavior of course is that it seems to fly in the face of the selfishness demanded by the struggle for existence. If you are off helping others – at your own expense – you can hardly expect to be one of the fitter progenitors of future generations. Indeed, what makes this problem so terribly challenging is that famously the social insects tend to be divided by castes and some, the workers, are sterile, devoting all of their attentions to the welfare of others. How can this be? In the *Origin*, Darwin solved this problem by thinking of the nest of the social insect as a kind of superorganism (Richards and Ruse 2015). Thus, workers and others, like the queen and the drones (the inactive males), are to be thought of as parts of the whole, rather than individuals in their own right. Just as let us say the hand and the eye are to be thought of as part of the body and not competing against the leg and the nose, so the workers are part of the whole and not competing against the queen and the drones. Inasmuch as the body succeeds, so the hand and the eye succeed. Inasmuch as the nest or hive succeeds, so the workers succeed. Darwin followed much the same line when addressing humans in the *Descent*. This was particularly so of our moral sense, which is the essence of sociality. Of course things are rather more complex, due to our powers of reasoning; but in the end it is a question of all for one and one for all, the tribe taking the place of the nest or hive (Darwin 1871: 1, 166).

It is significant that Darwin's evolutionary worldview was already pervasive by the time sociology and biology began coalescing into distinct professional guilds in late-nineteenth-century America (Calhoun 2007; Rainger, Benson, and Maienschein 1988). No less significant, the first generation of self-identifying sociologists insisted that *they*, rather than biologists, were best equipped to answer questions about the nature of sociality. They offered several explanations for the origin of the social impulse, including "social forces" (Hayes 1911) and "consciousness of kind" (Giddings 1896), but none of these explanations ever gained serious traction among academics. The most prolific writers paid terrific lip service to the evolutionary perspective, but they never seemed to grasp its full implications. For scholars like Lester F. Ward and Franklin Giddings, sociality remained an exclusively human phenomenon. They accepted that the evolutionary perspective implied physical and even psychical continuity among living things, but they could not shake the anthropocentric assumption that the mental and social lives of other species paled when compared to the mental and social lives of humans (Ward 1894).

Not everyone accepted such a limited view of sociality, though. Some Americans insisted that it was ludicrous to deny consciousness or cognition to other species, especially since the evolutionary perspective, and simple observation, suggested that other livings might have minds of their own. These researchers began promoting a new field of studies known as "comparative psychology," which, as its name suggests, compared mental phenomena in nonhumans in an effort to learn more about mental phenomena in humans. Sociologists like Edward Lee Thorndike (1898), John Watson (1903), and Albion Small (1905) were among the earliest in the United States to establish laboratories dedicated to experimental comparative psychology. These efforts notwithstanding, it was increasingly biologists and not sociologists who probed the origin of the social impulse. American biologists examined social and psychological behavior across a variety of taxa during the first few decades of the twentieth century. Herbert Spencer Jennings famously explored sociality at the microscopic level in his landmark book, *Behavior of the Lower Organisms* (1906). Meanwhile, after displaying an early interest in the mental lives of rodents (1907), Robert Yerkes devoted the remainder of his career to studying "psychobiological" behavior in nonhuman apes (Yerkes 1916, 1925, 1929).

Given Darwin's interest in the evolutionary peculiarities of social insects, it should come as no surprise that many of the United States' earliest and most ardent sociobiologists were entomologists. Among them, myrmecologist William Morton Wheeler proved the most influential. He was already the nation's foremost expert on social insects when he was named professor of economic entomology at Harvard in 1907, and he soon used his influence and his visibility to begin promoting a particular vision of nature. During a now-famous address at Woods Hole in 1910, Wheeler explained to his assembled colleagues why the ant colony was an organism. Not a metaphorical organism, he clarified, but rather a "true organism" (Wheeler 1911: 131). He reasoned that, like all organisms, ant colonies were reproductively differentiated, and that they developed along ontogenetic and phylogenetic lines. This vision inspired an entire generation of aspiring naturalists. Alfred Emerson was still just a grad student with a nascent interest in insects when he chanced to spend the summer of 1919 conducting fieldwork in British Guyana under Wheeler's tutelage. The experience had a profound influence on Emerson, who landed at the University of Chicago, and who resolved to study insect sociality the rest of his days. Although Emerson focused his attention on termites rather than ants, he was similarly convinced that the social-insect colony was a true organism in every sense (1939).

Emerson found a group of enthusiastic colleagues at the University of Chicago, and Warder Clyde Allee was perhaps the most influential among them. Allee graduated from the University of Chicago in 1912, and, nine years later, he returned to his alma mater as a professor. Over the next 30 years (1921 to 1950), Allee established his reputation as one of the most influential American ecologists of all time. He wrote extensively on the social dynamics of disparate species. *Animal Aggregations* (1931) provides a particularly illustrative summary of his research. Drawing on a wealth of research at both the microscopic and macroscopic level, Allee insisted that an individual's evolutionary success invariably depended on the presence of other individuals, and there was no such thing as a truly solitary organism. He wrote that the spatial arrangement of individuals in a restricted space, and the nature of their integration therein, could help explain the origin of "proto-cooperation." The Allee effect, as his observation became known, recognized that undercrowding could be just as lethal as overcrowding.

The 1930s also saw the coming of Mendelian genetics (a decade or two later to morph into molecular genetics) and in major respects this really did add the missing causal components needed to complete the Darwinian revolution (Provine 1971). Now, building on the theoretical insights of Ronald Fisher and J.B.S. Haldane in England and Sewall Wright in America, and thanks to the work of people like E.B. Ford in England and Theodosius Dobzhansky in the United States, natural selection could come into its own and much work was done showing its importance as the most significant causal factor in evolutionary change, at the same time boosting Darwin's insight that the key characteristic of organic life is it adaptedness (Ruse 2013a). The natural theologians were right – the world is as if designed. The only question is whether it is a hands-on Designer or the effects of ongoing unbroken law.

For fairly obvious reasons, the area of evolutionary biology covering social behavior rather lagged – it is hugely more difficult to measure the mating behavior of fruit flies than their wing size. However on the continent a number of behavioral scientists – labeled the "ethologists" – were starting to take the evolution of behavior very seriously, and with the move of one of the leaders (Niko Tinbergen) to Oxford expectedly social behavior was going to come more and more within the view of (what were now generally called) the neo-Darwinian evolutionists. In an almost paradoxical way, it was one of the general assumptions of the ethologists, particularly of the leader Konrad Lorenz, that was to spark much of the discussion. Lorenz along with many others (including all of the Harvard and Chicago biologists) calmly assumed that natural selection can work at many levels – favoring the individual on one occasion, and the group up to and including the species on another occasion (Lorenz 1966). This set up a reaction, because increasingly neo-Darwinians felt that "group selection" so-called is at best very rare. The trouble with it is that it lays itself too open to cheating. An organism with adaptations for helping itself is going to be at a selective advantage over an organism with adaptations for helping the group. In this new age of genetics, where it could be seen how each individual has biological interests in furthering its own gene replication, Darwin's ploy of treating nests and hives as super-individuals was judged inadequate.

Establishing the Principles of Sociobiology

Undoubtedly the most significant breakthrough came from a young English graduate student, William Hamilton (1964a, b). He argued that hymenopteran (ants, bees, and wasps) sterility can be explained as a function of reproduction by proxy as it were. In evolution, from a neo-Darwinian perspective, it matters not how genes are passed on, just that they are passed on in greater proportions than those of competitors. Hamilton pointed out that in the hymenoptera, because of a funny reproductive system – females have both mothers and fathers whereas males have only mothers – sisters are more closely related than mothers and daughters. Hence it is in the biological interests of a female to raise fertile sisters rather than fertile daughters! Those sterile workers, always female, are not so very "altruistic." They are furthering their reproductive ends by looking to those females in the next generation, rather than by selflessly serving the nest.

This process, which became generalized and was christened "kin selection," was obviously a paradigmatic example of selection working for the individual rather than the group, something that in the next decade in a popular account, the Oxford biologist Richard Dawkins (1976) was to label the "selfish-gene" approach. At the same time, independently, it was bolstered by the work of a young American ichthyologist, George C. Williams (1966). Having spent time as a postdoctoral fellow at the University of Chicago, he became disgusted at what he took to be the unwarranted flabby thinking of his seniors, particularly Allee. He like Hamilton turned to an uncompromisingly individual-selection approach over group selection and in 1966 published a very influential and fiery polemic, *Adaptation and Natural Selection*. Like everyone, he did not want to deny absolutely the possibility of group selection, but like everyone he thought its prospects very dim.

Conceptually the major American contribution to the subject came at the beginning of the next decade, from another graduate student, Robert Trivers. In a stunning series of papers, Trivers turned among other topics to that of "reciprocal altruism," where organisms help each other in expectation of help to be returned (Trivers 1971); to "parental investment," where parents put effort into their children's' well-being because they are now carrying the genes of the next generation (Trivers 1972); and to "parent–offspring conflict," where the biological interests of the parent (with several children) may conflict with the biological interests of the child (whose first concern is with self) (Trivers 1974). Whether or not these ideas were in the air – in the *Descent* Darwin certainly articulates the principle of reciprocal altruism – it was Trivers who gave them explicit articulation. (Perhaps a harbinger of what was to come, Hamilton was always an outsider and had trouble getting his dissertation to examination, and Trivers failed to get tenure at Harvard.)

Empirical work was now getting underway, and it is perhaps therefore no great surprise that Edward O. Wilson, on the faculty at Harvard, by the early 1970s the world authority on the ants (he was a student of Frank M. Carpenter who in turn was a student of William Morton Wheeler) and one who was doing major studies on pheromones (the chemicals used by insects for communication), should have had the idea and the ambition to draw all together. Thus was born *Sociobiology: The New Synthesis* (1975). (The first use of the term apparently dates from the 1940s, but it was Wilson who popularized it.) A massive work, copiously illustrated, it covers the theoretical work like that of Hamilton and Trivers, and then has a huge survey of the animal world, from the sponges, through the social insects, up through the vertebrates, ending with the great apes and finally with humankind. (The work was consciously modeled on an earlier survey of neo-Darwinism generally, *Evolution: The Modern Synthesis*, which appeared in 1942, authored by Thomas Henry Huxley's grandson, Julian Huxley.)

If 1859, the year of the publication of Darwin's *Origin of Species*, is the year when modern evolutionary thinking began, then 1975, when Edward O. Wilson published *Sociobiology*, marks the beginning of modern thinking about social behavior and its application to humankind, evolutionary psychology. The analogy may not be exact, however. Charles Darwin was not the first evolutionist (his grandfather, Erasmus, had endorsed the idea long before Charles was even born), but Charles Darwin did bring about a revolution. No one had gathered up the information as he had done and no one had made natural selection the main force for change. By comparison, the very subtitle of Wilson's book (*The New Synthesis*) indicates that he was primarily bringing together work, theoretical and empirical, that had already appeared elsewhere. Perhaps by making such a statement, one can think of the work as revolutionary, but even this in respect needs to be qualified, because one might with reason argue that Wilson and his coworkers were truly furthering the Darwinian revolution.

In any event, Wilson's *Sociobiology*, as well as Richard Dawkins' popular work, *The Selfish Gene*, which appeared a year later, caused massive controversy. Perhaps expectedly, many social scientists did not much like it. They expressed concern that biologists were ignoring or downplaying discoveries and clarifications that they had labored to make over the past century. Rather less expectedly, Wilson ran into great opposition from many of his fellow biologists, particularly people of the left, who felt that he was simply perpetuating vile myths about women and blacks and other traditionally downtrodden groups, myths that he was now trying to justify with the respectable cloak of neo-Darwinism (Allen et al. 1975).

Leading the charge were two of Wilson's colleagues in the organismic biology department at Harvard, the geneticist Richard Lewontin and the paleontologist Stephen Jay

Gould, the latter already on his way to becoming one of the best-known, popular-science figures in the world (Ruse 1979). Not entirely consistently, they accused Wilson of producing work that was both false and unfalsifiable! They argued that he was a racist, seeing certain groups of humans – no prizes for guessing which – as inherently inferior to others. They saw him as a sexist of the grossest kind, no matter that one of his students Sarah Hrdy (1978) was to do definitive work on the power of female monkeys in troops. Given the incandescent level of the critiques – at one meeting of the American Association for the Advancement of Science Wilson actually had water poured over his head ("You really are all wet Professor Wilson") – one suspects that there were emotional factors driving these charges. One certainly was ongoing rumblings in the aftermath of the Vietnam War. Several of Wilson's most prominent critics were politicized during that dreadful conflagration, and from thenceforth considered their bounden duty to oppose all kinds of oppression, real and apparent. Another was that many of Wilson's opponents were Jews and saw the kind of work he was doing as the thin end of a wedge that could only end in the kinds of events begun in the Third Reich of the 1930s and 1940s. Gould was much disturbed in this way and wrote a book, *The Mismeasure of Man* (1981),that made his feelings very clear.

Parenthetically, it is interesting to note that the philosophical community, with very few exceptions, from the start showed unremitting hostility to sociobiology, especially inasmuch as it applies to humans (see for instance Kitcher 1985.) This is perhaps not unconnected to the already-noted general fear that philosophers have of evolutionary theory, feeling that its successful application to our species would spell the end of the vaulted position of humankind on which so much of their work is predicated. Without denying the nonbelief of the average philosopher, it is interesting how they share the conviction with the evangelical Christian that we are special because we are made in the image of God, whether or not He exists.

Bloody but unbowed, Wilson followed his massive work with another explicitly on our species, *On Human Nature,* in 1978. He would have been less than human had he not taken a certain triumphant satisfaction in being awarded the Pulitzer Prize for this work. But less and less did he play a leading role in the science, starting to write books of a more popular and (as he would judge them) more philosophical nature. Turning therefore to others, what then can we say of the fate of sociobiology in the following four decades down to the present?

It is fair to say that, in the animal world, things settled down very quickly. Indeed, before Wilson, thanks to the new models of Hamilton and Trivers, together with some other innovative thinking – notably the use of game theory by the British biologist John Maynard Smith (1982) – evolutionary biologists had been doing both theoretical and empirical work on animal social behavior, and this continued full force in the years following. As indeed it continues today. The word "sociobiology" was considered contaminated by some and there were moves to find alternatives – "behavioral ecology" was one and others continued with the older term of "ethology" – but generally no one has felt the need for much of a name anyway, especially given that the whole point is that this work is and should be as being within the Darwinian paradigm (Ruse 2012). Other notable work would include the studies by Edward A. Herre and his associates in Central America on fig wasps (Herre, Machado, and West 2001; Ruse 2006).

In the human realm it is perhaps best to distinguish between studies about the past and studies about the present. The former belongs perhaps most prominently to the field of anthropology, particularly to paleoanthropology, and here we find physical anthropologists (or as they often style themselves nowadays evolutionary anthropologists) making very significant forward strides. It is well known that DNA studies have recently produced some quite astounding findings, for instance that beyond doubt Europeans carry some (about 4%) of Neanderthal genes (Krings et al. 1997; Lohse and Frantz 2014). Behavioral studies are obviously a lot trickier, but here too advances are being made through studies of tool use and diet and so forth.

One major factor in the evolution of humans is obviously the much-increased brain size and since big brains require large amounts of protein, we can say that our ancestors must have had access to such protein, and this could have come only from the bodies of other animals. Twenty-first-century graduate students may adopt a vegan diet but this was not an option for *Australopithecus afarensis* or *Homo habilis.* Much study has gone into the ways in which such protein was obtained, whether through hunting or scavenging or what. The same holds true of other aspects of human nature, for instance our linguistic abilities. Obviously one cannot take a tape recorder back a million years, but one can do such things as look at the nature of skulls and the indentations inside as a guide to which parts of the brain evolved first and most significantly (Falk 2004).

Developments in Evolutionary Psychology

As far as the present is concerned, generally students in this field have adopted the term "evolutionary psychology," and in fact one as often as not finds that research is going on in departments of psychology rather than departments of biology. In part, this obviously reflects the fact that psychology itself has moved in many respects closer to biology, with much attention now being paid to the physical underpinnings of our psychology as to the thought processes and behaviors of human beings. For all of the controversies of the 1970s, many saw bright prospects in the application of Darwinian principles to human nature and before long major studies were being performed and reported on (Daly and Wilson 1998; Hrdy 1999).

Some of the most interesting work applicable to all humans – John Toby and Leida Cosmides are prominent here – has been done on reasoning processes (Cosmides 1989; Cosmides and Toby 1996). There are some fairly standard psychological illusion tricks one can play on people, where if they are given real-life examples – young people in a bar drinking or not drinking alcohol – they solve the problems easily, but where if they are given the same problems in a theoretical form – turn over a card with an even number or not and that sort of thing – most people fail them almost all of the time. The evolutionary psychologists argue that evolution has fitted us to solve real-life problems and that it has done nothing on our abilities for solving theoretical problems that would not have been encountered in the Stone Age.

This has all led to major controversy about the extent to which our minds are still caught in the past (Ruse 2012). All agree that the mind/brain is a product of natural selection and that we have our features because they are adaptive, but does this mean that they are adaptive now or that they were adaptive once but now are not necessarily adaptive? A popular example is our fear of snakes but our too-ready acceptance of hand guns. In America today far more people are killed each year by hand guns than by snakes. Does this mean that we are out of adaptive kilter? We have adaptations to avoid snakes but not guns? Clearly culture comes into this somewhat, but how much? Would it be fair to say that most features today are adaptive, if not all? After all, we know that human evolution can take place very rapidly. There is no more obvious characteristic separating humans than skin color, but it seems that it is a very recent phenomenon. Perhaps it took only 15,000 years to develop. Likewise lactose tolerance; most adult humans cannot tolerate milk products, but with the coming of agriculture milk products became readily available and there was strong selection pressure to produce humans who could digest and use them.

Despite the opposition of philosophers, some of the most significant recent work has been on the evolution of morality (Ruse 2009). It is clear that our moral sense is an important adaptation. Humans are highly social beings with many of the features needed for such an adaptive strategy – for instance, our immunity to many diseases and parasites. Morality is clearly part of our tool bag. If we have to interact with people many times each day, then we need ways of getting along. More than that, we need quick and dirty ways of getting along. We cannot calculate self-interest every time we have an interaction with a fellow human. We need abilities to make quick decisions, even if later we may come to rue them. Morality helps here – I feel I should help you so I do, and later you feel you should help me and so you do. No arguments needed. Much recent work has gone into tracing the roots and nature of morality, looking at the great apes, looking at our past, looking at our current psychology and much more. Whether this all solves the problems of moral philosophers, whether for instance it tells us much about the foundations of morality, are much-discussed questions (Ruse and Richards 2016). But everyone seems to agree that it does tell us something.

Finally, we must not forget culture. Obviously a major distinguishing feature of human beings is our rich culture, whether or not you think it unique or that other animals also have their cultures. How do biology and culture interact? One much-respected hypothesis makes much of imitation. When in Rome, do as the Romans do, or, more precisely, do as the successful Romans do. "By imitating the successful, you have a chance of acquiring the behaviors that cause success, even if you do not know anything about which characteristics of the successful are responsible for their success," write Richerson and Boyd (2005: 120). You can see how bad ideas can hitchhike as parasites on the back of a system like this. Whether or not you think religion a bad idea, Richard Dawkins (2006) has made much of the ways in which religious beliefs come about through imitation and how they get themselves thoroughly engrained before anyone realizes how bad and dangerous they are. Perhaps here is a point where one might need to distinguish ideas that are bad in the sense of being false and ideas that are bad in the sense of being counter-reproductive in a Darwinian sense. It is by no means obvious that a true idea will necessarily lead to larger families than a false idea. Think of some of the silly things said about contraception.

There are still controversies. Almost all scientists today accept the predominance of individual selection over group selection. Interestingly and almost paradoxically, one who goes against the flow is none other than Edward O Wilson! Recently he has started arguing strongly for a more holistic approach to biological mechanisms including endorsing group selection (Wilson and Wilson 2007). Expectedly he brought on himself the wrath of the biological community, with no fewer than 150 people signing letters to *Nature* in disagreement. Perhaps, at this point, we have come full circle. Wilson was the student of Carpenter, who in turn was the student of Wheeler. Wheeler, along with his fellow scientists at Harvard – notably L.J. Henderson – always had a weakness for the thinking of Herbert Spencer, a man who notably endorsed an organicist view of society (Ruse 2013b). Are we hearing today faint echoes of those ideas that so engrossed thinkers 150 years ago when evolutionary thinking was just coming into its own?

Bibliographic Essay

Thanks to the far-reaching influence British naturalist Charles Darwin and British philosopher Herbert Spencer, a majority of American scientists had already converted to an evolutionary worldview by the late nineteenth century. As a result, the first generation of self-identifying sociologists and psychologists accepted that mental activity, human or otherwise, was a product of the evolutionary process. Early practitioners like Edward Lee Thorndike (1898) at Columbia University, John B. Watson (1903) at the University of Chicago, and Albion Small (1905) at Clark University all placed an early emphasis on the value of experimental studies. For more information on the development of American sociology in the late nineteenth and early twentieth centuries, consult Craig Calhoun's authoritative history (Calhoun 2007).

During the early twentieth century, biologists began to cast a much wider net, studying increasingly diverse kingdoms of life in an effort to understand the laws of social evolution. To learn more about Johns Hopkins biologist Herbert Spencer Jennings and his use of unicellular organisms in psychological research (Jennings 1906), one should consult the excellent work by Judy Johns Schloegel and Henning Schmidgen (2002). To learn more about Harvard biologist William Morton Wheeler's "superorganism" hypothesis (1911) and its influence on American sociobiology, read Charlotte Sleigh's masterful work on the topic (Sleigh 2002, 2007). Anyone interested in learning about Warder Clyde Allee's research on proto-cooperation (1931) and Alfred Emerson's unique marriage of natural selection and "emergent evolution" (1939) should read Greg Mitman's *State of Nature* (1992).

Support for holistic concepts like "emergence" and "superorganisms" evaporated in the wake of World War II. Biologists like W.D. Hamilton (1964a,b), John Maynard Smith (1964), and George C. Williams (1966) disputed the notion that natural selection could operate on groups, insisting that selection instead acted solely on individuals or the genes of which they are comprised. For an excellent summary of this contentious history, read Mark Borrello's *Evolutionary Restraints* (2010). One should also read *The Price of Altruism* (2010), Oren Harman's stirring biography of George Price, the American theoretician whose ideas on altruism changed the course of sociobiological thought on two continents.

Darwin's *Origin* notwithstanding, there is more perhaps no text in the history of sociobiological thought that has aroused as much passion as Edward O. Wilson's *Sociobiology* (1975). With breathtaking scope, Wilson examined the origins and nature of sociality across all branches of life and determined that four systems qualified as "pinnacles" of sociality: colonial invertebrates, social insects, nonhuman mammals, and, ahem, humans. Wilson's overt attempt to "biologicize" the social sciences touched off a major controversy across the academic spectrum (Alcock 2001; Allen et al. 1975; Lewontin 1978; Ruse 1979, Segerstrale 2000). Owing in part to this controversy, many researchers who were interested in mental and social evolution found it prudent to labor under a different banner, evolutionary psychology. To learn more, read *The Adapted Mind* (1992), by Jerome H. Barkow, Leda Cosmides, and John Tooby, as well as *The Moral Animal*, by Robert Wright (1994). And finally, it is altogether fitting that E.O. Wilson continues to rock the boat. While he placed kin selection at the center of sociobiology in 1975, he has since very publicly revoked that endorsement, insisting instead that natural selection, and not kin selection, provides the initial impetus for sociality (Wilson 2012, 2014; Wilson and Wilson 2007). His reversal has sent shockwaves across the academic landscape, indicating that the field of sociobiology remains as controversial, and as vibrant, as ever

Chapter Twenty-One

SOCIOLOGY

Sebastián Gil-Riaño

This chapter examines how historians of the social sciences and sociologists alike have interpreted the rapid growth and international ascendancy of American sociology during the interwar and particularly post-World War II period. First, I examine traditional narratives concerning the rise of American sociology, which emphasize its role as a professional academic project that emerged to understand the massive societal transformations brought about by the industrialization and urbanization of the late nineteenth century and to direct societal change in ways consistent with ideologies of American exceptionalism. These narratives emphasize the US nation-state as the most important context for understanding the coming into being and growth of sociology as a science and professional discipline. However, more recent approaches have sought to decenter the United States as an all-encompassing context in favor of a "global" or "transnational" context for the emergence of modern sociology. From this vantage point, the history of American sociology and the discipline of sociology more broadly, appear as inextricably entangled with the imperatives of the imperial projects of the North Atlantic. At stake in sociology's historiography are the epistemological moorings of the discipline itself: Is an objective and value-neutral science of society possible? Can sociology be both reflexive and objective? Does sociology bear the indelible stain of imperialism?

By surveying the historiography of sociology in the United States and its global turn, this chapter seeks to link the history of American sociology with postcolonial approaches to science studies more broadly. Over the past two decades, historians of science and science studies scholars have increasingly taken up studies that go beyond the nation-state and that situate technoscience in spaces beyond European and North American urban centers. A basic concern in global and postcolonial studies of science is to bring greater symmetry to the way scholars analyze and narrate the development and spread of technoscience in the modern era by bringing metropole and postcolony,

A Companion to the History of American Science, First Edition.
Edited by Georgina M. Montgomery and Mark A. Largent.
© 2016 John Wiley & Sons, Ltd. Published 2020 by John Wiley & Sons, Ltd.

center and periphery, north and south under a single analytic frame (Anderson 2002b: 646). Although some of this literature has taken inspiration from the work of postcolonial theorists such as Edward Said, Homi Bhaba, and Gayatri Spivak, it has also sought to move past postcolonial theory's emphasis on textual and literary analysis by engaging material processes and the production of rationality through scientific discourses as crucial domains for analyzing the ongoing consequences of the history of Western colonialism. Postcolonial studies of technoscience have mostly focused on the biomedical sciences. However, as sciences that have laid claim to knowledge of objects such as the human psyche, cultural systems, social structures, and economic rationality the social sciences are perhaps the most closely implicated in anthropologist David Scott's characterization of colonialism as a project characterized by "the concerted attempt to alter the political and social worlds of the colonized, an attempt to transform and redefine the very conditions of the desiring subject" (Scott 1995: 55). With Scott's definition of colonialism in mind, this chapter will examine the key historiographical frames that have been used to explain how sociology rose to prominence in the United States and the ways that these conventional interpretations are being reframed through the adoption of postcolonial and transnational lenses.

Narrating the Rise of Sociology as a Profession

Historians and sociologists often describe the emergence of professional sociology and social science in the United States during the latter decades of the nineteenth century as a response by liberal elites to an excessive democratization of intellectual authority and to a changing society that threatened to veer away from its republican and capitalist ideals. The transformations that took place during the Gilded Age threatened the place of the genteel scholars who had played a leading role in defining the republican ideals that characterized American political culture. The rapid industrialization of the United States during this period attracted millions of immigrants from Europe and Asia to cities such as New York, Chicago, San Francisco, and Los Angeles, and the implementation of Jim Crow laws in the South prompted thousands of African Americans to move North in search of better opportunities. As a result of these changes, American society shifted from a primarily rural, white, and pastoral society to an increasingly urban, cosmopolitan, and industrial society. At the same time, the rapid industrialization of the Gilded Age was also characterized by unfettered capitalism, which generated massive social inequalities and concentrations of wealth in the hands of a few industrialists. The urban centers that rapidly expanded in this period gave rise to growing numbers of poor industrial workers whose labor fuelled the economic development of this period but who did not enjoy the same improvement in quality of life accrued to the middle and upper classes. Consequently, this period was also pivotal in the formation of strong labor unions that questioned capitalism's efficacy as a mode of production and experimented with communist and socialist values (Ross 1991: 87).

   One of the central interpretations of the emergence of sociology in the United States, is that it has been a science concerned with directing societal change in ways consistent with the liberal and egalitarian social values of America's founders. Indeed, a common argument concerning the rise of professional sociology in the United States is that it arose part and parcel with the modernizing impulse of the late nineteenth century and with the need to administer and integrate increasingly diverse populations into the national state. Historian Thomas Haskell, for example, argues that genteel scholars created the American Social Science Association (ASSA) – the first professional social science association in the United States – in 1885 in an attempt to regain some of their vanishing intellectual authority in a rapidly industrializing society. Haskell argues that like the formation of professional bodies for physicians that were designed to distinguish between charlatans and competent practitioners, the ASSA was created by anxious liberal elites and intellectuals (predominantly white men from New England) who sought to bolster their authority by demarcating competent judgment and opinion on social affairs from incompetent speculation, and who were eager to devise more scientific approaches to social philanthropy – a project they dubbed "social science" (Haskell 2000: 66–7).

However, because of the highly challenging and causally complex problems generated by an increasingly interdependent industrial society, the ASSA's eclectic and unfocused approach instead prompted the demarcation of specific social science disciplines that could provide an adequate framework for individuals to devote their entire careers to the study of social problems. As the ASSA declined in influence in the 1890s, a new generation of scholars, deeply influenced by the theories of Auguste Comte and Herbert Spencer, and including Albion Small at the University of Chicago, Franklin Giddings at Columbia University, Graham Sumner, and Lester Ward decided that sociology could act as a broad synthetic social science and pushed for the creation of sociology departments in universities (Haskell 2000: 190–210).

Like Haskell, intellectual historian Dorothy Ross situates the origins of American social science in the social crises of the late nineteenth century. However, whereas Haskell sees the professionalization of social science as a necessary response to an increasingly fraught social landscape, Ross interprets the emergence of professional social science as a political attempt by middle-class liberals to preserve and further inculcate an exceptionalist vision of the United States as a democratic and capitalist nation, and thus as a liberal project of social control. Her interpretation of the first generation of sociologists (and particularly William Graham Sumner and Lester Ward) contends that, under the influence of Spencerian evolutionism and in response to Marxist critiques of capitalism and conservative critiques of democracy, their overarching concern was to decipher the laws of historical change and the means by which society could be manipulated to allow for their liberal vision of societal progress. Indeed, Ross points out how these early sociologists conceptualized society as analogous to physical nature and as thus open to a kind of mechanical manipulation (Ross 1991: 92).

Sociologists of the Progressive Era (which Ross periodizes as 1896–1914) retained this project of attempting to preserve a liberal identity for the US in the face of political and industrial upheavals. However, instead of a focus on the laws of progress, Progressive Era sociologists enthusiastically adopted a vision of science concerned with "prediction and control." Indeed, Progressive Era sociologists also adopted statistical and survey methods to a much greater extent and viewed these as methods that distinguished the scientific work of sociology from that of social workers and social reformers. This emphasis on rigorous statistical methods oriented toward social control allowed sociologists to imagine themselves as technocrats charged with maintaining social cohesion and minimizing social unrest as the United States transitioned to a fully industrialized society. It was during this period, Ross argues, that "social control inscribed and was intended to inscribe a normative vision of modern capitalist society at a time when that vision was seriously contested" (Ross 1991: 247).

After World War I, sociologists remained committed to statistical rigor and social control but abandoned the grand theories of Comte and Spencer and instead adopted a highly empirical approach to the study of social phenomena guided by more circumscribed concepts derived from ecology such as natural cycles and cyclical processes. Emblematic of this shift was the rise of the sociology department at the University of Chicago as the leading center for sociological training during the interwar period. With the leadership of the journalist turned sociologist Robert E. Park and with W.I. Thomas's 1918 publication of *The Polish Peasant* as a crucial reference point, Chicago blossomed into the leading center for the sociological study of urban problems, cultural contacts, race relations, and immigration studies. Under Park's leadership, Chicago's sociology department secured large funds from the Rockefeller Foundation for major research projects and attracted enthusiastic graduate students who were eager to solve social problems. As a mentor, Park sought to transform "the reform sentimentality of these students, often brutally, into an equally intense commitment to a distinctive curiosity" (Turner and Turner 1990: 47). Equipped with this curiosity as well as a commitment to scientific detachment, Park's students eschewed the grand theorizing of previous generations of sociologists and instead focused on detailed empirical studies of specific social problems arising in urban centers. Indeed, Chicago sociologists adopted an approach that privileged intricate understanding of the psychological profiles of individuals from different racial and ethnic groups and the ways that these reflect relations of competitive cooperation inherent to a liberal society. Park and other Chicago sociologists likened the different kinds of competitive economic relations they observed between ethnic groups as akin to the competitive relations between different species in an ecosystem and characterized their approach as "natural history" or "urban ecology." By gaining a better empirical understanding of these relations and the way they were manifest in the attitudes of individuals, Chicago sociologists hoped to render the process of cultural and ethnic assimilation more manageable and open to social engineering. Park's "race relations cycle" and Thomas's cycle of organization–disorganization–reorganization are examples of the kinds of social processes that Chicago sociologists identified as objects of study and as phenomena open to scrutiny and manipulation (Saint-Arnaud and Feldstein 2009: 50–1).

Chicago sociologists were very successful in attracting funds from private philanthropic organizations such as the Rockefeller and Carnegie Foundations, and played leading roles in the creation of some of sociology's key journals and professional societies. The success of Chicago sociologists in attracting funds and building up sociology's institutional framework, coupled with their adoption of a scientific ethos of detachment, have been touted as evidence of sociology's professional maturation during the interwar period. In conventional histories of sociology in the United States, the rise of Chicago sociology during the interwar period is seen to be the culmination of a process of professionalization that began in the 1890s. By linking the rise of professional sociology to the social upheavals of the Gilded Age and Progressive Era, historians have argued that sociology emerged as a discipline concerned with maintaining social order and defending liberal democracy and capitalism in the face of potential political alternatives as well as vast demographic changes. Thus, a crucial themes in these accounts is how professional sociology emerges as a discipline torn between the desire to ameliorate and improve society and the compulsion to gain intellectual legitimacy by adhering to the standards of impartiality and objectivity of the natural sciences. Narratives concerning the professionalization of sociology in the 1890s and its culmination during the interwar period thus privilege the social and political struggles of a modernizing US as a crucial explanatory context for understanding how sociology as a formal discipline and community of inquiry came into being. As we will see in the next section, histories concerning sociology in the post-World War II period expand this geographic focus and bring sociology's connections to the Global South into the fold.

Cold War Social Science and the Rise of Modernization Theory

Whereas histories of sociology describe the interwar period as one in which professional sociology cemented its place in the academic landscape, the Cold War era has been described as American sociology's "Golden Period," where sociology and the social sciences enjoyed unprecedented levels of funding and support thanks to the linkages they developed with federal agencies during the war and to the US government's desire to maintain a competitive scientific edge with the USSR in the wake of *Sputnik* (Turner and Turner 1990: 133–78). From roughly 1940–1970, American sociology enjoyed a dramatic rise in funding from federal agencies and this, coupled with the number of baby-boomers entering universities, spurred major growth in the number of sociology degrees that were granted on an annual basis, a proliferation of new journals and specialty associations, and a huge rise in membership in the American Sociology Association (ASA) (Turner and Turner 1990). At the same time, sociological ambitions in this period grew to unprecedented levels as leading sociologists such as Harvard's Talcott Parsons announced that they were on the cusp of a scientific revolution that would unify the social sciences and place them on an epistemological foundation equivalent to the natural sciences (Gilman 2003). The rapid growth and shifting epistemological priorities of sociology in this period had lasting consequences for its identity as a discipline. Not all sociologists supported the push toward theoretical unification and quantification in the late 1960s, which generated a backlash against what came to be known as "mainstream sociology" and the formation of alternative sociological traditions that sought to counter the perceived hegemony of Talcott Parsons and Robert Merton. At the same time, public controversies over American counterinsurgency measures in postcolonial nations and the role of social scientists in such projects generated critical scrutiny of social science's entanglements with government and military, and raised profound questions about the objectivity of social science.

The intellectual transformation of sociology in the postwar period is often described as a shift in where and how leading research was produced. Whereas the ethnographic and ecological studies of Chicago sociologists dominated the interwar period, by the postwar era Chicago sociology was eclipsed by Talcott Parsons and his colleagues at Harvard's department of social relations and the highly statistical methods and middle range theories developed by Robert Merton and Paul Lazarsfeld at Columbia University (Abbot and Sparrow 2007). This geographic shift was partly due to the retirement of Park and dispersal of some of the key Chicago faculty members, but perhaps more importantly to the rapprochements between East Coast sociologists with military agencies during the war. During World War II, many sociologists lent their research skills to the war effort and worked closely with war agencies such as the Research Branch of the War Department. At an institutional level, the ASA established a committee to facilitate collaboration with the government and military in 1940 and hundreds of sociologists enlisted with the National Roster of Scientific and Technical Personnel which served as a roster where federal agencies could recruit for war departments and the civil service (Abbot and Sparrow 2007). Harvard and Columbia sociologists were particularly active in this rapprochement with the government and military and as a result enjoyed great success in the postwar era when academic research and the social sciences experienced a great windfall in funding from federal agencies.

The linkages sociologists developed with the military and government were also complimented by the rise of large collaborative studies that deployed survey and other statistical methods with an eye to describing and attempting to alter the psychological dispositions underlying social problems. An example of this kind of study is the Swedish economist Gunnar Myrdal's multiyear study of American race relations, which was sponsored by the Carnegie Corporation of New York and enlisted the input and collaboration of many leading social scientists including Franz Boas, W.E.B. Dubois, Ruth Benedict, Otto Klineberg, Horace Cayton, Ralph Bunche, John Dollard, and E. Franklin Frazier and resulted in the nearly 1500 page-long *An American Dilemma* (1944) (Jackson 1990; Southern 1987). Another example was the widely read study of soldier attitudes and morale, *The American Soldier* (1949), which was written by the Harvard sociologist Samuel Stouffer and based on surveys of more than half a million GIs conducted by Stouffer and a team of social scientists in the War Department during the war.

The turn toward an increased use of statistics and large collaborative studies during the 1940s inspired leading sociologists to attempt to bring theoretical unity and coherence to the social sciences which eventually gave rise to modernization theory. As the historian Michael Latham has shown, many of the ideas that underpinned modernization theory first appeared in the work of Cold War sociologists (Latham 2000: 30). Indeed, two of the intellectual architects of modernization theory were the Harvard sociologists Talcott Parsons and Edward Shils, both of whom were dissatisfied with the epistemological foundations of sociology and sought to articulate a general theory of society and action that could be applied across the social sciences. A crucial theme in the theories of Parsons and Shils, often referred to as structural functionalism, was the theorization of the social institutions that maintained social order in both "primitive" and "modern" societies. For Parsons, one of the crucial differences between these two types of society was that in "primitive" societies social order was maintained through community and family institutions, whereas in modern societies with capitalist markets, social order was maintained through formal legal systems and democratic processes. For Parsons, the legal and democratic systems of "modern" societies thus represented highly specialized and dynamic institutions that allow industrial societies to maintain social order in the face of the transformations brought about by societal change. Indeed, one of the key components of Parson's theories was the characterization of societies as highly integrated and interdependent systems where economic, political, and social institutions are interrelated.

Parsons and Shils articulated theoretical models that were often highly abstract and that sought to describe the basic structures underlying all societies. In so doing, they offered social scientists a conceptual framework for classifying societies where the United States figured as a paradigm of modernity and acted as a measuring stick against which all other societies would be measured. Social scientists from other disciplines such as the economist Walt Rostow and the political scientist Lucian Pye were quick to incorporate the tenets of structural functionalism into their analyses and eventually transformed it into what came to be known as modernization theory. Although many scholars offered different takes on modernization, their theories shared some core assumptions including the notion that (1) there is a sharp dichotomy between traditional and modern societies, (2) the economic, social, and political changes in a society are interdependent, (3) there is a common and linear developmental path toward the modern state in all societies and (4) the progress of developing nations toward modernity can be accelerated through contact and technical support from modern nations (Latham 2000: 4). In the 1960s, as Europe's hold over its colonies began to disintegrate and communism emerged as an appealing ideology for anticolonial movements throughout the so-called "third world," modernization theorists found a welcome audience among military and government officials, and their ideas about development became infused into the foreign policy of the Kennedy administration and projects like the Alliance for Progress, the Peace Corps, and the Strategic Hamlet project in Vietnam (a counterinsurgency program that sough to contain communist influence) (Latham 2000).

However, the mid-century infatuation with modernization theory was short-lived as public disenchantment with the Vietnam War brought concerned citizens to scrutinize the multiple entanglements between the social sciences and the military and to question the purported objectivity of the social sciences. A focal point for discussion was Project Camelot, a 1960s military sponsored counterinsurgency project that enlisted social scientists to study postcolonial nations in an attempt to identify the causes of social unrest and anticipate potential revolutionary outbreaks. The project's projected cost of six million dollars would have made it the largest social science project in American history, but the project was cancelled after it was publicly denounced by Latin American intellectuals who accused it of being a Trojan horse for American imperialism and thus a threat to their national sovereignty. For historian of social science Mark Solovey, the project's cancellation can be seen as a pivotal moment in an epistemological revolution that unfolded during the 1960s and precipitated significant contestations of the orthodox understanding of "social science as an objective, value-neutral scholarly enterprise immune to extra-scientific (and especially political or ideological influences" (Solovey 2001: 173).

From the vantage point of the history of professional sociology that this chapter has examined, the 1960s can thus be seen as a crucial moment of rupture when the project of professionalization that began in the 1890s, and relied upon the invocation of objectivity as a means of legitimating the authority of social science, began to unravel. Intellectual challenges to modernization theory and structural functionalism coupled with mounting criticisms about government and military funding of social science brought the links between professional social science and liberal conceptions of the United States as a model of modernity and progress to the forefront and provoked a more reflexive engagement with sociology's history and aims. As a consequence of the debates that took place in the 1960s, sociology emerged as a much more contested field with a plurality of subfields and methodological approaches. The fracturing of sociology during the 1960s in turn provoked sociologists to reconsider the history of their discipline with an aim to recuperating sociological traditions and approaches that had been concealed through the construction of a sociological canon.

Decentering the United States: Postcolonial Approaches to the History of Sociology

Is it possible that the imperial gaze embedded in modernization theory is not a Cold War product but rather one of the foundational concerns of professional sociology? If so, what consequence does this have for how to narrate sociology's history? In recent years, sociologists have begun addressing such questions by turning their attention to the imperial history of their discipline and attempting to ponder the possibility of a "postcolonial sociology." Similarly, historians have begun to situate early American sociology within imperial contexts and to introduce postcolonial lenses into the analysis of sociology's emergence in the United States. This emerging historiography both revises and compliments the historical accounts of American sociology described in this chapter. Like the conventional historiography, postcolonial historiography recognizes social science as a form of situated knowledge concerned with social governance and social control. However, whereas standard histories of sociology have privileged the domestic struggles of a modernizing United States as the most crucial context for understanding sociology's emergence, postcolonial accounts of sociology's history locate its emergence at the high point of European and North American imperialism and argue that the management and control of peoples in the metropole and colony has been a constitutive and abiding sociological concern.

Postcolonial histories of sociology have been particularly concerned with revising historical understandings of sociology's rise. In recent years, for example, social theorist Raewyn Connell has sought to persuade sociologists to move past the emphasis on a classical sociological canon centered around the theories of Northern theorists and to instead engage much more deeply with sociological traditions from the Global South. Crucially, Connell's arguments rely on a revised historical interpretation of the sociological canon that situates the theories of classical sociologists such as Emile Durkheim, Max Weber, Franklin Giddings, and Lester Ward within the context of Western global expansion and colonization and argues that their theories are infused with a grand ethnographic gaze afforded by the data produced by imperial conquest (Connell 1997). Connell notes, for example, that the early sociological traditions that emerged in Europe and North America at the end of the nineteenth century were characterized by an eclectic mix of topics that included things as diverse as polyandry in Ceylon, mining in California, holy war in ancient Israel, Buddhism in India, Malay magic, and conceptions of kinship in Australian aborigines. According to Connell, this broad perspective on human history was organized by a central concern with elucidating the "difference between the civilization of the metropole and other cultures whose main feature was their primitiveness" (Connell 2007). Inherent to this project was a conception of time where colonized societies were seen as historical antecedents to the modernized metropole and thus became an object against which notions of historical progress were narrated. As such, the sociological traditions that emerged in the late nineteenth century were "formed within the culture of imperialism, and embodied an intellectual response to the colonized world" (Connell 2007).

By situating the rise of sociology within the context of late nineteenth-century imperialism Connell suggests that the notions of progress and difference, typical of late-nineteenth theorists, were not peripheral sociological concerns but rather central problem spaces that have occupied the intellectual core of the discipline since its emergence. At the same time, Connell uses empire as a framework for explaining why sociology flourished in the United States during the interwar period while deteriorating in Europe. Like the historian Michael Adas, who argues that World War I was a crucial turning point that undermined the intellectual underpinnings of European notions of superiority (Adas 2004); Connell argues that the war indexes a shift in imperial dynamics where the metropole moved from Europe to the United States and thus also a shift in the center of sociological production. Concomitant with the flourishing of American sociology during the interwar period was a shift in sociological methods and objects of study. Rather than a concern with grand theories that would make sense of difference on a global scale, Connell argues that American sociologists instead became concerned with understanding difference within the metropole through statistical techniques and with developing methods of surveillance that would serve the purpose of "social control."

Building on Connell's insights, historical sociologists such as Julian Go and George Steinmetz have similarly sought to re-examine the history of the discipline and its imperial entanglements as well as to deploy sociological frameworks for understanding the history of imperialism (Go 2011, 2013a; Steinmetz 2013). The projects embraced by historically minded sociologists relate to a broader project of applying the critical insights of postcolonial theory to sociology in an attempt to move past the analytic bifurcations and Eurocentric tendencies of the discipline and attempt to fashion a postcolonial sociology (Go 2013b).

Historians have also reassessed the history of American sociology and social science from a geographic perspective that highlights the significance of the Global South. For example, Andrew Zimmerman, a historian of German colonialism, argues that the sociological theories of "race," assimilation, and cultural difference that emerged in Chicago during the interwar period were linked to a transnational politics of improvement that "left room for reformist corrections to the perceived failings of supposedly inferior races" (Zimmerman 2010: 206). However, whereas typical histories of Chicago sociology situate this reformist impulse as a response to the urban problems of the United States, Zimmerman argues that Chicago sociologists' ideas about race, civilization, and economic progress were part of a much broader transnational dialogue about how to manage free labor in the wake of the Civil War and in the context of German colonization in Europe and Africa at the end of the nineteenth century. To link these various regional histories, Zimmerman examines the infamous 1901 Tuskegee Expedition to Togo: an expedition that brought together agricultural experts from the Tuskegee Institute in Alabama with German colonial officials who sought to reconstruct the system of labor being used in the New South in the German colony of Togo. Zimmerman demonstrates how German and US sociologists provided the intellectual underpinnings for both the Tuskegee Institute in Alabama and German colonization schemes and how sociologists from both sides of the Atlantic came to view the American South as a "model colonial society for white liberals" (Zimmerman 2010: 185). In Zimmerman's work we see notions of modernization and development crystallizing out of the colonial projects of the early twentieth century as opposed to the Cold War contexts with which they have often been linked.

Whereas Zimmerman's work offers a postcolonial account of Chicago sociology by linking it with broader imperialist projects, historian Henry Yu's *Thinking Orientals* reinterprets the legacy of Chicago sociologists by demonstrating the role they played in fashioning an *orientalist* discourse. Focusing on how Chicago sociologists constructed the so-called Oriental Problem of the American Pacific Coast during the interwar period, Yu shows how the Oriental Problem was primarily a construct of white protestant social scientists and social reformers who viewed so-called Orientals as exotic and culturally opposite to the West and thus not easily assimilated into American culture. In framing the study of Orientals in this way, Chicago sociologists created an institutional structure in which Asian Americans were valued for their knowledge of the exotic but also expected to perform in ways consistent with their perceived racial identity. Indeed, Yu notes that Chicago sociology's tendency to theorize race and culture as primarily a question of consciousness had the effect of rendering racial identity into an aesthetic object and of producing an "institutional structure that valued the exotic nature of the knowledge possessed by Orientals and other 'ethnics'" (Yu 2001: 9). From this vantage point the social identities used by sociologists to describe themselves and others appear as hybrid and malleable categories open to revision and contestation.

By situating the emergence of professional sociology within the context of turn-of-the-century European and North American imperialism, these recent reappraisals of sociology's history extend the theme of social science as an instrument of "social control" and "social engineering" prominent in earlier historiographical approaches. However, whereas the historiography of American sociology has tended to associate sociologists' ambitions to rebuild the Global South in the image of a modern industrial West with the work of Cold War modernization theorists, postcolonial approaches to sociology's history suggest that modernization has deeper roots. Indeed, the postcolonial approaches surveyed here suggest that Cold War sociologists' infatuation with modernization theory does not represent a rupture with early twentieth-century social theory but rather a reformulated version of older themes. By invoking the history of global empires as a crucial context for understanding the emergence and development of sociology as a discipline, these recent approaches invite historians and historical sociologists to adopt a much wider lens that involves mapping sociology's transnational history as well as greater attention to the lines of continuity that can be observed between the social theories of the late nineteenth century and those of the post-World War II era.

Future histories of American sociology face many complex and potentially rewarding challenges. An ample body of literature on the emergence of the social sciences in the United States has now documented the relation between the formation of professional communities of sociologists in the United States and the societal problems produced by the rapid industrialization, urbanization, and demographic transitions that transformed the country at the outset of the twentieth century. Postcolonial approaches to sociology's history have expanded the analytic possibilities for narrating the history of American sociology by bringing the context of global empires and the shifting imperial dynamics of the twentieth century into our historiographic frame. Rather than viewing the formation of American sociology as a project primarily concerned with maintaining social order in a modernizing United States, transnational and postcolonial approaches invite scholars to consider how the project of establishing social order at "home" was also linked with exporting social order "abroad." Indeed postcolonial histories of sociology seek to decenter the US nation-state as the most critical nexus for understanding the production of sociological knowledge and instead offer a much broader contextualization that takes trans-Atlantic (and trans-imperial) dynamics, the Global South, histories of migration, and the commodification of racial identities as crucial analytic frames.

However, whereas postcolonial approaches suggest a need to decenter the United States as an all important frame of reference, the historiography concerning Cold War social science and the dramatic growth of the social sciences in the United States during the postwar era illustrates why historical analyses of social science cannot simply abandon the United States. Indeed, the military-sponsored expansion of the social sciences during the Cold War allowed US sociology to become a global leader and American sociologists to exert an overwhelming influence in the formation of international sociological associations (Backhouse and Fontaine 2010: 102–136). Future histories of sociology in the United States therefore face the challenge of decentering the United States and more adequately contextualizing American sociology within a global cartography, while at the same time recognizing American sociology's unrivalled and unprecedented global influence.

Bibliographic Essay

An excellent place to start research on the history of sociology in the United States is the collection of essays edited by sociologist Craig Calhoun and titled *Sociology in America: A History*. The collection consists of 21 essays that offer comprehensive coverage of the key personalities, institutions, patrons, and research paradigms that have defined the history of sociology in the United States as well as discussions of sociological traditions and subfields that have received much lesser historiographic attention such as feminist sociology and the sociological analyses of race produced by African American scholars (Calhoun 2007). The collection offers a thorough introduction to the range of issues at stake in the historiography of sociology and includes contributions from some of the most influential historical sociologists including Andrew Abbot, Stephen Turner, Patricia Hill Collins, Immanuel Wallerstein, and Howard Winant. The volume also offers two historiographic essays that provide overviews of the key works in the history of sociology. However, it should be noted that the volume's contributions come almost entirely from sociologists and tend to be focused on issues that are primarily of importance to practitioners of the discipline.

The early history of sociology was written primarily by intellectual historians and has been concerned with illustrating some of the ideological and social factors that compelled sociologists to professionalize as well as the often-frustrated efforts sociologists have made to maintain objectivity. Aside from the classic works of Thomas Haskell and Dorothy Ross discussed earlier in this chapter, readers should consult the works of Robert C. Bannister including *Sociology and Scientism* (1987) and *Social Darwinism* (1979) for accounts of sociology's struggles with objectivity and relation to the evolutionary thought of the Victorian era (Bannister 1979, 1987). For a classic interpretation of how social sciences like anthropology, psychology, and sociology sought to distinguish their work from typological and hereditarian theories of race, see Hamilton Cravens' *The Triumph of Evolution* (1978). For an "institutional analysis" of US sociology that seeks to make sense of the trends and developments of sociology in the first 100 years of its history by paying particular attention to its sources of funding, see Turner and Turner's *The Impossible Science* (1990).

Because it has such a prominent place in the historiography of sociology and has attracted so much attention from both sociologists and historians of social science, the Chicago school of sociology is both a crucial reference point and inevitable point of passage for those seeking to understand sociology's historical legacy in the United States. Andrew Abbott's collection of essays *Department and Discipline* (1999) offers a useful overview of the voluminous literature on the Chicago school, and argues that the historiography of the Chicago school has moved away from studies concerned with delimiting and defining its institutional and conceptual contours to revisionist studies that situate the Chicago school in broader historical currents and thus aim to question the school's status as a unique and distinct entity (Abbott 1999: 4–33). Among the most important of these revisionist narratives are the studies produced by Mary Jo Deegan and Kathryn Kish Sklar who were both concerned with demonstrating the profound influence of social reformers such as Jane Addams and Florence Kelly on the institutionalization of Chicago sociology and the development of survey methods through their work in the settlement house movement (Deegan 1988; Sklar 1991). By tracking this relatively concealed aspect of Chicago's history, Deegan and Sklar showed that Chicago sociology's intimate and ethnographic methods were born from a tradition of grassroots social reform and racial uplift as much as they were from a concern with greater scientific objectivity. In a similar vein, sociologist Pierre Saint-Arnaud and historian Mia Bay have examined the sociological studies of African-American sociologists such as W.E.B. Dubois, Edward Franklin Frazier, Charles Spurgeon Johnson, Horace Cayton, and Oliver Cox and demonstrated how they developed innovative methodological and theoretical frameworks that often went ignored by their white colleagues during the first half of the twentieth century (Bay 1998; Saint-Arnaud and Feldstein 2009). More recently, urban historian Davarian Baldwin has argued for the importance of understanding the Chicago school's intellectual legacy not as a project that was concerned with the construction of racial categories but rather as a theoretical corpus created in response to the "national anxieties about racial differences in urbanizing America" (Baldwin 2004). The voluminous literature on the formation, development, and legacy of the Chicago school provides an excellent place to engage the myriad historiographic approaches that have been used for the purposes of understanding the historical significance of sociology in the United States.

However, although Chicago sociology during the interwar period has already generated a large historiography, recent studies that place Chicago sociology in a much broader transnational context suggest that there is room for further study. As mentioned in the main section of this chapter, Andrew Zimmerman's *Alabama in Africa* offers many insights into the trans-Atlantic connections between Chicago and German sociologists and the modernization of the Global South. Like Zimmerman, the Latin American historian Anadelia Romo has recently examined Chicago sociologists' influence on Brazilian social science as part of her historical study of how the Northeastern state of Bahia came to be imagined as the cradle of Brazil's African roots. In her study titled *Brazil's Living Museum: Race, Reform, and Tradition in Bahia* (2010), Romo shows how Chicago-trained sociologists and anthropologists including Donald Pierson, Ruth Landes, and E. Franklin Frazier began to study Brazilian race relations in Bahia during the 1940s on the advice of Robert. E. Park. Under the influence of Park, who had a notable interest in racially mixed peoples and wanted to create a "miscegenation map of the world" (Anderson 2012: 105), this generation of Chicago social scientists came to see Brazil as relatively free of racial prejudice and played an important role in promoting the ideology of "racial democracy" (Romo 2010). Historians of the Pacific have also begun to document a similar trajectory for Chicago sociology in relation to Hawai'i. Recent historical studies of Hawai'i by Maile Arvin and Christine Manganaro have shown how Chicago-trained sociologists such as Romanzo Adams and Andrew Lind played important roles in constructing the notion that Hawai'i constituted a "racial paradise" free of racial prejudice and how this discourse was used to conceal the colonial history of the island (Arvin 2013; Manganaro 2012).

Those interested in understanding US sociology's global influence post-World War II, will find much of value in the growing historical literature on modernization theory. In addition to the studies by Latham and Gilman mentioned in this chapter, *Staging Growth: Modernization, Development, and the Global Cold War* edited by David Engerman and others offers a diverse collection of essays that seek to connect the rise and fall of modernization theory in the United States with a global history of the Cold War (Engerman 2003). Laura Brigg's *Reproducing Empire: Race, Sex, Science, and U.S. Imperialism in Puerto Rico* also examines how US sociologists and social scientists came to view Puerto Rico as a "social laboratory" during the 1940s and 1950s and as a testing ground for economic development policies centered around population control (Briggs 2002: 109–42).

In recent years, the Cold War era has attracted much attention from historians of the social sciences, intellectual historians, historical geographers, and diplomatic historians interested in understanding why the social sciences enjoyed such unprecedented growth during this time period and the kinds of epistemological challenges social scientists encountered as they sought to reconcile ideals of scientific objectivity with the services they lent to federal and military agencies. This literature on Cold War social science has provoked lively debates and discussions among historians of social science and raised profound questions about the degree to which scientific knowledge in this period was shaped by military imperatives, and about the legacy of Cold War social science for democracy in the United States. *Cold War Social Science: Knowledge Production, Liberal Democracy, and Human Nature*, edited by historians Mark Solovey and Hamilton Cravens, provides an excellent overview of the relations between power and knowledge in various fields of Cold War social science including area studies, linguistics, psychology, and anthropology (Solovey and Cravens 2012). Joel Isaac's recent study, *Working Knowledge: Making the Human Sciences from Parsons to Kuhn*, provides an interesting counterpoint to studies on Cold War social science insofar as it situates Harvard's department of social relations and Talcott Parson's structural theories within a primarily intellectual project concerned with establishing a common epistemological foundation to philosophy, natural science, and the human sciences centered on scientific practice (Isaac 2012).

Chapter Twenty-Two

SPACE AND PLANETARY SCIENCES

Erik M. Conway

For all intents and purposes, the activities we call space and planetary sciences did not exist prior to the twentieth century. Their ancestors, geophysics – itself a discipline of the early twentieth century – and the much older planetary astronomy, were sources of techniques, intellectual community, and, to a degree, institutional support. The American Geophysical Union and American Astronomical Society served as the intellectual homes for these new disciplines as they emerged. But the US Department of Defense, founded in 1949, and the National Aeronautics and Space Administration (NASA), founded in 1958, financed their foundation and growth into thriving intellectual pursuits.

David DeVorkin has chronicled the military origins of space science in his *Science with a Vengeance*. A handful of scientists associated with laboratories operated by the War and Navy Departments began using V-2 rockets recovered from Germany at the end of World War II for various kinds of research, including efforts to gather solar spectra, cosmic ray traces, and determine the structure of the upper atmosphere. This research was supported by the armed services because it was utilitarian. It was important to the military to understand the medium through which their missiles flew, and these scientists were willing to serve the services' needs in order to pursue their own scientific interests (DeVorkin 1992).

Central to DeVorkin's interpretation was the rocket, which served as the foundational research tool of space science. While there had been scientists pursuing the same research questions prior to World War II, none persevered. The V-2 brought in "a whole new set of workers, with talents and alignments quite distinct from those of the people traditionally active in upper atmospheric research" (DeVorkin 1992: 2). The skill set of the rocket researchers was different than that of their predecessors, as was their willingness to tolerate failure – most V-2 launches were unproductive in a scientific sense. The flights helped advance both rocket and instrument technology, without much new

A Companion to the History of American Science, First Edition.
Edited by Georgina M. Montgomery and Mark A. Largent.

scientific knowledge creation. Yet a new community of practitioners stabilized around the rocket, and this community formed the nucleus of early US space science.

The formation of the Department of Defense in 1949 only contributed further to the growth of the new field. Due to the first Soviet atomic bomb tests and later the Korean War, spending on missile technology by the new agency soared, carrying the rocket-friendly scientists along with it. While historians of science and technology have largely focused on ballistic missile technology, sensor development for infrared and radar guided missiles for anti-aircraft use also flourished during the 1950s, as did research into the details of electromagnetic transmission through the atmosphere. Both of these areas were financed for practical military reasons – more effective missile guidance – but had powerful influences on science years later.

By the time NASA was formed, space scientists had been actively pursuing rocket research for only about a decade. Their field was rather embryonic. The new agency's leaders had the opportunity to expand, and to redirect if desired, the course of the field. Because the space and planetary sciences were new, Paul Forman's argument that military patronage fundamentally altered the trajectory of physical research is not quite relevant (Forman 1987), There was no pre-existing trajectory to alter. Rather, the establishment of NASA provided an opportunity to a generation of physical scientists to develop new research fields.

Important for understanding the trajectories that the space and planetary sciences took, though, was the reality that NASA was formed not as a scientific agency, but as an engineering one. It was an engineering agency required by its founding legislation to "arrange for participation by the scientific community in planning scientific measurements and observations to be made through use of aeronautical and space vehicles, and conduct or arrange for the conduct of such measurements and observations" (Space Act 1957). This made the new agency a rather different animal than, for example, the slightly older National Science Foundation. This chapter will illuminate some of the ways NASA's early focus on engineering has influenced its evolution as a scientific enterprise.

NASA as an Engineering Agency

It is rarely stated so baldly, but NASA was created first to engineer civilian launch vehicles, and second to do useful things with them. That ordering of priority was intentional, and also happened to align well with the institutional cultures of the new agency's predecessors. NASA was not formed anew out of thin air. Rather, it was a hybrid composed of several different engineering facilities.

NASA's oldest component was its direct predecessor, the National Advisory Committee on Aeronautics. Formed in 1915, the NACA (*never* pronounced "Nacka") had a single purpose: "to supervise and direct the scientific study of the problems of flight, with a view to their practical solutions" (Bilstein 1989). It was a research organization, but not one whose subject was the natural world. Rather, its purpose was to learn about and improve a particular technology, the airplane. To do that, the NACA developed research facilities that included the ability to study the atmosphere, but ones limited to atmospheric qualities that mattered to airplanes. It was a research agency aimed at engineering better flying machines. Reflecting that orientation, the title of the head of the NACA's first research center, the Langley Memorial Aeronautical Laboratory, was Engineer in Charge (Hansen 1987).

By the time it was transformed into NASA, the NACA had grown to three major and two lesser research facilities. The Langley center, founded in 1918, had spun off a second aerodynamics-focused center in Sunnyvale, California, named for Joseph S. Ames, in 1939, to aid West Coast aircraft manufacturers, and in 1941 an engine research laboratory, built in Cleveland, Ohio and named for aviation pioneer George Lewis. Langley developed a small rocket test facility on Virginia's eastern shore in 1944 known as the Pilotless Aircraft Research Division (PARD), and a High Speed Flight Station at Muroc Dry Lake in California in 1946 (Dawson 1991; Hansen 1995). The NACA's researchers had branched out from aerodynamics to structural engineering, materials science, aerothermodynamics, and into gas turbines, cryogenic propellants, and telecommunications, all in the pursuit of "higher performance" for the agency's research artifact. But it had never become a scientific institution.

NASA acquired two facilities from the US Army, the Jet Propulsion Laboratory (JPL) in 1959 and the Army Ballistic Missile Agency's Redstone Arsenal facility in 1960. JPL had resulted from some very small-scale rocket experiments in 1939, and been incorporated as an Army Ordnance research facility in 1944 (Koppes 1982). Unlike most Army Ordnance facilities, it was operated by Caltech under a contract, not as a uniformed army base. Its purpose was the development of rocket technology, and during the 1940s and 1950s, it pursued both solid and liquid fueled rockets. It developed the Corporal short-range ballistic missile, which evolved from a liquid-fueled research rocket, in the early 1950s, and the Sergeant solid-fueled ballistic missile later in the decade. The Laboratory hired extensively from Caltech's physics and chemistry departments but despite the efforts of its cosmic-ray physicist director during the 1950s, William Pickering, was unable to gain permission to use the Laboratory's rockets for scientific research. It was, for all intents and purposes, a single-focus weapons laboratory.

JPL was joined by a rival, and sometime partner, in 1946 with the army's acquisition of Werhner von Braun's rocket team from Nazi Germany. Located first at White Sands Proving Ground in New Mexico and later relocated to the Redstone Arsenal in Huntsville, Alabama, this group's focus was on large, long-range ballistic missiles. While many of von Braun's people had scientific interests, just as JPL's did, their mandate was the design and production of big rockets.

In the early 1950s, JPL and the Huntsville rocketeers collaborated on a program to test the ability to bring warheads back from space safely, a capability key to the viability of truly intercontinental range missiles. This was the Re-Entry Test Vehicle program, carried out in deep secrecy. The two organizations built nine of what they called a "Jupiter C," and in 1956 and early 1957 flew three successfully. The remaining flight hardware was stored and became the basis of the United States' first successful orbital launch, JPL's Explorer 1 in February 1958 (Conway 2007).

With NASA's founding in October 1958, JPL decided to leave the army fold and its director, Pickering, lobbied in Washington to make the switch. JPL formally shifted its flag to NASA in January 1959. With the weapons business mostly behind it (the Laboratory retained an obligation to complete the Sergeant missile program for a few years), Laboratory leadership proposed to NASA an expansive and ambitious program of planetary and lunar exploration that JPL would undertake for the agency. Execution of that program became JPL's focus for the 1960s. Von Braun's group stayed with the army until 1960, unsure what their role would be in NASA, but then joined as the George C. Marshall Space Flight Center (MSFC). Their role remained the same as in the army. During the 1960s, they were responsible for the design and production of the Saturn I and V rockets for the Apollo program.

NASA founded two new centers shortly after opening for business. In 1960, it created what was informally known as the Beltsville Center prior to approval of its name, Goddard Space Flight Center. This was the only plausibly science-focused of the agency's collection of facilities, composed of a nucleus of scientists and engineers from the Naval Research Laboratory (NRL). This contingent also provided most of the agency's first generation of headquarters-based science managers (Newell 1980). The NRL had been a major participant in space science during the 1950s and had been responsible for the Vanguard program to develop a "civilian" scientific launch vehicle for the International Geophysical Year.

NASA's last major center, first known as the Manned Spacecraft Center in Houston, Texas, was formed by still another group of Langley Research Center engineers led by Robert Gilruth (Hansen 1995). This group had been formed under the name Space Task Group and was the core of the Mercury program that flew the first Americans in space. It moved to Texas in 1962 and grew rapidly afterwards, but remained the locus of NASA's human space flight programs.

NASA, then, was formed from a collection of engineering organizations. Yet beyond the formal requirement that it conduct scientific missions, its first generation of scientific leaders had personal commitments to science, too. Their challenge was how to build a scientific program within an engineering agency, and find a way to gain support from scientists outside NASA for the new scientific fields they sought to create.

Organizing Science in NASA

NASA's first chief scientist, NRL alumni Homer Newell, recounted in his memoir that in NASA's earliest days, it was a struggle to keep science on the agency's organization chart. For NASA's first year, the science program was subsumed under "Space Flight Development;" nearly two years after NASA's founding, under the agency's second administrator, space science became a directorate co-equal with aeronautics and human space flight functions (Newell 1980). While NASA's science directorate has had several names since (originally the Office of Space Science, currently the Science Mission Directorate), its purpose has remained essentially unchanged: it funds and manages robotic science missions, and occasionally funds experiments carried by crewed spacecraft as well.

Newell and his first generation of science managers – John Townsend, Nancy Roman, Robert Jastrow, Gerhardt Shilling, Robert Fellows, and others – also had to establish and maintain relations with the scientific community external to NASA. One component of that challenge was figuring out what to do with the National Academy of Sciences (now the National Academies). During the International Geophysical Year, a special Space Science Board of the National Academy's National Research Council had been empowered to choose the experiments that would be flown aboard NRL's rocket, and that body wished to continue the role of choosing space scientists (Naugle 1991; Newell 1980). That was not acceptable to Newell or his own superiors at NASA, and instead they implemented a policy that permitted the Space Science Board to make recommendations but left prioritization and selection of experiments to NASA.

Three more decisions had lasting impacts on the evolution of space and planetary sciences in the United States. Newell believed that maintaining communications with the Space Science Board would not be sufficient to sustain good relations with the external scientific community. He established a set of advisory committees on which both NASA scientists and university-based scientists were members as another means of gaining advice and maintaining community. While the structure of these committees has varied over time (as have the federal regulations governing such committees), advisory committees continue to exist at NASA.

More controversial were the remaining two decisions. One was to allow scientists employed by NASA centers to propose instruments and missions. The external scientific community was very opposed to this, believing that the agency's insiders would have an unfair advantage over academic scientists. If NASA were to have scientists, in the opinion of the outside community, they were to exist to support the academic community, not to compete with it. Their view was that NASA should provide engineering services to external scientists, not do science on its own. That view was unacceptable to Newell and his former NRL colleagues at Goddard Space Flight Center as well as to JPL's Pickering and his staff. They had scientific ambitions of their own. Thus this decision put Newell at odds with the community he needed to cultivate. Selection committees, in addition to the advisory committees, were created and tasked with ensuring that instrument selections for missions were made in an unbiased way (Naugle 1991).

The final decision involved who would be permitted to build instruments for NASA's space science program. Goddard Space Flight Center's emerging policy permitted academic scientists to build their own instruments and required their participation in integrating it with the spacecraft; JPL required that instruments be built by JPL even when the proposer was not from JPL. Neither policy proved more effective, but JPL's policy was widely disliked. During the early 1960s, Newell permitted the two organizations to go their own way, so to speak, but by the end of that decade, headquarters had imposed Goddard's policy agency wide (Naugle 1991).

In his Office of Space Science, Newell created three scientific divisions, representing, to his mind, three sets of related disciplines: geophysics and astronomy; lunar and planetary sciences; and life sciences. Each had a director and a deputy director. One was always to be a scientist, and the other an engineer. The divisions, in turn, had what Naugle called "program chiefs" (now program scientists), and program managers (now generally called program executives), who were engineers. Program scientists were responsible for their disciplines, while program executives were responsible for one or more flight projects. Naugle described the resulting parallel networks of scientists and engineers as "shadow networks" in his own memoir on the period. In his words, "the shadow science network handled the purely scientific issues and the shadow engineering network handled the purely engineering issues. Issues involving both science and engineering went through the formal line organization. These shadow networks served to speed the routine work, promote teamwork between scientists and engineers, and to make the most efficient use of payload space available on the NASA missions" (Naugle 1991).

This management process and structure was laid out formally in a publication known originally as Technical Management Instruction 37-1-1 (TMI 37-1-1), drafted in late

1959 and 1960. While that instruction has evolved, particularly in the area of risk management, the basic management structure of NASA's scientific enterprise has not changed much. The current *Science Mission Directorate Handbook* specifies a management triumvirate: program executive (an engineer), program scientist (a scientist), and a program resource analyst (financial manager), an evolution no doubt intended to improve financial management (NASA 2008). That management handbook, in turn, is an expansion of NASA Procedural Requirements (NPR) 7120.5D, NASA's *Space Flight Program and Project Management Requirement*, a document issued by the agency's Office of the Chief Engineer.

This headquarters-level management structure was reflected at the center level, where individual missions were administered. Space science missions had a project manager (engineer), a project scientist (scientist), and with the addition of what are called "competed missions" to the agency portfolio in the early 1990s, some missions had a "principal investigator" (scientist). Only in these competed, principal-investigator-led, missions was a scientist the final authority; in the other mission types, the project manager had this power (NASA 2008). This reform helped solve a particular set of problems deriving from NASA's need to satisfy a large community of scientists with each mission in order to gain support for them, which frequently led to large, complex, and ultimately unaffordable mission concepts (Conway 2015; Neufeld 2014). Making a single investigator responsible for a mission was expected to result in narrower, more focused scientific goals, less complex missions, and thus lower costs.

These shadow networks of engineers and scientists are often in tension. During JPL's Ranger missions to the Moon (1960–1965), James D. Burke, the project manager, was frustrated by NASA headquarters officials efforts to keep imposing additional instruments on the spacecraft when he saw his job as simply getting it to work reliably. Changing payload demands did not help. Ultimately, after repeated failures, he was removed along with most of the instruments (Hall 1977). The first successful Ranger mission, Ranger 7, carried only a set of television cameras. Similarly, the Surveyor lunar landers were intended to carry out a very extensive scientific program, for which NASA officials initially chose a 300 lb payload. For engineering reasons, the first successful Surveyor carried only a television camera (20 lb); the second the camera and a surface scoop for testing the strength of the lunar surface; the final three finally carried a mineralogy instrument prepared by the University of Chicago.

This was early in the space age, of course, and the stripping of instruments off JPL's lunar spacecraft reflected a poor understanding of how difficult engineering successful space vehicles would be. Yet the oversubscribing of scientific payloads continued through at least the late 1990s, despite tremendous improvement in the ability of engineers to predict spacecraft performance and to build in reliability (Conway 2015). This reflected the desire to maximize scientific return from each mission on the part of NASA's science managers, often to the frustration of engineering managers attempting to maintain cost control.

The engineering shadow network creates tensions within NASA as well, and ones that are complex. NASA engineers whose focus was on producing reliable spacecraft for scientific missions were often seen as excessively conservative technologically, refusing to employ new technology because its reliability in flight was unknown. In the 1990s, for example, a reform movement within the agency (known as "faster better cheaper") promoted this view (McCurdy 2001). Yet many NASA engineers often work on new technology and promote its use for new instrumentation and new spacecraft capability; on the Ranger missions, JPL's engineers used solar arrays even though the mission could have been carried out using batteries alone, wanting to prove the technology for interplanetary use; they developed and flew a digital camera and telecommunications system on their Mariner Mars 1964 mission when analog technology could have done the job; they developed semi-autonomous driving technology in the 1980s and 1990s using military and NASA advanced technology funding that enabled the first Mars rovers (Koppes 1982; Mishkin 2003; Nicks 1985; Westwick 2007).

Thus within NASA's shadow engineering network, tensions exist between a desire for innovation and the demand for reliability. Combined with the ambitions of NASA's science managers, and those of the external space science community, these tensions have often been resolved via the ascension of those mission concepts that are both technologically and scientifically ambitious over those that promise only scientific returns (Conway 2015). Ambitious missions resolve all these tensions favorably, save one: the desire for cost control. Rarely do the most ambitious missions arrive on time and budget.

The Space Sciences and Human Space Flight

From the very beginning, tensions have existed between the scientific community and advocates for flying humans in space. A great many scientists did not think astronauts added any scientific value, and their presence raised costs by orders of magnitude. Geophysicist Lloyd Berkner helped smooth these tensions over during the 1960s as chair of the National Research Council's Space Science Board, though he could not quell all dissent (Needell 2000). Deletion of most of the scientific instruments from the Surveyor lunar landers did not help his case that scientists should support human exploration.

Though Berkner did not live to see it, he was largely vindicated by the Apollo program's scientific experiments and lunar sample return. Geologist Eugene Shoemaker trained the astronauts in basic field research techniques that they used on the moon. Lunar Excursion Modules carried diverse experiment packages, including seismic, magnetic, heat flow, gravity, dust, micrometeorite, and cosmic ray investigations. Moon rocks returned to earth were deposited in a Lunar Receiving Laboratory built at the Manned Spaceflight Center in Houston (now the Johnson Space Center), and from there were distributed to university laboratories for analysis (Beattie 2001).

After Apollo, NASA's Skylab shifted the human program away from the moon into low-earth orbit, and to a different scientific focus. Skylab primarily fostered solar science via the "Apollo Telescope Mount," a set of solar telescopes and instruments. The short-lived laboratory also served to study the effects of longer term effects of the space environment on the human body, an aspect of space medicine this chapter cannot encompass (Compton and Benson 1983).

Skylab's successor, the Space Shuttle, shifted NASA focus again, this time toward the science that could be done from low-earth orbit. In particular, the Shuttle fostered earth remote sensing and optical astronomy. The Shuttle's limitations as a planetary mission launch vehicle also nearly led to termination of NASA's planetary exploration program (Logsdon 2013). Only three deep space missions were launched by Shuttles, ultimately: the Galileo mission to Jupiter; the Magellan mission to Venus; and the European Space

Agency's Ulysses solar mission. After the loss of Shuttle Challenger and its crew in 1986, NASA's policy changed to remove scientific payloads from the Shuttle fleet, and thereafter the links between space and planetary sciences and human space flight were largely broken. The Hubble Space Telescope servicing missions in 1993, 1997, 2002, and 2009 were the principal, and certainly most visible, exceptions.

NASA and the National Academy of Science

Homer Newell's effort to reduce the National Academy's role in selecting his new agency's scientific agenda was effective, but only temporarily. After 1965, NASA's budget began to shrink rapidly. Newell's program managers began to seek the Academy's blessing for their plans again. In turn, as NASA sought "national consensus" on its mission plans via the National Academy, the scope and complexity of its missions expanded, driving up costs and triggering demands for reform.

Outside NASA, there was already a brief history of organizing "decadal surveys" that were in part surveys of near-term prospects, and in part requests for funding of specific scientific infrastructure projects. A 1964 survey by the National Research Council's Panel on Astronomical Facilities, chaired by A.E. Whitford of the Lick Observatory, had called for construction of eight medium (60 to 84 inches) and three large optical telescopes (150 to 200 inches), and a design study for a giant telescope ("the largest feasible"), in addition to construction of a very high resolution radio telescope consisting of up to 100 steerable antennae and expansion of the existing Caltech-operated Owens Valley Radio Observatory. The 1964 survey had also carefully avoided levying any specific suggestions on NASA's domain of space astronomy, other than calling on NASA to fund ground-based astronomy too: "A move by the National Aeronautics and Space Administration to broaden its already strong interest in basic astronomical research so as to include ground-based stellar astronomy would be highly advantageous to the national scientific effort and to the National Aeronautics and Space Administration itself" (Panel of Astronomical Facilities 1964: iii).

Indeed, because Newell decided that the planetary exploration program his agency was pursuing needed greater access to planetary observations across the electromagnetic spectrum, he did start funding ground-based astronomy, though specifically directed at the planets (which were unpopular among astronomers) (Ruley 2013). Thus only a year later, it was no longer necessary to avoid NASA's territory. A 1965 summer study at the Academy's Woods Hole facility under the auspices of the Space Science Board (then chaired by Harry Hess) proposed a broad program of space astronomy, planetary, and lunar exploration, though it emphasized "planetary exploration as the most rewarding scientific objective for the 1970–1985 period" (Space Science Board 1966: iii). As a form of promotion, however, this report was unsuccessful. Dramatic budget cuts the same year led to a revisiting of the planetary exploration portion by a smaller committee led by Gordon J.F. MacDonald in 1968.

During the three years between Hess's study and MacDonald's, a long-brewing conflict within the scientific community boiled over into Washington politics. Newell had promoted a "next step" to Mars that would send Saturn V-class scientific laboratories to Mars as preparation for human expeditions there. This hugely expensive program had not been seen favorably by scientists, who believed it was premature and that it would result in diversion of funding from other worthwhile research areas (Ruley 2012). Their opposition was heard in Congress, which blocked it and forced the re-evaluation of the planetary science program that MacDonald's survey panel carried out.

MacDonald's group defended the priority of several smaller missions already in development, such as Mariner Mars 1971 (Mariner 8 and 9), promoted a "Mariner class Mars lander" for 1973 (which became the Viking mission of 1976), and also launched an initial salvo in favor of "small missions" to the inner planets. His committee also voiced frustration at the relatively low priority of planetary exploration within NASA: "In the coming fiscal year, planetary exploration will receive at most two percent of the total space budget. We believe that this amount is totally inadequate to take advantage of the opportunities available to us" (Space Science Board 1968: 4). Their appeal went unanswered; indeed, planetary exploration's priority fell within NASA for the next two decades.

The chair of the next astronomy and astrophysics survey, Caltech's Jesse Greenstein, also faced the challenge of having to craft an ambitious program in an era of shrinking resources. "At a time when a stronger attempt (was) being made to integrate NASA planning into the general national scientific effort," his task was to prioritize competing funding demands in what they perceived as an era of austerity, with NASA treated as just one of several agencies (Astronomy Survey Committee 1972: v). The report had been requested by "several federal agencies," and jointly funded by the National Science Foundation and by NASA. Like its predecessor, this survey recommended funding of the Very Large Array (this time successfully), but reduced its demands for large ground telescopes. It also cast its weight behind the idea of a Large Space Telescope, which became the Hubble Space Telescope many years later (Smith 1990).

Greenstein's survey was perhaps the most successful of these decadals, in that most of its top priorities were funded – though not necessarily completed within the specified decade. The austerity that had afflicted NASA in the 1970s, though, resulted in decreasingly successful efforts. A 1976 study by the Space Studies Board successfully defended three existing programs, the High-Energy Astronomy Observatory, Pioneer Venus, and Mariner Jupiter Saturn (renamed Voyager in 1977), and advocated for a new start for the Large Space Telescope. It also initiated a longstanding (40 years to date) effort to gain approval for a Mars sample return mission. The Large Space Telescope was funded; sample return was not. Only two years later, the Academy's Committee on Planetary and Lunar Exploration (COMPLEX) weighed in with another study, this time focused on the inner planets, and fundamentally different than its predecessors. Chaired by Caltech's Gerald Wasserburg, instead of proposing specific missions, it proposed "scientific objectives to be achieved," believing "that this approach provides greater flexibility for program planning and permits a perspective to be maintained despite the vagaries of year-to-year funding variations" (COMPLEX 1978: v). This approach, while very useful if one wished to understand what questions scientists thought worth pursuing at the time, was not very helpful for an agency that needed to defend its programmatic choices.

The first few years of the 1980s proved to be what John Logsdon has called the "survival crisis" for NASA's planetary science programs (Logsdon 2013). John Naugle, who became the agency's chief scientist in 1977, responded to the crisis by organizing a new study, this time under the auspices of the agency's Advisory Council. His planetary program manager, Geoffrey Briggs, served as the committee's executive secretary. Known as the Space Science and Exploration Committee, it drew primarily on academic scientists and was designed to convert the scientific goals laid out by Wasserberg's COMPLEX study into specific, defensible missions. Knowing that expensive missions would never gain support, Briggs pushed the committee to divide its program not by inner and outer planets, but by complexity. Relatively simple, low-cost missions went into a 1983 document known as the "core program," while more scientific, and technologically, ambitious missions were placed in an "augmented" program plan published in 1986 (SSEC, 1983, 1986). Some of the core program missions were ultimately funded, ending several years of drought for the planetary science community.

NASA's attitude toward the National Academy clearly changed over the years, and increasingly the agency's science managers worked to not only manage the relationship between the two organizations, but to gain Academy approval for specific plans. Direct engagement between them helped ensure that the Academy's panels understood NASA's positions, and that the agency's diverse scientific communities appeared to "speak with one voice" about priorities. While not always successful as tools of advocacy, these decadal surveys continue to be carried out. Most recently, for example, COMPLEX published *Visions and Voyages for Planetary Science in the Decade 2013–2022* (2011), again placing Mars' sample return at the community's highest priority.

In an essay on the history of solar physics in NASA, DeVorkin raised an obvious question about NASA's influence in the process of forging these consensus documents (DeVorkin 2004). He was concerned about the way the agency's coalition-building process in support of large missions like the Large Space Telescope/Hubble Space Telescope seemed to cause space science projects to balloon out of control (and sometimes out of existence). This phenomenon is also visible in my own work on the history of Mars exploration; missions have tended to expand to attract the largest possible constituency, and sometimes then been cancelled because they exceeded the available budget (Conway 2015). This is a known problem among planetary scientists, and led to a new competed mission paradigm that emerged in the 1990s. Geoff Briggs, for example, understood that the need to create large coalitions was making planetary exploration unaffordable by the early 1980s. The competed mission idea avoids the entire national consensus-building process by empowering individual scientists to assemble their own mission coalition, under an enforced cost cap. The Discovery program parented by Wesley T. Huntress in the early 1990s operates this way (Neufeld 2014).

Building New Specialist Communities

A final important aspect of the growth of the space sciences was the role of existing specialist societies in adapting to the demands imposed by rapid growth in membership and the opportunities afforded by new technologies and greatly expanded funding. New specialist communities needed venues in which to meet and share both results and future plans, as well as ones in which to publish. Traditionally that function resides not in government agencies (though the NACA had maintained highly respected internal publication venues that were widely distributed), but in independent scientific societies. Both the American Geophysical Union (AGU) and the American Astronomical Society (AAS) undertook reforms aimed at accommodating new specialties during the 1960s, with the younger AGU moving to do so first.

Robert Jastrow and Gordon J.F. MacDonald, astrophysics and geophysics, respectively, petitioned for creation of a "planetary physics" section within AGU in 1960. The Union's leadership then appointed NASA's own Homer Newell to chair a committee to study the issue. Unsurprisingly, Newell's panel argued for creation of a new section, but one named "planetary sciences." Not everyone in the Union's leadership favored that idea and a second committee revisited the subject. This panel did not achieve unanimity on the issue either and recommended that the Union's membership vote on the question. Held in November 1962, the vote passed, and Newell became the first president of AGU's Planetary Sciences Section (Tatarewicz 1990).

The AAS took longer to accommodate itself to the rapidly evolving scientific landscape of the 1960s. In part, that was because the Society's leaders tried very hard to avoid overspecialization, which they feared would happen if they even accommodated parallel sessions at their annual meeting (Cruickshank and Chamberlain 1999). They feared that it would no longer be possible for astronomers to stay broadly informed about the state of their profession if it were no longer possible to attend every presentation. The Society's tradition had been to accept nearly all papers and have them all read in a single session, but by the early 1960s even shrinking papers to 10 minutes, and having many read "by title only," they could no longer avoid parallel sessions.

But it was an incipient revolt by solar physicists, a community that had seen tremendous funding growth in the 1950s and 1960s from military, NSF, and NASA sources, that resulted in formal disciplinary sections being permitted within the AAS. NASA's solar physics program manager, Harry J. Smith, first suggested in 1965 that NASA-funded solar scientists begin meeting separately to discuss their research needs; one of them, Leo Goldberg, who happened to president of AAS at the time, suggested instead that specialist meetings organized under the Society's sponsorship be started. But the Society's council overall did not wish to encourage further fragmentation of the astronomy profession than was already being generated by parallel sessions. They would approve a single specialist meeting but go no further. In fact, two were held before the AAS's council acceded to the demand for routine, annual specialist meetings, and more: in August 1968, it approved the establishment of discipline-based "divisions" of the Society (DeVorkin 1999; Hufbauer 1991).

By then, three increasingly separate sub-specialties within astronomy sought this form of community building. Solar physics was the largest and best organized (Thomas 1999). More modest in size, and certainly held in lower esteem by traditional astronomers, was the community of planetary astronomers, who chose to follow AGU's lead in naming their nascent discipline planetary science. The Division for Planetary Sciences' initial membership was 135 (Cruickshank and Chamberlain 1999: 265; Tatarewicz 1990). The third, and smallest, group was high-energy astrophysics, which had not (yet) seen quite the scale of growth that the others had (Trimble 1999).

Growth in astronomy and astrophysics continued, though at varying rates over the years. In 1960, AAS had hosted around 1000 members; the 1991 astronomy and astrophysics decadal survey found that AAS had grown from 3000 members in 1981 to just over 4000 US members in 1989. AAS claimed about 5000 US based members in 2013 (Anderson and Ivie 2013; Astronomy and Astrophysics Survey Committee 1991). AAS' 2013 demographic survey indicated that about 19% of its members were primarily interested in planetary science; that year, 763 scientists attended the Division for Planetary Sciences' meeting in Denver (Division for Planetary Sciences 2014). The annual meeting of the Lunar and Planetary Institute, formed in 1968, had grown from 600 attendees in 1975 to a little over 1500 in 2007, though with a decline during the 1980s, when there were few planetary missions, to only 500 in 1984 (Conway 2010).

If planetary sciences had grown to become a substantial fraction of the AAS, solar physics, often called heliophysics by the early 2000s, was claimed as a primary interest by only 8% of AAS members in the AAS 2013 survey. It was no longer the dominant subspecialty of AAS (Solar Physics Division 2014). High-energy astrophysics, though, had grown from its initial tally of 85 in 1969 to over 900 in 2011 (HEAD 2011). Interestingly, one important factor about AAS remained unchanged despite the large shifts in disciplinary interests over the twentieth century. Stellar astronomy had been the primary interest of most members of the Society in the 1960s; while it no longer claimed a majority of AAS' members in 2013, at 30%, it was still the most common interest within the organization.

Shifts in disciplinary interest may reflect shifts in funding priorities within the agencies that support space and planetary sciences. The NSF's share of funding for astronomical research declined from 1980 through at least 2000, leading to community complaints that while the number of ground facilities had increased, the ability to use them was decreasing. NASA's funding of research grants had increased during the same period, leaving NASA the larger funder of astronomical research. Yet in a 2000 study of federal funding for astronomy, the National Research Council's Committee on Astronomy and Astrophysics commented that "Despite this significant funding shift toward a basically mission-oriented agency, the committee was unable to detect any dramatic changes in the distribution of the subfields of astronomy" (Committee on Astronomy and Astrophysics 2000: 52).

NASA's role as the dominant funder of astronomy came with a kind of risk that the community had not had to face prior to the agency's foundation. As an agency built around missions, and not facilities, failure of a mission meant loss of support to investigators associated with it. This became most obvious around the long-lived Hubble Space Telescope. In 2000, the aging telescope's grant program supported a quarter of all US astronomers; its failure would be catastrophic to the profession (Committee on Astronomy and Astrophysics 2000: 54). Failure of an instrument at a ground observatory would not have the same far reaching effects – only a single investigator's team would be harmed, and then likely only temporarily, as a ground-based instrument can be repaired.

A similar kind of risk pertained to the planetary sciences community that NASA forged. When priorities shifted away from planetary sciences, as they did in the 1970s and 1980s, the field suffered loss of both data to work with and funding, with the consequence that the field actually shrank. Investigators moved to other fields, particularly earth sciences (Conway 2010; Tatarewicz 1990). And, of course, if a major mission failed, so did careers attached to it, especially in cases when the researchers were graduate or recent postgraduate.

The space and planetary sciences' dependence on complex feats of engineering, and therefore upon engineers, mark an important difference between them and older fields of American science. Breakthrough science depends upon breakthrough engineering. Mission selection thus depends upon both the quality of the proposed science and that of the proposed technology. While astronomers were rightly afraid of losing control of their disciplines to science managers, the reality of public funding of the space sciences has been more complex. Scientists, science managers, engineers, and politicians all have voices in the process of choosing missions. Because missions may last many years – a decade is no longer unusual – choice of a specific mission, like replacing the optical Hubble Space Telescope with the infrared James Webb Space Telescope as planned late this decade, can define an entire community's research direction for a decade or two.

Finally, the increases in funds made available to scientific research by the early Cold War created large new scientific disciplines. The end of the Cold War in 1989/1990, though, did not reverse those gains. The heady growth in funds that had characterized the 1960s had already ended, with a kind of rough stability setting in; while the larger discipline of physics suffered at Cold War's end, the space and planetary sciences did not.

Bibliographic Essay

There are three starting points for developing an understanding of the early years of space and planetary sciences in the United States, DeVorkin's *Science with a Vengeance* (1992), Doel's *Solar System Astronomy in America* (1996), and Karl Hufbauer's *Exploring the Sun* (1991). DeVorkin addresses the military origins of the space sciences, while Doel provides much of the intellectual background to what become the planetary sciences. Hufbauer provides the same service to heliophysics.

A useful source of documents for understanding the evolution of NASA science is the *Exploring the Unknown* series of books (also available online), which compile key primary documents with explanatory essays. While the documents are useful, the essays must be used with care. Not all are by historians; instead, some are by former NASA managers. They contain errors of both omission and commission, and even fundamental redirections within NASA are neglected. NASA once had an Applications Directorate, for example, and the term Applications no longer even appears anywhere in its high-level organization charts (Conway 2014). Yet this abandonment of one of the agency's fundamental functions goes without mention in these volumes. That said, the memoirs of NASA managers are often very useful for the perspectives they provide, and I have used Newell and Naugle for that reason.

I have relied heavily on DeVorkin's edited centennial volume for the AAS, *The American Astronomical Society's First Century* (1999), to help explicate the evolution of one of the three major scientific societies that serve as intellectual homes for the space sciences. I could not locate similar volumes for the American Geophysical Union or the American Institute of Physics, though the latter maintains an important archive of oral histories relevant to the space sciences.

The NASA History Office in Washington maintains a publication program that has resulted in a number of books useful for understanding the evolution of the space and planetary sciences. In addition to the memoirs cited, Tatarewicz, Beatty, Compton and Benson, the Exploring the Unknown series, and all the NACA histories were supported at least partly by the NASA History program. Many, though not all, are available as electronic downloads at history.nasa.gov.

NOTE: All opinions and interpretations expressed in this work are those of the author and not of the National Aeronautics and Space Administration, the California Institute of Technology or of the Jet Propulsion Laboratory.

Part II

Topics

Chapter Twenty-Three

BIOTECHNOLOGY

Nathan Crowe

Biotechnology now occupies a significant place in our cultural, political, and social landscape. The word can be heard in the offices of your local doctor, in boardrooms on Wall Street, in congressional offices on Capitol Hill, and by farmers at the seed shops. It is an important part of the American economy, occupying a large sector of the workforce, and is one of the most important growth industries in the future. At the same time, biotechnology generates more than just revenue – it also spawns ethical debates that deal with fundamental questions about the nature of life, basic human rights, and the limits of science. Consequently, studying biotechnology not only offers insight into the nature of science and technology, it can also be a lens to analyze larger social and cultural values.

Given its importance in so many aspects of our world, it is helpful to understand the history of biotechnology in the United States. Despite its seemingly high-tech, even futuristic character, biotechnology has a much longer history than most people realize. Understanding how it has evolved over time, and how it has both been shaped by American society and helped shape it, can give us insight into how it has become so important in today's world, as well as insight into why it seems to be so contentious.

This chapter will begin by considering possible frameworks for defining biotechnology. Following that, I will discuss several of the major narratives in its history. Firstly, I will cover the history of genetically modified organisms, with emphasis both on the agricultural products that have been created and the animal and microbial organisms that have been become important parts of the biomedical arena. Secondly, I will articulate how the relationship between publically funded scientists and private companies has developed over the course of the twentieth century. Finally, I will highlight the growth of control and engineering ideologies in the biosciences and society's evolving ethical sensitivity to these philosophies. These narratives will be interwoven together chronologically; the first section will begin with discussions of nineteenth-century

A Companion to the History of American Science, First Edition.
Edited by Georgina M. Montgomery and Mark A. Largent.
© 2016 John Wiley & Sons, Ltd. Published 2020 by John Wiley & Sons, Ltd.

American agriculture, the second section will articulate developments in biotechnology from World War I through the 1960s, and the final section will highlight the development of recombinant DNA technologies and the rapid growth of the current biotechnology sector. Together, these historical threads will provide a robust picture of how current biotechnologies have become so prominent in American science and society.

Defining Biotechnology: Problems and Particulars

What constitutes a "biotechnology"? For some, products such as genetically modified organisms (GMOs), particularly crops such as corn and soybeans that have genes that make them immune to specific pesticides, are quintessential biotechnologies. Others might consider the very processes that develop GMOs, the cutting and pasting of genes, for instance, as the more appropriate example of biotechnology. Still others might think of the industrial vats of yeast and *E. coli* that are used to generate a myriad of different products, from vaccines and pharmaceuticals to industrial ingredients like alcohols and solvents. Others have even argued that the machines which are crucial to working with biological materials are biotechnologies themselves: the PCR and protein folding machines in laboratories, the paternity tests, and the DNA sequencing kits, for instance. And what about the integration of humans and technology in the form of prosthetics? It seems that biotechnology encompasses both products of science and technology and the processes that create them. Ultimately, it is a nebulous concept that has been defined, and redefined, by users, commentators, lawmakers, and futurists, not to mention those who have chosen to study the topic including sociologists, anthropologists, and historians. In effect, nearly every combination of the definitions that are behind the words "biology" and "technology" can produce some sort of confluence that could be, and have been, labeled as biotechnology.

In 1984, in an effort to provide some concrete definition to the word "biotechnology," the US Office of Technology Assessment (OTA) reported that "broadly defined, includes any technique that uses living organisms (or parts of organisms) to make or modify products, to improve plants or animals, or to develop microorganisms for specific uses" (OTA 1984). Though the OTA laid out an encompassing definition, the report concentrated on the new, powerful techniques of recombinant DNA that had arisen since the mid-1970s. These techniques had helped spur the dramatic growth of a new industrial sector focused on cutting and pasting genes for the purpose of creating pharmaceutical drugs and health treatments (more on this later). This new sector of biotechnology, particularly in the economic sense, seemed radically different than the traditional plant breeding and industrial microbe use that had defined biotechnology before that time. By defining it broadly, but focusing on the new technologies of recombinant DNA, the OTA report contributed to the branding of biotechnology as a new instance of private entrepreneurialism mixed with new methods of biological engineering. Many of the biotechnologies that people think of today fall into this category: the GMOs of Monsanto, or the promises of personalized medicine, for example. In many ways, this connotation of biotechnology has become the current mainstream definition, raising questions about how similar or dissimilar modern definitions of biotechnology are to earlier conceptions of the word.

Historians and other scholars of science and technology have approached the history of biotechnology in a variety of ways in an attempt to understand the relationship between the issues and concepts surrounding today's biotechnology compared to previous conceptions of the word. Some historians have focused on the development of specific biotechnologies, showing the continuity or discontinuity of issues surrounding biotechnologies today and agricultural processes of the past (Bud 1993; Kloppenburg 2004). Others have focused on the attitudes and goals of scientists regarding the need for increased control over life, as well as the cultural responses to this integration (Landecker 2007; Pauly 1987). Still others have looked at the political, social, and legal apparatuses that have empowered various groups such as academic biologists, venture capitalists, and entrepreneurs to come together to explore, exploit, and profit from the processes of life (Hughes 2011; Rasmussen 2014; Thackray 1998; S. Wright 1994). The consequence of this is that the concept of biotechnology has been dissected, enlarged, undermined, and reconstructed many times. For the more interested reader, see the bibliographic essay for descriptions of more focused histories.

Biotechnology before *Biotechnologie*

The etymology of the word "biotechnology" has been traced back to 1917 when a Hungarian named Karl Ereky (1878–1952) used it as a clarion call for the application of industrial methods to create, harvest, and distribute agricultural crops, particularly animals like pigs (Bud 1993). For Ereky, the word *Biotechnologie* described the new scientific theories, methods, and chemicals that could be, and were being, used to revolutionize traditional agricultural practices like animal husbandry. Ereky, for instance, touted his pig factory that could systematically turn plant products into meat through the application of scientific techniques on an industrial scale. Advances and research in biotechnologies, he argued, were the key to enhancing agricultural outputs and solving many of the problems surrounding food production and distribution, particularly in a post-World War I rebuilding process (Bud 1993).

Of course, the application of science and engineering to agriculture had taken place before Ereky's coining of the word biotechnology. In many ways, Ereky's choice of "*Biotechnologie*" was simply the creation of a word that brought together many different ways in which science and technology had been applied to agriculture over the past century (Bud 1993). Most notably is how farmers had long been selecting plants and animals in an effort to obtain, on a regular basis, desirable traits. For corn that might mean higher yields and bigger kernels; for animals, that might mean more milk production or leaner meat. In the nineteenth century, when the United States was still in large part an agrarian society, the selection of desirable traits in agricultural goods was key in helping establish the United States as a large and effective producer of food. In fact, the vast majority of grains, vegetables, and fruits that have become staples of the American agricultural industry were not indigenous to North America. Rather, in the 1820s and 1830s the US government sent out several international expeditions and relied on ambassador networks to add to the number of species to which farmers had access. It took decades, however, for farmers to cultivate varieties of these imported plants that worked well in North American conditions. For instance, the center of wheat production (a crop originally sourced from Eastern Europe and Asia Minor) began in Pennsylvania in the 1840s and gradually moved westward over the next 20 years as varieties were bred to take advantage of the environmental conditions of the Midwest. Similar stories can be told about corn (indigenous to Central and South America) and cotton (varieties originated in South and Central America as well as Central Asia) (Kloppenburg 2004). Overall, during the middle of the nineteenth century, the federal government supported and facilitated farmers in their quest to produce more and better crops by supplying them with seeds on a regular basis.

In the nineteenth century, it became increasingly important to the federal government to invest in the success of agriculture. One of the most important ways in which it did that was by creating land-grant colleges and agricultural experiment stations through the Morrill Land-Grant Acts of 1862 and 1890 and the Hatch Act of 1887. Agricultural experiment stations, attached to and under the direction of the newly established land-grant colleges, were intended to help farmers systematically apply scientific principles to agriculture. Traditional breeding methods of choosing the best from each generation had already produced a variety of domesticated animals and plants. However, there was hope that new scientific principles drawn from the recent success of experimental biology could substantially increase the success of farming initiatives. In the late nineteenth century the rise of experimental biology ushered in a new sense that through systematic experiments with fertilizers, plant breeding programs, and maintenance, farmers could routinely enhance existing varieties and produce new ones (Marcus 1985).

At the turn of the twentieth century, the application of scientific theories to agriculture became even more plausible with the rediscovery of Gregor Mendel's work on the patterns of inheritance. Mendel's laws of inheritance immediately piqued the interest of biologists, some of whom would go on to develop new fields like genetics, as well as agricultural specialists who saw the potential for moving plant and animal breeding beyond methods that relied mostly on trial and error. Using Mendelian calculations, farmers and agricultural experiment station scientists helped produced new agricultural varieties that held up better against droughts and disease and produced superior harvests. The success of experimental techniques across the biological fields also helped in the development of more effective fertilizers and methods of planting and care. Essentially, by the time that Karl Ereky had begun to use the word *Biotechnologie* to describe an emphasis on the application of scientific and industrial practices to agriculture, it was already becoming a major part of farming in the United States and throughout the world. As an example, E.M. East, a plant breeder at the Connecticut Experiment Station and a pioneering geneticist, wrote *The Relation of Certain Biological Principles to Plant Breeding* in 1907, which illustrates the type of thinking that already existed in agriculture in the United States by the time that Ereky began using the term *Biotechnologie* in 1917. In particular, the rise of genetics in the 1910s that grew out of the rediscovery of Mendel's law's began to shift the focus of agriculturalists toward thinking about the genes underlying the traits that previously had occupied their attention – a trend that would continue throughout the twentieth century.

Outside the agricultural breeding context, other industries were taking advantage of microbes to produce a variety of products. Zymotechnology, the technical word coined in the 1800s to describe the control of the fermentation process and its products, can be considered the oldest biotechnology, dating back at least to the ancient Babylonians and Egyptians who recorded their processes of a beer-like creation. The experimental approaches in the nineteenth century that helped advance agriculture also helped pave the way for people like France's Louis Pasteur to pioneer new theories and methods surrounding microbes and the fermentation process. Though much of the early work in fermentation occurred in Europe, some of the early pioneers in the field such as Robert Wahl and John Ewald Siebel founded institutes, journals, and consulting companies in the United States at the end of the nineteenth century that helped improve the science behind brewing and other fermentation practices closely related to agricultural industries (Bud 1993). Moreover, the successes of fermentation technologies during World War I to produce butanol and acetone, two heavily used products in the creation of synthetic rubber and explosives, advanced the status of engineering, chemistry, and biology studies surrounding microorganisms after the war. The Chemists' War, as it had become known, referred not only to the chemical warfare products used, but also the chemical and biological processes like synthetic rubber production, all of which made fermentation technologies seem destined to be an important part of American success in the twentieth century.

Overall, the development of biotechnologies in the United States at the end of the nineteenth and the beginning of the twentieth centuries had taken on a trajectory of ever increasing control over life. Scientists and industrialists alike viewed the continued application of scientific principles to plant and animal breeding and maintenance as the key to increasing production and cutting costs. And the profile of fermentation technologies had risen substantially after World War I as an industry that could successfully exploit organisms in a new way.

Though there were many examples of biologists and industrialists who were interested in gaining control over life's processes, one in particular came to embody the ideals of both biology and engineering. Jacques Loeb (1859–1924), a German-American biologist who spent his scientific career at the University of Chicago and the Rockefeller Institute, was a strong and provocative advocate of what historian Philip Pauly (1987) has called "the engineering standpoint" in biology. Loeb made it clear in the late 1800s that the goal of biologists should be to gain control over life. In other words, observations and experiments were not simply to answer questions about the world, but rather to gain specific knowledge about how to control life. In 1899, Loeb shocked the world when he announced his "invention" of artificial parthenogenesis. In his work, Loeb had experimentally manipulated a sea urchin egg to begin development without having to be fertilized by sperm. The experiment launched Loeb into stardom. A headline in the *Boston Herald* described it as "Creation of Life," *Harpers Weekly* put Loeb on the cover, and in 1901 Loeb was a finalist for the first Nobel Prize. By the early 1900s, long-form journalists would focus on Loeb and his work, drawing comparisons to Faust, Prometheus, and Frankenstein (Pauly 1987). At the time, Loeb was one of the few scientists that had such explicit goals about controlling life surrounding his work. By the 1950s, however, this would change.

Many of the previous examples may seem removed from contemporary notions of biotechnology and today's ethical discussions about them, but these early interactions between biology, technology, and society also evoked some ethical debates at the time. For instance, the celebration of science and technology at turn-of-the-century world's fairs can be countered with dystopian visions of science like H.G. Wells' *The Island of Doctor Moreau* (1896). Similarly, the continued scientific management and industrial integration into agriculture also engendered criticism that corresponded to the opprobrium that some thinkers had with America's rapid industrialization during the late nineteenth century (Kasson 1976).

Some of the evolutionary and breeding metaphors had begun to be applied to humans in the form of the eugenics movement, a movement fueled by the idea that scientific principles could be applied to controlling human reproduction in an effort to create healthier people and a better society. The eugenics movement had begun in England in the nineteenth century but gained a popular following in the early twentieth century in America, and remained mainstream until World War II (Paul 1995). During the early twentieth century there was little public opposition to eugenic ideas, even to some of the forced sterilization laws, particularly compared to the widespread condemnation that eugenics would eventually receive later on in the century. Yet, it is worth noting that the specter of eugenics from this time period would continue to haunt discussions of biotechnology throughout the next hundred years, especially in the 1960s and 1970s.

Biotechnology from World War I to Genetic Engineering

As the previous section illustrated, many of the important themes surrounding today's biotechnologies have roots in the nineteenth and early twentieth centuries: the increasing integration of scientific and technological principles into agriculture; the rise of genetics, which had little immediate impact before the 1920s, but would become an important aspect of biotechnologies later in the twentieth century; and the beginning of specific ideologies in biology that stressed increased control over life. Between World War I and the 1970s, many of these trends would continue. In addition, other concepts we associate with the biotechnology of the early twenty-first century would emerge during this period, particularly increased cooperation between academic life scientists and companies looking to capitalize on new advances. Also, after World War II, biologists, and eventually the newly minted bioethicists, would begin to voice hopes, skepticisms, and criticisms of new biotechnologies and their future impact on society, many of which would continue to reverberate in twenty-first century debates.

One of the most significant developments in agriculture in the United States following World War I was the creation and rise of hybrid corn. Through new breeding techniques that involved manually crossing two inbred lines, agricultural scientists found a way to create an extremely robust corn plant that exceeded the traditional open pollination breeding programs. In the 1920s and 1930s, the US Department of Agriculture (USDA) invested heavily in helping to create new strains of hybrid corn that significantly increased yield and allowed for further mechanization of farming. One repercussion of the proliferation of hybrid corn was that the corn seed produced could not be used to plant next year's crop since the second generation of hybrid corn resulted in a drastically inferior product. Consequently, farmers had to buy new seed each year instead of simply saving part of their crop to seed next year's soil. This situation created a new opportunity for the growth and proliferation of seed companies that would specialize in the creation, production, and marketing of hybrid seed. The success of hybrid corn and the new breeding and agricultural techniques that it helped usher in became particularly important for the Green Revolution of the 1960s and 1970s that saw the United States help increase food production around the world (Kloppenburg 2004).

The period after World War I also saw the rise of biochemistry and the increasing focus on reductionist perspectives in biology (Kohler 1982). These trends led to the discovery of many new biochemicals, particularly vitamins, and endocrines such as hormones and growth factors beginning in the early 1900s. Doctors and scientists in the first half of the twentieth century thought that these newly isolated chemicals held immense potential for biology and medicine since they seemed to be central to life's most basic functions. Many scientists immediately saw the potential for their work to be useful in medical treatments and sought to build connections with pharmaceutical companies. For instance, Edward Kendall, a biochemist at the Mayo Clinic in Minnesota, isolated and extracted a new biochemical he called "thyroxine." He assigned the intellectual property rights to the University of Minnesota, and in 1919 the university made an exclusive deal with the pharmaceutical company Squibb to produce and market the new drug. From this deal, Kendall received research funds from the sales revenue (Rasmussen 2002). Following in Kendall's footsteps, Harry Steenboch at the University of Wisconsin created the Wisconsin Alumni Research Foundation to help manage and distribute the funds of his patent on the discovery of vitamin D (Rasmussen 2001).

Many other chemicals between the 1920s and 1950s like insulin, estrogen, and steroids were discovered by life scientists and marketed and produced by pharmaceutical companies through a variety of legal combinations. Similar stories occurred in the agricultural realm with the discovery of plant hormones like Auxin, developed by Fritz Went at the California Institute of Technology (Caltech). Later in the 1930s, Went's student James Bonner, who also became a professor at Caltech, would enter into an exclusive relationship with Merck to develop further plant hormones. Historian Nicolas Rasmussen (2001) argues that the relationships between academic scientists and pharmaceutical companies that began in the first half of the twentieth century are not significantly different than those that emerged in the 1980s, which many have claimed is a phenomenon that manifested itself only after the development of recombinant DNA technologies. One of the biggest differences in the 1980s would be the sheer number of academics that became connected to private companies.

The fermentation technologies that had flourished in World War I continued to do so in the 1930s despite the onset of the Great Depression. During this period, the federal government committed resources to the continued development of fermentation research. The USDA set up regional research laboratories around the country to improve fermentation technologies using agricultural products. In the early 1940s, as World War II pushed federal and private resources into greater cooperation, the USDA fermentation facility in Peoria, Illinois became pivotal in the development of penicillin. Though the early penicillin research had occurred in England, British scientists had come to the United States in the hope of finding a way to increase the production of the antibiotic. The facility in Peoria quickly succeeded in increasing production by over thirty-fold, and the USDA secured working relationships with pharmaceutical companies like Merck, Squibb, and Pfizer to help further develop techniques and equipment to increase the production of what promised to be a miracle drug. In 1943, the War Production Board coerced many of these companies into producing penicillin using the established fermentation procedures, despite the fact that these companies had invested heavily in trying to develop processes to synthesize, rather than grow, the antibiotic (that would not occur until 1959) (Rasmussen 2001).

Penicillin had been discovered and developed by researchers in England, but it was scientists at the USDA fermentation facility in Peoria who determined how to produce penicillin on an industrial scale through fermentation technologies that ultimately ramped up the production of penicillin over hundred-fold compared to the techniques used in Britain several years earlier (Neushul 1993). As with other science-based war efforts, World War II instigated an unlikely, and formerly unknown, cooperation between government facilities, foreign scientists, and a private industry. This biotechnological success, and not just the success derived from government–industry–physicists relationships, contributed to the postwar push by political elites like Vannevar Bush to advocate for federally funded research in not just the physical sciences, but the biological ones as well. Thus, the success of fermentation technologies with penicillin helped pave the way for the explosion in funding that would come for the biosciences in the second half of the twentieth century.

Overall, fermentation technologies had become synonymous with biotechnologies by the middle of the twentieth century. Their association with microbes, industrial development, pharmaceuticals, and agricultural products had securely fit them in a place that connected "biology" and "technology." Microbiologists and chemical engineers dominated this area of biotechnology, and one could find a variety of partnerships between private companies, university biologists, and state-supported research centers. Between the miracle drugs and the important chemicals created through fermentation processes, this brand of biotechnology had a robust place in American culture, science, and industry by the 1950s. Scientists, both in academic settings and in concert with industries, had established a variety of processes and products that applied technological innovations to various aspects of life.

Biotechnology gained a different character in the 1950s, however. In 1953 Watson and Crick discovered the structure of DNA, the material component of genes and supposedly the blueprint of life. As many historians have shown, this was neither the birth of molecular biology nor did it radically transform the focus or practice of biology (Kay 1996). Rather, Watson and Crick's discovery exemplified the successes of an already burgeoning research area concentrating on the basic molecules of life, which included some of the research on vitamins and hormones discussed earlier. However, this "new biology," as it became known, inspired scientists to increasingly believe that they were close to gaining complete control over life that went beyond simply exploiting the functions of microbes as fermentation technologies already did.

Nobel Prize-winning geneticists like Joshua Lederberg, for instance, publically began to anticipate what major social issues the new biology might begin to solve, and what social problems it might create. Robert Bud (1993) argues that critical examinations of the pros and cons of technology, including those that would emerge from the biosciences, can be traced back to conversations led by British scientists such as J.B.S. Haldane and Julian Huxley (who perhaps inspired his cousin Aldous Huxley to write *Brave New World* (1932) in the process). However, for American biologists, the specter of the atomic bomb inspired much of this conversation. Biologists who were working on genetics and molecular biology in the 1950s and early 1960s believed that they were on the cusp of gaining unparalleled control over life. Yet the secretive creation of the atomic bomb had made some biologists fearful that biological sciences may have a similar potential to devastatingly alter society. In an effort to have a public debate about the potential of the new biology rather than sit silently in their labs, prominent

American biologists like Hermann J. Muller, winner of the 1946 Nobel Prize in physiology or medicine, and Joshua Lederberg, winner of the 1958 Nobel Prize in physiology or medicine, sought to anticipate where the field might take science and society. They often imagined how their work could improve the human race, through new and improved breeding programs or perhaps by tinkering with the human genome. Admittedly, they said, this would radically challenge traditional values, but in their view these were not simply possible futures, but inevitable ones, and they thought that society should discuss them now rather than wait to deal with the aftermath once science succeeded.

Lederberg was a part of a chorus of voices in the 1960s in discussing the potential for biology's future. Lederberg often played instigator in these discussions; for instance publishing pieces on the possibility of cloning through the use of nuclear transplantation, an embryological technique pioneered in the early 1950s by Robert Briggs and Thomas King that had never succeeded up to that point beyond amphibians (Crowe 2014). Lederberg's utopian visions of the future, which included routine cloning and biological manipulation, eventually elicited critical responses from theologians, philosophers, and other scientists. For instance, in the late 1960s, theologian Paul Ramsey articulated ethical arguments that denounced the perceived inevitability of what had become known by that point as genetic engineering, making claims that such interventions were an affront to human dignity and encroached into the space of "playing God" (Ramsey 1970). Furthermore, many of critics saw these scientific applications to humans as another iteration of eugenic ideals. Some scientists, including Leon Kass, who at the time worked at the National Institutes of Health (NIH), expressed similar viewpoints about the problems with cloning and modifying genes. Kass would refocus his work in the newly emerging field of bioethics, and would eventually become the chairman of the President's Council on Bioethics in the early twenty-first century.

Between World War I and the 1970s, biotechnologies became increasingly prevalent in American life. The stories of hormone, vitamin, and antibiotic discoveries and production illustrate the rising profile of biomedicine during this period. And though not common, new relationships between scientists, pharmaceutical companies, and federal funding emerged during this period as well, laying a foundation for the types of entrepreneurial associations that would become commonplace by the 1980s. Furthermore, the focus on the "master molecule," as Evelyn Fox Keller (2000) called it, by biologists during this period helped create a scientific optimism about the opportunity and possibility of controlling the basic functions of life. This not only led to new promises for what biomedicine might do in the future but also contributed to a growing critique of scientists, physicians, and their work. In the 1970s, both the scientific optimism for controlling life and the criticism of that project became even stronger as scientists developed new methods that allowed them to cut, paste, and rearrange the genes of living organisms.

Biotechnology as Bioengineering

In 1972, Paul Berg at Stanford University conducted a set of experiments that introduced transformative methods into biology. Berg brought together a number of techniques already used in molecular biology labs to cut out a particular chunk of DNA from a virus (SV40) genome and a microbe (*E. coli*) plasmid (a small, circular DNA genome) and fused two of the components together to create a plasmid that was part microbe, part virus. Theoretically, this technique would allow scientists to transfer a gene for a particular protein into a microbial genome, which would make the microbe produce that protein. The tools, both the physical enzymes required and theoretical understanding of how genes and DNA worked, were already known, and Berg brought them all together, opening the door for scientists to engineer life in a way that had only been imagined before. The methods would become known as recombinant DNA technologies, and, in effect, Berg's work fulfilled the prognostications of biologists like Lederberg in the 1950s and 1960s. The age of genetic engineering had arrived.

Berg quickly realized the potential for use and abuse of the new techniques. After studying some of the potential risks of creating these new hybrid-DNA organisms, Berg, along with several other prominent molecular biologists, published an article in *Science* in 1974 that articulated some of the potential biohazards of recombinant DNA technologies and called for a moratorium on the research until appropriate safety guidelines could be created. One of the most significant results of this moratorium was the Asilomar Conference held in Asilomar, California in February of 1975, which brought together over a hundred of the world's top molecular biologists and geneticists to articulate safety guidelines for this new controversial research. The NIH adopted the recommendations made during this conference, which allowed for continued research with limited oversight. Historians have come to see this as a moment in which the science community attempted to preempt the potential of public oversight, which scientists thought could end up being too restrictive, by instead policing themselves in a way that allowed them to continue with research and also appease public fears (S. Wright 1994).

During the recombinant DNA moratorium, however, two California biologists would continue their research into the concept. Using the same basic idea put forward by Berg, but with a slightly different technique, University of California San Francisco molecular biologist Herbert Boyer and Stanford University microbiologist Stanley Cohen transferred a frog gene into an *E. coli* genome. The intellectual property attorney at Stanford University quickly filed a patent for the process, which engendered criticism from the scientific community for what was perceived as an act contrary to the values of science since patents limited the circulation of knowledge and technologies (Hughes 2011).

In 1976, Boyer would team up with venture capitalist Robert Swanson and establish a new company called Genentech to commercialize recombinant DNA technologies. Swanson represented a rising investment culture centered in California that supplied the start-up funds needed for new businesses in exchange for significant equity in the new company. Sally Smith Hughes (2011) shows how Genentech navigated through the challenges of raising money, conducting research, building relationships with pharmaceutical companies and universities, and eventually achieving success with the creation of bacteria that could produce human insulin and human growth hormone. Some of these academic–industry partnerships echoed the relationships seen in the stories surrounding hormone development. However, Hughes (2011) argues that scientists like Boyer pioneered entrepreneurial attitudes unseen in previous periods.

In late 1980, Genentech would be the first company using the new recombinant DNA technologies to go public, making Boyer and Swanson, as well as a small cadre of scientists who had left academia to help them, immediate millionaires. Genentech's successes in the late 1970s inspired many other academics and entrepreneurs to start their own companies, creating a stream of new pharmaceutical products, diagnostic tests, and related technologies that continued throughout the 1990s and early 2000s (Hughes 2011; Rasmussen 2014).

Significant changes in 1980 to federal policy regarding biotechnologies increased the appeal for biotechnology innovation in both the private and the public sector. The first was the Bayh–Dole Act, which allowed for private patents on federally funded research. This allowed universities, for instance, to own the patents on the discoveries and inventions of their employees even if the research had been funded by federal grants, i.e. public money. Previously, researchers assigned the patent to the federal government if the research had been federally funded, an issue that became increasingly difficult to administer as federal funding for science dramatically increased after World War II, and with it many innovations and discoveries. This allowed universities and for-profit companies to greatly benefit from public funding going toward basic research such as the type of research that Berg, Boyer, and Cohen were doing in the early 1970s.

The second major policy change came with the Supreme Court decision in *Diamond v. Chakrabarty*, which stipulated that a living genetically modified organism could be patentable under intellectual property laws. This had profound effects on the burgeoning biotech industry, which had been eager to see if many of the patent applications from the 1970s regarding recombinant DNA, including those put forward by Cohen and Boyer, would hold up to judicial scrutiny. With a decision affirming the right to patent techniques associated with living functions, a new wave of investment and patents strengthened the prospects of the biotech industry, which was driven mostly by recombinant and molecular technologies.

In 1988, the *Chakrabarty* decision was extended to include the patenting of the first animal, the Harvard-developed OncoMouse^{TM}. Though heralded as an important tool in the quest to cure cancer, many scientists bemoaned the acquisition costs that now came with a Dupont-owned experimental organism. Like the patents on molecular techniques such as gene cloning and monoclonal antibodies, many criticized the conflict of interests that came from mixing scientific advancements with commercial interests. In essence, the capital-focused goals of private corporations seemed to undermine the free (or at least easy) exchange of information that scientists felt was necessary for advancement. Though the federal policies regarding patents had dramatically increased the amount of technology transfer between publically supported research and private enterprise, and with it the amount of entrepreneurial opportunities for researchers, there was a segment of the scientific community that questioned whether they had accepted a Faustian bargain in the process (Rasmussen 2014).

Federal policies regarding patents also had profound effects in agriculture during this period. There had been a longstanding tradition of offering patent rights to developers of new plant varieties. The 1930 Plant Patent Act, for instance, provided protections for developing novel asexually reproducing plant species. In 1970, a second law, the Plant Variety Protection Act, passed, which granted patent rights to companies or individuals that created new plant varieties through sexual reproduction, for example the hybrid varieties that had been developed in many popular crops (Kloppenburg 2004). Kloppenburg shows that the 1970 law generated a significant reorganization of the agricultural sector due to the new ability to patent sexually reproducing plants like hybrid corn, which resulted in mergers and takeovers that led to the large multinational agricultural companies that we see today (Kloppenburg 2004).

The 1980s also saw a continued focus on the genetic code and the promise of biomedical innovations related to it. The federally funded Human Genome Project (HGP) promised to gain enough knowledge about human health that a new age of personalized medicine would arrive upon its completion. Researchers at NIH completed the HGP in 2003, but few of its promises have come to fruition. However, new fields of genomics did emerge as it became clear that the hyper-focus on genes that had dominated most of the twentieth century could not capture the complex reality of gene–gene and gene–environment interactions. Also, as part of the massive federal project, Congress mandated that at least some of the funds went to research regarding the Ethical, Legal, and Social Implications (ELSI) of the project. Though controversial, ELSI did create greater awareness of the implications of new biotechnologies.

By the mid-1990s, when the rhetoric surrounding the HGP was at its zenith, some revelations concerning biotechnological advances quickly triggered large debates concerning the direction and values of science. In particular, the announcement in 1997 of the birth of a cloned sheep named Dolly sparked significant discussion in both the United Kingdom, where Dolly was born, and in the United States about the directions and intentions of science. Dolly brought together many aspects of biotechnology that had been debated over the previous decades; cloning and genetic engineering worries were paired with the implications of reproductive technologies like *in vitro* fertilization, a technique that had been developed in the late 1970s and had become a common infertility solution by the 1990s. Dolly's birth brought many of these issues together and prompted speculation that cloned humans could be next. Though national bioethics committees were convened in the United States, no major laws were created to disallow human cloning, only statements of severe condemnation. Biotechnologies related to cloning would continue to make headlines in the 2000s as cases of fraud dogged cloning research, and discoveries in the fields of stem cells and regenerative medicine would become more frequent.

Similarly, since the late 1990s the American public has increasingly become concerned with the amount of genetic engineering carried out on our food supply. The first genetically engineered crop that made its way to the public was Calgene's "Flavr Savr" tomato in 1994, and though it was pulled quickly due to backlash, it was a harbinger of things to come. For example, Monsanto introduced pesticide- and herbicide-resistant crops in the late 1990s, which now make up large shares of the most popularly grown crops in the United States (Kloppenburg 2004), and which are often used in many of the common processed foods.

The history of biotechnology in America since the nineteenth century highlights many themes that are important to the larger narratives of the history of science and American history. Just as agriculture dominated much of American life before World War I, so too did the management and breeding of agricultural products. Similarly, the rise of experimental methodology in biology and the application of evolution theories brought with it changes in perspective for agriculturalists and the general public that broke down plants and animals into collection of traits, and later genes, that could be manipulated through the application of science and technology. This played out not only in the creation of new plant varieties, but also in the invocation of eugenic ideologies. Today, scientists still apply the newest breeding science, but this time it often involves biotechnologies such as genetic engineering and *in vitro* fertilization, ultimately showing the deep connection between agriculture and science that has been a part of American society since the nineteenth century.

Throughout the twentieth century biotechnologies would be driven by many of the important events that shaped American history. World War II brought together scientists and industrialists in a quest to use fermentation technologies, at that time the most sophisticated biotechnologies available, to produce penicillin. The aftermath of World War II would not only generate President Eisenhower's military–industrial complex, but also lay the foundation for the explosion of the biomedical field beginning in the 1970s. Similarly, as federal regulations were continually re-evaluated throughout the twentieth century to give advantage to American businesses, science would become increasingly interconnected with companies seeking to commercially exploit the knowledge that publically funded research had developed.

Jacques Loeb's engineering ideals in early twentieth-century biology may have marked him as substantially different from most researchers at the time. However, by the end of the twentieth century his goal of advancing biological knowledge for the expressed intention to control nature for the good of humankind is a dominant philosophy. However, the ethical implications of this ideology have become increasingly apparent. Though one could trace a history of biotechnology in America as the development of science, technology, and federal policy, that narrative would be lacking one of the major reasons that biotechnologies have become an important part of American life – the social and cultural implications of those ideas. The rise of biotechnology is deeply connected to aims and visions of scientific actors, and its history should also include the ways that society reacts to not just the innovations, but to the perceived intentions behind them.

Today and in the future, biotechnology will continue to embody both savior and subverter. A history of the field should help guide biotechnology going forward, and understanding how American society has previously reacted to new biotechnologies should also help society manage fears and expectations and moderate political decisions when it comes to implementing regulation.

Bibliographic Essay

For a comprehensive history of biotechnology, see Robert Bud's *The Uses of Life* (1993), which chronicles how various aspects of biotechnology developed, highlighting the long history of the concept from its original connection in fermentation technologies to recombinant DNA technologies of the 1970s and 1980s. Edited collections that cover many aspects of biotechnology include Arnold Thackray, *Private Science: Biotechnology and the rise of the Molecular Sciences* (1998), which also contains a good treatment of the long history of biotechnology with another piece by Robert Bud, "Molecular Biology and the Long-Term History of Biotechnology." For a comprehensive encyclopedia account see Daniel Kevles, "Biotechnology" (2003).

There are many works on the development of specific biotechnologies. For the development of plant biotechnologies, see Jack Ralph Kloppenburg, *First the Seed: A Political Economy of Plant Biotechnology* (2004). On the development of two of the most common and important laboratory technologies used in biotechnology research, see Paul

Rabinow, *Making PCR* (1996), and Nicolas Rasmussen, *Picture Control: The Electron Microscope and the Transformation of Biology in America, 1940–1960* (1999). There are also several good treatments of the creation and development of model organisms as laboratory technologies. See Robert Kohler, *Lords of the Fly: Drosophila Genetics and the Experimental Life* (1994); Karen Rader, *Making Mice: Standardizing Animals for American Biomedical Research, 1900–1955* (2004); and Angela Creager, *The Life of a Virus: Tobacco Mosaic Virus as an Experimental Model, 1930–1965* (2001). Hannah Landecker articulated how cells themselves became technologies in the laboratory by focusing on the development of cell culturing techniques in *Culturing Life: How Cells Became Technologies* (2007). For discussions about reproductive technologies, see Adele Clarke, *Disciplining Reproduction: Modernity, American Life Sciences, and the Problems of Sex* (1998). For an account of the development of cloning technologies, see Nathan Crowe, "Cancer, Conflict, and the Development of Nuclear Transplantation Techniques" (2014). For recent scholarship on the relationship between biology and computing, see Miguel García-Sancho, *Biology, Computing, and the History of Molecular Sequencing: From Proteins to DNA, 1945–2000* (2012) and Hallam Stevens, *Life Out of Sequence: A Data-Driven History of Bioinformatics* (2013).

Many authors have specifically addressed the development of recombinant DNA technologies. On the effect that cancer research had on its development see Doogab Yi "Cancer, Viruses, and Mass Migration: Paul Berg's Venture into Eukaryotic Biology and the Advent of Recombinant DNA Research and Technology, 1967–1980" (2008). For specific accounts of the first biotechnology company that emerged from recombinant DNA technologies, see Sally Smith Hughes, *Genentech: The Beginnings of Biotech* (2011). Other useful accounts of this period include Stephen Hall, *Invisible Frontiers: The Race to Synthesize a Human Gene* (1987). For an important account of the rise of the biotechnology industry as a whole in the 1970s and 1980s see Nicolas Rasmussen, *Gene Jockeys: Life Science and the Rise of Biotech Enterprise* (2014).

For discussions of biotechnology and policy, good places to begin are Sheldon Krimskey, *Biotechnics and Society: The Rise of Industrial Genetics* (1991); Susan Wright, *Molecular Politics: Developing American and British Regulatory Policy for Genetic Engineering, 1972–1982* (1994); and Herbert Gottweiss, *Governing Molecules: The Discursive Politics of Genetic Engineering in Europe and the United States* (1998). For specific analysis of patent policy, see Daniel Kevles, "A History of Patenting Life in the United States with Comparative Attention to Europe and Canada" (2002a), and "Of Mice & Money: The Story of the World's First Animal Patent" (2002b). An important primary source in this is *Commercial Biotechnology: An International Analysis* by the Office of Technology Assessment (1984). On how federal education policy affected the development of agriculture, see Alan Marcus, *Agricultural Science and the Quest for Legitimacy: Farmers, Agricultural Colleges, and Experiment Stations, 1870–1890* (1985). On the impact of federal funding after World War II on biology, see Toby Appel, *Shaping Biology: The National Science Foundation and American Biological Research, 1945–1975* (2000). For a historiographic assessment of biotechnology, the history of molecular biology and policy, see Jean-Paul Gaudillière, "New Wine in Old Bottles? The Biotechnology Problem in the History of Molecular Biology" (2009). For analysis of the policies that helped create the Human Genome Project, see Robert Cook-Deegan, *The Gene Wars: Science, Politics, and the Human Genome* (1994).

There is an extensive literature on the relationship between biotechnology and private industry. For the interaction between universities and private industry, see Martin Kenney, *Biotechnology: The University–Industrial Complex* (1986). On the pharmaceutical industry's role in biotechnology, see Christer Nordlund, *Hormones of Life: Endocrinology, the Pharmaceutical Industry, and the Dream of a Remedy for Sterility, 1930–1970* (2011). Other good sources include Nicolas Rasmussen, "Biotechnology before the 'Biotech Revolution': Life Scientists, Chemists, and Product Development in 1930s–1940s America" (2001) and Jean-Paul Gaudillière, "Better Prepared than Synthesized: Adolf Butenandt, Schering Ag and the Transformation of Sex Steroids into Drugs (1930–1946)"(2005). For a discussion of the reflecting on the biotechnology–industry relationship, see Gaudillière, "The Pharmaceutical Industry in the Biotech Century: Toward a History of Science, Technology, and Business?" (2001). On the interaction between scientists, government, and industry during World War II, see Nicolas Rasmussen, "Steroids in Arms: Science, Government, Industry, and the Hormones of the Aadrenal Cortex in the United States, 1930–1950" (2002) and Peter Neushul, "Science, Government, and the Mass Production of Penicillin" (1993). For specific discussions about the relationship between modern biotechnology and capitalism, see Kaushik Sunder Rajan, *Biocapital: The Constitution of Postgenomic Life* (2006) and Melinda Cooper, *Life as Surplus: Biotechnology and Capitalism in the Neoliberal Era* (2008).

There have also been a number of important social criticisms written about biotechnology. For contemporary social commentaries and criticisms of biotechnology, see Ted Howard and Jeremy Rifkin, *Who Should Play God?* (1977) and Jeremy Rifkin, *Algeny* (1983). Rifkin again reiterated some of the dangers of biotechnology with *The Biotech Century* (1998). The perils of significant intervention into the human genome were outlined by Paul Ramsey, *Fabricated Man: The Ethics of Genetic Control* (1970). For feminist criticisms, see Donna Haraway, *Modest_Witness@Second_Millennium. Femaleman© Meets OncomouseTM: Feminism and Technoscience* (1997).

Two broad literatures that are useful in understanding the history of biotechnology in America include the history of technology and the history of twentieth-century genetics and molecular biology. A good place to start for useful histories of technology in the nineteenth century include John F. Kasson, *Civilizing the Machine: Technology and Republican Values in America, 1776–1900* (1976). For larger ideas about technology and culture, see Thomas Hughes, *Human-Built World: How to Think about Technology and Culture* (2004). Good book-length treatments of important trends in twentieth-century biology research include Robert Kohler, *From Medical Chemistry to Biochemistry: The Making of a Biomedical Discipline* (1982); Lily E. Kay, *The Molecular Vision of Life* (1996); and Evelyn Fox Keller, *The Century of the Gene* (2000).

Chapter Twenty-Four

Darwinism

Adam M. Goldstein

The phenomena answering to "Darwinism" are so diverse that there is good reason to believe that the term should be retired except as an entrée for scholars investigating the contexts in which it is used, which probably extend to all aspects of human endeavor. In this respect, Darwin should be regarded in the same class as Nietzsche, Freud, and Marx, each of whom Foucault (1990) describes as having created a conceptual framework that has been used as the basis for scientific practices integral to the organization of society and to the conception individuals have of themselves. Despite Darwin's pervasive influence on society as a whole, within a purely scientific context, Darwinians can be differentiated by the positions they take on a distinctive set of scientific issues, and different forms of Darwinism can be distinguished from one another on this same basis. The aim of this chapter is to give the reader a sense of what these scientific issues are, and how they have been addressed by key figures in the United States since the publication of Darwin's *Origin*.

Because Darwinism is so closely bound up with evolutionary biology, any comprehensive account of it must address controversies in that discipline. This chapter focuses on the scientific issues in order to give the reader a framework for understanding the interaction of Darwinism with broader social concerns such as religion, ethics, racism and slavery, the rise of research universities in the United States, and issues about education raised by creationism. While Darwinism serves as a flash point for broader social concerns, it would nonetheless be a mistake to believe that the fate of Darwinism is driven primarily by its relationship to these concerns as though it were an ideology, political position, or worldview competing with others, or as though its credibility and success in science depended on this relationship.

A Companion to the History of American Science, First Edition.
Edited by Georgina M. Montgomery and Mark A. Largent.
© 2016 John Wiley & Sons, Ltd. Published 2020 by John Wiley & Sons, Ltd.

What is Darwinism?

Darwinism is often the target of efforts to discredit evolutionary science. This is why it is critical to understand that, whatever the controversy about Darwin's intellectual and cultural legacy, there is conclusive scientific evidence that many of his most important claims are essentially accurate; that scientists view evolution as central to understanding all fields of biology; and that evolutionary science continues to grow in exciting and unanticipated ways beyond its nineteenth-century conception. Against a backdrop of diverse opinions, scientists of today agree, on the basis of sound scientific evidence, on two broad, central principles about evolution. The first concerns evolution itself: there exist relationships of descent among species, extending in unbroken sequence, to the origin of life on earth. The second is that biological adaptation arises and is maintained by an interplay between two kinds of processes: natural selection, on the one hand; and the processes that create and maintain heritable variation, including chance, on the other.

In addition to these principles, evolutionary biology since Darwin presupposes a conception of biological variation termed "population thinking" by Ernst Mayr (1995: 320). According to population thinking, a species' genetic endowment incorporates a range of variation, and these differences between organisms are considered biologically significant. Mayr contrasts population thinking with essentialism, which he sees as similar to Plato's idealism. According to Mayr's interpretation, essentialism treats differences among individuals as deviations from an average (mean) condition. This mean condition is seen as a biological reality, a property of the species that remains constant across generations, and that stands over and above individual organisms. In contrast, according to population thinking, traits particular to a lineage within a species can spread throughout the species, changing the average condition markedly. In the context of essentialism, this cannot happen: the average condition is the species' essence. This means that any significant alterations in it would result in a new species.

The scientific community has accepted evolution, the necessity of natural selection for adaptation, and population thinking. Nonetheless, since the publication of the *Origin* (Darwin 1859), there have been many scientists who accept evolutionary science, but reject Darwinism. The claim that one is a Darwinian or the ascription of Darwinism to someone else is therefore best interpreted as an attempt to establish a relationship with Darwin himself. Darwin is identified as the individual who bears the greatest degree of responsibility for formulating evolutionary science, introducing it to the scientific community and general public, and advancing scientific arguments in its defense at its origin. His views thus bear considerable prestige, and are regarded as having a presumption of evidence in their favor.

Consider more closely the kinds of relationships that can be established between a given scientist's work and Darwin's. First, a person might present his or her views as being essentially in accord with Darwin's, with the intention of being included in the Darwinian orthodoxy, so as to accrue prestige and credibility. For instance, someone might claim to have new evidence in favor of a position argued for by Darwin, or to have shown how one of Darwin's explanations of some organic phenomenon can be understood in terms of the best current science. Second, in contrast, someone might claim that, although his or her views are an alternative to Darwin's, they nonetheless maintain a close connection to Darwin or extend the spirit of Darwin's work. Here too the aim is to accrue prestige and credibility by aligning with a highly respected, orthodox view; but in this case, the aim is to advance the orthodoxy in a new way. Third, someone might identify him- or herself as anti-Darwinian or non-Darwinian, wishing to present his or her view as a highly original departure from the conservative establishment.

In addition to those who identify themselves as Darwinian, there are those who identify others as Darwinian. Here as well the aim is to establish a relationship to the orthodoxy in evolutionary science associated with Darwin. Taking the first case considered above, someone might want to identify someone else as Darwinian as a way of claiming that the person's views are especially credible. Taking the second case, the identification might be made as a way of indicating that the person's views extend the orthodoxy in an especially important way. Taking the third case, identifying someone as anti-Darwinian or non-Darwinian is often a way of indicating that the person's work lacks credibility, or to point to the person's work as evidence of sea changes in evolutionary science, whether credible or not.

The history of Darwinism is rich with controversy, because there is no one set of positions that can be used to distinguish Darwinians from non-Darwinians: over time, what it means to be a Darwinian has changed. The situation is made still more interesting because what is claimed by one person to be in tension with Darwinism can be seen from another point of view as being consonant with it. Controversies between Darwinians and non-Darwinians often amount to disputes over what the Darwinian orthodoxy is or should be. Four main issues are especially prominent.

The first of these concerns *acquired characteristics*. On the present-day view of biological inheritance, changes in the genetic material passed from parent to offspring are independent of the genetic material from which the parents' bodies develop. As a consequence of this, adaptations acquired during the lifetime of an organism are not generally passed on. A second consequence of this is that inherited variations are not generally correlated with what would be advantageous for their bearers.

The second issue concerns *biological hierarchy*. The genetic material in sperm and eggs is the basis for the development of cells, tissue, and organ systems in organisms; organisms are members of family groups and larger social groups; and those groups form the lowest level of biological taxa, species. Species are grouped in larger taxonomic groups, notably genera and phyla. Darwinians differ from one another and from non-Darwinians on several issues concerning the nature and role of these hierarchies: how are they produced in evolution, and how do they interact with one another in natural selection? The third issue is *the role of natural selection*. Biologists of the late nineteenth century understood Darwin's conception of natural selection, and in general, accepted that it occurs in nature. This leaves broad scope for disagreement. How widespread is natural selection? What is its relationship with other processes, such as development of the organism from an embryo, and chance? What is its relationship to progress and improvement in the history of life? What is its role in explaining human physiology and behavior? The fourth issue concerns *rates of evolution*: what is the relative timing in evolution of important events such as the divergence of species from one another or the spread of adaptation in a lineage?

The remainder of this chapter is devoted to identifying moments in the history of evolutionary biology in the United States in which Darwin's symbolic role as representative of scientific orthodoxy, understood in light of these four issues, has served to differentiate scientists from one another.

The First Four Decades of American Darwinism

As early as January 1860, just one month after the *Origin of Species* was first published in England and before it was widely available in the United States, many of the most influential American scientists established lasting impressions of Darwin's work. With the notable exception of Louis Agassiz, American scientists were generally favorably inclined toward Darwin's claims about common descent, but much less so toward his claims about natural selection as a primary source of adaptation and as a mechanism for creating species.

By the time Darwin's *Origin* appeared, the generation of American intellectuals responsible for creating the first research universities were building their careers. Harvard brought a critical mass of scientists to Boston, where the *Origin* was discussed vigorously in private and public. Asa Gray, a Harvard botanist, wrote a favorable review of the *Origin* in *The American Journal of Science* (1860); Louis Agassiz, also at Harvard, and among the most influential scientists in American history, did not accept evolution, and wrote against the *Origin* that same year in the same journal (1860). Despite a continuing record of achievement and mentoring relationships with students who would form a significant core of the next generation of American scientists, Agassiz faltered in a debate about evolution with William Barton Rogers, a geologist, later among the founders of Massachusetts Institute of Technology. The debate took place over four successive meetings of the Boston Society of Natural History, starting in early 1860. Agassiz's difficulty responding to Rogers' clear explanation and pointed defense of evolution exemplifies the attitude of American scientists: because of its explanatory and empirical richness, the evolutionary research program suggested by the *Origin* was seen as worth a significant commitment of time, attention, and scientific resources. Rogers evinces this attitude in his reply to Ralph Waldo Emerson's remark at the February 15, 1860 meeting that the *Origin* is primarily a statement of conclusions absent factual support.

> Mr. Darwin makes no pretension to an absolute demonstration, but, after an impartial survey of the facts ... and a candid appreciation of the opposing considerations, adopts the view set forth in his book, as ... a more rational and satisfactory explanation of the history of living nature than the hypothesis of innumerable successive creations. Prof. Rogers regarded the work as marked in an extraordinary degree by fairness in the statement of opposing as well as favorable arguments, by the absence of dogmatism, and by all other evidences of a truth-loving spirit, as well as by the extent and variety of its knowledge and breadth of its philosophical views (Rogers and Agassiz 1860: 234).

Against the backdrop of the generally favorable responses to the *Origin*, there emerged a distinctively American school of thought about evolution, the neo-Lamarckians, who developed a comprehensive theory about the history of life incorporating taxonomy, embryology, anatomy, and paleontology. Paleontologists Alpheus Hyatt and Edward D. Cope were the movement's earliest partisans, developing their views independently in the 1860s; Alpheus Packard, a Brown University zoologist, was attracted to the movement later, becoming its most vigorous advocate, starting in the 1880s. His contribution to the movement is exemplified by his 1901 translation and commentary on Lamarck's works on evolution. These three were joined by Henry Fairfield Osborne, who was a vertebrate paleontologist at Princeton and Columbia Universities, and who began his career as Cope's student. The mechanism claimed by the neo-Lamarckians by which adaptations are acquired and transmitted to new species is the hallmark of their views. The neo-Lamarckians were struck by the manner in which organisms' heritable adaptations seemed to appear as needed in response to threats against their survival in the environment, concluding that there is a mechanism by which organisms can originate heritable changes in morphology, changes significant enough to constitute a new species or even a new genus in their offspring. This is the theory of acquired characteristics. The neo-Lamarckians presented the theory of acquired characteristics in opposition to Darwin's claim that natural selection is the central mechanism by which adaptations become typical in a species, and were seen by their contemporaries as anti-Darwinian. They believed that Darwin mistakenly emphasized competition of organisms with their conspecifics as a factor in the evolution of adaptation; their view was that adaptations are created by the creative response of an organism to challenges it faces in its environment.

The theory of acquired characteristics was integrated with a sophisticated theory of embryology and developmental biology. Development in the embryo and in the juvenile organism is accelerated so that an adaptive characteristic acquired by an adult organism appears earlier in the lifecycle of its offspring. This was intended to explain long-term patterns of evolution seen in lineages such as cephalopods studied by Hyatt. At first, a variety of patterns is seen in adult organisms, some of which are adaptive; these appear in the juveniles in successive generations, which form new species, and continue to diverge in a similar manner. Because there are limits to the number and kinds of changes an organism can undergo in development, acquired characteristics, even those adaptive in their interactions with the environment, cause the lineage to deteriorate. On the neo-Lamarckian view, patterns of evolution in genera and orders resemble the life of an individual organism: like organisms in a juvenile state, species early in the history of the genus acquire adaptations from their parent species; like adult organisms, species in the middle and early late stages of the genus continue to adapt, and transmit their adaptations to new species; and, like elderly organisms entering a stage of senescence, species at the end of the genus' history exhibit bizarre, non-adaptive forms, before extinction. The neo-Lamarckians fell victim to changes in scientific opinion about inheritance. As Mendel's work was vindicated, scientific consensus turned against the view that adaptive, heritable variations could be acquired in response to an organism's interactions with the environment; according to Mendelians, the biological variation transformed by evolution into adaptation is inherited by means of sex cells, which do not change during a given generation in a way correlated with an organism's needs for survival.

The fate of the theory of acquired characteristics notwithstanding, the neo-Lamarckians made important contributions to documenting evolution in the fossil record, establishing traditions for the study of adaptation in ecological contexts, and developing concepts and vocabulary concerning the timing in development of the appearance of traits during an organism's lifecycle.

The Evolutionary Synthesis

In the middle third of the twentieth century, a movement to unify biology using an evolutionary theme began. This movement is known by a variety of names, including the neo-Darwinian Synthesis, the Evolutionary Synthesis, or simply the Synthesis, each derived from the title of British evolutionist Julian Huxley's *Evolution: The Modern Synthesis* (1942). The presumptions of the synthesis were that, one the one hand, because all aspects of the organic world result from historical processes of evolution, evolution should be expected to play a role in understanding any such aspect; and, on the other hand, that since the objects of study of every biological discipline influence evolution, each discipline should be expected to contribute to understanding its mechanisms.

For the Synthesis to take shape as a movement, it was required that the Mendelian theory of inheritance be reconciled with certain elements of Darwin's understanding of evolution connected with population thinking. On a Mendelian view, inherited variation is unitary: an organism has a given character state in a given trait only if the biological material (gene) responsible for that trait is in a given state; normally, the gene maintains its state in the sex cells, and is passed on unchanged. In the early years of the twentieth century, it was believed that, for each trait of central importance for the life of the organism, there is only one or a few genes, and that the entire organism requires only a small number in sum. This was believed to be incompatible with Darwin's claim that evolution proceeds by natural selection of small differences between organisms. Sewall Wright's "Evolution in Mendelian Populations" (1931), together with similar work by British evolutionists R.A. Fisher and J.B.S. Haldane at around the same time, convinced many scientists that this is not so.

Wright developed his understanding of Mendelian inheritance by working with the inheritance of coat patterns in guinea pigs and rats while a student at Harvard's Bussey Laboratory, and in his work on livestock pedigrees at the United States Department of Agriculture. In contrast with the Mendelians described above, Wright's view is that an organism's phenotype depends on a large number of genes, each of which exerts a small effect. There are two types of genes: structural genes are responsible for the enzymes or other reagents in a process essential to the functions of the organism; modifier genes are responsible for gene products that catalyze the action of the products of the structural genes. The modifier genes might be regulatory genes; in general, they control the degree, timing, or rate of expression of the phenotype resulting from the structural gene products. For a given phenotype, there are many modifier genes, even if there are few structural genes. The centerpiece of Wright's 1931 paper is a statistical theory about how the frequency of genes inherited according to Mendel's laws should be expected to change under the influence of various processes of evolution, on the understanding of gene expression just outlined. This theory implies a novel conception of evolution as a change in the frequency of a gene for a given character state of a given trait (an allele) in a population, across generations. Differences in evolutionary fitness between alleles are represented in the theory as different numerical values reflecting differences between the alleles' chances of being passed on to the next generation. Wright endeavored to describe, in a single mathematical expression, all processes that can result in changes in allele frequencies in this manner, including mutation, migration in and out of the population, and random processes.

In the 1931 paper and in a 1932 successor, Wright sketches his shifting balance theory of evolution, which he illustrates using a visual model he terms an "adaptive landscape," among the most distinctive and influential contributions of scientists in the United States to evolutionary biology, indeed, to the sciences considered as a whole. The shifting balance theory reflects Wright's experiences with livestock pedigrees. He discovered that breeding cattle is best accomplished by mating those with the desired characteristics in a small group, to concentrate genes for the desired traits in the lineage, and to outbreed the lineage in a limited extent with organisms from the general population, to avoid harmful gene combinations that can result from intensive inbreeding. This procedure is in contrast with the procedure of mating the animals with the desired characteristics with a large number of individuals in the general population with the intention of spreading the desired trait throughout. In nature, the shifting balance process, which Wright claims is the process by which the mean fitness of a species is most likely to increase, takes place in a population spread out over a generally uniform environment, but subdivided into groups, among which there is a limited exchange of individuals. Due to the statistical properties of Mendelian inheritance, there is a higher probability of novel gene combinations arising in smaller groups than in the population as a whole. This is the first stage of the process. If one of these novel combinations increases the fitness of their bearers in the small group, it will spread throughout the group by natural selection. This is the second stage of the process. In the third stage, the fitness of the species as a whole is increased by migrants from smaller groups bearing fitter gene combinations. Wright conceptualizes the shifting balance process using the visual metaphor of a topographical map, known as a fitness landscape or adaptive landscape. In a topographical map, altitude is represented by the density of lines indicating elevation: the closer together the lines, the steeper the slope. In an adaptive landscape, evolutionary fitness is represented in just this manner, and the organism's phenotype is represented by the x- and y-axes. A biological population's fitness corresponds to its "elevation" on the adaptive landscape. Natural selection has the effect of moving a population to a higher elevation on the map. Chance can also cause a population to move to a higher elevation on the map, but in most cases, chance events will cause a population to decrease in fitness. The chance occurrence of a favorable gene combination and its spread by natural selection in the first and second stages of the shifting balance process is one means by which a subpopulation can move to a "fitness peak" in the adaptive landscape; in the third stage, the population as a whole moves to the peak.

Against the theoretical backdrop concerning Mendelian genetics created by Wright, Fisher, and Haldane, a broad consensus about the process of evolution formed among supporters of the Synthesis. The "biological species concept," advocated most forcefully in the Synthesis period by Ernst Mayr, forms the core of this consensus. According to the biological species concept, a species is a group of organisms that can interbreed with one another. On this species concept, a species functions as a unified whole in evolution. Mating outside the group, if possible, results in sterile hybrids. Mixing among organisms in the species ("gene flow") maintains the species-typical phenotypes. Mutations are generally harmful; those that are beneficial spread throughout the population by natural selection. New species form by a combination of geographic isolation, a non-adaptive phase, and natural selection. The process is known, following Mayr, as "allopatric speciation." First, the incipient species population, a small fragment of the parent, finds itself in a new environment at the periphery of the parent species' range, and becomes separated from it. The incipient species is a random sample of organisms, which may differ from the parent in phenotype and genotype, and further differences are likely to arise due to inbreeding. If these differences are adaptive, the population will establish itself in the new environment to form a new species. In keeping with the biological species concept, the species forms a self-contained lineage in which favorable genes spread by natural selection whenever they occur.

Several key works, scientists, and institutions were central to the Synthesis. Chief among these is Theodosius Dobzhansky, an immigrant to the United States from Russia. Dobzhansky's field studies, published in a series of papers published through the 1970s but starting in the 1930s under the collective title *The Genetics of Natural Populations*, investigate genetic diversity in nature and its evolutionary explanation. Dobzhansky's work is informed by his collaboration with Wright, to whom Dobzhansky looked for mathematical expression of his insights into evolutionary theory; Dobzhansky's collaboration with geneticists Alfred Sturtevant at Caltech and T.H. Morgan at Columbia informed his insights into the physiology of gene expression and adaptation. Dobzhansky's *Genetics and the Origin of Species*, first published in 1937, and revised in 1941 and 1955, and its successor, *Genetics of the Evolutionary Process* (1971), each describe the general state of knowledge in evolutionary biology at the time of their publication. The Committee on Common Problems of Genetics, Paleontology, and Systematics, active in the 1940s, consolidated work across disciplines, resulting in the creation of the journal *Evolution* in 1947 and the Society for the Study of Evolution responsible for its publication. In addition to his role on the Committee and Society, in 1942, Mayr articulated the evolutionary synthesis from the point of view of biological taxonomy in his *Systematics and the Origin of Species* (1999). Paleontologist George Gaylord Simpson's *Tempo and Mode in Evolution* (1944) and *The Meaning of Evolution* (1949), and botanist G. Ledyard Stebbins' *Variation and Evolution in Plants* (1950), were also influential in articulating the evolutionary synthesis in the United States.

Molecular Evolution

The advent of molecular biology created a new field of investigation for evolutionary biologists. Darwin, his contemporaries, and early Mendelian geneticists, including Wright and his colleagues, were restricted to investigating the evolution of traits with easily detectable phenotypes, such as coat patterns in guinea pigs or lethal alleles. Changes in DNA in a species were presumed to occur by changes in rates of survivorship and differences in rates of reproductive success of organisms resulting from the interaction of their phenotypes with their environment and with one another. This presumption was tested when technologies for sequencing DNA and determining amino acid sequences in proteins provided direct access to the inherited material and the physiology of phenotype differences. In 1966, Lewontin and Hubby (Hubby and Lewontin 1966; Lewontin and Hubby 1966) published two papers of special importance for the study of molecular evolution (King and Jukes 1969: 788). Using the then-novel technique of gel electrophoresis, they determined the number of proteins produced at various gene loci in *Drosophila*. Their results were surprising because they suggested that there exists considerable variation at the molecular level not seen at the phenotypic level.

Lester King and Thomas Jukes, among the first to recognize the importance of studies such as Hubby's and Lewontin's, formulated a central principle of the study of molecular evolution: In their 1969 "Non-Darwinian Evolution," they state that "patterns of evolutionary change that have been observed at the phenotypic level do not necessarily apply at the genotypic and molecular levels. We need new rules in order to understand the patterns and dynamics of molecular evolution." As the striking title of their paper suggests, they believe that molecular evolution exhibits patterns and dynamics in tension with Darwinism. Their central aim is to show that there are patterns of diversity across species not resulting from differences in fitness among heritable variants, the variants in this case being alternative DNA base pair and amino acid sequences at homologous loci across the biosphere. This claim can be identified as what has come to be called the neutral theory of molecular evolution. Motoo Kimura reaches a similar conclusion in his "Evolutionary Rate at the Molecular Level" (1968), the first of many such contributions in Kimura's lifelong advocacy for the neutral theory, for which he is well known.

King and Jukes divide their argument for the existence of widespread neutral mutations into two types. First, they consider synonymous substitutions in base pairs in the DNA code. Each amino acid is encoded by a sequence of three nucleotide base pairs of DNA, the codon; but some codons which differ in structure do not differ in the amino acids for which they code. This means that exchanges of synonymous DNA nucleotides cannot be detected by observing phenotypes. King and Jukes review the literature on diverging lineages in mammals, finding that since the time of their last common ancestor, rates of divergence in proteins differ from rates of divergence at the DNA level. They conclude that this difference is independent of differences in the effectiveness (evolutionary fitness) of the protein products resulting from the DNA. The second kind of argument they make for the existence of molecular-level evolutionary change concerns proteins similar in function to one another. If an amino acid sequence at a particular site in a larger molecule can be exchanged for another with little or no effect on the function of the larger molecule in the organism's physiology, King and Jukes argue, it should not be expected that natural selection play a large role in determining which amino acid sequence is found at the site in question. Reviewing the literature on enzymes such as cytochrome c, insulin, and hemoglobin, which appear in many kinds of organisms, King and Jukes claim that there are significant rates of evolution at such sites.

Kimura is closely associated with the neutral theory, having spent the bulk of his career developing it, but his 1968 paper is much shorter than King and Jukes', and does not present the breadth of evidence they do. Accordingly, the latter carried more weight, initially. Kimura's argument is based on what is known as the cost of natural selection. Natural selection by means of differences in rates of survivorship among organisms with different phenotypes requires that some organisms die, of course, primarily those bearing the less-fit phenotype. Mortality due to natural selection is termed its cost. There is a limit to the cost of natural selection a population can bear: if there is too much natural selection, more organisms die than are born among the survivors, resulting in a net loss in population numbers that will lead to the population's extinction. Kimura's argument is that the amount of diversity seen by Hubby and Lewontin and others could not be maintained by natural selection, because the cost would be too high. King and Jukes read Kimura's paper while preparing theirs, affirming his conclusion – indeed, arguing that Kimura is too conservative in his estimate of the frequency of neutral sites.

Punctuated Equilibria

In 1972, Niles Eldredge and Stephen Jay Gould proposed a novel theory, the theory of punctuated equilibria, which concerns the distribution of rates of evolution in time: in general, a species takes on the characteristics that differentiate it from its parent species at the time of origin, remaining stable in those respects until extinction. The stable period is known as "stasis," the periods of equilibrium that are "punctuated" by rapid branching events resulting in new species. The resulting visual model of phylogeny is termed "rectangular." Suppose that time is represented on the vertical axis of a Cartesian grid, and that the magnitude of a character trait measurement is represented on the horizontal axis. At its origin, a daughter species is represented on a nearly horizontal plane as the character state of interest changes rapidly. After its origin, the daughter species is represented as a vertical line, because the character state does not change significantly during its lifetime.

The relationship of the theory of punctuated equilibria to Darwinism is especially complex. The theory is in tension with Darwin's own view of macroevolution. According to Darwin, who articulates his view on this issue in chapter 4 of the *Origin*, the origin of a species is not accompanied by any change in the rate of evolution. On Darwin's view, lineages continually evolve at a uniform rate, until they go extinct, or are transformed into a new species. With each generation, the lineage adapts to new ecological niches similar to those its recent ancestors occupied. If organisms close to one another in a lineage in geological time were fossilized, they would not appear very different from one another, because the adaptations required for life in their ecological niches are so similar. In contrast, organisms in the present day would be expected to differ greatly from their distant ancestors, because their environments would be so different. Since organisms are fossilized only rarely, it would not be expected to find fossils of organisms in a lineage close in geological time to one another. For this reason, Darwin did not consider the lack of fossils of such "transitional forms" to be evidence against evolution.

Eldredge and Gould claim that punctuated equilibria are better represented in the fossil record than Darwin's view, which they term "phyletic gradualism." They argue that the absence of fossils of transitional forms is not due to the rarity of their fossilization, but to the absence of such forms in the history of life generally. A new species is significantly different from its parent, and although time scales for divergence may be hundreds or thousands of generations, they originate in a time span effectively instantaneous in geological time. Moreover, in general, the appearance of a new species does not occur by transformation of its entire lineage. A new species is expected to coexist with its parent, not result in the parent's extinction by transformation. The theory of punctuated equilibria does not, like Darwin's view, imply that natural selection is the central process by which new species are formed. As discussed above in connection with the neo-Darwinian synthesis, on the allopatric model of speciation, which Eldredge and Gould claim would generate the punctuational pattern in the fossil record, chance plays an important role. The theory of punctuated equilibria has played an important role in stimulating development of ideas about biological hierarchies in evolution. The punctuational model implies that new species can diverge from their parents in directions in which the parents were not evolving, which has been understood to mean that speciation and extinction (macroevolution) have dynamics independent of evolution within species (microevolution). This does not imply that the causes of speciation and extinction are not causally dependent on physical interactions among organisms and their environments, or their genetics and physiology, but that there are patterns among those interactions, especially among emergent qualities of populations, that are of special importance for promoting branching or extinction. For instance, habitat fragmentation can promote extinction, because it can reduce the likelihood that organisms find mates, which makes it more probable that the species will go extinct, even if the population is adapting to the environment in other ways.

Whatever its relationship with Darwinism, it is important not to overestimate the significance of the punctuational model, which is sometimes interpreted as having metaphysical consequences incompatible with the theory of evolution. The existence of relationships of descent among species entails that there exist relationships of physical continuity, by means of biological inheritance, both between earlier and later generations within a species, on the one hand, and on the other hand, within higher taxa, among species that descend from one another. Anyone who affirms that evolution is characteristic of the organic world affirms this, as do Eldredge, Gould, and others who advocate the punctuational model. Controversy about the punctuational model concerns the nature of the biological processes connecting species to one another, especially the rate at which they occur relative to processes within species not leading to branching.

Sociobiology and Evolutionary Psychology

Sociobiology and evolutionary psychology are late twentieth-century research programs aiming to explain human culture using evolutionary principles, with an emphasis on natural selection of genes. Sociobiologists and evolutionary psychologists are motivated by two observations: first, human individuals successful in social interactions accrue considerable advantage for survival and reproductive success; and second, human culture and the capacities required for an individual's participation in it are as complex and stable as the most complex physiological traits. Notable among these are the capacity for language, including literary language and related capacities for aesthetic production and judgment; teaching and learning across generations; the extensive investment of parents in the rearing of their young in long-lived, stable family groups; and our capacity for moral reasoning and rule following. Arguing against those who believe that the human mind is a general-purpose information-processing system, evolutionary psychologists claim that it consists of modules, each of which has a specific purpose in human perception, cognition, or affective response, and that each evolved by natural selection in our ancestral environment in the Pleistocene era. Sociobiologists invoke W.D. Hamilton's (1964a, 1964b) theory of kin selection to explain cooperation, morality, and institutions aimed at creating fairness among individuals.

Sociobiology and evolutionary psychology have occasioned vocal criticism. Both emphasize adaptation of lower levels of biological hierarchy – biological inheritance at the genetic level. Many scientists reject this, because they believe that explaining human culture requires adaptation of social groups. Many also criticize evolutionary reasoning about human beings. Human behavior is subject to deliberation and choice, and development of capacities for cognition and affective response occur in social, political, and cultural contexts which themselves have developed over human history. Critics claim that this creates barriers to evolutionary claims: an individual's biological inheritance cannot constrain perception and behavior in a way that natural selection can act on, and there is no way to identify the alternatives upon which natural selection acted in ancestral populations in our species. Humanistic disciplines such as literary and art criticism, cultural anthropology, and sociology have been especially resistant to evolutionary psychology and sociobiology. Their practitioners view principles of criticism and interpretation as independent of their value for survival or reproductive success; and they believe that evolutionary psychology and sociobiology presuppose a mistaken and morally objectionable view of meaning and culture that overemphasizes competition between individuals.

Perspective: Unification and "Dys-synthesis" in American Darwinism

The neo-Darwinian Synthesis forms the watershed moment in American Darwinism. Prior to it, the neo-Lamarckians established a distinctively American approach to evolutionary biology, which they believed to be in tension with central elements of Darwin's views. Whether identified with the neo-Lamarckian school or not, American biologists in all disciplines pursued their work on the presupposition that the objects of their study bear historical relationships to one another. The proponents of the Synthesis explicitly identified their movement as Darwinian, intending to incorporate all biological disciplines in the Darwinian framework. Placing evolution in a formal, mathematical framework based on the insights of Sewall Wright and other population geneticists established a widely agreed-upon general theory highlighting the role of natural selection as the central process by which adaptation spreads throughout a species. The period since the 1960s has witnessed what Janis Antonovics (1987) terms the evolutionary "dys-synthesis." Innovations such as the theory of punctuated equilibrium and the neutral theory of molecular evolution illustrate ways in which the study of evolution has yielded insights quite different than what the proponents of the Synthesis envisioned, while those such as sociobiology and evolutionary psychology continue to develop priorities espoused by advocates of the Synthesis, controversies about the application of science in human life notwithstanding.

Bibliographic Essay

Edward J. Pfeifer's contribution to *The Comparative Reception of Darwinism* ("United States," 1972) is an excellent starting point for learning about the early years of Darwinism in the United States. Louis Agassiz, in his *Essay on Classification* (1962, reprint; edited by Edward Lurie), explains his non-Darwinian, non-evolutionary view of biodiversity, which he maintained until the end of his career; Lurie describes Agassiz's engagement with evolutionists such as Asa Gray in *Louis Agassiz: A Life in Science* (1988). Hunter Dupree offers a parallel treatment of Gray in *Asa Gray, 1810–1888*

(1959). Gray offers a striking argument for evolution and natural selection in an American context in "Sequoia and its History" (1872). Cope's "On the Origin of Genera" (1868), Hyatt's "The Evolution of the Cephalopoda" (1884) and Packard's "On Certain Factors of Evolution" (1888) are clear statements of the neo-Lamarckian view, which Pfeifer contextualizes in "The Genesis of American neo-Lamarckism" (1965). Toby A. Appel's "Jeffries Wyman, Philosophical Anatomy, and the Scientific Reception of Darwinism" (1988a) describes the American scientific community's adoption of evolutionary biology from the perspective of one of Darwin's earliest advocates, Jeffries Wyman.

William Provine has collected many of Sewall Wright's scientific papers in *Evolution* (1986a), a companion volume to Provine's biography of Wright, *Sewall Wright and Evolutionary Biology* (1986). Provine's *The Origins of Theoretical Population Genetics* (1971) explains the scientific context leading up to the period of the evolutionary synthesis. Provine, working with co-editor Ernst Mayr, is responsible for a collection of interpretative and historiographic essays on the Synthesis, including reflections by many of those who participated, *The Evolutionary Synthesis: Perspectives on the Unification of Biology* (1980). Some of the most important original documents of the Synthesis period include the volume by those associated with the Committee on Common Problems edited by Glenn Lowell Jepsen, *Genetics, Paleontology, and Evolution* (1949) and *Dobzhansky's Genetics of Natural Populations I-XLIII* (1981), edited by Richard C. Lewontin, which contains all of the papers in Dobzhansky's Genetics of Natural Populations series. Niles Eldredge's *Reinventing Darwin: The Great Debate at the High Table of Evolutionary Theory* (1995), Vassiliki Betty Smocovitis' *Unifying Biology: The Evolutionary Synthesis and Evolutionary Biology* (1996) and Massimo Pigliucci's and Gerd Müller's *The Extended Synthesis* (2010) presents comprehensive interpretations of the Synthesis movement.

John Beatty's "Chance and Natural Selection" (1984) outlines conceptual issues about genetic drift important for understanding non-Darwinian theories of molecular evolution, which he extends in his "Natural Selection and the Null Hypothesis" (1987). Michael Dietrich's "The Origins of the Neutral Theory of Molecular Evolution" (1994) describes the immediate context in evolutionary science in the years leading up to the Kimura and King and Jukes' papers. Kimura makes an extended argument for the neutral theory in *The Neutral Theory of Molecular Evolution* (1983). Scientific works assessing advances in molecular evolution include John C. Gillespie's *The Causes of Molecular Evolution* (1994) and Tomoko Ohta's "Mechanisms of Molecular Evolution" (2000). Ohta is a chief innovator in the field of molecular evolution.

George C. Williams, in *Adaptation and Natural Selection: A Critique of Some Current Evolutionary Thought* (1966) establishes the conceptual framework for arguments about genes in natural selection drawn upon by Jerome H. Barkow, Leda Cosmides, and John Tooby in the *locus classicus* of evolutionary psychology, *The Adapted Mind: Evolutionary Psychology and the Generation of Culture* (1992); consult Edward O. Wilson's *Sociobiology: The New Synthesis* (1975) for sociobiology. Stephen Jay Gould's and Richard C. Lewontin's "The Spandrels of San Marco and the Panglossian Paradigm: A Critique of the Adaptationist Programme" (1979) advances methodological criticisms of both evolutionary psychology and sociobiology, as do Niles Eldredge, in *Why We Do It: Rethinking Sex and the Selfish Gene* (2004) and Philip Kitcher, in *Vaulting Ambition: Sociobiology and the Quest for Human Nature* (1985). Peter J. Richerson's and Robert

Boyd's *Not by Genes Alone: How Culture Transformed Human Evolution* (2005) articulates a comprehensive theory of how family groups, social groups, human migration,
and institutions such as agriculture and religion interact in human evolution. Dennis
Dutton, in *The Art Instinct: Beauty, Pleasure, and Human Evolution* (2010) and Daniel
Dennett, in *Darwin's Dangerous Idea: Evolution and the Meanings of Life* (1995) make
the case for integrating evolutionary thought into thinking about meaning and culture.

Chapter Twenty-Five

SCIENCE EDUCATION

Adam R. Shapiro

Education is the way that scientists make new scientists. More than that, it is a form of knowledge transfer, of spread and growth and implied progress. "Science," like "America," is a modern invention. Both come from European traditions that expanded across the globe. To say that there was no science in other parts of the world prior to European contact is not to disparage the knowledge-making capacities of non-European cultures, but to place limitations on the definition of science itself and to say that science is not innate or universal. Science has a history, a social one, that started in a specific place and which migrated and spread from there just as did other human inventions. Education was one of the main mechanisms of that spread. Likewise, the Western conception of America, as a singular integrated place inflected with the historic tropes of naturalness and conquest, parallels the imagining of scientific progress as a cumulative mastery of nature. There were developed and diverse societies living across the Atlantic from Europe, but the invention of those spaces as "America" is an artifact of European colonialism. Much of the early history of the eventual United States can be understood as the process of making this ideological "America" coterminous with geographical America. To begin to understand the history of science education in the United States, one must begin with its role in American colonialism.

Science in the American Colonies

Education was one of several forms of early colonial engagement in the Americas; its presence was particularly notable in missionary contexts. In both New Spain and New France (including parts of these colonies later incorporated into the modern United States) Jesuit missionaries established schools where they taught native populations and attempted to catalogue, understand, and communicate indigenous forms of knowledge (Harris 2005). In the seventeenth century, it would be anachronistic to speak of *science*

A Companion to the History of American Science, First Edition.
Edited by Georgina M. Montgomery and Mark A. Largent.
© 2016 John Wiley & Sons, Ltd. Published 2020 by John Wiley & Sons, Ltd.

as an enterprise strongly distinct from philosophy or religion. Nonetheless, early modern natural history and natural philosophy – forerunners to what would become science – were sometimes part of the teaching content of these colonial encounters. In 1917, Herbert Bolton described the role of missionaries in Spanish California, noting that "it is quite true ... that they 'came not as scientists, as geographers, as school-masters, nor as philanthropists, eager to uplift the people in a worldly sense, to the exclusion or neglect of the religious duties pointed out by Christ'. But it is equally true, and greatly to their credit, that, incidentally from their own standpoint and designedly from that of the government, they were all these and more" (Bolton 1917: 47). Bolton further discussed the efforts of the priest Eusebio Francisco Kino who "helped to civilize" (52) California for the Spanish colonial rulers. Teaching rudiments of agriculture and animal husbandry is placed alongside counting scalps and facilitating military conquest. In New France, education in non-religious topics was also part of the "civilizing mission" of education, at least officially (Ronda 1972: 391–2). In both New Spain and New France, the abilities of the taught populations to learn and incorporate new knowledge was cited by Jesuits in evaluating the populations they encountered. As Peter A. Dorsey observed, the whole of the Jesuit pedagogical enterprise was informed by natural theology, the effort to discern evidence of divine being and character through experience of nature (Dorsey 1998: 400–1). In this view of pedagogy, education was not only a mechanism for imposition of a "civilizing mission;" the ability to demonstrate receptivity to education was a hallmark of native ability to become civilized.

Natural theology was even more influential on the religious life of the English-speaking American colonies along the Eastern coast of North America – especially among the social class who established the United States. English natural theology texts were intended to present religious truths, illustrations of the power, goodness, or wonder of a God seen by looking at the marvels of creation. Moreover natural theology invoked the natural world to justify answers to questions about morality and political theory. This use of natural theology to inform moral and political philosophy was perhaps more strongly pronounced in Britain than in other European nations. As John Hedley Brooke noted, natural theology was present in most European countries of the early modern era, but "it is, nevertheless, true that natural theology, as an aspect of scientific culture, was particularly prominent and persistent in Britain" (Brooke 1991: 197). This prominence and persistence crossed the Atlantic with the flow of texts and people that comprised British colonialism, and continued even after the rebellion against the English crown.

At least initially, natural theology's use in schools was intended for religious study, as illustration of evidences of Christianity and moral philosophy. In the universities and schools where natural theology was taught in the English American colonies, it was often incorporated either into clerical training or into a classical curriculum. Texts such as Samuel Clarke's *Discourse Concerning the Being and Attributes of God* (1704) and Joseph Butler's *Analogy of Religion* (1736) were frequently used in American universities, and also had a wide public readership, judging from their substantial presence in major colonial libraries (Lundberg and May 1976). Though these texts discuss themes like the perfection and the apparent design of plants and animals, neither give great botanical or zoological detail. John Ray's *Wisdom of God Manifested in the Works of the Creation* (1691) offered greater detail of the works of nature but was not used in American universities. These works drew from considerations of natural instinct and the apparent goodness of creation to testify to the character and nature of God and to examine the questions of virtue or moral obligation which arose from the idea of God as creator of the natural world.

Natural theology was often invoked to justify a conservative (or a gradualist progressive) social order, one in which revolution, or deviation from appointed roles of social class, gender or race, was seen as subversion of a divine plan. Claims of natural superiority or of natural affinities for certain occupations or social roles could be used to rebut calls for political reform. The idea that natural differences were part of a divine design was also used to justify colonialism itself (as well as the subjugation of Native Americans and of Africans held as slaves). These contexts gave a new urgency to the control of naturalistic knowledge, and created circumstances for science education to be a politically radical act. Promulgating ideas about laws of nature or facts about the natural capabilities of groups of people – contesting principles of inherent and immutable natural ability and the heritability of that ability – could undermine the natural theological claims used to justify the social order. In light of this, the American Declaration of Independence's invocation of "the Laws of Nature and Nature's God" is an assertion that the justifications for colonialism (of England over white Americans, at the very least) were contrary to nature itself – and God. The author of those words, Thomas Jefferson, has been regarded as one of the premier champions of education and of science in this era of the new American republic (Bedini 1990).

Teaching the Laws of Nature and Nature's God in the Early Republic

From the founding of the American republic, interpretations of the nature of equality, and the questions of what persons were admitted to have inalienable human rights were in sharp dispute. It is generally acknowledged that the morality and legality of slavery in the United States was one of the most hotly contested political issues from the beginning of the nation to the legal abolition of most forms of slavery in 1865. This debate strongly impacted on university education in the United States, and coincided with changes to the way that natural theology was taught that helped bring it closer in subject matter to a form of science education. Historians have noted that in this era natural theology began to be regarded less as religious argument and more a form of pious popular science.

Perhaps the most important single text in this transition of genre was the *Natural Theology* of English theologian William Paley (1802). The book was written toward the end of Paley's life, as part of a series of texts that included his *View of the Evidences of Christianity* (1794) and his *Moral and Political Philosophy* (1785). Paley was quite explicit in the introduction to the *Natural Theology* that the three books were written in the reverse order as they were intended to be read. In Paley's "system," natural theology began with evidence from nature, presumably comprehensible to anyone regardless of their inclinations toward religious faith. It used the evidence of nature to argue that a singular, benevolent deity both existed and had interest in human affairs. The *Evidences'* examination of the validity and interpretation of Scripture was predicated on the conclusions of natural theology. And the *Moral and Political Philosophy* rested upon the conclusions of both other books. It was the moral philosophy for which Paley was best known in his own lifetime, and was considered rather controversial for its expression of a new and early form of utilitarianism (unlike later forms of utilitarianism, Paley's principle of maximizing the utility or well-being of people also took into account their spiritual salvation). The books effectively provided a whole curriculum of instruction, and most American colleges in the early republic taught at least one and sometimes all three of these texts.

By separating the topics of natural theology, scripture, and moral and political philosophy into three books, readers and teachers of those texts could read them separately. Paley's *Natural Theology* reached its theistic conclusions from an exhaustive survey of the natural world – nearly two-thirds of the book's chapters are descriptions of natural history and natural theology – and theological argument occurs in chapters at the beginning and end. By the 1830s, an edition of Paley's *Natural Theology*, annotated by Harvard medical professor John Ware "to give this valuable work a more extended circulation in our colleges and high schools" was published in Boston (Paley 1831: ii). Nearly all of these annotations concerned the scientific details, particularly those of human anatomy. Ware's commentary refined and enhanced the presentation of natural history and human anatomy, which was what many students were examined on, even though improvements to the science did not typically address whether Paley's theological conclusions had to be adjusted.

At the same time that the natural history elements of natural theology became emphasized in print, leaders in American higher education began to consider changes to the teaching of moral philosophy in ways that broke apart Paley's "system." Much of the driving force for this concerned the fight over slavery. In 1835 Brown University President Francis Wayland wrote a moral philosophy textbook that supplanted Paley's. Wayland was opposed to slavery, though he also expressed caution about forcing abolition on the Southern states. Paley's moral philosophy was explicitly antislavery, but some Americans interpreted utilitarianism as legitimating the subjugation of the slave. The intellectual debate over abolition also tracked over the scriptural realm, with the Baptist Wayland engaged in a debate with South Carolina clergyman Richard Fuller over the scriptural warrant for American-style slavery (Fuller and Wayland 1845).

Unlike Paley, Wayland's moral and political philosophy was rooted in a strong anti-utilitarianism, and moreover, was not naturalistic. That is, while Wayland invoked some elements of common-sense and moral intuitions, it did not look to the natural world for the purposes of establishing or justifying what constituted moral action (Madden 1962). Others have interpreted the success of Wayland's moral philosophy as a triumph of evangelical religious sensibility over the more Enlightenment mindset of Paley's natural-theology-first approach (Frey 2002). In Wayland's system, natural theology was not a necessary prerequisite to moral philosophy; religiously, it was at best a further illustration of the power and goodness of God as seen in nature.

Outside of formal education, a similar shift was also occurring. As Bernard Lightman has shown in the British context, the emerging genre of popular science, presentations and explanations of natural phenomena written for a broad audience, incorporated some of the stylistic elements of natural theology, after the Bridgewater Treatises of the 1830s (Lightman 2007: 23–4). By presenting the public with descriptions of nature that were couched in the language of a creator or divine purpose, authors of popular science could pass as socially respectable, as reinforcing Victorian era cultural values, rather than be seen as advocates of political subversion. In effect, in this era, natural theology effected a transition from being read as a work of religion drawing from nature to a work of pious popular science (Topham 1992: 398). Its continued use in American higher education was one of the first formalized instances of science education. It is telling that many American libraries, including many school libraries, catalogued texts like the Bridgewater Treatises as works of science (even as some of them classified the older Paley as religion.)

By the 1840s, many Northern colleges had inverted their curricula, beginning with Wayland's moral philosophy (or a similar text) and then teaching natural theology and Christian evidences. Oddly, Paley's *Natural Theology* was very often kept in use, even as his other books were replaced. It increasingly became the case that natural theology's persistence in American higher education functioned as a form of science education: describing and explaining the natural world (albeit with reference to a divine plan). Higher education of this classical type mostly addressed a socially elite population. Wayland spearheaded a reform of the curriculum at Brown, though its success and influence at other institutions was initially limited. Wayland proposed eliminating the singular compulsory curriculum in favor of a more elective one, including more specialized tracts. With this came the prospect of the first specialized studies in science (Cohen 1998: 82). Around the same time, Harvard and Yale established scientific schools, and other colleges introduced scientific or engineering courses of study. These efforts were meant to attract more students, "however, they had few graduates and were struggling," according to a 1925 reflection on their history (Pegram 1925: 501). Despite the struggles of scientific schools, Stanley M. Guralnick argues that there was also an increased presence of science as part of the classical curriculum itself. "The colleges upgraded the level of those sciences already taught, such as mathematics, physics, and astronomy, and added many more, such as chemistry, geology, and biology." This led to an increase in scientific literacy among the college-educated class and to an expanded presence of science among college faculty. "More professors were hired to concentrate on more narrowly defined scientific areas" (Guralnick 1974: 354). Deborah Jean Warner described how in this era "schools for women placed a new emphasis on natural history and natural philosophy." This was done in the expectation that teaching science to women was intended to encourage them to "become 'cultivators' of science, not necessarily 'practitioners'," though some did (Warner 1978: 58).

The increased presence of school science around the 1830s and 1840s was paralleled by an increase in access to public education. In his case study of the second oldest public high school in the country, Central High School of Philadelphia (founded in 1836, began instruction in 1838), David F. Larabee describes a curriculum that "was explicitly practical. The school prepared students for direct entry into the city's commercial life rather than for college." With its primarily "proprietary middle-class" constituency, "students spent between zero and 15 percent of their time studying the classics and over 40 percent of their time studying modern languages and science" (Larabee 1986: 43–4).

Access to public education and the centrality of science in its curriculum in the late antebellum period demonstrates an ideology (at least in the North) of the use of science to promote the development of a middle class amid an environment of increased mechanical industrialization. As Larabee describes, the middle-class ideology that informed Central's founding and its curriculum was "concern about the disruptive effects of the expansion of capitalist social relations on the existing social order," which "sought to establish a new order that was both compatible with capitalism and congenial with middle-class interests" (48). This middle class was also the primary market for the sort of pious popular science that flourished in the mid-century.

Industrialization also led to increased need for technical specialization within the American labor population. Warren Dym relates a complex role for formal mining engineer training in the antebellum American West. "Mine managers and metallurgists in the field since the Forty-Niners only grudgingly employed young academics, and advanced a nativist notion of 'roughing it' that did not look kindly on theoretical traditions or academic knowledge of the sort promoted overseas" (Dym 2011: 299). While the hydrology, mineralogy, and geology of the American West posed unique challenges, the lack of an American school of mining created a dichotomy between the unschooled miners and those who studied in Europe. As Western settlement and territorial acquisition after the United States conquest of Northern Mexico became legitimated through invocations of manifest destiny and an assertion of American identity that largely excluded Chinese immigrants, conquered Mexicans, and supplanted Native Americans, science education became a marker of internal tensions within American identity. Simultaneously, miners in California and Nevada pointed to their self-educated credentials as a demonstration of their fitness, while others saw the need to create an American mining school as a way to ensure the Americanness of future industry. These issues came to a head during and immediately after the American Civil War.

Science Education and the American Civil War

There are a great many interpretations of the American Civil War, its causes and effects. In many accounts, the fundamental debate over the legality of slavery since the founding of the United States was politically intractable, and would only ever be solved by violent conflict. Others point to the war's roots in the legality of secession and the constantly evolving relationship between the USA and its constituent states. Other historians have pointed to the possibility of an economic clash between the interests of the increasingly capitalist and industrialized economy of the North and the agrarian plantation economy of the South, or the competition between Northern and Southern political forces for hegemony of the Western frontier (which intensified through the conquest of Northern Mexico and the discovery of mineral wealth in the American Southwest) (Foner 1974). As seen already, the implementation and interpretation of science education intersected several of these themes, from the natural and religious legitimation of slavery to the potential for economic technocracy. In at least some interpretations, the success of the North in winning the war could be attributed in part to its greater economic and technological resources and its embrace of an ideology that fused capitalist and middle-class interests (Donald 1960).

Even before the war ended, the political power of the North was enlisted to entrench and extend this ideology into the postwar period. Most notably, the Morrill Land-Grant Act of 1862 (passed while the Southern states were in rebellion and not participating in Congress) was a massive implementation of federal education policy that significantly expanded some forms of science education. The Act provided for federally owned lands to be granted to states for the purpose of developing university education. From the outset, universities created and supported by the federal land grants had a different mission than the private liberal arts institution. As Ray V. Herren and M. Craig Edwards write in their history of the Act: "The bill's passage marked the culmination of a five-year effort to bring about the creation of a university in every state that would serve the needs of common people and teach the practical skills required by an increasingly industrialized economy, including that portion comprising the agricultural sector" (Herren and Edwards 2002: 90). The first major effort by federal government to influence higher education was focused on creating institutions that would use science education as a tool for increased industrialization and scientific agriculture, in effect spreading and entrenching the economic models that had flourished in the North prior to the war. Other historians have seen the Land-Grant Act as part of a much longer dialogue about the disposition of federal lands in the American West, whose status could be more readily settled in the midst of war. That the land grants were devoted to education, particularly education "for the benefit of Agriculture and the Mechanic Arts" was, Scott Key argues, secondary to the economic benefits of putting the land to use (Key 1996: 211–12). Thus the foundation for widespread science education as a dominant trend in American universities, though already beginning in the antebellum years, was given a tremendous boost by the Civil War Congress's efforts to push through legislation ensuring Western settlement and resource expansion. The Morrill Act also led for increased calls to establish a national school of mining (Dym 2011: 300). Though this did not lead to a national school, Columbia University established its School of Mines in 1863, and several other schools followed (Pegram 1925).

A separate act of Congress in 1890 extended the provision of land grants to the Southern states which had comprised the Confederacy during the Civil War. But even earlier, the effects of the war brought a greater amount of science education to the South. Reconstruction-era political leaders in Southern states were largely unsuccessful in attempts to secure federal support for compulsory widespread schooling (Harris 1974; Scroggs 1961). Yet the idea that the North's economic and industrial advantages had played a decisive role in the outcome of the war would resurface in a later generation of Southern leaders' efforts to reform and modernize the "New South." In the earlier history of American slavery, slaves were sometimes given opportunities to learn, including science, but this was sharply curtailed after insurrections around 1835 (Woodson 1919: 2). The founding of colleges to serve the educational needs of emancipated slaves brought with it an emphasis on industrial and technical training, but this was often criticized as being insufficient. "The Negro college has not succeeded in establishing that great and guiding ideal of group development and leadership," wrote W.E.B. Du Bois in 1932. "Its vocational work has been confined to the so-called learned professions, with only a scant beginning of the imparting of the higher technique of industry and science" (Du Bois 1932: 69). While the science and industrial education of the colleges had contributed to some economic mobility for black Americans, according to Du Bois it had also entrenched them in a status of working for and service to the industrial capital economy. "The industrial school acted as bridge and buffer to lead us out of the bitterness of Reconstruction to the toleration of today. But it did not place our feet upon the sound economic foundation which makes our survival in America or in the modern world certain or probable" (70).

By the end of the nineteenth century, science education was somewhat more widespread, but as Du Bois had observed, it had often been co-opted in the form of technical education designed to prepare students for industrial application. Few of these instances were the kind of education intended to help create new scientists. Yet it was in the late nineteenth and early twentieth centuries that an American scientific community began to flourish, in part due to changes in American science education at the time.

Teaching Science at the Turn of the Twentieth Century

While science education had become available in some schools and in some parts of nineteenth-century America, expanding after the Civil War through the provision of more universities and schools, it was far from widespread, and rarely seen as either socially or legally necessary. The existence and even the kind of science education available varied widely across the country and across barriers of race, class, and gender. Even American states that had fairly large populations of teachers and students in the 1890s, such as New York and Pennsylvania, did not have statewide science standards in their schools. In Pennsylvania, teacher licensure was based on an exam that met state standards, but which was designed and implemented at the county level. The exam included tests of skills in arithmetic, grammar, and other subjects, but nothing approaching science (Woody 1954). This does not mean that there was no science instruction in most schools, only that it was not compulsory (even though professional associations of teachers had long proclaimed the value of teaching science to students). Back in 1873, Professor A.A. Breneman of the State Agricultural College (now Penn State) told the Pennsylvania State Teachers Association "even where physics had been introduced to the schools, they had been put in haphazard, as it were, and experiments conducted for the mere purpose of amusement or entertainment. We have had too much text-book and too little systematic practical teaching. Science should be introduced by contact with nature" (Jillson 1873: 75). Educators and scientists took this call for "contact with nature" to heart, as Sally Gregory Kohlstedt (2010) has described in her history of the "nature study movement" in this era.

As Kohlstedt notes early in her history, "throughout the nineteenth century most [American] children lived in close proximity to plants and animals" (11). The nature study movement was rooted in a belief in the value of bringing nature into the classroom, and taking the classroom pedagogy out to nature. But the nineteenth century proximity to plant and animal life began to give way in the twentieth century in an America that was becoming increasingly urban, and whose large cities were populated by an increasing diversity brought on by foreign immigration and by the migration of African Americans to Northern cities. This demographic shift exacerbated political and economic rivalries between urban-industrial and rural-agrarian populations of the United States. More often than not, this was tinged with elements of nativism and racism, especially as European immigration came increasingly drew from non-Protestant, non-English speaking populations who remained concentrated in cities.

The turn of the century, partly informed by developments in the psychology of education and in theories of management being incorporated into government reform, helped spur a major effort to modernize and reform the administration and content of American education. Instrumental in this was the founding of the National Education Association's Committee of Ten, which sought to bring school practices across the country into some amount of uniformity (United States Bureau of Education 1892). The late nineteenth and early twentieth centuries saw many states take a more active role in textbook adoption as a means to address perceived corruption among local school boards and the publishing industry (Shapiro 2010). The period also saw the widespread consolidation of small rural school districts and the shift in educational power away from small towns and counties to the state level. David R. Reynolds identifies this shift as part of a larger transformation in the structure of power in American society" in which "institutional and organizational consolidation was the means" to the political and economic ideology of early twentieth-century progressivism (Reynolds 2002: 3). Also central to the progressive modernist ideology was the incorporation of science and its methods to the study and improvement of society. The cultural importance of identifying practices (like management and salesmanship) as *scientific* had already become widespread in late nineteenth-century America. Such rhetoric was part of a larger trend of professionalization in American labor, that was itself connected both to workers unions and to increased regulatory authority of the state (Wilensky 1964). This is not restricted to the professionalization of science itself, but that the science of industrial practices was a cultural necessity for establishing professional status. This included the science of pedagogy, in part through the rise of teachers unions and through the creation of teacher training colleges, including many state-established normal schools.

Instrumental in making the case for teaching as a profession was the progressive ideology that professional teachers played an essential role in the functioning of society by teaching acculturation and preparation for modern life in addition to specific disciplinary subject matter. In the early twentieth century this was especially present in the large American cities, such as New York and Chicago. Both of these cities had large immigrant populations and often had those populations living in dense urban settings that were perceived as a social and public health concern. These were also centers of professional teacher training and of teacher unionization, places where new theories of pedagogy were readily communicated and where there was the political space to implement them. Schools were established and expanded and many addressed the mission of preparing the children of immigrants for participation in the vision of modern American life.

Central to this was an expansion of the mandate of science education. In New York, teachers restructured the life science curriculum, not only unifying botany and zoology under a common rubric of biology but also emphasizing the necessity of biological knowledge for the necessities of health and human progress (Pauly 1991). In complement to this "civic biology," in Chicago, the "general science" movement brought an effort to teach science as a general principle, and to do so to promote values of citizenship and public engagement along progressive lines. As John L. Rudolph recounts: "General science was … a subject with no corresponding academic field of study. It was a course fashioned *de novo* by a newly organized community of professional educators – created to provide the masses of new students streaming into the high schools with an appreciation of the value of science in modern society and the skills to apply scientific thinking in their daily lives" (Rudolph 2005b: 354).

Progressive Era science education was not only meant to substitute for the content of nature study, but also to recognize the needs of an urban population that relied on an industrial and professionalized economy to survive. Nativist presuppositions about immigrant communities informed the mentality that immigrant children needed to be instructed in hygiene and that biological instruction should include the teaching of eugenic principles of breeding and heredity. The perceived universality of science was idealized in the invention of a formulaic "scientific method" taught as part of general science (Rudolph 2005a). This helped to establish the ideal of a scientifically informed society as normative, desirable, and even natural.

These changes to the science pedagogy in the American high schools were largely developed with the urban student in mind. Students living in Lower East Side tenements or in the boarding houses of Bronzeville did not live "in close proximity to plants and animals" (with possible exceptions including molds, rats, and roaches). However, the normative claims made by the civic biology and general science movements in American science education did not confine their applicability to the Northern industrial metropolises in which they were conceived. By the 1920s, efforts to introduce these modes of science education into other parts of the country, especially in the rural South, met with substantive political resistance. While part of the objection was rooted in opposition to specific aspects of the curriculum, such as evolution or the inclusion of sex education, the primary motivation for political resistance was the perception that science education was a form of insidious cultural imperialism that imposed onto rural Southern communities Northern urban American values disguised as universal and ideologically neutral pedagogy (Keith 1995; Shapiro 2008). The successful implementation of such educational reforms in the South was in part facilitated by Southern politicians who embraced some aspects of modernist progressivism as a pathway to economic development in their states, but this adoption helped create backlashes like an organized anti-evolution movement, leading to the famous Scopes "Monkey" trial of 1925 in Dayton, Tennessee (Shapiro 2013).

In the first half of the twentieth century, American science education grew substantially, both in the availability of and enrollment in science education courses at both the high school and the university level. Though this effort was somewhat informed by scientists, many of the leaders in this movement were pedagogy experts and policymakers. The primary aim of the former was not, however, to necessarily create new scientists, but rather to create a professional labor culture and support for Progressive-Era policies that made use of science as a form of cultural authority. That authority was reinforced by the role that scientists played in the invention of the atomic bomb and the public perception of that development's role in American victory in World War II. This set the stage for a more direct and active role for scientists in education policy in the postwar era.

Science, Sputnik, and Cold War Education

On October 4, 1957 the Soviet Union launched Sputnik I, the first artificial satellite to successfully orbit the earth. A commonplace claim about this event is that the launch of Sputnik caused Americans to suddenly invest in science education as a way of closing a perceived gap in technological capacity during the early stages of the Cold War. A better interpretation of these events would be to say that the financial and intellectual investment in science education had already started before Sputnik, but that the public interpretation of Sputnik as a symbol of falling behind in military and economic competition served to foment public support for Cold War-era science education and facilitated its widespread adoption (Rudolph 2002). Federal investment in science had predated World War II, but had reached an unprecedented level of involvement in the

Manhattan Project. Postwar scientists sought to continue government investment but not all shared the military and political ambitions that continued to motivate federal science research. As Jessica Wang relates, postwar progressive-liberal scientists "struggled to articulate a politics of science that would reconcile the tensions between expert rule and democratic control, as well as avoid the dominance of science by Cold War military imperatives." In 1950, the federal government created the National Science Foundation (NSF), which was viewed by some of these scientists "explicitly as a non-military alternative for the organization of basic research" (Wang 1995: 140). A 1956 NSF grant established the Physical Science Study Committee (PSSC). The PSSC along with the Biological Sciences Curriculum Study (BSCS) and other programs in other science and math subjects had a tremendous impact on science education in the United States (Girard 1979; Phillips 2014). Perhaps more than any single change to curriculum, these NSF-led projects gave much greater input to practicing scientists, rather than pedagogy experts, and the curricula themselves seem to shift focus on the purpose of science education toward the recruitment and training of future scientists. This priority was already present at the university level, as David Kaiser has shown, many large American physics departments were principally concerned with an almost factory-like "production" of new physicists. "Most American physicists did not spend the bulk of their time working on weapons during the 1950s. Yet they received money from defense-related bureaus to create an 'elite reserve labor force' of potential weapons-makers in the ranks" (Kaiser 2002: 133). The funding, training, and production of science PhDs became one of the main priorities of government funding bodies and private defense contractors alike (Wolfe 2012: 46). American science education also became a tool in Cold War geopolitics, with curricular and education support being exported to developing nations in an effort to ensure influence and support (Wolfe 2012: 55–73). In this, the expectation that science education could be used as a tool of cultural and political hegemony was made explicit, as was the claim that, despite the USSR's technological successes, Soviet "science" was inherently compromised because of its pseudoscientific adherence to a Marxist ideology (Gordin 2012: 79-105).

Science, Schools, and the Next Generation

Little research has been done on the effect of the shifting priority of the Cold War era on those students who did not become scientists, but by the end of the twentieth century, at the end of the Cold War and beyond, science education in the United States no longer held the level of cultural authority it once had. There have been many studies in recent years concerning the effort to undermine the inherent credibility of scientific facts, and the ongoing competition between science and ignorance or pseudoscience (Gordin 2012; Oreskes and Conway 2010; Proctor and Schiebinger 2008). By the late twentieth century, efforts to contest the imposition of forms of science education and the content of that education had largely merged as political movements. In particular, resistance to the imposition of educational control at the federal level had escalated even before the PSSC/BSCS – in reaction to the US Supreme Court's 1954 invalidation of segregation laws in *Brown vs. Board of Education*, and with further oversight incorporated into higher education acts in the 1960s and 1970s. Perhaps the single most galvanizing event in the backlash against federal action in American education came with the 1983 US Supreme Court ruling that Bob Jones University could be disqualified from its tax-exempt status for its practices of racial discrimination. According to Randall Balmer, it was the Bob Jones case (not the legality of abortion) that gave rise to the "religious right" in American politics (Balmer 2006: 14–16). That political movement has since added anti-evolutionism and, more recently, opposition to the teaching of climate science to its portfolio.

In the twenty-first century, debates over science education in the United States have taken place within the context of American anxiety over the loss of economic, if not military, hegemony. This global context of political debate has had an uneasy intersection with the internal political tensions between federal and local control, and between the legitimacy and delegitimation of science as both a cultural enterprise and as a way of knowing. Unquestioned in these debates is the presumption that science education's benefit, indeed the benefit of any government action, consists primary in direct economic output. In recent years this has led to proposals to financially incentivize students majoring in STEM (Science, Technology, Engineering, and Mathematics) fields, and, in the case of state-administered universities, to curtail funding for non-STEM fields (Webley 2013).

Perhaps in anticipation of political opposition that might come from a federal initiative in the style of the Sputnik era, in 2013, the Next Generation Science Standards (NGSS) were published. Intended to be a national standard without being a federal standard, the NGSS trademark contains the adage "For States, by States." The standards were developed by "a consortium of 26 states" (NGSS 2013: iv) without any federal involvement or funding. Despite this explicit effort to avoid the taint of Washington DC, adoption of the standards by various states has seen a repeat of claims that the standards impose a cultural and ideological agenda (including their inclusion of topics like climate change, biological evolution, and human reproduction). At the same time the debate over the NGSS rages, decentralizing policies within states and widespread increase in the establishment of charter schools had made it easier for creationism to be taught in the United States (Kopplin 2014).

The dominant narrative in the history of American science has been one of development from fairly small resources as late as the early nineteenth century to a global rise beginning at the end of that century and continuing through most of the twentieth. It may be unsurprising that this arc parallels the growth of American military and economic power and the rise of education as a state and cultural institution in the United States. It may well be that the continued history of American science will continue to be determined by the cultural roles that science education plays in the twenty-first century.

Bibliographic Essay

In a 2006 article, Kathryn Olesko debunked a claim sometimes made by historians of science education that until recently the subject had largely been "neglected," that it fell between the disciplinary cracks between the history of science and educational history. "Not only is there a rich past, but it is also a past that is considerably more intricate than the current vogue to highlight the work of Thomas Kuhn and Michel Foucault" (Olesko 2006: 864). There's no question that Kuhn's 1962 *Structure of Scientific Revolutions* had a lasting impact and continues to inform many uses of science education in the history of science. Kuhn articulated a vision of scientific communities who not only used education to make new science, but who created forms and materials of education that entrenched successive revolutionary ways of thinking. Kuhn's claim concerned how scientific theories change, how revolutions occur; it was never his intent to give a comprehensive account of educational processes themselves. Yet there have been many studies that have taken his claim that new scientific paradigms succeed when they are reflected in new textbooks as a prescription to look to changes in textbooks as a way to track changes in science itself. As the history of science has now frequently been brought into synthesis with the social historical questions concerning the implementation, regulation, and political economy of education, the limitations of such strategies has become clearer.

There are many books that treat the history of specific parts of American science education: either particular scientific disciplines – such as David Kaiser's study of physics education (2005a) or Sally Gregory Kohlstedt's history of nature study (2010) – or that look at science education in a specific period of American history – such as John L. Rudolph's (2002) study of the Cold War era – or of particular themes – like Kimberly Tolley's survey of science education of American girls (2003). More sweeping surveys of formal science education (that centered around schools) have come from authors whose backgrounds stem more from education research or educational history (DeBoer 1991).

Part of the difficulty in a truly comprehensive treatment of the subject is the difficulty implicit in defining its boundaries. What about subjects about which there is historical disagreements over what counts as science (as opposed to pseudoscience)? Especially when such topics (like "creation science") are the source of explicit school controversy, it's reasonable to include them under the domain of the history of science education, but not all pseudoscience controversies have been school controversies (Gordin 2012; Proctor and Schiebinger 2008). Does research on the history of science popularization, or of science presentation on television and radio, count as part of science education (Bowler 2009; LaFollette 2008)? Does training in the use of technology or training in applied science deserve more consideration? What about the social sciences? Does instruction in "moral science" still count?

If there is a general trend in the scholarship on the history of American science education, it is that the grand narrative (as much as one exists) is held together by examinations of formal education, places where policy can be instantiated and where the kind of interaction among historical actors can be easily identified as one of teaching learning. Case studies on specific scientific topics or eras may build outward from this and situate formalized education within a wider network of social interaction. There is increased recognition of the role that informal education plays, and the roles that external agents (like tutors, political advocacy groups, or local literary clubs) serve in connecting formal and informal educational processes.

Chapter Twenty-Six

ENVIRONMENTAL SCIENCE

Daniel Zizzamia

Fueled by the complexity of environmental concerns, the field of environmental science emerged in the 1960s. Environmental science is an interdisciplinary field that combines a concern for the human role in the environment, the effect of the environment on human health, and the scientific study of interactions that constitute the animate and inanimate natural world (Pfafflin, Baham, and Gill 2008; Wyman and Stevenson 2007; Young 2005: xi). Due to its anthropocentric inclinations, the field is often paired with environmental engineering. Environmental science simultaneously represents a lament for the negative effects of industrialization and the faith in the power of human ingenuity in altering the natural world that created those very problems.

The history of environmental science can be stitched together by thinking through the threads of interconnection that constitute the scientific fields and cultural developments that led to the formation of a coherent field of study labeled "environmental science." American environmental science has its origins in the nineteenth-century fusion of Humboldtian and Baconian science that explorers and scientists used to survey the natural resource potential of the vast American West. The field's subsequent historical development arose from the energetic rise of an interconnected industrial United States, the specialization of American science, the emergence of the integrative science of ecology, and the growth of environmental concern that blended humanist and scientific ways of seeing the world. It was only in the 1960s, when these strains coalesced that the field of environmental science was born.

Humboldtian Science and American Western Exploration

While the field of environmental science has recent historical origins, the core of what this term would come to encapsulate in the twentieth century had its genesis in the

A Companion to the History of American Science, First Edition.
Edited by Georgina M. Montgomery and Mark A. Largent.
© 2016 John Wiley & Sons, Ltd. Published 2020 by John Wiley & Sons, Ltd.

exploration of Louisiana Territory by Meriwether Lewis and William Clark and the Humboldtian expeditions that followed. The Prussian aristocrat, explorer, and naturalist Alexander von Humboldt was *the* scientific celebrity of the nineteenth century who virtually all men of science sought to emulate. In writing the classic *Cosmos*, von Humboldt proclaimed his "principal impulse … was the earnest endeavor to comprehend the phenomena of physical objects in their general connection, and to represent nature as one great whole, moved and animated by internal forces" (1850: vii). Humboldt's holistic approach sought to capture the interconnections of all natural phenomena. He was also concerned with social justice and the human place in nature. Humboldtian science represented a transition from natural philosophy to a restructured and holistic natural science (Dettelbach 1996: 304). During this transition, American scientific men adopted his methodological approach and blended it with the Baconianism necessary for collecting data in the extreme expanses of the US territory. These visions laid the foundation for the execution of early explorations into the North American interior (Bruce 1987: 68–74; Cannon 1978; Goetzmann 1959).

The extensive geography of the western frontier required this characteristic blend of Humboldtianism and Baconianism. The Louisiana Purchase was a bargain, but it brought with it the intimidating prospect of assessing the actual worth of the acquired lands. Was the West a garden, an avenue to the Orient, or simply a barren and arid wasteland to fortify the western edge of the United States from foreign invaders? To determine its character, expeditions into this vast territory had to observe everything and document the nature and environment of the West (Allen 1991). Anxious to determine the utility of western lands and later to plan the route for the transcontinental railroad, the US government deployed military men and men of science to assess the region by collecting massive amounts of data and by viewing the land using the broad gaze of Humboldt. In the early to mid-nineteenth century, the US government was largely responsible for expeditions that resembled an approach to seeing the world as broadly and interconnected as the contemporary field of environmental science (Goetzmann 1959, 1966, 1987).

The US government's vital role in western exploration did not change after the Civil War. The bloody conflict tore the nation in two and temporarily brought a halt to the western surveys and upset the routines of American science (Bruce 1987: 271–312). After the Civil War, the fusion of Humboldtian science and Baconian data collection reached its zenith under the wing of the US government with explorer, naturalist, and entrepreneur Ferdinand Vandeveer Hayden (Cassidy 2000; Foster 1994). By the late 1870s, the immensity of the American West had been subdued and surveyed through military expeditions that marched across the continent. The completion of the first transcontinental railroad in 1869 and more lines relentlessly extending into remote regions altered time and space. The early "Great Reconnaissance" and "Great Surveys" of the West had accumulated a great quantity of data for scientists to examine from the comfortable eastern institutions of science for years to come. The railroads also brought ease of transportation across the nation to scientists who could now get out to remote points in the West and perform detailed studies (Vetter 2004). This enabled more focused and specialized analyses of the western environment. The space of the West seemed less imposing and no longer required the broad vision of the Humboldtian and the dedicated data collection of the Baconian.

Scientific Specialization

The growing pool of American scientists, often trained as members of these surveys, began to see western surveys as their province and seized their opportunity to wrest control from the military with the aid of a fiscally minded Congress (Goetzmann 1966; Rabbitt 1979). Assisted by the increasing industrialization of the United States and the railroad transportation revolution, this push was part of a larger program to regulate and institutionalize scientific work to make it mirror European science. Governmental and industry patronage had to be pursued because the United States did not possess the same European sociocultural environment from which science was born (Bruce 1987: 35, 42, 36, 251, 261; Daniels 1968: 12–13; Dupree 1986: 1–90, 379–81; Slotten 1994: 33). These patrons were primarily looking for practical and utilitarian science, not basic science. Science essentially had to be remade in the United States because the ideals of disinterestedness and no-strings-attached research had no practical place in society (Lucier 2009).

The practice of "boundary work" was necessary for wrestling with the contradictions inherent in interested parties funding an enterprise founded on ideals of disinterestedness (Slotten 1994). A field of open access to funding sources through the government and industry permitted a more democratic sphere that credentialed scientists believed they should police. Amateurs were crucial to the early practice and promotion of science while the public and US government financially supported the nascent American scientific establishment (Bruce 1987). As national societies such as the American Association for the Advancement of Science (AAAS) gained prominence in the mid to late nineteenth century the role of the amateur diminished and the era of the "professional" dawned in the United States (Kohlstedt, Sokal, and Lewenstein 1999; Reingold 1976; Veysey 1975). The predisciplinary environment that welcomed amateurs was displaced by disciplinary fragmentation controlled by credentialed experts. While fragmentation of industrial America temporarily devastated interdisciplinarity, it made room for an abiding faith in the capacity for experts to engineer, control, and enhance environments.

The Progressive Era: Environmental Health, Conservation, and Preservation

The fragmentation of American science through specialization coincided with the separation of the study of the human body from the natural environment. With the aid of the germ theory of disease, the human body was less likely to be seen as embedded in its environment but rather a bounded entity (Nash 2006; Valencius 2002). The holism that was characteristic of Humboldtian science faded as the United States entered into the Progressive Era. However, Humboldtian holism limped along in the shadows and inculcated itself into elements of Progressive ideology. The turn of the century saw the rise of a class of technocrats and engineers who were tasked with remedying the social and environmental ills inflicted upon American society as a result of the excesses of rapid industrialization (McGerr 2003). Elements of the holistic view of Humboldt and a deep faith in the human capacity to make adjustments to the operation of the natural world were evident in the rise of two elements of today's environmental science – industrial health and sanitation, and the philosophies of conservation and preservation.

The environmental consequences of an industrialized United States came to the growing cities with a vengeance. While the denizens of American cities were quickly recognizing the effect of the environment on their health, their realization was often overshadowed by a tendency to overlook the environmental and health consequences of technology (Tarr 1996). An overwhelming faith in technology overrode the potential application of many holistic solutions. Yet not all urbanites were consumed by the promises of technology. It was in the filth of the city that grassroots concerns about environmental health and safety had their origins. Urban environments were where women, minorities, and industrial workers nurtured the growth of environmentalism and environmental justice in the Progressive Era (Gottlieb 1993). Urban spaces made it harder to see the separation between social and environmental issues because they were so intertwined in the built landscape.

The Progressive Era's social, political, and cultural changes not only drew from the power of everyday people concerned about their environment, but also from the rise of technocrats who saw themselves as above and outside nature and therefore able to engineer a better world. The professional technocrat known as the sanitary engineer arose out of the terrible squalor in the cities and its negative effects on human health (Melosi 2005). The cities illustrated the paradoxical promises and pitfalls of industrialization. Industrial technologies and processes brought increasing power and control, but also a host of unintended environmental consequences. Mired in the waste of industry, the ideology of conservation emanated from the Progressive technocratic vision. Conservation was a scientific movement of natural resource rationalization and efficiency often aligned with industry to engineer a less wasteful world. The obverse environmental objective known as preservation grew up alongside conservation as its adherents sought to keep industry out of the wilds of the world (Hays 1959).

Aesthetic preservation and utilitarian conservation both had social consequences. In the case of preservation, the romanticized uninhabited wilderness of the national parks had to be manufactured through the dispossession of Native Americans (Spence 1999). Conservation and preservation also had the effect of legitimizing or criminalizing the use of western commons (Warren 1997). By creating new laws with the purpose of managing the natural world, state technocrats altered communities and created new crimes such as turning simple subsistence into poaching (Jacoby 2001). As these examples indicate, it was not social justice concerns but rather the powerful current of truth-telling science and transformative technologies that informed early environmental legislation and policy.

Ecology

The science of ecology also arose out of the technocratic temper of the Progressive Era and supported the applied science of conservation (Kingsland 2005). This ecological view reflected the technocratic ideals of the Progressive Era and did not directly involve the study of humans because humans were understood as technicians manipulating the machinery of nature. The dominating spirit did not exist in isolation and without opposition. An alternative ecological outlook saw a holistic, humble vision of humans as a part of the natural world. This dichotomy represented the fundamental tension within the field of ecology (Worster 1994). However, some scientists could hold both of these seemingly opposing views in their minds. For example, in Charles Darwin's "entangled bank," nature was simultaneously a balanced cooperative and a battlefield of warring parties (Hagen 1992b).

The balance achieved by early ecologists represented a unification of the technocratic *hubris* informing environmental management and the interdisciplinary holism that placed humans in the natural world and embraced wider systems of natural interconnections. Out of this creative tension, the basic framework for environmental science may have developed from the science of ecology in the 1920s and 1930s when ecologists were generating more sophisticated theories describing natural phenomena and had gained a degree of authority in environmental matters (Young 2005). It is crucial to recognize that while ecologists were balancing opposing visions of nature, social agendas shaped their science. Ecologists in the first half of the twentieth century often brought "biological understanding to problems confronting human society in what seemed to be an acutely troubled time" (Mitman 1992: 1). In this sense, ecology sought to explore the human place in nature. Yet, more often than not, ecologists used science to support cultural values and naturalized elements of social behavior using metaphors infused with cultural baggage.

The tension could not hold forever, and by the Cold War ecology began to fragment as many ecologists sought to align themselves with the "hard science" of the postwar period. They wanted to purge the field of the sociocultural elements that informed their science before World War II. Unfortunately for these scientists, this was a losing fight in the face of the rising tide of environmentalism that sought out expert testimony concerning a healthier natural world under siege by industrial capitalism. Ecology often became politicized and the tension inherent in ecological thought contributed to the work of environmental science and helped environmentalists frame their concerns in terms of ecosystems (Hagen 1992b).

The Human Place in Nature: Environmentalism and Humanists

The rise of environmentalism is also important to the field of environmental science because it reflects how humans view their impact on the natural world. However, as with any historical question of origins, there are multiple interpretations. In part this is due to the fact that there is no stable definition of environmentalism that is not anachronistic. Nevertheless, it is often treated as an umbrella term for conservation, preservation, and quality of life issues related to the environment. That being said, some historians trace environmentalism to colonial expansion in the tropics and its attendant contributions to advances in scientific knowledge on island environments (Grove 1995). Others credit individual scientists such as Alexander von Humboldt who was a "proto-environmentalist" operating within the colonial framework that was nevertheless committed to the environment and social justice, paving the way for future conservationist and preservationist sentiments (Sachs 2006). American naturalists from the period 1740–1840 have also been credited as proto-conservationists seeing ecological interconnections, bolstering the Romantic movement, and questioning the logic and practice of improvement (Judd 2009). Yet the practice of "improvement" according to another historian examining early American agriculture meant reclaiming land from the wilderness and making the natural potential of the land last, creating "stable agro-environments" (Stoll 2002: 21). In improvement, as practiced by agriculturalists, was therefore the seeds of utilitarian conservation and even of elements of the science of ecology.

Environmental concern has a long lineage linked to natural resource management and the consequences and externalities of American industrialization, but the modern social movement known as environmentalism surfaced in earnest after World War II. The Faustian bargains of consumerism and mass-produced homes in the period immediately after the war into the 1970s sparked a radical shift from viewing suburban development and mass-produced homes as demonstrations of progress, to viewing the sprawl of suburbia as an environmental disaster (Rome 2001). Conservation had given way to environmental concerns due to the rising tide of "quality of life" concerns (Hays and Hays 1987). In the 1970s, the general public espoused concerns related to environmentalism and consumption and even organized in mass gatherings such as Earth Day (Rome 2013). During the 1970s this "environmental culture" was given strength through popular works such as *Silent Spring, Our Synthetic Environment, The Limits to Growth, Blueprint for Survival*, and *Small is Beautiful*, which outlined a sentiment of a cultural and environmental crisis that was directed toward "doomsday." A growing middle class was beginning to "concern themselves more with what the sociologists were calling 'quality of life' ... than with 'standard of living'" (Sale and Foner 1993: 7, 30). Quality of life included amenities such as natural areas for outdoor recreation and aesthetic enjoyment, and a healthy and unpolluted environment.

The field of environmental science emerged from the freshly fashioned concern for ecology, environmental justice, and a "natural" conception of the environment developing during the 1960s and 1970s. The growing field of environmental science studied the place and practices of the human species within the larger context of the planet earth. Its origins were in the scientific imperative to comprehensively comprehend the natural world, and the environmentalist social atmosphere of the postwar period. Environmentalists often mustered environmental science for their own uses because technical expertise was required in assessing environmental impacts, making policy, and judging the veracity of concerns (Hays 2000: 137–54). Yet, environmental scientists frequently preferred a technocratic vision of environmental affairs.

Amidst the scientific considerations expanding as a result of the growth of environmental science, humanists have been engaged in a project to understand the place of human culture in defining the parameters and priorities of a scientific vision of the environment. With the rise in contentious environmental politics in the late twentieth century, it became important not to simply ask about the dynamic physical interaction between humans and the environment, but also the impact of human culture on that dynamic. This line of inquiry has been used to interrogate and affect changes to the policies and practices of environmental science.

The field of environmental history, which formed its first official academic society in the late 1970s, has contributed significantly to interrogating the dynamic interplay between human culture and the environment. Environmental historians' insistence on understanding the social impact on scientific knowledge production and how we define terms like "wilderness" and "nature" shaped how environmental science was practiced and environmental policy was constructed and implemented. A number of works of environmental history predate the founding of the society such as the scholarship of Roderick Frazier Nash. Nash's most celebrated work was an intellectual history of American attitudes and ideas toward wilderness in which he traced the story from wilderness as a howling, desolate abode of evil with origins in the Judeo-Christian tradition, to a sanctuary absent of human presence represented by the Wilderness Act of 1964. Nash's work showed that "wilderness" was not simply a scientific category, but rather a construction of human culture (Nash 1967).

As with wilderness, the cultural critic Raymond Williams rightly stated, "the idea of nature contains, though often unnoticed, an extraordinary amount of human history" (Williams 1980: 67). Historian William Cronon's edited volume *Uncommon Ground* (1995) probed that very idea. Central to this volume was the belief that ideas of nature are not natural and should be interrogated for their cultural components. The contributors agreed that there was an inherent problem in calling upon nature as moral authority to situate claims of how things ought to be. This is especially problematic, Cronon asserted, given the fact that many people with different moral universes create and cement concepts in the natural world to justify their worldview. In creating laws and definitions concerning nature we should always ask "Whose nature?" or risk devaluing the "contested terrain" inherent in selecting a single set of meanings (1995: 51, 52).

A direct critical response to *Uncommon Ground* came in the form of an edited volume by Michael Soulé and Gary Lease titled *Reinventing Nature?* (1995). The editors saw the deconstruction of the terms "nature" and "wilderness" executed by Cronon and his contributors as a threat to the goals of environmentalists and environmental scientists because they could be put to use to "justify further exploitive tinkering with what little remains of wilderness" (1995: xv). The focus of the text concerned the impact of this deconstruction on policy and the ability for scientists to make claims about the natural world that are not contested as purely culturally constructed. The authors took a materialist stance and "agree that we can gain dependable, scientific knowledge about this independent, natural world, in spite of differences among us in class, culture, gender, and historical perspective" (1995: xvi). Cronon had recognized the fears inherent in Soulé and Lease's argument and understood that nature is autonomous. But he believed that if we are to interpret nature, we can only use anthropocentric and faulty tools available to us to do so (Cronon 1995). Environmental scientists had to wrestle with their methods in order to reconcile them with an appreciation for autonomous nature and the imperatives of human culture.

* * *

The Cold War was a pivotal period for the emergence of environmental science because humanistic and scientific visions of the natural world simultaneously rose in prominence and found room to unite into a single field of inquiry. The Humboldtian tradition was in a sense revived in the interdisciplinary framework of the field which overcame the fragmentation begun in the late nineteenth century. With the rise of interest groups and grass-roots environmentalism there was also a return of amateur pleas for authority that coincided with a breakdown of disciplinary boundaries. By the 1970s formal university training in "environmental science" had developed (Weis 1990). Just as it was a partner in its historical inception surveying the West, the US government also strongly encouraged the rise of the field of environmental science. The field was put to multiple uses including managing environmental concerns and crises in the Environmental Protection Agency, weaponizing the natural world in the fight for global domination during the Cold War, and to study the earth's atmosphere in the Environmental Science Services Administration, which later became the National Oceanic and Atmospheric Administration (Fleming 2010; Hamblin 2013; Hays 1987; National Science Board 1971).

The history of environmental science in the United States is in its early stages and there is much work yet to be done. The literature that does exist places its conceptual origins in what historians call "Humboldtian science" practiced in the early nineteenth century. This period was in a sense predisciplinary and therefore resembled what is now considered interdisciplinary. With the rise of American industrialization after the Civil War, American science began to change character. As a result of rising wealth, the transcontinental railroads, and the exploitation of the vast natural and scientific resources of the West surveyed by early explorers, scientists were able to specialize and divide into discrete disciplines. Industrialization also produced environmental costs that had their effect on the bodies and minds of American citizens. During the Progressive Era, environmental health, conservation, and preservation concerns radiated from industrialization's impact. A faith in the human capacity to transform and manage the natural world was also fostered. Out of this context came the field of ecology which focused on relationships in the natural world and considered the human place in the environment. However, it took the development of modern environmentalism in the post-World War II period to coalesce these strains into the field of environmental science. Owing to the political circumstances surrounding its origins, it is a field that is tied directly into the social character of scientific knowledge production and is continually subject to contestation due to the immense scope of its gaze.

Today, environmental science is a thriving field that will continue to play a vital role in shaping public understanding of the natural world and potential solutions to the United States' environmental problems. For example, environmental science has proved its relevance in controversies surrounding anthropogenic climate change. Scholars have proposed that human society has exited the Holocene and is now in the "Anthropocene" geological epoch (Steffen, Crutzen, and McNeill 2007). Departure from natural cycles of climatology and ecology of the planet are due to the massive release of CO_2 through human activities since the 1800s. Humans have taken advantage of geological processes by using fossil fuels, and if prevailing climate science is reliable, have altered "natural" cycles of geological time. The interdisciplinary nature of environmental science has made it the perfect field to analyze the social and environmental impacts of climate change and find potential solutions. Even environmentalists such as Bill McKibben have used environmental science to lament this expansive human influence over the entire planet (1990). Using the products of deep time (fossil fuels) to power industrial machinery has encouraged the development of environmental science which has begun to use tools such as plate tectonics and atmospheric science to consider humanity's powerful impact on geological processes and the geological record (Oreskes and LeGrand 2001; Zalasiewicz, Waters, and Williams 2014).

Bibliographic Essay

Historians have not closely analyzed the field of environmental science. However, the British Society for the History of Science released two edited volumes that traced the history of the "environmental sciences." The first was titled *Images of the Earth: Essays in the History of the Environmental Sciences* (1979) and it focused on geology, but the editors L.J. Jordanova and Roy Porter saw the term "geology" as anachronistic and misleading preferring to view their subject as "the history of systematic understanding of the earth" (2). The second volume titled *Science and Nature: Essays in the History of the Environmental Sciences* (1993), by Michael Shortland addressed the impact of environmentalism on the history of the environmental sciences. Once again, "environmental science" was used as a category for historical analysis, but this time it was used to lump together the history of specific sciences like ecology and climatology that pertain to environmentalist concerns such as air pollution and conservation.

Historian Peter Bowler also contributed to the history of the environmental sciences in *The Earth Encompassed: A History of the Environmental Sciences* (2000). While admitting to the futility of covering the history of every science related to the environment, Bowler's study traced many of the constitutive developments of environmental science. Emphasizing the impact of culture on science, he covered geology, paleontology, meteorology, geography, physics, chemistry, natural history, ecology, and environmentalism from Ancient Greece to today. Bowler also recognized the anachronism in using the encompassing term "environmental sciences," which had recently been derived from modern environmental concerns. Because of his emphasis on the "environmental sciences" rather than "environmental science" he saw them as discrete specializations rather than a unified field of academic study. Bowler indicated that these specializations would have been alien or anathema to earlier scientists, but that the unity that once existed has been irretrievably lost in modern science through increasingly focused study.

Due to its recent origins as a field of study, and the relative lack of histories tracing the development of the field of environmental science, the best way to trace the history of environmental science is through histories that may have not used the term "environmental science" or "environmental sciences" but have nevertheless underwritten its history. Christian C. Young's *The Environment and Science: Social Impact and Interaction* (2005) has come closest to tracing the origins of the field of environmental science rather than the environmentalist-inflected historical category of the "environmental sciences."

The field of environmental science shares many elements consistent with Humboldtian science. The scientific and exploratory methods that have been ascribed to Alexander von Humboldt in the term "Humboldtian science" were developed by historian William Goetzmann in *Army Exploration in the American West* (1959), where he fused Humboldt's cosmic vision that "erect[ed] grand structures of classified data which eventually would enable man to know all of nature" with the broad gaze and appreciation of grandeur inherent in Romanticism (18). The historian of science Susan Faye Cannon later refined the category of Humboldtian science by emphasizing its theoretical elements and downplaying its Baconian character in *Science in Culture: The Early Victorian Period* (1978). Cannon sought to overturn the consistent characterization of American science as merely fact-gathering Baconianism by hitching American scientists to Humboldt's holistic theoretical approach. Recently, Laura Dassow Walls, *The Passage to Cosmos: Alexander von Humboldt and the Shaping of America* (2009) attributed wide-ranging scientific and cultural influence to Humboldt. However, his legacy has not always been viewed as positive, but rather wedded to colonial exploitation in works such as Mary Louise Pratt, *Imperial Eyes: Travel Writing and Transculturation* (1992).

The history of American western exploration is well-travelled terrain. Foundational texts in this canon include: Richard Bartlett, *Great Surveys of the American West* (1962) and William H. Goetzmann, *Exploration and Empire: The Explorer and the Scientist in the Winning of the American West* (1966). The history of American science and exploration overlap and complement each other since they were intimately linked in the nineteenth century. Histories of American science also reveal the movement toward increasing specialization that characterized the end of the century. The central texts in this canon include: Robert V. Bruce, *The Launching of Modern American Science, 1846–1876* (1987); George H. Daniels, *American Science in the Age of Jackson* (1968); Hunter A. Dupree, *Science in the Federal Government: A History of Policies and Activities* (1986); John C. Greene, *American Science in the Age of Jefferson* (1984); and Nathan Reingold, *Science, American Style* (1991).

As the era of exploration was coming to a close and American science was fragmenting into discrete disciplines, the Progressive Era arose from the human and environmental consequences of unchecked industrial capitalism. The history of the rise of the science of conservation within the context of the technocratic mood of this era was covered by Samuel Hays, *Conservation and the Gospel of Efficiency: The Progressive Conservation Movement, 1890–1920* (1959). Placing the science of ecology alongside this expert-informed science of resource management is Sharon Kingsland, *The Evolution of American Ecology, 1890–2000* (2005).

Environmental science's link to ecology is undeniable due to the emphasis both fields place on interconnections in the natural world. Essential reading in the history of ecology is Donald Worster, *Nature's Economy: A History of Ecological Ideas* (1994). Worster traced the history of ecology to the eighteenth century with socioeconomic ideas about the "economy of nature" and ended with the environmentalism of the post-World War II era or the period he termed the "Age of Ecology" when the first atomic bomb was tested and "ecology became a political movement." For a foundational feminist analysis of the evolution of ecology, see: Carolyn Merchant, *The Death of Nature: Women, Ecology, and the Scientific Revolution* (1989).

The origin of environmentalism is disputed historical terrain, but it is generally accepted that an environmental movement gained wide public acceptance in the post-World War II era and it co-opted elements of ecology. By tracing environmentalism to the nineteenth century, many histories miss the subtleties of how environmental concern has evolved over time to become the late twentieth-century's environmentalism. Bridging this gap, Neil Maher, *Nature's New Deal: The Civilian Conservation Corps and the Roots of the American Environmental Movement* (2008) answers the question "how did Americans get from Progressive Era conservation to post-World War II environmentalism" (6)? Nonetheless, most of the canonical sources for the history of environmentalism and environmental politics trace the birth of modern environmentalism to the post-World War II era. Examples include: Samuel Hays and Barbara Hays, *Beauty, Health, and Permanence: Environmental Politics in the United States, 1955–1985* (1987); Adam Rome, *Bulldozer in the Countryside: Suburban Sprawl and the Rise of American Environmentalism* (2001); and Hal Rothman, Gerald D. Nash, and Richard W. Etulain, *The Greening of a Nation? Environmentalism in the United States since 1945* (1998).

Another way to examine the evolution of environmental concerns is to examine the lives of three generations of activists like George Perkins Marsh (1801–1882), Aldo Leopold (1887–1948), and Rachel Carson (1907–1964) – David Lowenthal, *George Perkins Marsh, Prophet of Conservation* (2000); Curt Meine, *Aldo Leopold: His Life and Work* (1988); Linda J. Lear, *Rachel Carson: Witness for Nature* (1997).

Around the same time that environmental science was becoming a field of its own, the field of environmental history emerged from environmental concerns of the post-World War II era. Environmental history uses environmental science to examine human–nature dynamics as they pertain to the course of history. One of the first successful works to do so was William Cronon, *Changes in the Land: Indians, Colonists, and the Ecology of New England* (1983). Donald Worster is the environmental historian most often juxtaposed to Cronon. Worster's work is materialist and critical of capitalism's impact on the environment. For examples of his work, see: Donald Worster, "Transformations of the Earth: Toward an Agro-ecological Perspective in History" (1990); *Dust Bowl: The Southern Plains in the 1930s* (2004). For an article-length exploration of the relationship between environmental science and environmental history, see: Donald Worster, "The Two Cultures Revisited: Environmental History and the Environmental Sciences" (1996).

The field of geography emphasizes seeing humans as elements in an ecological framework and uses the full range of environmental sciences to examine that relationship. Because it fuses both humanist and scientific scholarship, geography is perhaps the field whose history most closely aligns with that of environmental science. A classic in this field is Clarence Glacken, *Traces on the Rhodian Shore: Nature and Culture in Western Thought from Ancient Times to the End of the Eighteenth Century* (1967) that examined three overlapping and enduring questions that have infused all thought about the relationship between humans and the natural world: Is earth a designed and purposeful creation? Has the environment effected and molded "the character and nature of human culture?" How have humans changed the earth from its original condition? His intellectual impact is evident in the work of Andrew Goudie who addressed Glacken's second question in *Environmental Change: Contemporary Problems in Geography* (1983) and the third question in *The Human Impact on the Natural Environment* (2000). For a survey of the history of the field of geography, see: David N. Livingstone, *The Geographical Tradition: Episodes in the History of a Contested Enterprise* (1993).

The work of political ecologists, considered by some as a subfield of geography, has also fed into the history of environmental science. Political ecology is often defined as the field of human–environmental research that concerns itself with social, economic, political, and environmental matters. One of the foundations of this field is abandoning anthropocentrism and embracing nonhuman nature as an equal partner in society. This belief in an extension of rights to the natural world has a history charted in Roderick Frazier Nash, *The Rights of Nature: A History of Environmental Ethics* (1989). Many scholars in the field of political ecology view this extension of morality to nonhuman nature as a crucial goal so that there may be a more equitable and sound relationship between humans and the natural world. The key to this is the use of environmental science to speak for the natural world in the political realm. For scholarship that analyzes this relationship, see: Stephen Bocking, *Nature's Experts: Science, Politics, and the Environment* (2004); Anna Bramwell, *Ecology in the Twentieth Century: A History* (1989);

Tim Hayward, *Ecological Thought: An Introduction* (1995); Bruno Latour, *Politics of Nature: How to Bring the Sciences into Democracy* (2004).

The history of environmental science is admittedly in its infancy, but there are a number of scholars who are contributing to its growth. In large part due to its political contentiousness as an avenue of environmental science research, one area where the field is blossoming is in histories of atmospheric science and climate change. For a history of global warming, see: Spencer R. Weart, *The Discovery of Global Warming* (2008). For an examination of how humans have attempted to control the weather, see: James Fleming, *Fixing the Sky: The Checkered History of Weather and Climate Control* (2010). For a recent history of the politics of global warming, see: Joshua P. Howe, *Behind the Curve: Science and the Politics of Global Warming* (2014). Also, for an analysis of the role of radioactive fallout in the post-World War II period in understanding human health and ecology and assisting the rise of the modern environmental movement, see: Jerry Jesse, "Radiation Ecologies: Bombs, Bodies, and Environment during the Atmospheric Nuclear Weapons Testing Period in the United States, 1942-1965," PhD dissertation (2013).

Chapter Twenty-Seven

THE AMERICAN EUGENICS MOVEMENT

Christine Neejer

When most Americans hear the word eugenics, they often envision the horrors of Nazi Germany; concentration camps and grotesque medical experiments in the name of racial purity and hatred. The association of eugenics with Nazism has fueled three major assumptions regarding the history of eugenics. First, we assume that eugenics was an idea and practice that existed largely outside the United States. Second, we assume that eugenics was a reactionary and fringe ideology, which went either unknown or unsupported by everyday people and leaders alike. Third, we too easily accept the incorrect notion that support for eugenics plummeted after World War II.

Since the 1990s, the American eugenics movement has become a subject of increasing interest among historians. In their research, historians have discovered that all three of these assumptions regarding eugenics are inaccurate. From the 1880s through the 1970s, the United States was home to a vibrant and popular eugenics movement; it was not isolated to Europe. The ideas and practices of this movement changed over the years. But what remained consistent was that intellectuals, physicians, scientists, politicians, and religious leaders established and promoted eugenics, working in highly respected educational, research, and government institutions. As such, the major projects of the historiography of the American eugenics movement have been to understand the ideas and practices of American eugenicists, situate eugenics within the broader contexts of American history, and perhaps most challenging of all, understand the popularity and longevity of eugenics in the United States.

Historian Wendy Kline defines eugenics as efforts "to strengthen family and civilization by regulating fertility" (Kline 2001: 1). Eugenicists believed that crime, poverty, addiction, sexual immorality, and certain diseases were linked to the genetic makeup of a person, and that a person could pass down these traits to their children. To eugenicists, the transmission of such traits fueled these social problems, encouraged dependence on charities and assistance programs, and weakened the country as a whole. Eugenicists

A Companion to the History of American Science, First Edition.
Edited by Georgina M. Montgomery and Mark A. Largent.
© 2016 John Wiley & Sons, Ltd. Published 2020 by John Wiley & Sons, Ltd.

believed reproduction should be a privilege, not a right, based on whether one's genetic dispositions would harm the overall population (Kline 2001).

There were closely related policy implications that fueled an acceptance of eugenics, via both positive and negative eugenics initiatives. Positive eugenics initiatives were efforts to increase reproduction among families that eugenicists deemed "fit." These efforts usually targeted married middle- to upper-class Anglo-Saxon Protestants. Forms of positive eugenics included physicians encouraging these patients to have large families and education programs funded by eugenic organizations. Eugenicists encouraged these potential parents to see childbearing not as an individual choice, but as a duty to their race and their country. Negative eugenics initiatives consisted of efforts to decrease the amount of "unfit" people by discouraging or limiting their reproduction. Sterilization was the most common method of negative eugenics. While some women did chose sterilization as a birth control method, sterilization was typically compulsory, meaning that the individual had no legal right to refuse. By the 1970s, approximately 63,000 Americans were sterilized without their consent. In the twentieth century, 27 states at one point had laws that legalized sterilization for eugenics purposes (Largent 2008). Both were grounded in the assumption that the general fitness of any particular set of parents predicted the fitness of their children.

Setting the Stage for Eugenics

The American eugenics movement emerged in the nineteenth century, an era in which profound changes left many Americans uneasy about the future. The Industrial Revolution, which had started in Europe in the 1760s, ushered in a new era of technology and manufacturing, which fueled the expansion of cities and migration from rural areas for work in factories. This process of urbanization was directly linked to another major change – mass immigration. Starting in the early nineteenth century, millions of people immigrated to the United States from Europe. Eighteenth-century immigrants were smaller in number and typically English-speaking and Protestant. New groups, beginning with the Irish in the 1820s and followed by Italians, Poles, Germans, Swedes as well as immigrants from Mexico and China, profoundly altered the ethnic, racial, and religious makeup of the United States over the course of the nineteenth century. Each group brought their own language, cultural traditions, food preferences, family structures, and religious beliefs with them, and often worked, lived, and socialized with other immigrants from their place of birth. These groups seemed increasingly strange to many Anglo-Saxon Americans, whose families had immigrated generations ago, fueling racist ideas, ethnic stereotypes, and widespread discrimination against immigrants. The combination of urbanization and immigration led many middle- and upper-class, white Americans to see groups such as the poor, unwed mothers, criminals, or people with illnesses or disabilities not as unlucky individuals worthy of charity but as inferiors who were ruining the country. Along with immigrants, upper-class whites increasingly viewed poor whites and African Americans as inferior groups unworthy of assistance as well (Kline 2001).

At the same time, medical and scientific elites increasingly understood life through a biological lens. Influenced by French scientist Jean-Baptiste Lamarck, who believed that a person could pass down traits they acquired in life, by the 1860s some researchers suggested that heredity, along with environment, shaped the likelihood of a person exhibiting immoral behavior. By the 1880s, Larmarck's ideas went out of favor, replaced by German researcher August Weismann. Unlike Lamarck, Weismann believed that heredity was not influenced by environment but based solely in reproduction (Carson 2011). American physicians and scientists used Weismann's research as evidence that the unwanted traits of degeneracy, which they already believed to be rooted in heredity, caused the "three Ds" of dependency, delinquency, and mental defect. As such, they argued these social problems could not be lessened via environmental interventions like charity (Gallagher 1999). Instead, the most effective intervention would be to decrease the biological chances that such traits would continue to be passed down generation after generation. By the 1890s, surgeons had developed safer, less invasive, sterilization methods that, unlike pre-Civil War castration methods, allowed the patient to stay sexually active and experience pleasure (Carson 2011). These medical advances, combined with anxieties regarding social programs and a new faith in biology, set the stage for American eugenics.

The Rise of Eugenic Ideology

Francis Galton – a British scholar, inventor, and explorer – believed that science could provide a solution to "race suicide," and in 1883 he coined the term "eugenics." In 1904, Galton outlined what became the major principles of eugenics in his essay "Eugenics: Its Definitions, Scopes, and Aims." Published in the *American Journal of Sociology*, Galton aimed to gain support of eugenics among academics, scientists, and other intellectuals. Using zoology as a reference, Galton argued, "all creatures would agree it is better to be healthy than sick" and just like animals, humans also needed to encourage the "best specimens" and "useful classes in the community to contribute *more*" (1904: 2, italics in original). Galton advocated that communities should encourage "thriving families" with fit lineages to reproduce and discourage "unsuitable" couples from reproducing (1904: 4). Galton believed that creating a new "national conscience" of eugenics was more humane than letting genetically inferior groups continue to spread unfit traits: "What nature does blindly, slowly, and ruthlessly, man may do providently, quickly, and kindly" (1904: 5). To Galton, humanity had a duty to advance civilization, and eugenics provided a path.

Even before Galton popularized the concept of eugenics, the United States had a longstanding history of using castration as a punishment, particularly for sex crimes. Thus in 1888, when Dr. Orpheus Everts suggested that castration could have therapeutic benefits, his peers found his ideas logical and ethical. Everts did not support castration as a punishment alone, but instead believed that castrating criminals with a history of sex crimes would ensure that the criminal could not reoffend and that he could not pass his predilection for sexual violence to any potential offspring. In the 1890s, more physicians began to support Evert's view of castration as a more humane alternative to execution or long-term imprisonment for sex criminals. Like many elite groups, physicians also believed that degeneracy was on the rise, including sex crimes, and looked to science and medicine for a solution (Largent 2008).

In 1901, Theodore Roosevelt popularized the new term "race suicide," a belief among some white, middle- and upper-class Americans that groups with predilections toward degeneracy, particularly immigrants, African Americans and poor whites, were having more children than "fit" parents like themselves. By the early 1900s, Americans had smaller families, typically only a few children compared to the larger farming families of the 1700s. As women advocated for more legal rights, educational opportunities, and access to the professions, many women postponed having children. This led to anxiety among a number of elite groups, who worried that families of "fit" Anglo-American stock would eventually die out. Roosevelt was one of those elites who viewed a white woman who had fewer children as a traitor to her race (Kline 2001).

Harry Clay Sharp, a young physician from Indiana, read new research on castration and eugenics with great enthusiasm. Upon graduating medical school, Sharp served as a physician at the Jefferson State Reformatory in Indiana. Struck by the foul living conditions and overcrowding, Sharp wondered if a vasectomy could provide a more efficient method to combat crime and degeneracy compared to long-term institutionalization. In 1899, Sharp sterilized a young man who was imprisoned for "excessive masturbation," which physicians commonly viewed as a sign of degeneracy. After the sterilization, Sharp observed a decline in the patient's rate of masturbation, and believed this proved that the vasectomy was an effective treatment of degeneracy. From 1900 to 1907, Sharp sterilized 500 to 600 inmates and published his results in highly respected medical journals, advocating for the vasectomy as a humane, cost-effective, and scientific approach to curbing "inherited defects" including criminal and sexual behaviors (Largent 2011: 30). In 1907, Indiana lawmakers passed the nation's first compulsory sterilization law largely due to Sharp's unwavering efforts to promote the sterilization of criminals. The law created a legal process in which either county courts or the board of an institution, such as a prison or hospital, could order the sterilization of "confirmed criminals, idiots, imbeciles and rapists" without the individual's consent (Stern 2007: 9). Historians consider this a "benchmark" law because other states used this law as a starting point when crafting their own (Carson 2011). In 1908, the president of the American Medical Association published an important editorial in the history of eugenics, arguing that coercive sterilizations were not simply a good idea, but that physicians had a civic duty to perform them (Largent 2011).

Not surprisingly, historians have studied eugenics in Indiana because in 1907 it became the first state to pass a sterilization law. But the history of eugenics in California has also garnered significant attention because it was the largest state in the country. In 1909, California became the second state to legalize coercive sterilization for eugenics purposes. From 1909 to 1964, when the law was repealed, over 20,000 Californians were sterilized without their consent, accounting for almost one-third of the national total. During this period, legislators repeatedly amended California law to expand the scope of who could be sterilized. The original 1909 law targeted prisoners, particularly sex offenders, and disabled residents of institutions, but legislators expanded the law to include people deemed "mentally deficient," a vague term which included individuals with schizophrenia, depression, epilepsy, addictions, or developmental disabilities such as Down syndrome, as well as those deemed sexually promiscuous or dependent on charity due to poverty. Working with physicians, legislators crafted the regulations so that a person deemed unfit by the state eugenics board had no legal recourse to challenge the order or request a second hearing, and the institution was not required to notify the patient's family of the procedure.

In 1908, psychologist Henry Goddard became the director of research at the Vineland Training School for Feebleminded Girls and Boys in New Jersey, which housed one of the first research centers on developmental disabilities. Like many physicians and scientists, Goddard believed that individuals with unfit traits were largely unable to improve and he worried that such groups reproduced at a greater rate than "fit" families. But Goddard disliked the term "feebleminded" – a catch-all term eugenicists used that lacked a formal definition or set of criteria. Who counted as feebleminded was vague and flexible, dependent entirely on the assessment of an individual physician who typically based their diagnosis on a person's race, income, sexual behavior, or criminal history as well as what contemporarily would be called signs of mental illness or disabilities. Goddard believed that the only way to understand inferior intelligence was to first determine what was intellectually average. Goddard created the concept of "mental normality" by studying the intellectual development of children in various age groups and determining the average intellectual skills for each age range. Using this data, in 1910 he proposed an assessment system in which a researcher matched a feebleminded adult's intellectual abilities to a specific age group of children, thus giving a person a specific "mental age." He then translated this mental age into an intelligence quotient, or IQ, which became the standard intelligence test for decades in the United States. Goddard developed terms for various levels of low intelligence, including moron, imbecile, and idiot, in declining order of ability. Goddard argued that morons, who had the mental age of an 8 to 12 year old, were the most significant threat to civilization. He believed that schools and institutions incorrectly treated this group as normal and assumed they had the ability to improve. Goddard believed that morons were trapped in a savage state of being, mirroring prehistoric humans before they learned morality. Thus, the most efficient solution was not medical care, punishment, rehabilitation, or education, but simply ensuring they could not reproduce via segregation and sterilization (Kline 2001).

In 1912, Goddard published *The Kallikak Family,* his best-known work. Goddard traced the family lineage of "Deborah Kallikak," a pseudonym for a young woman institutionalized at the Vineland Training School. Aiming to illustrate the inheritability of unfit traits, Goddard researched both sides of Kallikak's family back to the Revolutionary War. Goddard argued that one side was fit; family members were upstanding, successful, well educated, and moral. The other side showed classic systems of inferior traits, including criminality, low intelligence, poverty, and immoral behavior. Goddard believed these two genetic lines merged when Deborah's great-great-great grandfather, a solider of superior stock, impregnated a tavern girl with feebleminded traits on his way home from battle. Goddard framed this seemingly forgettable sex act as evidence of the gravity of unchecked reproduction, as it led to the proliferation of unfit traits generation after generation, filling jail cells, poor houses, and institutions. *The Kallikak Family* confirmed its readers' worse fears, that the unfit transmitted their inferior traits through reproduction, but it also offered a simple, scientific solution via sterilization (Kline 2001).

Along with institutions in Indiana, California, and New Jersey, historian Mark Largent credits the research of Charles Davenport, a prominent biologist, as key to the rise of eugenics. Davenport admired Francis Galton's eugenics ideas and aimed to use biological and mathematical research methods to study heredity; Davenport believed that eugenics required a foundation of solid scientific data (Largent 2011). In 1904,

Davenport joined the Cold Harbor Springs Laboratory in Long Island, New York. He engaged in a variety of heredity studies on animals and suggested applications in studying human heredity, linking the work of biologists, animal breeders, pro-eugenics physicians, and social reformers. In 1910, Davenport established the Eugenics Record Office (ERO) at Cold Harbor. The ERO had two main goals: to study human heredity and to educate policymakers and the public on the practical application of their research. ERO researchers and fieldworkers complied family lineages to study the heredity data of a variety of traits and behaviors, including addiction, crime, intelligence, disabilities, personality, miscegenation, disease, and sexual history as well as body measurements. One of Davenport's most well-known publications was his 1911 book *Heredity in Relation to Eugenics*. Used as a college textbook for decades, Davenport argued that heredity was key to understanding most human traits, everything from hair color, to musical ability, to likelihood of disease. Davenport believed that it was necessary for the public to understand heredity, so they could "fall in love intelligently" and make the best possible offspring to improve the human race as a whole (Largent 2011: 58). Davenport's research provided eugenics with an air of scientific credibility.

Putting Eugenic Ideas into Practice

American involvement in World War I marked a new era in the history of American eugenics. Historian Allan Brant argues that World War I was the first war in which professional groups identified sexually transmitted diseases as a threat to national security. When soldiers contracted illnesses such as syphilis from unprotected sex, they needed medical treatment. Physically weakened due to their symptoms, many soldiers became unable to fight. Military officials, social reformers, and physicians, the three professional groups most concerned with this new threat, believed soldiers were patriotic and innocent young men, who could easily be encouraged to engage in immoral conduct by sexually promiscuous women, including both prostitutes and women who socialized or worked in local bars and dance halls. In some towns, law enforcement imprisoned women who contracted syphilis while working as prostitutes and released them upon receiving mandatory medical treatment. The practice of blaming women, particularly working-class women and women who engaged in sexual behavior deemed immoral, for the high rates of sexually transmitted diseases among soldiers seemed logical to medical professions and social reformers. Brant argues that the cultural attitudes regarding sexually transmitted diseases during World War I were that of punishment and sin – physicians, military officials, and reformers alike believed that a woman contracted syphilis simply due to her sexual promiscuity. They viewed high rates of infections as proof of declining morality and social values, not a need for public health interventions (Brandt 1987).

Historian Wendy Kline, who was one of the first scholars to consider how race and gender shaped the eugenics movement, cites World War I as a turning point in American eugenics. The concept of blaming working-class women's sexuality for soldiers' syphilis rates fit well with increasingly popular eugenics ideas. New developments in scientific testing and labeling, especially the new term "moron," provided eugenicists with what they saw as a more scientific approach compared to the catch-all term of feeblemindedness. Academics and scientists agreed that sexual promiscuity was a sign of hereditary inferiority and mental defects, and that such inferior traits were more prevalent among working-class groups. Kline argues that eugenicists learned significant lessons from World War I, which shaped the eugenics movement for the next 30 years. First, eugenicists increasingly viewed women as the problem, believing particularly poor women were sexually unrestrained and this led to high reproduction rates among families deemed unfit. But they also saw women as the solution. They learned from World War I that segregating every unfit woman was impossible – even in areas in which prostitutes were imprisoned, more sexually promiscuous women proliferated soldiers' dance halls and bars. There were simply too many of them to institutionalize during their reproductive years. Instead, sterilization provided a simple and scientific solution (Kline 2001).

California is significant in the history of American eugenics not only for its early sterilization laws, but also the application of this new focus on sterilization after World War I. Kline identifies the Sonoma State Home, which was originally founded in 1884 to assist developmentally disabled children, as becoming a leading center of sterilizations under superintendent F.O. Butler in 1918. Butler transitioned his approach from segregation to sterilization, believing that medical professionals had a responsibility to protect the public from women who spread their inferior traits via reproduction. Butler believed that sexually promiscuous behavior was a symptom of a woman's deeper mental deficiency, and he categorized these patients as "female high-grade morons," expanding Goddard's first use of the term. Similar to Goddard, Butler believed that sterilization was simple, cost-effective, and the most efficient treatment for this new group of patients. This belief fueled a notably aggressive sterilization policy at the Sonoma State Home. Butler showcased sterilized patients as success stories, arguing that the procedures decreased their rates of immoral conduct. Only a few years after World War I, the Sonoma State Home was what scholars now call a "feeder institution," meaning that staff admitted patients to the institution for the sole purpose of sterilization, releasing them soon after. Butler promoted his practices so well that institutions across the country looked to Sonoma as a model, and Butler assisted state governments in developing eugenics laws and sterilization policies (Kline 2001).

Starting in the 1910s and 1920s, eugenics ideas expanded beyond circles of elites. This was in part due to specific campaigns by eugenicists, such as researchers at Cold Spring Harbor, who wanted to gain public support for eugenics. It was also due to middle-class professionals and reformers, especially women, who began to see the benefits of eugenics in the issues that mattered to them. It may seem surprising that some women supported eugenics given that eugenicists increasingly viewed women as the root of the problem. While there were some women, particularly those subjected to sterilization, who we can assume opposed eugenics ideas and practices, many middle- and upper-class white women wholeheartedly supported eugenics. Some women supported eugenics because they believed eugenics ideas recognized the importance of reproduction and women's health to the entire population, instead of viewing it as a "women's issue" with no relevance to men. Many women also shared ideas held by pro-eugenics men, who were also concerned about what they believed to be rising rates of inferior traits while "fit" families had fewer children. Numerous women also supported eugenics because it encouraged more control over reproduction, and some supported only voluntary sterilizations while others advocated for coercive sterilization (Hasian 1996). Margaret Sanger, a leading birth control advocate who founded Planned

Parenthood, was one of many women reformers who supported eugenics in part of larger efforts to control reproduction. Sanger advocated for the use of a variety of birth control forms, and encouraged women to choose sterilization to limit their families. Many women in fact wrote to Sanger, asking for information on sterilization. Yet, Sanger also supported coercive sterilization, believing that medical professionals and the state had a duty to ensure unfit families, particularly poor immigrants, limit their reproduction (Kline 2001). In the early twentieth century, women fought their way into more professions, and in some cases, such as social work and child health, formed new ones. After World War I, many state offices grew and became bureaucratized, offering new positions to professionals interested in studying and curbing social problems, such as isolation and illnesses among poor families.

The rise of eugenics in Vermont illustrates the central role of women professionals in this time period. Henry Perkins, a professor from a well-established Vermont family, pioneered the eugenics movement in the state (Gallagher 1999). Perkins had assumed that the low rates of immigration to the state had shielded it from social problems compared to urban areas. Perkins was shocked to discover that despite this heritage, in World War I Vermont had the second highest amount of men who draft boards deemed too weak to fight due to "defects." While some Vermonters believed these defects were the result of poverty and childhood diseases, Perkins argued that rural poverty, migration to cities, and the influx of French Canadians (who he believed to be inferior stock) fueled increasing rates of unfit traits in the state. Perkins aimed to merge eugenics ideas with the work of social reformers to study and combat the rise of these traits. Vermont was typical of many states in that during the 1920s women entered social work, child development, and social science research. They established new organizations, state offices, and research centers to study social problems of families they deemed unwholesome. These professionals engaged in a variety of research projects and fieldwork to collect data on thousands of Vermonters. One of the leaders of these projects was Harriet E. Abbott, a Vassar graduate who went on to study social work in Chicago, a center for the profession, and eugenics at the Eugenics Record Office in Cold Spring Harbor. She shaped the projects into social science research, collecting and studying all of the field data on specific traits and calculating the cost of such traits to the taxpayer. Perkins used the research of Abbott and other professional women from organizations such as the Vermont Children's Aid Society and the State Social Service Exchange to support campaigns for sterilization laws (Gallagher 1999).

As professional groups advocated for sterilization laws and established research centers, many also established programs to promote eugenics to the public. Eugenics was not a movement with grassroots origins, but starting in the 1910s and 1920s, many believed that promoting eugenics to ordinary, non-elite families through popular culture could only help their efforts. Some programs aimed to promote positive eugenics and reward "fit" families. Better baby contests at state fairs were one of the first eugenics programs in popular culture. Mary DeGormo, a teacher, was one of the first organizers of a better baby contest, which took place at the Louisiana State Fair in 1908. DeGormo worked with a local pediatrician to reformat animal breeders' score sheets to assess each baby's physical and mental fitness against an ideal score.

Within a few years, better baby contests became popular events at state fairs across the country. The winning baby was typically featured in the town's newspaper and the family earned prizes. Women reformers and professionals, including social workers and state officials working in emerging maternal health programs, saw better baby contests as one of the most successful ways to educate poor women who were usually out of their reach. They taught fairgoers basic child-rearing information, such as how to feed infants, as well as broader eugenics ideas of "race betterment" by rewarding the baby with the traits deemed most fit. Not surprisingly, contest organizers did not allow African American babies to enter (Stern 2002). Founders of better baby contests in Iowa established the fitter family competitions in Kansas, in which physicians and community leaders would assess the "fitness" of an entire family. They would score each family based on the presence of unfit traits such as mental illnesses, disabilities, and addictions and fit traits including attractiveness, personality, character, and family size. The winning families received medals and trophies, and soon the event spread to other states (Boudreau 2005). Such contests were only part of broader eugenics programs at fairs. At the 1915 San Francisco Exposition, Race Betterment Week was the most popular event at the fair, warning attendees of the "moron girl" and celebrating the fit "mother of tomorrow" (Kline 2001).

Fairs were not the only public forum for the promotion of eugenics ideas. Women's organizations and clubs throughout the country undertook eugenics educational programs. The Woman's Christian Temperance Union, one of the most popular women's organizations in the country that promoted temperance, women's suffrage, and many social issues, developed and conducted classes on eugenics ideals to poor women. Other eugenicists worked with national organizations such as the Boy Scouts and National Education Association to develop lesson plans on pedigree charts and Bible stories that promoted eugenics ideals (Hasian 1996). Historian Christine Rosen argues that Protestant religious officials, both in leadership positions and directors of small churches, were largely in support of eugenics ideas, believing that scientific developments like eugenics provided new ways to achieve godly ideals of perfection. As such, it was common for Sunday school lessons to promote religious justification for eugenics (Rosen 2004). Popular magazines, particularly women's magazines, often published columns dedicated to promoting eugenics ideas. One could read about better baby contests and concerns over rising rates of degeneracy not only in professional medical journals, but also in magazines like *Good Housekeeping* (Hasian 1996).

Legal and Legislative Victories

By the 1920s, the American eugenics movement was flourishing. Within a decade after the passage of the first sterilization law in Indiana, Alabama, Connecticut, Kansas, Michigan, Nebraska, New Hampshire, New York, North Dakota, Oregon, South Dakota, and Wisconsin all legalized coercive sterilization of individuals deemed unfit, modeling their laws after Indiana and California. In the 1920s, Delaware, Iowa, Maine, Minnesota, Mississippi, North Carolina, Utah, Virginia, West Virginia, and Washington also passed sterilization laws. As these laws went into effect, scientists, physicians, and researchers studied sterilization in depth, developing protocols, publishing research studies on sterilized patients, and establishing institutions for procedures. It is important to note that passage of eugenics laws meant that physicians recorded sterilizations as part of their case notes, which have been a key source for historians. Yet, it would be incorrect to assume that documented cases of sterilizations were the only ones that occurred. There is evidence that sterilizations occurred in states without formal laws or before such laws were passed. For example, Colorado never passed a sterilization law, yet journalist Harry Bruinius uncovered evidence that superintendents of Colorado institutions approved sterilizations of patients as a "routine operation" in this period (Bruinius 2006: 328). Researchers at the University of Vermont also found records of 270 sterilizations in Pennsylvania, another state that never legalized sterilization despite support from legislators and respected physicians there. Similarly, in many states, eugenicists were unsuccessful in their efforts to pass sterilization laws, but these states still had thriving state-funded eugenics programs supported by community leaders and professional groups. For example, Florida was one of many states which never legalized sterilization but still established isolated institutions called "colonies" which segregated women deemed unfit so they could not reproduce (Noll 1995).

Historians cite two specific events in the 1920s that solidified this rising popularity and power of the eugenics movement in the United States. The first was the passage of the Immigration Act of 1924, also known as the Johnson–Reed Act. This federal law significantly altered immigration to the United States because it created new quotas, or limits, on who could immigrate to the United States. The law put significant restrictions on immigrants from Southern and Eastern Europe and prohibited immigration from the Middle East and East Asia. While the law organized immigrants based on their national origin, it also specifically limited the number of Jews allowed to immigrate regardless of their citizenship. Eugenicists, including those from California institutions and the Cold Spring Harbor Laboratory, helped craft the law and they served as key experts during Congressional debates, using their research and clinical experience to justify the quotas. Central to the law was the concept of "old stock" versus "new stock." The law encouraged immigration of whites from Northern Europe, because eugenicists believed these groups had superior heredity that would maintain or even improve the fitness of the United States as a whole. The law drastically limited or prohibited immigration from regions where eugenicists believed defects and degeneracy flourished, including European countries such as Italy, Greece, Poland, and Czechoslovakia as well as China and India. As such, immigration from these eras plummeted after 1924 (Ngai 2004).

The process of immigration also became more medicalized and formalized after 1924. Upon arriving at an American port, immigrants underwent invasive medical exams to ensure they lacked signs of defects or degeneracy. Historian Amy Fairchild, who studies medical inspection of immigrants, argues that medical staff used these exams to assess if the person was physically and mentally healthy enough to work. This assessment was rooted in eugenics fears that immigrants who entered the country were often poverty-stricken, unhealthy, immoral, and thus depended on charity, becoming a burden to taxpayers. Staff deported immigrants with medical problems or any signs of degeneracy including disabilities, body abnormalities, low intelligence, and believing that these individuals would be unable to maintain employment (Fairchild 2003). Medical staff also deported immigrants who showed signs of sexual immorality, inappropriate gender expression, or evidence that they engaged in same-sex relationships, which they also viewed as a symptom of degeneracy (Canaday 2011). The 1924 law permanently altered the makeup of the United States, and it did so via eugenics. The law elevated eugenics to a new level of prominence, framing eugenicists as innovative researchers who could transform ideas into specific laws and practices that would safeguard the United States against the supposed threats of immigration.

The second major event that solidified the prominence and power of the eugenics movement was the 1927 Supreme Court case *Buck v. Bell*. In the 1920s, Harry Laughin, director of the Cold Spring Harbor Laboratory who replaced Charles Davenport, led efforts to design sterilization laws that would survive court challenges. In 1924, Virginia passed a sterilization law under Laughlin's direction. That year, Dr. Albert Priddy, the superintendent of the Virginia State Colony for Epileptics and Feebleminded, issued a request to sterilize an 18-year-old patient named Carrie Buck, believing that her history warranted sterilization and to test the law in court. Priddy believed that Carrie Buck's family had a long history of sexual immorality and low intelligence, and assessed Carrie as having a mental age of nine. Buck was adopted and attended school, but teachers and her adoptive family claimed that she exhibited uncontrollable behavior and she eventually became pregnant as a teenager. Her adoptive parents committed her to the Virginia State Colony in 1923 after she gave birth to her child. A fieldworker from the Eugenics Record Office reported that Buck's daughter, Vivian, showed signs of "backwardness" as well (Lombardo 2010).

Representatives for Buck fought the sterilization order in local and state courts, and eventually leading to the Supreme Court. Buck's attorney argued that sterilization against her consent violated her right to due process and the Equal Protection Clause in the fourteenth amendment. In an eight to one decision, the Supreme Court sided with the institution and ruled the Virginia sterilization law as constitutional. Justice Oliver Wendell Holmes, Jr., a respected intellectual and legal scholar, wrote the widely cited ruling. He justified the decision due to experts' statements that Buck's biological mother had a history of sexual promiscuity and feeblemindedness. Holmes argued that it was in the state's interest to limit reproduction of those with inferior traits because it would ultimately decrease the amount of crime, poverty, and disabilities. Holmes referred to Buck's sterilization as a "lesser sacrifice" to ensure the health of the country as a whole, and famously concluded "three generations of imbeciles are enough" (Largent 2011: 101, 103). Buck returned to the institution and was sterilized.

Paul Lombardo, one of the leading historians of eugenics, has conducted extensive research on *Buck v. Bell* and interviewed Buck before her death in 1983. Lombardo argues that the evidence that justified Buck's sterilization was largely unfounded. Lombardo discovered that Buck became pregnant after her adoptive mother's nephew raped her, and that her adoptive family aimed to institutionalize her in part to hide the shame of the incident. Buck, who worked as a domestic servant after her release from the Virginia State Colony, was an avid reader and showed little evidence of low intelligence. Lombardo places much of the blame on Buck's attorney, who made significant mistakes during the trial, such as failing to call key witnesses for the defense, and openly supporting eugenics ideals. Lombardo also uncovered Buck's daughter's school records, which placed her on the honor roll (Lombardo 2010).

Buck v. Bell was a major victory for eugenicists. States expanded their sterilization laws, using the Virginia law as a new model. Institutions and state health departments formalized sterilization procedures. Sterilizations drastically increased throughout the country, as physicians, institutional boards, and lawmakers were confident that they were immune to legal action and were doing their duty to the country (Lombardo 2010). In 1928, eugenicists E.S. Gosney and Paul Popenoe established the Human

Betterment Foundation in California. Researchers at the foundation conducted massive studies of heredity to advance eugenics ideals and expand sterilization laws, utilizing their access to patients' records at the Sonoma State Home and other large institutions in the state. Their results were used as the foundation for eugenics programs, lessons plans in schools, medical textbooks, and legislative campaigns across the United States, furthering California's position as a leader in eugenics ideology and practice (Kline 2001).

The Great Depression provided more fuel for the eugenics movement. It spurred new anxieties about the stability of the family, as Americans witnessed seemingly financially stable families lose their jobs and homes. Eugenicists increasingly viewed women as the crux of a family's success – they blamed women of unfit families for continuing to reproduce, and celebrated women of fit families for fulfilling their duty to reproduce children of superior stock. Americans increasingly believed that a stable family was central to the survival of the country as a whole, as institutions such as banks seemed less reliable (Kline 2001). During the Great Depression, eugenicists increasingly linked their work to the emerging public health movement, which blossomed during this period. Eugenics programs expanded with new funds from relief programs designed to curb the effects of the economic crisis. Minnesota was one of many states which legalized coercive sterilization in the 1920s, but the sterilization rates dramatically increased in the 1930s because New Deal funding provided more medical staff to issue and perform sterilizations (Ladd-Taylor 2011). Similarly, Lauren Briggs, a historian who studies Puerto Rico, argues that sterilization programs designed to limit poor Puerto Rican women's reproduction also flourished in the 1930s with funds from New Deal programs (Briggs 2002).

In the 1930s, eugenicists aimed to broaden their educational efforts to the public and not limit themselves to sterilization in state institutions, which seemed like a battle they had won. Similar to the fairs and education programs of the 1910s and 1920s, eugenicists used popular culture to foster cultural attitudes that supported eugenics. In the 1930s, eugenics-themed movies, plays, and books provided lessons to viewers that normalized eugenics ideas. Erskine Caldwell's 1932 novel *Tobacco Road*, which director John Ford transformed into a movie in 1941, was one of the most popular eugenics-themed stories. Caldwell's father was a minister who conducted research of poor, Southern farmers and published his work in leading eugenics journals. *Tobacco Road* tells a fictionalized account of a family who exemplifies all of the traits of the unfit; they are poor, uneducated, lazy, unhealthy, sexually promiscuous, and unable to cope with the modernization of farm life. Authors like Caldwell casually promoted eugenics ideas by framing such families as worthy of pity, but stuck in a vicious cycle they were unable or unwilling to escape (Nies 2006).

In 1936, the court battles of Ann Cooper Hewitt furthered this focus on the family. Hewitt was a 21-year-old wealthy heiress living in California. She was sterilized without her consent at the request of her mother, who believed Hewitt was sexually promiscuous. Hewitt sued her mother and the two physicians who performed the sterilization, which occurred in a private practice, not a state institution. California law did not specifically include private practices in sterilization laws, and it was unclear if the law would protect Hewitt's sterilization. The judge dismissed the case, speculating that California's sterilization law would not require the consent of a private patient if the patient's guardian consented. Hewitt's mother was her legal guardian, and thus she had no recourse. Given Hewitt's status, the ruling made headlines across the country. It also shaped popular attitudes regarding sterilization. The judged implied a sense of trust that informed family members should acknowledge their duty to sterilize a family member for the greater good of the country. It also expanded the notion of who counted as unfit and unworthy of having children. It ensured that women like Hewitt, who were apparently sexually promiscuous but showed no signs of mental illness, criminality, disability, or low intelligence, should still be considered for sterilization (Kline 2001).

Cracks in the System

In many states, the 1930s was the decade with the highest rates of sterilizations. Yet, slowly a few cracks emerged, which began to challenge the eugenics movement. The first was the rise of Nazism in Germany. Some American and German physicians were linked in professional circles, attending eugenics conferences and reading each other's research. Yet very few American eugenicists supported Nazism outright, and most eugenicists did not see similarities with their work and Nazism. Starting in 1936, a few eugenicists, including Gosney and Popenoe of the Human Betterment Foundation, became concerned that their practices and ideas would be associated with Nazi medical practices and government policies even though they themselves viewed them as unrelated. Gosney and Popenoe shifted the language of eugenics, to further distance their work from Nazism. They reiterated that they believed sterilization should be used very selectively, not on large demographic groups of the population. They specifically avoided using any racial terms, arguing that eugenics targeted individual abnormality and not racial or ethnic groups. Eugenicists used Gosney and Popenoe's shifts to rebrand their movement in the 1940s and 1950s. Eugenicists increasingly focused on positive eugenics, developing programs to encourage fit families to breed. Pro-eugenics physicians viewed themselves as a type of marriage counselor, who assessed the well-being of marriages and offered an assessment regarding optimal family size (Kline 2001). Eugenicists focused less on promoting sterilization efforts to the public, but this does not mean sterilization rates declined. In fact, in many states, particularly in the South which passed sterilization laws later than other regions, sterilization rates continued to rise after World War II. However, eugenicists were less able to frame work as prevention of "race suicide" as they had in the decades before World War II. In the 1942 case *Skinner v. Oklahoma*, the Supreme Court ruled that coercive sterilization could no longer be imposed as a punishment for crime. The number of convicted criminals sterilized was very small compared to other groups, and the ruling had no bearing on *Buck v. Bell*. In fact, roughly one-third of all coercive sterilizations performed in the United States, over 20,000, occurred after 1942. Yet, the ruling was significant because eugenicists realized they could no longer assume the Supreme Court would support their policies (Lantzer 2011).

By the 1950s, eugenicists became less able to distance themselves from Nazism, even though they viewed American eugenics as unrelated. In 1956, the *Georgetown Law Review* published one of the first articles that outlined similarities between American and Nazi eugenicists. American Catholics had consistently opposed coercive sterilization beginning in the early 1930s, viewing it as a form of birth control, although they did support segregation and limited marriage rights for individuals deemed unfit (Hasian

1996). Published out of Georgetown University, an elite Catholic university, the article secularized Catholics' anti-eugenics stances. Historian Mark Largent argues that this article was a turning point in the history of American eugenics, linking it with the atrocities of Nazism in a scholarly journal (Largent 2011). While historians agree that the American eugenics movement and Nazi use of eugenics were distinct movements, this connection only slowly gained attention in the 1950s as Americans became more aware of the specific atrocities of the Holocaust.

The 1960s and 1970s were the final decades of a popular and power eugenics movement in the United States. The Gay Rights and Women's Liberation Movements provided sterilized patients a framework to fight eugenics, as these activists challenged professional and popular assumptions that oppressed groups were unworthy of reproductive choice. Yet thousands of Americans were sterilized without their consent in these decades; while eugenics ideals declined, eugenics practices like coercive sterilization continued. Similarly, eugenicists used covert tactics to provide documentation that woman consented to the sterilization, and increasingly framed sterilization as voluntary. The sterilizations of Elaine Riddick and the Relf sisters provide a striking example of "consensual" sterilization in this period. In 1967, Elaine Riddick, a 14-year-old African American girl from North Carolina, was sexually assaulted and became pregnant. Riddick lived with her grandmother, who had limited reading abilities. Upon delivering her son, Troy, in 1968, a social worker who worked on behalf of the state eugenics board tricked Riddick's grandmother into signing a consent form authorizing Riddick's sterilization. The social worker only told Riddick's grandmother that signing the form ensured the state would not commit her granddaughter to an orphanage – she did not mention anything regarding sterilization. In 1973, the Southern Poverty Law Center exposed the case of Mary Alice and Minnie Lee Relf, 12- and 14-year-old African American sisters from Alabama whose family relied in part on welfare assistance. Nurses from a local public health agency funded by federal grants requested that the sisters' parents allow them to administer Depo-Provera shots, which at the time was only in experimental stages and linked to high cancer rates in lab tests. Their parents also lacked the reading and writing skills to fully understand the contract. Regardless, the nurses sterilized the Relf sisters without their family's full understanding of the procedure. Particularly in the South, it was also common for hospital staff to conduct what they called "Mississippi appendectomies" in which physicians sterilized women during unrelated operations, particularly appendix removals, without their consent (Kluchin 2009: 110).

Historians found evidence of similar tactics used against Native American and Latino women. Scholar Jane Lawrence uncovered that a quarter of all known sterilizations performed on Native American women occurred in the 1960s and 1970s, usually under the direction of the Indian Health Service (IHS) (Lawrence 2000). IHS staff sterilized young girls without their family's consent and failed to tell women key information about the sterilization. In 1977, 10 Latino women sued county hospitals in Los Angeles because they were coerced into signing "voluntary" sterilization consent forms during labor and staff only provided forms in English (Briggs 2002). Many of the first court cases that challenged coercive sterilization were unsuccessful. In 1970, hospital staff provided Norma Jean Serena, a Native American woman, a consent form to sign *after* her sterilization. Ultimately, the US District Court sided with the physician who claimed he discussed the procedure with Serena ahead of time (Kluchin 2009).

Regardless of the legal outcomes, these cases exposed the eugenics movement as the exact opposite of its ideals – it was not progressive, humane, and scientific but deceptive, poorly regulated, and lacked sound scientific justification. Professional groups slowly distanced themselves from eugenicists. Many women formed anti-sterilization organizations in which they publicly discussed their own sterilizations and protested the practice. The American Medical Association quietly reversed its pro-sterilization stance, and biologists discredited the research of Davenport and other eugenics researchers. This anti-eugenics sentiment gained more traction upon the public realization in 1972 of the atrocities associated with the Tuskegee Syphilis Experiment, a research project funded and directed by the US Public Health Service starting in 1932. In this experiment, physicians and researchers from the Public Health Service told African American men with syphilis they were receiving treatment, when in fact they received no health treatment, so researchers could study the disease untreated. Participants never received any treatment despite their belief otherwise. The Tuskegee Experiment led to new ethical guidelines in health care and medical research, including informed consent and accurate reporting. While not a sterilization experiment, Tuskegee shaped the declining eugenics movement by providing new checks on physicians and researchers. By the late 1970s, sterilization rates declined and most states had removed sterilization laws by the early 1980s.

Bibliographical Essay

After eugenics fell out of professional and popular favor in the 1970s, two significant trends have occurred as contemporary Americans grapple with the legacy of the eugenics movement. First, in the 1990s, historians began to study the American eugenics movement in depth. Before the 1990s, most historians ignored the American eugenics movement or viewed it as an outlining event lacking historical significance. Starting in the 1970s, scholars and activists established new fields of study, including women's studies, Holocaust studies, African American history, and social history, to incorporate the lives of oppressed groups in historical narratives. Historians of eugenics used these fields as frameworks to establish a new approach to the history of eugenics that forced readers to recognize the popularity and power of the movement. Many leading scholars used their research on eugenics as a way to hold their own schools and state government accountable for their participation in the eugenics movement. Paul Lombardo, a leading historian of eugenics, edited the essay collection *A Century of Eugenics*, which was published by the University of Indiana Press, as part of the Indiana Eugenics Legacy Project (2011a). One of the most useful resources on eugenics is a digital project created by students and faculty at the University of Vermont, which housed eugenics research in the early 1900s. "Eugenics: Compulsory Sterilization in 50 American States" provides in-depth data on the eugenics programs in every state, including sterilization rates, leading eugenicists, institutions, laws, and key primary source documents, all of which are open access and widely accessible online.

Along with a new dedication to studying eugenics, survivors of sterilization have increasingly told their stories to reporters, government officials, and academics. Many states have responded by publicly acknowledging and discrediting their eugenics history. Several governors, including in California and Virginia, offered formal apologies to survivors (Ingram 2003). The Indiana State Library displays a historical marker that outlines Indiana's leading role in the eugenics movement (Lombardo 2011a). North Carolina, which had one of the most aggressive sterilization programs, has been the only state to begin a compensation program for sterilization survivors, which as of August, 2014 is still processing claims and has yet to award any funds (Burns 2014). Historians and survivors alike ultimately hope that such efforts will educate the public on the history of eugenics in the United States and provide a framework to think critically about the ethical implications of science and medicine today.

Chapter Twenty-Eight

EVOLUTION AND CREATION DEBATES

Arthur Ward

The history of the debate between creationism and evolution in the United States is, in part, a story of controlling a cultural narrative. Is it primarily a struggle between science and religion? Between academic or religious liberty and governmental overreach? Or, evidence versus faith? Believers in evolution are more likely to see the debate as a fight for scientific enlightenment against the forces of ignorance, while self-identifying creationists are more likely to see the debate as a fight for moral righteousness against the forces of moral nihilism. Each side has found it important to guide the public narrative of the debate, starting with the unfair presumption that there are only two sides to begin with, when there are of course a great many positions in between the two poles. The debate in the United States has found philosophical, theological, and scientific inquiry tightly married to political and cultural platforms.

The Debate Crosses the Atlantic

Darwin's *On the Origin of Species* was a detailed, careful work of science that sought to support a theory of evolution with two central commitments: the common ancestry of organisms and the mechanism of natural selection as the primary (though not sole) cause of evolution. But Darwin himself was not a showman or debater. Following the publication of the *Origin* in England, it was not Darwin but other more outspoken allies who took up the fight against creationism. Though these allies did a great deal to successfully spread the popularity of evolution, they were often less careful when it came to the science. In these early days, Darwin's followers took mainly to spreading the thesis of common ancestry and largely avoided defending or advocating for the theory of natural selection, often deliberately undermining natural selection by propounding Lamarkianism or divine providence in order to explain common descent (Ruse 2000). Most colorful and powerful among Darwin's defenders was Thomas Huxley, a proud

A Companion to the History of American Science, First Edition.
Edited by Georgina M. Montgomery and Mark A. Largent.

"agnostic," a term he coined. Huxley was a fierce debater, naming himself "Darwin's bulldog," and delivered many charismatic lectures and debates, most famous among them against the bishop Samuel Wilberforce at Oxford in 1860. Infamous as the debate became afterward, however, we know very little of what was said (no transcript exists) and the most famous exchange (Wilberforce: "Are you descended from apes on your grandfather's or your grandmother's side?" Huxley: "I would rather be descended from an ape than from a bishop of the Church of England!") almost certainly didn't occur in those words or in such a dramatic fashion (though, a woman was said to have fainted afterwards) (Ruse 2000, 2005). Yet the event earned a reputation as a major clash, and it set the stage for future public spectacles in England as well as in the United States.

Early American exposure to evolution and its conflict with creationism came in the form of a series of debates between two Harvard colleagues: famed biologist and geologist Louis Agassiz and the lesser known botanist Asa Gray. Agassiz carried significant clout as a groundbreaking scientist (he discovered the existence of ice ages in Europe), and he strongly opposed Darwin's theory of evolution on the grounds that it conflicted with his own preferred essentialist understanding of the natural world. Gray, on the other hand, was a recent but avid supporter of Darwin's – so deep in Darwin's inner circle that he had seen and made comments on a draft of the *Origin* before its publication (Dobbs 2005). Unlike the showman Huxley in Britain, Gray took a quieter approach, having always felt more comfortable researching than lecturing. Agassiz was by far the better orator of the two. In 1859, months before the *Origin* was published, but after Darwin had started giving lectures on the contents of the book, Agassiz and Gray had several debates over the course of the year. Gray came away the victor every time (Dobbs 2005).

Agassiz's reputation as a talented scientist was well deserved, and he approached Darwin's theory with respect and care. From within a scientific framework, he found it lacking and less capable than belief in a creator when seeking to explain a number of phenomena. In his book *The Structure of Animal Life* (1866), for example, he writes detailed descriptions of homologous traits across species:

> Facts like this, I think, show the immediate working of mind in the construction of the animal kingdom. It is not a kind of work which is delegated to secondary agencies; it is not like that which is delegated to a law working its way uniformly; but is that kind of work which the engineer retains when he superintends and controls his machine while it is working. It is evidence of the existence of a Creator, constantly and thoughtfully working among the complicated structures that He has made. (Agassiz 1866: 131)

It was understandable that Agassiz would approach Darwinism with skepticism. For one thing, without knowing what we know now about genes and mutation, the causal mechanism of natural selection appeared to be rather weak. Moreover, it appeared that natural selection was too slow a process to align with the best estimates of the earth's age at the time (Ruse 2005: 87). The earth was too young, most thought, for natural selection alone to account for the wide array of speciation present. We now know the earth is a great deal older than nineteenth-century science believed, and Darwin has been vindicated yet again. Nevertheless, at the time, selectionists did not have the scientific tools and methods to persuade others that their view was accurate.

Three facts about Gray made the debates with Agassiz noteworthy. The first is that Gray was a scientist of lesser stature than Agassiz, both professionally and socially, and so the confrontation had a David and Goliath flavor for many. The second is that Gray was actually a very devout Christian, a great deal more so than Agassiz himself, and so he could not be harnessed with the anti-religious stance of Huxley (Dobbs 2005). Thirdly, Gray's theism led him to see evolution as consistent with God as a creator, a view that is still prevalent today, and will be discussed at length later in the chapter. In this passage, Gray draws an analogy to billiard balls that are struck, noting that the determined and predictable pattern of deflecting balls do not contradict the fact that they were struck by someone with a plan.

> And if they prove it on the supposition that the unseen operator acted *immediately* – i.e., that the player directly impelled the balls in the directions we see them moving, I insist that this proof is not impaired by our ascertaining that he acted *mediately* – i.e., that the present state or form of the plants or animals, like the present position of the billiard-balls, resulted from the collision of the individuals with one another, or with the surroundings. The original impulse, which we once supposed was in the line of the observed movement, only proves to have been in a different direction; but the series of movements took place with a series of results, each and all of them none the less determined, none the less designed. (Gray 1860)

This passage looks to reflect Deism, the view that God set natural processes in motion and does not continuously interfere with them. In fact, Gray believed in divine provi dence, akin to something that is now called theistic evolution, a view where God acts continuously in the world through natural processes, for instance controlling genetic drift upon which natural selection can act. Gray's embrace of divine providence as a patch for evolution upset Darwin (Ruse 2000: 95). However, it reflected a common sentiment among early supporters of evolution, which itself can be partly explained by widespread weaknesses in the science of natural selection at the time.

The early days of the Gray/Agassiz confrontation was a model of civil disagreement. Each man respected the other, and they were not just colleagues but friends and neigh-bors. They debated Darwin's ideas on a level playing field of scientific rigor in a manner now entirely lost. But as Gray began to get the better of Agassiz, their quarrel became meaner. The purity of their scientific disagreement inevitably became affected by power struggles within the Harvard academic community, and Gray became offended by Agas-siz's arguments for the natural inferiority of some races, which could not help but be swept up in the drama of American slavery and the impending civil war (Dobbs 2005).

The Debate Becomes a Culture War

As Gray won the debates with Agassiz, and belief in Darwinian evolution took hold in Europe thanks to Huxley and others, Darwin's theory of natural selection still lacked the scientific rigor it needed to gain traction in academic science departments (Ruse 2005: 96). But natural selection is a causal theory, and such details as causes were not needed to gain converts in the public arena. Huxley paid almost no attention to natural selec-tion, and Gray boosted it generously with divine providence. Darwin's theory would eventually become grounded in science once again in the twentieth century, but for a time in the late nineteenth and early twentieth centuries, it was more of a pop-culture phenomenon. A coincidental convergence of public education and public health needs due to mass immigration to the United States and industrialists looking to build legacies resulted in dozens of fine museums being built in major cities. These were places to showcase magnificent dinosaur skeletons that were being unearthed in the American West, all the while exposing the public to a watered-down version of Darwin's theory of evolution (Ruse 2005: 101).

There was another side-effect of the popularization of watered-down Darwinism, and this was the unintentional encouragement of latent racist and sexist sentiment among some in academia. Phrenology, the practice of predicting one's intellectual potential by measuring portions of one's skull, was already popular, despite the lack of scientific rigor. It is easy to see how Darwin's theory of the "survival of the fittest" itself fit into the broader narrative of anyone looking to justify the natural superiority of white males, given their relative "success" compared to other groups such as women and non-whites. Thus was born the movement only later known as "Social Darwinism" by its detractors. A pseudo-science built upon a thin understanding of Darwin's own work and a heavy dose of racism and sexism, Social Darwinism was stoked by Herbert Spencer, an early supporter of Darwin's, as well as Darwin's own cousin Francis Galton. It resulted in lazy, *post hoc* evolutionary justifications of entrenched social hierarchies with no interest in alternative explanations. In short, it turned the descriptive phrase "survival of the fittest" (itself a coinage of Spencer's rather than Darwin's) into a normative prescription. It was, sadly, quite influential even during the start of the twentieth century, where it was seen by some as lending bloodlust to World War I. Thus it is important to say that much of the opposition to evolution at the dawn of the twentieth century, opposition that would be made infamous in the Scopes "monkey trial" to come, stemmed from those who rightly felt they had the moral high ground. Darwin's theory had become the refuge, at least to a degree, of racist and sexist scoundrels, and Christians especially in the United States saw the entire enterprise as the natural result of wandering from biblical teachings (Ruse 2005).

Two Christian cultural movements emerged in the early twentieth century as a response to perceived rampant industrialization, poverty, war, and immorality: Fundamentalism and the Social Gospel. Christian "Fundamentalism" in the United States owes its name to a series of books published by the Bible Institute of Los Angeles between 1910 and 1915 that urged Americans to return to a biblical morality and look at the Bible as the literal revealed word of God (Marsden 2006). The Social Gospel was more of a left-wing movement that concerned itself with the plight of the poor and downtrodden who were suffering in the new era of industrialism (Marsden 2006). A figure who was pivotal in both movements was William Jennings Bryan, who would become a hero in the struggle against evolutionism in the infamous Scopes trial.

Though biblical literalism and creationism became more prevalent during the rise of Fundamentalism, "creationism" itself is not a unitary doctrine: in Christianity alone there are at least three distinct stances that are vigorously debated in seminaries worldwide. Going in order from least to most compatible with scientific understanding, the stances are *Young Earth Creationism*, *Old Earth Creationism*, and *Theistic Evolution*. Young Earth creationists believe that biblical creation happened in six 24-hour days and that the earth is approximately 6000 years old, finding themselves rejecting not just evolutionary biology, but cosmology, geology, and countless other areas of modern science (Ruse 2013a). Old Earth creationists loosen their interpretation of biblical creation, believing that the six "days" of creation are best thought of not as 24-hour days but as vast "periods" of time, a thesis that was scrutinized during the Scopes trial discussed in the next section. Unlike Young Earth creationism, Old Earth creationism is consistent with the geological record, carbon dating, and cosmic microwave background radiation. However, Old Earth creationists still believe in the miraculous creation of humans, and thus reject evolutionary biology. By contrast, theistic evolution is a theological position with little to no conflict at all with science, evolutionary or otherwise. Theistic evolutionists believe in an Old Earth and also that God created humans, but did so by means of evolution. One version of this theory adheres to Deism, the view that God created the initial conditions of the universe but does not intervene. The other option is that God continuously intervenes, guiding genetic drift and mutation, thereby stacking the deck before natural selection takes place (Sober 2008a). Asa Gray was a theistic evolutionist, and today this legacy finds a worthy spokesperson in Francis Collins (2006), a respected scientist who led the Human Genome Project and the US National Institutes of Health. Though not even the most zealous Christian would take every word of the Bible literally, books 1–3 of Genesis are given a privileged status because they describe creation, the fall from Eden, and the promise of a messiah. As such, many Christians are less willing to give them a metaphorical reading, feeling that failure to take these important passages literally would undermine very central aspects of the religion.

The Scopes Trial

The most famous public showcase for the debate between evolution and creationism was the trial of John Scopes in 1925. He was accused of teaching evolution science in Tennessee, a state that had made teaching such material illegal in state-funded schools only a few months earlier with a law called the Butler Act (Young and Largent 2007). The events surrounding the trial were dramatized in the play *Inherit the Wind*, which was made into a successful 1960 movie starring Spencer Tracy and Gene Kelly. The dramatized version told the story of a courageous but naive schoolteacher trapped in a closed-minded town, beset by fundamentalists, prosecuted by a zealous politician, reported on by a cynical journalist, and defended by an aging and quirky white knight. Perhaps surprisingly, almost all of the central features of the fictional rendering just described are accurate, but one big detail of the Scopes conflict was altered for the story. In reality, Scopes did not stumble naively into legal trouble; prompted by strategists in the American Civil Liberties Union (ACLU), Scopes sought out trouble with the intention of creating precisely the spectacle that ensued. And it was such a spectacle, it has influenced the way that Americans have looked at the debate between evolution and creationism ever since.

The Scopes trial was billed as a battle of Goliath versus Goliath. The prosecuting attorney was William Jennings Bryan, a three-time candidate for president well known for his moral stances in favor of women's suffrage, for prohibition of alcohol, pacifism during World War I, and against Darwinism. Though he failed in his presidential bids, he was widely admired by many, earning the nickname "the great commoner" by campaigning on a train that traveled the country so he could meet voters while most candidates rarely left their offices. Ready to defend Scopes against the powerful Bryan was Clarence Darrow, the most famous defense attorney in the country, known for his wit and staunch civil libertarian sensibilities as well as his atheism. He was fresh from defending notorious thrill-murderers Leopold and Loeb, and while they were found guilty, Darrow succeeded against great odds in saving them from the death penalty. There exist detailed transcripts from the Scopes trial, but most of the United States received its trial news from the nation's most famous journalist of the day, H.L. Mencken. What the world saw through his eyes was a backwards but charming town full of earnest and moral, if misguided, common-folk:

> Elsewhere, North or South, the combat would become bitter. Here it retains the lofty qualities of the *duello* ... Is it a mere coincidence that the town clergy have been very carefully kept out of it? There are several Baptist brothers here of such powerful gifts that when they begin belaboring sinners the very rats of the alleys flee to the hills. They preach dreadfully. But they are not heard from today. By some process to me unknown they have been induced to shut up – a far harder business, I venture, than knocking out a lion with a sandbag. (Mencken 1925)

What the Scopes trial revealed was that the days of academic debate between evolution and creationism were long over and the struggle now would play out like a circus act in newspapers, courtrooms, and other forums of public opinion. Each side became primarily obsessed with strategy, and the strategic winner of the Scopes trial, despite the guilty verdict, was the defense led by Darrow. His pivotal tactic, immortalized in the fictional retelling, was calling the prosecuting attorney Bryan on the stand as an expert witness on biblical creation. This move was forced in part by the court barring any expert testimony from scientists as to the legitimacy of evolution. After all, whether evolution was legitimate or not was entirely irrelevant to the narrow question of whether Scopes was guilty of teaching it in a classroom. So Bryan, in a moment of hubris, agreed to be interrogated by the finest interrogator of the time, and it was an unmitigated disaster from a public relations standpoint. Bryan's mistake was to put forth his own biblical views as a better explanation than evolutionary theory, thus allowing the trial to become a referendum on biblical literalism:

DARROW   Does the statement "The morning and the evening were the first day" and "The morning and the evening were the second day" mean anything to you?

BRYAN   I do not think it necessarily means a twenty-four hour day.

DARROW: You do not? .....

BRYAN: My impression is they were periods, but I would not attempt to argue as against anybody who wanted to believe in literal days.

DARROW   Have you any idea of the length of the periods?

BRYAN   No I don't.

DARROW   Do you think the sun was made on the fourth day?

BRYAN   Yes.

DARROW   And they had evening and morning without the sun?

BRYAN   I am simply saying it is a period. .....

DARROW	And they had the evening and the morning before that time for three days or three periods. All right, that settles it. Now, if you call those periods, they might have been a very long time.
BRYAN	They might have been.
DARROW	The creation might have been going on for a very long time?
BRYAN	It might have continued for millions of years. (*State v. Scopes*, 1925)

The Scopes trial was instrumental in shaping the narrative for each side for the next century. The creationists came to be seen as Mencken and Darrow saw them: backwards-thinking, ideological, and bumbling. And in fundamentalist circles, the reputation of evolutionists was cemented as glib, condescending, and stubborn. After the Scopes trial, creationists were more careful to focus their energies on making a negative argument against evolution rather than defending a positive argument for the plausibility of their own view. By the end of the twentieth century they would finally craft a strategy, called the *wedge strategy*, that bore some fruit. The goal of aligning American culture with Christian principles would remain the same, but there would be an indirect approach: the new focus of the debate would be over scientific pluralism and academic and religious freedom.

The Contemporary Legal Battles

Despite new cultural and legal strategies employed by creationists, the trajectory of legal disputes over the teaching of evolution in schools over the last half-century has bent in the favor of evolutionists. Since the Scopes trial and the Butler Act, until just a few decades ago, some states were banning the teaching of evolution and lawsuits were filed to knock down those restrictive laws. These days, the shoe is largely on the other foot, so to speak. Rather than evolutionists fighting to get their curriculum included at all, they are now fighting to exclude creationism from the science curriculum. Evolutionists see this as a continuation of the same mission: to keep the curriculum in schools reflective of the best science available and protect students from the encroachment of religious doctrine masquerading as science. Creationists see a different narrative: while it is true that past laws wrongly restricted the science curriculum by excluding evolutionary science, they think that contemporary laws are now wrongly restrictive in excluding alternatives to evolutionary science, namely what they call *creation science*, usually centered around the theory of *intelligent design*. They complain that this is now an issue of respecting a plurality of views about human and earth origins, as well as scientific methodologies. This section will explore four landmark legal decisions in recent decades. The next section will explore some of the deeper scientific and philosophical issues that lie beneath the legal story.

Clarence Darrow and the ACLU won the public relations battle in 1925 by ridiculing Bryan and the creationists who prosecuted him. But while Scopes was found guilty, he was let off on a technicality. Thus the opportunity was lost to appeal the decision to a higher court and create some further-reaching legal precedent (Ruse 2005). Arkansas legislators eventually repealed the Butler Act in 1967, but not before inspiring similar laws in other states, some of which survived longer than the Butler Act itself. Notable

among these was a 1928 Arkansas statute barring the teaching of evolution that met a
US Supreme Court challenge in 1968 (Ruse 2000). The case was *Epperson v. Arkansas*
and it allowed for the first major public legal confrontation between evolution and
creationism since the Scopes trial. It was really the final battle that Darrow and the
ACLU of 1925 had been denied. The Supreme Court overturned the restrictive law,
arguing that it violated the establishment clause of the first amendment.

After *Epperson*, it was no longer permitted for states or school boards to prohibit
the teaching of evolution. Instead, what some did was to mandate the teaching of
alternatives to evolution. At the time, the alternative being advocated was known as
"creation science" (Ruse 2000: 263). It was the project of a group called the *Institute
for Creation Research* and it mostly consisted of a well-organized explanation for how
current geological, biological, and fossil evidence was consistent with a literal Young
Earth interpretation of the Bible. The creationists ran the same risk they did with the
Scopes trial, however, for as soon as the details of their own alternative was put forward, it would be open to great scrutiny. What the movement pushed for in the 1970s
was "balanced treatment," resulting in an Arkansas statute that required "creation science" be taught alongside evolutionary science. In the case of *McLean v. Arkansas* a
federal court ruled that creation science was not in fact science, and therefore violated
the establishment clause (Ruse 2000). In 1987 a similar decision was reached by the
United States Supreme Court in *Edwards v. Aguillard* against a Louisiana law which
mandated that creation science always accompany the teaching of evolution. Both the
McLean and *Edwards* decisions depended heavily on expert testimony from philosophers of science such as Michael Ruse who testified that "creation science" failed to
qualify as a science because it was untestable, overly confident, and lacked a scientific
methodology. Some of these criteria for what qualifies as science came under criticism
by other philosophers of science (Laudan 1982), a discussion that will be covered in
the next section.

The 1990s saw the decline of "creation science" and the rise of the "intelligent
design" as its successor. This was an explicit element of the "wedge strategy," which
included, among other steps, sowing skepticism among evolutionists through the advocacy of intelligent design, as well as recommending that schools "teach the controversy"
("The Wedge Document" 2014) The document reads, "The attention, publicity, and
influence of design theory should draw scientific materialists into open debate with
design theorists, and we will be ready." Intelligent design was less explicitly religious
in tone than "creation science," but when a school board in Dover, Pennsylvania took
the familiar step of requiring that intelligent design be taught alongside evolution just
as "creation science" had been previously, the legal result was also a familiar one. In
Kitzmiller et al. v. Dover the US District Court ruling was that intelligent design too
failed to qualify as legitimate science and therefore could not be required teaching in
science classrooms. Again, the ruling made heavy use of philosophical testimony on the
nature of science, creationism, and intelligent design.

The Philosophical Debates

The most public philosophical debate related to evolutionary theory has to do with the
movement known as *intelligent design* and the corresponding theory of *irreducible complexity*. However, this debate mostly plays out in the arenas of courtrooms and school

boards. There is no serious academic debate about the validity of intelligent design: both scientists and philosophers overwhelmingly reject it as a plausible alternative to Darwinian natural selection. The thesis of irreducible complexity is significant in the history of the debate, though, including in the legal decisions mentioned earlier, whatever its scientific and philosophical shortcomings.

To understand intelligent design, it helps to look at some historical background of theistic design and creation. In 1802 William Paley published *Natural Theology*, the focal point of which is an argument by analogy. Paley imagines coming across a pocket watch while on a walk, and upon observing its intricate and harmonious gears as well as its functional nature he concludes that the only rational conclusion would be that it was the product of a designer. Then, turning his attention to organic traits and mechanisms such as the wondrous human eye, he argues that since the same qualities of harmony, intricacy, and function are present, the same inference to a designer (God) is appropriate. This argument contains two steps: one premise is that certain traits are so impressively complex and functional that they must have been designed, and the other premise is that the concept of *design* implies that a designer is responsible. The conceptual link between *design* and *designer* was not thought to be the problematic step in the argument, so after Darwin the pushback against Paley's argument was against the first step of inferring design from complexity and function (Dennett 1995). More recently, some notable theorists such as Richard Dawkins (1996) and Daniel Dennett (1995) have argued that Paley is actually correct about the inference to design, but incorrect that *design* necessitates a *designer*. Evolution by natural selection is essentially a design process, they argue, and Darwin's great achievement was showing that design could be achieved incrementally without a designer. Other evolutionists (Davies 2003; Lewontin 1978; Ward 2012) continue to hold the more traditional resistance to Paley, arguing that all the complexity, harmony, and functionality in the world does not necessarily yield *design*.

Intelligent design is a theory of biological origins first developed by the biochemist Michael Behe and supported by the mathematician William Dembski and a few (but not many) other academics. Simply put, intelligent design claims that evolutionary biology is an insufficient explanation of biological origins and that instead, there are scientific reasons to believe that life was created and designed by an intelligent being. To support this theory, Behe argues that some biological traits cannot plausibly have evolved incrementally as prevailing evolutionary science would have us believe. He calls these traits *irreducibly complex*, by which he means "a single system composed of several well-matched, interacting parts that contribute to the basic function, wherein the removal of any one of the parts causes the system to effectively cease functioning" (Behe 1996). Such a system cannot have evolved incrementally, Behe argues, because any primitive ancestral version of the trait that lacked one of the parts would not function at all, thus not providing a fitness advantage, precluding the possibility of natural selection. At first glance, many traits appear to be irreducibly complex to the casual observer. The repellent spray of a skunk deters predators, but at any more elementary stage that was less repellent in scent, how would it deter predators and be any advantage at all to the skunk? A Venus flytrap is quick enough to close on an unsuspecting insect, but any slower and the insect would easily escape, so how could this trait have adapted slowly over time? Without its full functionality, it looks like it has zero functionality. Other examples that Behe uses are the bacterial flagellum and the blood-clotting cascade in vertebrate animals.

As tempting as these examples have been to some, the thesis of irreducible complexity suffers from at least two serious shortcomings. One problem is that it does not follow from the fact that a system currently requires all of its parts to function, that it could not have been evolved incrementally. A common metaphor used to illustrate this point is the assembly of a dry stone arch (Schneider 2000, 2014). It meets the criteria of irreducible complexity as Behe defines it above: if you take any one of the stones away, the whole arch collapses. So how could it have been assembled incrementally? The answer, of course, is with a scaffold! To build a dry stone arch, one first builds a support scaffold, which can be as simple as a pile of dirt, and then one rests all the stones, incrementally, over the scaffold. Once the scaffold is removed, the completed arch stands without any support needed. We can generalize from this example: systems that are seemingly irreducibly complex can be assembled incrementally if it is done with the help of an extra "scaffold" part that is then lost or removed. A biological example of this can be seen with the Venus flytrap. Biologists believe that ancestral Venus flytraps had a sticky mucous inside that would entrap a fly. The closing-action trait then evolved after the stickiness trait and provided a fitness advantage because it resulted in fewer flies escaping the mucous (Dunkelberg 2014). Once the closing mechanism became fast enough, the mucous became redundant and over time disappeared (as unnecessary traits are well known to do after a while). So, the mucous was in a sense a "scaffold" for the fly trap, allowing for an easy evolutionary explanation for a system that otherwise might seem mysterious.

Another way of explaining seemingly irreducibly complex traits is by discovering that ancestral versions of the modern trait served a wholly different function. Look at Behe's own example of a spring-loaded mouse trap, a mechanism he thinks is a paradigmatic example of irreducible complexity. Take away even one part of the mouse trap, Behe says, and the whole thing ceases to function. Yet while partial mouse traps cannot function to catch mice, they can do many other things. Kenneth Miller (2008) notes that partial mouse traps (that is, combinations of just one, or a few of the parts) can function as a tie clip, keychain, or spit-ball launcher to name a few possibilities. Evolution is full of examples of one trait changing function over time, and it is a mistake to infer from current functionality and project that backwards historically. What looks like a thumb on the panda is really a wrist bone that slowly evolved and changed function over time. Whale fins and flukes were once arms and legs and changed function gradually. Given the innumerable examples of functional change we know about, when confronted by an as-yet unsolved puzzle like the spray of the skunk, it would seem to be a grave mistake to stop the inquiry and conclude that intelligent design must be at work. On the contrary, investigating what other functions the spray (or a less smelly spray) could have served in ancestral skunks seems a promising research agenda.

At root, intelligent design's challenge to modern evolutionary theory is that it is not obviously false, and it cannot be ruled out as an explanation of biological origins. However, as a theoretical model for gaining information about the natural world, it has had nearly no success. Further, it has a history of failed claims that evolutionary theory has no explanation for some trait that is then nevertheless explained all the same. Many have thought that the primary sin of intelligent design is that it violates the methodological naturalism of science. This is the stance that science, as a system of gaining knowledge, only uses tools and methods capable of detecting spatiotemporal entities, thus ignoring supernatural explanations of phenomena (Sober 2007). Compare this to metaphysical naturalism, which denies the existence of supernatural entities and explanations to begin with. Methodological naturalism does not deny the existence of the supernatural, for instance God's existence, it ignores it. If this becomes the focus of the criticism of intelligent design, many of the other arguments over what science requires (for instance verifiability, falsifiability, testability, and tentativeness) are unnecessary. Methodological naturalism has proved its value as a commitment of science. Without it, science grinds to a halt: inserting "God did it" has shown itself to be a highly unsuccessful method for gaining knowledge of the natural world. Adhering to methodological naturalism, scientific theories of evolution do not rule out the possibility that creationism or intelligent design is true, but insofar as they are supernatural explanations, they cannot be counted as *scientific* theories. This was the finding in *Dover* that gave a victory to opponents of intelligent design (Young and Largent 2007).

Public and Private Debates

In February, 2014 there was a well-publicized debate between popular-science television host Bill Nye and Ken Ham, a leader in the creationist movement and founder of an organization called *Answers in Genesis*. The debate was hosted at the Creation Museum, a tourist attraction run by *Answers in Genesis* that presents visitors with an interpretation of natural history that is consistent with a Young Earth interpretation of the Bible. Nye delivered standardly accepted scientific evidence for the slow evolution of organisms and the billions of years-old age of the earth. Ham argued that the standardly accepted scientific evidence was fallible and asserted that only the Bible was a source of infallible information about the distant past because it was written by someone who was there, namely God (Nye and Ham 2014). The Ham–Nye debate did not unearth any novel or interesting arguments, and as with most public debates, both sides declared themselves victorious. However, it is an interesting cultural artifact because of how it compares to the Gray–Agassiz debates nearly 150 years earlier, or the Scopes trial nearly 90 years earlier. For while the science of evolutionary biology has advanced in leaps and bounds since Darwin, with overwhelmingly more confirming discoveries than disconfirming, creationism as an alternative to evolutionary science has not advanced.

This imbalance of progress, as well as revelations from the Discovery Institute's "wedge strategy" of provoking confrontation has dramatically shifted the willingness of scientists to even enter into debate. While Gray and Agassiz approached each other as academic peers (even if Agassiz was massively better known and better funded), many contemporary prominent academics and scientists were unhappy that Nye was debating a creationist in the first place. Their feeling was that by giving an absurd view a public airing, it lent an undeserved legitimacy to the creationist cause, thus playing right into the wedge strategy. The biologist Richard Dawkins stated in an interview with characteristic sharpness: "Just as I wouldn't expect a gynecologist to have a debate with somebody who believes in the stork theory of reproduction, I won't do debates with Young Earth creationists " (Andrews 2013). This sentiment fits into the greater narrative of the creation/evolution struggle in two ways. Firstly, it further feeds the "academic freedom" narrative of the creationists, whereby they can claim that their views are being unfairly excluded. Secondly Dawkins' comment mirrors the public position of many scientists and philosophers who argue that unscientific challenges to Darwinian evolution are a general threat to academia and must not be given room to breathe. It is therefore apparent that any new challenge to evolution cannot avoid the traffic signals and taboos of 150 years of cultural and legal conflicts.

Bibliographic Essay

For an excellent compilation of primary source documents that document this debate, readers should look at *Evolution and Creationism: A Documentary and Reference Guide* (2007) by Christian Young and Mark Largent. It includes bite-sized passages from the important documents behind this topic, spanning from before Darwin through modern day. Any further reading into the debate between evolution and creationism should begin with one of the many fine books by Michael Ruse, who is the dominant historian of the topic. He has also been an activist against creationism and intelligent design, and while that may threaten his objectivity as an observer, it makes his work all the more important to be familiar with for interested parties on either side of the debate. Ruse has two excellent surveys of the debate in the last 150 years, *The Evolution–Creation Struggle* (2005) and the slightly more accessible *The Evolution Wars: A Guide to the Debates* (2000), which is enriched with many illustrations and photographs. In both of these texts, Ruse acknowledges his own identity as an interested party in the debate without letting his history become advocacy, and readers can trust the history to be fair and accurate.

For a deeper dive into the opposition between Agassiz and Gray, readers should look at *Reef Madness: Charles Darwin, Alexander Agassiz, and the Meaning of Coral* (2005) by David Dobbs. For more on the rise of Fundamentalism in the twentieth century, see George Marsden's *Fundamentalism in American Culture* (2006). Some well-known books on the Scopes trial are *Summer for the Gods: The Scopes Trial and America's Continuing Debate Over Science and Religion* (1997) by Edward J. Larson and *When All the Gods Trembled: Darwinism, Scopes, and American Intellectuals* (1998) by Paul Conkin.

The philosophical literature on creationism and intelligent design varies widely in quality and accessibility. A good starting point is the online Stanford Encyclopedia of Philosophy entry on "Creationism" written by Michael Ruse (2014). Richard Dawkins' *The Blind Watchmaker* (1986) is an influential but not a careful treatment of the topic, while Daniel Dennett wrote the much more philosophically rigorous *Darwin's Dangerous Idea* (1995). Together, Dawkins and Dennett put forth an optimistic vision of biology ripe with "natural design" and "purpose." For dissent to this perspective see Richard Lewontin's "Adaptation" (1978), Paul Sheldon Davies' *Norms of Nature: Naturalism and the Nature of Functions* (2003), or Arthur Ward's essay directly disagreeing with Dawkins and Dennett, "Trouble for Natural Design" (2012). As with most issues in the philosophy of science, the most technical and authoritative work has been written by Elliott Sober, for instance his article "What is Wrong with Intelligent Design?" (2007), or his book, *Evidence and Evolution: The Logic Behind the Science* (2008a).

Intelligent design is the brainchild of Michael Behe, and a defense of the idea can be found in his book *Darwin's Black Box: The Biochemical Challenge to Evolution* (1996). The issue is further discussed in a volume co-edited by William Dembski and Michael Ruse called *Debating Design: From Darwin to DNA* (2004). Counterarguments to intelligent design and irreducible complexity are plentiful, and many can be found in the Dembski–Ruse volume above. Since irreducible complexity is such a new idea, and the need to respond quickly seemed pressing, much of the debate surrounding the concept occurred on websites. Two very accessible websites are run by prominent intelligent design opponents, Pete Dunkelberg (http://www.talkdesign.org/faqs/icdmyst/ICDmyst.html#venus) and Thomas Schneider (http://schneider.ncifcrf.gov/paper/ev/behe/). For a good scholarly scientific article by Schneider, see "Evolution of Biological Information" (2000).

There are two other internet resources of note. The debate between Ken Ham and Bill Nye (https://www.youtube.com/watch?v=z6kgvhG3AkI) was billed as a "YouTube debate" for a reason. The full affair is worth watching to get a sense of the major sticking points from the two polarized sides, however, it is worth noting that both parties ignore a substantial array of positions in the middle ground. The other significant internet resource for serious students of the topic is the "Wedge Document"(http://ncse.com/creationism/general/wedge-document). This outlines the strategy of fundamentalists to insert intelligent design into school curricula, and it only exists online because it was never published in print and is typically not retained as a public resource by fundamentalist organizations. It exists instead preserved as a resource held by groups opposed to creationism.

Chapter Twenty-Nine

FIELD AND LABORATORY

Jeremy Vetter

The rising status and prominence of American science from the mid-nineteenth century onwards owed at least as much to practice and place as to ideas, theories, and disciplines. Nowhere was this more true than in the two key co-defining places of modern science – lab and field – and in the borderlands that emerged between them. Lab and field may be examined separately, in relation to one another, or in relation to other places of science, such as the museum, observatory, and garden. From their emergence as central to the practice of science in the second half of the nineteenth century, through their twentieth-century transformation and dramatic scaling up, both lab and field as places have formed the material basis upon which American scientific power has been built. In turn, the expansion and growth of lab and field have been grounded in material changes in the capitalist economy and industrial society.

As categories of place-based practice, lab and field offer novel and productive conceptual frameworks for narrating the history of American science, transcending the conventional disciplinary categories. Commonalities across disciplines in lab and field practice have not only been important historically but have enabled rich sharing of historians' own approaches and methods across specialties ranging from the physical and earth sciences to the life and human sciences. Moreover, just as economic historians have cited the vast natural resource endowments of the North American continent – in combination with the organization of capital and labor – in seeking explanations for the country's growth and development, so too have material resources played a key role in the rising status of science in the United States during the industrial age. Historians have drawn productive analogies between laboratory and factory, on the one hand, and between the field and natural resource extraction or agriculture, on the other.

This chapter will explore the rise of the laboratory and the field in late nineteenth- and early twentieth-century American science, followed by an examination of the borderlands between them, and finally the myriad transformations in both lab and field

A Companion to the History of American Science, First Edition.
Edited by Georgina M. Montgomery and Mark A. Largent.
© 2016 John Wiley & Sons, Ltd. Published 2020 by John Wiley & Sons, Ltd.

during the twentieth century, as American science came to be at the very center of world science. Along the way, recurring themes, sometimes indicating similarities and at other times vast differences between lab and field, will include environmental settings, tools and technologies, the organization of work, and the relationships with lay people, amateurs, and the public at large.

Rise of the Laboratory

Histories of the rise of the modern laboratory have tended to focus on Europe – mainly Britain, France, and Germany – with only brief mention of the United States, as a site of emulation. One recent survey of the history of chemistry, for example, makes only brief mention of the United States in its account of the rise of the laboratory for teaching and research (Levere 2001: 122). American laboratories make a greater, if still modest, appearance in a leading survey synthesizing the history of nineteenth-century physics, mainly through examples of emulation of European laboratories, such as Joseph Henry at Princeton and Henry Rowland at Johns Hopkins. In both Europe and the United States, as laboratories proliferated during the second half of the nineteenth century, they became "places of precision" that were "increasingly part of an industrial culture that depended on disciplined regimes of accuracy and exactitude" (Morus 2005: 227, 235). This focus on European origins in lab history, while understandable, should not fool us into missing the vast expansion in US laboratories during the second half of the nineteenth century.

Although their origins may be traced to the experimental, especially chemical, laboratories of early modern Europe, often located in private homes, modern labs emerging in both Europe and the United States became closely connected not just to science research but also to the transformation of science teaching practices. As Larry Owens (1985) points out in a seminal early article on American lab history, when the laboratory first emerged at research universities modeled after their German counterparts – at Johns Hopkins University in the 1870s, for example – they were one of several alternatives, alongside Yale's playing fields and Harvard's gymnasium, for rationalizing and reforming the American university. Soon enough, however, the laboratory spread across the American higher education landscape, so that by the end of the nineteenth century, new laboratories were a highly visible feature of major collegiate and university building projects. Labs were becoming ubiquitous, not just at the many institutions of the East Coast and the Midwestern industrial belt, but also in the newly settled Western states, such as Nebraska (Cahan and Rudd 2000: 40–6, 87–91). Indeed, the westward expansion of the laboratory was nearly coextensive with the spread of colleges and universities themselves.

Though everyone agrees that laboratories became central to the practice of science during the industrial age, and the foundation upon which much of modern science has been built in the United States and elsewhere, the lab itself surprisingly remains something of "a neglected subject," without any "comprehensive surveys of labs for particular regions or periods (how many, where, when, what for)" (Kohler 2008: 761–2). We are especially lacking comprehensive surveys of labs in the United States. However, a composite portrait based on the scattered accounts in the secondary literature, along with sampled primary source evidence, is possible. The core feature of the modern lab is its constructed placelessness – that is, the determined attempt to situate it nowhere in particular, so that the knowledge produced inside a lab will be valid everywhere. Labs are places designed for strict control of the material environment and purposeful manipulation of variables. Yet, despite the attempted erasure of local environmental conditions, labs are dependent on a particular, human-constructed environment: the modern urban infrastructure of water, electricity, gas, and other utilities, together with indoor shelter and protection from unintended disturbance. Access is restricted to authorized personnel, who themselves are dependent on the urban (or suburban) built environment for their housing and subsistence. Almost always, labs have been organized around tools, machines, and instruments. Their work organization – before the rise of "big science" later in the twentieth century, as will be discussed below – has been typified by a fairly standard hierarchy and division of labor consisting of a lead scientist who directs the laboratory, scientific assistants who may themselves eventually direct their own labs, and lab technicians, whose roles in science have also attracted scholarly attention during the turn to practice (Barley and Bechky 1994; Mukerji 1989: 135–45).

Despite their close association with academic settings, however, laboratories were never restricted to the ivory tower. On the contrary, both by analogy and direct interconnection, the lab and the industrial factory rose to dominance together. In both their teaching and research aims, laboratories vastly expanded productive output, fostered precision and standardization, and inculcated industrial values of rationality, efficiency, and control. "Like Victorian factories," writes Iwan Rhys Morus (2005: 236), "physics laboratories could be depended upon to produce a steady stream of diligently designed, mass-produced, and standardized products." The emergence of science labs at colleges and universities created a place for practices, including tools and techniques (not to mention the students trained to deploy them), that moved easily between academic and industrial worlds. Moreover, around the turn of the twentieth century, the research laboratory itself crossed over into industry, in the form of industrial research labs whose aims extended beyond the immediate production process to larger scientific problems. While, in one sense, the industrial research lab represented the culmination of a shift in the process of invention from the craft machine shop of the early to mid-nineteenth century to the corporate lab in such practical domains as telegraphy (Israel 1992), in another sense, the industrial research labs at large business corporations such as General Electric, AT&T (Bell Labs), and DuPont represented something new altogether (Hounshell and Smith 1988; Reich 1985; Wise 1985).

Modes of Field Practice

If the labs across the American landscape in the late nineteenth century had their origins in the chemical laboratories and "houses of experiment" of the early modern period, then the lineage of fieldwork may be traced to the long, intertwined histories of natural history, and exploration. Indeed, exploration itself constitutes one important enduring category of field practice in the history of American science, ranging from the exploration of the North American continent through the nineteenth century (Carter 1999; Goetzmann 1966; Orsi 2014), to exploration of remote spaces beyond US borders into the early twentieth century, such as polar exploration (Robinson 2006), or even space exploration. Exploration may be properly seen as one important mode of field practice, changing over time but with certain distinguishing features largely enduring, such as a strong association with rugged masculinity and the heroic risking of the explorer's own body – a characteristic that was arguably not unique to exploration in some periods, since it was also shared with other scientific practices during the Gilded Age and Progressive Era, including self-sacrifice in lab science (Herzig 2005a).

But exploration has been only one of many distinct modes of field practice that have proliferated over time. Alongside the rise of the laboratory, the industrial age also saw the further development of other modes of practice in the field. Transportation technology deployed across the landscape was pivotal in the transformation of field practice, especially the nineteenth-century industrial age's most iconic and consequential conveyance, the railroad (Vetter 2004). As rail systems expanded their tentacles across North America, at first east of the Mississippi River in the antebellum period and then across the western half of the continent beginning in the late 1860s, after the Civil War, the vast natural resource base of North America could be tapped with much greater ease and at much lower cost, as accessible to the field scientist as for capitalist economic development. The railroad made possible not only the movement of field scientists themselves but also the materials they collected, which by the late nineteenth and early twentieth centuries included some of the world's most significant vertebrate paleontological quarries, located in the American West (Brinkman 2010; Jaffe 2000; Rea 2001; Rieppel 2012; Vetter 2008; Wallace 1999). As dinosaurs and large mammals made their way to the museums of Pittsburgh, Chicago, and New York, they undergirded some of the era's most spectacular public displays of science, as well as a key research domain of American strength and leadership in world science (Rainger 1991).

Even more closely related to exploration was the survey mode, in which a more systematic coverage of geographical territory enabled field scientists to map the natural and human world with far greater granularity and precision. Surveying emerged at first out of cadastral surveying for property boundaries, but in the antebellum period, much of American field science revolved around geological surveys in East Coast states (Cohen 2006; Spanagel 2014), as well as the US Coast Survey (Slotten 1994). The rapid westward expansion of survey science across the Mississippi River after the Civil War produced the "great surveys" of the late 1860s and 1870s (Bartlett 1962; Smith 1987), whose ample historiography has been enriched more recently by biographical accounts of well-known survey leaders such as Ferdinand Hayden and John Wesley Powell with substantial coverage of their field practices (e.g., Cassidy 2000; Worster 2001). With the founding of the US Geological Survey in 1879, federal government geological and topographical survey fieldwork was brought under a single institutional tent, but its reach and size only increased (Manning 1967). Moreover, throughout the nineteenth century, we should not forget that geological fieldwork, especially when related to fossil fuel extraction – increasingly important and lucrative in the industrial era – often occurred through private consultants as much as through publicly funded science (Lucier 2008). Survey practices may have been most visible in nineteenth-century geology, but they also spread to other science fields, such as the human and life sciences through ethnological and natural history surveys. The late nineteenth- and early twentieth-century period was in fact a golden age for survey expeditions by museums and government agencies into the "inner frontiers" of North America to secure biological specimens for research and public display (Kohler 2006). Another new kind of field survey work that took off around the turn of the twentieth century was the soil survey, led by the US Department of Agriculture's Bureau of Soils in cooperation with states.

Whether at individual field sites such as paleontological quarries or through surveys that covered more territory, emerging patterns of work organization in the field often included a hierarchical division of labor comparable to that in a laboratory, if sometimes more fluid. A typical field party, whether in geology, soils, zoology, botany, or ethnology, had a scientific leader, often one or more scientific assistants, and one or more hired laborers such as cooks, packers, and guides. On the North American continent, at least, these were typically much smaller parties than the grand field expeditions of the early to mid-nineteenth century, deployed strategically using the system of railroads to move in and out of the field. Field parties almost always answered to a supervisor based at a museum, university, or government agency, and a copious traffic in communication through the mail, and occasionally the telegraph, maintained lines of authority but also allowed for some flexibility in decision making for those in the field. Through the early twentieth century, field parties were still mostly male dominated, although there were significant exceptions, such as animal collector Martha Maxwell, anthropologist Alice Fletcher, amateur naturalist Annie Alexander, and others (Benson 1986; Bonta 1991; Mark 1988; Stein 2001). Special-purpose expeditions, such as those to watch eclipses, might be larger in size and incorporated women through a gendered division of labor (Pang 1996). Over time, some kinds of fieldwork, such as in eugenics, were even defined as "women's work" (Bix 1997). Other kinds of fieldwork – geology and soil science, for example – remained almost exclusively male dominated long into the twentieth century.

As the presence of cooks, packers, and guides in field parties suggests, the field as a place for science was significantly more permeable to the lay public than the laboratory. For their part, these "technicians of the field" sometimes served for multiple years and thereby developed special skills in serving scientific field parties, though they tended to work seasonally and to be recruited from local populations. Moreover, they frequently possessed experiential knowledge of places that visiting field scientists did not have, at least initially. Other members of the lay public encountered field parties without becoming attached to them. Whether in short- or long-term relationships with field scientists, locals often exerted significant leverage over fieldwork by virtue of their distinctive knowledge and control over land and resources – even if they could not ultimately upend the overall structure of power relations (Schneider 2000; Vetter 2008).

Another distinct mode of practice for field science was not to send out a party at all, but rather to establish a geographically extensive network of lay people, or in some cases of amateurs, who might collect or observe for scientists based in some distant metropole. Two early American examples from the mid-nineteenth century were Spencer Baird's network of naturalist correspondents and the weather observers connected by telegraph, both based at the Smithsonian Institution (Fleming 1990; Goldstein 1994), and there were even earlier precedents in networks of agricultural improvers in the early to mid-nineteenth century (Pawley 2010). By the late nineteenth and early twentieth centuries, networks linking experts with lay or amateur collaborators became ubiquitous in the field sciences, ranging from archaeology to meteorology (Burns 2008; Vetter 2011). Moreover, historians have fruitfully examined the conflict and competition that often occurred between professionalizing field scientists and amateur or vernacular authorities, in environmental domains as diverse as plants, birds, oysters, and oil (Barrow 1998; Frehner 2011; Keeney 1992; Keiner 2010). Finally, historians have also traced the widely popular "nature study" movement around the turn of the twentieth century to re-engage members of the public, especially children and youth, in knowing nature outdoors and not just in the lab (Armitage 2009; Kohlstedt 2010). In sum, the direct roles of the lay and amateur public in field science were unusually diverse and wide-ranging.

Lab–Field Borderlands

Around the beginning of the twentieth century, then, both lab and field had emerged as key places for the practice of American science. They were not always isolated from one another, however. Not only has there often been movement between lab and field, but there have been significant developments in what may be called the "lab–field border-lands," by analogy with the historical geography of differing human ways of life, such as agricultural settlement and nomadism (Kohler 2002). In the earth and physical sciences, the metaphor of the "borderland," though with disciplinary traditions as proxies for lab and field, has been applied to the shift from field and museum to laboratory experiment in the founding of the Carnegie Institution of Washington's Geophysical Laboratory (Servos 1983). More broadly, in the life sciences, the notion of a "lab–field border-lands" as an analytical framework has gained significant attention in recent years. As biologists became dissatisfied with the perceived excesses of the late nineteenth-century trend toward laboratory microscopy, they sought ways to bring features of lab practice – tools, techniques, quantification, and experimental design – back out in the field. From the first quarter to the second quarter of the twentieth century, however, field biologists came to be more secure: not just borrowing strictly from the lab, but instead growing more confident innovating their own borderlands practices. They created "a culture that resembled those of both laboratory and field but was distinctly different from either one." In so doing, "they perpetuated the traditional aims and methods of natural history by giving it some of the analytical force of laboratory precision and causal analysis. They were of the field but met laboratory standards of evidence and inference" (Kohler 2002: 218, 256). Thus, as a framework for studying place and practice, this notion of a lab–field borderlands is quite distinct from the earlier postulation of a naturalist–experimentalist dichotomy (Allen 1979), which generated a lively and engaging debate in the 1980s about the methodological and intellectual orientations of American biologists.

Many noteworthy and influential places for the practice of American science have been situated somewhere within the lab–field borderlands, even if the traffic between lab and field has not always been so heavy or fraught with anxiety. Certain pro-tected areas have been designated as "natural laboratories," with determined efforts at "preservation for science" by restricting human use and intervention in order to observe the workings of nature – for instance, at Glacier Bay, Alaska from the 1920s onwards (Rumore 2012). More generally, field stations as a mode of practice have been located in the lab–field borderlands. These "labs in the field" combined the buildings, instruments, and precision of the laboratory with easy access to the nat-ural environment around the station. Late nineteenth- and early twentieth-century field stations differed significantly in where exactly they were located in the border-lands between lab and field, with East Coast marine stations such as the Marine Biological Laboratory at Woods Hole, Massachusetts, exhibiting the closest approach to the lab ideal (Pauly 1988) and Rocky Mountain biological field stations being especially close to the field (Vetter 2012), but they were all hybrid institutions. Moreover, US field stations drew on both a European research model and a more distinctly American summer teaching model (Benson 1988a). This practical, utilitarian mixture met its match in the even greater proliferation of agricultural field stations across the American landscape, which were hybrids in their own way, ranging from the very lab-like central stations that most states operated, supported by federal funds from the Hatch Act of 1887, to their numerous branch stations in the field, dispersed geographically to encompass the diverse environmental conditions found within their territories.

Lab-field hybridity has continued to characterize places of American science throughout the twentieth century, in fields as diverse as plant biology (e.g., Kingsland 2009) and urban studies in the human sciences (Gieryn 2006). The examples are too numerous to catalog here, and in the widest sense the terrain might be extended to include all manner of sites that combine aspects of lab and field together, including some whose roots go back much earlier, such as botanical gardens and observatories. The observatory, in particular, was a place with its own deep historical lineage that influenced the development of the laboratory, yet at the same time it was situated spatially in the natural environment much like a field site. Emphasizing commonalities in both space and techniques, one recent collection of essays on the observatory sciences (Aubin, Big, and Sibum 2010), including an American case study, argues for significant borrowing from observatory to laboratory. Indeed, both types of sites were expanding massively in the late nineteenth-century United States. "Americans built observatories enthusiastically and on an emphatic scale," notes Richard Staley (2010: 226) – 40 by the mid-1880s, which was more than either Britain (32) or Germany (26) – with a distinctive emphasis on large telescopes. Even apart from their role in exploring outer space as a place for science, observatories have demonstrated the difficulty of categorically separating lab and field (Lane 2010).

Other places of science, such as museums, might be viewed as perpetuating a third analytical pole of science practice that has opened border zones with both field and lab during the twentieth century, exemplified in the heavy traffic between museums and the field (Sunderland 2012) or the revitalization of practices associated with natural history museums in the lab setting (Strasser 2010). In all these cases, the analytical payoff of examining lab and field (and museum) is even greater when the practices are so intermingled, bringing together the placelessness and control of the lab with the rich context and embeddedness of the field. The apotheosis of lab–field hybridity in recent times has perhaps been the Biosphere 2 project, located near Tucson, Arizona (Reider 2009), which garnered extraordinary international attention, and sometimes intense criticism, for the attempt of its creators in the late 1980s and early 1990s to construct a completely sealed and enclosed mini-world holistically replicating several diverse biomes of the earth (i.e. "Biosphere 1"), including eight "biospherians," or human inhabitants. From the earliest seaside summer stations to Biosphere 2, hybrids of lab and field practice have been a fount of innovation in twentieth-century American scientific life.

Lab and Field: Twentieth-Century Transformations

Besides the emergence of places for American science in the borderlands between them, both lab and field have also undergone further transformations that have undergirded the increasingly central status of American science in the world and have, in turn, been produced by the rising economic and cultural power of the United States. In the lab, one especially visible shift in practice has been the development of ever more expensive, elaborate instrumentation, linked with scaling up laboratory practice to larger scale coordination and division of labor (Galison 1997). Indeed, the dominance of "big science" and the construction of its key exemplars are almost exactly coterminous with the centrality of American science itself by the middle of the twentieth century. A seminal collection of historical work on "big science" with many US case studies (Galison and Hevly 1992) makes this point clear. The proliferation of high-profile labs during the World War II and Cold War periods has included the pioneering Lawrence Berkeley Laboratory (Heilbron and Seidel 1989), Los Alamos (Hunner 2004), Argonne (Holl 1997), Brookhaven (Crease 1999), and national labs in general (Westwick 2003), as well as on-campus sites as diverse as Ames Laboratory at Iowa State (Goldman 2000) and Lincoln Labs at MIT (Slayton 2012), and finally ever larger particle accelerators such as Fermilab (Hoddeson, Kolb, and Westfall 2008) and the Stanford Linear Accelerator, or SLAC (Z. Wang 1995). As universities adjusted to the massive federal funding for national security and defense-related research that persisted from World War II into the Cold War period (Dennis 1994), entire campus laboratory ecosystems were created accordingly at such research powerhouses as MIT and Stanford (Leslie 1993).

More broadly, key instruments, and the communities of practitioners around them, have received special attention in studies of twentieth-century American laboratory science, including the electron microscope in biology (Rasmussen 1997), nuclear magnetic resonance spectroscopy and mass spectroscopy (Reinhardt 2006), the polymerase chain reaction (Rabinow 1996), and probe microscopy in materials science (Mody 2011). Within the biological sciences, moreover, living organisms have crossed the threshold into the laboratory and found new homes there. Historical studies of influential model organisms, or "model systems," have included many globally significant American examples, such the early twentieth-century breakthroughs in genetics – and in the moral economy of scientific practice – using the fruit fly (Kohler 1994), mid-twentieth-century advances using tobacco mosaic virus as a "model system" (Creager 2002), and mice, which were transformed into standardized lab creatures between 1900 and 1955, not just in practical terms but also as symbols of the lab for the public (Rader 2004). Even more recently, American laboratory design and practice have shifted to a newer ideal based on more direct engagement with that public, through creating architectural openness and visibility, while at the same time restricting access and maintaining some degree of control (Gieryn 2008).

American field science practice, too, has undergone great transformation during the twentieth century. In the human sciences, statistical data collection practices for such key aggregate indicators as unemployment were "industrialized" during the New Deal of the 1930s (Didier 2011), and highly visible examples of gauging the mass American public in the mid-twentieth century, for instance, through opinion polling and sexual behavior surveys, have also offered intriguing evidence about ground-level practice (Igo 2008). In the environmental field sciences, such as land use mapping and meteorology, the advent of aerial technologies around the 1920s and beyond has transformed practice, both to bring about a new way of seeing and classifying the land (Checkovich 2004) and the wholesale emergence of a new exemplar of "infrastructural science" (Turner 2010). These new practices for seeing the objects of science in larger aggregates – either the mass human public or the natural landscape and atmosphere – have been matched on the ground with the deployment of more deliberately designed experiments and protocols for observation in the field. Such refined practices have proliferated not only in domains directly relevant to capitalist economic development such as forestry (Brock 2004), but also in conservation biology (Davis 2007) and primatology (Montgomery 2005). In all these cases, American research was becoming increasingly central to innovations in scientific practice by the middle decades of the twentieth century. Even in less nationally focused histories, such as that of Amanda Rees (2009) on the "field workers' regress" in primatology – by analogy with the classic experimenter's regress in the laboratory – American science became a significant part of the story. Not just in scientific research practice itself, but in popular culture – for example, American films about animal behavior (Mitman 1999) – the displayed field site has loomed ever larger in discussions about the relationship of humans to the natural world.

As a place for science, the field has been reconceptualized as existing at every conceivable scale, from the local to the global. Within ecology, in particular, the role of local field sites in both practice and the generation of important ideas has remained important throughout the century (Cittadino 1993; Way 2011; Young 1998), even as some field sites, which have often been built on Cold War nuclear technologies and the specific sites of "big science" of the lab discussed above, were reconfigured for more intensive manipulation and control (Bocking 1997; Kirsch 2007). American field science has related not just to local place and practice, but also to the larger regional politics of place, with examples ranging from habitat-based re-envisioning of endangered species in California (Alagona 2013) to attempts to control fire ant infestation in the American South (Buhs 2004). Moreover, in the mid-twentieth century, surveillance technologies in the field such as radiotelemetry remade field sites into "wired wilderness" (Benson 2010).

Field surveillance practices were applied around the world in the context of US global dominance during the Cold War era, including on the Indian subcontinent (Lewis 2004). More broadly, American field science moved aggressively beyond the nation's borders into the Latin American tropics (Christen 2002; Raby 2012), and global oceans (Burnett 2012; Doel, Levin, and Marker 2006; Kroll 2008; Rozwadowski 2005). More recently, the surveillance of the field has scaled up to encompass the entire globe and its shared common atmosphere (Edwards 2010), though the language of "the field" has often tended to disappear even as it was reconceptualized at a much larger scale. Yet throughout the twentieth century, the field sciences have faced perceptions of relative decline compared to lab science – and later also to computer modeling – which has been felt not just in such fields as biology and physics with their strong laboratory orientations but even in the earth sciences (Oreskes 1999: 288–91). The lay public has continued to play a role in, and contest the authority of, field scientists to speak for nature. Field practices at the global level, for example, are notorious for the lack of epistemic closure in the wider public, even for such overwhelming problems as climate change. At the same time, at the very local level, field scientists have faced opposition from users and collaborators, such as growers in California who have insisted on the irreducible particularity of conditions on their farms, when interacting with agricultural experts conducting field trials (Henke 2000). From the global to the local, the inherent messiness and lack of full control in the practice of fieldwork, has perhaps only been matched by the enduring relevance of that very same fieldwork to the problems of the material, environmental world and human interaction with it.

Lab and field are places of scientific practice that have been central to the development of American science, including its rise to unparalleled global prominence by the middle of the twentieth century. As John Servos (1986: 612) pointed out three decades ago, when studies focused on lab and field practices were just beginning, "the specialties in which America developed rapidly were typically field or laboratory sciences, heavily dependent on observational or experimental evidence or techniques." Indeed, in many ways the promise of constructivist history of practice in lab and field has still not been fully realized, as historians of American science continue to orient themselves predominantly around disciplines and theories, even as they have broadened out to consider a wider range of social, cultural, and political issues.

Lately, three robust historiographical developments in US history seem to offer great promise for further enriching the historical study of lab and field: (1) environmental history, which is enabling historians to integrate the abstract, conceptual "place" of lab or field with the natural, if culturally constructed, "place" of the tangible, physical, material world in which the lab or field is situated, at scales ranging from local and regional to global; (2) the history of American capitalism, which raises questions at the heart of both lab and field practice in their relations with industrial and resource development – and to which the history of American science seems poised to make a vital contribution, since knowledge practices have been so closely tied to the capitalist valuation and manipulation of nature; and (3) the United States in the world, which promises to more broadly contextualize lab and field practices in order to capture their transnational, comparative, and global character. As places, lab and field remain not only relevant, but essential, to writing the history of American science.

Bibliographic Essay

The historiography of lab and field as places emerged during the last two decades of the twentieth century, during a lengthy period of cross-fertilization between the history of science and interdisciplinary scholarship in STS, a term which refers either to "science and technology studies" or "science, technology, and society." Two key developments in the writing of both history and STS during this period, which have influenced the study of lab and field, have been (1) constructivism – arguing that scientific knowledge is strongly shaped by the particular settings in which it is produced – and (2) the shift toward practice. Reflecting the widespread belief that the United States had become the center for world science in the second half of the twentieth century, the pioneering constructivist laboratory ethnographies that appeared at roughly the same time were based in California (Knorr-Cetina 1981; Latour and Woolgar 1986; Lynch 1985; see also Traweek 1988). Shortly thereafter, the study of science practices and tools took off in the early 1990s, beginning with influential collections of essays in STS (Clarke and Fujimura 1992; Pickering 1992).

Around the same time, sets of historical case studies – some of them based on US examples – also began to appear, including seminal collections on both lab (James 1989) and field (Kuklick and Kohler 1996). Since then, while some of the initial momentum and excitement has subsided, wide-ranging historical work in both lab and field history has continued to appear (for overviews, see Gooday 2008; Vetter 2010). As conceptual frameworks for analyzing place and practice in the history of science that have been applied to a wide range of case studies across national boundaries, lab and field can be found to varying degrees throughout the secondary literature cited throughout the above essay. Very few key texts exist, and none that provide a synthesis across all the lab or field sciences in the United States. The best entry point for the lab–field borderlands, and a common reference point for historical debates about lab and field, is Robert E. Kohler, *Landscapes and Labscapes* (2002). A more fully developed history of American scientific practice told through the lenses of lab and field remains on the horizon.

Chapter Thirty

Gender and Science

Donald L. Opitz

A broad, multidisciplinary field of inquiry, the study of gender and science encompasses multiple agendas concerned with gender, sexuality, women, and men within, and in relation to, the sciences, technology, and medicine. Its pursuit by historians of American science mirrors the broader field's plurality of approaches and agendas. The phrase owes its origins to feminist critiques of the sciences and, particularly, scientific and medical perspectives on sexuality and sex differences. On the one hand, *gender* as a term connoting socially constructed identities in relation to material bodily features and functions, as well as the relationship between those identities, emerged within American developmental psychological debates of the 1950s and 1960s. Yet, its specific conjunction with *science* occurred in the late 1970s amid the burgeoning feminist critiques of scientific constructions of sex and gender. Keller (1978) coined the label for this field in her article bearing the very title, "Gender and Science." This chapter will survey the most important bodies of gender and science scholarship touching on the history of American science, with special attention to the historiographical trends and emergent areas of recent research.

The Scientific Origins of "Gender"

Notwithstanding the eighteenth-century French usage of the term *genre* within sociopolitical contexts, as highlighted by Offen (2006), most historians locate the emergence of the modern meaning of *gender* within the field of American developmental psychology and its subsequent adaptation by feminists and sociologists. As chronicled by Irvine (1990), Hausman (1995: 95–106), Meyerowitz (2002: 98–129), and others, psychologists who studied intersexuality and sex differences in the 1950s and

A Companion to the History of American Science, First Edition.
Edited by Georgina M. Montgomery and Mark A. Largent.
© 2016 John Wiley & Sons, Ltd. Published 2020 by John Wiley & Sons, Ltd.

1960s adapted the terminology of gender, until then primarily residing within linguistics, and provided theoretical definitions of gender concepts like "gender role" and "gender identity." Earlier in the century, American psychologists sometimes employed *gender* as a synonym for *sex* in the strict, biological sense, as did F.H. Saunders and G. Stanley Hall (1900: 546; see also DeLuzio 2007: 90–132). But others paved the way for applying "gender" as a means to classify men's and women's socially constructed roles. In her influential anthropological study, Margaret Mead (1949) elaborated on her observations of the socialized differences between males and females, but she retained the popular terminology of "the sexes" and "sex roles." Nevertheless, one of Mead's reviewers noted how she "informs the reader upon 'gender' as well as upon 'sex,' upon masculine and feminine rôles as well as upon male and female and their reproductive functions" – in other words, recognizing "gender" as a way to differentiate social roles from biological functions (Bentley 1950: 312). This reviewer – a Cornell psychologist who advocated for a brand of dynamic functionalism in opposition to structuralism – earlier defined his notion of gender as the "socialized obverse of sex," signifying "social matters" that "separate the boys from the girls" among humans but not animals, "where there is sex but no gender" (Bentley 1945: 228; on Bentley, see Devonis 2012).

As David Haig's (2004) impressively large-scale survey demonstrated, such early "nonjocular" uses of gender were extremely rare among titles of indexed scholarly articles, until Johns Hopkins psychologist John Money (1955: 254) decidedly coined "gender role": "The term *gender role* is used to signify all those things that a person says or does to disclose himself or herself as having the status of boy or man, girl or woman, respectively." Money and his followers advanced an "interactionist" position on the question of how individuals acquire gender roles, arguing that, as with language, biology primes individuals for the acquisition but the environment determines which specific gender roles are imprinted. Moreover, Money held that individuals acquire gender roles very early in their development and, once acquired, those roles are resistant to change (Haig 2004: 93; on Money, see Goldie 2014). As Rebecca Herzig (2005b: 200) perceptively pointed out, the terminology's birth within a clinical context concerned with medical interventions like hormonal manipulations and reconstructions of genitalia made gender "part and parcel of shifting technological practice," suggesting a more broadly technoscience context to the term's origins and development.

At the International Psycho-Analytical Congress held in Stockholm in 1963, California psychoanalysts Ralph Greenson and Richard Stoller, who presented in a symposium on homosexuality jointly introduced the term *gender identity*, distinct from *gender role*, to emphasize a more subjective experience, i.e., "one's sense of being a member of a particular sex ... the awareness of being a man or male in distinction to being a woman or female" (Greenson 1964: 217; see also Stoller 1964: 220). Stoller later elaborated these views and emphasized gender's distinctness from biological sex: "*gender* is a term that has psychological or cultural rather than biological connotations" (1968: 9). Thereafter, into the 1970s, the psychosocial character of gender gained traction among psychologists and sociologists, even as debates ensued over the extent to which individuals' acquisition of gender is primarily due to nature (biological) or nurture (environmental). By the mid-1970s, however, more widespread discourse around the meanings of *gender* percolated in disciplinary spaces outside psychology, especially among critiques by feminists trained in the natural and social sciences.

Gender as a Feminist Category

The rise of "gender and science" as a concern for science studies consisted of the debates that established, elaborated, and destabilized gender theory and its forerunners: sex differences, sex roles, mental sex, and psychological sex (Meyerowitz 2002: 98–129). Early critiques of scientific theories about sex differences, which typically naturalized women's (usually subaltern) social roles with respect to men's according to biological differences – also known as biological determinism – swelled during the post-Darwinian debates of the late nineteenth century. Noteworthy American advocates for women's rights, like Congregationalist minister Antoinette Brown Blackwell (1875) and schoolteacher Eliza Burt Gamble (1894), challenged interpretations of sex differences but nonetheless upheld the view that they are ascribable to innate causes. Other liberal critics challenged this paradigm, particularly reform writer Charlotte Gilman Perkins (1898), who argued against the immutability of "sex-distinction" and called for reform of women's economic status through social and political means. As Cynthia Eagle Russett (1989) argued, in the nineteenth-century context, men of science like Darwin and Spencer spoke in terms of sex, not gender, and contributed to a "genuine scientific consensus" across disciplines that explained Victorian womanhood in terms of "arrested development," employing physical and biological principles like the conservation of energy, sexual selection, and the physiological division of labor (x, 10-11). According to Russett, this scientific "construction of Victorian womanhood" formed a "masculine power play" (206), and parallel lessons might be drawn about modern research on sex differences in fields like endocrinology, neuroscience, and sociobiology (14).

Such lessons were already being urged among a resurgence of feminist critiques that took its cues from the 1970s women's movement, particularly by female scientists who raised the "woman question" in science, or variants of "what is to be done about the situation of women in science?" (Harding 1986: 9). Although not a new question, women's liberation provided a new context and indeed the inspiration of a global movement. Moreover, sexologists like Money and Stoller, who introduced the terminology of gender, equipped feminist critics with a new vocabulary, despite its early problematic and inconsistent usage (see Gould and Kern-Daniels 1977). Haig (2004: 94) assessed the period up to the late 1970s as "small beginnings" in feminists' uses of "gender" in view of the continued popularity of "sex" and "sex roles." To the extent writers began employing the new language of gender, they often used "gender" interchangeably with "sex;" but, gradually, the terms were bifurcated with gender increasingly mapped to the social and sex to the biological. Sociologist Ann Oakley (1972: 16) elucidated the "crucial distinction" of these terms along such lines in order to clarify the arguments surrounding sex differences: "'Sex' is a word that refers to the biological differences between male and female: the visible difference in genitalia, the related difference in procreative function. 'Gender' however is a matter of culture: it refers to the social classification into 'masculine' and 'feminine.'" Gayle Rubin (1975: 159) advanced a similar view in terms of a "sex/gender system." Nevertheless, slippery usages persisted. In her debut feminist critique of science, in which she extrapolated from her personal struggle in the male-dominated field of physics, Evelyn Fox Keller used "sex" and "gender" virtually interchangeably. She, noted, for example, how an infant born with ambiguous genitalia could be doomed to "an apparent *sexual identity* at odds with his or her genotype" as a result of "fallacious sex assignment at birth" and differential treatment reinforcing that assignment; this posed evidence for sexologists' argument that "*gender identity* appears to be established, primarily on the basis of parental treatment, by the age of eighteen months" (Keller 1974: 16, emphasis added).

While other feminist writers launched critiques of biological determinism from the humanities and social sciences (Griffin 1978; Ortner 1972), Keller arose among the earliest voices directly addressing the issues connected with what she would soon dub "gender and science." Keller (1995a) retrospectively dated her coinage to the title of her 1978 essay, but elements of the feminist perspective underlying this phrase already appeared in 1974 when she wrote, "the differential performance of men and women in science, the apparent differences between conceptual styles of men and women everywhere, are the result, not so much of innate differences between the sexes, but rather of the myth that prevails throughout our culture identifying certain kinds of thinking as male and others as female" (19). The power of myths, and language more generally, in shaping scientific thinking – what Keller, along with others, understood primarily in social psychological terms consistent with Kuhn's (1962) paradigms – became a recurring, dominant theme in the critiques (Longino 1990: 10–11). Thus, in "Gender and Science," Keller asserted, "the association of masculinity with scientific thought has the status of a myth which either cannot or should not be examined seriously," with the implication being that such a serious examination is precisely what was then needed (Keller 1978: 187). The scrutiny soon followed, focusing primarily on biology and medicine, with Bleier (1984), Fausto-Sterling (1985), Keller (1985), Birke (1986), Harding (1986), Martin (1987), Haraway (1989), Longino (1990), and Hubbard (1990) among the leading critics. Despite the nuances among their individual arguments they nonetheless collectively contributed to a feminist consensus that demonstrated science as value-laden in ways that sustained a masculine mythology, culture, and epistemology that in turn sidelined women's agency in practicing science and shaping scientific discourse.

The History of Women in American Science

In addition to the theoretically driven critiques, further work advancing "gender and science" tended to cluster around two primary historical agendas. Like Keller, some feminist historians took up the project of exposing the overriding, mythical association of science with masculinity and the implications of that association for the stratification of men's and women's roles within science. These scholars have focused on deconstructing and overthrowing prevailing scientific perspectives on masculinity and femininity and their usage in arguments against women's capacity to do science – masculinity conventionally associated with agency, rationality, and objectivity, whereas femininity with passivity, sentimentality, and impressionability. As Marina Benjamin (1991: 12) noted, a central concern in this literature is the dialectical relationship between male culture and female nature, and thus the tendency of scientific (i.e., cultured) men symbolically relegating women, stuck in nature, among their objects of study; put another way, the dialectic metaphorically equated "woman with nature as that which is known, as opposed to that which is capable of knowing" (Benjamin 1991: 3). In parallel to this scholarship, another burgeoning literature attended to compensating for the historical neglect of women's participation and contributions in science, and historians therefore worked toward correcting the prevailing fallacy that demographically the sciences had been, nearly exclusively, gendered male. These two bodies of literature often overlapped in their explanations for the historical gendering of science and, reciprocally, the scientific gendering of human history and society, and they share in their vision for changing the status quo. Together, the studies have contributed to efforts in "righting the record," "removing barriers," making science "female-friendly," and working "towards a feminist transformation of the sciences" as suggested by a few of the representative titles (Bystydzienski and Bird 2006; Kass-Simon and Farnes 1990; Rose 1994; Rosser 1990).

There has been plenty within American science to fuel both historical projects, but as Russett (1989: 11) quipped, "Science, however, does not observe national boundaries," and as such some of the most influential works, historiographically speaking, examined Western scientific debates transpiring on both sides of – and indeed traversing – the Atlantic. Catalyzing historical critiques of scientific constructions of men's and women's social roles, Elizabeth Fee's (1973) presentation at the inaugural "Big Berks" Conference of Women Historians analyzed Victorian American and British ethnologists' construction of an "evolutionary tale" that she argued "confirmed their own social order," one in which male domination and "the thorough domestication of women" constituted a "quintessential characteristic of civilization" (38–9). Fee afterwards probed similar arguments in her essays on biology, physiology, and psychology (1976), as well as craniology (1979). By this time Carolyn Merchant was immersed in her profoundly influential book-length study (1990) on the shift of scientific thinking during the seventeenth century from primarily an organicist to mechanistic paradigm, with science now masculinized and nature feminized in a sexualized way: to be unclothed, penetrated, and dominated. Thus by the early 1980s, the historical works of Fee, Keller and Merchant, along with others, began to form a collective project concerned with, as Keller (1995a: 82) retrospectively summarized, "exploring the force of gender and gender norms not only in the making of men and women but also as silent organizers of the cognitive and discursive maps of the social and natural worlds that we, as humans, simultaneously inhabit and construct."

Much of the focus of this early scholarship on gender fell specifically on masculinity, or perhaps more precisely, "masculinist" science, and throughout the 1980s scholars added to this work. Brian Easlea carried a similar argument as Carolyn Merchant's beyond the context of early modern Europe and associated science's durable masculinist features with psychosocial concerns such as "gender and penile insecurity greatly exacerbated by the existence of social classes" (Easlea 1981: 55). He then applied this logic to explain the aggression behind the nuclear arms race as the inevitable outcome of an "irrational male behaviour" that "stems from men's oppression of women" (Easlea 1983: 5). Further attending to masculinity and patriarchy, Russett (1989), as already noted, critically analyzed Victorian "sexual science" while Donna Haraway (1989) examined American primatology as a form of "sexualized discourse" (11). Such scrutiny of the masculinist logic, metaphors, and narrative forms of science widened to include studies of science's masculinist imagery (Jordanova 1989), institutional culture (Noble 1992), as well as the intermingling of science and medicine with other cultural sources of masculine codes of honor (Nye 1993).

Despite this literature's emphasis on the pervasiveness of masculinity in science, another line of inquiry worked to debunk the myth that science was, according to David

Noble's (1992) title, a "world without women." To many of the writers who picked up this project, it was a deeply personal one. Through a mix of curiosity, serendipity, and perseverance, feminist scholars began revealing the sheer prevalence of women scientists, both historical and contemporary, as well as the conditions that either curtailed their careers or rendered them invisible within the historical record. Among the few pioneers in the early 1970s (for example, Solomon 1973; Wilson 1973), Margaret Rossiter, who had abandoned majoring in chemistry in favor of history of science, was singular in launching a career-long inquiry into the history of American women scientists. She started this in 1972 upon discovering, to her "astonishment," biographies of a few women in the first edition of *American Men of Science* (Rossiter 1982: xi, 2002: 62). This resulted in an article for the *American Scientist* (1974), which generated over 200 reprint requests, particularly among women scientists and their relatives who cited further cases (Rossiter 2002: 63). The project snowballed. Within a few years Rossiter determined patterns to "women's work" in the sciences resulting in women scientists' "hierarchical" and "territorial" segregations, the latter especially illustrated in the case of home economics (Rossiter 1980). With too much material for one book, she planned multiple volumes and published the first dealing with a period extending from the late nineteenth century up to 1940 (Rossiter 1982). Although followed by two further volumes that carried the project through the recent past (Rossiter 1995, 2012), this first volume provided the field with an enduring framework and set of terms through which to understand and talk about the trends, for example, the idea of a "Madame Curie effect," or the imposition of unattainable expectations based on the anomalism of an exceptional case (Rossiter 1982: 122–8; see also Des Jardins 2010). Rossiter (1993) latter added "the Matilda effect," or the systematic undervaluation of women's contributions in science, to the field's growing vocabulary.

Alongside Rossiter's monumental undertaking, other scholars similarly collected evidence for women scientists' historical experiences – or, as Rossiter's (1982) subtitle emphasized, their "struggles and strategies." Writers' approaches included the autobiographical (for example, Hynes 1982; Keller 1977), biographical (Abir-Am and Outram 1987; Keller 1983; Kohlstedt 1978a; Ogilvie 1986), and, like Rossiter, more broadly contextual (Kohlstedt 1978b; Morantz-Sanchez 1985; Rosenberg 1982; Verbrugge 1988). Much of the early work within this area comprised what Gerda Lerner (1986: 13) termed "compensatory history," yet it laid the groundwork for expanding the field in at least three directions: toward more specialized, contextualized studies focusing on women; accounts integrating women into broader narratives; and efforts to more richly historicize the interrelation of science and gender with other analytical categories like class and race, following cues from influential, poststructuralist perspectives like Foucault (1976), Scott (1986), Haraway (1988a), and Butler (1990). Toby Appel's (1994) close examination of the "female subculture" of late nineteenth-and early twentieth-century physiology among American women's colleges illustrated the richness of greater specialization. Elizabeth Keeney's (1992) seamlessly interweaving of women practitioners in her book on nineteenth-century American botany demonstrated the potential for integrative history. Paving the way for a new historiography, Haraway's *Primate Visions* (1989) scrutinized the influences of primatologists' social positions as determined by the contextual, local convergences of gender, class, and race, in the shaping of their scientific narratives about primates. Elizabeth Lunbeck's (1995) study of early twentieth-century psychiatry at the Boston Psychopathic Hospital was "loosely organized around a Foucauldian notion of discipline" (6), and her analysis considered the gendered construction of psychiatric ideas, professional conventions, and demographics based on norms for masculinity and femininity.

As this literature branched out and grew, assessments of the potential represented in the burgeoning feminist historiographies were both optimistic and cautionary. On the one hand, some scholars concluded that the international literature offered evidence and approaches with great potential for "re-mapping the historical world of science" (Christie 1990: 108–9) and radically changing "the most deep-rooted beliefs about our scientific history" (Benjamin 1991: 17). Yet, in her comprehensive review, Kohlstedt (1995: 55–6) cautioned against too much optimism. As evidence of the continuing neglect of the role of women in synthetic narratives about the sciences, she highlighted the case of biochemistry, according to Rossiter (1982: 289) a "moderately feminized" area: among the new histories, Kohlstedt noted, women biochemists failed to attract serious attention. Haraway (1997: 24–9) echoed this caution by underlining the absence of gendered analyses within the Robert Boyle industry that memorialized him as an exemplar of the new experimental way of life associated with the scientific revolution of sixteenth-century Europe. As she summarized the difficulty: "the blind spot of seeing gender as women instead of as a relationship got in the way of the analysis" (28). Elizabeth Potter (2001), however, had soon offered a partial corrective.

The Maturation of a Feminist Science Studies Field

Despite the prognostications and cautions, as the new millennium approached at least three trends signaled the maturing of "gender and science" (and its various permutations) as an interdisciplinary field of study that had come into its own. First, the proliferation of women's and gender studies courses across the disciplines included topics and seminars devoted to women, gender, and science (as well as medicine and technology), and faculty coming from various disciplines recognized the need for a number of readers consisting of the field's "classic" essays as well as updated topical surveys of the robust literature (of which examples will be cited below). Also around this time, and given the field's maturity, feminist scholars asked whether their efforts collectively made any impact in changing American science, whether in terms of its epistemology, institutional culture, imagery, gender demographics, or markers of gender equity like pay levels and career attainments. Given that gender inequities differed by scientific discipline, to some degree scholars' interests gravitated toward those areas with the worst problems, especially the physical and mathematical sciences and allied technical fields. Finally, the reality of continued wholesale ignorance of – if not outright backlash against – feminists' concerns precipitated a turning point and renewed sense of collective effort, in turn reshaping the agendas and approaches of feminist science studies scholars.

As the field matured, anthologies and special journal issues – often organized around critical themes and questions – multivocally opened new lines of inquiry. Among the pioneers were volumes devoted to critiques of research on sex differences (Teitelbaum 1976), to personal narratives of struggle (Ruddick and Daniels 1977), and to themes like "Women, Science, and Society," as highlighted in a special issue of *Signs* (Stimpson and Burstyn 1978). Workshops, conferences, and symposia – notably the Genes and Gender Collective first held at the American Museum of Natural History in 1976

– added further anthologies and monograph series to the feminist "collective" genres. From the mid-1980s onwards, individual scholars with noteworthy repertoires took advantage of the anthology format for collecting, updating, and adding to their acclaimed essays (for examples, see Keller 1985 and Haraway 1991). By the late 1990s, the widespread need for collations of exemplars from the rich, heterogeneous literature provoked a new generation of readers and guides, particularly as academic departments developed new courses devoted to feminist science studies.

Paving the way, further collections of reprints from the journals *Hypatia*, *Signs*, and *Isis* appeared in succession, offering representative theoretical approaches as well as empirically rich historical studies (Kohlstedt 1999; Laslett et al. 1996; Tuana 1989). Marketed directly to students in women's studies and the history of science, these guides also pushed the agenda for greater inclusion of the feminist themes "yet to be captured in traditional textbooks" (Kohlstedt 1999: 7). Yet the disciplinary filters of these retrospective collections left unmet the broader needs of multidisciplinary courses designed not only to engage students in critical examinations of the historical and sociological factors shaping scientific research, but also to expose students "to emerging and controversial research" in the very sciences under examination (Wyer et al. 2001: xiii–xvi). Further retrospective collections, as well as anthologies showcasing new directions in the research, thus proliferated, demonstrating not only the wealth of scholarship in a maturing, and expanding, field, but also the multiplicity of perspectives guiding the volumes' assemblies (Lederman and Bartsch 2001; Mayberry, Subramaniam, and Weasel 2001; Wyer et al. 2001). Additionally, Suzanne Le-May Sheffield (2004) offered an exemplary, introductory synthesis of research on "women and science" that effectively launched a new genre of textbooks.

In parallel to these endeavors, feminist scholars repeatedly asked a most fundamental question about their work, encapsulated in the title of Londa Schiebinger's book, *Has Feminism Changed Science?* (1999). Capstone publications based on major conferences held at the University of Minnesota in 1995 and Princeton University in 1998 added to the collective scholarly responses that again expressed ambivalence about the impact of feminist research as well as the need to widen the field of "gender and science" to include technology and medicine (Creager, Lunbeck and Schiebinger 2001; Kohlstedt and Longino 1997). As Kohlstedt (2006: 38) noted, optimism characterized the responses owing to "numerical and statistical increases in women across most fields." At the same time, as one group of editors stressed, "a new generation" of feminist narratives emphasized the "disciplinary heterogeneity of feminist science studies" and imagined "bold and brave new worlds" (Mayberry, Subramaniam, and Weasel 2001: 3).

The emphasis on inclusion of technology under the rubric "gender and science" signaled an underlying reconceptualization of this area, provoked both by a growing recognition of the difficulty in demarcating modern science from technology (as well as the historical malleability of their interrelation) – heralded by Haraway's (1997) use of "technoscience" – as well as concern over the persistence of gender inequities in the technological fields, particularly computer science and engineering. Clearly, the focus of the feminist studies agenda required further attenuation. Calling attention to the slow progress of interventions to reverse the underrepresentation of women in science, technology, engineering, and mathematics (STEM) fields in the United States, Bystydzienski and Bird (2006: 1) underlined a new configuration of Rossiter's territorial segregation: "Thus, while women are currently slightly overrepresented among all high school and college graduates, and constitute the majority of undergraduates and master's-level graduates in the biological sciences, in areas such as physics, computer science, and engineering women continue to obtain only a small proportion of the college degrees." They also observed the persistence of segregation hierarchically, as female departmental chairs and full professors in engineering, mathematics and statistics, earth sciences, chemistry, and physics and astronomy, combined, amounted to a very small percentage of the total: "It is still the rule that the higher the educational level, the fewer the women, and especially so in STEM fields." The so-called "pipeline" that analogized the pathway from early education to leading research positions remained "leaky," as confirmed by the results of a major study disseminated by a joint Committee on Maximizing the Potential of Women in Academic Science and Engineering (2007), formed by the National Academy of Science, National Academy of Engineering, and Institute of Medicine.

With the heightened STEM equity concerns, scholars who traditionally examined gender in relation to science and technology, separately, increasingly aligned their efforts and collaborated in works addressing both. The volumes by Creager, Lunbeck, and Schiebinger (2001) and Wyer et al. (2001) exemplified this very confluence of interests. Meanwhile a number of important historical works focusing on the gendering of technology and women in technology appeared from the mid-1990s onwards, further urging the addition of technology to the feminist science studies agenda. These works included examinations of women and gender in relation to technological invention and machinery (Oldenziel 1999; Stanley 1995), computer science (Abbate 2003), and engineering (Canel, Oldenziel, and Zachmann 2000). Within the same context, Amy Bug (2000) took on physics and Margaret Murray (2000) mathematics. Rossiter (2003) added her historiographical overview of this budding literature.

Provoking a turning point in feminist scholars' attention to these issues, the president of Harvard University, Lawrence Summers, gave a controversial speech on the question of women's status in STEM at a conference on Diversifying the Science & Engineering Workforce hosted by the National Bureau of Economic Research in Cambridge, Massachusetts in January 2005. In considering the underrepresentation of women among tenured positions in science and engineering at leading research institutions, he highlighted the availability of smaller pools of females qualified for such positions, which he argued was a consequence of statistical differences, by sex, in mathematical and scientific aptitudes, as shown through recent research on schoolchildren's standardized test scores. Summers' comments resurfaced arguments of biological determinism that fueled a media frenzy and provoked outcries by feminist scholars. In her contribution to Londa Schiebinger's (2008) volume assessing "gendered innovations in science and engineering," Charis Thompson observed how, following Summers' speech, "higher education entered a new phase of the politics of gender and science" and suggested that the time was ripe to "bring the richness of the field of gender and science into policy" (2008: 109–10). Women scientists joined the feminist studies "backlash," as framed by environmental toxicologist Emily Monosson (2008) in her introduction to a volume that reinserted the problematic of motherhood, or more generally "the pull of family versus career," in post-Summers gender and science discourse (12–14). Clearly Summers' inflammatory speech re-energized the feminist science studies agenda while serving as a new reference point in the battle against biological determinism.

Emergent Historiographical Themes

Alongside feminist sciences studies scholars' heightened efforts to bring their work into the policy realm for achieving greater gender equity in STEM, the historiographical shift of a "new generation" of scholarship proceeded to unfold. Among the trends, a resurgence within the broader history of science discipline of book-length biographies that offered more nuanced analyses of scientists' private lives beckoned for gendered analyses and more narratives about women. Although not concerning an American scientist, chemist Ruth Lewin Sime's (1996) admirable biography of physics Nobelist Lise Meitner, decades in the making, paved the way for feminist writers to contribute to this genre's renaissance. In American science, noteworthy outcomes of this trend included biographies of Rachel Carson (Lear 1997), Grace Murray Hopper (Williams 2004), and Mary Putnam Jacobi (Bittel 2009). Meanwhile, scholars critically re-evaluated narratives of women's passivity, demonstrating instead women's agency, in areas like home economics (Goldstein 2012; Stage and Vincenti 1997), women's writing on nature (Gianquitto 2007), science and health education (Kohlstedt 2010; Verbrugge 2012), scientific illustration (Shteir and Lightman 2006), and domestic research spaces (Opitz, Bergwik and Van Tiggelen 2015). Moreover, important works carried the baton forward in analyzing recent research on sex in the disciplines of endocrinology, psychology, neurobiology, and genetics. These scholars inserted the material role of "hormonal bodies" into gender (Roberts 2007); argued for "sexual fluidity" as opposed to fixity (Diamond 2008); challenged the gender/sex dichotomy through, for example, posing sex as an "effect of gender" (Jordan-Young 2010: 17); and urged an analytical approach of "modeling gender in science" as opposed to excising it (Richardson 2013: 226).

With the field's early emphasis on "righting the record," and despite the call to view gender as a relationship as opposed to a category synonymous with women, detailed gendered analyses on men's roles and masculinity have, with few exceptions, only recently gained traction. Feminists had earlier explored how masculine norms infused the practices, cultures, and epistemologies of science, or how science has been masculinist, but the nuances of how science contributed to changing constructions of masculinity and how gender mattered to the identities of men of science still needed addressing. Pioneering this shift in the agenda for masculinity was Nye's (1993) account concerning medicine in modern France, from which he extrapolated lessons for other contexts, including that of American science (Nye 1997). Adding to Nye's perspective, Naomi Oreskes (1996) challenged the received feminist wisdom that objectivity overshadowed other values traditionally used to discredit women, arguing instead that women tended to work in the most objective of occupations but, alternatively, not those considered most "heroic." Further scholars subsequently subjected scientific heroism to gendered analysis (Robinson 2006), as well as how gender played out in fieldwork (Pang 1996). In a similar vein, masculine technoscience subculture became a historical focus, as in Roger Horowitz's (2001) edited volume of essays that considered technological "toys" for "boys" and Nathan Ensmenger's (2010a) study of computing as a field overtaken by male antisocial, nerdy types (nicely complemented by Abbate's (2012) focus on women). Expanding greatly on these themes, contributors to Milam and Nye's (2015) *Osiris* volume addresses areas ranging from high-altitude physiology to penile plethysmography. At the time of this writing, the "masculinity" dimension to gender and science is still very much an emerging one.

Even so, a continuing area of neglect within the wider literature is focused inquiry into the scientific constructions of sexual orientation and sexual identity, as well as the particular issues faced by scientists whose "non-heteronormativity" or "queerness" clashed with science's professional cultures (on the terminology, see Warner 1991). A surge of studies emerged in the mid-1990s following national media sensations over scientific reports purporting evidence for the existence of genetic and anatomical causes for homosexuality, and the ethical concerns posed by the hypothetical possibility that such discoveries could lead to its "cure." Neuroanatomist Simon LeVay (1996) summarized the developments and the ethical issues from a scientific insider's point of view, and historians followed with analyses demonstrating how the terms of the modern debates had century-old precedents (Rosario 1997; also Rosario 2002). Much in the spirit of the reclamation project of feminist history of science, some scholars sought to reclaim the lives of scientists considered to be gay, lesbian, bisexual, transgender, and queer, though in the fields of science, medicine, and technology, a range of additional layers had to be penetrated to access hidden subjects (Hansen 2002; Opitz 2012: 246–51). Meanwhile, good histories of the scientific research on intersexuality, homosexuality, and transsexuality came out in regular succession (Dreger 1998; Kessler 1998; Meyerowitz 2002; Terry 1999). Joan Roughgarden (2004) controversially critiqued the theory of sexual selection and argued for a greater diversity of animal and human sexuality than previously portrayed by science. More recently, Anne Fausto-Sterling (2012) considered the ethical and political dimensions to queer "sexual science," but she also brought the subject back home to the context of biological research on sex and gender. Singular in drawing attention to the issues associated with sexual orientation, in addition to gender, within laboratory culture is the personal account offered by neuroendocrinologist Neena Schwartz (2010). Although queer theory is a field that had come into its own (Warner 1991), its full potential for the history of American science is yet to be realized.

Bibliographic Essay

The classic text providing an introduction to the primary issues in the field remains Keller (1985). Subsequent works providing updated assessments include Harding (1991), which contains an important essay on race; Bystydzienski and Bird (2006); Creager, Lunbeck and Schiebinger (2001); Jordanova (1993); Keller (1992); Keller (1995a); Keller and Longino (1996); Kohlstedt (1995); Kohlstedt and Longino (1997); Schiebinger (2008); Sheffield (2004) and, focusing on masculinity, Milam and Nye (2015). Rosario (2002) provides an introduction to the literature concerning homosexuality. Singular in opening the field to a historiographical focus on the body is Lacqueur (1990). Although focusing on European women, Schiebinger (1989) is pathbreaking in analyzing women's contributions, as well as the gender issues surrounding them, between the sixteenth and twentieth centuries. Unparalleled in comprehensively analyzing the history of women scientists in the United States are Rossiter's three volumes (1982, 1995, 2012). A valuable biographical resource is Ogilvie and Harvey (2000). Anthologies exploring cross-gendered and same-gendered scientific collaborations within families and households are Pycior, Slack, and Abir-Am (1996) and Lykknes, Opitz, and Van Tiggelen (2012). Green and LaDuke (2009) importantly adds to the literature on women in mathematics. Impressive in breadth, Schiebinger (2014) reprints a sampling of the field's representative works. With some overlap, the updated edition by Wyer et al. (2014) includes further works indicative of the field's emergent trends. For a complementary introduction to the gender and technology literature, see Herzig (2005a). On medicine, still unsurpassed is Apple (1990).

Chapter Thirty-One

THE GERM THEORY

Jacob Steere-Williams

In a 1902 article in *Popular Science Monthly*, Providence, Rhode Island's Health Officer – later President of the American Public Health Association – Charles Chapin, triumphantly declared "the end of the filth theory of disease." Its replacement, Chapin maintained, the germ theory, "had ceased to be a mere theory" (Chapin 1902: 235). The germ theory of disease is the scientific principle that explains how specific microorganisms are the cause of specific infectious diseases. Although some naturalists and medical thinkers in the early modern period had entertained hypotheses about the relationship between minute living organisms and disease, the central scientific findings that ushered in the germ theory were developed between the short historical time frame between the mid-nineteenth and early twentieth centuries. The phrase itself did not even appear in Anglo-American medical texts until the early 1870s. Debates about germs in the late nineteenth century were historically the most contentious, and it is more historically correct to think about a multiplicity of germ theories, rather than a single, unified, ontological germ theory.

Historians have long assumed that European scientists played the leading role in establishing the germ theory, with figures such as the French chemist Louis Pasteur, the British surgeon Joseph Lister, and the German bacteriologist Robert Koch, having reached heroic status in the history of medical science (Winslow 1943: Ch. 14). Yet, the germ theory – the veritable basis of scientific medicine – was debated, cultivated, and constructed across international, if not global lines. The laboratory-based science of bacteriology has also been granted the privileged status as the prized tool of how scientists studied germs in the past (Rosen 1993: Ch.7). This view was no better culturally enshrined than in Paul de Kruif's 1926 romanticized medical history, *The Microbe Hunters*, or in his close friend and collaborator Sinclair Lewis' 1925 classic *Arrowsmith*. Yet, as explained in this chapter, debates over germs in the nineteenth and early twentieth centuriesy came from a multiplicity of scientific and medical practices. In this way

A Companion to the History of American Science, First Edition.
Edited by Georgina M. Montgomery and Mark A. Largent.
© 2016 John Wiley & Sons, Ltd. Published 2020 by John Wiley & Sons, Ltd.

we can better understand how late nineteenth-century epidemiologists pioneered the deductive spatial and statistical evidence of the germ theory, even before bacteriological confirmation of the causative organism of cholera or diphtheria, for example. Pathologists, botanists, experimental physiologists, chemists, and veterinary scientists, too, all played pivotal roles in the complex development of the germ theory. It is important to remember that the formation of scientific disciplines that exist today were only in the process of professionalizing in the Victorian period, not the fixed disciplines that they were to become in the twentieth century. We should not, therefore, place too much emphasis on the successes of bench scientists isolating specific microbes.

Less contentious has been the way in which the germ theory served as a historical watershed between traditional, Hippocratic–Galenic medicine, and modern biomedical science. The germ theory provided the rationale – and led to the institutional and professional context – for large-scale public health intervention in the late nineteenth century, one that helped to fundamentally change the health landscape of the Western world and lead to what scholars call the Mortality (or Epidemiological) Transition (Condran 1987; McKeown 1976). The epistemological role of the germ theory, moreover, in modern Western history, was more than redirecting clinical medicine, laboratory science, and preventive medicine; the tenets of the germ theory served as a paradigmatic shift in the reconceptualization of the natural world. The theory directed not only how scientists practiced, in other words, but how everyday people behaved, lived, and worked.

The construction of the germ theory and the ascendency of its "mythical power," to quote leading medical historians Nancy Tomes and John Harley Warner, was nowhere more curious than in late nineteenth- and early twentieth-century America (Tomes and Warner 1997). Historians once thought that Americans were both passive and late in accepting the germ theory when compared to Europeans, in part due to the belief that nineteenth-century America was a backwater for scientific and medical research. Yet, it is still an interesting and thought-provoking historical question to contemplate how the United States became a leading node in a complex web of the study and control of infectious diseases by the mid-twentieth century.

The advocates of the germ theory – its scientific practitioners and its public health disciples – played a significant role in transforming American medical science as a destination, rather than a point of departure, for studying germs. In what follows, this chapter unpacks this transition in three sections. The first part examines Victorian debates over germs and infectious diseases, clarifying how a multiplicity of germ theories across Europe and the United States were debated alongside the longstanding filth theory of disease and belief in spontaneous generation. Part two turns from theoretical debates to three late nineteenth- and early twentieth-century sites where the germ theory gained acceptance in actual practice; in American laboratories, hospitals, and homes. This section charts both the rise of American universities such as the Johns Hopkins University, the University of Chicago, and the University of Michigan as leading centers for bacteriological training and studying disease. The final part of this chapter transitions from a focus on bacteria to the early twentieth-century research on viruses and immunology, where American medical scientists fully demonstrated the global reach of their research, especially through the Rockefeller Institute for Medical Research. Finally, the conclusion reframes the history of the germ theory, shedding light on its previous European and bacteriological focuses, and on its relevance to current scientific debates.

Germs in Theory: Victorian Debates and the Architects of the Germ Theory

Before the twentieth century, it is ahistorical to discuss the germ theory as a single, unified idea that dominated medical science. The most heated and controversial debates over the role of microorganisms in causing or contributing to the incidence of infectious diseases in humans or animals occurred in the last three decades of the nineteenth century.

Before the mid-nineteenth century, the vast majority of medical thinkers believed that diseases were caused by rotting organic filth – human, animal, or plant – that escaped into the air and produced disease-causing emanations, known as *miasmas*. The filth, or miasmatic theory of disease causation had a longstanding historical gestation in the Western world, and can be evidenced in very early Greek writings attributed to Hippocrates, such as *Airs, Waters, Places*. By the nineteenth century it was an all-encompassing physiological theory that both helped to spur public health activism – cleaning up bad smells in the environment might prevent disease – but also to encourage moralistic and religious condemnation for individuals who were stricken by disease – sinners or sexual miscreants were seen to be corrupting their internal constitutions and thus more likely to be effected by bad airs. "All smell, is," the famed British barrister turned sanitarian Edwin Chadwick exclaimed in 1846, "if it be intense, immediate acute disease" (Chadwick 1846). The medico-moral theory was best elucidated by historian Charles Rosenberg, who in his widely influential *Cholera Years*, noted that, "cholera was an inevitable and inescapable judgment ... a scourge not of mankind but of the sinner" (Rosenberg 1962: 40).

The miasmatic theory was supported by what was seen at the time as solid science and experiential evidence. What sustained miasma theory scientifically was the well-used theory of spontaneous generation, which held – since at least ancient Rome – that certain forms of life could, under specific conditions, arise from nonliving matter. The miasmatic theory, being predominately environmental within the Hippocratic–Galen framework, was particularly attractive to nineteenth-century Americans.

The epidemiological profiles of the wide array of infectious diseases that struck nineteenth-century America provided an arena for sorting out complex etiological debates. Besides the common lot of deadly childhood diseases such as measles, scarlet fever, and diphtheria, Americans faced tuberculosis (known as phthisis or consumption at the time), smallpox, typhoid fever, and typhus, all with great ferocity. The two endemic fevers to the United States, malaria and yellow fever, though considered environmental, were at times considered noncontagious, others due to a specific poison which travelled with people or ships (Humphreys 1999). As Margaret Humphreys has recently shown, surgeons during the American Civil War (1861-1865) may also have been prepared to accept proto-germ theories based on their experience with hospital gangrene, and especially widespread practices of disinfection (Humphreys 2013: 305).

American physicians and budding medical scientists were undoubtedly introduced to proto-germ theories by the early research of Pasteur, on fermentation, and Lister, on putrefaction, in the 1860s and 1870s. In a contest sponsored by the French Academy of Sciences in 1859, Pasteur devised a simple experiment to disprove spontaneous generation, where he heated a microbe-laden culture in a long, s-shaped swan neck flask. The heating destroyed the microbes present, and the shape of the flask prevented air-borne microbes from entering. If the flask was broken, Pasteur demonstrated, air-borne particles could fall into the flask and colonies of germs would proliferate (Geison 1995). He had successfully elucidated the distinction between aerobic microbes – those that require oxygen – and anaerobic microbes – those that thrive in its absence. In the years that followed, Pasteur articulated an early, chemically based version of a germ-based theory (Latour 1988).

Beginning in the 1860s, British surgeon Joseph Lister extrapolated Pasteur's chemical research to the problem of postsurgical infections, known as "hospitalism." By the 1870s, Lister had developed a rigorous system of antiseptic surgery, employing carbolic acid sprays to kill microbes present on wounds, and elaborate dressing techniques for preventing germs from entering wounds on the mend. Pasteur and Lister had far from solidified a single germ theory, but many Europeans and North Americans saw their views coalesced into a biologically based theory of disease causation (Brieger 1966).

Some Americans even tried to replicate European experiments or pioneer their own. The American physician James H. Salisbury is a prominent example. Salisbury, working in New York, boasted in the 1860s to have discovered the fungal origin (palmella) of a host of fevers, particularly malaria and typhoid, through experiments exposing volunteers to palmella-soaked soils that they supposedly breathed in. Ultimately Salisbury's theory was disproved, as other researchers found palmella in ordinary offices and homes where no fever existed. Despite few original contributions, Americans were abreast of European developments. Historian Nancy Tomes has shown through an interesting analysis of medical student theses from the 1870s that American physicians were well aware of debates over germ theories even before the craze of laboratory-based bacteriology first brought American students to Europe to study such methods (Tomes 1999: 26).

By the late 1870s even the phrase "the germ theory of disease" embraced two hypotheses. The first was the botanically inclined theory of the noted English microscopist Lionel Beale, who argued that disease-causing microbes were units of bioplasm produced by and discarded from the body. The developing hypothesis from Pasteur and to a lesser extent Lister, meanwhile, suggested that disease-causing microorganisms were living agents. The latter theory was furthered by English scientist John Tyndall, whose research into microbes suspended in the air helped to further demonstrate Pasteur's initial findings.

Many scientists disavowed biologically based germ theories. Some favored chemically inclined theories, popular etiological stances often encompassed by the term zymotic theory, which stated that the microbes responsible for causing disease were the byproducts of chemical processes similar to fermentation, combustion, or even organic poison. In an age of etiological uncertainty, metaphors dominated the rhetorical landscape; advocates of biological, botanically based germ theories applied what many scholars refer to as a "seed and soil" metaphor of disease, which helped to explain why some members of a community might get sick and others were spared (Worboys 2000). Particularly interesting is the way that numerous camps – even those who still maintained a belief in spontaneous generation – incorporated contentious theories of evolution, though by far the biggest advocates of evolutionary theory, as we might expect, followed a biological view of disease.

A prominent example of the multiplicity and complexity of germ theories was the belief in the dangers of sewer gas. While initially consistent with the miasmatic theory, by the 1870s and 1880s the fear of sewer gas was renewed by advocates of the germ theory, who claimed that specific germs lurked in sewers and as the gases wafted into American homes, they could be breathed in and cause specific diseases. Such hybrid views meshed with the theories of the British physician Charles Murchison and the German physician Max von Pettenkofer, who believed that specific germs needed to undergo a fermentation-like process in the soil before they could be infective to individuals.

The historian John Harley Warner has successfully argued that before the 1880s most American physicians believed that knowledge about disease was best gained from the bedside of patients. This reflects the larger importance of the debate over diseases etiological in the chaos of the 1870s. More broadly speaking, at stake was the relative merit and scientific authority of knowledge about disease gained from the bedside – the traditional bastion of knowledge making on disease – or the field or the laboratory. Americans, then, perhaps more than their European counterparts who were interested in the living microbial world, were initially hesitant to value evidence gleaned from experimental laboratory studies that were initially more prominent in Europe.

Bacteriological skepticism on the part of American physicians was in part due to the failure of early experimental studies on germs. Early microscopists, in Europe and North America, after all, were largely unsuccessful in providing *causal* proof of the existence of a single microbe that was responsible for a specific disease (Cassedy and Warner 1995). As Joseph Richardson, Professor of Hygiene at the University of Pennsylvania and germ theory advocate noted in an address before the Philadelphia Social Science Association in 1878, "in spite of the bold assertions of certain enthusiast and savants ... no really skillful microscopist will at present maintain that minute vegetable organisms, found in connection with contagious maladies, are as yet proved to have any definite relation to them as causes of disease" (Richardson 1878: 4).

Early investigators seemed to find microbes everywhere, in air, water, foodstuffs, and in and on the human body. They needed, and were later equipped with, a set of laboratory guidelines for proving causality between particular microbes and specific diseases, what became known as Koch's postulates (Gradmann 2009). Up to the 1870s, epidemiological methods – referring to studies of the communication of disease in populations – were far more successful in furthering germ theories than laboratory methods. Such reticence toward accepting laboratory-based research began to change in the 1880s.

The pivotal moment in the acceptance of bacteriology as a prominent tool of studying germs was Koch's research on anthrax in the mid- to late 1870s, where the German bacteriologist identified *Bacillus anthracis*, a large bacillus that he cultured in an aqueous solution of cattle eyes. Koch further demonstrated that the bacillus only existed in diseased animals; when injected into healthy ones, it consistently produced the disease. This was the kind of causative proof that had been missing from earlier experimental studies, and fuelled the rapid approval of the biologically based germ theory from the 1880s in America. An often-cited recognition was Abraham Jacobi's well-known address before the New York Academy of Medicine in 1885. There Jacobi, a German-Jewish immigrant and pioneer in American pediatrics, warned against what he dubbed "bacteriomania," the quick pronouncement by amateurish bacteriologists that they had "discovered" the specific microbe responsible for a specific disease. Jacobi's indictment was not, however, waged against well-trained experts such as Koch, and he noted that the majority of American physicians "have readily accepted the new gospel with but few exceptions" (Jacobi 1886: 163; Tomes 1999: 42).

While the multiplicity of germ theories began to unite into the 1880s, there were nonetheless many detractors. The critiques of proto-germ theories were many and plentiful in Europe and North America. The most famous was perhaps was the esteemed German sanitarian Pettenkofer, who in 1892 hoped to refute Koch's 1883 discovery of the cholera bacillus and the etiological link between it and the clinical condition, by swallowing a "cholera cocktail," made from the rice-water discharges of a sufferer of the disease. To Koch's surprise – and no doubt ours today – Pettenkofer lived with only an upset stomach.

Detractors aside, by the 1880s the landscape of American public health attitudes toward germ theories radically changed, as Americans enthusiastically took up the study and prevention of germs. No longer were Americans content with reading European literature; instead, they travelled to Europe for hands-on training in the new laboratory sciences.

Germs in Practice: The Laboratory, the Hospital, and the Home

Mid-nineteenth-century American medical institutions lacked the rigorous physiological, pathological, and chemical laboratories, to say nothing of the state of advanced medical training, which were by then a staple in many European universities. Such was the reason some wealthy Americans, as Warner has shown, travelled to the Edinburgh lecture halls of William Cullen or the Paris clinics of Xavier Bichat during the first half of the nineteenth century. By the 1880s, however, the center of training in new bacteriological techniques was Germany. With newfound interest in germs, many Americans travelled to Europe in hopes of training under figures such as Koch.

One of the architects of American public health, William H. Welch, provides an illustrative example. Born in Virginia in 1850, Welch attended Yale University and obtained an MD from the College of Physicians and Surgeons in New York City. Welch first travelled to Germany in the mid-1870s for additional training in pathology and experimental physiology, first with Carl Ludwig in Leipzig, then with Julius Cohnheim in Breslau. The lure of Germany, Thomas Bonner has argued, was the affiliation of laboratories within the structure of university medical schools (Bonner 1995). Upon his return in 1879, Welch opened the first laboratory course in America, at Bellevue Medical College in New York. Viewed as one of the rising stars of American public health, he was offered the professorship of pathology at Johns Hopkins University. In the mid-1880s he returned to Germany, visiting the renowned laboratories of both Pasteur and Koch. Welch was part of cadre of American experts – along with William Osler, William Henry Halsted, and Howard Kelly – that led Johns Hopkins University School of Medicine to be a foremost institution for studying disease (Fee 1987). Other Americans, notably T. Mitchell Prudden, who also studied at Koch's laboratory, helped to bring staining and culturing techniques back to the United States. Prudden, the first Chair of the pathology department at what became Columbia University, founded one of the first courses in bacteriology in New York.

Victor Vaughan and Frederick Novy at the University of Michigan also helped to stimulate the field. The two spent an entire summer in the mid-1880s at Koch's laboratory studying the new methods in bacteriology. Back in Ann Arbor, they pioneered research into ptomaines, a bacteriological and chemical theory that sought to show how certain microbes produced chemical toxins that caused disease. Although Vaughan is remembered as a more prolific leader in American public health, Novy had done the heavy lifting, training more American students how to identify, culture, and isolate bacteria than any other American medical scientist. The celebrated De Kruif, for example, originally trained under Novy.

Theobald Smith, who received his medical degree from Albany Medical College in 1883, exemplified the keen interest between human and animal diseases in this period. Smith, working at the Bureau of Animal Industry in Washington, DC alongside D.E. Salmon, helped to discover the etiological transmission and cause of Texas cattle fever, and the two pioneered research on the salmonellas. Smith's epidemiological and bacteriological work on tuberculosis was equally prolific; he established the first formal department of bacteriology at Columbian University (now George Washington University) in the late 1880s.

Armed with European techniques, the study of bacteriology blossomed in late nineteenth-century America. It was, however, institutionalized unevenly across the country, with leading medical centers at the University of Michigan (under Novy and Vaughan), the University of Pennsylvania (under Alexander C. Abbott), and Harvard (under George Whipple and Milton J. Rosenau). George Sternberg, famous US army surgeon, early bacteriologist, and authority on yellow fever, published one of the early American manuals in 1892, *The Manual of Bacteriology*, which was unrivaled until William T. Sedgwick, who first worked at the Lawrence Experiment Station of the Massachusetts Board of Health, and later at Massachusetts Institute of Technology, published *Principles of Sanitary Science* in 1901.

The isolation of many of the most common and dangerous infectious diseases provided some assuagement to the sanitary revolution which, as Martin Melosi has argued, brought widespread improvements to sewerage and water systems, and food protection across the United States (Melosi 2000). That said, public health departments still relied on epidemiologists to trace outbreaks in populations, and chemists to study bacterial counts in food and water supplies, particularly milk.

With institutional networks of bacteriology solidified throughout the United States by the 1890s, how were ideas about germs actually put into practice? Historians of American medicine generally refer to the late nineteenth-century zeal for widespread sanitary improvement based on the principles of the germ theory as the "new public health" (Rogers, 1989). Armed with knowledge of many microorganisms responsible for the spread of infectious diseases, American public health leaders sought to follow three commands: identify, isolate, and disinfect. Whereas the older, nineteenth-century version of public health focused on cleansing the environment, the new public health of the early twentieth century largely focused on stopping diseased individuals from spreading specific infections, though epidemiologists continued to think in terms of population dynamics.

The most visible sign of the "golden age of bacteriology," as early twentieth-century American public health leader Charles-Edward Armory Winslow noted, was the isolation hospital, established in cities and states across America. There individuals with – first clinically, later bacteriologically – confirmed cases of specific diseases would be held for a period of time in hopes of stopping chains of infection. Such isolation hospitals – sometimes wards within hospitals – depended a great deal on municipal bacteriological laboratories. Chapin had established the first of such institutions in 1888, and by the early twentieth century they proliferated across the country, to a greater extent than anywhere in the world.

As bacteriology was put into practice, however, new questions arose with regards to scientific authority, physician autonomy, and individual liberty. In the half-century between 1880 and the mid-1930s, when sulfonamides entered the Western pharmaceutical arsenal, bacteriologists in America could only count a few direct successes in therapeutics.

One notable example of the germ theory-inspired breed of public health in the United States was the campaign to control diphtheria in New York City from 1890–1930. As historian Evelynn Maxine Hammonds has shown, the city pioneered a bacteriologically inspired system of diphtheria identification, detection, and therapy (Hammonds 1999a). Armed with William Park's secure culture kit, Herman Biggs, the director of the city's laboratory, devised a thorough system diagnosing mild forms of the disease, distributing the diphtheria antitoxin, and testing the latter's efficacy in providing immunity. Though an important example, even here bacteriology did not straightforwardly lead to public health intervention until at least the 1930s. Instead, the entire period leading up to the development of sulfa-drugs demonstrates the protracted, and especially political nature of implementing knowledge of germs.

Implementing American bacteriology at times also elucidated the tenuous relationship between public health necessities and individual rights. The prominent example is the infamous case of Mary Mallon, known colloquially as "Typhoid Mary." Mallon, an Irish immigrant cook in New York City, was a healthy carrier of typhoid fever. After scrupulous case-tracing by George Soper, Mallon was found to have high concentrations of typhoid bacilli, leading New York City public health officials to isolate her on North Brother Island. She remained there from 1907 to 1909, but later returned for the same offense in 1915. Mallon died on the island in 1938, having spent 26 years in confinement. As Judith Walzer Leavitt maintains, the case shook the public health system of early twentieth-century America, and subsequently forced Americans to ask new questions about individual liberty, the state, and public health (Leavitt 1996). The quarantining of San Francisco's Chinatown, offset by Joseph Kinyoun's bubonic plague research, health inspection for typhus and glaucoma on Ellis Island, and the Tuskegee syphilis experiments from the 1930s all provide striking examples of the discriminatory nature of early American public health history that followed from the uneven implementation of the germ theory (Kraut 1994; Markell 2004; Reverby 2009; Risse 2012).

The first 50 years of the operation of the germ theory in America, then, was defined by a passionate quest to gain knowledge of effective bacteriological techniques (Duffy 1990: Ch. 13). As Americans travelled to European laboratories, they also began to organize and institutionalize bacteriological teaching and practice across the United States. In the following 50 years, however, there was tension between the promises of bacteriology and the actual successes on the ground.

Beyond Germs: From Bacteria to Viruses and Beyond

What the germ theory initially accomplished was a reorganization of medical science, particularly a reorientation in the valuation of experimental laboratory training and research. As explored above, Johns Hopkins University is perhaps the best example of this shift in American medicine. The popular and cultural acceptance of the germ theory, however, was in spite of the direct translation of laboratory-based bacteriology to effective therapeutics. The bacteriological revolution, historians now agree, did not single-handedly lead to immediate change in the health landscape of Western society, though it did result in the flourishing of the cultural authority of medical science and the redirection of preventive medicine (Cunningham and Williams 1992).

In the first decade of the twentieth century, the only successful therapeutics to come directly out of germ-based bacteriology was the diphtheria antitoxin, a rabies vaccine, and to a lesser extent, the improvement of the smallpox vaccine. Less successful were experimental vaccines against bubonic plague, cholera, and typhoid fever, though the latter showed some success in the Spanish–American War. The first half of the twentieth century saw the maturation of research on microorganisms, and in the avenues of prevention and treatment. This was due in part to Progressive Era private philanthropy, notably the Rockefeller Foundation, and to the increase in federal support for disease-based research.

The Rockefeller Institute for Medical Research, established in 1901, and later the Rockefeller Sanitary Commission, founded in 1909, provided an unprecedented model for privately funded medical research. The 1909 commission had the express goal of reducing the incidence of hookworm disease in the American South, and in bolstering local public health services. It later grew in size and productivity, particularly the International Health Division (Farley 2003; Palmer 2010).

An early forerunner for the National Institutes of Health, the federal Hygienic Laboratory was directed by the energetic Joseph Kinyoun (1860–1919) and was established in 1887. Initially designed as a diagnostic laboratory to support federal quarantine laws against cholera, bubonic plague, yellow fever, and smallpox, the Hygienic Laboratory, with first Kinyoun and after 1899 Milton Rosenau, expanded in scope to include more advanced research in microbiology. Kinyoun alone advanced experimental immune serum for smallpox, and early vaccines for pneumococcus and streptococcus. In 1930 the Ransdell Act transformed the growing Hygienic Laboratory into the National Institutes of Health, the leading center for microbiological research in the decades to follow. In addition to the increase in private and federal funding for disease research, professional backing by the American Public Health Association and especially the Society of American Bacteriologists provided the institutional nexus for furthering American microbiology.

While there were obvious successes in American public health in the early twentieth century – Joseph Goldberger's 1915 finding that pellagra was due to a dietary deficiency speaks to this trend – it was not until the 1930s that bacteriologists pushed passed a major obstacle in both conceptualizing and treating disease. As early as 1896 Martinus Willem Beijerinck claimed that microbes smaller than bacteria, able to pass through filters and lie undetected by microscopes, must also cause disease. The development of the electron microscope in the 1930s provided the technical tool for launching virology, but the process was already well underway, particularly in plant molecular virology.

Development of virology in 1930s provides an interesting case for pushing the paradigm of the traditional tenets of American bacteriology. High-profile cases such as President Franklin D. Roosevelt's bout with polio in the 1930s led to massive public mobilization, and virologists increasingly took the spotlight from bacteriologists. Jonas Salk, at the University of Pittsburgh, became first to discover an effective vaccine

(Oshinsky 2006). Furthermore, the role of the American pharmaceutical industry in the development of penicillin in the 1940s and 1950s seemingly marked the fulfillment that the germ theory had promised in the late nineteenth century. John F. Mahoney, of the Public Health Service, for example, triumphantly found in 1943 that penicillin was more than 90% effective against syphilis.

By the middle of the twentieth century the disease landscape had fundamentally changed. While infectious diseases of the nineteenth century had gradually declined, the steady increase in chronic disease, especially cardiovascular disorders and cancers became the leading causes of biomedical concern. Yet, some scholars have identified the way in which the germ theory limited initial research. Bacteriological reductionism, which had germ hunters searching for the "cancer germ," for example, may have in fact impeded biomedical research. As the twentieth century came to a close, emerging and re-emerging infectious diseases, among them Legionnaire's disease, Ebola, antibiotic-resistant tuberculosis, and Rift Valley fever, were arenas where the germ theory has still guided basic microbiological research, yet the discovery of prions has fuelled a reassessment of the tenets of causality provided by Koch's postulates.

Recently some scholars have gone as far as to call for a "new" germ theory of disease. Evolutionary biologist Paul Ewald has led the way, combining microbiology and evolutionary theory to suggest that many of the diseases thought to be environmental or genetic may in fact be caused by infections (Ewald 2002). This provides an interesting parallel to the historical narrative of the germ theory, and implies that a multiplicity of germ theories is perhaps still the best frame with which to understand ideas about the causation of disease.

In modern history, the germ theory stands at the center of several fundamental shifts, including the mortality transition and the rise in the cultural authority of science. American medical scientists entered debates about germs only tenuously in the 1860s and 1870s, though they followed such developments prodigiously. A cadre of medical scientists then institutionalized laboratory practices on germs at home in America. By the mid-twentieth century the United States was the model for a country with a well-funded commitment to basic and experimental research on disease.

A historical sketch of the germ theory in the United States reveals several important conclusions. The first has to do with the complexities in the fabric of American science, which historians describe as maturation from the mid-nineteenth to the early twentieth century. This was in part due to the institutional and professional frameworks which supported biological research, and in this way microbiology overlaps with several scientific fields in its development. The more important trend the historical narrative of the germ theory reveals has to do with scientific authority. That there even is an essay on "the germ theory" in the early twentieth century is indicative of the epistemological purchasing power the concept holds in Western popular and scientific imagination. Ideas and theories about germs, after all, have been constructed in particular times and in particular places, and given value-laden power unevenly over time.

Bibliographic Essay

Historical interest in the early reception of the germ theory in America began in earnest in the mid-twentieth century. The first wave of historians to examine the period between

1860 and 1880 argued that American physicians and scientists were slow in accepting the germ theory of disease, which in turn was seen largely as a product of Europe. Phyllis Allen Richmond's 1954 paper, "American Attitudes toward the Germ Theory of Disease, 1860–1880," is indicative of such a position. Some historians, such as Lloyd Stevenson (1955) and Richard Shryock (1972), went as far as to contend that American sanitarians were hostile to new experimental approaches.

By the 1990s, as the history of medicine began to take a decidedly social and cultural turn, historians began to revise the older narratives of American exceptionalism. Nancy Tomes' work, in a series of articles and in her celebrated *Gospel of Germs* (1999), has analyzed the power of the germ theory in transforming the lives of everyday Americans, thus reframing the history of the germ theory. Tomes argued that while Americans made few original contributions toward the germ theory, they were mindful of European developments. She went as far as to call Richmond's findings the "myth of American backwardness" (Tomes and Warner 1997: 22).

From the 1990s, historians started to ask new questions about the ontological stability of the germ theory, most notably Michael Worboys, whose 2000 account *Spreading Germs* unpacked the complexity of germ theories, rather than the march of a single unified theory. As a counterbalance, readers might also consult K. Codell Carter's 2003 *The Rise of Causal Concepts of Disease*, which undermines the social history of the germ theory from a philosophical perspective, focusing on the power of the germ theory vis-à-vis its ability to provide causal proof.

Because historians tended to assume that the production of the germ theory was a European-inspired development, the majority of early historical research tended to be whiggish. Nonetheless, those interested in Pasteur should consult both Gerald Geison's 1995 work, which examines Pasteur's laboratory notebooks, and Bruno Latour's 1993 book, which, in a philosophically reflexive way, penetrates the knowledge production of Pasteur's theories among French scientists and the public. Christoph Gradmann (2009) has produced a study on Koch and German bacteriology that is attuned to new approaches of treating science as a cultural process. Worboys' 2000 account should be consulted for those interested in Lister, but Jerry Gaw's 1999 monograph is a full description of the diffusion of Listerism in Britain. Some American scholars, such as Thomas Gariepy (1994), have even examined the slow and uneven spread of Listerism in the United States.

As of late the more interesting research has focused on germ practices, rather than simply armchair debates over germ theories. There are also many individual case studies that exemplify American public health approaches toward specific diseases. Hammonds' 1999 monograph on diphtheria in New York City, and Leavitt's 1996 work on "Typhoid Mary" have been mentioned, but others include Alan Brandt's 1987 work on venereal disease, Katherine Ott's 1999 book on tuberculosis, David Oshinsky's 2006 monograph on poliomyelitis, and Margaret Humphrey's excellent 1999 book on yellow fever.

For those interested in the application of the germ theory within the Rockefeller Foundation, the best recent examples are Stephen Palmer's 2010 *Launching Global Health*, and Warwick Anderson's 2006 case study of American tropical medicine in the Philippines, titled, *Colonial Pathologies*.

Chapter Thirty-Two

INSTRUMENTATION

Sara J. Schechner

The study of scientific instrumentation in the United States is a relatively new field of historical research, even though mathematical instruments were carried onboard the European ships that brought explorers to the coasts of North and South America 500 years ago. This chapter will trace the historiographical development of the field, describe current topics of research, list resources available (particularly in the form of material culture), and make suggestions for future projects.

Defining Scientific Instrumentation

It is necessary to begin with three questions: What is an instrument? What makes it scientific? And where does one begin and end? As scholars have pointed out, there is ambiguity in the term, "scientific instrument" (Baird 2004; Taub 2009, 2011; Van Helden and Hankins 1994; Warner 1990). The *Oxford English Dictionary* (*OED*) defines "instrument" as "a material thing designed or used for the accomplishment of some mechanical or other physical effect." By this definition, a scientific instrument could be a microscope that magnifies the image of tiny things, a Bunsen burner to heat a flask of water, or a surgical knife used to open a body. The *OED* goes on to say that the word may "also [be] applied to devices whose primary function is to respond to a physical quantity or phenomenon, esp. by registering or measuring it." This adds thermometers and magnetic compasses, spectrometers and acoustical resonators, pulse height recorders and timers. But the *OED* suggests boundaries to these definitions, saying that the former are mechanical contrivances that are usually "portable, of simple construction, and wielded or operated by the hand" and the latter "may function with little direct human intervention and be of complicated design and construction." These qualifiers seem too restrictive for modern scientific instruments, for an environmental test chamber or a particle accelerator are rarely portable or simple, and a wind vane,

A Companion to the History of American Science, First Edition.
Edited by Georgina M. Montgomery and Mark A. Largent.

meter stick, or electrometer need not be complicated. But the *OED* concedes that an instrument is "distinguished from a *machine*, as being simpler, having less mechanism, and doing less work of itself; but the terms overlap." It also acknowledges that the user's social and professional status influences the choice of terminology: An "instrument" is "now usually distinguished from a tool, as being used for more delicate work or for artistic or scientific purposes: a workman or artisan has his *tools*, a draughtsman, surgeon, dentist, astronomical observer, his *instruments*."

A broader definition of an "instrument" is given by a recent encyclopedia devoted to "Instruments of Science" (Bud and Warner 1998). It includes test equipment for industry and health; drawing instruments; model organisms such as mice and *Drosophila*; cameras and projection lanterns; models such as globes, planetaria, and orreries; calculating tools such as sectors, slide rules, and computers; pocket sundials; and sextants – in short, the principal apparatus of astronomy, biology, chemistry, geology, mathematics, medicine, physics, psychology, navigation, surveying, and horology. As collections are assembled, they can be even more expansive in their interpretation. The Harvard Collection of Historical Scientific Instruments, for example, holds telegraph, telephone, radio, and radar equipment, phonographs and early pressings, balloons and kites, test tubes and chemical glassware, mineral specimens, metabolism cages, zoetropes and kaleidoscopes, anatomical models, tuning forks, an organ, vacuum tubes, a rocket nose cone, and a whiffle ball. Thus the term "scientific instrument" applies to objects employed by scientists not only for experiment and measurement but also for teaching, calculating, modeling, and communication. It applies to the apparatus of land-based, practical disciplines related to astronomy – i.e., time finding, navigation, and surveying. It also applies to the material culture of science that finds its way into the hands of nonscientists for their daily use – e.g., pocket sundials and pocket calculators, almanacs, magnifying glasses, dunking birds, and pregnancy test kits. The categorization owes much to the commercial interests of manufacturers and retailers, patent offices, and mechanics' fairs in the nineteenth century, as well as to the classifications employed by museum curators and collectors (Taub 2009; Warner 1990).

As Ulrich et al. (2015) have shown, the classification of all forms of material culture is unstable. A Pyrex pie dish bought at a local supermarket becomes a scientific instrument when placed underneath a metabolism cage to catch the lab rat's waste to be analyzed. A tuna fish can swapped for a Fisher Scientific feeding bowl inside the cage is also part of the instrument. Human teeth pulled from the mouths of Bostonians at Harvard Dental School circa 1900 were anatomical specimens first, carefully suspended from wires inside glass vials. Sixty years later they became another type of scientific instrument when they were ground up and analyzed at the Harvard School of Public Health in order to establish a baseline for environmental levels of polonium to be compared to the polonium absorbed from tobacco smoke by Bostonians. In practice, a scientific instrument can be anything used for a scientific purpose.

To complicate matters further, Warner (1990) has pointed out that the term "scientific instrument" was not adopted until the mid-nineteenth century. Before that time, the instruments were labeled "mathematical," "optical," and "philosophical." Mathematical instruments included all the tools used by mathematical practitioners, who worked in the sixteenth through nineteenth centuries in a variety of applied disciplines in which arithmetic and geometry were employed to solve real-world problems. These disciplines included astronomy, surveying, navigation, fortification, gunnery, dialing, cartography, and computing. The instruments were distinguished by having engraved divisions against which measurements were taken. Examples included sundials, theodolites, sectors, quadrants, and astrolabes. Drawing instruments were usually classified under mathematical instruments, because etuis included protractors, measuring rules, sectors, and dividers, which were employed together for calculations, plus pens, pencils, and compasses used to make maps and charts. Late in the nineteenth century, such instruments came to be called engineering instruments. Optical instruments included lenses, prisms, and mirrors, plus telescopes, microscopes, refractometers, and such which incorporated these components, as well as spectacles and reading glasses. Philosophical instruments took their name from the study of nature known in the early modern period as "natural or experimental philosophy." Philosophical instruments included air pumps, electrical machines, chemical apparatus, and pedagogical devices. In historical discussions, scholars frequently resort to the terminology of the day, but they also pragmatically employ the label "scientific instrument," knowing full well that it is anachronistic but too useful to do without.

Historiography

Most writing about scientific instrumentation has been done by Europeans about European objects but the historiographic arc is instructive and applicable in many ways to the American scene. For most of the twentieth century, scholars were divided into two camps. On one hand, there were historians of science who were biased towards text-based sources. On the other, there were individuals who worked closely with surviving apparatus – e.g., curators of museum collections, collectors, and dealers.

The founder of the discipline of the history of science, George Sarton, emigrated in 1915 to the United States and spent most of his career at Harvard. Sarton was interested in the broad sweep of scientific ideas across civilizations and centuries, and paid little attention to scientific instruments and the details of experiments. Another influential visitor to the United States was Alexandre Koyré. He saw the history of science as part of intellectual history. In "Galileo and Plato" (1943), Koyré argued that observation and experience got in the way of good science. Galileo, he said, dispensed with real-world experiments in favor of philosophical thought experiments unclouded by the messiness of real-world equipment. For Koyré, the heroes of the Scientific Revolution were not craftsmen-scholars, but philosophers who rarely built more than a theory. Sarton, Koyré, and their students, consequently focused attention on the logic of scientific ideas, the influence of philosophical beliefs, and other matters "internal" to the ship of science. By contrast, so-called "externalists" followed the course set in the 1930s by Boris Hessen in *The Social and Economic Roots of Newton's "Principia"* (1931) and Robert K. Merton in *Science, Technology and Society in Seventeenth Century England* (1938), exploring how that ship was tossed about by waves of religion, politics, and economics. By the late third and fourth quarters of the twentieth century, scholars were merging these two streams, with case studies of how religious and political beliefs strongly colored scientific ideas and practice (Shapin and Schaffer 1989).

During this same period there were very few specialized museums of science. Old instruments were kept with decorative arts in art museums, and with technology and industrial arts in national museums. They were also scattered among rooms in historic houses, left in closets of retired apparatus in institutions of higher learning, or beloved in private collections. For example, from 1923 to 1938, the Museum of Fine Arts, Boston, exhibited a private collection of portable sundials dating from 1600 to 1900, which had belonged to Harold C. Ernst, MD, professor of bacteriology at Harvard Medical School. At nearly the same time (1927–1936), the Metropolitan Museum of Art in New York put on display the sundial collection of the late John C. Tomlinson, Sr., a prominent New York attorney. These loaned objects complemented the museum's own holdings of sundials, clocks, and other scientific instruments, which were kept in its decorative arts department and were overseen by curators of Western European art. After the closure of the Philadelphia Centennial Exposition, 1876, the Smithsonian Institution relocated 60 boxcars worth of exhibits on American history, art, zoology, geology, anthropology, medicine, and the technologies of metallurgy, printing, transportation, textiles, fisheries, and agriculture to a new US National Museum. When the building opened in 1881, exhibits on scientific instruments created by firms like Keuffel & Esser (mathematical, drawing), Codman & Shurtleff (surgical, dental), Joseph Zentmayer (microscopes), and W. & L.E. Gurley (surveying), were part of the mix. The Adams National Historical Park in Quincy, Massachusetts, has always shown off globes and telescopes belonging to President John Quincy Adams in their original home setting, and Thomas A. Edison's apparatus remains on view in his laboratory in West Orange, New Jersey, now part of the Edison National Historical Park. Colleges and universities frequently preserve many laboratory instruments in formal or informal collections. Research equipment of one generation often becomes the teaching instruments of the next. Examples of such collections are found at Harvard, Dartmouth, Transylvania University, and the University of Mississippi, to name just a few. Medical schools and hospitals also preserve anatomical specimens, microscopes, and surgical instruments in special collections – for instance, in Harvard's Warren Anatomical Museum in the Countway Library of Medicine or in the Medical Museum of the Armed Forces Institute of Pathology (now part of the National Museum of Health and Medicine). The East India Marine Society of Salem, established in 1799 by sea captains, still exhibits maritime instruments amidst natural and artificial curiosities brought back from voyages around either the Cape of Good Hope or Cape Horn.

In the first half of the twentieth century, those who took care of this material culture of science were usually not formally trained in the history of science, being art historians or senior scientists, and they tended to take an antiquarian approach. They appreciated the instruments for their artistry and craftsmanship, or they sentimentally valued them because of a famous former owner. If the instruments were displayed at all, they were displayed in isolation as artistic productions or library furnishings, which often made the circumstances of their prior scientific use or social context hard to understand. Publications by collectors and scientists concentrated on the development of a particular class of instrument, often including a chronology of related devices and their makers. These early works read like catalogs, but their value should not be dismissed. They laid down important scaffolding for locating the objects and makers in time and place. Examples include Clay and Court (1932) on microscopes, King (1955) on telescopes, and Mayall and Mayall (1938) on sundials.

In 1947, an International Union of the History of Science (IUHS) was established under the banner of the International Council of Scientific Unions and the endorsement of UNESCO with the sponsorship of the Académie Internationale d'Histoire des Sciences. One of its first commissions, established in 1952, was a Commission des Instruments Scientifiques (today known as the Scientific Instrument Commission of the International Union of the History and Philosophy of Science, Division of History and Technology). Its first major project was the establishment in 1956 of a committee to produce a worldwide inventory of historical scientific instruments. The creation of a society of curators, historians, and scientists interested in the preservation and study of early scientific instruments was an important step in the development of the field. Key players in those early years were Henri Michel, Francis Maddison, Maurice Daumas, and in the United States, I. Bernard Cohen, Derek de Solla Price, and Silvio A. Bedini. The Scientific Instrument Commission is still active today, but the field has been augmented by specialized societies for astrolabes, sundials, telescopes, slide rules, balances, and maps, each holding its own meetings and publishing its own journals. Organizations based in the United States, but with international membership, include the North American Sundial Society, the Antique Telescope Society, the Oughtred Society, and the International Society of Antique Scale Collectors.

At a time when few scholars looked at scientific apparatus as more than window dressing, two individuals played an important role in bringing their cultural value to light in the United States: David P. Wheatland and I. Bernard Cohen (Schechner 2012). A 1922 graduate of Harvard College with a bachelor of science degree, Wheatland went to work in 1928 for the Harvard Physics Department, first as a technical assistant to a faculty member, then as department secretary, and in 1940, as the assistant director of the Cruft Research Laboratory of Physics. He oversaw the building of the Mark I computer and was its first civilian operator. Wheatland's duties led to numerous encounters with obsolete instrumentation often discarded in stairwells and attics of the science buildings on campus. He was already a collector of rare books on electricity and magnetism, and he was astonished to see the apparatus depicted in engravings in those books reified in the castoff instruments. He understood that these objects represented an important part of local scientific heritage, but he feared that they were in physical danger due to neglect as well as the propensity of faculty and students to cannibalize them for spare parts. Since the Physics Department did not then see any value in the instruments, Wheatland took them into his office for safe-keeping. He often reunited parts that had long been separated, and he cleaned and repaired the apparatus. When his small office became filled to overflowing with "foundlings," Wheatland sought a new space for the assemblage.

One person taken with Wheatland's cause was I. Bernard Cohen, since 1942 an assistant professor of the history of science, Sarton's student, and the first person to get a PhD in history of science in the United States. Cohen was investigating similarities in the work of Benjamin Franklin and Isaac Newton, and learned that Franklin in the 1760s had personally selected in London many of the rediscovered instruments now in Wheatland's Harvard office (Cohen 1941, 1956). Cohen helped Wheatland to recover other primary documents in the university archives related to the acquisition and use of the apparatus for research and teaching. Together they set up the first exhibition of the instruments in February 1949 (Wheatland and Cohen 1949). This project led to the formal establishment of the Collection of Historical Scientific Instruments with

Wheatland as curator. Cohen's findings about experimental philosophy at Harvard were published in *Some Early Tools of American Science* (1950).

Wheatland continued to collect with context in mind as he rescued instruments from dusty corners and dumpsters at Harvard. He had a real knack for knowing what would be of fundamental historical importance long before anyone else thought to save it. His charming book (Wheatland 1968) stood out because it was neither a technical description of the featured instruments nor a family tree on which he located them like some sort of evolving species. Rather it contained stories about each instrument: what it cost to buy and repair, how professors and students interacted with it, expeditions it went on, and so forth.

If this group of early instrument devotees shared anything in common with the historians of science of their day, it was an "externalist" method. They believed that to understand science – much less its apparatus–one could not simply look at scientific theories. Those who worked with scientific instruments argued for the importance of technical developments, artisanship, commercial practices, cultural aesthetics, social hierarchies, and consumption of instrumentation in understanding the scientific enterprise. During the second half of the twentieth century, more and more members of the Scientific Instrument Commission were trained in the history and philosophy of science, and their publications situated the instrumentation into larger themes in the history and philosophy of science. Topics included mathematical practitioners and the London instrument trade, the material culture of astronomy in people's daily lives, iconography of scientific instruments, how scientific theory and social values influence instrument design, popular science and the spectacle of experiments, courtly patronage and competition, outfitting research expeditions, technology transfer, and relationships between master and apprentice, scientist and craftsman (Anderson, Bennett, and Ryan 1993; Grob and Hooijmaijers 2006; Morrison-Low 2007; Schechner 2001).

Back in the other camp, some historians of science not associated with instrument collections became interested in the 1980s in the connections between theory, experiments, and laboratory culture (e.g., Gooding, Pinch, and Schaffer 1989). Instruments were part of these histories, but they were viewed as unproblematic in and of themselves. In 1994, the History of Science Society published an issue of *Osiris* devoted to the topic of *Instruments* (van Helden and Hankins 1994). While regarded as a "coming of age" for instrument studies, contributors to this volume relied almost exclusively on textual sources in their papers, and only one author had any hands-on experience with actual museum objects.

Some who had hands-on experience found this irksome, particularly when artist David Hockney made the news and garnered a lot of support from the public and scientists for his book, *Secret Knowledge: Rediscovering the Lost Techniques of the Old Masters* (Dupré 2005). Hockney claimed that Renaissance artists could not have painted with such exquisite realism without the help of optical instruments on the sly. One problem with broad claims (such as those made by Hockney and his defenders) about the use and performance of particular instruments was that these claims were based solely on what natural philosophers (scientists) or mathematical practitioners (engineers and technicians) had written about their own inventions or procedures. Such reports are notoriously unreliable and idealized. For instance, during the time of the Old Masters, many published descriptions of instruments were no better than science fiction, describing an imaginary device that was never made or never could work given the quality of materials available and the craft skills of the period (Schechner 2007). Although the hands-on scholars were glad to see what Jim Bennett, the keeper of the Museum of the History of Science in Oxford, called the "current vogue for instrument studies" (Bennett 2003), they still felt marginalized by mainstream history of science where few scholars thought it advantageous, much less necessary, to examine surviving examples or use reconstructions. To address these concerns, Bennett helped to organize a conference, "Do Collections Matter to Instrument Studies?" which was sponsored jointly by the Scientific Instrument Commission of the International Union of the History and Philosophy of Science and the British Society for the History of Science, and held in Oxford in June 2002. In the United States, Sara Schechner, the curator of Harvard's Collection of Historical Scientific Instruments, organized "The Material World of Science, Art, Books, and Body Parts," which was the opening plenary session for the annual meeting of the History of Science Society in Milwaukee in November 2002. The papers delivered at these meetings argued strenuously that the material culture of science offered historians rich evidence that could not be gleaned from textual sources alone. This point continues to be made by museum-based scholars (Morris and Staubermann 2010), but only recently has it been given space in journals such as *Studies in the History and Philosophy of Science* (Taub 2009) and *Isis* (Taub 2011) in the form of small assemblages of papers that take tangible scientific instruments as their starting points.

Bibliographic Essay

The first instruments to make it to the shores of the New World were navigational, surveying, time-finding, and time-keeping instruments, which were needed to explore the coastlines and establish working colonies. The inventories of expeditions such as those of Martin Frobisher in 1576–1578, reports by Thomas Harriot and others, and archaeological digs at sites like Jamestown give a fair picture of these early instruments: They included mariners' compasses, azimuth compasses, nautical charts, dividers, lodestones, sand glasses, logs and lines, sounding leads, cross staffs, backstaffs, mariners' astrolabes, quadrants, sundials, nocturnals, surveyors' theodolites, plane tables and plane table compasses, variation compasses and dip needles, globes, armillary spheres, and mechanical clocks. A pocket sundial played a central role in the famous story of Captain John Smith being rescued by Pocahontas (Schechner 2007). And chemical apparatus for assaying ores was employed in Jamestown. An excellent introduction to the instruments of colonization from an American perspective can be found in a special issue of *Rittenhouse* (Hicks 2007). For the instrumentation set within global contexts, see *The World of 1607*, the catalog of an exhibition celebrating the 400th anniversary of the Jamestown settlement (Jamestown-Yorktown Foundation 2007).

Mathematical instruments also had the distinction of being the first scientific instruments to be sold, made, and repaired in the American territories. Crude slate sundials were made locally at colonies like Avalon, established in 1621 in Newfoundland (Schechner 2004). Otherwise, the earliest known, native-built instruments were eighteenth-century wooden instruments for navigation and quirky wooden copies of brass surveying instruments. A London-made surveyor's compass was fashioned of brass with a silvered magnetic compass whose wind rose was exquisitely engraved and surrounded by a raised ring divided into degrees. Its American-made counterpart was rough-hewn wood with a printed paper wind rose and a divided circle of pewter. American manufacturers turned to wood because the local supply of metals was small, and there were few skilled artisans who knew how to work with it before the second quarter of the nineteenth century. Instrument makers could melt down broken brass implements to fashion new parts, but impurities in the metal made such brass unsuitable for instruments with magnetic compasses. Given the expense and trouble of importing quality brass in ingots and sheets, makers turned to wood, or simply skipped manufacturing all together and sold ready-made imported mathematical instruments (Bedini 1975: 191–6).

To learn more about the makers of surveying instruments in the eighteenth and nineteenth centuries, Smart (1962–1967) offers a starting point in the form of a catalog of makers, but some of the information is dated. Bedini's *Early American Scientific Instruments and Their Makers* (1964) is also devoted to surveying instruments. He divides makers into those who worked primarily in wood, and those who worked in brass, and an appendix divides practitioners into geographical regions and the types of instruments they sold. A problem with this book, however, is a concern that Bedini often took at face value advertisements by persons claiming to be instrument *makers*, when now it is more clear that many were *sellers* and *repairers* of goods imported from England (Schechner 2009). A rare, London-trained mathematical instrument maker in the American colonies was Anthony Lamb; his story is told by Bedini (1984). Bedini's *With Compass and Chain* (2001) is a more nuanced examination of the work of cartographers, surveyors, and instrument makers in the American colonies and new republic, and his *Thinkers and Tinkers* (Bedini 1975) includes the work of navigators and their instrument makers.

Telescopes arrived in the New World not long after their invention in 1608 in Holland. A spyglass was employed for military purposes on a Portuguese ship off Brazil in 1614, a Dutch ship off Peru in 1615, and by the English governor of Bermuda in 1620, but the first telescope for astronomical use was owned by John Winthrop, Jr., the governor of the colony of Connecticut, circa 1657 (Schechner 2014). In the eighteenth century early newspapers and diaries show public interest in stargazing. Colonial astronomers needed high-quality instruments for eclipse expeditions, observations of the Transits of Venus in 1761 and 1769, and geodetical surveying. Ships' captains found spyglasses useful at sea, and telescopes were in demand by officers during the American Revolution. But except for the very rare, occasional homemade instrument, all telescopes were imported from Europe. As with the navigational and surveying instruments, the reason was the want of brass for the tubes of reflecting telescopes. The other problem was the glass. There were no local manufactories for optical glass. Indeed, a close examination of the advertisements of opticians and spectacle makers until the end of the nineteenth century shows that the component lenses, mirrors, and prisms were imported. Even Bausch and Lomb imported its lenses until 1870, making a name for itself on the frames that held the lenses. It was more cost-effective to buy ready-made from Europe than grind one's own (Schechner 2009). When America's first major telescope makers came on the scene in the antebellum period – Amasa Holcomb, Henry Fitz, and Alvan Clark – and supplied the observatories that were popping up like mushrooms around the United States, they also imported the optical glass that they ground for their instruments (Multhauf et al. 1962; Warner and Ariail 1995). Only after World

War I disrupted the imports from Europe, especially from Germany, did the US government make native glass production a high priority (United States Army, Ordnance Department 1921).

Much has been published about the 400-year history of the telescope, including American innovations in wide-field photographic telescopes (astrographs) for land-based meteor studies, space telescopes for deep sky work, radio telescopes, and spectroscopic instruments applied to telescopes (Brandl, Stuik, and Katgert-Merkelijn 2010; Morrison-Low et al. 2012). Nevertheless, little has been written about the development of the optical instrument industry in the United States. *Artists and Optics* (Warner and Ariail 1995), a book devoted to the firm of Alvan Clark and Sons, the telescope makers, is primarily an expanded catalog of all the known instruments sold by the firm, but extremely useful. So are the many articles about lesser-known telescope makers and American observatories published in the *Journal of the Antique Telescope Society*. Nonetheless, much more work is needed to understand how Clark went from being a portrait artist to the maker of the world's largest refractors. The careers and workshop practices of his rivals (such as Henry Fitz and John Brashear) are also worthy of future study.

A similar story of imports making up for a dearth of American innovation and skill is told for microscopes until the mid- to late nineteenth century. Warner (1985) has suggested that a lack of collaboration between engineers, artisans, businessmen, scientists, and government was to blame for the absence of a precision optical industry in Philadelphia. This situation deserves further examination and comparison with the conditions that aided the development of optical firms like the Spencer Lens Company in upstate New York and the work of Robert B. Tolles of Boston Optical Works in Massachusetts.

The lack of scientific glass and brass before the second quarter of the nineteenth century also delayed the manufacture of philosophical apparatus in America, because these materials were central to instruments such as air pumps, electrical machines, hydrostatic and mechanical apparatus, thermometers, and barometers (Schechner 2006). Consequently the study of philosophical instrumentation in the American hemisphere is a study of imports from the parent colonies (England, France, Spain, and Portugal) until the mid-nineteenth century. Many books catalog the cabinets of apparatus at colleges during this period (e.g., Cohen 1950; Granato and Lourenço 2014; Pantalony, Kremer, and Manasek 2005), and Schechner (1982) describes how many items passed through the hands of the Reverend John Prince of Salem, Massachusetts. Prince not only served as an intermediary between the academies and the London instrument makers but also did repair work (Schechner 1996, 2006). Laboratory exercises in the period 1880–1920 are addressed in Heering and Wittje (2011).

The apparatus for teaching mathematics in the United States from 1800 to the present – blackboards and projectors, slide rules and blocks, protractors and graph paper, geometric models and calculators – is well explored in Kidwell, Ackerberg-Hastings, and Roberts (2008). Another good resource is *Rittenhouse: Journal of the American Scientific Enterprise* (1986–2012), a print journal devoted to the production and use of scientific instruments in North America. This is a good place to learn about the instrument makers and retailers from the nineteenth century onwards who supplied everything from school apparatus to telegraph equipment, research quality laboratory microscopes to home medical devices, ship chronometers and experimental psychological apparatus. The journal has now evolved into an online platform, *eRittenhouse* (2013–) and broadened its coverage to all of the Americas. In addition to the manufacture, sale, and use of instruments in the Americas, it also looks at their social impact.

Clocks and watches are generally considered a specialty of antiquarian horologists, but time finding employs observations of the sun and stars with sundials and other astronomical instruments, and these in turn are used to set clocks, watches, sand glasses, and various timing devices. Moreover, clocks not only regulate our lives but also our scientific instruments. America's relationship to clocks is the subject of Stephens (2002).

To extend human life and improve its quality, people turn to healers, physicians, and surgeons. The material culture of health and medicine is the object of study by curators in the Medical Museums Association and related organizations devoted to the history of pharmacology. A history of the American surgical instrument trade is Edmonson (1997).

For research questions on any topic related to scientific instruments the website of the Scientific Instrument Commission (http://iuhps.org/) maintains a database of publications and a cumulative bibliography, as well as links to online scientific instrument trade catalogs, videos, and other resources.

Lastly, major resources for studies of scientific instruments are the instruments themselves, and these can be found in museums, colleges and universities, astronomical observatories, research laboratories, libraries, industrial factories, hospitals, and even the local city hall. Notable North American collections and their specialties include:

- *Adler Planetarium* (Chicago, IL): astronomy, navigation, surveying, time finding, time keeping, mathematics, and cartography, featuring many instruments from the Middle Ages and Renaissance.
- *Bakken Library and Museum* (Minneapolis, MN): electrical phenomena in the life sciences and medicine.
- *Canada Science and Technology Museums Corporation* (Ottawa, ON) supports three museums – the *Canada Science and Technology Museum,* the *Canada Aviation and Space Museum,* and the *Canada Agriculture and Food Museum*: Material culture illustrating how science and technology have shaped Canadian culture; instruments of astronomy, surveying, meteorology, physics, and industries used by departments of the Government of Canada, such as Natural Resources and the National Research Council.
- *Case Western Reserve University* (Cleveland, OH), *Dittrick Medical History Center:* diagnostic, therapeutic, and surgical instruments.
- *Chemical Heritage Foundation* (Philadelphia, PA): alchemical and chemical instruments and artifacts related to the chemical industry.
- *Dartmouth College* (Hanover, NH), *Hood Museum of Art, Dartmouth Collection of Scientific Instruments:* items dating back to the founding of the college in 1769 as well as more recent research.
- *Harvard University* (Cambridge, MA), *Collection of Historical Scientific Instruments:* teaching and research instruments from a broad range of scientific disciplines, many with a Harvard history.

- *Harvard University* (Boston, MA), *Warren Anatomical Museum in the Francis A. Countway Library of Medicine*: medical and surgical instruments, anatomical models and specimens.
- *Huntington Library* (Pasadena, CA), *Burndy Library Collection*: electrical instruments, rare early light bulbs, and miscellanea.
- *Manitoba Museum* (Winnipeg, MB): navigational instruments associated with the Hudson's Bay Company.
- *Mariner's Museum* (Newport News, VA): navigational, oceanographic, and maritime instruments from the sixteenth to twentieth centuries, as well as communication equipment such as lights, buoys, and radios.
- *Maritime Museum of the Atlantic* (Halifax, NS): nautical instruments, nineteenth century onward.
- *Massachusetts Institute of Technology* (Cambridge, MA), *MIT Museum*: artifacts documenting MIT's scientific and engineering work.
- *McGill University* (Montreal, QC) is home to three significant collections of instruments: The *Rutherford Museum* has apparatus used by Ernest Rutherford at McGill, 1898–1907. The *McPherson Collection* holds physics instruments from the mid-nineteenth century to about 1920. The *McCord Museum*, formerly administered by McGill, but now a private institution, has material culture related to Montreal and Canada.
- *Musée Stewart au Fort de l'île Sainte-Hélène* (Montreal, QC): globes, sundials, mathematical, surveying, and medical instruments, and the philosophical instruments used by Abbé Nollet and his student Sigaud de Lafond.
- *Musées de la Civilization* (Quebec City, QC) is a consortium of four museums whose focus is the history of French-speaking culture in North America but also includes ethnographic collections related to First Nations. The *Musée de la Civilization* has one of the largest collections of French scientific instruments outside of Paris, including late eighteenth- and nineteenth-century laboratory apparatus, which are part of the important *Séminaire de Québec Collection*.
- *Mütter Museum of the College of Physicians of Philadelphia* (Philadelphia, PA): from suture needles to iron lungs, medical and surgical instruments of all kinds, as well as anatomical models and specimens.
- *Mystic Seaport* (Mystic, CT): navigational and maritime instruments, maps and charts, and watercraft reflecting America's relationship with the sea and inland waterways since 1530.
- *National Air and Space Museum, Smithsonian Institution* (Washington, DC): astronomy, space astronomy, and aeronautics from the modern period, with a focus on US history.
- *National Museum of American History, Smithsonian Institution* (Washington, DC): instruments of astronomy, physics, chemistry, horology, surveying, and navigation, many of which are of European origin, although American artifacts are featured.
- *National Museum of Health and Medicine, Smithsonian Institution* (Washington, DC): material culture related to the history and practice of American medicine and military medicine, including anatomical models and specimens, surgical, dental, diagnostic, and therapeutic instruments, and the *Billings Microscope Collection*.
- *South Carolina State Museum* (Columbia, SC), *Robert B. Ariail Collection of Historical Astronomy:* telescopes and astronomical apparatus, many American-made.

- *Transylvania University* (Lexington, KY), *Monroe Moosnick Medical and Science Museum*: natural philosophy teaching apparatus, 1820–1850.
- *University of Mississippi Museum* (Oxford, MS), *Millington-Barnard Collection of Scientific Instruments*: nineteenth-century apparatus for teaching astronomy, physics, and natural philosophy.
- *University of Toronto Scientific Instruments Collection* (Toronto, ON): astronomy, chemistry, computing, psychology, and physics, primarily nineteenth and twentieth century.
- *Vancouver Maritime Museum* (Vancouver, BC): navigation, fishing, boating, naval affairs, with focus on the Pacific Northwest.
- *Yale University* (New Haven, CT), *Yale Peabody Museum of Natural History, Division of Historical Scientific Instruments*: containing apparatus primarily related to physics, medicine, and anthropology.

Today research in scientific instrumentation is more expansive than it has ever been and more integrated into historical and philosophical approaches. And yet, publications about the development and use of instruments in an American context are scarce. So many topics beg to be studied – e.g., the transition from imports to native fabrications (Max Kohl of Chemnitz to CENCO of Chicago); the impact of economics and trade tariffs, war and politics; why the American System of Production was put to work in the Waltham Watch factory but not the workshops of major suppliers of telescopes and microscopes; how instruments were chosen for expeditions (such as the survey of Mason and Dixon or the westward exploration of Lewis and Clark); the place of instruments in industrial settings, such as breweries and pharmaceutical firms; how "new" materials (such as hard rubber, plastics, aluminum, anodized metals) become part of instruments; the relationships between instruments and art; and whether "big science" buildings such as observatories or accelerators are instruments in and of themselves.

Chapter Thirty-Three

SCIENCE AND LITERATURE

Stephen Rachman

The dynamic relationship between science and literature in the West has been colored in the last 50 years by a sense of separation, at least since British scientist and novelist C.P. Snow's 1959 lecture on "The Two Cultures" highlighted a lack of conversance and dialogue between the sciences and the humanities (Snow 1959: 3). The impact of Snow's work has been to spur dialogue between these two branches of knowledge or modes of thought, and a general inquiry into the relationships between science and literature. In the decades following Snow, a familiar kind of cultural contestation became normative as humanists asserted the position that science and scientific knowledge, far from being a value-neutral form of "objective" investigation of reality, was not a privileged practice but socially constructed. The rise of the disciplines of history of science and interdisciplinary areas such as science and technologies studies, medical humanities, literary genres such as science fiction, and most recently neuro-literary studies are just a few outgrowths of these developments. This chapter will trace the origins of these developments through the nineteenth and into the twentieth centuries in the American context.

As prescient as Snow's comments might have seemed in the late 1950s (even though delivered in England they were taken up with vigor in the United States), the interdisciplinary relationships between science and literature have a long and complicated history in the United States, of which their perceived separation is just a part. This history is perhaps best understood as a series of parallel developments with, in the words of Robert J. Scholnick (1992), "permeable boundaries."

The modern sense of the word "science" with its restriction to the culturally neutral study of the natural or physical world (Otis 2002: xvii), skeptical protocols, methods, and emphasis on experimentation began to gain currency in the English-speaking world during the 1830s and 1840s, and the word "scientist" in the 1870s (Williams 1983: 279). But, even as this distinction was beginning to be called for, there was no sense

A Companion to the History of American Science, First Edition.
Edited by Georgina M. Montgomery and Mark A. Largent.
© 2016 John Wiley & Sons, Ltd. Published 2020 by John Wiley & Sons, Ltd.

that the literary and the scientific were two separate realms of understanding with their own specialized terminology and conceptual apparatuses. For much of the nineteenth century, in the words of Laura Otis, "science was in effect a variety of literature" (2002: xvii). For example, when the popular science journal, *Scientific American*, began publication in the mid-1840s, it featured patents and inventions, even carrying poems on topics such as the nature of life and the triumph of justice ("What is Life?" 1848: 1). By the 1940s, *Scientific American*, while retaining its orientation toward popular science and engineering, had shifted its focus to the more familiar "hard sciences" of chemistry, biology, and physics. Advances in technology, social transformation by way of industrialization, changes in educational curriculum, and discoveries of natural phenomena contributed to a growing sense of separation. What began to drive the distinction between science and literature in the period between the years before the Civil War and the advent of World War II was a large epistemological shift. It was not merely that scientific knowledge communicated in print was no longer a form of literature, but rather that its goals, contexts, and status as a form of knowledge had come to appear fundamentally different in kind than other forms of writing. As one observer in 1874 put it, a scientist "would not mean one who 'possesses knowledge in general,' so much as one who rejects all but knowledge for the foundations of hypotheses, and therefore constructs only with such materials as he already 'knows'" (Anonymous 1874: 321). The sense of exclusion and specialization became synonymous with scientific practice, as method "was specialized to one kind of method," especially one involving a "neutral methodical observer and an external object of study" (Williams 1983: 279).

But if science has come to be seen as specialized in language, protocols, and operations since the nineteenth century, then the same could be said of literature, criticism, and philosophy. If the lay public feels closed off from the sciences, then they feel much the same with respect to, as George Levine has remarked, "the increasingly arcane operations of literary criticism. The distinction is one of degree, not of kind: science is no more exempt from the constraints of nonspecialist culture than literature is; nor has it ever been" (1987: 3). During the eighteenth and nineteenth centuries, the concept of literature evolved from all kinds of polite learning into specialized senses pertaining to and reflecting the cultural distinctions of nations (e.g., English and the American literature) and particular forms of imaginative and creative writing (e.g., poetry, fiction, and drama), the latter notion becoming particularly pronounced in the Romantic movement (1780s–1850s) (Williams 1983: 185–6). In this way a curious development of parallel specialization has contributed to the sense of separation, if not antagonism, between science and literature. These parallel transitions toward specialization are one indication of the ways in which the histories of science and literature have always been in dialogue with one another, each field making its claim on and against the other. To the extent that science and literature have diverged, these two paths have come to represent competing forms of authority. The epistemological authority of science to inform our understanding of the world – even how we imagine it; its claims on disinterested observation and sensory evidence; its driving role in technological innovations that shape our lives, command respect and attention and the literary world has and must take cognizance of and be influenced by them. But for all of science's power and authority, the literary critique of it maintains a deep discomfort with discourses and practices that ignore, reduce, or abstract to the point of nonrelevance the full amplitude and variety of human experience in its social, moral, and aesthetic dimensions.

Colonial Era and Gothic Fictions of Science

To trace the divergence of science and literature, we need to consider it stemming from, not two cultures, but a common broadly shared Western cultural development. While we do not generally think of it as such, the cultural and political development of the United States from its earliest English colonization in the late sixteenth and early seventeenth centuries through the revolutionary period of the eighteenth century and the early republic of the nineteenth century coincided with what came to be known retrospectively as the "Scientific Revolution" and the Enlightenment. The long historical arc of the development of the sciences in the seventeenth and eighteenth centuries on the North American continent was one in which the literary and the scientific were regularly intertwined. Of the five senses of science given in Samuel Johnson's *Dictionary of the English Language* (1755) only one ("Certainty grounded on demonstration") conveys anything remotely resembling the modern sense; the others range from "knowledge" to "one of the seven liberal arts" (Aldridge 1992: 40). When Noah Webster published the first *American Dictionary of the English Language* (1828), the definition of science had not shifted from Johnson's. The definition is essentially the same. "Authors have not always been careful to use the terms art and science with due discrimination and precision," Webster pointed out in a note but his distinction concerns itself with theory and practice. "Music [because it has theory based on abstract principles and can be played] is an art as well as a science" (Webster).

A number of figures prominent in the founding of the United States blended scientific and literary activity in ways that mutually informed both aspects of their work. Benjamin Franklin's diverse interests as an inventor (lightning rod, stove) and investigator of natural phenomena (the electrical properties of lightning, the pattern of the Gulf Stream), his work as a printer, celebrated publisher of *Poor Richard's Almanack* (1733), and author of the *Autobiography* (1791), make him the foremost figure of eighteenth-century literary–scientific cultural synergy. Thomas Jefferson's work as author of a work of natural history, *Notes on the State of Virginia* (1787), the Declaration of Independence (1776), sponsorship of Lewis and Clark's voyage of discovery, and his founding of the University of Virginia make a strong case for his shared interests alongside Franklin. But, despite the American founding fathers' intellectual balance, one finds in their examples a decided lack of interest in poetry and fiction, the two literary genres through which the literary would distinguish itself most aggressively in contradistinction to the scientific as these domains would begin to diverge.

With the exception of his *Autobiography* (the fictional elements of which are more self-promotional than novelistic), Franklin was more of a printer, publisher, and collector of witticisms than an author of creative works, and this reflected the colonial society he inhabited. His nonfictional publications sold much better than the few novels he promoted. He was the publisher of the first novel in the colonies, Samuel Richardson's *Pamela* in 1842–3 – a literary landmark often considered the first English novel and a bestseller in London, but in Philadelphia Franklin's reprint met with little interest. It would be 25 years before another novel would be published in what was to become the United States. Only in the 1790s were novels published with any regularity, and just a few of those were written by American authors (Green and Stallybrass 2006).

While the Enlightenment was most certainly known for scientific and philosophical inquiry into nature, the distinction between literature and science was not of particular philosophical relevance, and Franklin, alive to all things philosophical, was largely indifferent to the aesthetics of novels. In the case of Jefferson, there was a distinctly anti-literary bias. "A great obstacle to good education," he opined,

> is the ordinate passion prevalent for novels, and the time lost in that reading which should be instructively employed. When this poison infects the mind, it destroys its tone and revolts it against wholesome reading. Reason and fact, plain and unadorned, are rejected. Nothing can engage attention unless dressed in all the figments of fancy, and nothing so bedecked comes amiss. The result is a bloated imagination, sickly judgment, and disgust towards all the real businesses of life. (Quoted in Spahn 20111: 209)

The tension between fiction and reality expressed in Jefferson's concerns (and he was directing these remarks toward the issue of female education) anticipated the divisions that would come to mark the divisions of the literary and the scientific in the twentieth century.

Ironically, Charles Brockden Brown (1771–1810), a Philadelphia novelist of the 1790s who attempted to fuse the dark world of the gothic with the natural sciences and medicine, would dedicate one of his gothic novels, *Wieland: Or, The Transformation: An American Tale* (1798), to Jefferson. Over the course of a series of novels written at the turn of the nineteenth century, Brown developed plots that turned upon liminal natural phenomena: somnambulism (*Edgar Huntly*), spontaneous combustion, ventriloquial psychosis (*Wieland*), and yellow fever (*Arthur Mervyn, Ormond*). Situated as cultural pathologies and quasi-clinical conditions, they function as triggers for his plots and as engines of sociocultural critique. In terms of forging American gothic aesthetics, Brown saw his own literary operations as an innovation in European narrative traditions by dispensing with, "[p]uerile superstitions and exploded manners, gothic castles and chimeras" (Brown 1805). Using a series of fascinating yet, at the time, vaguely understood physiological phenomena that he conceived of as both mysteries of the human organism and cultural pathologies, he Americanized the gothic novel, politicizing it, transferring the balance away from patently superstitious sources of irrational terror toward ones, if not exactly more rational, then rooted in fears based upon mysterious phenomena found in contemporaneous discourses of nature and reason. In an 1805 article on "Terrific Novels," Brown elaborated on this distinction. Gothic terror in its debased form attempted to "keep the reader in a constant state of tumult and horror by the powerful engines of trapdoors, back stairs, black robes, and pale faces: but the solution of the enigma is ever too near at hand, to permit the indulgence of supernatural appearances" (Brown 1805). His objection was not to terror per se but that the terrifying devices of standard gothic fiction amounted to little more than jumpiness induced by scary atmospherics and costumes. For Brown, gothic anxiety's proper métier was not the trapdoor-riddled monk's castle or the black robe but the mental conditions aroused by fears of madness, disease, bankruptcy, politics, the ordinary violence of human passion. Brown's friendship with Dr. Elihu Hubbard Smith (and connections with other members of the New York Friendly Club, especially Drs. Edward Miller and S.L. Mitchill) were forged in large part through a shared enthusiasm for medicine, nature, politics and literature, and an educational vision which could synthesize and integrate these branches of knowledge. His experience of the yellow fever epidemics of 1793 and 1798 (in which

Smith died and Brown suffered a mild case) were signal events in his life and, as William Dunlap and all subsequent biographers have claimed, fundamental sources of his fictional productions. The difficulties involved in Brown's shifting of gothic stimulus away from the vocabulary of conventional fear toward more legitimate sources of anxiety are immediately present to any reader of his fiction. In Brown's gothic, when one begins to travel beyond fear one encounters science, and more often then not, medical science. Brown's novels are freighted with complex back stories and explanatory passages that invest the most fantastic and phantasmagorical aspects of his work with an imaginative intensity that maintains a realistic density – a tension between science and literature, a type of patho-psychophysiological fiction that would foreshadow the Romantic era.

The line of gothic intersections with science has had a long-lasting impact on the literature and the image of science in the popular imagination. Mary Shelley's *Frankenstein* (1818), a cautionary tale of Promethean creation, remains the dominant emblem of the antagonism between science and literature, both in the desires science inspires and the fears that remain the wages of these desires. Variants of this framework include the works of Edgar Allan Poe (see below), Robert Louis Stevenson, *The Strange Case of Dr. Jekyll and Mr. Hyde* (1886), H.G. Wells, *The Island of Dr. Moreau* (1896), H.P. Lovecraft, "Cthulhu" stories (ca. 1920s), Michael Crichton, *The Andromeda Strain* (1969) and *Jurassic Park* (1990), and Neal Stephenson, *Snow Crash* (1992).

Nineteenth-Century American Literature

With the nineteenth century, American literature and literary culture (magazines, journals, publishing houses, distribution networks, authorship, copyright law) developed at a rapid pace. This was also the period in which the concept of literature and the literary was evolving. In the United States, the first major developments are a literary culture marked by its romanticism and a strong sense of literary nationalism. The most important early figure in American literary engagement with science is Edgar Allan Poe (1809–1849). Like Brockden Brown and Shelley, Poe worked primarily in a gothic tradition. He is known as the originator of detective fiction ("Murders in the Rue Morgue," 1841), a popular genre that uses investigative and deductive methods analogous to those found in the sciences. Never credulous of the fictive elements in his own work, he attributed their apparent novelty to their "*air* of method" (Rachman 2010: 18). He was also a pioneering figure in the development of science fiction most notably through the fantastic sea voyage in *The Narrative of Arthur Gordon Pym of Nantucket* (1838); the moon voyage "The Unparalleled Adventure of One Hans Pfaall" (1835); tales of mesmerism "The Facts in the Case of M. Valdemar" (1845) and alchemy "Von Kempelen and His Discovery" (1849). His influence would be particularly strong on the French author of the 1860s, Jules Verne, and the American H.P. Lovecraft in the twentieth century, but his various works express the full range of complexity with the respect to the sciences.

Paradoxically, Poe was also a figure of anti-science romanticism, the reputation for which rests largely on his early "Sonnet—To Science" (1829), which begins

Science! true daughter of Old Time thou art!
  Who alterest all things with thy peering eyes.
  Why preyest thou thus upon the poet's heart,
  Vulture, whose wings are dull realities?

Echoing, John Keats' "Lamia," Poe's sonnet offers up a version of science as fundamentally anti-literary, forcing the poet and the poetic tradition out of its mythic habits of expression and perception. "How should he love thee," the poet asks of science, "or deem thee wise?" The epistemological challenge of "science" and its assault on myth confronts the poet with the sense of a fall from an Edenic state as he is left torn from "The summer dream beneath the tamarind tree" (Poe 2000: 91). The poem's final image of dispossession caused by science or, as John Limon has suggested, more accurately, "American science philosophy" is a fundamentally destructive force. Limon argues that "in the intellectual world Poe grew up in, American science philosophy was pedestrian . . . but science . . . was revolutionary and rich with imaginative hypotheses. Poe played with this disparity throughout his career" (Limon 1990: 21). Poe frequently engaged science and scientific questions in order to "poeticize" it by subjecting it to a series of imaginative operations.

Therefore, despite the forcefulness of Poe's sonnet, it would be misleading to assume that it is a direct statement of Poe's final position on science. Indeed, Poe's "Sonnet— To Science" served as the preamble to "Al Aaraaf" (1829), a poem that deals with a collection of spirits who inhabit a star – a famous supernova sighted near the constellation Cassiopeia by Danish astronomer, Tycho Brahe, in November 1572. Poe was capable of lamenting the conditions under which science placed poets while at the same time using scientific discoveries for poetic purposes. Throughout his career, Poe would transmute scientific conditions of knowledge into aesthetic categories. This propensity can be seen from his earliest poetry through to his cosmological speculations on the finitude of the universe in *Eureka* (1848).

In Poe, one finds a theme that runs through much of the dialogue between science and literature as it would take shape in the nineteenth and twentieth centuries. As Hazard Adams described it, Romantic critics "came to characterize literature as concrete and specific and science as general and abstract" (Adams 1970: 164). While this is surely best considered as a tendency rather than a strict opposition (there is, of course, a good deal of abstraction in literature and a good deal of concrete specifics in scientific writing), the complexity of Poe's interactions with scientific thought, philosophy, and practice proceeded necessarily from the play of this opposition. If, as Italo Calvino has proposed, scientific writing tends toward "a purely formal and mathematical language based on an abstract language indifferent to its content" and the literary tends "to construct a system of values in which every word, every sign, is a value for the sole reason that it has been chosen and fixed on the page" (Calvino 1986: 37), then in Poe we often find a confounding mixture of these two modes of writing. As a consequence we find a challenge and engagement between them. This collision between the aims of abstract and concrete writing is one of the driving strategies of Poe's work, how it straddles the scientific and the literary.

Another pioneering figure of American literature who takes up themes of science, Nathaniel Hawthorne (1804–1864), is of note. His works feature botanists, chemists, physicists, inventors, and doctors who exhibit a range of creative and destructive powers. In "Dr. Heidegger's Experiment" (1837), the title figure, a scientist, works in his laboratory to develop an elixir of life, a theme that would return in last unfinished work, *The Dolliver Romance* (1864). In "The Birthmark" (1843), a scientist kills his wife in an attempt to perfect her by removing what he perceives to be a sole blemish on her otherwise faultless body. In "Rappaccini's Daughter" (1844), the title figure, an Italian botanist, seeks, through his study of poisonous plants, a method of "immunizing" his daughter from afflictions, but succeeds in making his daughter poisonous to others. In "The Artist of the Beautiful" (1844), Owen Warland, a watchmaker and inventor of delicate impractical mechanisms, sacrifices a life of love and children in order to dedicate himself to the invention of an automaton butterfly, so realistic that it cannot be easily distinguished from a living specimen. Through all of Hawthorne's stories of scientific-technical elites, the theme of human perfectibility through scientific or mechanical means is countered by a chastening sense of mortality, fragility, or limitations. Like Poe and Shelley, he shares a concern for the creative and destructive potentialities of science and scientific culture.

The use of science for purposes of social critique also informed the scientific interests of the Transcendentalist movement, especially the work of the poet, essayist, and philosopher Ralph Waldo Emerson (1803–1882) and his chief disciple Henry David Thoreau (1817–1862). Emersonian Transcendentalism, through his seminal essays, "Nature" (1836), "The American Scholar" (1837), "Self-Reliance" (1841), and "Experience" (1844) expressed a profound commitment to original, individual thought and natural inquiry. This movement shared with science a skepticism about received learning, but this was mixed with a heterodox spirituality and a pervasive romantic idealism. As a naturalist, Thoreau had the deepest commitments to science of the Transcendentalist movement and *Walden* (1854), and his later essays reflect the complexity of shifting investments in language and science. As Robert D. Richardson, Jr. has shown, Thoreau sustained "a steady, indeed a constantly deepening, interest in fact or data" in his investigations of the natural world. At the same time, like other American writers of the period, he was "dismayed at the prospect of the passive or mindless accumulation of data" (Richardson 1992: 114). For Thoreau, data without a proper interpretive framework was useless. As he would devote a chapter of *Walden* to "Higher Laws," Thoreau saw derivation of law as the proper aim of data gathering and, even more crucially, the experiential was the proof of knowledge. "Much is said about the progress of science," he observed. "I should say that the useful results of science had accumulated, but that there had been no accumulation of knowledge." For Thoreau, science could only become knowledge through direct experience (Richardson 1992: 114).

In 1847, Thoreau collected local specimens for Louis Agassiz, the Harvard biologist responsible for the professionalization of science in mid-nineteenth-century America. From this point forward, Thoreau would deepen his own scientific activity culminating in his studies of Cape Cod and the Maine woods, and his posthumously published, "Dispersion of Seeds." Yet, as the work progressed from problems of human freedom to questions of the flora and fauna of New England, he grew skeptical of the uses to which the study of the natural world ought best be put. "He is the richest," he wrote in May, 1853, "who has most use for nature as raw material of tropes and symbols with which to describe his life" (quoted in Richardson 1992: 121). In Richardson's assessment, over time, Thoreau became skeptical of new scientific methods, terminology, and discovery. Rather, Thoreau "asserts that scientific language actually gets in the way of our understanding how the world relates to us" (Richardson 1992: 122). This insistence on the subjective value of science would run counter to the empiricism and objectivity that would mark emergent scientific method; it would mark the difference between the biologist and the naturalist. By the 1860s, while Thoreau, who eagerly absorbed Darwin's *Origin of Species*, would begin to incorporate the principles of natural selection into his thinking, he retained an insistence on the power of language to describe the world in terms of the self, not independent of it.

Elements of the Transcendentalist approach to science persist throughout the nineteenth century, most prominently in the poetry of Walt Whitman (1819–1892), whose *Leaves of Grass* (1855–92) was notable for introducing an unprecedented graphic depiction of the animality of humanity by way of a radical departure from traditional English prosody. His Emerson-inspired free verse experimentation, with its frank depictions of sexuality and unabashed nationalism, scandalized and revolutionized American poetry. Less widely known but central to his work was an abiding interest in science. For example, "This Compost" (1856) was inspired by reading the pioneering German organic chemist Justus von Liebig's *Chemistry, in its Application to Agriculture and Physiology* (1846). Whitman's review of this work paid close attention to the chemistry of fermentation and putrefaction.

> What chemistry!
> That the winds are really not infectious,
>
> . . . .
> That when I recline on the grass I do not catch any disease,
> Though probably every spear of grass rises out of what was once a catching disease.
> Now I am terrified at the Earth, it is that calm and patient,
> It grows such sweet things out of such corruptions,
> It turns harmless and stainless on its axis, with such endless successions of diseas'd corpses,
> It distills such exquisite winds out of such infused fetor,
> It renews with such unwitting looks its prodigal, annual, sumptuous crops,
> It gives such divine materials to men, and accepts such leavings from them at last.
> (Whitman 1996: 496)

Whitman's journalism in the *Brooklyn Daily Times* from the late 1850s endorsed the miasma theory of contagious disease – the notion that disease was spread by noxious effluvia of decaying matter – expressed in "This Compost" (Aspiz 1980: 63–4). Whitman's interest in science is a direct application of the problem Thoreau identified with scientific language – the interference it seems to run between human understanding and the world. The speaker positions himself in a place of wonder; wonder at the natural processes of distillation and renewal that rely on upon decay and disease. Through the contemplation or perhaps recognition of these processes, Whitman transforms the pathological concept into a creative impasse at the threshold of culture, the place where science meets literature.

In a 1914 essay entitled "Science and Literature," the naturalist and champion of Whitman, John Burroughs (1837–1921), attempted a synthesis of this romantic tradition.

Science is the critic and doctor of life, but never its inspirer. It enlarges the field of literature, but its aims are unliterary. The scientific evolution of the great problems—life, mind,
consciousness—seem strangely inadequate; they are like the scientific definition of light as
vibrations or electric oscillations in the ether of space, which would not give a blind man
much idea of light. The scientific method is supreme in its own sphere, but that sphere is
not commensurate with the whole of human life. Life flowers in the subjective world of our
sentiments, emotions, and aspirations, and to this world literature, art, and religion alone
have the key. (Burroughs 1914: 424)

Early in the twentieth century Burroughs expressed a vision of separate but conjoined
spheres of science and literature. The supreme power and strange inadequacy of science
in relation to subjective experience would define a principal axis of relation through
which the two disciplines would continue to inform and question each other.

Medicine as a Site of Literary–Scientific Intersection

In mid-nineteenth-century America, the figure of Oliver Wendell Holmes, Sr. (1809–
1894) embodied a literary–scientific nexus. His varied roles as physician, occasional
poet, professor of anatomy and dean of Harvard Medical School, "Autocrat of the
Breakfast-Table," novelist, promoter of the stereoscope, coiner of the word anesthe-
sia, and New England Brahmin savant amount to an interdisciplinary project in and of
themselves. The spheres of medicine and literature that Holmes participated in developed dramatically as professions during the nineteenth century. More than any other
literary figure in this era, Holmes' applied a new scientifically informed understanding
of disease to his social and literary vision. His landmark epidemiological study of the
contagiousness of puerperal fever shaped his socio-literary vision in his self-described
"medicated novel," *Elsie Venner: A Romance of Destiny* (1861).

Elsie Venner involves the story of the strange and ophidian Elsie whose abnormal
behavior was caused by prenatal poisoning (Elsie's mother died in childbirth after having been bitten by a rattlesnake). Holmes used this novel as a platform from which to
attack theological conceptions of original sin. For Holmes, the "medicated" element
of his novel involved an attempt to challenge and replace the religious concept of sin
with a medical concept of inherited constitutional weakness. In its most crucial scenes
the novel is a forum for Dr. Kittredge, Elsie's family physician, to explain why the practitioner's vision is more adequate and appropriate for understanding the contemporary
human condition. "We have nothing but compassion," Kittredge explains, "for a large
class of persons condemned as sinners by theologians, but considered by us as invalids.
We have constant reasons for noticing the transmission of qualities from parents to offspring, and we find it hard to hold a child accountable in any moral point of view"
(Holmes 1861: 286). In *Elsie Venner*, psychological and behavioral problems are ultimately physiological and Holmes used literary forms to extend medical authority to
broader areas of religion and culture.

The intersection of medical science and social authority that Holmes pioneered
became a regular literary forum. Holmes wrote two more "medicated" novels, *The
Guardian Angel* (1867) and *A Mortal Antipathy* (1885). Silas Weir Mitchell (1829–
1914), a prominent Philadelphia physician and novelist, continued in this tradition with

"The Case of George Dedlow" (1866). See also, Sinclair Lewis, *Arrowsmith* (1925) and Albert Camus, *The Plague* (1947). A feminist critique of this form of medical authority can be found in Charlotte Perkins Gilman, *The Yellow Wallpaper* (1892), and Sylvia Plath, *The Bell Jar* (1963). The impact of this kind of writing on realism and naturalism was also extensive as authors such as Frank Norris, *McTeague* (1899) and Jack London, *The Call of the Wild* (1903) sought to apply Social Darwinism to American society with an eye toward the kind of biological determinism that shaped Holmes' thought. The expansion of medico-scientific authority as a form of cultural authority remains an important site through which one can observe the shifting dialogue between science and literature. It is indeed a permeable and ever-shifting boundary as we continue to debate the nature of reality and the discursive nature of scientific inquiry.

Bibliographic Essay

Students interested in tracing the divergence of science and literature in broad cultural terms would do well to consult the entries under these terms in Raymond Williams' *Keywords: A Vocabulary of Culture and Society* (1983). Related entries in this work on art, imagination, nature, and technology helpfully clarify the difficulties that arise in understanding the shifts that have occurred in the meanings of these words over the last three centuries.

Because of the legacy of the Darwinian debates in the second half of the nineteenth century, the relationship between science and literature as a field of study has always been more robust in Great Britain than in the United States. C.P. Snow's *The Two Cultures* (1959) was in some respects a continuation of the public debates about science, literature, and culture between Matthew Arnold and Thomas Henry Huxley in the 1880s. This emphasis is also reflected in the preponderance of British materials in Laura Otis's anthology, *Literature and Science in the Nineteenth Century* (2002). George Levine's edited collection *One Culture: Essays in Science and Literature* (1987) nicely consolidates the major threads of this work and includes an essay by Gillian Beer whose influential *Darwin's Plots: Evolutionary Narrative in Darwin, George Eliot and Nineteenth-Century Fiction* (Cambridge: Cambridge UP, 1983) detailed the strong connections between Victorian literature and evolution. Levine's *Darwin and the Novelists: Patterns of Science in Victorian Fiction* (Cambridge: Harvard UP, 1988) extends the line of Beer's research to the major English novelists of the nineteenth century.

As a field, the American context, while certainly influenced by the British one, has tended to focus on technology and invention rather than science per se. While this chapter has focused on science and literature, a more technologically oriented essay would need to address the influential study of Leo Marx, whose *The Machine in the Garden: Technology and the Pastoral Ideal in America* (New York: Oxford UP, 1964) traced the ways in which technological advances impacted America's sense of itself as a "pastoral" nation. While scientific advancement is inevitably linked with technological progress, Marx pays scant attention to science per se. Marx's work has spawned its own line of technologically oriented cultural analysis in both American and non-American contexts, notably Herbert Sussman, *Victorian Technology: Invention, Innovation, and the Rise of the Machine* (2009), Cecilia Tichi, *Shifting Gears: Technology, Literature, Culture in Modernist America* (1987); Klaus Benesch, *Romantic Cyborgs: Authorship and*

Technology in the American Renaissance (2002); Joel Dinerstein, *Swinging the Machine: Technology, Modernity, and African American Culture Between the World Wars* (2003); and J. Adam Johns, *The Assault on Progress: Technology and Time in American Literature* (2008).

For scholars interested in the intersection of science and literature in the United States, Robert J. Scholnick's edited collection *American Literature and Science.* (1992) offers a good sampling of essays on a range of American authors from colonial times through the twentieth century. It also features an essay by N. Katherine Hayles, whose *The Cosmic Web: Scientific Field Models and Literary Strategies in the Twentieth Century* (1984) and *Chaos Bound: Orderly Disorder in Contemporary Literature and Science* (1990) have explored the relationships between field models of cosmological physics, thermodynamics and modernism and postmodernism. John Limon's *The Place of Fiction in the Time of Science: A Disciplinary History of American Writing* (1990) discusses a range of nineteenth- and twentieth-century authors and works in terms of the philosophical concerns of science and literary attitudes toward the cultural politics of science. These works discuss the relationships between science and literature covering various literary genres from romanticism, realism, naturalism, modernism, and postmodernism.

The intersection of literature and medical science has received a great deal of scholarly attention and there are numerous articles and monographs well beyond the scope of this chapter. Much of the energy of this scholarship can be traced to Michel Foucault's *The Birth of the Clinic: An Archaeology of Medical Perception* (1975). Foucault most forcefully articulated the emergence of the transparent medical gaze in the eighteenth-century French clinic as a cultural construct. Susan Sontag's *Illness as Metaphor and AIDS and its Metaphors* (1990), perhaps the most widely-read essay on medicine and literature, explores the literal and metaphoric aspects of disease throughout the nineteenth and twentieth centuries. Charles E. Rosenberg's and Janet Golden's edited collection *Framing Disease: Studies in Cultural History* (1992) provides a fine overview of the cultural intersections of the history of medical science and its relation to cultural, social, and literary contexts. Kathryn Montgomery Hunter's *Doctors' Stories: The Narrative Structure of Medical Knowledge* (1991) and Howard Brody's *Stories of Sickness* (1987) offer excellent surveys of the intersections between medicine and literature with an emphasis on narrative and narrative structures.

Finally, the history of science fiction as a genre has grown dramatically in recent years. For handy overviews of the developments in science fiction and its relationship to science, see Adam Roberts, *Science Fiction (The New Critical Idiom)* (2006) and *The History of Science Fiction* (2007).

Chapter Thirty-Four

MUSEUMS

Amy Kohout

Last December I found myself face to face with the Hyde Park Mastodon. Discovered during backyard pond excavating in 1999, the bones of this imposing creature were carefully removed, cleaned, identified, modeled, and pieced back together for display at Ithaca's Museum of the Earth (Paleontology Research Institution 2003). The mastodon stands directly in front of a painted landscape so that if you're looking at it head-on, it appears to be part of a generic "natural" scene. But stand to its side, and the background is instead a plain, white wall. From this angle, the mastodon is a scientific specimen, bones and tusks gleaming against an empty canvas. I appreciated the opportunity to view the Hyde Park Mastodon against both backdrops. While museums like this one display "science," that is not all they display. On my visit to the Museum of the Earth, I encountered both art and nature; I tried my hand at scientific fieldwork; I learned about the history of the museum's collections; and I was asked to think about the future through an exhibit on climate change. Though sometimes seen as simply a piece of pre-laboratory scientific practice, museums and their history have far more to offer historians of science.

This chapter will explore the history of museums and their relationship to American science. First, it will examine museums in the early American republic. Second, it will trace the growth of these institutions throughout the nineteenth century. Third, it will consider the display strategies used by curators from the Victorian age to the present. Major shifts in the exhibition of natural and scientific material occurred in the twentieth century, and looking at these display decisions and techniques can help us to understand changing goals and audiences for these exhibits. Finally, I will conclude by returning to what the museum – as a place and as an idea – can do for our understanding of the history of American science. While we often consider the museum as a physical structure, a site with a set of materials that matter to us, the museum also offers us

A Companion to the History of American Science, First Edition.
Edited by Georgina M. Montgomery and Mark A. Largent.
© 2016 John Wiley & Sons, Ltd. Published 2020 by John Wiley & Sons, Ltd.

an opportunity to study and rethink practices – practices that may have originated in museums or been central to their function, but have moved beyond their walls.

From Peale to Barnum

Charles Willson Peale, curator of what is often called the nation's first museum (established 1786), also had a mastodon, which he excavated in 1801 near Newburgh, New York and brought to Philadelphia. In 1804, when Alexander von Humboldt visited Peale's museum, he would have seen the great beast featured in the Mammoth Room at Philosophical Hall. Other skeletons, displayed alongside it, offered the viewer a sense of the mastodon's scale. By 1807, the Mammoth Room featured a painted background as well as Peale's *Exhumation of the Mastodon* (1806), detailing the innovative excavation processes required to recover the specimen (Sellers 1980: 161–2, 203). Though the mastodon was the focal point, Peale's collection contained far more. Insects, birds, fish, mammal skins, skeletons, antiquities, and ethnological artifacts: Peale's museum had it all. And Peale was not simply the curator of his museum; he was collector and preparator as well.

Charles Sellers has suggested that Peale's Philadelphia Museum anticipated later museum exhibits: habitat grouping, lifelike taxidermy mounts, naturalistic displays. Although eventually overtaken by the pull of the strange, the odd, and the fantastical (an approach made most famous by Phineas Taylor Barnum), Peale and his museum demonstrated what Sellers described as "enjoyment while learning." Peale envisioned a representative collection of natural history specimens: animals, vegetables, minerals. In a lecture delivered in 1800 Peale enthusiastically asked, "Can the imagination conceive any thing more interesting than such a Museum?" (Peale, Peale, and Hawkins 1800: 35).

Peale's imagination certainly could not, but he grounded his arguments for a national museum in his commitment to the new republic. European powers had national museums; should not the United States also have a national institution committed to the pursuit of scientific knowledge? (Peale, Peale, and Hawkins 1800: 17–31) Peale envisioned a museum for the people. He linked knowledge with citizenship. Furthermore, an American museum could help countermand the claims of certain naturalists who dismissed American species as inferior to those on the European continent. The American *incognitum*, as the mastodon was called, came to represent the "natural antiquity" of the United States: a rich heritage different from, but not inferior to, the ruins and culture of the Old World (Semonin 2000: 4, 276).

Elected to the American Philosophical Society shortly after opening his museum in 1786, Peale maintained relationships with the learned societies of Philadelphia and beyond; members of these groups often utilized Peale's collections for their scientific pursuits (Kohlstedt 1992). But museums like Peale's were not the only institutions with natural history collections. Private collectors and learned societies built "cabinets" of specimens to be studied, and in 1826, the Academy of Natural Sciences, a research-oriented membership organization, opened its own taxonomically organized museum to the public (Conn 1998; Kohsltedt 1992).

Peale advocated tirelessly for a national museum, and he hoped that his museum (including his mastodon) could form its root: "And since I have subjects, a sufficient number of every class to make a brilliant display in a large building, why not with my labours make a beginning?" (Peale, Peale, and Hawkins 1800: 36). That never happened. Peale died in 1827, and his sons, Titian and Rembrandt, took over the museum. Titian's work as a naturalist extended beyond the museum to include official service with the United States Exploring Expedition (the U.S. Ex. Ex.) from 1838–42. Shortly after Titian returned, financial trouble necessitated closing the museum and selling off its holdings. P.T. Barnum purchased a large portion of the collection.

Though Barnum's American Museum contained many scientific specimens, it was the strange and fantastical artifacts that drew the public through the museum's doors. One of Barnum's most famous exhibits was the "Fejee Mermaid," supposedly captured in the islands and carefully preserved. In reality, Barnum's "mermaid" was an unnatural "hybrid" specimen: the tail of a fish sewn to the upper half of an orangutan (Fabian 2010: 156-7). While most famous for the art of "humbug," P.T. Barnum was more than a hoaxer. Though not exactly what Peale envisioned a museum should be, Barnum's museum should not be dismissed as all spectacle. Both men – and their museums – utilized the tools of science and art, education and entertainment (in varying degree), to engage with the people of the early republic.

When the members of the U.S. Ex. Ex. (Titian Peale among them) returned home with their extensive collection of natural specimens and cultural artifacts, they also brought with them a Fiji man, Veidovi, who died days after arriving in New York's harbor. In *The Skull Collectors: Race, Science, and America's Unburied Dead* (2010), Ann Fabian described how Veidovi's head was cut off, studied, and displayed by the National Institute for the Promotion of Science in Washington, DC. Barnum even used public interest in Veidovi's head to generate attention (and ticket sales) for his "Fejee Mermaid" (Fabian 2010: 125, 128, 156–7).

The National Institute displayed the materials collected by the U.S. Ex. Ex. alongside the items that accompanied a special (and large) bequest made to the United States by Joseph Smithson to be used for "the increase and diffusion of knowledge" (quoted in Henson 1999: S249). The Smithsonian Institution was chartered in 1846, and Joseph Henry was named its first secretary. Though Henry selected Spencer Baird to be the Smithsonian's assistant secretary, he did consider Titian Peale for the position (Orosz 1990: 162).

The United States National Museum

Even after the formal establishment of the Smithsonian Institution, its actual purpose and focus remained unclear. Though initially against the idea, Henry ultimately supported a museum as part of the Smithsonian, especially after witnessing the draw of the Institution's collections for visitors to DC, many of them brought to the capital by the Civil War (Orosz 1990: 201–12). Additionally, the Smithsonian was the designated repository for items collected on federal expeditions like the U.S. Ex. Ex. The middle decades of the nineteenth century brought increased federal exploration of the western portion of the continent, and with it, animal, botanical, and geological specimens for study and display.

Though these expeditions were not primarily concerned with growing the Smithsonian's holdings, the surveys led by the likes of Hayden, King, Powell, and Wheeler did much to introduce the American people to the territory of the West. Their reports encouraged both settlement and commercial investment, and the materials they gathered generated excitement and further study. We cannot examine the relationships between American science and American museums without an understanding of where all of the artifacts and specimens studied and displayed came from. And though more formal biological surveys eventually replaced the mixed military and civilian groups that pursued multiple objectives across territory unknown to them, these earlier expeditions highlight the complexity of the pathways objects take to reach museum storage facilities and museum displays, as well as the ways that institutions like the Smithsonian are tangled in the history of American exploration, expansion, and empire (Goetzmann 1966; Kohler 2006; Sachs 2006).

What happened to these artifacts after collection? For animal and botanical specimens, as much preparation as possible occurred in the field. Skins were stuffed, parts were pickled, plants were pressed; all were measured and labeled. Once in Washington, they were studied. Individual specimens were described, compared with other items in the collections, and classified as something known or something new. Natural history work at the Smithsonian expanded collective knowledge of the natural world, and much of this found its way into reports, scientific papers, and published taxonomic lists (Lewis 2012).

The material collected by federal expeditions was not only biological; it was also ethnological, and this was nothing new. After all, the earliest formal expedition for the exploration of the Louisiana Purchase, led by Meriwether Lewis and William Clark in 1803, was under instructions from Thomas Jefferson to follow the Missouri to the Pacific, and to learn as much as possible about the Native people they encountered (many of the items collected by Lewis and Clark made their way back to Peale's Philadelphia Museum) (Jefferson 1803). Government expeditions later in the century gathered a significant number of ethnological artifacts, and these materials, in addition to existing work in archaeology and linguistics, ensured the Smithsonian's commitment to anthropology, even if the field and its practices were both new and varied (Hinsley 1981).

The Civil War, too, generated specimens, many of them medical. The Army Medical Museum was created to advance medical study, and as such, army doctors were charged with sending to the Surgeon General's office "all specimens of morbid anatomy, surgical and medical, which may be regarded as valuable" (quoted in Fabian 2010: 168). Soliciting specimens and artifacts for display was an active process. Museum curators cultivated relationships with scientists, collectors, dealers, doctors, and enthusiasts in order to grow their institutional holdings. While earlier museum men, including both Peale and Barnum, developed relationships with people who could help them acquire specimens, the correspondence network Baird built while assistant secretary of the Smithsonian Institution was unsurpassed – nearly 5500 letters per year near the end of his tenure (Goldstein 1994: 576). Who was Baird writing to? Baird's network extended beyond scientists to include naturalists with a range of occupations: farmers, teachers, doctors, homemakers. From this network, the Smithsonian received specimens to grow the collections and assistance in raising public awareness of American scientific work. When Baird became secretary in 1878, his letter writing tapered off. This was due to the shift in his responsibilities, but also to a shift in the scientific landscape: the proliferation of universities, local scientific societies, and regional resources created other possibilities for professional and amateur scientists and naturalists at the same time that they served to distance the Smithsonian from individual correspondents (Goldstein 1994: 596–8).

And, perhaps most significant for our consideration of museums, Baird, who had been committed to the idea from the start, was able to use the 1876 Centennial Exposition in Philadelphia to make a case for a permanent national museum. With the help of George Brown Goode, Baird coordinated a wildly successful Smithsonian exhibit, and negotiated to acquire many of the exposition's exhibits at the fair's end. The outcome was approval for the construction of the United States National Museum (Yochelson 1985: 15–16). When Baird became secretary in 1878, Goode became the museum's director. He overhauled its organizational framework, and established new display practices that stressed the educational importance of the museum's exhibits. Goode wrote extensively about his vision for museums and museum administration, and his ideas are regularly referenced in discussions of museum work today. He famously said that "a finished museum is a dead museum, and a dead museum is a useless museum" (Goode and Kohlstedt 1991: 347).

Other factors shaped the place of museums in the practice of American science – and other institutions. Though smaller in scale and scope than the Smithsonian, the Museum of Comparative Zoology (MCZ), founded by Louis Agassiz in 1859, is significant for several reasons. First, Agassiz. Louis Agassiz was a prominent European zoologist who came to the United States in 1846 to deliver a popular lecture series at the Lowell Institute titled, "The Plan of Creation as Shown in the Animal Kingdom." He stayed on to take a position at Harvard University, and to establish the MCZ. The fact that it was situated at Harvard highlights the museum's importance as "a training ground for a new generation of professional zoologists." Mary Winsor wrote, "It was not enough for Agassiz to have a vast quantity of specimens, he must also have students, for they had a central role in his museum, an intimate, symbiotic function in his plan" (Winsor 1991: 9). Furthermore, many of the students who moved through Agassiz's museum went on to become key figures in the development of institutions important for our understanding of American science.

Not only important as a physical space linking these scientists, the MCZ offers a window onto the history of biology, a field that underwent significant transformation in the mid-nineteenth century, thanks to Charles Darwin's *On the Origin of Species* (1859). Darwin's theory of natural selection had serious implications for Agassiz's work – and for natural history more broadly. The central focus of Agassiz's comparative zoology was locating, identifying, and describing distinct species; natural selection, adaptation, and evolution blurred the dividing lines between species (Pauly 2000; Winsor 1991). In challenging the notion of discrete divine creation of species, these theoretical developments also challenged the primacy of museums like the MCZ as the site for the scientific work of compiling definitive lists and complete specimen collections. Indeed, by the late 1870s, Alexander Agassiz, whose significant wealth allowed him to continue his father's work at the MCZ, was advocating for field stations and laboratories, not museums (Tonn 2015; Winsor 1991).

The history of Agassiz's museum illustrates the challenges of maintaining and displaying ever-increasing collections. Agassiz had always planned on an organized display of the entire collection, but its sheer size meant that at any given time, most of it was in "barrels and boxes and jars" (Winsor 1991: 120). Some credit Agassiz with the idea of "dual arrangement," but it was Asa Gray who articulated the different interests of the research community and general public (Winsor 1991). Researchers needed access to a full collection, while visitors to the museum might benefit most from a few interesting and educational examples. Eventually, the MCZ embraced separate collections for research and for display, and soon other major museums followed suit.

The late nineteenth century was a period of American institution building, and museums were no exception. Alongside industrialization and urbanization and the growth of hospitals and schools, museums were growing too, both in terms of the size of their collections and their place in civic life. During the 1880s, over four million people visited the exhibits of the United States National Museum (Pauly 2000: 69). And alongside what Robert Wiebe has called "the search for order" in the late nineteenth century, the natural sciences experienced a great degree of professionalization (Wiebe 1967). Steven Conn's narrative of this period includes a "struggle for authority" between museums and universities (Conn 2010: 17), and in the relationship he described between the Academy of Natural Sciences in Philadelphia and the University of Pennsylvania, struggle does seem to be the right word (Conn 2010: Chapter 2). But Philip Pauly painted this period of professionalization, of the separation of museums from scientific research, as more contingent and fragmented. In *Biologists and the Promise of American Life: From Meriwether Lewis to Alfred Kinsey* (2000), Pauly took readers on a hypothetical tour of people and institutions involved in the life sciences at the end of the nineteenth century, revealing less coordination, but plenty of "possible audiences and patrons for life science initiatives" (Pauly 2000: 124). Museums do not figure heavily into the next part of the story; Pauly followed university efforts to establish programs in biology, and identified the Woods Hole Marine Biological Laboratory as a site for important research and wide-ranging conversations among the participants, many of whom would become key players in the direction of the field and its importance for American life more broadly.

At the turn of the twentieth century, academic biology was increasingly focused on the laboratory – and the microscope. Embryology and reproduction, in particular, captured biologists' attention. This shift has been characterized by some historians as a "revolt from morphology," a radical shift away from natural history as it had been practiced in the nineteenth century – observing, collecting, and describing specimens and species – and a move toward a more experimental approach based in the laboratory. Recent work has softened this "revolt" to more of a shift in emphasis; though natural history work and practices continued, they were no longer at the center of the life sciences (Maienschein, Rainger, and Benson 1981). And museums, long central to the practice of American natural history, played a smaller role in the broader category of biology.

Henry Osborn, a paleontologist who helped to shape the Columbia University department of biology in the 1890s (and even fostered ties with the laboratory at Woods Hole), served concurrently as a curator at the American Museum of Natural History (AMNH). In 1908 he was named the museum's president, and during his tenure at the museum, he built a premier paleontology research program at the same time that he prioritized "the public dimension of paleontology" (Rainger 1991: 88). The AMNH was itself a public institution. It received funding through New York City's Department of Public Parks, but the museum's trustees – some of the city's wealthiest men – handled the rest (Rainger 1991: 55).

Osborn, himself white, educated, and elite, espoused traditional values. He used paleontology, and the museum more broadly, to advocate – and fundraise for – programs and exhibits consistent with the kind of social uplift that was characteristic of elite society during the Progressive Era. Museum exhibits promoted nature conservation and nature study. (Some of Osborn's views were quite ugly – racism, eugenics, anti-immigration – and elements of these ideas were on display as well.) Osborn's leadership – and his ability to attract patronage, press, and public interest (100,000 visitors each year by the 1920s) – carved out concurrent spaces for both research and public education at the AMNH in the early twentieth century (Rainger 1991: 180-182).

In recent decades, scholars interested in American museums have turned their attention inward, to the objects and exhibits inside American museums. Focusing on specimens and artifacts – and their preparation and display – allows us to explore not only how museums used and understood their holdings, but also how they engaged the public.

Curation and Display

Looking at what is happening inside the museum necessarily means considering it as a physical space. Architectural historian Carla Yanni has examined the intersection of museum buildings and the production of knowledge about nineteenth- and twentieth-century nature on display inside these structures. Though focused on British museums, Yanni's work helps us to understand what Steven Conn has called "object-based epistemology." Objects, especially if displayed properly, could reveal their meaning to viewers; when arranged, the objects together offered a narrative, and in the late nineteenth century, that narrative was often a "positivist, progressive and hierarchical view of the world" (Conn 1998: 5; Yanni 1999). Can you picture it? Imagine lots of specimens, arranged taxonomically and under glass. The Wagner Free Institute of Science currently offers visitors an opportunity to see what a systematically organized late nineteenth-century museum looked like: rows of long cases filled with objects carefully labeled. And plenty of taxidermy. These displays were laid out to instruct – by viewing specimens in order, students and visitors learned the arrangement of the natural world. Ordered, hierarchical display strategies were also central to the structure of world's fairs, where the globe's finest achievements were supposedly exhibited. Robert Rydell has argued that by organizing the world's peoples, cultures, and commodities along an axis of "progress," American exposition planners displayed an interconnected set of arguments about race and empire, arguments that resonated with the natural science and physical anthropology of the day (Rydell 1984).

The power of these narratives, bolstered by the display of artifacts and specimens to illustrate them, feels, at times, totalizing. Indeed, cultural historians have explored the depth and reach of these narratives in American life, and museum studies scholars have drawn on Foucault to examine the museum as a site of control (Conn 1998: 11–12). But beginning with the objects themselves, as Samuel Alberti has suggested, foregrounds the relationships between objects, curators, and museumgoers. In "Objects and the Museum" (2005), Alberti advocated for approaching the museum through object biographies. Though Alberti was careful not to ascribe agency to objects – it is the people who interact with them that make meanings – objects are central because they "channeled and enabled a series of relationships" (Alberti 2005: 571). Elsewhere, Alberti has explored the language of museum objects. Museum processes – preserving, skinning, stuffing, assembling skeletons – are what transform once-alive objects into specimens. Not yet artifacts at the moment of their collection, they become material culture once humans begin to remake them for display (Alberti 2008: 81–2). There is a rich and growing literature on taxidermy, paralleling, perhaps, an increased interest in the practice itself.

Examined against the backdrop of narratives of the professionalization of science, taxidermy brings into relief a complex network undergirding the more visible labor of scientists and curators. Taxidermists did messy work: they constructed lifelike mounts out of once-alive animals. To do so well meant having an impressive arsenal of knowledge about one's subject: skeleton, musculature, habitat, behavior. Susan Leigh Star traced this "highly elaborate and expensive artisanal skill" through its failed attempts at professionalization, and examined the relationship between the relegation of taxidermy as a craft and the formalization of biology as a profession (Star 1992: 261). While taxidermy focused on representing particular specimens "naturalistically," turn-of-the-century biology was moving toward formalization, standardization, and industrialization (Star 1992: 272). But rather than place taxidermy's decline inside a much too simple shift from old natural history to new biology, Star argued that taxidermy – the invisible work of constructing "orderly, beautiful, collectible" nature – was not only important for understanding modern biology, but also that examining taxidermy practices illuminates a larger point: "Recovering the material basis of science by looking very directly at the stuff it uses and the stuff it leaves behind is one way to begin restoring the links, and reclaiming the mess" (Star 1992: 282).

While Star focused on the work of taxidermy, Rachel Poliquin focused on its cultural history. From early modern curiosity cabinets to postmodern art installations, *The Breathless Zoo: Taxidermy and the Cultures of Longing* examines motivations for and reactions to taxidermy – what do we do with these animals, "dead but not gone"? What do they mean? (Poliquin 2012: 9). Poliquin presents taxidermy as "a primary technology for making creaturely life visible" (Poliquin 2012: 136). It is an aesthetic practice. Taxidermy is about display – which means it is also about looking. Taxidermy was – and is – exhibited beyond the boundaries of the natural history museum: Walter Potter's anthropomorphized animal scenes, "unnatural" hybrids, discarded or degraded specimens repurposed for art. Considering these examples alongside more traditional specimens helps us to look beneath the realism and consider the arguments implicit in these displays (Poliquin 2012).

Scholarly analysis of this cultural work has centered on the habitat diorama, a display strategy that situated specimens in landscape scenes appropriate to the species represented (Wonders 1993). Perhaps the most famous habitat dioramas came to life in the AMNH Hall of African Mammals. The Hall housed gorillas, elephants, lions, and more, all collected and prepared under the supervision of the museum's taxidermist, Carl Akeley, and it is the subject of Donna Haraway's "Teddy Bear Patriarchy: Taxidermy in the Garden of Eden, New York City, 1908–1936" (1984). Haraway read the Hall of African Mammals in the context of American notions of manhood and nature. She used Akeley's work, work he was using to offer a story of Africa, a story of conservation, to instead narrate the world he was part of, and to make visible "the commerce of power and knowledge in white and male supremacist monopoly capitalism" (Haraway 1984: 21). Akeley's gorillas reveal less about Africa than they do about American culture. Habitat dioramas were a way for museumgoers to engage with representations of nature, with once-alive artifacts of a wild world. They offered opportunities to experience (in close proximity) creatures arranged in ways approximating life, and as such, they were popular. And for a time, they were considered the best way to educate the public about American (and other) nature. But even after natural history was displaced by biology, dioramas persisted because of their place in broader systems of museum patronage. Donors liked dioramas, and museums liked them too for the ways they enabled collecting expeditions (Rader and Cain 2008: 157). Eventually even the dioramas began to feel static and outdated.

Mid-twentieth-century shifts in curatorial strategy bring into sharp relief a central question that museum curators and scientists at museums had been struggling with since the beginning: what – and who – are museums for? Karen Rader and Victoria Cain have examined the shift from "natural history" to "science" in museums in the middle twentieth century. Historians have overlooked these decades, focusing instead on "museology bookends": the shift away from rows of specimen-filled glass cases (and toward realist, visual displays like dioramas) in the early twentieth century, and the move toward less linear and more interactive exhibits later in the twentieth century (Rader and Cain 2008). Rader and Cain identified a complicated process of adaptation and negotiation both within museums and between museums and scientists as reduced budgets in the 1930s and 1940s illuminated tensions between the research and educational aims of many institutions. At the Smithsonian, scientist-curators and exhibition staff disagreed about how to represent museum collections in their Hall of Marine Life: "would it present a series of separate and specialized research findings or a broader thematic narrative?" (Rader and Cain 2008: 159). The resulting exhibit was a compromise. It featured a large whale sculpture, and signaled a shift toward "immersion-style exhibits designed around scientifically-mediated, lifelike art, rather than nature's own specimens" (Rader and Cain 2008: 161).

The AMNH had its own whale – or rather, model of a whale. First displayed in 1907, the whale model was celebrated as an authentic representation of the real thing. Michael Rossi examined the construction of this whale – the "measuring, photographing, casting, and sculpting" used to make an accurate model – in order to understand how curators negotiated questions of authenticity (Rossi 2010: 339). Could a model reach taxidermy's "real-ness"? A key justification for the realness of the AMNH whale was that it reflected the measurements of a particular whale. This was not an aggregation of generalized whale characteristics; this whale was the physical double of a once-alive animal that had been seen and studied by scientists. Still, some had doubts, and the debates over the accuracy – and the value – of a whale that was not actually a whale highlight the stakes of museum displays. What were they for, if not to display "real" nature, real artifacts, to viewers?

They were for creating an experience of the natural world. After World War II, the AMNH shifted to a new exhibit strategy, even more centered on the experience of museumgoers. The original whale model was replaced by a fiberglass version in 1969, the central piece in the new Hall of Ocean Life. Museum curators, in focusing less on specimens and more on scientific ideas and concepts, "increasingly tended to view scientific authority governed not by objects, but by the experiences of viewers as participants" (Rossi 2010: 359).

This shift in focus was accompanied by a shift in audience. The move toward immersive exhibits was also a move toward children, toward engagement with public school students and their teachers. Though certainly connected to changes in the site and practice of biological sciences, this shift is also reflective of mid-twentieth-century ideas about public education. Albert Parr, the man who led the AMNH during this period, championed the link between museums and schools, and worked to expand the museum's educational programming (Conn 2010: 144–51).

Bruce Lewenstein and Steven Allison-Bunnell have demonstrated that the research and exhibit goals of museums were not always in conflict. Sometimes exhibition needs create possibilities for substantive scientific work (Lewenstein and Allison-Bunnell 2000). They pointed to the rainforest exhibit at the Smithsonian's National Museum of Natural History, which began "as a traditional hall of botany in the early 1960s" and was later "mounted as a thematic story about environmental degradation" (Lewenstein and Allison-Bunnell 2000: 194). Museum staff took a collecting trip to South America. Though the purpose of the trip was tied to the rainforest display, members collected botanical samples for research alongside the photographs they took to help with the display. This example highlighted the interplay between research and exhibit goals, and furthermore, that "museums use their need to create public spaces as tools and opportunities for creating new scientific knowledge" (Lewenstein and Allison-Bunnell 2000: 195).

Although research and exhibition projects can coexist productively, it is also true that museums are less equipped to conduct research in the sciences outside the fields of natural history. In institutions where exhibits shifted away from natural history and toward displays focused on scientific processes, there was less likelihood of research and exhibit synergy. (Some fields, like paleontology, still lend themselves to museum-based work.) Furthermore, since the middle twentieth century, the science museum landscape has broadened significantly. And museums of technology invited visitors to do more than look. Institutions such as the Chicago Museum of Science and Industry (1933), the Franklin Institute (opened to the public in 1934), and the Exploratorium (established 1969) focused on engaging visitors with exhibits designed to teach about how things work in a hands-on way. The Exploratorium's founder desired a "laboratory atmosphere"; visitors could conduct experiments as part of the exhibits, which were purposefully arranged in a nonlinear way (Rader and Cain 2008: 163). I have vivid memories of "participating" in a static electricity "experiment" at the Ontario Science Centre as a child. This trend toward interactivity continues, further enabled by technology that can be used to simulate the distant past. As part of "Pterosaurs: Flight in the Age of Dinosaurs," which opened at the AMNH in 2014, visitors could flap their arms and fly, hunt, and fish as part of the exhibit's "virtual flight lab" (AMNH 2014). Flying like a dinosaur is a far cry from looking up at Peale's mastodon.

Objects Today

Steven Conn asked a question that seems to undergird the shifts that occurred in twentieth-century museum exhibits: "Do museums still need objects?" In his volume of the same name (2010), Conn explored "the disappearance of objects" and the evolving relationships between institutions and their publics accompanying this change in curatorial practice. While not focused on natural history or science museums, Conn acknowledged the particular power of specimens in the late nineteenth and early twentieth centuries, and suggested the possibilities these collections hold for the place of museums in future ecological research and education – species that were once collected and displayed for their "representativeness" are now appreciated and studied for their "rarity" (Conn 2010: 52).

This observation – that natural history collections may become a significant resource to scientists interested in biodiversity and climate change – reveals that although scientific practices shifted away from the museum in the twentieth century, putting more distance between scientists and curators, there are still plenty of ways they interact. This idea that museum collections will become scientifically valuable again, not just because of what those specimens can reveal about environmental questions, but also because we've developed new methods of analysis, is where many essays on museums and science end. But recent work has moved in a different direction – away from the objects themselves, and toward how we use them.

Many scholars of both museums and American science have focused on exploring the sites of knowledge production; an additional debate in the history of biology has centered on the nature of this production. The shift from museums to universities (and their laboratories) corresponded to a shift toward a biology broader than natural history, toward experiment, toward the microscope, toward cells and genes. But this move, from description, taxonomy, systematics, and comparison to a more micro and more experimental biology, was not a simple replacement; natural history research and practices are with us still. In fact, recent work by Bruno Strasser has further complicated the standard narrative of the history of the life sciences. Rather than stress breaks or new paradigms, Strasser has used collecting practices as a way to see continuities between nineteenth- and twentieth-century life sciences, as well as hybridity in twenty-first-century work with collections (databases) of protein sequences. Alongside experimental work, biologists continue to perform the kind of comparative study that nineteenth-century museums and their collections supported (Strasser 2011, 2012).

This emphasis on practice is also visible in museums. While historians are exploring museum practices in scientific contexts, contemporary science museums are emphasizing scientific practices in their displays and exhibits. From offering visitors the chance to try scientific fieldwork (unearthing a fossil) or participate in an experiment (experiencing static electricity), museums are drawing attention to processes. Some even put research staff, working behind glass, on display! Other museums offer commentary on the history of both scientific practice and museum curation through their displays: for example, the Wagner Free Institute of Science has preserved its Victorian-era glass cases filled with specimens, and viewers can experience the collection as it was at the turn of the twentieth century. In recent decades, historians have begun returning to museums, to examine them not simply as repositories of old science, but as sites for re-examining museum practice, and by extension, scientific practice as well.

Bibliographic Essay

An interest in museums brings together scholars from several fields: history, science and technology studies, anthropology, and museum studies, as well as practitioners interested in museum collections, from scientists to curators. I found my way to natural history museums through environmental history when I learned that an army surgeon

I was researching collected specimens for the Smithsonian everywhere he was stationed. In exploring his life and work, I began tracing the pathways of the birds he collected and prepared (Kohout 2013). I share this to provide some context for my own exploration of museum history, as well as to identify areas of potential reader interest that my essay does not cover in great detail.

First, my essay is American-centric in scope, as befits this volume. There is an extensive literature focused on the origins of the modern museum in the European context. This material enriches our understanding of the history of collecting practices, the history of private collections and cabinets of curiosity, as well as the history of European learned societies and museums, many of which pre-date their American counterparts. Good starting places are *Cultures of Natural History*, edited by Nicholas Jardine, James A. Secord, and E.C. Spary (1996) and Paula Findlen, *Possessing Nature: Museums, Collecting, and Scientific Culture in Early Modern Italy* (1994). Also, see the historiographical essays written by Mary Winsor in the *Cambridge History of Science* (2009) and Sally Gregory Kohlstedt in the December 2005 *Isis* "Focus" section dedicated to museums.

I make reference to the imperial underpinnings of American collecting expeditions, but there is far more to be said on the subject. On collecting practices and imperialism, see Jim Endersby, *Imperial Nature: Joseph Hooker and the Practices of Victorian Science* (2008) and Daniela Bleichmar, *Visible Empire: Botanical Expeditions & Visual Culture in the Hispanic Enlightenment* (2012). On colonial museums, see *Cathedrals of Science: The Development of Colonial Natural History Museums during the Late Nineteenth Century* (1988) by Susan Sheets-Pyenson. I have only briefly mentioned world's fairs; the best place to start remains Robert Rydell's *All the World's a Fair: Visions of Empire at American International Expositions, 1876–1916* (1984).

My essay has focused primarily on natural history, but there is also work on the history of anthropology. For an overview of debates over ethnographic artifacts, see Conn (2010), chapter 2. See also Douglas Sackman's *Wild Men: Ishi and Kroeber in the Wilderness of Modern America* (2010), which examines the relationship between Ishi and Albert Kroeber. I only discuss habitat dioramas, but Ronald Rainger's *An Agenda For Antiquity* (1991) includes coverage of human history dioramas at the AMNH.

Museums of science and technology also warrant more coverage – and more scholarship. There is not much written on the Franklin Institute; see Conn (2010), chapter 4. On the Exploratorium, see Hilda Hein, *The Exploratorium: The Museum as Laboratory* (1990). Sharon Macdonald's edited volume, *The Politics of Display: Museums, Science, Culture* (1998b), includes essays on power, race, interactivity, and curation. *Museums of Modern Science* (2000), edited by Svante Lindqvist, is a collection of essays from academics and curators on issues facing science museums past and present. I would also direct readers to *Life on Display: Revolutionizing U.S. Museums of Science and Natural History in the Twentieth Century* (2014), a new book by Karen Rader and Victoria Cain.

For a broader theorizing of the museum, I would direct readers to museum studies. Tony Bennett's *The Birth of the Museum: History, Theory, Politics* (1995) takes a Foucaultian approach, and *Museums in Motion: An Introduction to the History and Functions of Museums* (1996), by Edward Alexander, offers an overview from a practitioner's perspective. *A Companion to Museum Studies* (1998a), edited by Sharon Macdonald, offers several directions for further study.

It has been suggested that we are living in a golden age of museums; in 2014, there were 35,000 museums in the United States (IMLS 2014). I'd like to close by highlighting a few that are doing particularly interesting things, especially within the context of the history of American science. First, there is the Museum of Jurassic Technology, which seems to be arguing for a return to museums as cabinets of curiosity, rather than sites of order and authority (see Weschler 1995). The newly opened Morbid Anatomy Museum in Brooklyn is committed to the "exhibition of artifacts, histories and ideas which fall between the cracks of high and low culture, death and beauty, and disciplinary divides" (Morbid Anatomy Museum 2014). In addition to hosting temporary exhibits and researchers interested in its books and artifacts, the Morbid Anatomy Museum serves as a gathering space for lantern slide lectures, workshops with their taxidermist-in-residence, and vintage film screenings. And lastly, the Lost Museum at Brown University is an exhibit that re-creates portions of a natural history museum that used to exist on campus in the late nineteenth century (Fountain 2014). This project speaks both to the impermanence and the enduring power of natural history objects, arranged just so.

Chapter Thirty-Five

NATURAL HISTORY

Pamela M. Henson

Natural history is the study of organisms in their environment, and the relationships between plants and animals and the physical world around them. Natural history includes the observation, collection, description, systematic study, history, and classification of natural objects or organisms. Study of natural history can be found in Greek, Roman, and Arabic cultures, with attempts to name, describe, and organize the natural world. Natural history was passed down through the Middle Ages by such works as Aristotle's *History of Animals* and Pliny the Elder's *Naturalis Historia*. Renaissance naturalists reinvigorated the discovery, description, and naming of plants and animals. The eighteenth and nineteenth centuries saw wide exploration of the natural world, and the development of new systems for the description and classification of natural organisms and objects.

The seventeenth and eighteenth centuries were very active periods for the field of natural history. John Ray (1627–1705), father of English natural history, developed a classification system that illustrated the glory of God through knowledge of the creation. Carl von Linné, better known as Linnaeus (1707–1778), a Swedish naturalist, known as the father of modern taxonomy, advanced taxonomic practice and theory through his *Systema Naturae* published in 1735. Gilbert White of Selbourne (1720–1793) was regarded as England's first ecologist who encouraged respect for nature. Best known for his *Natural History and Antiquities of Selborne* (1813), he advocated phenology – the study of periodic plant and animal lifecycle events and the influence of climate and habitat. Natural historians such as Ray and Linnaeus established classification systems for all living creatures, from single-celled organisms to plants and animals. "Cabinets of curiosities" gave way to well-organized museums that displayed the diversity of the natural world to the general public as well as naturalists.

The discovery and exploration of the New World included the discovery of many plants and animals similar to those of the Old World, but also organisms never seen

A Companion to the History of American Science, First Edition.
Edited by Georgina M. Montgomery and Mark A. Largent.
© 2016 John Wiley & Sons, Ltd. Published 2020 by John Wiley & Sons, Ltd.

before. Naturalists sought to explain where these strange creatures fit in God's natural order. There is limited documentation of Native American knowledge of natural history, primarily from Central American codices, compilations of knowledge recorded by missionaries, and images of nature captured in material culture, such as paintings and totems. After contact in 1492, Europeans viewed American natural history as pristine and untouched by Native Americans, but recent scholarship has demonstrated that Native Americans altered the landscape significantly, developing agriculture sites by centuries of harvesting, tilling, sowing, and burning (M. Anderson 2005; Denevan 1992).

American Natural History

European settlers initiated natural history studies of the New World using the theories and methods of European science. Colonists brought attitudes toward nature that were quite different from Native Americans', including a different sense of property and the values of a capitalist economy, and these led to complex long-term changes in the natural world in New England (Cronon 1983). Colonists' ideas about nature included that God created the natural world for human use; the Great Chain of Being established a human place above the natural world; and the natural world was a dangerous place that should be domesticated. European naturalists were curious about the plants, animals, and geology of the Americas and established networks to collect the raw materials of natural history for European intellectual centers, where they were named and classified.

New World collections were sought by many Old World naturalists. In 1748, Linnaeus' student Pehr Kalm spent two years collecting the flora and fauna of Pennsylvania, New York, New Jersey, and Canada. Kalm solicited plant lore from Native Americans as well as the colonists, and the new species were classified within Linnaeus' *Systema Naturae*. Sir Hans Sloane (1660–1753) was a British physician and collector, notable for bequeathing his collection to found the British Museum. Peter Collinson (1694–1768), a cloth merchant, used his commercial network to develop a market for New World seeds and plants in England. He financed the American travels of John Bartram and Mark Catesby, and was a patron of the fledgling American Philosophical Society and the Library Company of Philadelphia (O'Neill and Mclean 2008). Mark Catesby (ca. 1682/83–1749), an English naturalist, collected from Virginia south to the Bahamas from 1712–1719. His *Natural History* (1731–1743) introduced Europe to the variety and beauty of American bird life (Evans 1993). Colonial naturalists were occasionally made fellows of such organizations as the Royal Society of London, further cementing the ties between Old World and New (Stearns 1948).

The flora and fauna of the New World were sometimes quite different from their European counterparts. In 1712, when Puritan divine Cotton Mather (1663–1728) described a mastodon, its huge size attracted notice on both sides of the Atlantic. Since extinction was not thought possible, mastodons defied explanation. Cadwallader Colden's (1688–1776) flora near his New York home was published by Linnaeus. John Bartram (1699–1777), a Pennsylvania Quaker, combined a reverence for nature with great curiosity. Bartram established a garden with the most varied collection of North American plants in the world, and a network of plant exchanges with the London merchant Collinson that reached collectors throughout Europe. He was one of the first practicing Linnaean botanists in North America (Bartram 1766). Following in his father's footsteps, William Bartram (1739–1823) continued to discover and classify native American plants. William Bartram's *Travels Through North & South Carolina, Georgia, East & West Florida, the Cherokee Country* (1794) was well received in the Old World for its beautiful drawings and meticulous observations.

Alexander Wilson (1766–1813), a Scottish immigrant, was introduced to natural history by William Bartram. In 1808–1814, Wilson published his nine-volume *American Ornithology* and also conducted the first breeding bird census in Bartram's garden. His 1810 meeting with Audubon probably inspired Audubon to publish his own books on birds. John James Audubon (1785–1851), American naturalist and painter, was noted for his magnificent *Birds of America* (1840–1844), considered one of the finest bird studies ever completed. He traveled widely to document American birds with detailed illustrations of birds in their natural habitats.

Natural theology still dominated scientific inquiry of nature after the United States was founded in 1776; however, American naturalists now wanted to study and name their own species. Educated citizens were familiar with William Paley's 1802 work, *Natural Theology, or Evidences of the Existence and Attributes of the Deity collected from the Appearances of Nature*, with its Watchmaker Analogy to argue the design of human beings and the cosmos demonstrated the existence of a divine designer, and it was adopted by Christians and Deists alike. The Bridgewater Treatises published from 1833 to 1840 also spread their message that the power, wisdom, and goodness of God were manifested in the Creation. But in the new democracy, Deism continued to grow, with its rejection of the supernatural, such as miracles. Rational knowledge was also championed by Thomas Paine's influential 1795 work, *The Age of Reason*.

Thomas Jefferson (1743–1826) best exemplifies the New World naturalist. Fascinated by the natural world, Jefferson sought knowledge of nature as intrinsically valuable, as well as useful in the development of the US economy. Jefferson engaged in a public dispute with Count Georges-Louis Leclerc, Comte de Buffon, curator of the King's Natural History Cabinet in France. In his 1766 *Histoire Naturelle*, Buffon claimed that New World species were degenerate forms of Old World species. Since North America was a cold and wet clime, *all* species found there were weak, shriveled, and diminished. If Americans tried to raise domesticated species, they would produce feeble offspring. Buffon's ideas provided a popular scientific justification for Old World superiority. In Jefferson's *Notes on the State of Virginia* (1782), he argued there was no reason that differences between the New World and Old should translate into degeneracy in the former and attacked Buffon's data with tables enumerating the weights of animals from Europe and America. After Jefferson sent a large moose specimen, Buffon was forced to relinquish his theory. The moose paled in comparison to the mastodon species discovered near Newburgh, New York, in 1801.

The new United States established its place in the world order through practical and scientific exploration of the globe. The US Exploring Expedition (USEE) circumnavigated the globe from 1838 to 1842 to establish the United States as a world power and acquire knowledge about far-flung regions of the globe. Joel Poinsett (1779–1851), Secretary of War with an interest in the natural world, ensured that natural history collecting was in the USEE's mandate. Upon its return, collections were displayed at the US Patent Office and later became foundation collections at the Smithsonian's National Museum (Philbrick 2003).

For economic and scientific purposes, many states initiated natural history surveys. The New York State Geological and Natural History Survey was formed in 1836 to document the mineral wealth of the state. In 1870, it was reorganized as the New York State Museum of Natural History. By the 1850s, East Coast natural history had been explored extensively, and the government supported expeditions to the American West to facilitate settlement and assess economic possibilities (Goetzmann 1966).

Among the most influential naturalists of the mid-nineteenth-century period was Louis Agassiz (1807–1873), a Swiss-born biologist and geologist. He was appointed professor of zoology and geology at Harvard in 1847, founding its Museum of Comparative Zoology. Noted for his belief that all organisms were based on four ideal body types, he made significant contributions to fish classification and the study of glaciation and geology. But Agassiz's impact was diminished by his writings on special creation, that is, that God had created each species perfectly adapted to its environment, as well as his resistance to Charles Darwin's new theory of evolution by natural selection (Winsor 1991).

When Darwin published *On the Origin of Species* in 1859, it was disseminated rapidly across the United States. The North American natural history community soon split along followers of ideal morphology, such as Agassiz of Harvard, and proponents of Darwin's theory, led by Harvard botanist Asa Gray and Spencer Baird of the Smithsonian. Agassiz's colleague Asa Gray (1810–1888) was the most important American botanist of his era, as he unified taxonomic knowledge of North American plants. Gray's most popular work was his *Manual of the Botany of the Northern United States* (1864). Gray arranged the first US edition of *On the Origin of Species* in 1859, and argued for conciliation between Darwinian evolution and the tenets of theism, at a time when both sides perceived the two as mutually exclusive. Gray also pioneered the study of geographical distribution of organisms, with his disjunction theory. At Cornell University, entomologist John Henry Comstock (1849–1931) revised taxonomic practice to incorporate Darwinian theory and taught a generation of evolutionary taxonomists. The "natural" classification of organisms was now seen as reflecting their evolutionary history, as organisms changed and diversified in response to natural selection (Henson 1993). As the United States and Canada expanded across the continent, naturalists continued to discover, describe, and classify a wealth of new organisms. Naturalists also continued to explore the globe, finding organisms that fit within existing groups, as well as exotic new creatures that challenged existing classification systems. Natural history collections grew so rapidly in size and complexity that "systematists" – those who classified into a natural system reflecting the evolution of life – began to specialize by organism type, such as birds or insects, and then even more narrowly by butterflies or ants or beetles.

By mid-century, women participated more actively in science, making considerable headway in the field of botany; indeed in 1887, J.F.A. Adams published an essay in *Science* titled, "Is Botany a Suitable Subject for Young Men?" expressing concerns over the feminization of botany. The new US Department of Agriculture provided employment to many young women educated at the newly founded women's colleges.

By the late nineteenth century, conventions for the naming of organisms varied by intellectual school, geographic region, and type of organism, making comparative studies quite difficult. In the 1890s, European and New World zoologists sought to establish commonly accepted international rules for all disciplines and countries to replace unwritten rules. The "International Rules on Zoological Nomenclature" were officially published in 1905, and amendments were subsequently passed by zoological congresses. A new version of the nomenclatural rules was published as the first edition of the *ICZN Code* in 1961 (Johnson 2009).

By the late nineteenth century, with the development of tools like the microscope and fields like chemistry, laboratory biology began to overtake the earlier emphasis on natural history and organismal biology. Cytology, embryology, physiology, and genetic studies attracted attention, students, and funding, as natural history fell out of vogue. The field was reinvigorated at mid-twentieth century by the evolutionary synthesis that combined ideas from Darwinian evolution, systematics, and genetics, which was developed by three major figures. Ernst Mayr (1904–2005), taxonomist and historian of science at Harvard, was a leading evolutionary biologist whose work contributed to the synthesis and development of the biological species concept. George Gaylord Simpson (1902–1984), an American paleontologist at Columbia University and the American Museum of Natural History, was the most influential paleontologist of the twentieth century, and a major participant in the synthesis, contributing *Tempo and Mode in Evolution* (1944) and *The Meaning of Evolution* (1949). An expert on extinct mammals and their intercontinental migrations, he anticipated such concepts as punctuated equilibrium. George Ledyard Stebbins, Jr., (1906–2000) was an American botanist and geneticist at the University of California whose work on the genetic evolution of plant species led him to develop a comprehensive synthesis of plant evolution incorporating genetics. His *Variation and Evolution in Plants* (1950) provided an alternative to Mayr's biological species concept that accounted for the very different modes of plant reproduction (Cain and Ruse 2009; Cravens 1978; Mayr and Provine 1980).

As the western prairies underwent rapid change through settlement, naturalists became aware of the delicate balance of natural environments, spurring the rise of the conservation movement and study of ecology. With the West fully settled, the study of nature focused on using natural resources and plowing the great prairie lands. The near disappearance of bison and extinction of the passenger pigeon gave impetus to the budding environmental movement. In addition, the 1930s Dust Bowl and Great Depression taught Americans that a deeper understanding of the natural world was needed even for agriculture. The environmental movement grew as it became clear that each part affected the whole, and America's natural abundance could be finite. Ecology became well established as a holistic branch of biology that attempted to define not only the species and physical environment but also the complex set of interactions between all the parts. Natural history studies of a region served as the baseline for ecological studies, identifying the organisms that lived in that environment as well as subsequent changes (Sutter 2002).

Aldo Leopold (1887–1948), an American author, ecologist, and environmentalist, was best known for his *A Sand County Almanac* (1949), which influenced modern environmental ethics and the movement for wilderness conservation. His ecocentric or holistic ethics of wildlife preservation emphasized biodiversity. As a founder of the science of wildlife management, he advocated creating "wilderness" regions, and rejected the utilitarianism of conservationists such as Gifford Pinchot and Theodore Roosevelt. Roosevelt (1858–1919) began studying natural history while just a boy. He loved hunting and adventure in the open West and considered a career as a naturalist. As president, he supported natural history studies, but along with Pinchot, had a far more utilitarian view of nature than Leopold. He oversaw acquisition of the Panama Canal, an enormous environmental disruption, but also supported a baseline biological survey of the region prior to the opening of the Canal (Nash 1967).

As animals and plants faced rapid extinction, new emphasis was placed on environmental conservation. Rachel Carson's seminal volume, *Silent Spring*, published in 1962, found a receptive audience in the counterculture and led to the creation of the annual Earth Day commemoration in 1970. The Endangered Species Act of 1973 provided tools for species conservation as well as funding for basic research on threatened organisms (Winston 2011). Continued research in the New World tropics made naturalists aware of the significant biodiversity found in those climes, and single species studies gave way to analysis of biological communities, diversity, distribution, and interactions with the physical and biological environment. The tropics were also found to have a far greater degree of seasonal climate variation than previously understood. By the close of the twentieth century, the effects of global warming refocused studies from individual species to entire ecosystems.

By the late twentieth century, advances in technology were incorporated into natural history and provided new sources of evidence of biological processes, from cell sorting to DNA analysis. DNA sequencing added to the characters used by systematists to classify organisms and hypothesize rates of evolutionary divergence of organisms. Museum natural history collections allowed new forensic techniques to assist with analysis of human remains in crimes and identify species after bird strikes in the increasingly crowded skies. The divide between laboratory biology and organismal biology was bridged by application of molecular techniques to the study of organisms. New research tools, such as the scanning electron microscope, allowed systematists to incorporate characters formerly not visible to the human eye into traditional taxonomic analyses. As DNA sequencing became more affordable and accessible, systematists began to routinely incorporate characters from the genome in their classifications.

Systematic biology had absorbed the evolutionary synthesis and biological species concept, but taxonomists looked for additional analytical tools to determine evolutionary relationships. Cladistics, from the Greek for branch, grouped organisms together based on their shared unique characteristics inherited from the group's last common ancestor but not present in more distant ancestors. It was based on the work of Willi Hennig, a German entomologist, who named it phylogenetic systematics. Analysis of large sets of characters was facilitated by the development of computers that could process ever larger datasets. Phenetics was an alternative method of computational phylogenetics during the 1950s through the 1980s that used overall similarity in morphology or other observable traits. Pheneticists built dendrograms from similarity data, rather than reconstruct evolutionary trees. Interest in phenetics waned when it became clear that it was unable to provide reliable information about evolutionary relationships among taxa (Hamilton and Wheeler 2008; Hull 1998; Vernon 2001).

The evolutionary history of organisms was also advanced by the theories of continental drift and plate tectonics and had major implications for the field of biogeography which studies how species and ecosystems are distributed through geographic space and geological time. Underwater exploration by Bruce Heezen, Marie Tharp, and others in the 1950s revealed seafloor spreading, the process of new crust formation between two plates that are moving apart. Tharp represented the growing role of professional women biologists educated at land-grant colleges and women's colleges, moving from the peripheries of science into the mainstream (Creager, Lunbeck, and Schiebinger 2001). Studies of the ocean floor also led to discovery of seafloor hydrothermal vents where life forms existed without energy produced by sunlight. Previous biology teachings had emphasized photosynthesis as the source of all energy for multicellular organisms. Chemical synthesis by bacteria living inside the rift tube worms, clams, and other organisms demonstrated a previously unknown energy chain for maintaining life and leading some to posit a chemical synthesis origin of life (Ballard 2000).

As laboratory biology increased in importance by the mid-twentieth century, many universities abandoned organismal biology and systematic/evolutionary studies in favor of molecular biology. Studies in the 1970s addressed the crisis in systematic biology and sought to ensure that university collections survived. The Post Report, Michener Report, Steere Report, and Belmont Report all sought to address the decline in organismal biology and ensure preservation of collections languishing in basements of universities (Capshew and Rader 1992).

As the world grew smaller with improved communications, distinctly American science gave way to international teams of collaborative researchers. By the second half of the twentieth century, large-scale expeditions to remote areas had given way to targeted research trips to collect specific organisms. Postcolonial restrictions on collecting limited specimens that visiting researchers could collect themselves, which then led to more exchange relationships between natural history institutions. Expeditions gave way to field stations where researchers could observe organisms and their interactions in their natural environment over an extended period of time.

By the late twentieth century evolutionary biologists such as Geerat Vermeij (1946) brought new insights to evolutionary theory with his concept of an evolutionary arms race between mollusks and their predators. He documented a dynamic interactive process where predator and prey continually changed in response to changes in the other. The study of evolutionary developmental biology (or evo-devo) compares the developmental processes of different organisms to determine ancestral relationships and uncover the evolution of developmental processes. Using techniques of molecular biology, evo-devo focuses on the origin and evolution of embryonic development; the relationship of modifications of development and developmental processes to the production of novel features; the role of developmental plasticity in evolution; how ecological changes impact development and evolutionary change; and the developmental basis of homoplasy and homology – shared novelties vs. novelties that evolved separately. Especially important is the study of how genes regulate embryonic development in model organisms (Amundson 2005; Laubichler and Maienschein 2007).

Today the sum of natural history knowledge can be easily found by anyone across the globe at a single web portal. The Encyclopedia of Life (EOL) began in 2007 to provide a webpage for every species. EOL compiles trusted information from resources across the world such as museums, learned societies, and expert scientists, into one massive database and online portal at EOL.org. The site provides basic descriptions, classification of organisms, life histories, and behaviors, for each organism. Accessible to amateurs as well as professionals, EOL fostered the resurgence of Citizen Science, voluntary natural history work by non-specialists that contributes to the larger scientific enterprise.

Natural History Institutions

The study of natural history has relied on the establishment of organizations to care for collections and share ideas about the natural world. In the colonial era, Boston and Philadelphia were the intellectual centers of the New World, and educated citizens formed voluntary organizations similar to those of Europe to advance knowledge. The Library Company of Philadelphia was founded in 1731 by Benjamin Franklin so intellectuals could pool resources and share access to stored knowledge. The New World's first learned society, the American Philosophical Society, was founded in 1743 in Philadelphia to promote useful knowledge in the sciences and humanities, including natural history. Other societies soon followed such as the American Academy of Arts and Sciences founded in Cambridge, Massachusetts, in 1780, to foster scholarship, civil dialogue, and useful knowledge (Wheeler 1983). New Yorkers were equally interested in the natural world but with a more practical cast and less influence from European intellectual centers (Gronim 2007).

After independence, many new organizations were established. The Academy of Natural Sciences of Philadelphia was founded in 1812 to cultivate science and useful learning. When the Academy opened its doors to the public in 1828, even women such as Graceanna Lewis (1821–1912) were permitted to study the Academy's books and specimens (Oleson and Brown 1976). The Linnaean Society of New England (1814–1822) was established in Boston to promote natural history through a museum, lectures, and excursions. Its collections were eventually donated to the Boston Society of Natural History. The Boston Society (1830–1948) was dedicated to the study and promotion of natural history through a scholarly journal and museum. By 1838, its collections numbered 20,000 specimens, and its library of 600 volumes was open to the public.

Several scholarly institutions were established in the nation's capital in the early nineteenth century. The Metropolitan Society, later named Columbian Institute for the Promotion of Arts and Sciences (1816–1838), was the first "learned society" in Washington. The Washington City Museum, curated by John Varden, existed from 1829 to 1841. Other groups lasted short periods of time and were absorbed in the National Institute for the Promotion of Arts and Sciences, founded in 1840, hoping to acquire the bequest of James Smithson. The Institute took responsibility for the nation's natural history collections in the Patent Office gallery, but it foundered and its collections were eventually donated to the Smithsonian Institution (Daniels 1994).

Science, including natural history, also became the province of the new national government, with the creation of the US Coast Survey in 1807. Most government-supported science was conducted to encourage economic development in the new democracy. In 1836, a federal Patent Office was created to stimulate new and useful scientific research and inventions that could be patented and would spur economic growth. The US Department of Agriculture, created in 1862, focused on the economic development of agriculture through research, education, and regulation. The federal government also financed western exploration and natural resource surveys to help develop the American economy as new regions were settled (Dupree 1957). State natural history surveys also supported research into the natural resources of each region (Welch 1998). Government funding for applied biology was also supplemented by foundations created by the new American industrialists such as the Carnegie Institution and Rockefeller Foundation (Kohler 1991; Sarton 1950).

The Migratory Bird Treaty Act of 1918 reflected a new awareness that migratory species, such as song birds, lived their lives across national boundaries, from the Arctic to southern South America, and that species conservation required cooperation across broad regions. Initially focused on the United States and Canada, the treaty later extended to Mesoamerica. The Biological Survey of the Panama Canal Zone in 1910–1912 led to the formation of a permanent tropic biology research station in the canal watershed. Other tropical stations soon followed, reflecting a growing interest in the tremendous diversity of life in the tropics. Tropical field studies became an important part of basic training for naturalists (Sterling 1989).

As natural history collections grew too large for small societies to manage, museums for research and public display were established. Charles Willson Peale (1741–1827), a noted naturalist and painter, founded the first museum in the United States in 1786. Peale's Museum in Independence Hall in Philadelphia made collections widely accessible for study and for popular education. Peale also organized the first US scientific expedition in 1801 and a system of bird exchanges with England. His museum was among the first to adopt Linnaean taxonomy, in contrast to competitors who presented their objects as random oddities of nature (Porter 1986).

The Smithsonian Institution was established in 1846 as a trust instrumentality of the United States to carry out the bequest of Englishman James Smithson. The first national scientific research organization, the Smithsonian's enabling act provided for a museum of natural history. Spencer F. Baird (1823–1887), the first curator, built the US National Museum as the premier American natural history collection. The creation of the Smithsonian reflected the American desire to equal Europe's intellectual and cultural achievements. The Smithsonian initiated a publication program for US scholars, who found it difficult to publish abroad, and a publications exchange program to ensure that American publications reached libraries across the globe and that Americans had access to European scientific literature (Henson 2004). The US National Museum housed the government's collections, as well as its own collecting efforts.

Agassiz's Museum of Comparative Zoology, established at Harvard in 1859, was the first of many college museums that were established in the late nineteenth and early twentieth centuries to maintain collections for study by professors and students. The American Museum of Natural History in New York was founded in 1869, with a broad public education as well as scholarly mandate. From 1881, the museum launched a golden age of exploration that lasted until 1930, with expeditions that discovered the North Pole; explored Siberia; traversed Outer Mongolia and the great Gobi; and penetrated the jungles of the Congo, taking museum representatives to every continent on the globe. These explorations fired the public imagination and reached a broad audience through such organizations as the National Geographic Society, founded in 1888. *National Geographic* magazine and films transported the American public to exotic societies, the depths of the oceans and the highest peaks on earth, stimulating public interest in the natural world.

As natural history itself specialized by type of organisms, so did professional societies, such as the American Society of Ichthyologists and Herpetologists founded in 1913 and the American Society of Mammalogists in 1919. But an organization that covered all of systematics and its theories, methods, and problems, was also needed. The Society of Systematic Biologists (SSB) started as the Society of Systematic Zoology in 1947 and revised its name to Biologists in 1971. The SSB was created to advance the science of systematic biology across various organismal groups. Members of the organization and its publication, *Systematic Biology*, work on the theory, principles, methodology, and practice of systematics.

Specialized societies also reflected increasing professionalization into scientific disciplines, often limiting the role of amateurs, who in turn created new organizations to meet their needs. The American Nature Association, headquartered in Washington, DC, published *Nature Magazine*, an illustrated monthly focused on popular articles about nature from 1923 to 1959. The magazine reflected the early twentieth-century popular interest in recreational study of the natural world. With leisure time on Americans' hands, the study of nature became a popular pastime, encouraged by the Nature Study Movement, Scouting programs, and the Chautauqua movement.

Educational Institutions

Educational institutions in colonial America were modeled on European ones, focusing on classics and religious instruction, with no attention to the sciences. The Boston Latin School, established in April 1635 to educate the Boston elite, followed a curriculum based on the classics. The oldest American institution of higher learning, Harvard University, was established in 1636 by the Massachusetts legislature, but by the nineteenth century had developed a strong science program. The second-oldest college, the College of William & Mary, was chartered in 1693 by English King William III and Queen Mary II to study divinity, philosophy, languages, and arts as well as sciences. Thomas Jefferson studied there and later reorganized the curriculum to include the law of nature and nations, anatomy and medicine. Yale University was created in 1701 by Connecticut clergymen to train ministers and lay leadership. Presidents Thomas Clap and Ezra Stiles were instrumental in developing a scientific curriculum.

Higher education became more accessible to American citizens after the Morrill Acts of 1862 and 1890 provided for creation of "land-grant" colleges that focused on teaching practical agriculture, science, military science, and engineering – a far different mandate than the traditional classical curriculum. Iowa State University, Kansas State University, and Rutgers University were immediately founded. With an emphasis on scientific agriculture, these colleges focused on both classical and practical scientific research on the natural history of agricultural areas. Private schools, such as Cornell University, were able to develop large scientific schools with land-grant funds. Land-grant funded biology did lead to a bifurcation of some organismal fields such as entomology into systematic entomology that emphasized basic research and classification versus economic entomology that focused more on managing pests for agriculture and medicine (Pauly 1984).

Science also entered the primary and secondary school curriculum in the nineteenth century, replacing rote memorization with laboratory experiments and dissections. In contrast to these junior versions of real science, the nature study movement, a popular education movement in the late nineteenth and early twentieth centuries, taught students to know and respect the natural world. In another chapter in this volume, Kohlstedt details the role nature study played in American interactions with the natural world.

The post-World War II growth of community colleges and new programs for adult, life-long learning engaged a wide audience of adults in natural history education, as did avocational societies such as the National Wildlife Federation. At America's zoos, entertainment had to be supplemented by conservation of endangered species and animal research, as investment in a diverse world replaced curiosity at exotic animals (Hanson 2002). Educational television programs, such as the Sunrise Semester of the 1960s, brought natural history education into the home, complemented by the rise of public television programs focused on wildlife studies. Along with the environmental movement, popular interest in avocational natural history studies continued to grow. The Cornell Bird Project exemplified the growth of Citizen Science with a large online database built on bird observations contributed by amateurs and specialists worldwide. In the 1990s, the development of the internet and Encyclopedia of Life ensured access to natural history knowledge to scientists, teachers, students, and amateurs across the globe.

Bibliographic Essay

Studies of American natural history were late coming to the history of science. Physical sciences received far more attention than the biological sciences, and American science was not deemed valuable until the mid-twentieth century. Most early studies of the history of American natural history were written by practitioners in the field, such as Joseph Ewan's writing on botany, *A Short History of Botany in the United States* (1969) and G. Brown Goode on natural history, *The Origins of Natural Science in America: The Essays of George Brown Goode* (1991), edited and with an introduction by Sally Gregory Kohlstedt. Pioneer works by historians of American science include Hunter Dupree, *Science in the Federal Government: A History of Policies and Activities to 1940* (1957) and *Science and the Emergence of Modern America, 1865–1916* (1963), and Nathan Reingold, *Science, American Style* (1991), Hamilton Cravens, "American Science Comes of Age: An Institutional Perspective, 1850–1930" (1976), and John Greene, *American Science in the Age of Jefferson* (1984), focusing on biology. Major figures in the field have always attracted biographers, with later works placing their subject in social as well as intellectual context. Examples are biographies of Louis Agassiz by Mary Winsor, *Reading the Shape of Nature: Comparative Zoology at the Agassiz Museum* (1991) and Thomas Jefferson by Keith Thomson, *Jefferson's Shadow: The Story of His Science* (2012). Studies of special groups, such as federal science, museum science, university science, and scientific institutions offered insights of the infrastructure that supported science, as in the works of Hugh Slotten, *Patronage, Practice, and the Culture of Science: Alexander Dallas Bache and the U.S. Coast Survey* (1994), Mary Winsor, *Reading the Shape of Nature* (1991), and Sally Kohlstedt, *The Formation of the American Scientific Community: The American Association for the Advancement of Science, 1848–1860* (1976). By the 1980s, interest in laboratory biology shunted natural history aside, much as laboratory biology had replaced natural history at the universities, seen in the volumes edited by Ronald Rainger, Keith Benson, and Jane Maienschein, *The American Development of Biology* (1988) and *The Expansion of American Biology* (1991), Robert Kohler, *Landscapes and Labscapes: Exploring the Lab–Field Border in Biology* (2002), and *All Creatures: Naturalists, Collectors, and Biodiversity, 1850–1950*

(2006). General studies of natural history gave way to specialized fields, mirroring the specialization in the field, notably botany by Elizabeth Keeney, *The Botanizers: Amateur Scientists in Nineteenth-Century America* (1992), ornithology by Mark Barrow, *A Passion for Birds: American Ornithology after Audubon* (1998), and oceanography by Helen Rozwadowski, *The Sea Knows No Boundaries: A Century of Marine Sciences under ICES* (2002). The 1980s also saw interest in the role of women and minorities in science and the role of gender in scientific practice, pioneered by Margaret Rossiter, *Women Scientists in America: Struggles and Strategies to 1940* (1982), *Women Scientists in America: Before Affirmative Action, 1940–1972* (1995), and *Women Scientists in America: Forging a New World Since 1972* (2012), Londa Schiebinger, *The Mind Has No Sex? Women in the Origins of Modern Science* (1989), and Sandra Harding, "Women's Standpoints on Nature: What Makes Them Possible?" (1997). Taxonomic methods and the impact of Darwin have been important themes, seen in the work of David Hull, *Science as a Process: An Evolutionary Account of the Social and Conceptual Development of Science* (1988), and Pamela Henson, "Comstock's Research School at Cornell University, 1874–1930" (1995). Teaching practices and institutions of natural history have been addressed by Sally Gregory Kohlstedt in *Teaching Children Science: Hands-on Nature Study in North America, 1890–1930* (2010).

Chapter Thirty-Six

NATURE STUDY

Sally Gregory Kohlstedt

The phrase, nature study, was given to a core curriculum that emerged in the late 1880s and that, within a generation, became incorporated into public schools across the United States and subsequently found expression in Britain and several Commonwealth countries as well. Not entirely unprecedented, the program built on traditions and techniques for studying nature reflected in natural history publications from ancient times. Curiosity and engagement broadened as the diversity of living things and unexpected topography were continuously revealed through Western worldwide exploration and imperial expansion starting in the sixteenth century and intensifying in the eighteenth and nineteenth centuries. That sense of discovery was paralleled by romantic trends that celebrated local landscapes especially in Britain and Germany. Such continuities and influences were important to the educational phenomena named nature study, but the program was more immediately influenced by Progressive Era theories and practices that undergird a new pedagogy.

The progressive outlook reflected increasing ambivalence about an urbanizing society and at the same time aligned with the ambitions of an increasingly academic scientific community (Lears 1981). In the case of nature study, scientific leaders, educational reformers, conservation activists, and a community of writers and artists identified an opportunity to counter the angst about what could be viewed as the negatives in their culture and create through children an outlook that would accommodate the new knowledge constructively. Thus nature study gained its strength and certain limitations from the multiple real and potential alliances that undergird it and which pulled its practitioners in varied and sometimes inconsistent directions. In the early years, however, strong and articulate advocacy came from among a well-educated cohort of educational theorists and administrators affiliated with public and private normal schools (subsequently known as teachers' colleges); they, in turn, relied on the

A Companion to the History of American Science, First Edition.
Edited by Georgina M. Montgomery and Mark A. Largent.
© 2016 John Wiley & Sons, Ltd. Published 2020 by John Wiley & Sons, Ltd.

persistence and details of implementation of individual teachers, many of whom had a significant interest in science.

The idea that children should learn more about the natural world around them is longstanding, and indeed many nineteenth-century educators believed that children were born with an innate curiosity for the natural world. Already by the early nineteenth century, as more printed primers, periodicals, and pamphlets were designed for children, topics and objects from nature held a prominent place. Whether these texts were designed to demonstrate examples for parents and tutors or for teachers in classrooms, they often used a familiar style intended to engage children as well (Shteir 1996). In the United States some of the earliest books were imported from Britain, and their authors experimented with conversational modes, using dialogue between pupil and teacher or other techniques to build a capacity to ask questions and engage in tentative answers. Soon American authors produced their own materials and mixed styles that could be narrative, didactic, illustrative, or familiar. Among the most popular and engaging best sellers were those that introduced the natural world to children and were intended to be used in conjunction with field experiences in a nation still largely rural and with easy access to the countryside. Botanical books were particularly common, and *Familiar Lectures on Botany* (1829) by Almira Hart Lincoln Phelps remained in print for decades. Both she and her sister, Emma Willard, were staunch proponents of education for girls and curricular innovators, with Emma producing maps and texts on historical geography including her *Geography for Beginners* (1826). These were used primarily in academies and as auxiliary material to the more standard assignments in reading, writing, and mathematics (Kohlstedt 1990). The introduction of geography into public schools at mid-century was an innovation that also emphasized the importance of place and space in thinking about both the topography of the globe and the impact of rivers, mountains, fertile soil, and other features as they affected human mobility and productivity (Schulten 2001).

Historians of education note that the expansion of education and rethinking of curriculum at every level at the end of the nineteenth century led to increasing commentary on appropriate pedagogy. While presidents Charles Eliot of Harvard University, M. Carey Thomas of Bryn Mawr College, and David Starr Jordan at Stanford University emphasized new disciplinary expertise and a curriculum designed to be more flexible at their respective higher education institutions, they were also public intellectuals who spoke out about the ways in which pre-collegiate education needed to change to provide better preparation for advanced work (Cremin 1961). Later in the century and expanding on pre-Civil War initiatives, normal schools became the fundamental mechanism for training teachers effectively by providing not only content knowledge but also pedagogy appropriate for particular subjects. Some states moved swiftly to build impressive structures that dramatized the importance of and commitment to public education, while a few of the newer private universities created their own departments or colleges to educate both teachers and administrators equipped to run ever larger public schools and school systems. Leadership in these institutions collaborated with both academic colleagues and leading public citizens as they defined a curriculum that they viewed as progressive (Ogren 2005).

Alignment of these capacities and educational aspirations made it possible for advocates of additional science in schools to push forward new curricula. The approach to science for younger children was to be conducted through observation gained from direct and intimate engagement with the natural world around them, reflecting similar engagements of adults in bird watching, hiking, and other outdoor activities (Barrow 1998). While historian Peter Schmitt (1990) suggested popular enthusiasm about nature was a reaction to changes brought by technology and urbanization, others like Roderick Nash (1967) argued that attention to "the wild" (identified as its most authentic in the American West) was stirring the American imagination, which led to investigating and in some cases preserving untamed or reclaimed "nature" closer to home. As natural history museums, landscape art, and new capacities in photography and reproduction contributed to the cultural fascination with nature both far and near, educators themselves reflected and in turn generated capacity for teaching nature with special attention to the very material ecology close at hand.

Many of the instructors who advocated for nature study had a background in science and brought specific skills which allowed them to lead the summer institute lecture programs, to become faculty in normal schools, and to write textbooks that enabled even established teachers to learn new approaches and for pupils to engage in unfamiliar topics. A number of projects, including one sponsored by the Boston Society of Natural History for over two decades in the late nineteenth century, experimented with ways to better equip teachers to teach science. Several of its active members had been inspired by naturalist Louis Agassiz and his short-lived summer school on Penikese Island, where he encouraged teachers to gather sea urchins and other natural objects and to learn by studying them intently. Scientists joined a drumbeat of more science for children in the early grades by talking among themselves at meetings of the American Association for the Advancement of Science, writing articles for *Science* and *Popular Science Monthly*, and advocating in popular journals like *Lippincott's Magazine* (Kohlstedt 2005a). There were also informal programs that encouraged individual children to create small hobby collections of shells, butterflies, or other local specimens or to join Agassiz Clubs, where they could meet regularly with others for field trips and sharing experiences. All of these activities meant that when nature study as a curriculum was introduced, it found fertile ground that anticipated both the readiness and interest of even young pupils.

The Nature Study Approach

The nature study curriculum for the public schools, which was formally introduced in the 1890s in multiple settings and especially in major cities and statewide in the Midwest and the West Coast, emphasized the importance of experiential learning. An educational approach that contrasted with what a generation of educational activists argued was rote learning would depend on some of the first formal studies of child development. Imported theories from European educators that emphasized tactile and direct experience as most effective with young pupils, including those of Heinrich Pestalozzi and Maria Montessori, formed the heart of this new approach to learning, and these became linked by the end of the century to new ideas about biology as well (Nyhart 2009). These imported ideas had a general rather than an explicit influence as they were translated into the American context by instructors at normal schools who were determined to teach both content and pedagogy. In fact, teacher training institutions became the cauldron in which these ideas were reworked and reformulated to produce fresh thinking about both what subjects should be taught and the most effective methods for teaching them in ways that engaged pupils and encouraged their enthusiastic learning.

There is some ambiguity in dating the origins of nature study, but the intention to teach children about geography, botany, and other aspects of natural history was evident in private academies by the mid-nineteenth century. Boston was particularly notable in the decades after the Civil War because teachers in the public schools were invited to take classes at the Boston Society of Natural History and learn from its museum specimens. Their participation was voluntary as they attended after school and Saturday classes taught by faculty from Harvard and the Massachusetts Institute of Technology; they also made frequent reference to the well-known naturalist Louis Agassiz who had occasionally lectured to prospective teachers at Framingham Normal School. Leadership for the Boston project came from one of his auditors there, Lucretia Crocker, and from Ellen Swallow Richards, later known for her pioneering work in creating the field of home economics; both women had a significant interest in the natural sciences themselves (Kohlstedt 2005).

The method of using tactile objects had precedence, and a renewed interest in the approach can be traced to European educational theorists like Pestalozzi, Montessori, and Friedrich Froebel. In the United States, the concept of object lessons had been publicized in the post-Civil War years by Edward Sheldon at Oswego Normal School in upstate New York. In the latter decades of the nineteenth century, German research and writing on child development in psychology also began to influence educational thinking as Americans went abroad to join university seminars or take advanced degrees. Among those going abroad to learn research techniques and understand current movements were psychologist G. Stanley Hall and educators Charles McMurry and Frank McMurry. After taking PhDs in different German universities, the brothers joined the Pedagogical Seminar of William Rein at Jena and eventually returned in the 1890s to Illinois, where they were leaders in forming the National Herbart Society as a way to continue to study and promote the ideas of Johann Friedrich Herbart about correlated or interdisciplinary learning; it later became the National Society for the Scientific Study of Education. Their sister-in-law by marriage, Lida Brown McMurry, initially at Illinois Normal School taught and supervised nature study teachers but subsequently moved with Charles to Northern Illinois Normal School in DeKalb and wrote textbooks directed at both teachers and their pupils. This integration of educational theory and classroom practice served the movement well as major private universities also began to hire faculty equipped to teach educational philosophy and psychology.

At least two individuals are typically credited for articulating the possibilities for nature teaching as a broad model for the country and both were progressive educators working in normal schools. One was an instructor in Massachusetts who worked primarily to educate prospective teachers in smaller towns and country schools, Arthur G. Boynton, whose personal interest in science and leadership propelled him to the position of principal at Bridgewater State Normal School. Influenced by Agassiz and his summer school for teachers not far from present-day Woods Hole, Massachusetts, Boynton introduced natural science courses for teachers by taking advantage of his location to incorporate fieldwork and seaside experiences into his program so that the future teachers learned techniques for presenting nature studies lessons first hand. He found it particularly effective to use this approach at regional summer institutes for teachers throughout the state in the late 1880s and early 1890s. Education through such institutes had sprung up in a number of states, including New York and Illinois, but these were informal and often ephemeral in their offerings even as they attracted the most ambitious educators.

A second initiative was urban and established by Wilbur Jackman under Colonel (the name derived from Civil War service) Francis Parker. Parker had become nationally known as a leader in progressive education, largely because he was an effective spokesman for the innovative staff and programs which he supervised as principal at the Cook County Normal School, which served Chicago teachers. In the late 1880s, Parker hired a high school science teacher from Pittsburgh, Jackman, who had an idea about introducing what he would call "nature study" into the curriculum. Jackman, who had taken a BS degree from Harvard, had been working on a program that would teach children science largely through hands-on learning in and beyond the classroom. Under Parker, he quickly developed his "Outlines in Science Education" for teachers to use at the practice school affiliated with Cook County Normal, and in 1891 published these as a textbook entitled *Nature Study for the Common Schools*, thus effectively giving a clear name to the approach being introduced to Chicago-area teachers.

Both the new University of Chicago and the well-established Columbia University were interested in building the areas of psychology and educational philosophy, persuaded that these emerging social science disciplines were coming into prominence and that they were likely to be important because education played such a prominent role in progressive politics. The university leadership hired academic faculty interested in research whose work with educators would influence the next generation of informed citizens. The president of the University of Chicago, William Rainey Harper, persuaded Parker to move from his normal school in order to lead a new more comprehensive Chicago Institute on the edge of campus, complete with a practice school. Here Jackman would have even more space for innovation on nature study curricular development, using a side yard for a school garden. Full faculty cooperation, however, was never realized and John Dewey, a leading figure on the Chicago faculty, found himself at odds with Chicago's leadership and left for a position at Columbia University. During the 1890s, under President Seth Low, Columbia rapidly changed from its traditional men's liberal arts model to one closer to other emerging research universities, and Low made a number of strategic faculty hires. In 1897 James Earl Russell, who was brought to New York City to address the ambiguous status of a normal school loosely affiliated with the university, formed what became the well-known Teachers College, with a stronger link to the university through its plan for research on education. In turn, Russell, who had studied at Cornell University, hired instructors who could implement this evidently emerging curriculum into the Teachers College curriculum and its experimental schools.

With flagship universities thus engaged, nature study gained credibility among educational leaders and reformers in major school systems across the country. Clifton Hodge, then professor at the newly established Clark University under President G. Stanley Hall, also sought ways to make nature accessible to urban children and worked with local teachers in Worcester, Massachusetts to use their classrooms as models. Hodge's *Nature Study and Life* (1902) became one of the leading texts among those teaching in schools in other growing cities like Cleveland, Ohio, Lincoln, Nebraska, Denver, Colorado, and Los Angeles, California. He provided detailed and concrete suggestions for teachers that encouraged them to have their students consider what they might learn about animal life from pets, what fascinating insects and other creatures and special plants were harbored in city parks, and how they might use abandoned city lots for gardens.

While these higher education initiatives necessarily attended to the ways that nature study might be taught in larger urban settings, a concern about rural life motivated a parallel effort to train and work with teachers who were often isolated in small country schools. Here, too, the goals was to expand the experience of pupils who, although surrounded by nature, might never have appreciated some of its beauty or analyzed the information to be gained by careful analysis of all living things that might better inform the routine practices of farming and country life. In New York State, where rural depopulation seemed to many a serious problem in the late nineteenth century, the legislature sought ways to promote agriculture and had found success in its active extension program coordinated through the Agricultural Experiment Station. Anna Botsford Comstock, a former teacher and wife of Cornell University's prominent entomologist, served on a state committee that included "nature study" in its authorization of funds to Cornell (Henson 1996). By that time Comstock had already developed a course of study used in the Westchester County school district and began to offer a correspondence course for interested teachers using a pamphlet series entitled "Home Nature-Study Course." It included 12 intensive lessons that required field observation and written assignments submitted to Comstock and her staff that were intended to make teachers themselves more informed observers, even as it demonstrated activities that could engage their pupils. She worked closely with a faculty colleague (she was not granted full faculty status until 1913), Liberty Hyde Bailey, whose strong reputation in horticulture and eventual deanship of the College of Agriculture positioned him to be an effective proponent of the nature study curriculum. Together they built a program with nature study leaflets and summer institutes for teachers that would be influential and emulated across the country from Maine to California. Bailey's frequently republished *The Nature-Study Idea* was a reflective, inspirational set of essays that emphasized the moral outlook and aesthetic values of having children directly engaged with their local environment; from his point of view this would build an appreciation for country life and lead, as well, to responsible behavior toward living things. Comstock was practical as she worked with Cornell faculty to produce small pamphlets that would eventually be compiled into an 800-page volume, *Handbook of Nature-Study* (1911) that remains in print.

Nature study was a malleable educational program and was well designed to work within the context of a progressive educational movement and at a time when American interest in the natural environment was evident everywhere. On the one hand, it responded to the growing emphasis on teaching science in the public schools with the goal of a more literate public able to understand both the complexity and the interrelatedness of the natural world. Students in New York State, for example, might focus on the well-watered forests and shade-loving plants that harbored an array of insects and small animal life. Their projects, based on topography and living things in their locale, would be very distinct from those in sand-blown landscapes with sparse flora and fauna and concentrated irrigated agricultural crops typical of southern California. On the other hand, it was intended to draw students into study by encouraging their curiosity and engagement with the particular characteristics of their environment in order for them to generalize to the more universal biological concepts of growing things, animal–plant relationships, and geological terrain. Many nature study programs, influenced by Herbartian ideas, emphasized ways in which nature study correlated with other subjects in the curriculum.

Given the evident malleability of nature study, some textbook authors contributed to its expansion beyond the identification and explanatory aspects of nature study to present an aesthetic sensibility or a conservationist perspective, or both. As Kevin Armitage (2009) points out, science and a more open-ended approach that started with curiosity were not always easily compatible elements, but, in fact, variance in content and approach could be accommodated under the nomenclature of nature study. Innovative teachers wrote about the ways they linked this work to other classroom lessons or to special opportunities beyond the school that were eagerly anticipated by their pupils. Nature study teachers used literature, some from eighteenth-century poets like Robert Burns, while others drew on contemporary authors like environmentalists John Muir and John Burroughs, to evoke a sensibility about nature that complemented more analytical assignments. Anna Comstock was herself an artist and her *Handbook* recommended drawing as a skill that built powers of close observation and an appreciation for the sophistication of nature even as children traced leaves, outlined rudimentary maps, and sketched what they saw on field trips. By the 1910s, although diverse and multifaceted, the nature study curriculum was a visible and integral part of public schooling.

Framing a Movement

As nature study became embedded as a subject of instruction in many leading normal schools and thus could be incorporated into recommended curricula at the state level or mandated at the local level, school administrators, nature study supervisors, and teachers sought acknowledgment of their professional achievement and opportunities to enhance the field. Echoing the trend toward certification and national organizations evident in the progressive spirit at the turn of the century, the emerging experts presented their programs and methods in regional and national educational meetings and, while there, discussed additional ways to advance nature study. From basic curricular outlines in the early years, discussion advanced to examples of dedicated nature study rooms in schools, organized trips to museums as well as parks, and school gardens in cities. Textbooks and supplementary materials became big business, although much of what teachers did built on resources that they supplied to their students or gathered with them. The proliferation led at least one grumpy textbook author to complain that there were more textbooks than effective teachers, a comment that reflected the complicated challenge of bringing experienced as well as entering teachers into compliance with a genuinely new curriculum.

The earliest nature study advocates had relied on publicity and persuasion as they introduced this curriculum at regional and national educational meetings, provided models for practice during summer teachers' institutes, and demonstrated possibilities through the daily course of study at the practice or experimental schools affiliated with major normal schools. These activities were of critical importance in defining this curriculum and making it an integral part of expanding school systems that were growing to accommodate general population growth, as well as pupils now required to attend for a longer period. Advocates were determined to establish that science education was appropriate from primary school through high school and that nature study with its emphasis on observation was the most effective way to introduce younger pupils to not only biology and geology but also meteorology, astronomy, and physics.

By the 1910s, nature study was an established and mandated curriculum in large cities like New York, Chicago, and Los Angeles. Access to resources could be complicated, so nature study supervisors were hired to help teachers acquire natural materials, take advantage of local parks and recreation centers, and schedule visits to natural history museums and zoological parks. They created links with local horticultural or Audubon societies and found space for school gardens, many of which were extended during World War I when many children joined the School Garden Army in order to help off-set food rationing. Rural school programs were more dependent on individual teachers who gathered advice and resources through state-sponsored summer institutes and collaboration with local farm associations (Lavender 1997). Typically rural programs reflected the model developed at Cornell, where Bailey and Comstock worked within the College of Agriculture and utilized the community connections of the New York State Experiment Station.

A number of editors attempted to publish a journal on nature study at the turn of the century, but none lasted more than three or four years. In 1905 Maurice Bigelow, whose appointment as a faculty member at Teachers College where he wrote his PhD thesis on nature study in its Horace Mann School positioned him to take the lead, coordinated an effort with several other leading advocates including Bailey, Hodge, and his colleague John Woodhull, to establish *Nature-Study Review*. With an initial cadre of subscribers from across the country and with attention to diversity — early membership extended to George Washington Carver at Tuskegee Institute in Alabama – the *Review* soon provided a forum for showcasing and quite deliberately evaluating methods used in nature study. Bigelow was aware of criticism by his psychologist colleague, Edward Thorndike, and used the journal as a way of responding. Thorndike, reacting to the outlook in Bailey's popular book, *The Nature-Study Idea*, and, impatient with the seeming informality of teaching that relied on initial observation of students, denounced nature study as too sentimental and lacking scientific rigor. These challenges were openly discussed in the new journal but were also contradicted by projects and more positive outlooks that documented successes by instructors in the normal schools as well as the growing number of mostly women who became supervisors of nature study in large public school systems. Such supervisors helped individual teachers gain skills, find resources, and establish nature study reference resource rooms in their schools. Bigelow launched a subscription campaign for the *Review* but found that even a modest charge was too much to recruit ordinary teachers so that the majority of subscribers remained normal school instructors of nature study, textbook authors, and those whose major responsibility focused on teaching children science. The *Review* also helped launch a Nature-Study Society, which met annually, most often with the peripatetic American Association for the Advancement of Science but occasionally with the National Education Association. The journal and society became linked and, for a time, there were also regional sections of the society at sites where the curriculum became standard, notably New York City, St. Louis, Missouri, Rockford, Illinois, and other cities. Annual meetings were typically mixed presentations that explored activities and innovative approaches to teach nature study, such as use of aquariums or successful assignments affiliated with field trips alongside reflective pieces on the value of nature study or even debates about whether nature study was indeed a science.

While nature study texts typically emphasized subjects representing botany, entomology, zoology, geology, astronomy, and meteorology, the approach was to start by eliciting responses from the pupils and then build to a more scientific vocabulary by using comparisons and identifying taxonomic parts important for later definitions. By the 1920s, changing philosophies in education, the introduction of standardized tests, and a new generation of normal school leaders questioned the effectiveness of nature study. The tension about whether child development ideas about engaging the imagination and interest of youngsters and encouraging questions would lead to a basic understanding of science had been evident from the start. Many leaders like Clifton Hodge, who suggested that there was a place for systematic studies even in the early years, nonetheless argued that starting with local and familiar objects would lead to the most effective learning outcomes. For those who wanted to have certain basic principles of science taught in a didactic way that would ensure that students would arrive at high school and then college with the fundamentals needed for advanced study, nature study seemed to lack rigor. The criticism surfaced in journals like *Science* and commentators such as psychologist Edward Thorndike were harsh, using terms like sentimentalism and namby-pambyism to describe the teaching observed in the Horace Mann School at Teachers College. Moreover, as Thorndike and some contemporary scientists warned, those who grew up identifying with the natural world would not learn that nature was "red in tooth and claw" and might resist laboratory work with animals (Ruse 1999).

The most significant shift was initiated in New York City, ironically a place where nature study was very well established in the public schools. In the 1920s, Gerald Craig, a former World War I veteran and experienced teacher, returned to Teachers College to take a PhD and wrote his dissertation comparing nature study and elementary education courses. Although he had been a member of the American Nature-Study Society, he moved quickly to produce a new curriculum for the Horace Mann School with the telling title "Tentative Course of Study in Elementary Science" and, with his advisor, established the National Association for Research in Science Teaching. Having taught nature study briefly after service in World War I, Craig initially did not directly challenge nature study but rather emphasized the importance of have a more systematic approach to science. To scientists impatient with nature study, Craig's formulation of an elementary science curriculum was more straightforward in that teachers followed clear assignments that took less time and imagination than nature study. By the 1930s, New York State adopted a recommended curriculum on elementary science in a shift that was happening elsewhere in the country. A close look at the new curriculum indicates that some of the elements of nature study persisted because gathering examples from nature or having plants and small animals in the classroom were effective teaching devices. As Craig developed a very profitable textbook series, "Pathways to Science," he became increasingly disparaging toward nature study suggesting that it had been ad hoc and allowed for "anthropomorphism, sentimentalism, animism, emotionalism, prejudice and unfounded opinion" (Craig 1940: 17, 27). The general educational trend was away from progressive methods, but the vehemence of the dismissal of nature study helps account for the fact that the older approach was ignored or treated with distain by historians in subsequent decades.

Impact of Nature Study

Nature study flourished for essentially one generation, from the 1890s through the 1910s, supported by a cohort that had trained in the activist and often inspiring normal schools throughout the country, where innovative techniques were valued and attention to child-centered pedagogy dominated curricular planning by instructors and their graduates. Well embedded in many states and school systems, nature study books and curricular materials would persist well after those dates in classrooms where teachers found the approach effective. When at its most influential, nature study was taught in private as well as public schools, introduced to those in special schools for the blind and the feebleminded, and present on the agenda of most state and national association meeting programs. As an export to Britain, it had some followers in Ireland, Scotland, and England, but enthusiasm for and the vocabulary of nature study may not have been as widespread there as it was in the Commonwealth countries of Australia and New Zealand; in the latter country it continued to be a well-respected approach even after World War II (Jenkins 1981; Kohlstedt 1997; Wylie 2012).

By the 1920s, too, nature study had diffused into multiple other informal educational settings. Nature study leaders were among the founders of the 4-H movement in rural areas, while in more urban areas served by the Boy Scouts and Girl Scouts, badges in nature study could be earned by activities like those used in classrooms. Student-initiated projects became regularized as science clubs and other co-curricular activities supplemented regular school work, and science proved an area of interest and one that gained community and corporate sponsorship (Terzian 2013). Natural history museums became collaborators with many school systems and, in fact, developed educational programs for children that engaged them after school and in summer programs (Kohlstedt 2013). The nature study movement, perhaps because of its chronological coincidence with the national park and conservation movements, was also directly responsible for encouraging the development of the smaller nature preserves established near towns and major cities. Even after the nature study curriculum was no longer described in school curricula, most of these sites remained as informal tributes to the early twentieth-century enthusiasm for nature study and, indeed, serve a cohort of students now coming through a more ecological curriculum in the twenty-first century. There has also been an explicit revival of the use of the term nature study evident particularly in specialized charter schools, in home schooling, and private schools, some of whom look to the English advocate for nature study in the early twentieth century, Charlotte Mason, for inspiration or to the writings of Richard Louv (2005).

Bibliographical Essay

Historical accounts of nature study have approached the movement from a number of viewpoints, and the estimation of its importance and effectiveness vary just as they did among contemporaries. In particular, its influence has been noted by not only educational historians but also by those interested in literature, conservation, and other aspects of American culture in the early twentieth century. In these works the attention is on the ways in which nature study seemed to serve other movements.

Two lines of argument threaded through early assessments of the nature study movement. Ellwood Cubberley, recognized for his early histories of education in the United States, taught at the young Stanford University. His influential book, *Changing Conceptions of Education* (1909), and subsequent publications viewed nature study as evidence of fruitful new directions that incorporated science and art in stimulating children to learn. Carrying the positive outlook of progressive educators, Cubberley and his view would prevail among historians of education into the 1950s. However, a succeeding set of academic historians, influenced by their contemporaries who promoted an approach that would lay the fundamentals of biology, physics, and chemistry in earliest grades, posited a different evaluation of nature study. Their strong opinions were reflected in Orra Erwin Underhill's account in *The Origin and Development of Elementary School Science* (1941). He treated the nature study as an educational phase through which topics of natural history had been introduced but which had proven inadequate for genuine scientific education. The introduction to his book echoed Thorndike's criticism with the comment that nature study teachers had been "emotional rather than scholarly about natural phenomena" (quoted in Kohlstedt 2010: 225). Most twentieth-century historians of education tended to privilege theory or political influence over curriculum and practice, but where nature study was discussed it was usually with a dismissive tone that implied it was ineffectual and sentimental. A few dissertations and master's theses on nature study, almost all produced in educational colleges, stressed nature study's positive local influence or its relationship to conservation efforts.

By the late twentieth century, however, the situation changed. As educational leaders focused on effective pedagogy for younger pupils, new attention to environmental education led to interest in the antecedents of such schooling. A new emphasis by literary scholars on nature writers in the late nineteenth century discovered that some of them were popular with teachers who taught nature study and, indeed, that authors like John Burroughs, John Muir, and Olive Thorne Miller (Harriet Mann Miller) were delighted to learn that their work would be used by teachers and pupils (Gates 2002; Philippon 2005). Thus, for example, Eric Lupfer in his "Reading Nature Writing: Houghton Mifflin Company" (2002) as part of his longer project of describing the emergence of "naturism" as a literary genre, discovered that teachers were an important market for such nonfiction, formed local reading groups to discuss such work, and found it useful for their teaching. Even as adult readership on natural history and related topics grew, so too did both fiction and nonfiction for children. As Donna Varga (2009) suggested in her review of children's book through the end of the nineteenth century, there were increasing numbers of books that could indeed take "Babes in the Woods."

Also attentive to nature writers, Kevin C. Armitage's book, *The Nature Study Movement: The Forgotten Popularizers of America's Conservation Ethic* (2009), looked closely at the particular ways that advocates for nature study and the environmental movement found common ground. Building on the insights of other environmental history scholars like Peter J. Schmitt but taking a more positive outlook, Armitage pointed out the very specific ways in which ideas about nature study were linked by authors, photographers, and activists to immediate concerns about loss of wildscapes, abuses of land and water, and particularly immediate danger of extinction especially after the apparently last carrier pigeon died in a Cincinnati zoo in 1914. Other authors also noted, often tangentially, the ways in which adult hobbyists, especially those in gardening and ornithological groups, worked with the nature study movement (Dunlap 2011). Much of the nature study education, however, was done with rocks, plants, animals, meteorological measurements and other material objects close at hand, as Ellen Doris (2002) found in her careful analysis of the content of *Nature-Study Review* and reports of what teachers were doing with their classes.

The most comprehensive account of the nature study educational movement is Sally Gregory Kohlstedt's *Teaching Children Science* (2010), which traced the movement from its informal origins through its introduction across the United States as well as the challenges that led to its retrenchment by the 1930s. Framing her work through professional and institutional developments, she explored the multiple ways in which nature study education was influenced by pedagogy theory imported from Germany, by the growth of both informal amateur interest in birds and other species, and by the quite specific engagement of leading science education advocates. Unlike most previous studies, her book was attentive to the role of teachers and supervisors and particularly the role of women whose interest in science helped them create an important career niche through nature study. Another book that picks up on the theme of women's involvement both as pupils as well as teachers was Kim Tolley's *The Science Education of American Girls* (2004), which underscores the fact that nature study could build on the interest of women whose interest in science led to few other occupational paths, with the result that much of the supervision of nature study in urban school systems was done by women.

Chapter Thirty-Seven

SCIENCE AND POLICY

Kevin C. Elliott

The historical relationships between science and policy in the United States encompass a wide range of topics and issues. This chapter uses Harvey Brooks' (1964) famous distinction between *policy for science* (decisions about how to fund or structure knowledge production) and *science for policy* (the use of knowledge in decision making) as an organizing framework. The first three sections examine *policy for science* and are divided into three important historical periods: (i) 1789–1940; (ii) 1940–1980; and (iii) 1980 to the present. These sections highlight the very different relationships that developed between government agencies, scientists, universities, corporations, and private foundations in American history.

The latter half of the chapter turns to *science for policy*. Both in the United States and internationally, the role of science in public decision making has undergone extensive scrutiny throughout the latter half of the twentieth century and the beginning of the twenty-first century. The final two sections of the chapter examine two of the most important themes that have emerged from these discussions. First, a series of cases in which private interests generated questionable science in order to influence public policy left scientists and policymakers scrambling to find ways to produce credible, trustworthy, policy-relevant science. Second, models for incorporating members of the public in science policy have undergone a dramatic transformation in recent years, shifting from the view that non-scientists should be primarily passive recipients of information to the view that non-scientists can play an important role alongside experts in science policy-making. These and other ongoing issues provide opportunities for historians of science and other science studies scholars to promote more productive relationships between science and policy.

A Companion to the History of American Science, First Edition.
Edited by Georgina M. Montgomery and Mark A. Largent.
© 2016 John Wiley & Sons, Ltd. Published 2020 by John Wiley & Sons, Ltd.

Policy for Science: 1789–1940

The role of the federal government as a supporter of American science became particularly significant during and after World War II. In contrast, national support for science was ambivalent at best in the wake of the country's founding in the late eighteenth century. While early presidents like George Washington, Thomas Jefferson, and John Quincy Adams enthusiastically supported proposals for scientific projects like the creation of a national observatory, a national university, and a coastal survey, Congress questioned the constitutionality of spending federal money on such pursuits (Dupree 1957). The Lewis and Clark expedition was one of the few early federal projects that included a major scientific component, and its impact was still limited because there were no established scientific institutions for classifying and reporting on the expedition's botanical and zoological findings (Dupree 1957: 27). Even the creation of the Smithsonian Institution, which was funded by a private bequest by the Englishman James Smithson, was delayed for a decade because of debates in Congress over the appropriateness of the institution and the form it should take (Dupree 1957; Greenberg 1968).

Despite early difficulties, federal involvement in science slowly increased over the next century. The US Constitution included the word "science" once, in its clause that granted Congress the power to grant patents. Following the initial Patent Act of 1790, Thomas Jefferson himself helped to evaluate patent applications in his capacity as Secretary of State. The Patent Act of 1836 established a more permanent Patent Office with a commissioner and staff that was put in place just in time for a significant rise in technological innovation and patent applications from the 1840s to the 1860s (Dupree 1957: 47). During the first half of the nineteenth century, the War Department also supported science through Western surveying expeditions and naval explorations, as well as through the creation of a Corps of Engineers and the training of engineering students at the US Military Academy. In 1863, legislation was passed to create the National Academy of Sciences (NAS), although the advisory role of the NAS was initially fairly limited (Douglas 2009: 23).

The government's support for scientific research accelerated significantly during and after the Civil War, as Southern politicians who had attempted to restrict the federal government's power lost their influence. Congressman Justin S. Morrill of Vermont successfully sponsored the Land-Grant College Act of 1862, which provided federal funds to each state for supporting colleges devoted especially to education in agricultural and practical skills. The Department of Agriculture was also established in 1862, and in 1887 the Hatch Act strengthened the land-grant colleges by providing federal funding for state agricultural experiment stations. After multiple and sometimes clashing efforts to survey Western lands, the US Geological Survey (USGS) was officially created within the Department of the Interior in 1879. A host of other science-intensive bureaus and agencies were established soon afterwards: the Reclamation Service, the Bureau of Mines, the Forest Service, the National Bureau of Standards, the Bureau of the Census, and the Public Health Service (Dupree 1957: 290–1). One of the most significant creations was the National Advisory Committee for Aeronautics (NACA), which had its own lab but which also contracted out some research to universities (Douglas 2009: 26; Greenberg 1968: 59).

Two themes stand out in this pre-World War II period of science policy. First, the federal government's support for science focused almost exclusively on solving practical problems, and it is doubtful that legislators would have supported the goal of promoting or subsidizing scientific research in general (Mirowski 2011: 96). Second, with small exceptions like NACA, federal agencies did not contract with universities to produce research. The nation's elite universities began to focus more of their attention on scientific research in the early twentieth century, but the financial support for these projects came primarily from industry and private foundations rather than from the government (Mirowski 2011). Thus, most support for science in the pre-war years focused on applied topics. As science journalist Daniel Greenberg claims, "[W]hile basic science existed more or less as an orphan on the American landscape, applied science and technology were accorded an esteemed place and rather generous support" (Greenberg 1968: 52). But neither government officials nor scientists were inclined to change America's dominant approach to science policy. The official view in 1903 of President Theodore Roosevelt's Committee of Organization of Government Scientific Work was that science "on the part of the Government should be limited nearly to utilitarian purposes evidently for the general welfare" (Dupree 1957: 296), and a parallel National Academy of Sciences report produced soon afterwards did not challenge this conclusion (Greenberg 1968: 62; see also Smith 1990: 30–2).

Policy for Science: 1940–1980

Approaches to science policy began to change during the Great Depression, and these changes accelerated during and after World War II. In an effort to help struggling universities, Herbert Hoover tried to create a National Research Fund to support scientific research with private contributions, but the attempt failed (Dupree 1957: 342). Efforts during the Great Depression to use money from the Public Works Administration and the National Bureau of Standards to fund university research were also unsuccessful (Smith 1990: 32–3). By the end of the 1930s, however, the federal government had begun dispensing grants to universities through the new National Cancer Institute (founded in 1937) and the Works Progress Administration (Smith 1990: 33), and the country's leaders were expressing more openness toward funding scientific research outside of government agencies (Dupree 1957: 360–6). During World War II, these preliminary steps toward supporting university research with federal money took a dramatic step forward and ushered in a new era of science policy.

A seminal event in this history took place in 1939, when Albert Einstein wrote a letter to President Franklin D. Roosevelt at the urging of Leo Szilard, Eugene Wigner, and Paul Teller (three eminent émigré scientists from Hungary). Their goal was to draw Roosevelt's attention to the ramifications of recent scientific developments in physics, especially the possibility for using nuclear reactions to generate power and possibly weapons. Roosevelt responded by creating an Advisory Committee on Uranium that was chaired by Lyman J. Briggs, director of the National Bureau of Standards (Greenberg 1968: 74).

The Briggs committee moved slowly and cautiously, with limited funding, and was soon superseded by the efforts of a collaboration of American scientific leaders:

Vannevar Bush (president of the Carnegie Institution of Washington), James B. Conant (president of Harvard), Karl Compton (president of Massachusetts Institute of Technology (MIT)), and Frank B. Jewett (president of the National Academy of Sciences and Bell Telephone Laboratories) (Greenberg 1968: 78). Bush was an electrical engineer who had served as the vice president of MIT before becoming the leader of the Carnegie Institution. He had experience working in a number of different institutional contexts, including serving as a member and chair of NACA and doing applied research for the military as a naval reserve officer. Using his personal connections, he arranged a meeting with President Roosevelt in June 1940 to call for a major new effort to fund scientific research in support of military objectives (Greenberg 1968: 74).

Based on Bush's guidance, Roosevelt released an executive order on June 27, 1940, which Daniel Greenberg called the "birthday of the great postwar involvement of science and government" (1968: 79). Roosevelt established the National Defense Research Committee (NDRC), which became the Office of Scientific Research and Development (OSRD) a year later. Bush led the OSRD with James Conant as second-in-command. The distinctive feature of the OSRD was that it gave scientists freedom from direct military control while tasking them with projects of relevance to the military. Civilian scientists dominated OSRD, which built on NACA's precedent of contracting out research to university labs. According to Greenberg, the creation of OSRD was a "political landmark for the nation's scientific enterprise. For the first time in the nation's history, substantial federal funds were going to university laboratories. Furthermore, while the contract [i.e., the contract system developed by the OSRD for funding university research] was intended to reconcile freedom with accountability, it was clearly weighted toward freedom" (1968: 80). The funding system developed by OSRD became a very influential precedent, because it generated extraordinary research successes during the war, including radar and the atomic bomb.

Toward the end of the war, Bush met with Roosevelt again, and the President tasked him with the job of proposing a postwar plan for funding scientific research. The resulting document, *Science: The Endless Frontier,* was one of the most influential documents in the history of American science policy. Its central vision was to create an agency (which became the National Science Foundation in 1950) that would provide government funding for university research while providing scientists with maximum autonomy. It also codified Bush's view that basic research was a crucial national resource that provides the fuel to generate subsequent innovations in support of national defense and economic growth. But this was not the only plan offered. Senator Harley Kilgore of West Virginia also wanted to create a government agency for funding science, but he did not share Bush's focus on preserving the autonomy of scientists to engage in whatever research seemed most important to them. As Kilgore put it, the motivation for federal research was not "building up theoretical science just to build it up. The purpose is what has been the purpose of scientific research all the war through, and what the incentive for scientists is, to do something for the betterment of humanity" (quoted in Greenberg 1968: 103).

Ironically, while it was initially difficult to pass Bush's plan through Congress, members of the military were so impressed by the contributions of science to the war effort that they enacted much of his vision in their own agencies. For example, in the 1940s a group of naval officers founded the Office of Naval Research (ONR) with the goal of funding basic research in universities. Other agencies adopted similar policies. By 1954, the federal government was spending about $120 million per year on basic research, with almost $80 million coming from various branches of the military: the Atomic Energy Commission, the Department of Defense, and NACA (Greenberg 1968). Moreover, this money provided for basic research was only a small fraction of the rapidly expanding pot of money that the federal government provided for research and development after World War II. In 1949, federal expenditures on research and development passed $1 billion. By 1957, they reached $4.4 billion, and by 1964, they had reached $14.8 billion (Greenberg 1968: 158). After the Soviet Union successfully orbited Sputnik in 1957, fears that the USSR could attain dominance in science and technology provided ongoing incentives for aggressively supporting science.

Policy for Science: 1980 to the Present

The funding regime developed during and after World War II began to lose support in the late 1960s and 1970s, as figures across the political spectrum began to question whether it provided the benefits it promised (Smith 1990: 73–5). But most figures point to the 1980s as a particularly important turning point when a variety of social forces coalesced to promote a new era of American science policy focused on privatized science developed for commercial purposes. For example, a number of new bills passed in response to worries that the United States suffered from economic malaise. The Bayh–Dole Act of 1980, which gave universities and small businesses the ability to patent inventions made with federal funding, is typically held up as the symbol of this new era. But this act was just one piece of legislation among a variety of others: the Stevenson–Wydler Act of 1980 (which opened up commercialization of research from national labs), the National Cooperative Research Act of 1984 (which limited antitrust prosecution for corporations involved in joint research projects), and the National Technologies Transfer Act of 1989 (which allowed federally sponsored research facilities to spin off classified research to private firms) (Mirowski 2011).

Even this set of legislation was only one component of the broader social forces that contributed to increased privatization and globalization of research after 1980. The development of recombinant DNA technology by Stanley Cohen and Herman Boyer in the 1970s launched a great deal of high-tech molecular biology research. More and more of this work became patentable after the Supreme Court decision of *Diamond v. Chakrabarty* in 1980, which allowed the patenting of living organisms as long as they were classified as human-made rather than products of nature (Biddle 2014). This case was one of a number of court decisions, laws, and policies that strengthened intellectual property protection and weakened antitrust regulations (Mirowski 2011). The increasingly restrictive system of intellectual property in the United States then influenced the rest of the world via the Uruguay Round negotiations of the General Agreement on Tariffs and Trade, which created the World Trade Organization in 1995.

As a result of these and other factors, trends in US research and development (R&D) have changed dramatically. Whereas the federal government was funding roughly two-thirds of US R&D in the 1960s, the situation was reversed by the year 2000, so that industry is now funding two-thirds of US R&D (Mirowski 2011). Universities have experienced increasing pressure and incentives to develop collaborations with industry and to engage in research that promotes economic development (Slaughter and Rhoades 2004; Washburn 2005). Historians and sociologists have used the terms "Mode 2" and "Triple Helix" to describe the resulting forms of research, which are characterized by industry–academic–government collaborations, transdisciplinarity, and an emphasis on producing knowledge that is geared toward application (e.g., Etzkowitz 2002; Nowotny, Scott and Gibbons 2001). Even within industry, research practices have changed to reflect new trends toward globalization. Large pharmaceutical and agricultural biotechnology companies are now increasingly exporting projects to contract research organizations (CROs), which can operate wherever in the world the costs and regulatory environments are most favorable.

This new era of science policy poses a variety of questions and challenges for policymakers, universities, and corporations. Policymakers face increasing questions about whether new, restrictive intellectual property regimes truly promote innovation and social welfare (Biddle 2014; Sterckx 2010). Meanwhile, universities are struggling to balance their multiple social roles. Sheldon Krimsky (2003) has argued that universities have historically had at least four roles: (i) training well-educated and thoughtful citizens; (ii) supporting national defense; (iii) promoting economic growth; and (iv) generating "public-interest" research about social and environmental problems. Krimsky worries that recent pressures to focus on promoting economic growth are leaving universities less capable of playing their public-interest role to address important social problems. Universities are also facing significant uncertainty about future research funding. For example, after federal funds for the National Institutes of Health (NIH) doubled in the first years of the twenty-first century, they have largely remained flat, leaving the biomedical research enterprise in a state of "malaise," as a flood of new trainees finds it increasingly difficult to find promising jobs (Alberts et al. 2014). Finally, there is currently a great deal of public skepticism about whether current research in fields like agriculture and medicine is reliable and socially helpful or whether it is directed primarily toward promoting a narrow range of private interests.

Science for Policy: Private-Interest Science

The potential for private-interest groups to generate scientific research directed primarily toward promoting their own agenda has become a major science policy issue. Historians of science have played an important role in raising public awareness about the potential for questionable research practices in science produced for policy purposes. Perhaps the most famous case is that of tobacco, which Robert Proctor (2012) has studied in detail. Beginning in the 1950s, the tobacco industry worked with public relations expert John Hill to plot a campaign for protecting their products against growing evidence that cigarette smoking was associated with lung cancer. Proctor summed up the approach of the industry:

> ... its goal at many points was to *generate* ignorance—or sometimes false knowledge—concerning tobacco's impact on health. The industry was trebly active in this sphere, feigning its *own* ignorance of hazards, while simultaneously affirming the *absence of definite proof*

in the scientific community, while also doing all it could to *manufacture ignorance on the part of the smoking public.* (Proctor 2008: 13-14, italics in original)

Partly based on his examination of this case, Proctor coined the term "agnotology" for the study of culturally induced ignorance or doubt. The ignorance-generating strategies employed by the tobacco industry included funding research projects designed to produce preferred results, withholding worrisome research findings, re-analyzing studies that appeared to show harm from tobacco, harassing opposing scientists, and developing PR and lobbying campaigns to promote preferred messages.

Other historians have identified very similar agnotological strategies in other public- and environmental-health cases (Elliott forthcoming a). For example, Gerald Markowitz and David Rosner (2002) have documented how the lead and vinyl industries engaged in the same efforts to withhold information, re-analyze studies, marginalize opposing scientists, and lobby aggressively on behalf of their products. Mark Largent (2015) shows how the aspirin industry engaged in similar efforts to cast doubt on the relationship between aspirin and Reye's syndrome. In an effort to derail mandated warning labels on aspirin bottles, they enlisted the assistance of the powerful head of the US Office of Information and Regulatory Affairs (OIRA), Jim Tozzi, who later founded a major Philip Morris-funded nonprofit organization dedicated to challenging regulations.

All these strategies, including the founding of think tanks and nonprofits dedicated to questioning the science behind regulatory initiatives, are now playing a prominent role in contemporary efforts to deny anthropogenic climate change (McCright and Dunlap 2010). In their book *Merchants of Doubt* (2010), Naomi Oreskes and Erik Conway examine how an industry-oriented think tank called the Marshall Institute used the same sorts of tactics to halt momentum for addressing climate change during the first Bush administration. They note that these sorts of think tanks have become a powerful force for challenging action on environmental issues; one study found that 92% of the 56 "environmentally skeptical" books published in the 1990s were linked to right-wing foundations, along with 100% of the 13 skeptical books published in the 1980s (Oreskes and Conway 2010: 236). Sociologist Robert Brulle (2014) recently raised the additional concern that the funding sources for these think tanks and foundations are becoming less transparent. He notes that major "climate denial" organizations like Koch Affiliated Foundations and ExxonMobil are increasingly routing their gifts through "donor-directed" foundations that allow donors to designate the ultimate beneficiaries of their gifts without making that information public.

Given this context, a major challenge for contemporary US science policy is to develop strategies for addressing these efforts at generating "private-interest" science. While cases like tobacco, lead, and climate change are particularly striking, similar problems are at least as pervasive in the medical field. For example, scholars have identified a "funding effect," according to which pharmaceutical studies funded by those with a financial stake in the outcome are up to four times more likely to generate results favorable to the funders than independently funded studies (Bekelman, Lee and Gross 2003; Sismondo 2008). Moreover, pharmaceutical companies pay medical education and communication companies (MECCs) that specialize in generating journal articles with these favorable results, finding prominent academic scientists to include as authors, and using key opinion leaders (KOLs) to communicate the results to other physicians

(Goldacre 2012; Sismondo 2009). In response to these trends, a number of scholars have called for more independent funding of biomedical and environmental-health research through government agencies like the NSF and NIH (see e.g., Elliott 2011; Shrader-Frechette 2007), but the vast majority of studies performed for regulatory purposes are still funded by industry (Conrad and Becker 2012).

One of the most common solutions to these problems is to require that the authors of journal articles disclose their sources of funding and their roles in the research. This is an important first step, given the evidence that pharmaceutical companies commonly pay prominent academic authors to put their names on journal articles that are "ghost written" by private medical communications companies (Healy and Catell 2003; Sismondo 2009). Nevertheless, while disclosure can increase transparency, it is unlikely to prevent abuses of policy-relevant science (Elliott 2008, 2011). Other scholars have suggested developing criteria for evaluating the credibility of studies and developing standardized study protocols so that it is more difficult for private-interest groups to manipulate science produced for regulatory purposes (Conrad and Becker 2012). Unfortunately, these strategies are also likely to be limited in their effectiveness (Elliott 2014, forthcoming b). Recent innovative proposals for generating more trustworthy science for policy have focused either on sequestration (i.e., taking regulatory studies out of the hands of industry and performing them in government institutes) or collaboration (i.e., creating research initiatives that bring together representatives of industry, government, and citizen groups) (Elliott 2014). Proposals for collaborative research initiatives fit particularly well with another major theme in contemporary science for policy: public engagement.

Science for Policy: Public Engagement

Even in research contexts in which financial conflicts of interest are not a significant issue, the late twentieth century witnessed an increasing emphasis on finding ways to promote public engagement in the development of science and technology. This emphasis is the result of several factors. One factor is an increasing skepticism among politicians, scholars, and members of the public about the "social contract" model for the relationship between science and society. This model was developed by figures like Vannevar Bush (1945) and Michael Polanyi (1962) during and after World War II. They argued that there is (or should be) a tacit contract between science and society, such that society gives scientists autonomy to pursue projects that fit their own disciplinary interests, and science in turn will flourish and generate extensive social benefits. This view of science is closely related to what is often called the "linear model," which posits that basic research leads to applied science, which in turn leads to technological developments and social benefits.

The social contract and linear models were challenged on multiple fronts during the later decades of the twentieth century. Scholars like Donald Stokes (1997) pointed out that the linear model's distinction between basic research and applied science is more tenuous than it appears. For example, it is not clear that scientists need to be given maximal freedom to investigate whatever they want in order to function effectively, because research can be guided by practical considerations *and* simultaneously generate fundamental understandings of nature. Moreover, it is not clear that technological advances typically develop in a straightforward linear fashion from basic research;

sometimes they can develop autonomously and then stimulate subsequent research in an effort to improve or understand them (see Pielke 2005, 2007). Meanwhile, growing concerns about environmental problems during the latter half of the twentieth century illustrated that scientific advances can generate significant risks as well as benefits (Beck 1992). As a result of all these considerations, the assumption that funding more basic research will inevitably generate social benefits became increasingly dubious (Sarewitz 1996). At the same time, allegations of falsification and fabrication by scientists raised doubts about the ability of the scientific community to govern its own activities in the absence of external constraints (Broad and Wade 1982; Kevles 1998). Together, these factors generated increasing skepticism about the post-World War II model of providing extensive funding for basic scientific research while shielding the scientific community from democratic control.

Another set of factors that generated increased support for public involvement in the development of science and technology was the demise of the "deficit" or "public understanding of science" (PUS) model of public science literacy (Toumey 2006). In 1985, the British Royal Society published a report called *The Public Understanding of Science*, which suggested that individual and social decision making could be improved if the public were better educated about science, which in turn could be accomplished if scientists learned to communicate information more effectively. While this was a British report, many experts in both the United States and Britain (and around the world) have subscribed to the basic idea that public opposition to new scientific and technological developments is typically based on a *deficit* in their knowledge of the science and that the key to resolving social conflicts is to increase *public understanding of science* (Toumey 2006; Wynne 1992). In the wake of the Royal Society report, a number of social scientists and historians have argued that public opposition to scientific and technological developments is often caused not primarily by public ignorance but rather by other social concerns and values (see e.g., Largent 2012; Wynne 2005). Other challenges to the deficit model have stemmed from the realization that contemporary policy problems stretch the limits of science; thus, experts are prone to making errors and inaccurate predictions. Therefore, citizens have the right to decide how to respond to the resulting uncertainties, and they sometimes have "local" expertise that can improve both science and policymaking (Collins and Evans 2007; Elliott 2009; Epstein 1998; Irwin 1995).

In recent years, a number of different mechanisms have been employed for promoting public engagement in research. One approach is community-based participatory research, in which scientists work with citizens to perform research projects that meet community needs and concerns (e.g., Couvet et al. 2008; Shirk et al. 2012). Another approach is to develop citizen panels or consensus conferences, which enable small groups of citizens to be educated about a scientific or technological issue so that they can write an informed report summarizing their views about the major social concerns that it raises (e.g., Philbrick and Barandiaran 2009). Other approaches do not provide such direct opportunities for public engagement but still try to promote broader thinking about the social ramifications of science and technology. For example, the NSF now has a policy of including the "broader impacts" of research on science and society as part of their funding decisions. Funding agencies have also started to provide a percentage of money for investigating the ethical, legal, and social issues (ELSI) associated with major developments such as the Human Genome Project and the National Nanotechnology Initiative. Scholars have also experimented with efforts to "embed" social scientists and humanists in scientific labs in the hope that it will promote more careful thinking about the social ramifications of research (Schuurbiers and Fisher 2009).

The relationships between science and government have developed significantly since 1789, but it is clear that a great deal of work is still needed to address the major science policy issues that we are currently facing. In fact, the two major issues discussed at the end of this chapter exacerbate one another, because it is difficult to incorporate "upstream" public engagement in research when it is controlled by private entities. A number of other issues that could not be covered here also merit attention. These issues include the history of presidential science advising (see Golden 1988; Herken 2000), the story of the Office of Technology Assessment (OTA) and its demise in the 1990s (Bimber 1996), the recurring debates over the creation of a federal Department of Science (Dupree 1957; Smith 1990), policy issues related to the social sciences (Herman 1995; Solovey and Cravens 2012), and recent efforts to create a science of science policy (Fealing et al. 2011). The history told in this chapter also focused primarily on policy issues related to the *funding* of science, and it touched only briefly on scientific issues related to *regulatory* policy and *patent* policy (see Jasanoff 1990; Sell 2003). Another important contemporary issue is the increase in partisan wrangling over science, with each political party declaring that the other has abused or ignored science in the policy arena (e.g., Berezow and Campbell 2012; Mooney 2005). Finally, the overarching analytical framework of this chapter deserves further development. While it is undoubtedly helpful to examine policy decisions made *about* scientific research and also to examine the *use* of science in making policy decisions, the last sections of the chapter indicate that we ultimately need to bring these two projects together to form a *policy for science for policy* – public policies that help generate robust forms of science that inform decision making effectively (Pielke 2007; Pielke and Betsill 1997).

Bibliographical Essay

For the history of science policy between 1789 and 1940, Dupree (1957) provides a classic overview. Bruce (1987) and Tobey (1971) provide more limited discussions of the time periods surrounding the Civil War and after World War I, respectively. Douglas (2009) provides a brief historical overview of science advising during the nineteenth and twentieth centuries. Mirowski (2011) provides a valuable overview of the influences of government, private foundations, and industry on university science from the late nineteenth century until the present.

For the period between 1940 and 1980, Greenberg (1968) provides a detailed account of science policy between 1940 and 1950, with a more general overview of trends before World War II and during the early decades of the Cold War. He focuses on the development of a relatively unrestrictive system for providing federal grants to university researchers. Kevles (1971) provides a very influential history of the American physics community and its role in developing the atomic bomb during World War II. Smith (1990) examines science policy throughout American history but with special focus on the period between 1945 and 1990. Leslie (1993) examines American science during the Cold War. For a focus on policy issues surrounding the social sciences during the Cold War, see Solovey and Cravens (2012), and for an emphasis on psychology in particular during this time period, see Herman (1995).

For the time period between 1980 and the present, Mirowski (2011) provides a masterful overview of the recent trends in public policies, corporate structures, and universities that have contributed to the new era of globalized, privatized science. A number of other books have also chronicled the trend toward increasing privatization and industry–university relationships, including Etzkowitz (2002), Krimsky (2003), Slaughter and Rhoades (2004), and Washburn (2005).

For historical studies of private-interest science, see Proctor (2012) for a comprehensive overview of the tobacco industry, Markowitz and Rosner (2002) for analyses of the lead and vinyl chloride industries, McCright and Dunlap (2010) for discussion of climate change, and Elliott (forthcoming a) and Oreskes and Conway (2010) for overviews of multiple environmental issues. McGarity and Wagner (2008) and Michaels (2008) cover public health research, and Angell (2004), Goldacre (2012), and Sismondo (2008, 2009) discuss concerns about the pharmaceutical industry. Krimsky (2003) examines financial conflicts of interest in university research as well as potential solutions. Elliott (2008, 2011, 2014, forthcoming b) explores a range of potential solutions to financial conflicts of interest in research.

For analysis of public engagement in research, Epstein (1998) provides an important study of citizen contributions to AIDS research. Collins and Evans (2007), Irwin (1995), and Wynne (1992, 2005) examine the nature of expertise and show how non-scientists can have important forms of expertise that strengthen science policy in some contexts. Their work is part of the field of science and technology studies (STS), which includes a great deal of literature that highlights the complexity of policy-relevant research and the resulting benefits of public engagement; Hackett et al. (2008) and Sismondo (2004) offer important overviews of the STS field. Pielke (2005, 2007) provides an accessible introduction to the distinction between basic and applied research, as well as an analysis of the linear model. Guston (2000) and Sarewitz (1996) offer influential discussions of the breakdown of the social contract model for the relationship between science and society. For information about the scandals that contributed to this breakdown, Broad and Wade (1982) provide an influential overview, and Kevles (1998) analyzes the Baltimore affair, which was one of the most significant scientific scandals of the late twentieth century.

Chapter Thirty-Eight

POPULARIZING SCIENCE

Constance Areson Clark

Before the 1830s and 1840s, few Americans earned their livings as scientists. It was only after professional opportunities began to arise for them that natural historians or natural philosophers began gradually to adopt William Whewell's 1834 term "scientist" to describe themselves. Significantly, the terms "popularizing" and "pseudoscientist" began to appear in just the same period. As the process of professionalization began to create a new category of "scientists" set apart from non-scientists, scientists began to see a need to popularize their work, which meant, to them, education of the public *about* science *by* scientists. But scientists were never the only people writing about, speaking about, broadcasting, or otherwise "popularizing" science. In patterns that ebbed and flowed, tensions arose over such questions as: who had the authority to speak for and of science? How was not-science or pseudoscience to be distinguished from "real" science, and by whom? And who counted as the public for science? As early as 1858, the educator Horace Bushnell complained that people were beginning to act "as if nothing could be true, save as it is proved by the scientific method" (Thurs 2007: 79). Bushnell's complaint suggests an early concern about claims to authority for "scientific method" as a form of knowledge presumed to be distinct from and superior to other ways of knowing – a single arbiter of truth over all other forms of social knowledge.

It is not that there was no popular science in the United States before the 1840s. Enthusiastic audiences attended lectures demonstrating the "amazing wonders" of electrical phenomena in the eighteenth century (Delbourgo 2006: 87–128); and much evidence points to public interest in natural history in that period. But historians have noted a marked increase in the occurrence of lectures and publications about science in the early to middle nineteenth century, due in part to new markets for science popularization, and also to scientists' changing perceptions about the state of public knowledge about science (Burnham 1987: 131–46).

A Companion to the History of American Science, First Edition.
Edited by Georgina M. Montgomery and Mark A. Largent.
© 2016 John Wiley & Sons, Ltd. Published 2020 by John Wiley & Sons, Ltd.

The American economy grew in the early nineteenth century, especially after the end of the War of 1812, and along with the republican enthusiasms of the Jacksonian era, spurred educational reform and increased rates of literacy. During this time, many people associated science, democracy, and progress. But tensions arose with the professionalization of the sciences and increases in its authority. Technological advances in printing, binding, and reproduction of images stimulated an explosion in publication of new magazines, inexpensive newspapers, and books. Later in the century, fairs, expositions, and museums celebrated advances in science and technology. American audiences gained, and took advantage of, new opportunities to learn about new things. It is safe to say that science and technology stood out among the "new things" that fascinated them. So there was public interest. Science and natural history were popular.

But "popularization" was not simply the growth of dissemination of a popular subject. Rather, the notion of "popularization" grew up with the development of science as a newly professionalized activity, and of the "scientist" as a category of professionals claiming a specially demarcated place – and new authority – for their work. Science, in this process, became a thing separate from the rest of the culture, a thing to popularize. This was not a simple process. To complicate this, no public – and certainly not the American public – has ever been monolithic or homogeneous. Further, the ways in which the "non-scientist" public encountered science – the forms in which it was available – evolved along with the changing place of science in American society. Many of the settings for this popularizing, such as zoos, museums, and nature study, are explored in other chapters of this book, as are some of the scientific disciplines and movements that appeared as the subjects of popularization; this chapter will therefore attempt a road map of some of the larger patterns in the history of American popularizing.

"Men of Science" and the Lecture Circuit

The professionalization of science and the fragmentation of the sciences into disciplines, each with its own specialized professional society, networks, and journals, culminated in the late nineteenth century and made possible impressive advances in knowledge about the natural world; however, it also redefined science in a way that created the idea of a public of outsiders in need of education. For people like Thomas Henry Huxley, John Tyndall, and other members of the British coterie calling itself the X Club, or Alexander Dallas Bache, Joseph Henry, and colleagues in the American group called the "Lazzaroni," popularization was part of the strategy of establishing the identity and authority of, as many of them revealingly called themselves, "men of science" (Bruce 1987: 218–20; Gregory and Miller 1998: 22; Lightman 2011). Scholars have sometimes referred to a sort of "religion of science" among late nineteenth-century popularizers. Science, for many popularizers, represented – even caused – progress, including progress toward a more rational and less superstitious society. As scientists became more removed – or thought of themselves as removed – from the general public, they saw science as the realm of objectivity. And they often saw the public as superstitious. It became common to complain of the ignorance or gullibility of an imagined public cast as vulnerable to charlatanism, spectacle, and "humbuggery."

Yet Americans evinced a strong interest in science and natural history. An energetic lyceum movement grew up in the early nineteenth century, especially in New England, offering lectures, including lecture demonstrations on scientific topics (Bruce 1987: 116; Rossiter 1971: 602). Audiences attended lectures demonstrating principles and phenomena of physics and chemistry delivered by well-known scientists and itinerant lecturers (Burnham 1987: 136–7). People living in or near Boston could go to the Lowell lectures, which drew distinguished speakers such as the notably famous (and famously charismatic) Benjamin Silliman and Louis Agassiz, in part because an endowment by the textile manufacturer John Lowell made it possible to pay prominent scientists fees which could significantly improve their incomes (Rossiter 1971: 610). Lowell Institute lectures were so popular that some of them had to be repeated in order to accommodate overflowing crowds, and occasionally, people demanding admission to already-full lectures rioted (Burnham 1987: 137–38; Gregory and Miller 1998: 22; Kevles 1978: 14–15; Pandora 2009a: 354). The movement spread to other parts of the country. Newspapers published notices of lectures by Thomas Henry Huxley, Asa Gray, and Louis Agassiz, and in 1872, a special edition of the *New York Tribune* featuring the text of Tyndall's lectures sold 50,000 copies (Gregory and Miller 1998: 25; Kevles 1978: 14). Foreign visitors remarked on Americans' willingness to attend lectures on scientific subjects (Burnham1987: 35–6).

Republicanism and Cheap Print

In the context of Jacksonian America, much of this interest in science took a republican flavor. An expanding economy, educational reforms, and high rates of literacy (except among slaves, legally prohibited from learning to read), and a belief in upward mobility fueled interest in self-improvement. Rapid technological change and announcements of new discoveries and theories in science fired imaginations. So much changed radically in the 1830s and 1840s that the rapidity of change itself seemed astonishing, and many associated the change with technology and science. Americans witnessed a revolution in communications at this time. Newspapers and magazines proliferated, making the nineteenth century, as Burnham has put it, "the century of the magazine" (Burnham 1987: 130).

Magazine and newspaper publishing were growing businesses. Publishers found that they could create markets, and that science stories helped them to do that. Magazines specifically devoted to science and technology appeared. For example, *Scientific American* (started in 1845) focused on engineering and new inventions. General interest magazines, such as *North American Review* (1815), *Harper's Monthly Magazine* (1850), and other new magazines regularly included science stories (Burnham 1987: 36; Kohlstedt, Sokal and Lewenstein 1999: 35). In 1872, *Popular Science Monthly* began, with editor Edward Youmans evangelizing for science (Gregory and Miller 1998: 25). It was successful from the start. While not all of these magazines survived, Americans gained access to a wealth of new publications about science.

The historian of science Katherine Pandora has argued, in comparing the American and British contexts, that in the early nineteenth century, many Americans, steeped in the language of republicanism, saw science as something available to everyone. In a study of books for children, Pandora found authors addressing children with the ideals of a young democracy in mind: books encouraged children to investigate nature for themselves, not deferring too much to authority. Children should examine the real phenomena of nature, and they did not (in contrast to some British books for children) need access to expensive instruments. They might enlist their parents' help in building herbaria, but they should not be worried to discover that their parents did not know everything: people young and old could investigate nature together. In one book series, a boy employed as a hired laborer was especially adept at teaching other children. People of all classes could participate in this community of learners. Furthermore, these books were deliberately written in "plain language," part of a traditional American discursive style (Pandora 2009a, 2009b).

The Graphic Revolution and the Celebrity Scientist

Publishers also found that pictures sold magazines and newspapers. In addition to the proliferation of books and articles explaining recent scientific discoveries, the nineteenth century marked the beginning of what Walter Lippmann would later call a graphic revolution. Photography and the new technologies of reproduction of pictures made visual images widely available in a variety of forms and at newly affordable rates. Some popular magazines and newspapers included pictures of scientific apparatus and diagrams illustrating scientific theories, photographs and caricatures of scientists, and cartoons satirizing science. For example, evolution themes appeared regularly in cartoons, as did Charles Darwin himself. Janet Browne has suggested that Darwin's presence after publication of *On the Origin of Species* in 1859, in photographs, caricatures, and cartoons meant that he became identified with his theory – so much so that he became the very embodiment of the theory (Browne 2001, 2003). Pictures of prominent scientists introduced a novel form of scientific celebrity, auguring a new kind of relationship of scientists to the public while also perhaps creating a hierarchy of famous and less-famous scientists.

Popularization would increasingly involve celebration of scientists as individuals, often at the expense of coverage of the science that made them famous. Emphasis on the personalities of such figures as Darwin and later Marie Curie and Albert Einstein sold magazines. Not all scientists sought or welcomed this kind of fame, but it could be put to use. Sometimes scientists or inventors took advantage of publicity quite adroitly. The most striking example was Thomas Edison, deftly cultivating journalists and exercising considerable control over the journalistic construction of his public persona as "the wizard of Menlo Park" (Bazerman 2002: 28–9; LaFollette 1990: 98–100).

Media accounts in the 1920s of what was most often called the "Einstein theory" illuminate the growing prominence of the science celebrity. Interest took off after confirmation of the theory during the 1919 eclipse, amplified by Einstein's highly publicized visit to the United States. A movie attempting to explain relativity drew large audiences. Daniel Thurs has told the story of an elderly woman "having a riotous time" over several days at the New York Public Library reading about the Einstein theory (Thurs 2007: 98). Newspapers and magazines probably referred so often to the theory of relativity as the "Einstein theory" because it was much easier to focus on the man than to explain the theory. A 1921 visit by Marie Curie provoked similar interest, and one of the most notable facets of journalistic accounts of science in the American media in this decade of the takeoff of movie fandom was the treatment of both Einstein and Curie as science celebrities. Increasingly, too, newspapers and magazines regularly asked scientists and inventors to comment on things well beyond their own fields of expertise.

Science, Religion, and the Mythology of Warfare

Controversies over evolution gave rise to a metaphor that turned out to have a lasting legacy. Early in the nineteenth century, naturalists in the United States had followed the developing British conversation with interest, and like many British naturalists were equipped with a secure conviction of the argument from design – the idea that nature offered myriad examples of the goodness of the Creator. After publication of *On the Origin of Species* in 1859, some of Darwin's advocates, most notably Thomas Henry Huxley and John Tyndall, began constructing a picture of religion as the adversary of science, and two books published in the 1870s gave that idea a lot of purchase. John William Draper's *History of the Conflict Between Religion and Science* (1874) and Andrew White's *History of the Warfare of Science with Theology in Christendom* (1876) framed popular discourse about evolution, and created a mythology about the relationship of science and religion that has lasted.

A number of historians of science have challenged this mythology of warfare (Burnham 1987: 162–3; Larson 1997: 21–2; Thurs 2007: 69, 83). Historians have traced the increasing use of the term "scientific method" as part of the creation of a boundary setting science off from other forms of knowing. The "universality of the scientific method" (Burnham 1987: 30) meant that science could be defined broadly to include authority in all sorts of areas, including economics, politics, and history. But the reification of "scientific method" as a single thing amounted to a mystification of science, obscuring more than it clarified (Gieryn 1983; Shapin 1990; Thurs 2007, 2011).

Daniel Thurs and Ronald Numbers have noted that the term "pseudoscience" appeared at the same time as the terms "scientist," "popularization," and "scientific method" within the popular vernacular. These terms can all be seen as part of a single process of differentiation, of setting science apart from not-science (Burnham 1987: 167; Thurs 2011; Thurs and Numbers 2011). However, as Bernard Lightman has pointed out "the change did not take place without a fight" (Lightman 2011: 339).

Scientists, (Women) Popularizers, and Mythologies of the Public

Lightman has examined the work of British Victorian popularizers who were not professional scientists but who wrote highly influential books. These writers were able to write about natural history in ways that did not set science off from other ways of finding meaning in the world. Many of these writers were women and clergymen – among the groups especially excluded from the circles of professional science – yet their books enjoyed impressive sales. Further, Lightman suggests, their books may have been the way many Victorians learned about the natural world (Lightman 2011: 356–7).

Not all those with an interest in science were willing to cede authority to the credentialed professionals, self-styled "men of science," to speak of science, write about it, or follow its implications. People who were not themselves professional scientists claimed the right to interpret Darwin, in particular, for themselves. Historian Kimberly

Hamlin has analyzed the work of a group of American feminist women who argued that evolutionary theory, specifically Darwin's theory of sexual selection, affirmed their understanding of women's rights (Hamlin 2014).

Though women were able to contribute to the literature of popular science, images of women in popular media excluded them in symbolic ways. Even later in the twentieth century, when more women managed to enter the realms of professional science and succeed there, popular and mass media images of women scientists often portrayed them as oddities (LaFollette 1990: 78–96). Dorothy Nelkin has described startling accounts in women's magazines in the 1960s of women who had won Nobel Prizes in science. For example, a reporter for *McCall's* gushed about the physicist Maria Mayer, "a brilliant scientist, ... and she was so darn pretty...." Under the headline, "She Cooks, She Cleans, She Wins the Nobel Prize," a 1977 magazine referred to Roslyn Yalow as "a Bronx Housewife" (Nelkin 1995: 18–19). "In the twentieth century," LaFollette commented wryly, "many battles have been lost in the popular press" (LaFollette 1990: 95). The processes of exclusion that began in the nineteenth century had lasting legacies.

By the early twentieth century, many American scientists, perceiving – or assuming – general public ignorance and apathy about science, began to call for campaigns of public education. But fewer of them actually participated in such efforts than lodged complaints about the need for them. Science stories in print media in the United States saw a falling-off early in the twentieth century (Burnham 1987: 171; Tobey 1971: 3–12). After the end of World War I, however, a number of factors converged to create a new peak of science popularizing both by scientists and by newly professionalizing science journalists.

Modern Democracy and the Professional Popularizer

Warnings of a widening gap between the scientific expert and the layperson appeared as an increasingly common theme. In 1919 the *New York Times*, for example, published a series of editorials lamenting the ignorance of the general public on matters scientific and warning of dire implications for democracy (Thurs 2007: 104). This theme, that modern democracy depended upon a scientifically literate general public resounded. Concerns of this kind were instrumental in the establishment of an influential organization intended to help educate the public about science, the Science Service. Set up by the newspaper magnate Edwin W. Scripps with help from zoologist William Ritter and support from the American Association for the Advancement of Science (AAAS), the National Academy of Sciences (NAS), and the National Research Council (NRC), the Science Service was the first syndicated news service specifically for news about science. As Dorothy Nelkin put it, Scripps "believed that science was the basis of the democratic way of life." He also believed that "science news would sell" (Nelkin 1995: 81). It seems that he was right. The Science Service sold articles to more than 100 newspapers in the 1920s, reaching an estimated seven million people, one-fifth of the total circulation of the American press (Nelkin 1995: 82). Scripps hired chemist and popular science writer Edwin Slosson to run the Science Service, and historians have recognized Slosson's influence as significant and lasting: Nelkin says that he "created a market for science news and a pattern for the emerging profession of science journalism (Gregory and Miller 1998; Kevles 1978: 171–2; Nelkin 1995: 83; Tobey 1971: 62–95).

By the 1920s, in addition, the relatively new discipline of science journalism had begun to flourish. And just in time, it seemed. The number of newspaper and magazine articles, and especially the number of books about science, surged in the 1920s (Burnham 1987: 171; LaFollette 2009: 248–53). As the role of science journalists grew, they, like scientists before them, began to form a professional network. The size of the National Association of Science Writers, founded in the 1930s, would swell during the 1950s. These science writers remained, in the 1950s, as they had been in the 1920s, advocates for science, and they generally thought of themselves in that way (Gregory and Miller 1998: 38).

And there was radio. The number of radio listeners grew exponentially in the 1920s, from thousands to millions. Early in the decade, many people, including some science popularizers, saw radio as a medium with great possibilities for democracy and education (LaFollette 2008: 12). The Smithsonian Institution, the American Museum of Natural History in New York, and the Field Museum in Chicago offered radio programs on science, as did the Science Service. Some scientists were skeptical of radio as a sensationalizing medium, but Edwin Slosson pitched the idea that radio broadcasts could project the natural dignity of scientists. Austin Clark, organizing radio talks offered by the Smithsonian, insisted that lecturers arrive "in tuxedos and in a mental attitude to match" (LaFollette 2008: 19). Clark and Slosson both insisted that in order to preserve the dignity of science radio, talks should avoid controversial topics. It was a balancing act, though: Clark also favored "jazzy" topics that would attract listeners, such as "Animal Terrors of Ages Past" (LaFollette 2008: 16–17). Jazzy, but not controversial. Clark and Slosson both feared that broadcasts covering the Scopes evolution trial – which was broadcast over the radio by Chicago's WGN – would be too sensational for scientific dignity.

Radio soon became, however, a commercial medium. Initially, science series were broadcast locally, but with the advent of national radio networks, local access became more difficult, and scientists' control over their radio broadcasts began to evaporate. Outlets were limited, and programs that broadcasters did not see as sufficiently "commercial" could not compete. Networks pressured popularizers to emphasize personalities over facts. Scientists disagreed about the degree to which they should support popularization as a commercial enterprise. The early promise of radio as a site for meaningful science communication was not realized, according to Marcel LaFollette, because of attitudes of both broadcasters and scientists. Ultimately, LaFollette concluded, "In the competitive world of commercial broadcasting science was dramatized, personalized, and eventually marginalized" (2008: 1). A similar pattern would characterize the appearance of science on early television in the late 1940s and 1950s. Eventually, LaFollette has shown, popularized science on television emphasized the romantic, anthropomorphic, and sensational – and "facticity" – at the expense of the actualities of scientific practice. In television dramas ostensibly grounded in science, such as *CSI* "science was *referenced* as much as represented" (LaFollette 2013:, 224).

Drama in Every Test Tube: Selling the Romance of Science

While Slosson insisted that the Science Service should avoid controversy and sensationalism, he also used colorful language in advertising the service: Science offered "drama and romance . . . interwoven with wondrous facts" according to one of his advertisements; another claimed that "drama lurks in every test tube" (Nelkin 1995: 82). In newspaper and magazine articles and on the radio, the Science Service cast scientists as pioneers and discoverers.

The Science Service was not alone. Chemists, and the professional organizations that represented them, concerned about possible alarm over the use of chemical weapons in World War I, organized efforts to educate the public about the usefulness – and romance – of chemistry. In 1925 the Chemical Foundation took out a full-page ad in the *New York Times* offering six books on "The Progress – the Romance – the Necessity of Chemistry!" Other chemical associations also subsidized popular books through the 1930s and provided numerous copies to libraries and schools (LaFollette 2009: 248).

The romance of chemistry and of the laboratory sold books, and helped popularizers to "sell" science. Paul de Kruif's bestselling 1925 *The Microbe Hunters*, Marcel LaFollette has remarked "turned laboratories into sites of adventure and drama" (LaFollette 2009: 254). Popularizers at the American Museum of Natural History in New York also touted the "romance" of natural history in its general audience magazine *Natural History*, and the charismatic Roy Chapman Andrews effectively promoted images of the romance of fieldwork in paleontology to solicit support for the museum's Mongolia Gobi Expeditions (Clark 2008: 36, 48).

Controversy, Democracy, and the Theater of Science

The 1920s science publicity boom was also fueled by the evolution controversy of that decade, culminating in the notorious 1925 trial of John Thomas Scopes, accused of breaking Tennessee's law prohibiting the teaching of evolution in public schools. The first trial ever broadcast on radio, it was featured in cinema newsreels and covered daily in newspapers across the country. It inspired museum exhibits, editorials, and cartoons making fun of both (or all) sides of the controversy (Clark 2001, 2009; Davis 2008). Publicity about evolution had also fueled a surge in media science coverage in the years after 1859 as it would during later evolution controversies in the United States. But though such controversies may seem to skew the pattern of public attention to science, they also reveal underlying tensions and preoccupations.

One of the most intriguing things illuminated by the Scopes trial has to do with American ideas about the relationship of science and democracy. In contrasting popular attitudes in the United States with patterns he analyzed in early twentieth-century British culture, the historian of science Peter Bowler commented: "The belief that ordinary people had as much right as anyone else to decide what should be taken as valid knowledge about nature was particularly strong in America . . . . [In Britain] there was no equivalent to William Jennings Bryan's direct appeals to the people to determine what should be taught in the schools about evolution" (Bowler 2009: 18). An important theme in recent studies of the trial has been the surfacing of conflicting definitions of democracy, including concerns about the relationship of science and democracy in an increasingly technical society. Bryan argued that his anti-evolution crusade grew out of a need to protect the priorities of the democratic majority. Journalists ridiculed Bryan, pointing out that scientific truth could not be determined by majority vote, and Bryan accused critics of elitism. The charge of elitism hardly occurred in a vacuum (Clark

2008: 64; Larson 1997: 201). Concerns about the relationship of science and democracy, and accusations of scientific elitism, continued to percolate throughout the decade and into the 1930s. They would resurface in a very different context after World War II.

Post World War II and the Public Understanding of Science

After the end of World War II, partly in response to lessons learned from the mobilization of science during the war, especially the Manhattan Project, the federal government established a financial infrastructure for support of science on an unprecedented scale. This conversion to what historians refer to as Big Science reshaped the practices of science, and for many people, it seemed also to alter fundamentally both the relationship of science with the public and the accountability of science.

In a famous 1945 address, "Science – the Endless Frontier," engineer Vannevar Bush borrowed the rhetoric of science as romance to argue for increased government support of science and for efforts by scientists to educate the public about the need for such support. More pleas for scientists to communicate with the general public followed. In the context of Cold War competition with the Soviet Union, it seemed imperative to cultivate positive public relations for science (Kevles 1978: 367–92). At the 1951 Arden House conference, AAAS president Warren Weaver called for a campaign by scientists to "increase public understanding and appreciation of the importance and promise of the methods of science in human progress." The words "and appreciation" and "progress" should not be neglected. The AAAS and other scientific societies set out to correct what they perceived as the failure of the public to understand and appreciate science. A statement from one of the scientists involved in these efforts contrasted the "*knowledge*" of scientists with the "*opinions* of pressure groups and selfish politicians" (Kohlstedt, Sokal and Lewenstein 1999: 109–11).

Big Science created "a heightened sense of orthodoxy" (Thurs and Numbers 2011: 296) so that anything unorthodox would be seen not as heterodox science, but as not-science, or even worse, pseudoscience. Thurs and Numbers have suggested that "an explosion of talk about pseudoscience in the last third of the twentieth century" reflected another period of boundary drawing, excluding outsiders from the realm of sanctioned science, and by implication, confining popularization to the realm of the science "other." As Michael Gordin has observed "As is surely obvious, 'pseudoscience' is a term of abuse, an epithet attached to certain points of view to discredit those ideas." It is a word and an idea, Gordin continued, that "performs active work in the world, separating off certain doctrines from those deemed to be science proper" (Gordin 2012: 1).

Just as scientists seemed to exclude outsiders more rigorously than ever, Americans found themselves confronted with a more science-mediated world. The science before them was both more powerful and more inscrutable, and the issues more urgent. Atomic references saturated media discourse of the 1950s from books and lectures to movies, science fiction, cartoons, television programs, and even children's literature. Frankenstein made a comeback, and the term "mutant" became familiar. "Popularizations" of science appeared in the context of a new kind of mass culture and in the context of increasing public concern about the role of science in American policies.

Daniel Kevles has pointed out that President Eisenhower's warning as he left office concerning the dangers of a military–industrial complex would resonate with increasing skepticism about science in the 1960s. A counterculture that sometimes conflated science and technology viewed science as complicit in the Vietnam War, the atom bomb, and the dangerous pesticides described in Rachel Carson's best-selling *Silent Spring* (Kevles 1978: 393–403). The old charges of the elitism of science took on darker implications: The consequences of a sequestered and inaccessible science Establishment seemed very worrisome to many people. Science journalists, many of them the same generation that had extolled the virtues of science in the 1920s and 1930s, continued to play the role of science enthusiasts, but in the decades of the 1960s and 1970s, science popularization ebbed. Controversial theories of biological determinism, especially with publication of E.O. Wilson's 1975 book *Sociobiology* and the protests that it inspired caused something of a crescendo, and seemed to polarize discourse about science. Much of the controversy that swirled around the sociobiology debate had to do with the drawing of boundaries around the domain of science. The arguments were partly about who had authority and who did not (Nelkin 1995; Pandora and Rader 2008: 353; Sheldon 2014; Weidman 2011).

By the early 1980s, however, another boom in popular science was underway (Lewenstein 1987). Magazine publishers found a new strategy in niche marketing, and science magazines turned out to be a relatively lucrative niche. In addition, scientists turned to the writing of books ostensibly meant for a "popular" audience; but in this time of fragmented scientific specialties, the public for a popular science book often included scientists in disciplines other than the author's own. The conventions of scientific journal articles had so hardened that some scientists who wanted to engage in "big picture" thinking found in books the appropriate place to do it. Scientists also used books as vehicles for positioning themselves, their theories, and their disciplines. In a study of the use of books by paleobiologists in defining their field and making a case for its importance in evolutionary theory, for example, David Sepkoski has shown that paleontologists who also wrote popularizations, such as Stephen Jay Gould, Niles Eldridge, and colleagues used general audience books to establish a place for their science at – in John Maynard Smith's celebrated phrase – the "high table" of science (Gregory and Miller 1998: 85; Sepkoski 2014; Smocovitis 2014).

A "public understanding of science movement," beginning with scientists' postwar calls to increase the "public understanding and appreciation" of science became, as Gregory and Miller put it, something of an industry, spawning journals, professional meetings, and books. Critics have charged that public understanding literature has sometimes taken a top-down view of popularization as truth emanating from scientific experts to a supposedly passive public; more recently science studies scholars have sought stories that transcend the science/not-science dichotomy, and expand our understanding of the varied publics for science (Broks 2006: 118–24; Cooter and Pumfrey 1994: 238–9; Gregory and Miller 1998: 7–8, 88–99; Irwin and Wynne 1996).

The legacies in popularizations of science of the image of the scientist as hero, of the mythology of scientific method as a single thing, and of science as an exclusive club and the single arbiter of a simple truth have sometimes exacerbated public distrust of scientists and even of science (Gregory and Miller 1998: 90; Lewenstein 2001: 243; Thurs 2011). This distrust is sometimes quite reasonable. In *Merchants of Doubt* (2010), Naomi Oreskes and Erik Conway revealed that propagandists for the tobacco industry in the 1960s and more recently for the oil industry have assumed the authority of science to effectively cast doubt on widely accepted scientific conclusions through savvy and well-funded use of the media. Oreskes and Conway argue that images of science disseminated in mass media are misleading: Science is not the thing that it looks like in journalistic accounts. Journalists adhere to an idea of objectivity that insists on airing both sides of any controversy, using the associations of "scientific" with "objective" in a way that often obfuscates. In recent years, historians and sociologists of science have challenged the ideas that there is a monolithic public; that this public is passive; and that science operates in a rarified realm separate from the rest of society. We need more studies of particular publics and their participation in the cultures of science and technology. We need more studies that locate the ways in which "popularizing" is not a one-way transaction. More than ever, we need studies of science as a part of its larger context, and as part of a dialogue.

Bibliographic Essay

The terms "popularizing science" and "popular science" have become problematic for historians and sociologists of science in recent years, so much so that some of the most insightful historians working in the field have suggested abandoning them altogether. In part this has happened precisely because science studies scholars have arrived at much more nuanced understandings of practices of inclusion and exclusion in the history of science and a greater appreciation of the complexity of the publics for science. The term "popularizing" has been associated with what Stephen Hilgartner (1990) has called "the dominant view" and others call "the diffusionist model" of science communication, which implies or assumes a one-way transmission of scientific truth to a passive audience. Recent studies have challenged the diffusionist model. Although it raises – or because it raises – frustrating questions, it is a fruitful debate.

Much of the historiography addressing these questions has grown out of innovative work by historians of British Victorian science. The history of popular science had long been neglected, but beginning around 1980, it began to draw the attention of a number of historians who have changed the way we think about the history of science. In 1994 Roger Cooter and Stephen Pumfrey noted in "Separate Spheres and Public Places" that "surprisingly little has been written on science generally in popular culture" (238). Criticizing the public understanding of science movement for "largely draw[ing] upon traditional historical and sociological formulations of the 'popular' as passive lay consumption of learned products" (239), Cooter and Pumfrey laid the groundwork for investigation into "the ways in which science was actually practiced in popular culture" (243).

Even for those working primarily in the history of American popular science, work by historians of British Victorian popular science is essential both because comparisons between the American and British contexts are so relevant and because historians of British Victorian science have been pioneers, exploring new ways of thinking about not only who the popularizers were but also how various publics have participated in what is often called, following the title of an influential 2004 paper by James Secord, "Knowledge in Transit." Some of the most influential works in this category include (among others) Anne Secord's path-breaking 1994 essay, "Science in the Pub;" James

Secord's *Victorian Sensation* (2000); Bernard Lightman's *Victorian Popularizers of Science* (2007); the papers in the collection edited by Lightman, *Victorian Science in Context* (1997); essays in the collection edited by Aileen Fyfe and Bernard Lightman, *Science in the Marketplace* (2007). Peter Bowler's *Science for All* (2009) extends the time frame of this work. The work of Steven Shapin ranges widely and extends deeply: His essay "Science and the Public" (1990) is indispensable for anyone pursuing this topic. A 2009 Focus Section in *Isis*, "Historicizing 'Popular Science'," is also indispensable, especially the introduction by Jonathan Topham, and for those exploring the history of popular science in the American context, see Katherine Pandora's "Popular Science in National and Transnational Perspective: Suggestions from the American Context" (2009b). Pandora's work in general is important for a perspective on the American context, and her 2008 essay co-authored with Karen Rader, "Science in the Everyday World," is a valuable study of the field including suggestions for new directions, and an excellent review of the literature and the historiography.

Work by Stephen Hilgartner, Bruce Lewenstein, Alan Irwin and Brian Wynne, and Richard Whitley have included important critiques of the implications of top-down diffusionist models of popularization and suggested new directions for research. The collection of papers from a 1991 AAAS workshop edited by Lewenstein aims to provide an audience-centered approach, as does the collection edited by Irwin and Wynne, *Misunderstanding Science* (1996).

Studies by Marcel LaFollette, John C. Burnham, and Ronald Tobey have provided thorough and important analyses of the patterns of popular science in the United States in the nineteenth and twentieth centuries. Jane Gregory and Steve Miller's *Science in Public* offered an overview of both the history and the historiography up to 1998, and Peter Broks has published a useful overview, *Understanding Popular Science* (2006).

For more on the details of science popularized in a variety of media, see Greg Mitman's (1992) work on science in film and Christopher Frayling's *Mad, Bad and Dangerous* (2005). Bert Hansen's *Picturing Medical Progress from Pasteur to Polio* (2009) is a fine study of the history of popularizing medicine that includes, in particular, analyses of images of medicine in visual media. Robert Rydell's work, beginning with *All the World's a Fair* (1984), incorporates efforts of scientists and science advocates to integrate ideologies of progress, science, race, and empire in the medium of world's fairs and expositions. And finally, Bruce Lewenstein has posted a useful bibliography on his website: http://lewenstein.comm.cornell.edu/science-communication-resources/

Chapter Thirty-Nine

SCIENCE AND POSTCOLONIALISM

Banu Subramaniam

The new social movements in the latter half of the twentieth century in the United States were important not only for their political and social transformations but also for their impact on academia and the production of academic knowledge. Largely identity based, these movements pushed for institutional transformations in academia. Students fought for the hiring of a more diverse faculty, the creation of new programs and depart ments, and the production of knowledge that attended to the histories of domination (Abelove 1993; Boxer 1998; Delgado and Stefancic 2012). The civil rights movement inspired students to fight for the creation of ethnic studies programs; the women's movement motivated women's and gender studies programs; lesbian, gay, bisexual, and transgender (LGBT) groups facilitated the emergence of LGBT and queer stud ies programs. While the transformations resulted in creating new academic programs based on identity categories, their impact has been considerable in both interdisci plinary and disciplinary contexts. Scholars challenged academic knowledge as biased, favoring the interests of the powerful: rich, Western, white, straight, able-bodied men, and ushered in new lines of research that put race, gender, sexuality, class, and ability at the center of analysis (Abelove 1993; Boxer 1998; Delgado and Stefancic 2012). To the list of identity-based movements, efforts by scholars and activists from the third world and formerly colonized nations have brought colonial and postcolonial issues to the fore as an important omission in traditional academic work (Seth, Gandhi, and Dutton 1998).

Writing about science and postcolonialism for a volume on the history of American science is an exercise in irony, as it raises questions about the very framing of the field of science and postcolonialism. Both terms – science and postcolonialism – have been critically and thoroughly interrogated. What do we mean by the term "science"? Main stream representations of "science" refer to a cohesive body of disciplines in the natural sciences that emerged in the West and focused on studying the natural world with a

A Companion to the History of American Science, First Edition.
Edited by Georgina M. Montgomery and Mark A. Largent.
© 2016 John Wiley & Sons, Ltd. Published 2020 by John Wiley & Sons, Ltd.

unique and objective "scientific method." However, starting with Thomas Kuhn (1962: 42) who argued that science was a "rather ramshackle structure with little coherence among its various parts," scholars have increasingly presented science as a heterogeneous body of work created less by coherence and more by historical contingencies. By now scholars generally agree that there is no single enterprise of "science" but rather a large variety of "special studies" that have been grouped together under this label (Dear 2012). Given that the history of colonialism and empire have shaped the hagiography of science as something born and cultivated in the West, we also need to attend to the problems with this history and its many exclusions. Rather than uniquely Western in origin, historians have demonstrated science's embrace and appropriation of knowledges from other civilizations across the globe (Harding 1991, 2012; Schiebinger 2005, 2008; Teresi 2001). There is nothing exclusively "Western" about Western science or its genealogy (Harding 1991, 2008, 2009; Schiebinger 2004, 2008; Teresi 2001). How then should we understand this body of work we call "Western science" and what about other knowledge systems? What do we include and what do we exclude? Why? These are some of the key questions in the postcolonial studies of science.

The other term "postcolonialism" has its own troubles. The "post" in postcolonialism suggests that the term refers to all formerly colonized countries that have gained independence from their colonists. Yet "postcolonial" refers to only some postcolonies. There is an important distinction made between "settler colonies" such as Canada, North America, Australia, and New Zealand and other places where indigenous populations were colonized and subsequently gained independence (Wolfe 1999). It is the economic development of the settler cultures such as the United States that developed new market structures, new political, economic and legal infrastructures where science and technology played a central role, that has shaped global geopolitics and economics (Merson 2001). The term "postcolonial" in the literature thus only refers to the latter. Adding to the confusion, postcolonial is often loosely used simply to mean "non-Western" or the "third world."

Is colonialism something of the past? In the context of the history of the United States, the continued subjugation of Native peoples suggests an ongoing colonialism, or at best, an ambivalent postcolonialism (Cheyfitz 2002). The United States as a strong and active world superpower has been active in many and varied military interventions and "occupations." The political and economic might of the United States has long controlled the world stage creating a highly unequal and globalized world. Scholars suggest that a term like "neocolonial" is better suited to explain the world rather than postcolonial, which implies a degree of political and economic independence for the formerly colonized nations. Indeed, the developing world is if anything more dependent on science and technology of the developed world in these "neocolonial" times than they were in colonial times (Merson 2001).

Despite these complexities, both terms, postcolonialism and science, have come to be widely used. While there is much less work exploring the intersections of science and postcolonialism (in contrast to science and race or gender), but in the last decade, we have seen a rising interest in such work. Several special issues of journals have been devoted to science and postcolonialism (Anderson 2002b; MacLeod 2000; McNeil 2005; Phalkey 2013; Schiebinger 2005; Seth 2009). We have the first edited anthology on the topic (Harding 2012). The *Handbook of Science and Technology Studies* (STS) now includes an entry on science and colonialism (Anderson and Adams 2007). The inclusion of this topic in this current volume also demonstrates the growing importance and increasing attention to the relationship of science and postcolonialism.

How should we understand the relationship of science and postcolonialism? Science and technology have been "the jewels in the crown of modernity" (Harding 2012: 2), central to the expansion of empire and critical to the contemporary world. In many ways postcolonial science studies is a burgeoning field still in formation. There is little agreement among scholars on what we mean by postcolonial science (Abraham 2013), or for that matter, by "colonial science" (Phalkey 2013), the legacies of which profoundly shape postcolonial nations today. While exact definitions may be elusive, most scholars agree that science and technology were critical to the growth of colonialism, as they continue to be in the development of postcolonial countries.

The Legacy of Colonial Science

In tracing the history of contemporary configurations of science and postcolonialism, scholars suggest multiple and different genealogies (Brydon 2000; Harding 2009). However, most agree that we have moved considerably from earlier linear diffusion models of science. In early theorizing about the spread of science, George Basalla (1967) argued that science developed in the West and subsequently "diffused" to the rest of the world. In this model, once science diffused to a nation, it was incorporated in a multi-phased model, each phase helping the nation evolve toward a more modern society. This model has come under severe attack for several reasons (Prasad 2008; Raina 1999). It evokes a center/periphery model where the West remains the center and the originary home of science from which science diffused to the peripheries the world (Raina 1996). In such models, science, rationality, progress, and enlightenment ideals always rest in Europe or the West (Chambers and Gillespie 2001). As Dipesh Chakrabarty (1992) has argued "Europe remains the sovereign theoretical subject of all histories, including the ones we call 'Indian,' 'Chinese,' 'Kenyan,' and so on." Moving away from such Eurocentric histories, most scholars now see this linear diffusion model of science as a gross misrepresentation of colonialism. The Eurocentric narrative where science and technology has only diffused from the West to the rest of the world has also made invisible the many innovative science and technological developments that traveled to Europe and America *from* the peripheries (Arnold 2005; Goonatilake 1998). Rather than understanding the mobility of science as one from the center to the periphery, scholars argue that science and colonialism have been co-constituted by the center *and* the periphery through the transnational circulations of science. We need to attend to simultaneously trace the relationship of colonizers and colonized, of empire and colonized nations to understand the emergence and evolution of science. Colonial science was not a hegemonic European enterprise that "diffused" to the rest of the world, nor a purely multicultural enterprise that assimilated indigenous knowledges from across the world into the European canon. Science was not merely a tool of empire, but constitutive of empire, inextricably connected to the workings and ambitions of empire. Rather, there was a "complex reciprocity" between the colonial powers and colonized lands. Scientific practices were not developed outside of politics but were themselves at the heart of colonialism's political ideologies and institutional structures (Baber 1996; Prakash 1999). In such a reframing, one cannot tell the story of colonialism without science or narrate a history of Western science without colonialism (Baber 1996; Kumar 1995; Raj 2001). Furthermore, in more contemporary histories, the colonized nations and natives are no longer passive victims in the story of colonialism, but ones with agency, active engagement, and resistance. While the colonists presented science as a symbol of "modernity," natives at times resisted such symbols and at other times appropriated and translated science into local contexts in new and interesting ways (Prakash 1999). Ambivalence and hybridity were the hallmark of such engagement. What emerges in these new histories is an entangled history of science and colonialism, making it impossible to tell any neat story of science or colonialism or create a "pure" genealogy of either.

Science is thus best understood as co-constituted with colonialism (Seth 2009), and historians have documented that many sciences such as botany, zoology, ecology, bacteriology, anthropology, psychology, medicine, and mathematics, among others developed alongside the needs of colonial expansion. Science and technology helped colonial conquest through developing more innovative modes of travel and navigation, resource extraction, economic botany, infectious diseases, manufacturing technology, population management and so on (Alexander 2002; Anderson 1995, 2002a, 2006; Schiebinger and Swan 2005). The multiple and varied needs of empire created uneven development of the various scientific disciplines (Arnold 1996; Lemaine et al. 1976; Seth 2009). Sciences should therefore be understood as "sciences of empire" (Schiebinger 2008); indeed almost all modern science should be understood as "science in a colonial context" (Seth 2009). Theories of science were also central to the development of colonial policies and governance. For example, when an outbreak of cholera emerged, it was critical to characterize the nature of the problem. Was cholera a product of local conditions in the colonies and therefore best studied in the colonies or was it caused by a bacterial pathogen and therefore better studied centrally and more thoroughly in the metropole, and the knowledge then transferred to all the colonies? The science of the characterization of cholera was thus central to the development of imperial public policy and colonial management (Arnold 1993, 1996; Harrison 1996, Seth 2009). This reciprocal and co-constituted relationship of science and colonialism is key to understanding colonialism.

A central feature of the colonial enterprise was establishing a logic of biological difference – the ideology that colonists were justified in colonizing another people. Through ideologies of the superiority of Western technology (Adas 1989) and biology (Arnold 1993, 1996; Stepan 1982), colonizers scientifically rationalized colonization. Thus colonial power and policies were fundamentally biopolitical in nature, producing different bodies and keeping them in their "proper" places (Sylvester 2006). The many elisions and contradictions of science are apparent in the literature. In countries like India where the colonial enterprise depended heavily on the involvement of natives for governance, colonizers simultaneously maintained that the natives were worthy objects for their civilizing mission, yet could never be actually civilized completely (Prakash 1999). Natives of mixed blood were always seen as "fixed" and static while the European character was never fixed and was open to growth and change (Stoler 2002). Theories of biological superiority and inferiority produced ideologies of difference – of sex, race, sexuality, class, nation – deeming which bodies were capable of colonizing or of being colonized (Hammonds 1994; Harding 1993; Stepan 1982). The irrationality of the natives, their "hot" temperaments, and promiscuous sexualities always shaped the logic of difference and thus colonial subjugation. The deep misogyny that shaped male colonists is reproduced in policies of citizenship, governance, public and private rights, and of the embodied subjectivities of colonizers and colonized (Anderson 2002a, 2006; Stepan 1982; Harding 2008, 2012). Deeply entrenched in theories of the natural sciences, the legacies of these colonial ideologies of difference continue to haunt contemporary biological theories of difference (Subramaniam 2014).

Science and the Postcolonial

One of the achievements of postcolonial studies as Stuart Hall (1996) argues is that the "post" in postcolonial is "under erasure" making the colonial visible in the contemporary world. Postcolonial studies of science demonstrate how technoscience has been "rephrased within the framework of globalization," making visible colonial ontologies that live on in other guises and names (Anderson and Adams 2007). Understanding science in postcolonial contexts and unearthing the legacies of colonial science and its transformations in postcolonial contexts have been at the heart of postcolonial science studies. Postcolonial studies of science thus offer us "flexible and contingent frameworks" to better understand the travels and multiple "contact zones" of technoscience and empire across laboratories, cultures, societies, and nations, where these encounters across difference have engendered new cultural forms (Anderson 2009; Pratt 1992).

One of the reasons for the immense and enduring power of colonialism is that the elites of postcolonial states embraced science and technology as a central mode of modernity and development (Abraham 1998, 2006; Droney 2014; Roy 2007). There is a continuity between the last colonial civilizing process and contemporary international development projects (Anderson 2009; Escobar 1995). Therefore, contemporary scientific theories and practices continue to show the deep legacies of colonialism (Harding 2012; Traweek 1988). Contemporary science education in postcolonial countries also reinforces models of "Western science" as taught in the West, replicating Eurocentric models of knowledge in postcolonial contexts, naturalizing science as "Western science" (Mutua and Swadener 2004). Early third-world critics of science placed the failure of postcolonial states squarely at the feet of science and technology as a mode of postcolonial national development (Alvares 1992; Nandy 1988; Shiva 1989; Viswanathan 1997). In this more radical critique, postcolonial nations embraced science and technology as the "reason of state" (Nandy 1988), and with it they embraced the "violence" of Western science (Alvares 1992). Arguing that science and technology enabled colonialism, they contend that the third world has been poorly served by modern science and technology philosophies, policies, and practices and has resulted in the economic, political, social, and cultural regress for formerly colonized countries (Harding 2009). An embrace of science and technology is but an extension of the colonial enterprise. As such, postcolonialism continued colonialism rather than ushered in independence or decolonization. The postcolonial had thoroughly internalized the logic of colonialism, so that technoscientific research in the postcolonial state continues to be haunted by colonial and Eurocentric practices (Prasad 2008). As a result, they argue, it should not surprise us that even after independence, postcolonial nations remain marginal to the world of Western science, being the site of little innovation or agency. The center of science remains solely in and under the control of the West, and with only "minor variations" emerging from the third world (Goonatilake 1984). These critics contend that postcolonial nations need to retreat from the hegemonic structures of science in order to truly "decolonize" formerly colonized worlds and their knowledge systems.

Decolonization is important, these scholars argue because it will liberate and develop indigenous, alternate, and local knowledge systems. After all, there remains a rich tradition of alternate and local science and technologies which emerged from ancient civilizations far older than Western science. Here, scholars propose a wealth of local knowledge systems that could replace western science to create a truly liberated postcolonial "decolonized" state. A different set of postcolonial critics argue that this more radical position in many ways reinforces the binaries of center/periphery, West/East, and the notion that Western science and technology is always destructive and oppressive while local knowledge systems are always good and liberatory. Critiquing such simplistic renditions of bad West/good third world, they argue for a more nuanced and complicated reading of the relationship of science and the postcolonial state (Phalkey 2013). The challenge is to move the vocabulary beyond the binaries of West/East, West/rest, colonizers/colonized to understand their co-constitution in postcolonial states, just as in the colonial. We must understand the center and periphery as co-constituted, where the periphery continually makes the "center" (Philip 2004; Raina 1996). Rather than locate the postcolonial in the third world, they have stressed the circulation of an increasing transnational science and the creation of new complex subjects and science across the world that resist easy categorization (Fujimura 2000; Prasad 2014).

"Locating" Science

Joseph Needham has argued that the success of modern science was because while it was uniquely Western in origin, it was at the same time culturally universal and capable of being embraced and used across the globe (Needham 1969). Scholars have since shown that the universal nature of science and technology is an illusion. The illusion of universality is produced because the complex practices of scientific knowledge production and the "sticky materiality of practical encounters" are removed from their contextual understandings and represented as occupying a space of contextless universality (Tsing 2005: 1). As historians developed more textured readings of sciences across the globe, cross-cultural explorations continually demonstrated that science had adapted differently to different contexts (Habib and Raina 1989; Traweek 1988). Arguments about science's universality and homogeneity across the globe have thus been thoroughly critiqued, but geography and place continue to be a centrally important consideration in contemporary postcolonial science studies. Why and how does geography and location matter? These debates get to the heart of the issues in postcolonial science studies.

In making claims for different articulations of science across the globe, some scholars warn that such claims of unique local sciences are dangerous as they suggest that each particular locality produces unique local ontologies linked to knowledge that is singular and essential (Abraham 2006; Anderson 2009). Claims of an "Indian science" or "Hindu science," or "Islamic science," or "Christian science" dissolve into a relativistic and epistemological morass that is untenable. Understanding all knowledge claims as sciences misses something important about the spread of what we call "Western science," and the relationship of power and knowledge (Foucault 1980). Conversely, there is also danger in only claiming a "global" science without attention to the local since all practices are always acting in the local even when they claim to be "doing global" work (Verran 2002). In contemporary times, there is an increasing fetishizing and celebration of the "global," while ignoring the ravages of colonial and neocolonial exploitation for so much of the world. We need to understand the practices of "Western science" as a co-production of global and transnational networks and local laboratories and contexts, while keeping unequal power relations firmly in mind. "Context" has become central to much of contemporary postcolonial studies. Science cannot be reduced to the laboratory as a "fetishized" object unconnected to modes of its production and its social relations entirely erased (Anderson and Adams 2007). Yet at the same time, we need to understand that the sciences emerged within particular contexts for particular purposes and that these knowledges traveled across the globe. While we need a better account of how science and medicine worked in different colonials settings, we also need an account of what is colonial about science and medicine (Anderson 1998).

There is much mention of the local, and especially the indigenous in the postcolonial literature. It is important to note that the indigenous is not the same as local. Rather, the "local" is always continually produced and reinvented (Raj 2006). As Cori Hayden argues, we should understand modern bioprospecting in Mexico as not merely "a channel" for travel, local knowledge, biodiversity, community, and corporate interests, but as sites where bioprospecting contracts produce, invoke, and give shape to new subjects, objects, and interests (Hayden 2003, Seth 2006).

Despite these debates, one coherent aspect about the "postcolonial" that has emerged in the literature is that any such exploration of the postcolonial must begin from the instabilities of the master terms "west" and "rest," and analyses should foreground the co-constitution of west/rest in an unequal but simultaneous global formation (Abraham 2013). In understanding the co-constitution of science, scholars have moved from the center/periphery model to track the movement of science, whether it is through "contact zones" (Pratt 1992), "science in motion" (Prasad 2008), "moving metropolis," (Macleod 2000) or "science as circulation" (Raj 2013) – all signal the dynamism in scientific practice. There are many and varied modernities; not everyone is modern or scientific in the same way, and the us/them framing misses the profound movement and "circulation" of science (Raj 2006). While the majority of colonial and postcolonial scholars track the circulation within colonial circuits, recent work reminds us that we also need to pay attention to "inter imperialities" (Doyle 2014). Contemporary historians have rewritten longstanding narratives of science, religion, finance, trade, labor, and the arts, not as narratives of individual empires but as ones with multidirectional interactions across imperial powers. Such inter-imperialities are dynamic, uneven, and systemic and accretive, and important for understanding the formation of modern histories. Ultimately, we know science travels, but less on why and how (Anderson and Adams 2007). We need to better understand these circulations.

Heterogeneities and Multiplicities

A central contribution of postcolonial studies is showing how dominant Western epistemic practices have privileged certain ways of knowing as the sole and valid basis for the production of knowledge. Yet the relationship of colonizers and colonized nations was more than just dominance and subjugation, but rather one of co-production albeit through grossly unequal power relations (Abraham 2006). Postcolonial developments in their varied theories and practices are important sites to understand the many post-colonialities that emerged, the many modernities they have spawned, even while others suggest that the West has never been modern (Latour 1993)! Indeed the literature is filled with suggestions of multiple modernities (Harding 2008), and this allows us to recognize the tremendous heterogeneity of postcolonial modernity (Appadurai 2001; Sahlins 1999; Strathern 1999; Subramaniam 2000). "Modernity" has become a much deployed and appropriated term in the postcolonial literature. In part this is because dominant "Western" epistemic modes of expression have been embraced across the disciplines, from the sciences to the arts, public policy and governance (Bharadwaj 2013; Raina 2003; Sharp 2009). Challenging such a unitary and universalizing epis-teme, postcolonial scholars have developed an ambivalent relationship to the enlight-enment (Harding 2009), arguing that even "Western" science is not universal or partic-ularly homogeneous. This ambivalence can be seen in the range of responses to the role of science in postcolonial nations. Some wish to "delink" Western sciences to promote local traditions and sciences (Alvares 1992; Nandy 1988; Shiva 1988, 1989); others embrace Western science as part of a postcolonial progressive project (Nanda 1998, 2001). Yet others seek a middle ground in seeking for epistemic pluralism in integrat-ing local knowledges into modern science (Harding 2009).

Ultimately, postcolonial science studies challenges scholars to think of science itself as a "vernacular language," as a form of knowledge "produced through social alliances and contingencies of practical method" (Tsing 2009: 389). It would seem that there are many discontinuities and multiplicities within Western science, as well as local practices across the globe. These discontinuities are not the exceptions, but rather a central and recognizable part of the history of science, colonialism, and postcolonialism. Tracing the continuing and discontinuities of a circulating global and transnational science is the project of contemporary postcolonial science studies.

"De"colonizing Knowledge

While there may be many disagreements within postcolonial science studies, one thing is clear – postcolonial does not constitute decolonialization. Rather in many ways post-colonial is an extension of colonialism. Secondly, postcolonialism has at many times been far from liberatory, instead it has presented its own troubled histories of hierarchy and difference (Abraham 2006; Sur 2011). Neither postcolonial nor the indigenous as categories are homogeneous, just as not all nations or peoples share the same goals or interests (Hayden 2006; Reardon 2012).

Just as studies of colonial science unearthed the profound logic of difference that kept colonial hierarchies of race, gender, and sexuality in place, postcolonial science reveals how these logics of difference continue through the power structures in postcolonial states. Hierarchies of race, class, caste, sexuality, and nation are continually reproduced transnationally as well as in postcolonial nations (Hammonds 1999; Hammonds and Herzig 2009). Scholars have powerfully documented how difference is continually produced through scientific experimentation and the commodification of bodies and body parts, rendering postcolonial bodies as "disposable" commodities. The bodies that inhabit postcolonial nations continue to be spaces that are in the front lines of scientific research in a range of fields including pharmaceuticals, genomics, environmental pollution, nuclear development, reproductive technologies, and population control. The globalized character of transnational science and policymaking finds a whole host of actors across the first and third worlds who enable such circulations and exploitations. These transactions, however, function through the uneven grounds of politics across the globe and through differential power structures in international negotiations. Ultimately postcolonial states and local actors are bound by the universalizing logic of science and capitalism, creating very unequal partners and results (Adams 2002).

The contributions of postcolonial (also feminist science studies and critical race studies) in revealing the oppressive legacies of colonialism and science have also become embroiled in other fault lines of contemporary politics. In critiquing the unitary and universal power of science and in elevating indigenous knowledges as legitimate or even alternate sciences, new sites of power open up. Religious fundamentalisms and nationalisms have sought to occupy these sites as equivalent to science. In the United States, we have creation scientists, climate deniers, anti-choice activists who claim a "scientific" basis to their claims. Thus some social activists have critiqued postcolonial and feminist scholars for creating a scapegoat in science and opening up new spaces for assaults on liberty and freedom. Giving up science as a site of reason, they argue, fails to tap into the liberatory power of science for progressive social movements, suggesting that postcolonial science studies has done a disservice to the third world (Nanda 1998, 2001). Such arguments ignore the oppressive history and legacies of colonial and postcolonial science.

Despite the relatively recent focus on postcolonialism in science studies, and the still nascent field, and the deep theoretical debates that I have enumerated, scholars and activists have played an important role in exposing the entrenched racism and Eurocentrism that continue to shape the contours of academic science. They have also been important in pointing to the long histories of oppression that continue to shape postcolonial nations through the continued scientific production of biological difference – of sex, gender, race, caste, sexuality (Hammonds 1994; Harding 1993; Stepan 1982). Activists and scholars also point to the continued exploitation of the third world through the commodification of biological material, patenting of biological life, and as sites of experimentation and exploitation. Recent work in postcolonial science studies has also moved beyond expanding mainstream academic theories and their Eurocentric focus to incorporate postcolonial issues as centrally important considerations of knowledge and as sites that generate new questions, approaches, and theories (Philip, Irani, and Dourish 2010). While the work in postcolonial studies has largely been human-centered, some work in the environmental literature especially, has documented the profound ways in which non-human biota have been caught in the imperial travels and postcolonial migrations of humans in a globalized world (Crosby 1986; Grove 1995; Haraway 2008). These new narratives of non-human life promise to be an important and productive site to understand the porous nature of world borders. It is this decolonizing potential of postcolonial studies that continually challenges the Eurocentrism of the West, as well as the elites of postcolonial nations that lends hope to the liberatory potential of postcolonial science studies and a more just world.

Bibliographic Essay

The terms postcolonialism and science have been open to multiple genealogies and histories, but both are now widely used by scholars. However, the scholarship in each signaled the importance of the other. Science was key to colonial expansion, and became centrally important to the development of postcolonial nations. Similarly, the development of science was shaped profoundly by colonialism and postcolonialism. In recent years, we have seen a rise in work exploring the intersections of postcolonialism and science. Several special issues of journals have been devoted to the theme: While there is much less work exploring the intersections of science and postcolonialism (in contrast to science and race or gender), in the last decade, we have seen a rising interest in work relating to science and postcolonialism. Sandra Harding's *The Postcolonial Science and Technology Studies Reader* (2012) would be a good beginning to some of the current issues that energize the field. Several special issues of journals have been devoted to science and postcolonialism, each with useful introductions framing the field: Roy MacLeod's "Nature and Empire" in *Osiris* (2000), Warwick Anderson's "Postcolonial Technoscience" in *Social Studies of Science* (2002); Maureen McNeil's "Postcolonial Technoscience" in *Science as Culture* (2005); Londa Schiebinger's "The European Colonial Science Complex" in *Isis* (2005b); Itty Abraham's "The Contradictory Spaces of Postcolonial Technoscience" in *Economic and Political Weekly* (2006); Suman Seth's "Putting Knowledge in its Place" in *Postcolonial Studies* (2009); Jahnavi Phalkey's "Science, History, and Modern India," in *Isis* (2013a). The Society for the Social Studies of Science produces a *Science and Technology Studies Handbook* that now includes entries on science and postcolonialism (Anderson and Adams 2007). Also influential have been Susantha Goonatilake's *Aborted Discovery* (1984), Michael Adas' *Machines as the Measure of Men* (1989), Zaheer Baber's *The Science of Empire* (1996), Kavita Philip's *Civilizing Natures*, and Sandra Harding's *Sciences from Below* (2008)

In addition to the broad framing of the issues that concern postcolonialism and science, the field is also based on explorations of individual countries, since the issues that emerge seem very dependent on colonial histories and regional geopolitics. The scholarship from India is particularly rich and well developed. For Asia, please see the works of Itty Abraham, Warwick Anderson, David Arnold, Susantha Goonatilake, Irfan Habib, Deepak Kumar, Ashis Nandy, Jahnavi Phalkey, Kavita Philip, Amit Prasad, Dhruv Raina, Kapil Raj, and Anna Tsing; for the Americas, see Laura Briggs, Arturo Escobar, Evelynn Hammonds, Cori Hayden, Valerie Kuletz, Elizabeth Roberts, Lindsay Smith, and Kim Tallbear; for Africa, see Ruha Benjamin, Phillip Curtain, Laura Foster, Gabrielle Hecht, Anne Pollock, Ann Stoler, Richard Rottenburg, and Helen Verran.

Just as colonialism was predicated on the production of biological differences of different groups, hierarchies of difference persist in postcolonial nations. Questions of race, class, caste, sexuality, and nation continue to haunt postcolonial nations and scientific knowledge production (Hammonds 1999; Hammonds and Herzig 2009). A range of analyses document how questions of difference are produced in the different sciences. Scholars have powerfully documented how science under postcolonialism continues to produce bodily differences rendering some bodies "disposable" and commodifiable. Thus, marginal populations continue to be the sites for "experimental" bodies for new pharmaceutical markets (Foster 2012; Kahn 2005; Pollock 2014; Roberts 2012; Sunder Rajan 2006), body parts (Cohen 2005; Reardon 2013; Scheper-Hughes 2000), experimentation (Bharadwaj 2013; Briggs 2002; Petryna 2009, Sunder Rajan 2006, Takeshita 2011), environmental toxicity (Fortun 2001; Kuletz 1998; Petrayna 2013), nuclear development (Hecht 2002), genomics (Benjamin 2009; Egorova 2013; Nash 2012; Reardon and Tallbear 2012; Tallbear 2013) and population control (Bharadwaj 2012; Hartmann 1999; Towghi and Vora 2014).

Chapter Forty

RACIAL SCIENCE

Robert Bernasconi

In 1864, the French philosopher and anthropologist Georges Pouchet wrote that "today France and England walk entirely on the scientific path opened by the American school" (Pouchet 1864: 10). He was referring to the work of Louis Agassiz, Samuel George Morton, and Josiah Nott, and their promotion of polygenesis, the theory that the different races had different origins. Nevertheless, only six years earlier, in an earlier edition of the same book Pouchet had written: "in France and in England the old ideas are still dominant" (Pouchet 1858: 9). A similar shift in the status of polygenesis had been noticed 20 years earlier in the United States. In 1842, *The United States Magazine and Democratic Review* had affirmed without question the unity of the human race as a single species, but only eight years later an anonymous reviewer of W.F. Van Amringe's mammoth *Theories of the Natural History of Man* in the same journal declared that "Few or none now seriously adhere to the theory of the unity of races" (Anonymous 1850: 328; Horsman 1981: 145–7). Even allowing for some exaggeration on the part of these reports, and the fact that it neglects dissenting voices (for example, Waitz 1859: 430), the question inevitably arises as to how the American school of polygenesis came to hold such dominance, first at home and then internationally.

The international success of American polygenism is easier to explain than its domestic success. American natural historians had long been able to claim a first-hand knowledge of racial diversity that most of their European colleagues often lacked (Smith 1810: 210). They were also more deeply invested in racial issues. Anténor Firmin, a Haitian who was a member of the Société d'Anthropologie de Paris and so knew the French scene at first hand, speculated that the American school of polygenesis was motivated by a desire to maintain slavery at all costs, whereas French scientists were drawn to advocate polygenesis because they wanted to separate science from religion and were suspicious that in previous generations many scientists had argued for monogenesis more on biblical than scientific grounds (Firmin 1885: 51).

A Companion to the History of American Science, First Edition.
Edited by Georgina M. Montgomery and Mark A. Largent.
© 2016 John Wiley & Sons, Ltd. Published 2020 by John Wiley & Sons, Ltd.

Although it is tempting to see the defense of slavery as the driving force behind American polygenism, and this seems to have been the case for Josiah Nott, neither Agassiz nor Morton fit this model. Indeed, it is hard to find references in Morton's writings to the political problems of the day (Stanton 1960: 223 n4). Most authors stayed well away from the issue in their scientific writings, addressing it only in other contexts. There was a strong tendency for polygenists to favor slavery, in large part because they were likely to think of the racial characteristics of the so-called inferior races as permanent, but there was no straightforward correlation between what a given scientist believed about the unity of the human races and whether that same scientist supported slavery. So both Reverend James Bachman of Charleston, the foremost champion of monogenesis in the United State in mid-century, and James Cabell, professor of comparative anatomy at the University of Virginia, argued at length against Agassiz and Nott in favor of the unity of humanity, but they still defended slavery. Beginning in the mid-1830s those scientific writings were culled by advocates on both sides of the debate on abolition, but what was perhaps the first sustained appeal to science to support slavery was an import. In 1837 Guenebault published in translation selections from the second edition of Julien-Joseph Virey's *History of Mankind*. Virey was a polygenist who supported slavery and by choosing to extract passages from Virey's book that favored both causes, Guenebault was making an intervention in the debate. To be sure, the arguments for slavery that Guenebault found there were not new, but it was clearly meant for them to have a greater weight simply because their author was a renowned scientist (Guenebault 1837: 120–7; Virey 1824: 65–83). In short, slavery as an institution determined what American natural historians saw of blacks, but by and large its defense did not drive their research.

Nevertheless, although biblical quotations generally carried more weight than scientific citations during the debate on slavery, the fact that both sides were increasingly ready to appeal to the authority of science reflects the growing prestige of science, however crude. Slavery had not been challenged in the United States at the end of the eighteenth century to the degree that it had been in Britain and France, but racial science was at that time in its infancy and so played a smaller role in the debates. Science was a latecomer to the debate about the status of Negroes and the Native Americans. Nobody felt the need to appeal to it before alleging that both of these races were inferior. Indeed, the dominant sense of race in Europe in the nineteenth century would be better described as historical rather than biological; that is to say, it was based on the history of nations. It should be remembered that anthropology was still in its infancy and a strict division between cultural and physical anthropology was a product of the twentieth century so that it is anachronistic to impose it on an earlier time. But it was the politicians, not the scientists, who had the most to say about the meaning of racial identities. The legal definitions of who counted as white, black, or Native American was frequently changing so that racial identification was usually more about the borderline between races than about racial essences.

The scientific issue that natural historians could not avoid was not slavery, but survival. The population decline of the Native Americans meant that there was a need to find a justification for the genocidal policies directed against them. The survival of African Americans was also debated, although at first it might have been no more than a variation on Thomas Jefferson's fantasy that there might at some indeterminate future time be a way of removing all blacks "to other parts of the world" (Jefferson 1984:

264). In 1854, Frederick Douglass was responding to such ideas when in "The Claims of the Negro Ethnologically Considered" he declared that African Americans could not be colonized or exterminated and that they would not die out: "His tawny brother, the Indian, dies, under the flashing glance of the Anglo-Saxon. *Not* so the negro; civilization cannot kill him" (Douglass 1982: 524). Some even tried to defend slavery as a means to save African Americans from their potential destruction. Cabell suggested that "the actual bondage of the blacks in America was not intended, in the merciful and wise providence of God, as the only means of extricating them from their otherwise inevitable 'destiny:'" it was a way of "bringing them under the tutelage of a superior race without danger of becoming 'extinct before' such higher race" (Cabell 1858: 267n).

Miscegenation and Sexual Repugnance

At the beginning of the nineteenth century, American natural historians were very much on the margins of the racial discourse that had began in the previous century by Buffon, Kant, and Blumenbach, but on one central issue Americans could speak with greater authority. It was the question of adaptation, the impact of moving from one environment to another. Three contributions stand out when one looks back on the early years of race science in North America. First, John Mitchell wrote a groundbreaking essay on skin color in 1744 that concluded with a defense of the unity of the human species based on the idea that skin color in North America was a consequence of whatever was "most suitable for the Preservation of Health and the Ease and Convenience of Mankind in these Climes and Ways of Living" (Mitchell 1744: 146). Secondly, Williams Charles Wells, who at the end of his life lived in England but was originally from Charleston and who has been seen as a forerunner of Charles Darwin, wrote an essay in which he argued that accidental varieties of man that were well suited to combat disease in a given location would prosper there and that this could be used to explain the black skin and wooly hair of Negroes (Wells 1818: 432–9).

Finally, there was Samuel Stanhope Smith, who was professor of moral philosophy of the College of New Jersey, which would soon become Princeton College. He was the most influential of the three. In 1787 Smith published a defense of the idea of the unity of human species against Henry Home, Lord Kames, who had argued that the different races were "fitted by nature for different climates" as could be seen from the fact that in Charlestown [sic] Europeans "die so fast that they have not time to degenerate" (Home 1774: vol. 1, 11). Although the thought that each race somehow belonged in its own place was revived in 1850 by the Scot, Robert Knox (1850: 75–99), it was not one designed to appeal to the colonists. Smith found an answer in the idea that "nature has given such pliancy to the human constitution as to enable it to adapt itself to every clime" (Smith 1787: 6). However, Smith went beyond attempts to explain human diversity in terms of the effects of climate by adding a second instrument that he called "the state of society." He did not recoil from the conclusion that the descendants of "Anglo-Americans" who stayed long enough in North America would take on the same skin color as "Native Indians," but with the important proviso that this would happen only if they shared a similar "savage state" with them (Smith 1787: 59–60n). The case of Africans transported to North America proved more difficult to accommodate to the theory. Smith conceded that there was no evidence of their skin color lightening in North America, but he insisted that other physiological features of domestics slaves were in the process of being transformed to grow closer to those of their masters, whereas those of field slaves were not, thereby confirming the importance of the state of society as a factor in their adaptation (Smith 1787: 57–8). In 1810 Smith, who was by now president of Princeton, issued a heavily revised and expanded version of the book that took account both of criticisms and, more importantly for Smith's standing, of European authors like Blumenbach whom he had ignored in the first edition. While repeating his assertion that the black complexion of "American negroes" was not growing lighter, he now claimed that it was highly probable "that time will efface the black complexion in them" (Smith 1810: 255n).

Smith's argument seemed to point to the possibility that in the distant future the distinction between the races based on their alleged inequality would disappear. This refutes the widespread idea that in this time period a strong racial essentialism was universal. Throughout the century there were always a few, like Firmin (1885: 404–20), who held Smith's view. Others, like Frederick Douglass, thought the dissolution of the races and the formation of a new race could be accomplished by race mixing: "My strongest conviction as to the future of the negro therefore is, that he will not be expatriated nor annihilated, nor will he forever remain a separate and distinct race from the people around him, but that he will be absorbed, assimilated and will only appear finally . . . in the features of a blended race" (Douglass 1886: 438). By contrast, polygenesis clearly lent itself to a form of essentialism insofar as one of the more persuasive arguments in its favor was the idea that racial differences appeared to be permanent.

Charles Caldwell, a physician who went on to play a major role in establishing Kentucky as a center for medical education, was among the first to challenge in detail the evidence Smith used to support his environmentalist account of race. For example, Caldwell disputed Smith's observation that the fact the features of domestic slaves were more agreeable was to be explained by their proximity to the families for whom they worked. It was not a result of performing their duties that certain negroes had "become active, handsome, and pleasing domestics;" rather they had been selected to serve as domestic slaves because they were "previously agreeable both in feature and figure" (Caldwell 1814: 27). In 1830 and again in an expanded version in 1852 Caldwell moved beyond Smith to take on the English ethnologist, James Cowles Prichard, the foremost advocate of monogenesis as a scientific, rather than biblical, truth (Caldwell 1852: 17–50; Prichard 1826: 2, 584–95).

In 1839 Samuel George Morton came to Caldwell's assistance by using the evidence of his unsurpassed collection of skulls to show that neither caucasians nor negroes had changed in 3000 years. Relying on the widespread belief that it was not much more than 4000 years since the Creation, he concluded that there was not enough time for the caucasian and negro races to arise from a single origin given their differences (Morton 1839: 88). A similar argument was developed by Peter A. Browne who used samples of human hair from Egyptian mummies to confirm the permanence of racial characteristics to argue against the environmentalists with their claims about the power of adaptation (Stanton 1960: 149–54). If the races were unchanged over 3000 years or so, then the argument could also be made that the hierarchy that saw the white race at the top might be permanent, especially if the prevailing view that the ancient Egyptians were black could be toppled.

When the scientific concept of "race" was formulated in the eighteenth century, Buffon's rule for species identification, according to which two animals that can produce fertile offspring by procreating together belong to the same species, was in place and secured a scientific basis for monogenesis. Nevertheless, the number of possible counterexamples was mounting and as a result in 1847 Morton challenged Buffon's rule as offering "no proof of the unity of the human species" (Morton 1847: 211). He was thus able to embrace "the doctrine of *primeval diversities* among men – an original adaptation of the several races to those varied circumstances of climate and locality which, while congenial to the one are destructive to the other" (Morton 1847: 40n).

After Morton's death in 1851, Josiah Nott and George Gliddon prepared a volume dedicated to his memory, *Types of Mankind*. Part of its success was due to the inclusion of an essay by Louis Agassiz of Harvard University in which he declared with greater clarity than hitherto that "what are called human races, down to their specialization as nations, are distinct primordial forms of the type of man" (Agassiz 1854: lxxvi). Since his arrival in the United States from Switzerland in 1846 Agassiz had established a reputation as the foremost scientist in the country, and while it is something of a surprise that Agassiz so readily allowed himself to be used in this way by these two polemicists who wore their extra-scientific agenda on their sleeves, the fact that he did so enhanced their cause immeasurably.

Nott, a medical doctor in Mobile, Alabama, looked for arguments for polygenesis wherever he could find them and, unlike most of the figures involved, he was open about the stakes as he saw them. So in an essay that addresses the question of the race of the Egyptians, he announced: "I must show that the Caucasian or White and the Negro races were distinct at a very remote date, and that the Egyptians were Caucasians. Unless this point can be established the contest must be abandoned" (Nott 1844: 8). Nott did not on that occasion specify the "contest" but his audience would have known that he meant the perpetuation of the Southern way of life. Whether he was writing about the Jews, the ancient Egyptians, or the latest discoveries of Morton, his motivation was the same. One essay that proved especially influential was "The Mulatto a Hybrid – Probable Extermination of the Two Races if the Whites and Blacks are Allowed to Intermarry." First published in 1843, but with the main conclusion widely circulated in *Types of Mankind* (Nott and Gliddon 1854: 373), it was largely based on some dubious statistics and implausible observations, but the thesis announced in the title found a ready audience. He would elsewhere report the standard arguments against race mixing, such as, for example, that "the superior races ought to be kept free from all adulterations, otherwise they will retrograde, instead of advancing, in civilization" (Nott and Gliddon 1854: 405). Even Bachman suggested that in the United States "the admixture of a superior race with an inferior . . . in almost every case results in degradation and crime" (Bachman 1850: 106). But in "The Mulatto a Hybrid" Nott was one of the first to argue for racial purity on biological grounds. Until then, according to the highest authority, race mixing improved the stock (Prichard 1836: vol. 1, 148–9). Nott's basic idea was that to be mixed race was to be susceptible to chronic diseases and infertility to the point where it amounted to a disease in its own right (Nott 1844: 253). Agassiz gave support to Nott on this point from beyond the grave when his wife published a selection of his letters. He was quite specific that the very existence in the United States of "the half-breeds" was "only transient and that all legislation with them should be regulated with this view" (Agassiz 1885: 608), but at the same time he departed from the widespread view that all inferior races would disappear when confronted by a superior race. The "pure black" must be considered as permanent an inhabitant of North America as the white race (1885: 598–600). It was a mistake to be guided by the disappearance of the "American Indians" because that arose from their "peculiar character" (1885: 597). The new racism did not necessarily reject the idea of superior and inferior races, but the hybrid was now at the bottom of the hierarchy.

Nott's challenge to Buffon's rule may have been a great deal less insightful than Morton's but its impact was far-reaching. Count Oscar Reichenbach offered a modified version of Nott's argument, albeit without acknowledgment (Reichenbach 1864). Paul Broca, who as the founder of the Société d'Anthropologie de Paris and the author of numerous important works can genuinely be described as an eminent scientist, called Nott "one of the most eminent anthropologists in America" and quoted at length the conclusions of his 1843 article, as Jean Boudin, a French physician, had done earlier (Boudin 1857: II, 220; Broca 1860: 400–1). Broca extended Nott's argument to include the impact of immigration when he described how even though the Anglo-Saxons were still dominant in the United States, the influx of many other races raised the prospect that it might soon be inhabited by a hybrid race with all the attendant dangers (Broca 1859: 602). Soon such concerns were leading Americans to impose immigration controls, beginning with the Chinese, and here too the United States was in the vanguard.

Buffon's rule was in trouble even before Nott and Morton attacked it. Prichard had tried to uphold it by reducing it to a natural sexual repugnance between individuals of different kinds. This is what stopped wild animals from mixing across species thereby creating "a scene of confusion" (Prichard 1826: vol. 1, 97–8). Repugnance was nature's mechanism for avoiding hybridity and yet Prichard conceded that this mechanism could be circumvented when animals were no longer under natural conditions but had been domesticated. In his response to Prichard, Caldwell called into doubt Prichard's claims about repugnance (Caldwell 1852: 39–40), but Morton saw a way to turn this alleged sexual repugnance against monogenesis by using it as evidence of nature's intent (Morton 1847: 211). If the mutual repugnance that existed between Europeans and Africans had "only been partially overcome by centuries of proximity, and, above all other means, by the moral degradation consequent to the state of slavery," then this was further evidence in favor of polygenesis (Morton 1847: 210).

The idea of such a natural repugnance for sex across the racial divide gained traction in spite of successive generations of white masters raping their black slaves. In his monumental essay on inequality between the human races Joseph Arthur de Gobineau added a new twist. In addition to a law of repulsion "from the crossing of blood," he posited a law of attraction that the white races felt for the other races (Gobineau 1983: 167; Young 1995: 107). Indeed it was by acting on this sexual attraction that civilization arose (Gobineau 1983: 303, 342–4). Whereas Guenebault had not hesitated to include in his translation Virey's remarks that once their antipathy and disgust for negresses had worn off European men suffered from "a violent and almost morbid fascination" with them on account of their "great lasciviousness" (Guenebault 1837: 88), when Nott arranged for parts of Gobineau's book to be translated for an American audience, Gobineau's observations about the white man's attraction for women of other races were omitted (Gobineau 1856). Combined with Nott's conviction about the "the probable ultimate infertility of human hybrids of the mulatto type" the "aversion to hybridity" was nature's way of stopping the two races "destroying one another by amalgamation" (Wright 1860: 3, 13). Suitably cultivated this aversion could be employed to support efforts to "exclude blacks and mulattoes from all political privileges" (Wright 1860: 13). To defenders of the Southern way of life, aversion to racial amalgamation and the dire effects that followed from it when it did take place were providential. Louisa S. McCord of South Carolina in the context of a review of Knox's *The Races of Men*, declared extermination or slavery to be the destiny of the darker races as ordained by God, "who, for his own purposes, and in his impenetrable wisdom, has so formed the weaker race that they dwindle and die out by contact with the stronger; has so formed the stronger that they instinctively repel the thought of amalgamation with the weaker; has so formed both, that amalgamation leads to extinction" (McCord 1851: 414). But the idea of such a repugnance was not restricted to the polygenists. In 1850 Reverend Thomas Smyth, confirming Morton's claim that it was mutual, argued that it supported the idea of the equality and thus of the unity of the human races (Smyth 1910: 345).

Race mixing between whites and Native Americans was viewed differently. Roswell Haskins quoted Caldwell as arguing that Native Americans, whom he regarded as a branch of the Mongolian race, came close in cerebral development and general character to the degree of "white blood" they possessed. Mixing with whites was their only hope for survival: "The only efficient scheme to civilize the Indians is to cross the breed. Attempt any other and you will extinguish the race (Haskins 1839: 111). That was also Nott's view about Native Americans: "the full blood Indian" faced extermination with the approach of civilization and "whatever improvement exists in their condition is attributable to a mixture of races" (Nott 1844: 38). But the growing consensus was that neither mixing, nor anything else, could help the Native American (Kneeland 1851: 92), with the result that "in a few generations more the last of these Red men will be numbered among the dead" (Nott and Gliddon 1854: 69). Although Henry Morgan in his important study of the Iroquois complained about the treatment they had received and insisted on their potential for "civilization" (Morgan 1851: 33, 143), the longstanding view was that every "savage" stood in the way of "the perspective into futurity" by occupying the land that could be occupied by "five hundred of national animals, . . . the extinction of his race, and the progress of the arts which give rise to his distressing apprehensions, are for the increase of mankind, and of the promotion of the world's glory and happiness" (Sullivan 1795: 139).

Eugenics and Biopolitics

At the end of the nineteenth century American scientists of race would again be on the margins. The decline of the American school had a double cause. The most important was the publication of Charles Darwin's *On the Origin of Species* in 1859. Polygenesis derived its strength and its purpose from the idea that the races were fixed, but that was no longer sustainable within an evolutionary perspective, even though a certain polygenesis still survived, as did a certain racial essentialism (Stocking 1982: 42–68). The debate between monogenists and polygenists continued but it was no longer the "grand problem" Nott had declared it to be (Nott and Gliddon 1854: 50). Nevertheless, the question of the possible extinction of so-called inferior races gained in intensity and now the authority of science seemed to blend with divine authority as when

Herbert Spencer in a book on ethics announced: "if it be said that as the Hebrews thought themselves warranted in seizing the lands God promised to them, and in some cases exterminating the inhabitants, so we, to fulfill the 'manifest intention of Providence,' dispossess inferior races whenever we want their territories; it may be replied that we do not kill many more than seems needful, and tolerate the existence of those who submit" (Spencer 1879: 240). It is easy to forget how callous some of the preeminent minds of the time were.

At the same time the biopolitical racism that Nott had introduced declared that nature punished those individuals and races that did not follow its dictates. However, the scant or nonexistent evidence was now being argued with a greater pretense at rigor. The 1860 census was presented as confirming Nott's claims. The extinction of slavery would lead to increased "admixture" and this would impair "the colored race" without improving it morally. It could follow the same path as Native Americans and was "doomed to comparatively rapid absorption or extinction" (Kennedy 1864: xii). In the 1890s Frederick Hoffman, a statistician with the Providential Insurance Company, declared the negro in America "a vanishing race": "The Indian is on the verge of extinction, many tribes having entirely disappeared; and the African will surely follow him, for every race has suffered extinction whenever the Anglo-Saxon has permanently settled" (Hoffman 1892: 531). This essay and Hoffman's subsequent publications provoked Du Bois to write his classic essay "The Conservation of Races," a work which confirms that race thinking in the nineteenth century was as much a historical as a scientific category (Bernasconi 2009: 533–4; Du Bois 1897).

The new biopolitical racism was most visible in eugenics. The heyday of eugenics in the United States was the first four decades of the twentieth century, but it was already prepared for in the nineteenth. Gideon Lincecum, a Texas naturalist, was among the first in the United States to advocate sterilization as a way of improving the population when in 1849 he advocated the castration of criminals in place of execution: "The laws of hereditary transmission cannot be overruled" (Burkhalter 1965: 96). Some proposed castration as the solution to the perceived problem of the sexual crimes of "negroes" (Largent 2008: 25–7). Nevertheless, in the United States, whites were sufficiently confident in the boundaries that had been established by Jim Crow legislation: the expectation was that race mixing would end, especially given the tendency for the states to define the white race ever more narrowly in the direction of racial purity which was now equated with health. Once the barriers separating the races were established, the focus of eugenics was on identifying the obstacles to improving the white race. In a groundbreaking study, Robert Dugdale drew attention to increased instances of degeneracy among whites. In a study that drew praise from both Francis Galton, the founder of eugenics and Spencer, who favored the operation of the "survival of the fittest" over charity (Galton 1883: 63; Spencer 1884: 69n), he examined degeneracy across seven generations of a family to which he gave the name "the Jukes" and argued that although environment contributed to the problem, the "tendency of heredity is to produce an environment which perpetuates that heredity" (Dugdale 1877: 65).

In the early part of the twentieth century the center of racial science which had moved to France shifted back to Germany where Eugen Fischer became the first to apply Mendelian laws of inheritance to study the effects of race mixing. He concluded that the Boers and Hottentots should not mix further (Fischer 1913: 305–6). In time the Germans turned to such Americans as Harry H. Laughlin and Charles Davenport, both of them strong advocates of eugenics as well as strict immigration controls. However, increasingly the influence of Franz Boas and his students came to be felt. His insistence on a distinction between nature and culture led to a division between physical anthropology and cultural anthropology, which led to physical characteristics being separated from mental characteristics. Race became the preserve of biologists. Eventually this movement led to the UNESCO Statement on Race of 1950 written under the leadership of one of Boas' students, Ashley Montagu. When in the aftermath of the holocaust the world walked again on the scientific path opened by American scientists it was to deny the salience of the scientific concept of race.

Bibliographical Essay

The first version of Samuel Stanhope Smith's *An Essay on the Causes of the Variety of Complexion and Figure in the Human Species* was published in Philadelphia in 1787. This edition was reprinted in London in 1789, but in 1788 a corrected version was published in Edinburgh with the addition of some notes by "A Gentleman of the University of Edinburgh," who was in fact the American naturalist Benjamin Smith Barton. This edition was recently reprinted as volume 6 of *Concepts of Race in the Eighteenth Century* (8 volumes), edited by Robert Bernasconi (2001). The 1810 version was republished with an extensive introduction by Winthrop D. Jordan under the same title in 1965 by Harvard University Press. Some of the main works of the polygenists Samuel George Morton, Charles Hamilton Smith, and Nott and Gliddon (including *Types of Mankind*) have been reprinted together with Bachman's *The Doctrine of the Unity of the Human Race* and Hotz's translation of Gobineau with an appendix by Nott in a seven volume set under the title *American Theories of Polygenesis*, edited by Robert Bernasconi (2002). Various essays by Louis Agassiz, Josiah Nott, John Bachman, Frederick Hoffman, and W. W. Wright were republished in *Race, Hybridity, and Miscegenation*, three volumes, edited by Robert Bernasconi and Kristie Dotson (2003): Nott's "The Mulatto a Hybrid" is reprinted in the first volume, *Josiah Nott and the Question of Hybridity*. Louisa S. McCord's writings on race were reprinted in *Political and Social Essays*, edited by Richard C. Lounsbury (1995). Anténor Firmin's *De l'égalité des races humaines* was reprinted as volume eight of *Race and Anthropology*, edited by Robert Bernasconi (2003). It has been translated by Asselin Charles as *The Equality of the Human Races* (2000).

Charles Davenport's most important studies in this context are *Heredity in Relation to Eugenics* (1911) and, with Morris Steggerda, *Race Crossing in Jamaica* (1929). Harry H. Laughlin's most important books are *Eugenical Sterilization in the United States* (1922) and *Immigration and Conquest* (1939). Franz Boas republished many of his most important essays in 1940 under the title *Race, Language, and Culture*. Some of Boas' most important early writings were collected by George W. Stocking as *The Shaping of American Anthropology 1883–1911* (1974). In 1950 Earl W. Count of Hamilton College, who had had a strong connection with some Nazi theorists in the 1930s, edited an international collection of writings on race from Buffon and Kant to the 1940s: *This is Race* (1950). The collection is still useful in spite of its questionable origins. It includes selections from two significant figures: Earnest Albert Hooton and Carleton Coon. Nobody played a larger role in shaping the new postwar orthodoxy of the Boasian school than Ashley Montagu. His *Man's Most Dangerous Myth. The Fallacy of Race* (1942) went through six editions over 55 years and in the process grew from 216 pages to 499.

William Stanton's *The Leopard's Spots. Scientific Attitudes toward Race in America, 1815–59* (1960) introduced a whole generation of historians to the writings of Charles Caldwell, Samuel George Morton, George R. Gliddon, and others. Building upon it and almost equally valuable is Reginald Horsman's *Race and Manifest Destiny* (1981). Another survey that is useful is Thomas Gossett's *Race: the History of an Idea in America* (1995). A brief survey of racial classifications can be found in Gloria A. Marshall's "Racial Classifications" in *The 'Racial' Economy of Science*, edited by Sandra Harding (1993). On the relation of identifications of someone's race based on appearance as opposed to descent, especially in a legal context, see Ariela J. Gross, *What Blood Won't Tell* (2008). I have developed her idea that the margins of a category constitute the core in "Crossed Lines in the Racialization Process: Race as a Border Concept" (Bernasconi 2012).

The religious dimension of the discussion is covered by Colin Kidd in *The Forging of Races. Race and Scripture in the Protestant Atlantic World, 1600–2000* (2006). On Firmin, see Robert Bernasconi (2008), "A Haitian in Paris: Anténor Firmin as a Philosopher against Racism." The most scholarly survey of eugenics in the US and UK contexts is Daniel T. Kevles, *In the Name of Eugenics: Genetics and the Uses of Human Heredity* (1995). More sensational but informative about Davenport and Laughlin is Edwin Black's *War Against the Weak: Eugenics and America's Campaign to Create a Master Race* (2003).

For the early period Alden T. Vaughan's collection of essays throws light on the way Native Americans as well as African Americans were perceived: *Roots of American Racism: Essays on the Colonial Experience* (1995). The relation between racism and sexuality in the nineteenth century is explored by Robert J.C. Young in *Colonial Desire. Hybridity in Theory, Culture and Race* (1995). I have connected the concern with race mixing in the North American context to Foucault's conception of biopower in "The Policing of Race Mixing: The Place of Biopower within the History of Racisms" (2010). For the radical transformation that the concept of race underwent in the United States through the impact of Franz Boas and his students, see Elazar Barkan, *The Retreat of Scientific Racism* (1993). Milford Wolpoff and Rachel Caspari wrote an introductory but still useful book that includes extensive treatments of Earnest A. Hooton and Carleton Coon: *Race and Human Evolution* (1997). Hooton is discussed in a more scholarly fashion in Pat Shipman, *The Evolution of Racism: The Human Differences and the Use and Abuse of Science* (1994).

There are informative biographies of some of the main figures: Lester D. Stephens, *Science, Race, and Religion in the American South. John Bachman and the Charleston Circle of Naturalists 1815-1895* (2000); Edward Lurie, *Louis Agassiz* (1988); Reginald Horsman, *Josiah Nott of Mobile* (1987); Leigh Fought, *Southern Womanhood and Slavery. A Biography of Louisa S. McCord* (2003); F.J. Sypher, *Frederick L. Hoffman. His Life and Works* (2002).

Chapter Forty-One

RELATIVITY IN AMERICA

Daniel Kennefick

In the late nineteenth century many physicists regarded relativity theory with complacency. It was believed that this was a largely settled area of inquiry whose laws and rules were worth knowing not in order to question them, but as tools to be applied to other problems. A distinction was drawn between the relativity applied to particles and projectiles, in which what mattered was the relative velocities of physical particles and observers, and the relativity applied to waves and other phenomena associated with propagating media, in which the relevant velocities were those relating the state of motion of the observer to that of the medium, rather than to the physical source of the wave. In addition to this distinction one important limitation was understood to apply to the rules of relativity: they could only apply to inertial observers. Such observers were entitled to see themselves as at rest and apply the normal rules of physics to their observed systems. Observers in a state of accelerated motion were understood to be in motion with respect to the absolute reference frame of empty space and thus were not entitled to make any such assumption. One particular medium, the luminiferous ether, played a special role because it seemed, since it extended over all space, that its reference frame might be coterminous with the absolute frame of empty space itself. Therefore, in so far as interest in relativity was to be found in late nineteenth-century physics, it centered on just this question of performing experiments which could identify the state of the motion of the earth with respect to this reference frame, to find the absolute state of motion of our home in space. It so happened that the most expert work on this problem was performed in the United States in the 1880s.

Albert Michelson, the first American to receive the Nobel Prize in Physics, was a pioneer in American high-precision physics experimentation, a field which the United States would come to dominate. His specialty was the measurement of the speed of light. In 1881 most physicists believed that it should be possible to measure the speed of the earth with respect to the ether by careful measurement of the speed of light in

A Companion to the History of American Science, First Edition.
Edited by Georgina M. Montgomery and Mark A. Largent.
© 2016 John Wiley & Sons, Ltd. Published 2020 by John Wiley & Sons, Ltd.

different directions. For this purpose Michelson invented the interferometer, a device which compared the motion of light in two different directions by observing interference fringes. This device continues to play a leading role in relativity today as the method used to detect gravitational waves passing by the earth. In 1887, working with Edward Morley at Mount Wilson in California, Michelson conducted a very precise experiment which failed to find any evidence for the motion of the ether with respect to the earth. This result strongly influenced Hendrik Lorentz, one of the leading theorists of the day, in developing a theory of electrodynamics which anticipated (and helped inspire) Einsteinian special relativity.

Reception of Special Relativity

The reception of Einsteinian relativity in America must be seen in the context of this prehistory of experimental relativity. American physics in the late nineteenth century has been described by Dan Kevles (1987) as strongly emphasizing experiment over theory. Goldberg (1992) argues that in other countries experimenters were often initially skeptical of Einstein's special theory, while theorists were more accepting of his theory. Nevertheless, as Goldberg shows, in America the reception of relativity was not marked by much hostility, in spite of its weak theoretical tradition. Since American physicists could view Einstein's theory as a direct vindication of the experimental work of one of their country's most distinguished physicists, it was natural for them to set the theory in an experiment-friendly context. Goldberg's analysis of American textbooks up until the 1970s supports this picture, in which relativity is presented as a natural response to the empirical fact of the Michelson–Morley null result. This depiction of the origins of relativity theory has proved enormously influential, not just in the United States. Most recent students of physics probably have come away with the belief that the Michelson–Morley experiment played a key role in Einstein's path to relativity. The fact that Einstein never once referenced the paper in his publications naturally contradicts the validity of this discovery narrative, but its plausibility has served a useful pedagogical purpose. It is known from certain later accounts which Einstein gave, that he was aware of the Michelson–Morley experiment, but it does not seem to have featured strongly in his thinking on relativity. Einstein was a strong believer in the relativity principle, and therefore apparently regarded the Michelson–Morley experiment as proving something which should have been obvious. For him, quite well-known and standard results, such as the fact that either the magnet or the current-carrying coil may be moved to produce Faraday induction, were sufficient to convince him that the relativity principle applied to electrodynamics.

Nevertheless the narrative produced in American textbooks appeals strongly to the student being introduced to relativity. By raising the possibility of measuring the motion of the earth through space, and then showing how the Michelson–Morley experiment confounded such expectations, the textbook account dramatizes the principle of relativity in a compelling way. While the account is inaccurate in describing Einstein's path to relativity, it does give a rough and ready approximation of the viewpoint of those physicists who were the audience for relativity theory. In addition, the focus on Michelson–Morley as providing the empirical underpinning of special relativity had the advantage of drawing attention away from the scarcity of experimental results vindicating the theory. For the first two decades of the theory's life (dated from 1905) there were few tests which could be performed, and those were close enough to the limits of available techniques to leave room for debate. Over the succeeding decades an increasing number of measurements (many in atomic and nuclear physics) demonstrated that the theory was required to explain a variety of important effects. But these cases involved elaborate calculations which showed that a certain result was obtainable theoretically only if one included relativistic effects in the calculation. They were far from standalone tests which could provide a dramatic narrative for the textbook, however cumulatively convincing they may have been for contemporary physicists.

Reception of General Relativity

The reception of general relativity (published in 1915) contrasts with that of special relativity in a number of ways. Although tests of the theory were, in principle, even harder to come by (since the requirement of a strong gravitational field was added to the requirement of high velocities), astronomy did provide three classical tests whose early successful results provided a very dramatic reception narrative. Indeed, whereas the reception of special relativity took place in the obscurity of a highly technical part of the scientific world, the reception of the general theory took place in the full glare of worldwide publicity. While the theory enjoyed a significant and highly visible level of empirical success, it excited opposition in some quarters for its perceived complexity and its introduction of technically demanding mathematical tools, such as Riemannian geometry and tensor calculus, which were previously little known to many physicists. Private letters from the period, especially between astronomers who were called upon to test the theory, show that dismay was a common reaction to the "difficult" theory (Crelinsten 2006). By contrast, while special relativity had challenged certain foundational concepts of physics, it did not, at least as originally envisaged by Einstein, introduce novel mathematical techniques. On the other hand, while special relativity was relevant across a range of new fields of physics (atomic, nuclear, and particle physics), general relativity proved to be largely irrelevant to the new physics of the mid-twentieth century. Most physicists could afford to ignore it, unless they were interested in highly specialized topics such as cosmology. From the point of view of the United States, the reception of the general theory also dates to a time when theoretical physics was beginning to emerge from the shadows. Theorists were becoming more common as some of the new and much larger generation of American physicists after World War I chose to specialize in it. This was partly in response to exposure to Europe, where theorists were increasingly coming to the fore. Many bright young American physicists were exposed to the new breed of theorists either through spending time as postdocs in Europe, or through the policy of bringing the European approach to America, as in the case of Johns Hopkins University in Maryland, where John Wheeler, an American theorist who later founded an important school of relativity, was educated. Thus in the 1930s there were already a few American experts in general relativity and, as we shall see, the United States was to play a central role in the renaissance of relativity in the 1950s, when it emerged from a period of neglect and confinement to the mathematics faculties of European universities.

In one important respect the reception of general relativity in the United States followed a similar pattern to that of special relativity. American experimenters played major roles in confirming the theory, with Charles Edward St. John being the central figure in the solar redshift test (Earman and Glymour 1980b; Hentschel 1993), while William Wallace Campbell and his Lick Observatory provided the crucial second successful eclipse test of Einstein's light-bending prediction (Crelinsten 2006). Finally, it was the work of Edwin Hubble at the Mount Wilson Observatory who established the experimental evidence for an expanding universe, and it was this dramatic result which created the first really active field in which general relativity was relevant, that of modern cosmology.

Early American Relativity

The earliest American relativists were associated with the new universities of the West Coast. Richard Chase Tolman, a Caltech physical chemist, was primarily interested in the new theory's implications for cosmology. He played a significant role in the development of the concept of the expanding universe. His background in physical chemistry exposed him to special relativity because of his interest in electrons as the carriers of electric current (and thus he had an interest in fast-moving particles). He was therefore a relatively early adopter of general relativity and his tenure at Caltech placed him at the scene of Hubble's discovery of the expanding universe. Einstein himself came to Cal tech in 1931 to learn about Hubble's results. Although he had previously been involved in a controversy with Willem de Sitter and others, in which he opposed early cosmological models which featured universal expansion, Einstein quickly accepted the validity of Hubble's results and dropped his objection to solutions of his theory which accommodated it. He collaborated with Tolman in this early period and although Einstein's interests drifted away from cosmology, Tolman remained active in the field.

Another important early American cosmologist was a Caltech alumnus named Howard Percy Robertson. His role in cosmology was significant enough for his name to be attached to the standard cosmological model in use today (the Friedmann–Robertson–Walker metric). Both Robertson's and Tolman's interest in general relativity went beyond cosmology and they can legitimately be viewed as the first American relativists, in the sense of being physicists with a clear commitment to research in general relativity. The other important contributor to general relativity in this period also associated with Caltech was J. Robert Oppenheimer (who spent considerable time at Caltech even after his appointment at Berkeley). He was unusual in this period in examining the topic of gravitational collapse and collapsed objects in general relativity. Building on work by Tolman he collaborated with two of his students in a study of the physics of collapse, which was considerably ahead of its time.

While America had played an important role in the development of the expanding universe model of cosmology, it was the chief birthplace of the Big Bang theory which came to be the dominant paradigm of cosmology in the postwar period. During his work on the 200-inch telescope at Mount Palomar, near San Diego, Walter Baade confirmed and corrected Hubble's work, preparing the way for a detailed Big Bang theory which would build on the prewar work by the Belgian Georges Lemaître. A major breakthrough was made by the Russian émigré to the United States, George Gamow.

He had studied in Russia under Friedmann, one of the authors of the original general relativistic solution for an expanding universe, and it was he, together with his student Ralph Alpher (the son of immigrants), who first showed how the hot, dense early universe predicted by the Big Bang could explain the observed abundances of chemical elements (specifically hydrogen and helium) in the universe today. The marriage of "native" American experimental and observation prowess (Hubble and Baade's student, Allan Sandage) and theoretical contributions from immigrants (Gamow) was certainly a feature of postwar American science. Not all those making theoretical contributions in astronomy were immigrants, however, and the prediction of a cosmic microwave background was made by Robert Dicke and Jim Peebles at Princeton in the early 1960s (though similar ideas had been discussed earlier by Alpher and his collaborators). Dicke, an experimentalist as well as a leading theorist, intended to search for a cosmic background radiation, but it was found serendipitously by a team at Bell Labs, Arno Penzias and Robert Wilson. This triumph of a theory largely made in America came at the expense of its rival, the steady state theory, conceived by a group of European theorists in England (Hermann Bondi and Tommy Gold, both Austrian émigrés, and Fred Hoyle). Of course, there were advocates of these rival theories in all countries, but the prominence of American-based scientists in the landmark discoveries associated with the triumph of the Big Bang theory is fairly typical of postwar science. Such was the extent of this dominance that English came quite quickly in the postwar period to dominate the papers produced in international journals of physics, and American journals soon came to be leaders in the discipline.

American Anti-Relativism

American scientists had a largely favorable reception of relativity theory, and the American public had an overwhelmingly favorable view of Einstein the man, as they encountered him during several visits before he settled permanently there. Nevertheless there did exist antagonists of the man and his theory in American life. His political views, especially his pacifism and his interest in socialism, made him anathema to some on the right of American political life. His brave public stances in favor of civil rights for African Americans and against red scare attacks on communists and like-minded people only confirmed the general admiration of his character and the bitter dislike from some conservatives. Nevertheless, in contrast to his experience in Germany, Einstein did not have to fear any adverse consequences of this opposition, which was largely confined to FBI surveillance resulting in a voluminous file, filled with a great many inaccuracies about his life (Jerome 2003).

Equally, the American anti-relativity scene has been active since the early years of relativity, but without making any measurable impact upon American science. The majority of anti-relativists were not professional physicists, though quite a few had a technical background in some branch of science or engineering. To counteract their intellectual isolation, non-professional opponents of relativity have made some effort to found organizations and journals to provide themselves a platform for their views. Examples include the Academy of Nations founded by a coalition of German and American anti-relativists, including one of the leading early opponents, Arvid Reuterdahl of St. Paul, Minnesota, an engineering professor (Wazeck 2014); and (much later) the journal

Galilean Electrodynamics by another engineering professor, the Czech immigrant Petr Beckmann. The Academy, which was to give an imprimatur to the credentials of leading anti-relativists, by awarding them prizes and other honors, has long since vanished, but the journal still exists (http://home.comcast.net/~adring/).

Although there is some evidence from private letters that many astronomers tended to be skeptical of the theory (Crelinsten 2006), most professional scientists who harbored doubts did not publically contest its acceptance. Public opponents included Charles Lane Poor, an astronomer, and Dayton Miller, an experimental physicist and sometime collaborator of Edward Morley (earlier Michelson's collaborator). Miller continued working on Michelson–Morley experiments for many years, often claiming to have found results contradicting the null result of the original experiment. However, similar experiments performed by other experimenters did not support Miller's claims.

Einstein's School

As with theoretical physics in general, relativity in America advanced in the mid-twentieth century both by the emergence of homegrown theorists and through the immigration of distinguished Europeans. In relativity theory there could be no greater or more distinguished European physicist than Einstein himself, who settled in American in 1933 at the Institute for Advanced Study in Princeton. As Langevin remarked "The Pope of physics has moved, and the United States will now become the center of the natural sciences" (quoted in Pais 1997). Shortly after his arrival Einstein embarked on an intensive period of research on relativity with his assistants (in contrast to his preference for unified field theory work throughout most of his later career). Thus, although he never had a graduate student, he did have lead, for a few years, an American-based school of relativity, which played a large role later in the postwar renaissance of the subject. Working with his first American-born assistant Nathan Rosen, Einstein developed the concept of the Einstein–Rosen bridge (later called a wormhole) and also addressed the topic of gravitational waves in 1936.

At first he and Rosen thought they had come up with a proof of the non-existence of gravitational waves and they sent off a paper to the *Physical Review* (then beginning to emerge as a major physics journal and already the leading journal in the United States) entitled "Do Gravitational Waves Exist?" The paper was refereed by the best-known American relativist of the time, Howard Percy Robertson, and he spotted its flaw, that the singularity in the spacetime of plane gravitational waves was not the proof of non-existence Einstein and Rosen imagined it to be. He sent back a 10-page referee report which caused Einstein to angrily withdraw the paper and resubmit it unchanged to the *Journal of the Franklin Institute*. Fortunately Robertson was at Princeton, close to Einstein. He had been on sabbatical while working on the referee report, but upon returning to Princeton he struck up an acquaintance with Einstein's new assistant, the Polish physicist Leopold Infeld. He persuaded Infeld that Einstein was wrong and the much altered paper which eventually appeared acknowledged Robertson's role in changing the geometry of the waves to a cylindrical one with the singularity becoming a source at the central axis. After the war it was shown that the singularity was, in any case, only a coordinate one (Kennefick 2007).

Working with Infeld, Einstein than tackled the problem of motion in relativity in an influential paper, the Einstein–Infeld–Hoffmann (EIH) paper. His next student, Peter Bergmann, like Infeld and Rosen, went on to found an important school of relativity after the war, in his case (unlike those of Infeld and Rosen) in the United States, at Syracuse University. Thus Einstein's role in the development of relativity goes beyond his discovery of the theory to the more institutional one of planting seeds from which some of the major schools in the field emerged.

The Renaissance of Relativity

Until 1955 relativity barely existed as an active field of research in physics (Eisenstaedt 1989a). What work was done in the field was often accomplished by mathematicians working in mathematics departments of European and North American universities. Physicists paid the subject very little attention. Lacking journals, conferences, or other forms of institutional support, the field was very fragmented. It was not uncommon for important contributions to be made, only to be lost for years or decades because papers were not referenced or read by anyone interested in the topic. In 1955 a conference held in Bern, Switzerland, to commemorate the fiftieth anniversary of special relativity (and the fortieth of the general theory) had the effect, not so much of marking a long gone historical movement, but of giving birth to a new one. The conference attracted a considerable number of people interested in working in general relativity and they were all pleasantly surprised to find others with similar interests. A couple of years later Cecile and Bryce De Witt organized a conference at Chapel Hill in North Carolina which was specifically aimed to bring together researchers in this area. The success of this conference inspired the creation of a permanent series of international conferences on relativity, known as the general relativity series, with Chapel Hill being retroactively awarded the status of GR1 and the Bern conference often being referred to as GR0. After Chapel Hill an organization known as the International Society on General Relativity and Gravitation was set up to facilitate the conferences. This society also set up a first dedicated journal for the field, *General Relativity and Gravitation*.

A number of groups based in the United States played important roles in the rebirth of general relativity, especially that of Peter Bergmann at Syracuse University. As was the case in Europe, some of the active groups at this time were based in mathematics departments, an example in the United States would be the research group led by Václav Hlavatý, a Czech mathematician, at Indiana University. But in terms of its influence as a school of relativity, Bergmann's group would only be matched later by the group of John Wheeler at Princeton. Students of Bergmann and Wheeler were very prominent in founding other important groups throughout the United States.

Undoubtedly the major initial impetus for interest in general relativity was the common belief in the need for a quantum theory of gravity. This outlook encouraged interest in topics such as gravitational waves, since radiation seemed a natural place to search for evidence of quantum behavior, in analogy with the experience from the quantization of the electromagnetic field. Others saw gravitational radiation as a place where the theory strongly diverged from Newtonian gravity and thus pointed the way toward the most interesting new science. However, gravitational waves proved difficult to detect (see below) and the search for quantum gravity continues without immediate sign of a conclusion even today. Thus it is likely that interest in relativity would not have continued to grow but for the fortunate emergence of the topic of gravitational collapse, beginning with the discovery of quasars.

Quasars were initially studied as radio loud objects in the early days of radio astronomy, until Maarten Schmidt and John Beverley Oke at Caltech identified an optical counterpart to one such object (whose position had been accurately determined by Cyril Hazard in England from lunar occultations) and measured its redshift at 0.16, an unheard of value which suggested it was an extremely distant extragalactic object. An object with a redshift of 0.16 is, according to Hubble's law, nearly 750 megaparsecs from earth, which is to say two and half billion light years. To be visible at such an enormous remove the object must be incredibly bright, probably about as bright as a whole galaxy. Yet quasars (quasi-stellar objects or QSOs), as their name implies, are point sources, like stars. How can one squeeze the energy source of hundreds of billions of stars into a tiny region of space (one we now know to be closer in size to the solar system than to the galaxy)? It quickly became apparent to astronomers that only an advanced state of gravitational collapse could plausibly provide the required energy output. But the intense gravitational fields required to power quasars meant that only general relativity would be able to model such systems. Newton's theory could not be relied upon. Overnight a new field of relativistic astrophysics was founded. The name itself was probably coined for the first Texas Symposium on Relativistic Astrophysics, held in Dallas in 1963. The conference aimed to bring together astrophysicists and relativists to study this problem, and the importance of this turn of events for relativists is captured by the jocular conference-ending remarks of Tommy Gold: "It was, I believe, chiefly [Fred] Hoyle's genius which produced the extremely attractive idea that here we have a case that allowed one to suggest that the relativists with their sophisticated work were not only magnificent cultural ornaments but might actually be useful to science! Everyone is pleased: the relativists who feel they are being appreciated, who are suddenly experts in a field they hardly knew existed; the astrophysicists for having enlarged their domain, their empire, by the annexation of another subject – general relativity. It is all very pleasing, so let us all hope that it is right. What a shame it would be if we had to go and dismiss all the relativists again" (Schucking 1989)

American Funding of Relativity

One major reason for the renaissance of the 1950s was increased government funding for gravitational physics, and here the US government played by far the leading role. In the immediate postwar era the leading source of funding was the US Air Force (USAF). It funded important meetings, such as the Chapel Hill Conference of 1957, travel to conferences and between research groups, and directly funded research through the Aeronautical Research Laboratories (ARL) at Wright-Patterson Air Force Base near Dayton, Ohio. The USAF considered it appropriate, if it was to fund research in this area, to have in-house experts, so it hired Joshua Goldberg, a former student of Peter Bergmann at Syracuse. Goldberg and others recall that the USAF was able to provide transport for conferences, such as the one at Chapel Hill, using MATS (the Military Air Transport Service), an arm of the US military devoted to transport of personnel, cargo, and mail between different US bases around the world (Goldberg 1992). Since MATS only operated between NATO countries it could not be used when such conferences were held in the Soviet Union, or its allied countries. Similarly USAF money was available to some groups in Europe such as the King's College group in London, but could not be used by them to further their collaboration with the Warsaw group. Goldberg recalls that political considerations could be occasionally frustrating, since he was unable to offer support to a group based in Brussels, but the USAF was quite happy to support the German group of Pascual Jordan, despite his past Nazi-party membership.

Not only was Goldberg in a position to fund other groups from ARL, he also had resources to develop a group of his own. Roy Kerr held a position there immediately before his move to Austin, where he developed his famous solution, which came to represent rotating black holes in the theory. For much of the 1960s, indeed, the group had several permanent members, in addition to temporary positions akin to postdocs. There can be little doubt that USAF funding of relativity played a significant role in the postwar renaissance of the subject. Many of those active in the subject at the time have stressed the role of international conferences in strengthening the field, since up until 1955 many of those with an ambition to work in the subject were unaware of the existence of other scientists who shared their interest in it. The postwar paradigm of small research groups with one to three permanent members, a similar number of postdocs, and several graduate students was a very expensive proposition, by the standards of prewar physics. USAF funding made the enhanced level of activity associated with such mini groups possible, not only in the United States but also in some other NATO countries. It is worth noting that in many areas of physics it was the US Navy which led the way in military funding of basic research, but in relativity the US Air Force undoubtedly came to play the leading role.

By the end of the 1960s there was beginning to be some political discomfort at the close relationship between the military and academic research. In 1969 Senator Mike Mansfield of Montana, a Democrat who was then the Senate Majority leader, proposed an amendment to the Military Authorization Act which prohibited the military from supporting basic research which did not have a direct relevance to military needs. An aim of the legislation seems to have been to increase the role of the National Science Foundation (NSF) in funding pure scientific research, though in practice the NSF's budget did not increase by large amounts and most military-based funding continued, as it focused on applied science. Nevertheless the Mansfield Amendment did have its intended effect as far as funding of relativity research was concerned. No one could claim that the type of research which the ARL was funding counted as applied in any way and USAF funding of relativity came to an end. A great many of the groups were able to secure NSF funding instead (though obviously the NSF does not directly support research outside of the United States as the ARL once did). An element of continuity was achieved when the NSF hired Richard Isaacson, one of Wheeler's students, who had worked at the ARL, as their program director for gravitational research. Isaacson was able to ensure the continued advancement of the field and ultimately to inaugurate the era of Big Science in relativity, when he helped obtain funding for the LIGO project (see below), the largest project ever funded by the NSF (most big science projects are funded by other government agencies, most typically the Department of Energy and Department of Defense). Thus if the ARL helped relativity grow from the single isolated professor paradigm of the prewar period into the era of the small group, it was the NSF that took it to the next level.

Experimental Tests of General Relativity

While American theorists have certainly played a prominent role in the development of postwar relativity, it is in the experimental realm that American science has played its most outsized role. Undoubtedly the greater resources of American science, compared to other nations, was a major factor in this dominance. Up until the 1950s the three classical tests of general relativity remained the only tests of the theory available. The perihelion advance of Mercury had been quite settled by nineteenth-century astronomers and was not altered by their twentieth-century counterparts in any meaningful way. It was pointed out by Robert Dicke of Princeton that a large solar oblateness would change the amount of the observed anomaly and thus throw out the precise agreement with Einstein's theory. This would have been to Dicke's advantage, as a co-author of a rival theory, with Carl Brans, which possessed a free parameter to vary, so as to agree with whatever the new result happened to be. However, despite efforts by Dicke and others, no convincing evidence of such a solar oblateness has turned up.

The deflection of star light by the sun had been addressed by American astronomers in the early days, as mentioned above. Over the decades numerous teams from many countries unsuccessfully endeavored to improve upon the limited accuracy of those earlier efforts. In the end, the field came to be completely dominated by radio telescopes, when it was discovered that several radio-loud quasars are regularly eclipsed by the sun. Since the sun is comparatively radio-quiet, there was no need to wait for an eclipse and the quasar's position could be monitored right up to the limb of the sun, permitting far more precise agreement with the predictions of the theory. These observations were carried out by the Green Bank Interferometer at the National Radio Astronomy Observatory (NRAO) in West Virginia.

High-precision experimental gravitational physics was reborn in the 1950s as a result of the new experimental techniques uncovered during the atomic, nuclear, and particle physics revolution which physics had undergone in mid-twentieth century. The discovery of the Mössbauer effect provided a means of measuring wavelengths of light with unprecedented accuracy. This enabled Robert Pound and his graduate student Glen Rebka to do a laboratory experiment to confirm Einstein's gravitational redshift prediction. This had always been the least well-attested of the three classical tests of relativity, owing to the complexity of the solar environment, the main astrophysical testing ground. Overnight the Pound–Rebka experiment put general relativity on a far more secure footing than had ever been possible before.

In the era of the space age, a new test to augment the traditional three (light deflection, Mercury perihelion advance, and gravitational redshift) was suggested in 1964 by Irwin Shapiro of MIT, the gravitational time delay of light. Similar to the light-bending experiment, in that gravitational time delay is strongest in the solar system when light passes close to the sun, this effect depended on a change in the travel time (or speed) of the light as it passed through a gravitational field. The experiment demanded precise clocks and the availability of radar to enable a round-trip travel of the light beam (so that the same clock could measure the emission and reception of the radar). In early tests radar was reflected off the inner planets Mercury and Venus as they passed behind the sun as viewed from earth to demonstrate the effect. Later a series of spacecraft (including Mariners 6 and 7 and Voyager 2, all NASA missions) could be used to transpond a signal on reception of one from earth.

As the discovery of the quasar had great implications for relativity, so did the discovery of pulsars in the late 1960s. It quickly became apparent that here was another instance of a gravitationally collapsed object, most likely a neutron star. The discovery of the first binary pulsar at the Arecibo radio telescope in Puerto Rico by Joseph Taylor and Russell Hulse in 1975 provided the first strong field testing ground for general relativity. The painstaking work of Taylor and his collaborators ultimately produced strong evidence that the binary pulsar was decaying in its orbit as a result of gravitational wave emission, thus providing the first observational evidence that gravitational waves are real. Taylor and Hulse were awarded the Nobel Prize for Physics for their work in 1993.

On the theoretical side, the parameterized post-Newtonian formalism, developed principally by Ken Nordtvedt and Clifford Will, was a great theoretical boon to experimenters in the 1970s (Will 1993b). Inspired by the Brans–Dicke theory, a number of metric theories of gravity had been developed as rivals to general relativity from the 1960s on, and the plethora of theories, while serving as an opportunity for theory testing, challenged the experimenter because of the difficulty of accurately determining what the experimental claims of a particular theory actually were. The parameterized post-Newtonian (ppN) formalism reduced the complexity by permitting theorists to put the results of a test in this parameterized form. As long as the 10 or so parameters were known or had been worked out for a particular theory, it was a simple matter of plugging them into the relevant formula to learn its prediction for a given test.

Two particularly ambitious tests envisaged in the 1950s were the plan to conduct a gyroscope experiment in space, which eventually resulted decades later in the Gravity Probe B mission; and the attempt by Joe Weber to detect gravitational waves on earth. Weber's work became extremely controversial in the 1970s when his claims to have detected gravitational waves were discounted by those who attempted to replicate his success. Nevertheless, he succeeded in creating a field which continues to this day in the effort to detect gravitational waves (Collins 2004).

The United States: Birthplace of the Black Hole

Awareness of gravitational collapse as a source of vast energies was commonplace by the mid-nineteenth century and the discovery of white dwarf stars focused attention on the possibility of collapsed objects, which led to the birth of the idea of neutron stars in the 1930s. Although several lines of thought pointed in the direction of neutron stars as a possible end point of gravitational collapse, it was still a considerable surprise when the discovery of pulsars announced their actual existence to the world. Subrahmanyan Chandrasekhar discovered the upper limit on the mass of white dwarfs in the early 1930s. It was the counterintuitive fact of such possibly unconstrained gravitational collapse (with the body producing ever more heat energy as it shrank, and being ever less able to radiate it away because of its increasingly small size) that caused a backlash, articulated by Arthur Stanley Eddington, against Chandrasekhar's idea. This backlash was especially effective in inhibiting many astronomers from focusing on the implications of his work because Eddington was the founder of the basic theory of stellar structure upon which Chandrasekhar's work was based. Chandrasekhar himself withdrew from the topic, moving to America (Chicago) and to other areas of astrophysics. Eventually, when the subject had matured, he returned to the topic of gravitational collapse during the golden age of relativity.

Astronomers did, however, remain interested in the idea of gravitational collapse as a power source. The difficulty was that in traditional astronomy very luminous sources were rarely observed. Walter Baade and Fritz Zwicky, two European émigrés at Caltech, pointed out in 1934 that there was considerable evidence that supernova were as bright as the galaxies they occurred in, and were probably produced by individual stars. They therefore concluded that only gravitational collapse could produce such amounts of energy in a short time and that one should expect a supernova remnant to be a collapsed object like a neutron star. But since Baade and Zwicky could only point to a bright supernova (in Andromeda) from 1885 and a probable one in our galaxy only in 1572 (observed by Tycho Brahe), it was clear that there was little with which observational astronomers could work in trying to follow up their ideas.

Even if observational evidence was sparse, physicists could address the question implicit in Chandrasekhar's limit. It was Eddington himself who originally drew attention to the puzzle of white dwarfs, which were known to be small but hot stars with extraordinarily high densities, much greater than that of normal matter. The high temperature of the white dwarfs (known from their color) implied that the electrons within them were moving very rapidly, but yet must be packed very closely to the nuclei of the atoms (to explain the high density). It was Ralph H. Fowler who showed that this odd behavior made sense if one treated the electrons as a gas obeying the laws of quantum statistics. Nowadays we say that the electrons produce a degeneracy pressure as a result of their large velocities, which are the result of their desire not to overlap their wave functions while in the same quantum state (the famous Pauli exclusion principle). Fowler's student, Chandrasekhar, pointed out that there is a limit to the speed with which the electron can move, the relativistic limit of the speed of light. Thus, beyond a certain pressure, electron degeneracy pressure is no longer effective at resisting collapse. It was this result which so bothered Eddington. With the discovery of the neutron, came the realization from Lev Landau and from Baade and Zwicky, that a neutron star, where neutron degeneracy pressure played the role that electrons do in a white dwarf, might be able to resist even higher pressures, and therefore have greater masses. Landau referred to neutron cores, rather than neutron stars, thinking of high density cores of otherwise normal stars.

Landau's work interested Oppenheimer enough to consider the question of the limiting mass of neutron stars. Although severe uncertainties of the nature of nuclear matter limited the reliability of his results (with a student of his, George Volkoff) it seemed that the limiting mass was of the order of the sun's mass. This implied that many stars might have sufficient mass to one day collapse even beyond the nuclear densities of a putative neutron star. Thus, the neutron star did not solve the problem raised by the Chandrasekhar limit, which was: What did general relativity say would happen to a star collapsing, very likely in free fall, with nothing to slow its collapse? Oppenheimer tackled this problem with another student, Hartland Snyder. They showed that the star would collapse until a distant observer saw it approach its Schwarzschild radius more and more closely, doing so very slowly because of the gravitational time dilation effect. But they also noted that an observer freely falling with the collapsing stellar surface would experience no such slowing. It would fall through the Schwarzschild radius in quite a short time. Any further communication with the outside would then be impossible, as Oppenheimer and Snyder noted, which is why the Schwarzschild radius is now called the event horizon of the black hole. Thus, Oppenheimer and Snyder were the first to give us the essentials of the picture of a black hole. Although previous relativists had pointed out that different coordinate systems gave radically different results for measurements of space and time in the Schwarzschild metric, they had mostly taken this to be nothing more than a paradox which cast doubt on the validity of the theory. No one had properly addressed this metric as the endpoint of gravitational collapse and asked what was seen by a freely falling observer as opposed to a distant observer (Eisenstaedt 1989b).

The lack of much observational relevance meant that only a handful of people were much interested in gravitational collapse in the 1930s and 1940s. To further hamper progress, these few people came from different fields and did not necessarily know of each other's work. They came at the problem from different directions. The work of Landau and Oppenheimer was not, for many years, influential with relativists, many of whom were mathematicians probably not especially interested in the problem of gravitational collapse. Similarly the ideas of neutron stars must have seemed somewhat esoteric to most mid-twentieth-century astronomers.

What changed everything was the introduction of radio astronomy after World War II. As discussed above, this quickly led to the discovery of quasars, the first example of a highly luminous source which was on all the time and could thus be examined whenever one felt like it. Bringing together the relativists with the astrophysicists, as at the first Texas Symposium, was the first step toward solving this problem by creating the new field of relativistic astrophysics, which could focus attention upon it and foster communication. The relativists did rise to the occasion. Roy Kerr, while at the University of Texas, discovered the Kerr solution of the Einstein equations, which is like the Schwarzschild solution but with rotation. This was the most important breakthrough because it came to be interpreted as a rotating black hole, which we now believe to be the engine that powers quasars. Relativists developed, or rediscovered, a variety of coordinate systems which could be used to describe the Schwarzschild or Kerr metrics from the point of view of different observers. Until about the time of the first Texas Symposium the focus had been on the viewpoint of the distant observer (whose viewpoint was the canonical one, given by the Schwarzschild coordinates) and which encouraged discussion of frozen stars, as collapsed objects trapped forever on the verge of falling inside their own event horizons. A string of papers presented at that conference, and built on in the next few years, resulted in a radically altered viewpoint on these two metrics. This new outlook was encapsulated in the famous phrase, black hole, coined by John Wheeler. Wheeler observed that the hole was truly black, because the frozen star quickly became invisible as a result of essentially infinite gravitational redshifting. It was also truly a hole, since if one went down to try and interact with it, one would find that one did fall inside in a very finite period of time.

It is worth observing that both of these points had been made by Oppenheimer and Snyder in 1939, but it had taken time for the correctness of their approach to be fully appreciated. Undoubtedly this was at least partly because of the lack of any institutional infrastructure for the field in 1939. Wheeler's insight was made possible by the major contribution of students in his group at Princeton. This group and Bergmann's, the two main "schools" of relativity in the United States, produced a large number of distinguished relativists, who led the charge in driving the field of general relativity to new heights.

The Golden Age of Relativity

At the beginning of the relativity renaissance the focus of researchers was on the problem of quantum gravity. By contrast to the slow progress on this topic, the study of gravitational collapse and collapsed objects was blessed with almost instant success. It gave rise, indeed, to the golden age of relativity, identified by Kip Thorne as the period from the mid-1960s to the mid-1970s (Thorne 2003). This period was blessed by a rich seam of discoveries in the area of gravitational collapse and, simultaneously, in cosmology. Observationally new discoveries continued to foster an interest in relativistic astrophysics, including the discovery of pulsars and then binary pulsars, as well as the birth of x-ray astronomy and the discovery of what became gradually accepted as the best candidate black holes (such as Cygnus X-1). These discoveries were all rewarded with Nobel prizes, which were among the first prizes to be awarded in astronomy, and also highly unusual in being prizes awarded, at least in part, for the relevance of the research to relativity.

By the end of the golden age, relativists were moving from being published only in their own journals, which they had set up a couple of decades previously, to being published in *Physical Review*, an American journal which, over the same period, had become the world's most prestigious physics journal. As astrophysics began to go through its own golden age, with an avalanche of new techniques and big telescopes, ideas from relativity's past, which had been dismissed as impractical, made a dramatic reappearance. A particular example is gravitational lensing, predicted by Einstein in the 1930s, but dismissed by himself as unobservable. In the later part of the twentieth century not only was the Einstein ring produced by lensing observed, lensing also became an enormous tool of astronomy, both to study galaxies at extravagantly great distances from ours and to find planets around stars in our local galactic neighborhood (among many other uses). A double Einstein ring has even been observed (in which the light from two distant galaxies is bent around an intervening galaxy). The arrival of space-based telescopes and techniques such as adaptive optics permitted the imaging of stars, and other objects, at the center of our own galaxy, and other nearby galaxies so that the presence of supermassive black holes at the centers of normal galaxies could be confirmed. By about 1970 this had been a prediction made by Donald Lynden-Bell on the basis of the then accepted model of quasars as being active black holes (black holes with a hot accretion disk of material which had fallen into their gravitational wells). Since so many quasars were already known to exist at large redshifts by that time, it seemed clear that each galaxy we see near us today must be a former quasar. Therefore they must possess a quiescent black hole at their centers.

Relativity in America Today

Obviously the growth of relativity in the United States to the point where there are active groups at dozens of universities was not achieved without critical institutional support. Most of these relativity groups were based in physics departments and many veterans of relativity in the 1970s (and even the 1980s) recall how it was not uncommon to hear the opinion expressed by many physicists that relativity was not really "physics." Obviously, given this level of institutional resistance, relativists would hardly have gained these new footholds if they had not been able to receive research funding. The loss of USAF funding was, at least in the case of this field, made up for by increased NSF funding. As we have seen, with NSF backing, relativity was able to progress into the area of Big Science with support for the Laser Interferometer Gravitational-Wave Observatory (LIGO) project, which required hundreds of millions of dollars worth of funding (Bartusiak 2000; Collins 2004, 2014). Funding on a similar scale was also available from NASA, for instance with the Gravity Probe B project. Such was the general success of the field that funding continued to increase in spite of setbacks, such as the failure of the first incarnation of LIGO to make a detection, and the difficulties encountered by Gravity Probe B which greatly complicated its data analysis and reduced the accuracy with which it could claim to have established the reality of frame dragging (another prediction of relativity once claimed by Einstein to be beyond the reach of experiment).

On the other hand relativity also grew large enough to experience the downs as well as the ups of government funding. The notion of making a space-based gravitational wave observatory (a LIGO in space) was first proposed in the 1970s and grew into the Laser Interferometer Space Antenna (LISA) project which was selected as a joint NASA–ESA (European Space Agency) project in the Beyond Einstein program in the first decade of the twenty-first century. However, first the reorienting of NASA priorities toward manned space flight, and then budget cuts across the board, forced NASA to eventually terminate its participation in LISA. At the present time ESA has decided to continue with a scaled-down version of the project (sometimes referred to as eLISA), but it is not scheduled to fly until the second quarter of the twenty-first century. Few projects can be as potentially useful to astrophysics and cosmology as LISA plans to be, given its ability to detect signals from the supermassive black holes at the centers of galaxies, but the failure to secure funding shows the limits of astrophysics as a draw on the exchequer. Nevertheless the subject is larger and more active, both theoretically and experimentally, than it has ever been, and America continues its leading role in it, most notably with the Indigo project, to build a LIGO-style interferometer to detect gravitational waves in India, largely with American funds.

Bibliographical Essay

For Einstein's work and life in America one may consult the many biographies, especially that by Pais (2005) for its unrivaled discussion of Einstein's science, or the most recent one by Isaacson (2008). Studies of relativity post-Einstein are not common and, where they deal with activities in America, are mostly referenced in the text. One book which gives a first-hand account of most of the important developments in American relativity is Thorne (1995). Two particularly important topics for the history of relativity are not treated extensively here as they are the subject of book-length studies: gravitational wave theory is covered in Kennefick (2007) and experimental efforts in Collins (2004, 2014) and, in less detail, Bartusiak (2000); the history of quantum gravity is told in Rickles (2015). The history of cosmology has attracted more attention, see Kragh (1996) for the debate between the supporters of the Big Bang and steady state theories, and Way and Hunter (2013) for the discovery of the expanding universe. For personal recollections of this history, there is Lightman and Brawer (1992). An interesting book related to high precision experimental gravity is Franklin (1993). For Einstein's Unification program, see van Dongen (2011). For the development of black holes, see Eisenstaedt (1989b) and Melia (2009).

Chapter Forty-Two

SCIENCE AND RELIGION

Mark A. Waddell

The modern relationship between science and religion in the United States has been one of both harmony and contention. For most of the eighteenth and nineteenth centuries, American philosophers, scientists, and theologians found much common ground between the study of nature and the veneration of God. Even the advent of Darwinism in the latter decades of the nineteenth century did little to upset the careful balance that had been established between these two institutions, at least at first. The increasingly secular character of American intellectual life, however, eventually came into conflict with the populist Christian fundamentalism that had reached social and political prominence in the United States in the early years of the twentieth century. The hard-won reconciliations and accommodations established between religion and science slowly disintegrated, and by the beginning of the twenty-first century one could find numerous examples of an ideological and seemingly intractable conflict. Common ground still exists, however, in spite of the vociferous and fractious debates waged for the benefit of an increasingly polarized public.

For much of the history discussed here, American scientists and theologians were influenced heavily by their European counterparts. That influence really only waned in the wake of two world wars and the rise of American supremacy in science and technology, but until then most American intellectuals took at least some cues from the scientific and theological ideas of their contemporaries across the Atlantic. Part of what makes the United States unique in its relationship between science and religion, however, is the fact that its history has been punctuated by periods of intense and populist religious fervor, known as the "Great Awakenings." Typically characterized by increased attendance at evangelical gatherings and a general spike in popular interest in religion, these revivals tended to encourage a return to biblical literalism and attacks on "materialism" – eventually connected with mainstream science – as the source of contemporary social and political ills. While historians have debated whether the Great Awakenings

A Companion to the History of American Science, First Edition.
Edited by Georgina M. Montgomery and Mark A. Largent.
© 2016 John Wiley & Sons, Ltd. Published 2020 by John Wiley & Sons, Ltd.

were idiosyncratically American, it is undeniable that the populist and evangelical Christianity that emerged in the United States in the nineteenth and twentieth centuries had a profound impact on social and political attitudes toward science.

This chapter sketches the outlines of what is an enormously complex historical relationship; inevitably, there will be gaps in what is discussed. Those seeking more detail are encouraged to consult the bibliographic material at the end of the chapter as well as other chapters in this volume.

The Eighteenth Century: Finding God in Nature

In the decades before independence, most American colonists held to ideas that had predominated in pre-modern Europe, particularly the conviction that one could come to know God through the study of nature. Early modern philosophers and theologians had inherited from their medieval forbears the notion of the "two books" composed by God: on the one hand, His word, as recorded in Scripture, and on the other, His works, or the created world. (Grant 1996: chapter 5) Thus, to understand God one must consult both of these books, lending an important religious impetus to the investigation of nature. The founding members of the Royal Society of London, particularly men like Robert Boyle (1627–1691) and Thomas Sprat (1635–1713), believed wholeheartedly in this harmonious relationship between nature and God. Boyle went so far as to endow a series of public lectures or sermons – the Boyle Lectures – that have persisted up to the present day. The main goal of these lectures was demonstrating the truth of Christianity by means of natural philosophical and scientific inquiry.

Many natural philosophers in colonial America looked to figures like Boyle and his fellows in the Royal Society for inspiration, and the Puritan divine Cotton Mather (1663–1728) is an excellent example of the early American predilection for seeking consensus between religion and the study of nature. He wrote to the Royal Society in the early years of the eighteenth century, presenting what he called his "Curiosa Americana" in an attempt to ingratiate himself and secure a fellowship in the Society. These "curiosities" ranged from descriptions of New World plants and animals to huge bones that had been excavated in Claverack, New York, in 1705. Many in the colonies speculated that these bones were the remains of the Nephilim mentioned in Scripture, giants that had once walked the earth, and Mather used his letters to the Royal Society to argue that "their *Dead Bones* are *Lively Proofs* of the *Mosaic* [i.e., biblical] History" (Levin 1998: 75).

For many in the eighteenth century, nature held the keys to understanding and venerating God. If proof of Christianity was required, Mather argued, the natural world offered it in abundance. Thus, what emerged over the course of the century was a robust and widespread *natural theology*, which saw the study of the natural world as the best and most certain way to comprehend the existence of God. Natural theology was in many respects an inevitable response to the increasingly empirical and fact-centered philosophy that emerged in Europe toward the end of the seventeenth century and that became deeply entrenched in European and American thought during the Enlightenment of the eighteenth century. The careful mathematical science of Isaac Newton (1642–1727), intended by Newton himself to testify to the existence of God, encouraged a wide range of thinkers in the eighteenth century to apply the evolving tools of inductive reasoning and rigorous empiricism to theological as well as philosophical problems.

The ascent of reason during the Enlightenment cemented this way of thinking. Of special importance here was the religious movement known as *Deism*, which emerged during the Enlightenment as a response both to the perceived dogmatism of existing religious institutions (particularly Catholicism), and to the perennial and destructive problem of religious strife that had fragmented Europe in previous centuries (Outram 2013: chapter 9). Whereas earlier natural theologians had tended to believe that there were multiple paths to a knowledge of God, with the study of nature being foremost among them, Deists believed that the exercise of reason and the empirical study of the natural world were all that one needed to prove the existence of God. They rejected divine revelation of all kinds, including that found in Scripture. Many of the American Founding Fathers, Thomas Jefferson and Benjamin Franklin chief among them, were Deists, and Deism flourished in American intellectual circles throughout the eighteenth century and into the nineteenth. Thomas Paine's widely read *The Age of Reason*, which appeared in the United States in 1794, was a mature – if radical – articulation of Deistic ideas, and was met by a storm of protest from contemporary American evangelicals, who saw in Paine's rational reductionism a serious threat not only to organized religion but to public morals as well.

Deism began to disappear in the early years of the nineteenth century, replaced to a certain extent by movements that espoused a similar ethos, such as Unitarianism, or challenged by Christian revival movements such as Methodism and Pietism (Outram 2013: chapter 9). While Deism declined, however, the broader movement of natural theology continued to thrive in the United States for at least another century. Unlike Deism, which had adopted a thoroughly reason-centered approach to the divine, the simple idea that one could look to nature for knowledge about God was relatively uncontroversial and could be accommodated to a wide range of Christian beliefs.

1800–1859: Seeking Harmony

Natural theology remained a powerful force in most aspects of American culture throughout the nineteenth century, including science. As Herbert Hovenkamp has characterized it, "For [American Protestants], doing theology was a scientific activity. It employed the inductive method … drew its data from the world of nature … its conclusions could be demonstrated empirically" (Hovenkamp 1978: 37). The early decades of the nineteenth century, however, saw American natural theologians inspired by two particular sources, both English in origin: the work of William Paley (1743–1805), a clergyman who published his *Natural Theology, Or, Evidences of the Existence and Attributes of the Deity* in 1802; and the Bridgewater Treatises, a collection of essays by leading intellectuals of the early nineteenth century that sought to demonstrate "the Power, Wisdom, and Goodness of God" evident in the created world (Hovenkamp 1978: 25–6). William Paley in particular would have a lasting influence on the relationship between science and religion in America. He presented a teleological argument for the existence of a divine creator, pointing to the complexity evident in nature as proof for conscious and purposeful design. His most famous argument was the "watchmaker analogy," which continues to inspire American creationists today.

In the opening pages of his *Natural Theology*, Paley presented a hypothetical situation that found him wandering across a deserted heath and tripping over a rock. The question of how the rock arrived there seemed, to Paley, of little consequence. If he had come across a watch lying on the ground, however, he would have confronted a more difficult and important problem. The watch, with all of its inherent complexity, could not simply have arrived there by chance. Its myriad gears and other parts worked together so perfectly and harmoniously that it was absurd to imagine that they had just happened to arrange themselves in such a fashion. Thus, Paley concluded, "the machine which we are inspecting demonstrates, by its construction, contrivance and design. Contrivance must have had a contriver; design, a designer" (Paley 1802: 14). Moreover, even if that watch had been produced by another, earlier watch through some feat of mechanical replication, this earlier watch must also carry the tell-tale signs of contrivance and design, and so on back through the many "generations" of watches until one arrived, inevitably, at the first Designer.

If this watch must have a designer, Paley argued, how much more likely was the existence of a designer for the innumerable structures one could find throughout nature – parts of organisms that evinced far more complexity than a simple mechanical device? The argument had a profound impact on subsequent attempts to link God with the natural world. Charles Darwin (1809–1882) read Paley's work with great interest while he was composing his *On the Origin of Species*, though he would go on to argue that the presence of complexity in nature need not presuppose the existence of intelligent or purposeful design. For many American Protestants, Paley's ideas inspired an even greater reliance on empiricism; after all, if complexity and design were to be the cornerstones of a powerful natural theology, the pious naturalist depended upon a range of empirical methodologies in order to find and investigate traces of this design in nature. Hovenkamp has argued that this led American natural theologians, in the first half of the nineteenth century, to embrace the philosophy of the English experimentalist Francis Bacon (1561–1626), who had advocated an empirical and methodical study of nature that also allowed for the existence of a divine creator.

Thus, in the first half of the nineteenth century there existed a largely harmonious relationship between science and religion among American philosophers and theologians. This began to change, however, around the middle of the century – the 1850s in particular became a watershed for American natural theology. Widely embraced by most Christian denominations in some form or another up to this point, natural theology met its most significant challenges first in the increasingly popular philosophy of Immanuel Kant (1724–1804), and then in the wake of Darwin's *On the Origin of Species*. Kant was one of the great skeptics of the Enlightenment, and he had attacked metaphysics in general and the question of the First Cause in particular. His posthumous influence on American academic circles led to a more skeptical approach to metaphysics at the same time that the natural sciences were finding their professional footing and becoming increasingly independent of Protestant theology (Hovenkamp 1978: 44–6).

Even before natural theology began to run into difficulties, however, a surge of interest in geology set the stage for a series of scientific challenges to biblical authority that would occupy the rest of the nineteenth century. The 1830s and 1840s witnessed a number of geological discoveries, in England as well as in the United States, that called into question the biblical account of creation and suggested that the age of the earth was much greater than anything recorded by Scripture. Charles Lyell (1797–1875) published the first volume of his *Principles of Geology* in 1830, and its claims that modern geological structures had been shaped by slow-moving forces over hundreds of millions of years went on to inspire not only Charles Darwin's evolutionary theory but also a growing tide of biblical criticism. Bishop Ussher's crude attempts in the seventeenth century to approximate the age of the earth by counting the human generations recorded in the Bible finally disintegrated beneath the onslaught of an increasingly precise and plausible science.

In spite of these challenges, earnest attempts to accommodate these new scientific discoveries with Christian belief persisted throughout the middle of the nineteenth century. Leading American geologists like James Dwight Dana (1813–1895) at Yale and Edward Hitchcock (1793–1864), the president of Amherst College, tried to reconcile the new geological science with their Christian beliefs by interpreting the six days of Genesis as geological epochs, each potentially millions of years old. Others posited a "gap" in the Genesis account, with many millennia of geological time intervening between the original six days of Creation and the recorded events in the Garden of Eden. Such efforts to harmonize Christianity and geology were relatively uncontroversial, embraced even by many fundamentalists. (Larson 1997: 16) Nonetheless, by 1859 the relationship between science and religion in the United States was undergoing a slow but inexorable change, one that was only accelerated by the evolutionary theory of Charles Darwin.

1860–1925: Darwin Arrives in the United States

Darwin's *Origin of Species* reached the United States in 1859, at a critical juncture of events: the outbreak of the Civil War was looming, and American Protestantism was foundering. Its numerous denominations were divided among themselves, despite the fact that all of them faced the growing social problems posed by increasing urbanization, rapid industrialization, and the strident tones of a public and powerful scientific establishment (Russett 1976: 25). The biblical literalism embraced by many fundamentalist Protestant sects came to be regarded as untenable by more mainstream theologians, and as we have already seen the rise of an increasingly skeptical philosophy in academic circles threatened both natural theology and the wider study of metaphysics among American intellectuals. Thus, when the *Origin of Species* arrived in the United States it was destined from the very start to have a powerful and indelible impact on religious belief.

The advent of Darwinism called into question a central tenet of natural theology, namely, the argument from design. Darwin showed that complexity could be explained by natural selection rather than by providential or deliberate intervention from a higher power, and his theory of evolution explicitly rejected the teleology that was so important to natural theology – the idea, in other words, that there are goals or ends toward which change progresses. For Darwin, the only goals that mattered were those that ensured the survival of the individual; anything beyond that was mere conjecture and thus outside the realm of the natural sciences.

For thinkers on both sides of the Darwinian debate, to deny the existence of teleology in nature was to deny the very existence of God. This view was embraced in England with characteristic fervor by "Darwin's bulldog," Thomas Henry Huxley (1825–1895), who believed that harmony between science and religion was impossible, but also in the United States by Charles Hodge (1797–1878), one of the most influential American theologians of the time. Hodge conflated Darwinism with atheism precisely because it left no room for teleology or purpose in nature, and his views on Darwin's theory would go on to have a profound effect on evangelical Protestants well into the twentieth century (Russett 1976: 26).

Between the extremes presented by Huxley and Hodge, however, most orthodox American Protestants sought a middle way. Chief among them was one of Darwin's most influential and effective advocates in the United States, the Harvard botanist Asa Gray (1810–1888). Gray had written the first American review of *On the Origin of Species* shortly after its publication in 1859, and he argued that Darwinism did not necessarily spell an end to the argument from design. Instead, Gray believed that natural selection was a mere mechanism, a means whereby adaptation and change came about, and as such it could still be seen as operating in service to a larger design (Russett 1976: 34). When Gray wrote to Darwin with this idea, however, the latter responded with a gentle but firm rebuttal: "I had no intention to write atheistically. But I own that I cannot see as plainly as others do, and as I should wish to do, evidence of design and beneficence on all sides of us. There seems to me too much misery in the world. I cannot persuade myself that a beneficent and omnipotent God would have designedly created the Ichneumonidae with the express intention of their feeding within the living body of Caterpillars, or that a cat should play with mice" (Larson 1997: 17). Ichneumon wasps are parasites that lay their eggs inside the body of a living caterpillar, and whose young then hatch and devour the unfortunate creature from the inside out. It was an example of the cruelty that the young Darwin had come to see throughout nature, and that convinced him that this could not be the handiwork of a benign and providential God (Dixon 2008: 62). Darwin did not reject altogether Gray's notion of a Creator-God who took an active role in the development of life, however. His theory of natural selection depended upon the operation of natural laws, but he admitted, "I can see no reason why a man, or other animals, may not have been aboriginally produced by other laws, and that all these laws may have been expressly designed by an omniscient Creator, who foresaw every future event and consequence."

Whatever Darwin's own views, the theory of evolution by natural selection did not necessarily pose significant problems for most moderate Christians in nineteenth-century America. Many, like Asa Gray, were content to see evolution as the clever means whereby God had designed Creation. More significant problems were posed by the appearance of Darwin's *Descent of Man* in 1871, which argued that humans were animals, subject to the same pressures and forces that ruled selection in other species. By this time, the theory of natural selection had become especially troublesome for pious Christians. Its advocates claimed that adaptations or variations that were beneficial to the individual were the result of natural forces – or worse still, simply random – rather than examples of divine beneficence. At the same time, natural selection highlighted the arbitrary cruelty and indifference that seemed rampant in the world described by Darwin: "nature red in tooth and claw," as Tennyson later put it. With a majority of moderate American Protestants seeking to reconcile Darwinism with Christianity, it is unsurprising that they should ultimately separate natural selection from the larger idea of evolution. Over the latter decades of the nineteenth century, natural selection slowly faded from discussions of evolution, replaced instead by other mechanisms that one could accommodate more easily to orthodox Protestant belief – mechanisms, in other words, that could be accommodated to a teleological vision of natural change and progress. Of particular interest in the United States were the ideas of Jean-Baptiste Lamarck (1744–1829), who had proposed the first coherent theory of evolution in the early years of the nineteenth century.

Lamarck had argued for the inheritance of acquired characteristics: he believed that organisms passed on to their offspring particular characteristics that they had acquired through the use or disuse of various parts of their bodies. An oft-cited example is that of an animal that must stretch its neck upwards in search of edible leaves, and which then produces an offspring with a slightly longer neck. Lamarck's ideas were known to Darwin, who largely rejected them, but they assumed an important role in American academic circles in the latter decades of the nineteenth century. "Neo-Lamarckism" became a widely accepted alternative to Darwinian natural selection because it could easily accommodate a teleological or purpose-driven interpretation of evolution. Rather than the caprice and cruelty of chance that appeared central to natural selection, Lamarckism implied direction: organisms moved steadily toward better-adapted types, perhaps with the aid of a providential God. Thus, while the concept of evolution was virtually unquestioned by American scientists by the end of the nineteenth century, the theory of natural selection fell out of favor and all but disappeared. It would be revived only in the first half of the twentieth century, after the full integration of Mendelian genetics and the rise of the modern neo-Darwinian theory of evolution that persists today.

Darwinism had the potential to cause tremendous difficulties between science and religion, but in fact a majority of American scientists and theologians in the latter half of the nineteenth century were eager to chart a middle road between the extremes of atheism and fundamentalism. Ironically, it was those who were neither scientists nor theologians who, late in the century, began to point to a fundamental conflict between science and religion. Most prominent were John William Draper (1811–1882), who published his *History of the Conflict Between Religion and Science* in 1874, and Andrew Dickson White (1832–1918) whose *History of the Warfare of Science with Theology in Christendom* appeared in 1896. Draper's work was inspired by the First Vatican Council, convoked by Pope Pius IX in 1869 and formally adjourned the following year. Draper was particularly disturbed by the doctrine of papal infallibility that was adopted during the Council, and his *History of the Conflict Between Religion and Science* argued that religion in general and Catholicism in particular were, and always had been, incommensurable with the free pursuit of science (White 1966: 26). Like Draper, Andrew Dickson White also sought to free science from what he saw as the medieval and retrograde institution of religion, but unlike Draper he did not advocate for an end to religion. Instead, he believed that a reconciliation between the two was both possible and inevitable, precipitated by the widespread adoption of the doctrine of evolution, which he saw as "the master key of the universe" whose universal truths would fashion an unbreakable consensus between science and religion in the modern age (White 1966: 34–6).

The Scopes Trial and its Aftermath

Beginning around the beginning of the twentieth century, the tenor of the relationship between science and religion began, slowly, to change. Increasing numbers of Americans – some, like Draper, hostile to religion's influence on science, others steeped in the populist brand of American evangelicalism that was taking hold across broad swathes of the country – began to emphasize obvious points of conflict and friction between these two institutions. The hard-won accommodations that had been established between Darwinism and American Christianity began to fall apart, and reached their ultimate crisis in the widely publicized Scopes trial of 1925.

In May of that year, the American Civil Liberties Union (ACLU) placed an advertisement in the *Chattanooga Times* that read, "We are looking for a Tennessee teacher who is willing to accept our services in testing this law in the courts" (Larson 1997: 83). The law in question was the Butler Act, which the governor of Tennessee had recently signed into existence. It prohibited the teaching of human evolution in any school that received state funding, and reflected a growing unease with Darwinian evolution on the part of the American public, particularly in the conservative and evangelical bastions of the South. Civic leaders in Dayton seized on the brewing controversy as a way to promote their tiny, struggling town, and a 24-year old general science teacher named John Scopes was conscripted to teach human evolution to his high school biology students and thereby contravene the prohibitions of the Butler Act. The ACLU had its case, and Dayton soon had the attention of the entire country.

The Scopes trial – known popularly at the time as the Scopes monkey trial – came to represent a watershed moment for the relationship between science and religion in America. We have already seen that many Christian evangelicals had expressed reservations about Darwin's theory of evolution after 1860, particularly as it came to be applied to humans. The Scopes trial, however, marks the moment when these debates left the rarefied atmosphere of academic discourse and entered the general consciousness of the American public in a profound and lasting way. Surprisingly little has changed in the intervening 90 years. The same ideologies – even the same arguments – articulated within and around the trial remain entrenched in modern American culture today.

The trial became a national sensation, and was the first to be broadcast across the country on the radio. William Jennings Bryan (1860–1925), a well-known politician and three-time Democratic candidate for the presidency, took charge of the prosecution. He was a noted anti-evolutionist who had spoken to audiences across the country about the evils of Darwinism, especially as applied to social problems. His opponent was Clarence Darrow (1857–1938), a famous defense attorney who took the side of John Scopes and who relished the opportunity to ridicule Bryan's fundamentalist Christianity when he put the latter on the stand late in the trial. Bryan and Darrow represented the increasingly polarized views of the American public on the subject of evolution, and though Bryan died shortly after the trial ended his arguments lived on in the rhetoric of contemporary evangelicals.

At the conclusion of the trial, John Scopes was found guilty – a verdict that surprised no one – and assessed a fine of $100 by the judge, John T. Raulston. His conviction was ultimately overturned on a technicality, however, because in Tennessee only a jury, not a judge, could set a fine of more than $50. The Butler Act remained in force until

1967, when it was repealed, but the reverberations of the Scopes trial outlasted the law that sparked it. Immediately following the trial, the governor of Texas ordered that those pages that discussed the theory of evolution should be excised – literally, with scissors – from all biology textbooks in the state (Moore 2001: 792). Other textbooks quietly dropped the topic of evolution altogether, with the result that Darwinism, and evolution more broadly, were not taught in most American schools for almost thirty years after the Scopes trial.

Even once evolution reappeared in public curricula, however, it faced a wide range of religious challenges that continue to this day. In the early 1980s, for example, almost sixty years after the Scopes trial, several states tried to pass laws that would either restrict the teaching of evolution in public schools or that mandated that those schools also teach "creation science." One such law was the "Balanced Treatment for Creation-Science and Evolution-Science Act" passed by the Louisiana legislature in 1981, which did not require that school districts teach creationism, but which forced those districts to teach "creation-science" if they also taught evolution. After a series of legal challenges the law was eventually considered by the Supreme Court in *Edwards v. Aguillard,* and at the conclusion of the trial in 1987 the Court ruled that the Louisiana law violated the Establishment clause of the Constitution because it sought to advance a particular religion or religious belief.

When the Louisiana legislature passed their "Balanced Treatment" law in 1981, a sizeable percentage of American adults did not believe in human evolution, and public opinion has changed little since then. According to a 1982 Gallup poll, 44% of American adults believed that God created humans in their present form and that human evolution has never occurred. Another 38% believed that humans evolved into their present form with the guidance of God, with only 9% believing that human evolution had occurred without divine assistance. These numbers have changed little in more than thirty years: as of May 2014, 42% of those polled still believe that humans were created in their present form, with 31% attributing human evolution to some form of divine guidance and 19% believing that human evolution has proceeded without supernatural assistance (Gallup Poll 2014).

Given these poll numbers, it is unsurprising that in the decades since *Edwards v. Aguillard* American creationists have continued their attempts to integrate their beliefs into mainstream culture and, particularly, public education. The defeat suffered by creationists in 1987 convinced many to try a different approach, and so the theory of "intelligent design" was born. Championed in recent years by individuals like Michael Behe (1952–) and Steve Fuller (1959–), "intelligent design" proposes that many observed phenomena in nature bear the hallmarks of purposeful and deliberate design. Behe, a biochemist, has coined the phrase "irreducible complexity" to describe structures that he believes are too complex to have evolved in stages over time, an idea descended almost entirely from William Paley's writings of 200 years ago. A commonly cited example of this argument is the molecular motor that drives the flagella of some bacteria, which, according to Behe, could not have evolved by means of natural selection. Behe's argument is founded on the (erroneous) belief that any less-complex or simpler antecedent to the observed flagellar motor would serve no obvious function, and so would not be acted upon by natural selection to produce the more complex structure seen today. Along with other examples used to argue for intelligent design, this idea has been debunked and invalidated by a wide range of scientists (Pallen and Gophna 2007).

There is widespread consensus, even among moderate theologians, that intelligent design is little more than creationism dressed up in the guise of modern scientific analysis. This was the conclusion reached at the end of another prominent trial concerning the teaching of intelligent design in public schools, the 2005 *Kitzmiller v. Dover Area School District* case, which was sparked by an attempt by members of the Dover, Pennsylvania school board to mandate the teaching of intelligent design in high school biology classes. A number of local parents filed suit against the school board, and the case went to federal court in Pennsylvania. Much as the Scopes trial in 1925 had been used by the ACLU as a test-case for laws that blocked the teaching of evolution in public schools, *Kitzmiller v. Dover Area School District* was a constitutional test-case for the theory of intelligent design. Both Michael Behe and Steve Fuller testified on behalf of the defendants, but ultimately the judge in the case ruled that the actions of the Dover school board were unconstitutional. He wrote in his decision that intelligent design "is a religious view, a mere re-labeling of creationism, and not a scientific theory," and that any attempt to mandate its inclusion in public school curricula would violate the Establishment Clause by using government support to promote or advance a particular religious belief.

The Situation Today

Today, the United States remains a deeply religious country. Whereas the vast majority of American religious history has centered around Protestantism, the modern United States has become more pluralistic and diverse, giving rise to increasing cooperation and dialogue between different faiths, as well as a strong and public dialogue between those disparate faiths and facets of the surrounding culture. One such dialogue is that between science and religion, where deep divisions still persist. Evangelical Christianity and biblical literalism pervade many parts of the country, and public hostility to science remains a stubbornly intractable issue in many quarters.

For most of its history, the relationship between science and religion in the United States was characterized by intellectual debates that took place among a highly educated coterie of theologians, philosophers, and scientists. Over the past eighty years, however, that situation has changed dramatically. Thinkers on all sides are now reaching out to the American public, trying to gain converts to one position or another, but for the most part they are engaging with a public that understands relatively little of modern science (Sherkat 2011). This is a problem only exacerbated by the pseudoscience peddled by institutions like the Creation Museum in Petersburg, Kentucky or presented to general audiences in works like Michael Behe's *Darwin's Black Box* (1996). Nor is this tactic of public engagement something practiced only by creationists; Richard Dawkins (1941–), the evolutionary biologist and outspoken atheist, has used polemical works like his book *The God Delusion* (2006) to try and sway public opinion in the direction of his own views.

Religious debates about evolution, the age of the earth, and the reality of the Big Bang have much to do with the persistence of biblical literalism in some segments of American religious culture. Such beliefs do not necessarily belong to a majority of American evangelicals, but nor are they held only by fringe elements. In spite of the existence of such beliefs, however, it is not the case that evangelical Christianity is incompatible with modern science. On some issues, American evangelicals have found themselves working alongside scientists, as in the collaborations that have sprung up around the issue of the natural environment.

With the growth of the environmental movement in the United States in the latter decades of the twentieth century, some religious denominations – both Christian and otherwise – found themselves on the front lines of the culture wars. Strong support for environmental management and conservation was a central tenet in the many varieties of New Age faith that proliferated in the United States after the cultural revolution of the 1960s, and such groups continue to provide a religious perspective on problems that had traditionally been the purview of environmental scientists and activists (Taylor 2001a, 2001b). Likewise, some evangelical Christian denominations entered the fray in the late twentieth century, finding in Genesis clear justification for their self-appointed role as environmental stewards and protectors (Hitzhusen and Tucker 2013). This so-called "green Christianity" forged lasting collaborations between some Christian denominations and environmental conservationists, activists, and scientists, but also put many of these evangelical congregations at political odds with the religious right.

This particular example nicely highlights the complexity of the modern relationship between science, politics, and religion in the United States and focuses our attention on a central dimension of their longstanding historical relationship; namely, cooperation. The optimism expressed by Cotton Mather over 200 years ago, his belief that one can come to know God through the study of nature, has never faded from the American consciousness. There remain points of serious conflict, and they will probably persist for some time. What really characterizes the American relationship between science and religion, however, are the many ingenious and heartfelt attempts to reconcile the two.

Bibliographic Essay

The historical relationship between science and religion in the United States is unsurprisingly complex. In order to understand the American context, however, one must first understand the wider relationship between these two institutions, particularly in Europe, as this shaped the ideas and opinions of American intellectuals. Edward Grant's *Foundations of Modern Science in the Middle Ages* (1996), though removed by several centuries from the material in this chapter, contains an eye-opening discussion of early clashes between religion and natural philosophy that set the stage for the largely harmonious relationship that persisted in Europe for hundreds of years. Broad but useful overviews of that relationship include John Hedley Brooke's *Science and Religion: Some Historical Perspectives* (1991); the more recent *Science and Religion: New Historical Perspectives* (2010) edited by Thomas Dixon, Geoffrey Cantor, and Stephen Pumfrey; and *Science and Religion: A Historical Introduction*, edited by Gary Ferngren (2002) . A more concise introduction to the relationship between science and religion can be found in Thomas Dixon's *Science and Religion: A Very Short Introduction* (2008).

This chapter has focused, by necessity, on Christianity, but there are many works that provide a broader and, in some cases, comparative perspective. Examples include *Science and Religion Around the World* (2011), edited by John Hedley Brooke and Ronald L. Numbers; *Science and Religion: Christian and Muslim Perspectives* (2012), edited by David Marshall; and *Buddhism and Science: Breaking New Ground* (2013), edited by B. Alan Wallace.

While still broad in scope, other works focus more on recent decades. These include Benjamin E. Zeller's *Prophets and Protons: New Religious Movements and Science in Late Twentieth-Century America* (2010); *Science and Religion: One Planet, Many Possibilities*, edited by Lucas F. Johnston and Whitney A. Bauman (2014); and Alvin Plantinga's *Where the Conflict Really Lies: Science, Religion, and Naturalism* (2011). For an interesting perspective on attitudes held by scientists working today, see Elaine Howard Ecklund's *Science vs. Religion: What Scientists Really Think* (2010); Michael J. Reiss' "The Relationship Between Evolutionary Biology and Religion" (2009); and Edward J. Larson and Larry Witham's "Scientists and Religion in America" (1999).

Some of the best and most useful scholarship on religion and science in American history takes the form of overviews such as Edward A. White's *Science and Religion in American Thought: The Impact of Naturalism* (1968); Herbert Hovenkamp's *Science and Religion in America, 1800–1860* (1978); and Cynthia Eagle Russett's highly influential *Darwin in America: The Intellectual Response, 1865–1912* (1976). For a unique and very readable take on the complex relationship between religion, science, and popular culture in nineteenth- and twentieth-century America, see Fred Nadis' *Wonder Shows: Performing Science, Magic, and Religion in America* (2005).

Valuable and accessible primary sources relating to the material in this chapter include William Paley's *Natural Theology* (reprinted by Oxford University Press in 2008), John William Draper's *History of the Conflict Between Religion and Science*, which has been reprinted as recently as 2013, and Andrew Dickson White's *History of the Warfare of Science with Theology in Christendom*, reprinted by Prometheus Books in 1993.

There are numerous works that discuss the Scopes trial. One of the most comprehensive is Edward J. Larson's *Summer for the Gods: The Scopes Trial and America's Continuing Debate Over Science and Religion* (1997). Others include Adam R. Shapiro's *Trying Biology: The Scopes Trial, Textbooks, and the Antievolution Movement in American Schools* (2013); Adam Laats' *Fundamentalism and Education in the Scopes Era: God, Darwin, and the Roots of America's Culture Wars* (2010); Paul K. Conkin, *When All the Gods Trembled: Darwinism, Scopes, and American Intellectuals* (1998); Constance Areson Clark's "Evolution for John Doe: Pictures, the Public, and the Scopes Trial Debate" (2001); and Randy Moore's "The Lingering Impact of the Scopes Trial on High School Biology Textbooks" (2001).

Likewise, the ongoing debates between evolution and creationism in the United States have inspired numerous studies, including Edward J. Larson's *Trial and Error: The American Controversy Over Creation and Evolution* (1989) and Ronald Numbers' *The Creationists: From Scientific Creationism to Intelligent Design* (2006). *God and Design: The Teleological Argument and Modern Science*, edited by Neil A. Manson (2003), provides a rich collection of essays written by a range of thinkers on all sides of the current debates, from confirmed atheists to proponents of creationism and intelligent design. More comprehensive examples of creationist thinking can be found in Michael Behe's *Darwin's Black Box: The Biochemical Challenge to Evolution* (1996) and Stephen Fuller's *Dissent Over Descent: Intelligent Design's Challenge to Darwinism* (2008); for an introduction to the other side of the debate, see Richard Dawkins' acerbic but readable *The God Delusion* (2006) and his earlier, less combative *The Blind Watchmaker: Why the Evidence of Evolution Reveals a Universe Without Design* (1996).

Those who wish to learn more of the ties between environmentalism, earth stewardship, and modern religious movements in United States are encouraged to consult the following works: Evan Berry's "Religious Environmentalism and Environmental Religion in America" (2013); Gregory E. Hitzhusen and Mary Evelyn Tucker's "The Potential of Religion for Earth Stewardship" (2013); Bron Taylor's two-part analysis of the interactions between the modern environmental movement and a range of religious faiths in his "Earth and Nature-Based Spirituality" (2001); and Catherine Albanese's comprehensive *Nature Religion in America: From the Algonkian Indians to the New Age* (1990).

Chapter Forty-Three

SEX AND SCIENCE

Miriam G. Reumann

The purview of scientific communities developing in the United States at the turn of the twentieth century included collecting and disseminating information about human and animal sexuality; theorizing what kinds of sexual behaviors and identities counted as appropriate or inappropriate; designing and overseeing treatments for those whose sexualities were deemed in need of repair; and contributing to national discussions of appropriate social and legal policy regarding sex. Decisions about what kinds of bodies and behaviors counted as normal or deviant, acceptable or in need of intervention, have driven research in the life and social sciences even as some with marginalized or stigmatized sexualized identities have simultaneously been excluded from contributing to scientific expertise. Scientific work on sex has, therefore, been an essential component of the larger processes of how modern science constituted itself as an authority as scientific models and forms of knowledge expanded into everyday Americans' lives.

The scientific disciplines that were most engaged in investigations of human sexuality from the late nineteenth century through the twentieth included general biology, endocrinology, psychiatry, and medicine (especially gynecology and, later, urology). Sexology, which soon emerged as the primary discipline of sexual science, stands at the center of this topic area, but this chapter will also focus on the reproductive sciences, medicine and psychiatry, and technologies of birth control. Sexual science sought legitimacy and authority through exploring beliefs, behaviors and bodies, defining the normal and abnormal, and authorizing how and when to intervene in both, while often navigating concerns about being overly controversial and insufficiently scientific.

Long before the emergence of modern American scientific disciplines and institutions, models of bodily sexual difference had posited men and women as separate kinds of beings, though these models reflected changing understandings of each gender's sexual function and capacity. Thomas Laqueur has traced a shift from a one-body model, in which women's and men's reproductive and sexual anatomy were seen as essentially

A Companion to the History of American Science, First Edition.
Edited by Georgina M. Montgomery and Mark A. Largent.
© 2016 John Wiley & Sons, Ltd. Published 2020 by John Wiley & Sons, Ltd.

similar but reversed, to a two-body model in which the two were viewed as utterly different (Laqueur 1990). In this conception, sexed bodies were seen not only as genitally distinct, but as saturated with masculinity or femininity at all levels. Against this backdrop, as modern science examined gendered and sexualized bodies, practitioners drew on a complex legacy of assumptions and associations.

Although the term was most likely coined by American reformer Elizabeth Willard in her 1867 tract *Sexology as the Philosophy of Life*, the discipline that would become sexology emerged out of the nineteenth-century ferment of new scientific ideas and emerging social norms in the Western world to become the central site for investigations of human sexual behavior. The late nineteenth century saw the rise of a new view of sex as reflecting not merely sets of behaviors, but inborn and essential characteristics of an individual shared across members of their social group. The assumption that sexual acts led to pathology (with, for example, masturbation implicated in insanity) gradually shifted to a different belief, that acts were instead symptoms of underlying, essential conditions or identities. Clinicians and theorists of this era created written taxonomies based on case histories, offering new vocabularies of sex. This new scientific and popular understanding of sexuality moved from a primary focus on reproduction to one on sexual variation and entitlement to pleasure. A number of late-nineteenth- and early twentieth-century European founders of the discipline produced works that expanded popular knowledge of sex, especially of sexual minorities, and often encouraged greater understanding and acceptance of them. Richard von Krafft-Ebing, whose 1886 *Psychopathia Sexualis* was among the most influential such works, expanded the category of normal by emphasizing that – as Harry Oosterhuis described it – "sexual behavior could be abnormal without being perverse," meaning that some varieties of unusual behaviors or fantasies (sadistic or masochistic, for example) could be accepted as normal as long as they were ultimately aimed at heterosexual coitus (Oosterhuis 2000: 47). Swiss psychiatrist Auguste Forel's 1905 *The Sexual Question* urged gender equality and the end of most laws regulating nonviolent sexual behavior, and the German sex researcher Magnus Hirschfeld, who founded the first institute devoted to sexology in Berlin in 1912, would encourage acceptance of homosexuality, transvestism, and fetishism as harmless variations of sexual instinct rather than dangerous pathologies (Forel 1905).

The concerns and approaches of these European researchers varied, but they shared a new terminology and understanding of "sexuality" that combined sexual impulses, emotions, and preferences, and not only located them (sometimes) within the body but also saw them as embedded in and reflective of an individual or group's essential self, whether conceptualized as the brain or the personality. This legacy would be crucial for Americans as the nation developed its own sexual science.

The Rise of Sexology in the United States

At the turn of the twentieth century, a growing number of scientists and a lay audience welcomed these new investigations, which were increasingly available to middle-class and educated Americans. Embarking on a campaign to incorporate attention to sex into their respective disciplinary programs, they laid the groundwork for modern American sexual science. The sweeping social changes attendant on urbanization and the rise of a consumer culture brought often-bewildering transformations in courtship and dating even as World War I focused public and governmental attention on issues of sexual hygiene and disease. Declining family size brought changing ideals of marriage and family life and use of contraception into greater public view, while discussion of sexual ethics and social policies, which expanded throughout the Progressive Era along with state interventions into public health, contributed to more open debates about sexual behavior. The self-conscious questioning of "Victorian" values of female innocence, modesty, and a double standard fostered greater discussions about the nature and meaning of human sexuality, as did changing gender norms and the emergence of a new therapeutic culture that promoted self-knowledge, happiness, and pleasure.

From its inception, sexual science in the United States had clear differences from its European counterpart. As researchers from a range of different disciplines worked to create a new field here, many of them sought to make their version of sexology distinctively American, often seeing the European version as too decadent, too focused on marginalized sexualities, and insufficiently modern or scientific, due in part to its reliance on the individual case study. American wartime discourse on sex had often focused on the differences between US and European mores, exemplified by the French military's generous offer to share their brothels with the United States Expeditionary Forces, and reactions to continental sexology followed this pattern.

Another crucial difference between American and European sexology was the role played by race. In the years after the Civil War, when American sexology began to develop, the policing of bodies – especially those on the border between black and white, male and female, normal or abnormal – fit into larger national trends toward both establishing professional authority among scientists and creating rules and norms for a rapidly growing and often splintering nation. Therefore racial and sexual differences would often be seen as intertwined, underwriting one another (Somerville 2002; Terry 1999).

Sex and Scientific Professionalization

All of these factors worked together to make scientific claims valuable capital. Lily Kay has described how, as the biological scientists were gaining credibility, funding, and new institutional power, "reformers seeking to develop and implement programs of social control increasingly capitalized on the resources and the rising authority of the social, biological, and physical sciences." Science was increasingly viewed during the Progressive Era as "a symbol of reason and efficiency, the fountainhead of objective knowledge and industrial prowess, a euphemism for technological and social progress" (Kay 1993: 24).

One crucial factor in the rise of sexual science in the United States was its navigation of disciplinary boundaries at a time of rapid growth and professionalization. In the first decades of the twentieth century, a variety of fields – embryology, genetics, and physiology, among others – were increasingly being grouped together as the "life sciences," a designation that emphasized the ways in which they collectively demonstrated the values of modern scientific practice. These included an emphasis on experimentation over observation, a focus on the laboratory rather than the field, and the use of new technologies, including an embrace of quantification and statistics (Cunningham and Williams 1992). These developments helped to further greater credibility and status among scientists, especially as biology expanded dramatically into public school courses, university programs, research institutes, and medical schools (Benson 1991).

The subset of disciplines that dealt with reproduction developed later than other branches of biomedicine that studied major organ systems, taking shape only in the early twentieth century in part because the male and female reproductive systems were seen as a less respectable or legitimate area of study than others. Endocrinology, in particular, grew rapidly from the 1920s on, becoming a recognized specialty. It benefited from a turn-of-the-century rise in both popular interest and scholarly research in hormones. In an older model, nerves – along with the quality and quantity of one's fat and/or blood – had been seen as governing functions linked to gender and sexuality. Nerves became increasingly displaced by a model privileging the role of chemicals, especially the substances dubbed "sex hormones," in regulating such bodily processes.

One of the chief ways in which sexual scientists could prove their modern scientific bona fides was through researching hormones. Spurred by the work of gynecologists and physiologists – as well as the interests of pharmaceutical companies hoping for profitable treatments – researchers and the public, as Nelly Oudshoorn phrased it, "came under the spell of the gonads" in the early twentieth century (Oudshoorn 1994: 20). Initial discussions of hormones reflected persistent confusion over the workings of gendered bodies. First viewed as "antagonistic" substances, due to the common assumption that men's and women's bodies produced distinctly different substances, many hormones instead turned out to be shared by both sexes, a finding that undercut any totally dualistic model of sexual difference. Hormones were viewed, as Lara Marks describes in her study of contraceptive research, as a "new magic bullet" promising all kinds of rejuvenation and Oudshoorn notes that "the introduction of sex hormones generated a revolutionary change in the study of sex" (Marks 2010: 43; Oudshoorn 1994: 37).

This development initiated a shift from anatomical to chemical definitions of sex, with many possible variations and degrees as "sex endocrinologists introduced a quantitative theory of sex and the body" in which "male" and "female" hormones were both present in men and women, in varying amounts (Oudshooorn 1994: 38). Many sexologists, as sociologist of science Adele Clarke argues, gained credibility for their field by focusing on the more "scientific" area of endocrinology rather than on human sexuality or direct contraceptive research (Clarke 1998). This association was powerful enough that it dominated the agenda of funding programs, as well as attracting attention through media coverage of dramatic stories of rejuvenation. Thus, a clear division emerged between human sex research based on surveys, case studies, and personal narratives and that which drew from laboratory-based reproductive science, with the latter viewed as more scientifically credible and less suspect. This split helped to shape the kinds of research that were proposed and funded.

Ongoing divisions within sexual science were also interwoven with transformations in medicine and psychiatry. During and after the Progressive Era, many physicians in the United States who were struggling to elevate their professional and economic standing also took sex into account as they positioned themselves as scientific authorities whose education and skills rendered them central to addressing the problems of a new age. Regular physicians staked their claim to expertise not only by denigrating the skills of other kinds of practitioners but also by leveraging public faith in science. Harry

Marks has pointed out that the "modernity of twentieth-century medicine consists in its reliance on the physical and biological sciences" as "it was generally believed that establishing medicine as a science meant grounding medical practices in one or more of the laboratory-based disciplines which study the functioning of biological organisms – biochemistry, physiology, genetics, immunology – disciplines which in turn were to be based on the sciences of physics and chemistry" (Marks 1997: 1). Many MDs saw themselves as the best and most obvious choice of experts to handle social problems such as sexual deviance or violence, often drawing on the metaphor of curing a diseased body politic to support their disciplinary claims over the management of bodies, but endocrinology as a field aimed to treat some of the same presenting issues as medicine and psychiatry, including inappropriate gender presentation, sexual unhappiness, and infertility. Its advocates maintained that prioritizing bodily rather than psychological models of disease and cure rendered their approach more modern and scientific.

These tensions beset American psychiatry at a time when the profession was similarly expanding, creating competition over which fields could best lay laid claim to authority over sexuality. As Americans familiar with Freud – or, at least, with a transplanted version of his ideas – turned to therapeutic experts, sexuality was central in the gradual transformation of psychiatry from a low-status job, concerned largely with custodial care in asylums, to a culturally significant and higher-status profession concerned with intimate problems of everyday life. This process often incorporated a break between strict notions of normality and pathology.

Psychiatrists and social workers in the early twentieth century based much of their practice on individual patients and case studies but increasingly incorporated quantitative methodologies such as mental testing precisely because they seemed to offer a way to approach problems of daily life in a manner that was scientific rather than moral, and objective rather than subjective. Not surprisingly, this period saw the creation of new disease concepts and expansion of older ones, as well as an increasing interrogation of the "normal." Claiming ownership over sexual classification – which offered a host of diagnostic problems and new routes to authority for the human sciences –was, for psychiatric theorists and practitioners, crucial to the process of professionalization (Lunbeck 1995: 115).

Birth control advocates were another group who posed problems for professionalizing sexual scientists. Some chose to ally with them, albeit cautiously: the British Havelock Ellis, who along with Krafft-Ebing was the European authority most likely to be familiar to Americans, often worked with sex radicals and birth control advocates, and American physician and sexologist Robert Latou Dickinson led the National Committee on Maternal Health to promote use of contraception in the early twentieth century by, as Dickinson saw it, replacing suspect "propaganda" with trustworthy "science" (Reed 1978: 181–2). Most sexual scientists, however – especially those situated in academic departments rather than in private practice or with independent funding – avoided such potentially awkward associations.

If most scientists avoided close contacts with birth control activists, they were far more enthusiastic in embracing a different aspect of reproductive control: the tidal wave of eugenics. There was great interest among scientists and much of the general public in programs of positive eugenics (often known as "racial betterment," encouraging reproduction among those deemed most fit) as well as "negative" eugenics (efforts to restrict reproduction among those seen as unfit, including state-level sterilization programs). Many institutions shifted from segregating and incarcerating women deemed mentally deficient to focusing on sterilization, transforming "feeblemindedness" from a "treatable disease to a sexually-loaded, gender-specific permanent condition" (Kline 2001). The post-World War I United States saw a surge both in legislative activity relating to sterilization and in actual procedures. By 1926, 23 states had sterilization laws, and many of those had active programs, mostly in institutions. The 1924 case of *Buck v. Bell*, in which Virginia's sterilization law was declared constitutional by the Supreme Court, was followed by a flood of new statutes: by 1931, 28 states enabled sterilization of the mentally or morally unfit (Reilly 1991: 67–8). Sterilization enjoyed great popular and political approval, especially given the economic constraints of the 1930s, and the role of scientific and medical endorsement in its adoption and general acceptance offers an indication of how the authority of sexual science was on the rise.

Popular and Scholarly Sexology after World War I

By the late 1920s, popular demand for usable sexual information both complicated and helped to drive the field of sexual science in the United States. A significant number of sexual surveys and case studies, driven by greater interest in and open discussion of both heterosexual marriage and same-sex relationships, gained increasing public attention during the 1920s. In the context of expanding public roles for women, some re-envisioning of marriage, and a rising divorce rate, many moderns questioned aspects of Victorian culture cautiously expressing hope that sex research could help promote human happiness, along with stable marriages and societies. The costs of pursuing human sexology could be high: in the 1920s, several university-based sex surveys led to scandals, and such investigations could pose risks to one's reputation and job. Nevertheless, new research studies on humans were funded and published, carving out a new cultural space between popular titillation on the one hand and incomprehensible scientific jargon on the other.

American studies that existed before the 1920s tended to focus on commercialized vice and were often carried out by morals groups. That changed when a series of sizable and complex surveys were published between 1929 and 1932: penal reformer Katharine Bement Davis' *Factors in the Sex Life of Twenty Two Hundred Women* (1929); Gilbert Hamilton's *A Research in Marriage* (1929); and Lura Beam and Robert Latou Dickinson's *A Thousand Marriages: A Medical Study of Sex Adjustment* (1931) and *The Single Woman* (1934). These volumes were widely seen as something new, the first examples of a distinctly different mode of sexual investigation. Blending narrative case studies with statistical charts and social analysis, they inquired into the sexual behaviors of so-called "average" or "normal" populations rather than focusing on "deviant" ones, encouraging readers to draw implications for their own lives. Marking an important shift in which sexual desire and its fulfillment was increasingly often conceptualized as healthy for the individual and the social body, as well as necessary to good marriages, these works collectively helped to establish sexology as a more legitimate area of exploration and made its existence more familiar to a broader audience.

Each of these investigations had received funding from the same organization, the John D. Rockefeller, Jr.-backed Committee for Research in Problems of Sex, established in 1923. The committee's founding was a sign both of increasing interest in sex and society and of the funder's recognition that it was too shocking a topic for most sources to support. Most initial projects funded were clinical in nature, often animal rather than human studies; after helping to fund a spate of 1920s studies, the committee overwhelmingly favored animal and hormonal studies until it took an interest in Alfred Kinsey's work in the 1940s, perhaps viewing the former as more scientific and less controversial (Corner and Aberle 1953).

The advent of World War II and its aftermath, which promoted increasingly open discussion of issues of sexual behavior, saw further expansion of the trends that had prevailed during the interwar period: greater popular interest, more public discussion, and an insistence on large numbers and/or technological advances as the most accurate and scientific modes through which to study human sexuality. Increasingly, sex surveys would be seen as fuelling, not just documenting, sexual change.

The postwar science of sex was dominated by the 1948 and 1953 Kinsey Reports on male and female sexuality, making biologist Kinsey, previously known for his taxonomic work on gall wasps, a controversial household name. His major findings – that many more Americans than previously thought engaged in sex acts that were pre- and extra-marital, homosexual, or otherwise "deviant" – were used to underwrite both panicked jeremiads about the state of the nation and calls for greater tolerance and openness (Reumann 2005). The Reports demonstrated the continuing shift away from case studies in American sexual science; their focus on measurement was also quintessentially modern, reflecting the veneration of new and improved technologies and respect for quantification in general. Statistics appealed to investigators, funding agencies, and readers because their use blended long-accepted beliefs – that bodies and behaviors, in large numbers, reflected important data that could be accurately measured – with a new embrace of increasingly accurate and detailed measurement, exemplified by the Kinsey team's data collection and sorting via punch-card technology (Drucker 2014). In the twentieth-century United States, the large quantitative study largely replaced the case history as the privileged site for producing valid information about sex. Krafft-Ebing's *Psychopathia Sexualis*, for example, helped to establish the field in late nineteenth-century Europe by focusing on case histories dealing with masturbation, fantasy, homosexuality, and a wide range of non-normative sexualities (many contributed by the sexologist's own patients and correspondents), while Kinsey's voluminous studies more than a half-century later in the United States would also use self-reporting but would claim scientific authority through their massive sample size and use of statistical analysis, and Masters and Johnston's 1960s analyses would gain credibility and attention largely because of their use of modern technologies.

The bestselling status of the Reports helped bring scientific sexology to the attention of a mass public, a development on which future sexologists would build (Igo 2008; Irvine 1990; Reumann 2005; Terry 1999). Marriage education, which reached an increasing audience through both academic courses and popular advice literature, further popularized both the specific findings and recommendations of sexologists and their larger presence in a therapeutic and pro-marriage culture.

The post-World War II period also saw the continuation of a shift in the place of contraception and the role of birth control advocates, away from a more radical, sex-positive, activist-controlled model to an emphasis on contraception as a form of sex hygiene involving positive eugenics for the middle classes and promising greater happiness in marriage. Most reproductive scientists in this era distanced themselves from birth control activists, promoted medical control over reproduction rather than user control, and advocated for basic research rather than applied, since searching for human contraceptives carried major risks to reputation and funding.

The long "sexual revolution" and upheavals of the 1950s, 1960s, and 1970s would further cement the professional and popular links between sexual science and technology, as sexual surveys were largely replaced by pharmaceutical research and cultural narratives about sex became increasingly distant from scientific ones. Second-wave feminists, many of whom were deeply suspicious of professional science, often called for writing about sexuality that was personal, subjective, and individual, more like the nineteenth-century case study or confession than the more recent quantitative or clinical model of sexual science.

Theorizing Sexual Identities

A continuing theme in both sexual science and analyses of it is the role of such work in creating or shaping sexual identity over time. The emergence of various categories of sexual "perverts" is a classic example of what Ian Hacking has dubbed "making up people," a process in which "[n]ew slots were created in which to fit and enumerate people," and in which individuals "spontaneously come to fit their categories" (Hacking 1999: 161). Sexology's appearance as a discipline raises questions about the extent to which it reflected or created the kinds of identities it described: how did the appearance of the homosexual or the masochist as new taxonomic categories relate to the rise of groups organized around these new identities? The extent to which at least some informants engaged with the experts who popularized these categories – contributing their own life histories, practices, and slang, but also offering their own diagnoses, arguing back against some expert conclusions – suggests a complex relationship between subject and object (Oosterhuis 2000; Terry 1999). Ivan Dalley-Crozier has called attention to how subjects found in European sexological treatises "sometimes theorize their own condition" by offering suggestions or even authoritative declarations about the etiology and diagnosis of their sexual behaviors in autobiographical essays or other utterances, and Harry Oosterhuis' work similarly notes how many of Krafft-Ebing's clinical illustrations came from his own patients and from interested laypeople who corresponded with him and shared their own autobiographical narratives (Dalley-Crozier 2000: 114; Oosterhuis 2000). Oosterhuis characterizes many of these correspondences, some of which lasted for years, as mutual exchanges of information in which "perverts sought both explanations for their own longings and behaviors and validation from a respected professional, while in exchange Krafft-Ebing received valuable raw material" (Oosterhuis 2000: 186). Overall, as sexual science gained credibility and was successful in publicizing its findings, the spread of sexological ideas among educated Americans went hand in hand with an increasing scrutiny of the self and a heightened concern about the "normalcy" of one's own sexual actions and desires. For educated Americans, codes of sexual "normality" were increasingly often related to the teachings of sexual science, rather than religious or moral traditions.

Investigations of bodies deemed to show ambiguous genitalia offer a case in point. Sexual scientists often saw intersex bodies as a topic on which they could offer authoritative knowledge, laying claim to the correct determination of sex when called upon to adjudicate and treat cases of sexual ambiguity and change. The proper categorization and treatment of intersexual cases throughout much of the twentieth century was contested, as authorities debated how or whether to assign primacy to an individual's internal or external biological structures, score on psychological tests, or preferred gender identity. A number of historians link the scientific debates over intersex conditions to larger changes in the role of science, arguing that medico-scientific authority over and management of sex reached new heights with the rise of intersexed people. According to Alice Dreger, for Europeans the figure of the hermaphrodite aided in the "medical invention of sex," even as medical experts shifted over time between reliance on external genital morphology, hormones, gonadal structure, and psychology (Dreger 1998). In the United States, as Elizabeth Reis describes, Americans in the nineteenth and twentieth centuries moved from an earlier religious and moral framework for assessing intersexed bodies to a scientific one (Reis 2009). Bernice Hausman's analysis picks up at mid-century, arguing that post-World War II technology not only enabled transsexuals to express and make manifest their desires for new bodies; it actually created the category of the transsexual as a brand new and contingent sexual subjectivity (Hausman 1995: 2–3). As medical protocols and popular narratives of intersex continue to change in response to community activism, intersex provides an especially dramatic model of how the conventions of sexual science can alter.

As they developed a scientific community, American sex researchers who measured bodies, harvested organs, assayed hormones, and asked questions about fantasy and behavior from the late nineteenth century on shared a common assumption that they were participating in a common project. Some consistent trends unified sexual science over time, such as the recurring search for scientific magic bullets to increase desire and performance demonstrated both by the 1920s craze for glandular transplants and extracts and the more recent and ongoing embrace of pharmaceuticals like Viagra (Loe 2006). But more commonly, the tensions between their different approaches, which manifested in debates over funding support, methodology, and scientific merit, hampered the field as a whole. Sex scientists in many ways contributed to their own marginalization by their conflicts over how they wanted to proceed, laying claim to "science" by variously measuring bodies and secretions, embracing quantification, or turning to technology, while remaining ambivalent about social science methodologies and use of the case study. Historian Sharon Ullman notes in her study of early twentieth-century sexual behavior and culture that "[a]lthough sexuality had begun to emerge as the center of personal identity, the language for that concept did not yet exist – except within the rhetoric of vice and prostitution" (Ullman 1998: 104). Twentieth-century sexologists, in concert with some of their research subjects and against a background of increasing confidence in science, helped to formulate such language, creating models of proper, healthy sexual attitudes and behaviors that would influence generations to come.

Bibliographic Essay

Scholars interested in the intertwined histories of science and sex in the United States can draw from several subfields: the histories of sexuality, medicine and psychiatry, gender, and – to a lesser extent – studies of specific physical conditions, the biological and physical sciences, and institutions and individuals.

A useful grounding in the professional development of sexual science is provided by the voluminous literature on the history of the biological sciences; a few of the most relevant sources are Merrily Borell, *The Biological Sciences in the Twentieth Century* (1989), Andrew Cunningham and Perry Williams (eds.), *The Laboratory Revolution in Medicine* (1992) and Keith R. Benson, Jane Maienschen, and Ronald Rainger (eds.), *The Expansion of American Biology* (1991).

Studies of reproductive medicine are particularly relevant for the intersection of sex and science; these include Adele E. Clarke's *Disciplining Reproduction: Modernity, American Life Sciences, and the Problems of Sex* (1998) and Nelly Oudshoorn's *Beyond the Natural Body: An Archaeology of Sex Hormones* (1994). On scientific determinations of gender identity, see Marianne van den Wijngaard's *Reinventing the Sexes: The Biomedical Construction of Femininity and Masculinity* (1997). Caroline Herbst Lewis' *Prescription for Heterosexuality: Sexual Citizenship in the Cold War Era* (2013) provides a more focused case study by exploring how mid-century physicians, especially gynecologists, interpreted and enacted their roles as educators and translators of sexual science to patients. Among the many studies of homosexuality, those focusing most on medicine and science include Jennifer Terry, *An American Obsession: Science, Medicine, and Homosexuality in Modern Society* (1999) and Henry Minton, *Departing from Deviance* (2001).

In a large literature on sex, gender, and eugenics science, standout analyses include Laura Briggs, *Reproducing Empire: Race, Sex, Science, and U.S. Imperialism in Puerto Rico* (2002); Wendy Kline's *Building a Better Race: Gender, Sexuality, and Eugenics from the Turn of the Century to the Baby Boom* (2001), Alexandra Minna Stern's *Eugenic Nation: Faults and Frontiers of Better Breeding in Modern America* (2005) and Johanna Schoen's *Choice and Coercion: Birth Control, Sterilization, and Abortion in Public Health and Welfare* (2005).

A similarly rich literature has examined the history of contraception in the United States, including pioneering studies such as James Reed's *From Private Vice to Public Virtue: The Birth Control Movement and American Society since 1830* (1978) and Linda Gordon's *Woman's Body, Woman's Right: Birth Control in America* (1990), which examine the development and political fortunes of birth control, while Andrea Tone's *Devices and Desires: A History of Contraceptives in America* (2002) focuses on the technological and entrepreneurial sides of birth control manufacture and use. The development, reception, and use of the oral contraceptive pill has received by far the most attention from historians. Lara Marks' *Sexual Chemistry: A History of the Contraceptive Pill* (2010) focuses on the scientific networks that produced it, while Elizabeth Watkins' *On the Pill: Social History of Oral Contraceptives, 1950–1970* (1998) and Elaine Tyler May's *America and the Pill: A History of Promise, Peril, and Liberation* (2010) are more attentive to its cultural and social meanings.

In a growing literature on sexology as a discipline, George Corner and Sophie Eberle's mid-century history of the early years of the Committee for Research in Problems of Sex, *Twenty-Five Years of Sex Research: A History of the National Research Council Committee for Research in Problems of Sex, 1922–1947* (1953), contains little context for or analysis of the rise of American sex research but offers valuable primary source information on what kinds of projects weere solicited, submitted, and funded by the major funder of American sex research during its first quarter-century. One of the first serious examinations of Kinsey was Regina Markell Morantz's "The Scientist as Sex Crusader:

Alfred Kinsey and American Culture" (1977), which remains the most valuable brief overview. Julia A. Ericksen and Sally Steffen's *Kiss and Tell: Surveying Sex in the Twentieth Century* (2002) offers a general survey of American sexology with a sociological bent, focusing on the basic methodologies and findings of modern sexologists, while Janice Irvine goes further, both examining the context and reception of major sex studies, and analyzing the the development of the discipline's frameworks and assumptions in *Disorders of Desire: Sex and Gender in Modern American Sexology* (1990).

Another valuable resource for scholars interested in this area is studies of individual sex researchers and their work and social worlds. Alfred Kinsey has come in for particular attention in duelling biographies, James Jones' detailed but pathologizing *Alfred C. Kinsey: A Life* (2004) and Jonathan Gathorne-Hardy's more sympathetic *Kinsey: Sex the Measure of All Things* (2004). Sarah Igo's *The Averaged American: Surveys, Citizens, and the Making of a Mass Public* (2008) considers Kinsey's place among mass surveys of the twentieth century and Miriam Reumann's *American Sexual Character: Sex, Gender, and National Identity in the Kinsey Reports* (2005) examines how the Reports were taken up in national discussions of both politics and intimate life, while Donna Drucker's *The Classification of Sex: Alfred Kinsey and the Organization of Knowledge* (2014) explores Kinsey's approach to collecting, organizing, and tabulating data, a subject that preoccupied many twentieth-century sex researchers.

The history of intersex and of medical forms of sex change have elicited a great deal of attention, illuminating as they do the changing relationships between sexual identity, gender, and scientific intervention. Alice Domurat Dreger's *Hermaphrodites and the Medical Invention of Sex* (1998) focuses mostly on European ideas and practices, while Elizabeth Reis' *Bodies in Doubt: An American History of Intersex* (2009) examines significant American developments from the colonial era through the recent past. Additional valuable studies of the United States are Joanne Meyerowitz's *How Sex Changed: A History of Transsexuality in the United States* (2002), which incorporates patients' perspectives into her analysis of changes in the status of transsexuality and Berenice Hausman's *Changing Sex: Transsexualism, Technology, and the Idea of Gender* (1995), which concentrates more on changing technologies and the implications of sex reassignment for the meanings of gender and identity.

Despite the flood of work on sexuality in the last few decades, many aspects of scientific culture and practice have been overlooked in the field. Recent work on national and local patterns in birth control use and sex education has added to our picture of how people interacted with and negotiated scientific innovations, but much work remains to be done on the relations between professional and popular discourses of sexual science, such as how experts' writings on sexuality trickled down to or were reinterpreted for popular audiences, whether in the clinic, the classroom, or the pages of advice columns. Biographical studies of individuals could be built on and augmented with more work on larger institutional patterns of the impact of sexual science. We especially lack analyses of the role of sexuality in disciplines and studies that were not explicitly focused on gender difference or concerned with hormones, gonads, or sexual behavior. Much as analyses attentive to gender have illuminated the ways in which assumptions about masculinity and femininity have structured the broader culture and practice of science, analyses of other seemingly unrelated scientific disciplines and studies hold promise to illuminate how they may have reflected and shaped beliefs about sexuality in unintended ways. We lack work on specific subdisciplines and institutions beyond a few central sites: how did the sexual sciences work in far-flung or marginal locations? More broadly and speculatively, given that most work that brings together the histories of science and sexuality privilege one over the other, what would it look like to take both at once as central, and put sex and science in dialogue? This might mean looking for commonalities as well as differences in the mostly separate histories of homosexuality and heterosexuality; given that both categories and the identities they helped shape were constituted in part by scientific experts, what broader processes can we see at work when we look at them together?

Chapter Forty-Four

ZOOS AND AQUARIUMS

Christian C. Young

In the midst of World War II in Europe, a biologist working at the public zoo in Bern, Switzerland, began publishing a view of animals in captivity that reshaped the work of scientists, zoo administrators, and exhibit designers. Heini Hediger became known as the father of zoo biology. His studies suggested that all preceding efforts to keep wild animals on public display had neglected basic principles of animal behavior, and translations of his work soon began to influence zoo design around the world (Hediger 1942, 1969). Yet, this revolution – as it has sometimes been called – stands in contrast to a more contextualized narrative of the history of scientific work in zoos and aquariums. That narrative includes an understanding of natural history studies dating back to the eighteenth century. Hediger's revolution clearly had an impact on the way Americans and American scientists viewed the prospect of keeping wild animals in captivity, but framed against the backdrop of the American scientific enterprise, this appears as only one among many important influences on the scientific community.

In tracing the development of science in the United States, institutions that support scientific work provide benchmarks for comparison with developments in Europe and other parts of the world. Zoos and aquariums can serve as those benchmarks, recognizing that the emergence of public institutions for the display of animals began to take place in the nineteenth century, usually in larger cities, and often with support from existing scientific societies in those cities. Zoos of this kind were built first in Paris, Vienna, and London, with nearly 20 more built by the end of the century, as well as a number of short-lived public aquariums. In the United States, where fledgling scientific societies in newer cities confronted seemingly boundless opportunities for natural history studies, zoos claimed lower priority. Still, by the 1890s, many cities followed the lead of Philadelphia and New York and built their own zoos and aquariums. City leaders hoped such institutions would elevate them from quaint, frontier trading depots

A Companion to the History of American Science, First Edition.
Edited by Georgina M. Montgomery and Mark A. Largent.
© 2016 John Wiley & Sons, Ltd. Published 2020 by John Wiley & Sons, Ltd.

to estimable cultural centers. As such, the appearance of zoos in those cities suggests something of the vibrancy of scientific communities across the country.

To understand the growth and change that took place in science, an examination of zoos and aquariums reveals the opportunities and challenges of scientists associating themselves with those institutions. Scientists, then as now, rarely choose to study animals behind bars or glass. On one level, zoos and aquariums serve as a poor substitute for field and laboratory work. But these cultural institutions stand at an important intersection between field and laboratory, and between scientists and public exhibitions of scientific knowledge. By exploring zoos as a surrogate for the field, or considering aquariums as simulated habitats for laboratories, new questions about those institutions emerge even as old questions reemerge with added urgency. Moreover, the tension within the scientific community over maintaining these public spaces – expensive to an extreme and encumbered with popular notions of what animals are appropriate for display – exposes additional ways of asking questions of how scientists respond to broader interest in their studies. Zoos provide ready access for the public, while the locale for a field study is remote. Historians of science have long been interested in the study of field stations, but few accounts focus on zoos as research centers. Historians can learn more about zoos as research centers by drawing on the corollary of the field station. By contrast, the scientific lab may be nearby, but it is an isolated space, hidden from public view. Zoos and aquariums put scientific knowledge on display in a way that makes nature itself accessible to a larger audience. To varying degrees, these places put the processes of scientific discovery on display as well. From modest beginnings for such public displays, American zoos and aquariums reveal key aspects of the development of scientific communities.

Shifting Traditions of Natural History in the United States

Zoos and aquariums occupy a unique position in the history of natural history in that they generally resisted the increasing specialization that occurred in the fields of science that emerged in the late nineteenth century. Even as botany, zoology, ethology, ecology, genetics, anatomy, physiology, and other subdivisions arose with more narrowly focused practitioners, zoos and aquariums remained institutions where broad knowledge of natural history was needed. The methods of naturalists in these places also retained the important observational techniques of preceding generations. The ability to move out from the cage or tank and into the field in pursuit of additional specimens added, for many zoo and aquarium employees, important flexibility. Collecting in the wild remained at least a part of the practice of these keepers well into the 1930s, and for some institutions, well beyond that date (Hanson 2002).

In Europe, early zoos in large cities tended to have close ties with scientific societies and institutions in those locations. In general, the prior existence of those institutions provided an unsurprising basis for the establishment of zoos that would attach to scientific credentials. In the United States, where scientific societies emerged as cities grew, the pattern persisted. Self-trained naturalists, physicians, and collectors with wide-ranging interests gathered living and preserved artifacts from their surroundings. Philadelphia, home to the earliest American scientific community, became the first city to follow the European model of organizing a zoological collection. (The earlier establishment of a collection in Central Park in New York can only be seen as an early, successful, and continuous evolution from menagerie to zoo that was incomplete until the mid-twentieth century.) With physician William Camac as its chief promoter, Philadelphia's scientific community began planning its zoo in the 1850s. They hoped for permanent buildings and a full-time staff. These hopes went unrealized until after the Civil War, but in 1872, they hired an experienced scientific zoo man from the public zoological garden in Hamburg, Germany, and opened their collection to the public (Hanson 2002).

When the Smithsonian Institution embarked on its plans to develop its own zoo, it already had a small menagerie of animals behind the castle on the mall, just a few blocks from the US Capitol building. With a vibrant natural history museum and active cadre of scientific men and women, it seemed a logical institutional home for a prominent zoo. Establishing a zoological garden on 160 acres in Rock Creek Park became the task of William T. Hornaday in 1890. Instead of another small, cheap, smelly menagerie with limited scientific interest, the Smithsonian expected this public institution to serve as a national institution for scientific research (Hanson 2002).

The first zoo directors emerged from professional careers that represented the range of training of American naturalists. Several came into the profession from the US Bureau of Biological Survey, including Edmund Heller and John Alden Loring (Hanson 2002). Charles Haskell Townsend began his career with the US Fish Commission in 1883, studying salmon stocks in California. After assignments in the Aleutian Islands of Alaska, he moved east to become the director of the New York Zoological Society's aquarium in Battery Park. He remained with the aquarium when it reopened in Coney Island (Reidy, Kroll, and Conway 2007). These and other government biologists had become part of a network of specialists, some formally trained and others less so, but with years of experience in the field. The origins of that network date back to the middle of the nineteenth century, when Spencer Baird and C. Hart Merriam enlisted the assistance of college graduates and farm boys (Dupree 1957). While taxidermy served as an important hobby for young men who grew up hunting, most of the specimens collected in the nineteenth century were not suitable for museum display. Collecting for zoos, obviously, was even more demanding for the amateur.

Many zoos built on the natural history collections of the surrounding community. Most continued to accept donations of animals well into the middle of the twentieth century. Although the donation of exotic animals by wealthy philanthropists sometimes provided a zoo with credibility and established new excitement throughout the community, modest contributions of local animals could help to fill exhibits and attract crowds of visitors. The public might line up to see a baby badger donated by a resident of the city, even if it lived just a few days in captivity. During a single week in late 1931, the Milwaukee Zoo received donations of a bald eagle, a woodchuck, and a fox squirrel. Most zoos kept records of all such donations, and the lists contain a surprising diversity (Zoological Society of Milwaukee Archive).

· Commercial collectors served as another source of animals for exhibits. These suppliers took orders for sought after species. European zoos and their collectors provided animals to American zoos for decades, but American collectors moved into the market with varying degrees of commercial success. The Cincinnati Zoo, under the directorship of Sol Stephan, became an important holding center for collectors who had imported exotic animals in anticipation of sales to more far-flung institutions (Hanson 2002).

Independent collectors achieved success both in providing zoos with specimens and in providing the public with riveting accounts of their collecting adventures. A few published books on wild animal collection read like safari travelogues. While these accounts highlighted the popular interest of zoo work, many zoo collectors maintained a clear scientific agenda (Hanson 2002).

In cities across the country before 1890, short-lived menageries featuring deer and unusual livestock became common. After 1890, following the institutionalized examples in large eastern cities, the rapidly expanding mid-western cities of St. Louis, Milwaukee, St. Paul, and Omaha established zoos with more exotic animals. Smaller cities like Springfield, Missouri; Grand Rapids, Michigan; and Bloomington, Illinois followed suit. These public exhibits did not typically involve scientifically trained staff, nor did those cities support significant numbers of scientists interested in the functioning of zoos or the natural history of the animals there (Kisling 2001). The collections were, however, increasingly organized around the principles of zoos established by scientific communities in Europe.

The Influence of European Models of Science on American Institutions

During the nineteenth century, an impressive expansion of zoos took place in major European cities. Paris had public animal exhibits in the Jardin des Plantes before 1800, as did the royal Austrian zoo near Vienna. London's public zoo opened in 1828. Somewhat later in Germany, zoos built by scientific communities served as emblems of engagement with the larger world. Scientific work in zoos contributed to the care, presentation, and breeding of animals between the 1850s and 1870s, distinguishing zoos from other collections of animals and menageries. Zoos provided institutional support that employed naturalists. The professional demands of creating exhibits and caring for animals established an incentive for the scientific community to expand. Exhibits required imagination, entrepreneurial schemes, and meaningful connections to cultural iconography. While naturalists working in zoos may not have consistently led these efforts, they became involved. Their understanding of the food, shelter, and space requirements of animals, along with emerging awareness of breeding seasons and habits contributed to exhibit design. Zoo directors could arrange exhibits where the activities of animals – not their mere presence – would be on display (Nyhart 2009). Out of these connections came a growing awareness that nature protection also relied on a broader public appreciation of animal lives in more sophisticated exhibit settings.

Zoo directors and boosters sometimes exaggerated the cultural significance of their institutions, while naturalists worked to achieve more specific goals that included learning about the classification of peculiar and exotic species in their collections. To a large extent, the taxonomic groups they housed together became standard in both scientific and public understandings of animal families. This pattern was to recur at the end of the nineteenth century and well into the twentieth in the United States. Early efforts to acclimatize animals in zoos can be traced to the 1850s in Paris and the 1860s in Cologne (Osborne 1994). The efforts were short-lived as some planners wanted to provide more natural conditions in order to keep animals in their native state (Nyhart 2009). By exhibiting live animals, zoos could move beyond the limits of taxidermy and museums, with their dead animals and static collections. Naturalists involved in planning new exhibits began testing the limits of landscape design, expanding the biological perspective for an increasingly intrigued public.

The earliest public aquariums also addressed a desire among Europeans to view the inaccessible marine world. Small, private aquariums offered glimpses of plants and fish to wealthy hobbyists. The large-scale aquarium was an English innovation, reinvented in 1860 as an aquarium-salon built in Vienna, and an in-ground aquarium at the Hamburg zoo built in 1863 (Nyhart 2009). Taking on the challenge of size for a public exhibit of either salt- or freshwater aquariums, designers struggled to overcome technological obstacles and to maintain viable conditions. Those difficulties persisted in American attempts to create aquariums.

Standing at the intersection of European and American zoos at the end of the nineteenth and into the early twentieth centuries, Carl Hagenbeck influenced the way zoos obtained and displayed their collections around the world. Near Hamburg, Hagenbeck's revolutionary approach to zoo design illustrated first and foremost his entrepreneurial sense. Many of the improvements on display there reflected his commitment to allowing animals to roam in spaces that appeared open and unconfined. The grouping of animals around certain habitat themes was praised as more scientifically sophisticated. In the end, Hagenbeck succeeded in creating a cost-effective means of housing the many animals that moved through his park and on to other collections and zoos. The work of his staff to encourage the study of "acclimatization" among animals that might not otherwise be suited to the seasonal changes of northern Germany was best understood as a less expensive means of keeping a wider variety of animals than could be accomplished in climate-controlled buildings. Naturalists in zoos elsewhere expanded their attempts to that same end (Rothfels 2002).

In Europe and the United States, the increasing number of zoos and aquariums provided occupational opportunities for naturalists. Unlike careers as taxidermists or school teachers, which typically required longer training and specialization, zoo naturalists learned on the job and were able to advance in their work. Natural history existed within a network of scientists who did hands-on work in raising the level of appreciation of nature in society (Nyhart 2009).

Animal displays appeared in American cities as those cities developed into centers of commerce and culture. Animal displays to satisfy public fascination preceded scientific infrastructure, and yet, certain parallels in the work of collecting specimens for study and for display illustrated the accumulated work of naturalists in America. The Peale Museum in Philadelphia, founded in 1784, famously exhibited a rhinoceros from 1826 to 1829, delighting the public and illustrating the expansion of US exploration and collection in the early nineteenth century (Reidy, Kroll, and Conway 2007). In an effort to organize menagerie collections in multiple cities in the 1830s, owners of those collections formed the Zoological Institute. That one business organization lasted less than three years, however, and illustrates the lack of coherence in the broader goals of the animal display enterprise (Kisling 2001). In Philadelphia, where the Bartram Botanical Garden had been established in 1731 and the Peale Museum provided a clearer model for public exhibitions, the city's Zoological Society began planning a zoo shortly after its founding in 1859 (Hanson 2002).

While European zoos and aquariums provided important models for their American counterparts, the development of successful and enduring institutions in the United States depended on the support of scientific societies and tailoring exhibits to the

American public. In both of these senses, zoos had to be built in distinctively American ways.

The Role of Public Institutions in Supporting Scientific Work

The first call for a nationally sponsored collection of animals for the public came from Joel R. Poinsett in 1841. Writing for the National Institution for the Promotion of Science, Poinsett proposed a zoological garden, housed by the Smithsonian Institution, based on the model of the Jardin des Plantes in Paris. While the proposal circulated for five years in Washington, no planning took place on the ground. In Boston, James Cutting followed the example of a popular exhibit of tanks to establish a public Boston Aquarial Gardens in 1859. Three years later, P.T. Barnum bought the aquarium, but it closed shortly thereafter. On the near side of Cape Cod, the US Commission of Fish and Fisheries opened an aquarium in the mid-1870s. The Woods Hole Science Aquarium, still operated on the same site by the federal government, can be considered the oldest institution of its kind in the country. The great New York Aquarium opened in 1876, but soon closed. Another aquarium in that city would be decades in the planning.

Three of the earliest and continuously operating American zoos – in Philadelphia, Washington DC, and the Bronx – provide important illustrations of how the scientific community became involved in these institutions. The Philadelphia Zoological Society finally succeeded in opening its zoo in 1872, after more than a decade of delay due mostly to the unrest of the Civil War. The zoo benefited from the donation of animals from a popular but meager display assembled in cages on the mall near the Smithsonian castle. Public expectations for a menagerie in the capital city had been piqued, however, and Congress responded to calls for a national zoo over the next decade by setting aside land in Rock Creek Park. In 1888, permanent exhibits containing 225 animals were on display in the park. Famed naturalist and big-game hunter William T. Hornaday organized the zoo and served as its first director (Mann 1930).

Members of the New York Zoological Society, all civic leaders in New York City, decided it was time for them to open a world-class zoological park in their city. They enticed Hornaday to join their effort – he had left the fledgling National Zoo after differences of opinion placed his authority in question – and expand their vision for an even grander effort. Equal to the task, Hornaday combined his experience as a collector, taxidermist, exhibit designer, and naturalist with a growing passion for wildlife conservation to establish a coherent mission for the New York Zoological Park, the Bronx Zoo. Hornaday has influenced how zoos have been run ever since (Bechtel 2012). By the beginning of the twentieth century, 25 other cities had built zoos and public aquariums. By 1949, 77 zoos and 6 aquariums had been established.

American zoo designers in the early twentieth century began to create exhibits without bars, a practice that received increased attention after Carl Hagenbeck opened his wild animal park in Hamburg in 1907. Both the Bronx Zoo and the National Zoo featured large enclosures for groups of animals, a result of the greater acreage zoo designers could access in expanding cities. For other zoos, directors drew special attention to new and renovated spaces that resembled Hagenbeck's designs. The Denver Zoo, which had opened in 1896, introduced barless exhibits in 1918, followed by St. Louis the next year. The Detroit Zoo added new exhibits that replaced caged enclosures in the 1920s, as did Chicago and Cincinnati in the 1930s.

In 1901, the Philadelphia Zoo opened the Penrose Research Laboratory, the first such research lab for veterinary pathology and animal nutrition. Like some European zoos, workers in the Penrose lab explored the question of acclimatization of tropical animals in their temperate climate. In a notable advance, researchers used glass enclosures for primates in the hope of preventing the spread of disease from the humans who otherwise stood on the other side of bars in the open air, sharing pathogens that could infect gorillas, chimpanzees, and other apes. Treatment of tuberculosis in apes was a particular focus of Philadelphia scientists at the zoo (Kisling 2001).

At the Bronx Zoo, the New York Zoological Society established the scientific journal *Zoologica* in 1907, a publication produced continuously until 1973. As evidence of the scientific work done in conjunction with that zoo, *Zoologica* included detailed articles on collection expeditions both on land and at sea. These articles provided practical examples that could serve zoo naturalists as well as zoologists at other institutions. There were observations of exotic animals in their native habitat that could inform exhibit designers, as well as descriptions of nesting and mating behaviors that would prove useful to zoos and aquariums attempting early captive breeding programs. Most of the articles in this series represent the best of natural historical work that had been continued under the auspices of zoos and aquariums into the twentieth century.

Scientific work also expanded around the comparatively small number of aquariums. These public exhibits connected marine studies in tanks with oceanic research efforts, providing a vital network of new scientific information. By 1924, aquariums had opened and closed again in Miami Beach, Philadelphia, and Venice, California. In Chicago, the Shedd Aquarium opened in 1930 near the Field Museum of Natural History and Adler Planetarium on the city's lakefront. The Shedd Aquarium was the first inland aquarium to include a saltwater exhibit, filled by 20 rail cars that transported a million gallons of water and specimens from Key West. The exhibits of that institution became a standard attraction in the city for decades. In 1971, a research vessel in the Caribbean Sea operated by the Shedd expanded that institution's scientific prominence. The earliest public aquariums in other cities still operating include the Bell Isle Aquarium, which opened in Detroit in 1904, the Waikiki Aquarium in Honolulu that opened in the same year, and the San Francisco Aquarium, established in 1923. The Boston Aquarium closed, but was eventually reopened as the New England Aquarium in 1969 after nearly a decade of planning. Similarly, the New Orleans Aquarium shut down for what seemed like a permanent closure, but eventually housed the Aquarium of the Americas in 1990 (Kisling 2001). As with zoos, these American aquariums satisfied and stimulated public curiosity about ocean life. They also provided the nodes of a professional network of scientists working with specimens and answering questions about habitats, physiology, reproductive behaviors, and interactions among species.

The Connection Between Scientific Research, Education, and Conservation

Learning more about the animals on exhibit in zoos and aquariums, in addition to satisfying scientific compulsions, promoted the success of a broader mission for these institutions. Animals needed to live long enough to satisfy the expectations of zoo directors and the interested public. The display of animals in spaces that resembled their native habitat attracted notice. Zoo scientists and patrons – following the model Hornaday had established – saw the zoo as a center of conservation interests. More than ever, habitats for active and healthy animals beyond the zoo were threatened. As western North America became more settled, examples of how human activity destroyed habitat were increasingly visible. European and American zoo designers alike became attuned to this devastation and made reference to it in their exhibits with calls for preservation. With a scientific sophistication that went beyond Hornaday's urgent appeals earlier in the century, they could point to an emerging ecological sensibility, which suggested impending collapse of natural habitat when certain species became disturbed.

The tradition of conservation efforts at the Bronx Zoo was extended in the early 1940s. Newly elected as president of the New York Zoological Society, Fairfield Osborn focused the zoo's conservation program on a singular scientific component. He helped establish a wildlife refuge on land donated by the Rockefellers adjacent to Yellowstone National Park near Jackson Hole, Wyoming. Osborn expected this large-scale project to advance the cause of wildlife conservation and raise the profile of the Bronx Zoo. At the same time, some scientists bristled at the idea of turning a section of wilderness into an extension of a zoo exhibit. Prominent wildlife biologist Olaus Murie fought the plan as an attempt to intrude upon nature with the vision of a controlled menagerie. Murie hoped such landscapes could be retained without schemes contrived by millionaires and urban zoo directors. Osborn countered that any effort on behalf of wildlife conservation was justified at a time when war and technological prowess seemed to obliterate natural landscapes (Mitman 1996). At the Bronx Zoo, Osborn contributed extensively to the notion that zoos and conservation efforts should be intimately linked. Having served as a trustee of the New York Zoological Society since 1922 and its secretary from 1935 until he became president in 1940, his connections ran deep. Under his leadership in the decades that followed, more naturalistic exhibits were built, typically as a means of alerting the zoo visitor to the kinds of habitats that needed to be conserved in the wild (National Cyclopaedia of American Biography 1984). The differing perspectives between Osborn and other conservation-minded scientists, initially contentious, gave way to decades of collaboration on large-scale wildlife protection operations (Mitman 1996). Zoos increasingly sought to establish remote research centers that would complement their exhibits as well as provide space for animals that could not be quartered for the long term in existing facilities. San Diego's world-renowned zoo in Balboa Park, just north of downtown, built a sprawling "safari park" located (to the dismay of ill-prepared tourists) an hour or more – depending on traffic – from the city. These spaces promoted conservation, and they also provided institutional support for a growing and diverse community of professional researchers.

Prior to the 1950s, many zoo and aquarium employees found their way into jobs as keepers through patronage systems. Institutions run by cities or municipalities often hired individuals with connections to the officials who ran them. Since there were few specific qualifications established for zoo workers, and even fewer people with scientific training in biology or zoology who might apply, zoos tended to hire and promote people who liked animals and worked hard caring for them. As organizations, particularly the American Association of Zoological Parks and Aquariums (AAZPA), began to develop criteria for zoo accreditation, patronage jobs started to disappear. At the same time, zoos found a growing community of young, college-educated scientists to fill positions at institutions large and small. The growth of this community in the 1960s reinforced the scientific status of the AAZPA in developing more rigorous guidelines for specific exhibits and institutions in general (Donahue and Trump 2006).

With growing ranks of scientists involved directly in the running of these institutions, an agenda emphasizing broader conservation and educational goals grew ever stronger. This agenda coincided with increasing concern about environmental issues and habitat protection around the world. Near the end of his career at the Bronx Zoo, Osborn wrote an influential article published in the *New York Times* entitled, "Another Noah's Ark – For New York" in 1963. That same year, the AAZPA joined the International Union for Conservation of Nature and Natural Resources (IUCN), establishing an alliance that would promote collecting and breeding programs for zoos under strict scientific authority. This authority became a key feature of defending the importance of zoos to a public that became increasingly educated about the plight of animals. Many zoo patrons also became concerned about animal treatment and even skeptical of the relevance of zoos (Donahue and Trump 2006). Zoos and aquariums walked a fine line in promoting conservation and exhibiting endangered animals.

One research component of zoo and aquarium work that connected to the conservation mission and justified captivity focused on breeding studies. Historically, breeding programs contributed directly to the continuation and expansion of exhibits that began with small numbers of rare and exotic animals. Successful breeding meant that exhibits could be enlarged, and that exchanges with other institutions would enhance the overall collection. In cooperation with government wildlife agencies and nongovernmental organizations, research stations and breeding farms emerged. With an eye to international scientific cooperation, Ulysses Seal and Dale Makey, researchers in Minnesota, began in 1971 to develop an International Species Inventory System (ISIS). They envisioned a computerized database with many species coordinated across multiple institutions around the world. Support from the American Association of Zoological Veterinarians, the US Fish and Wildlife Service, and the AAZPA provided seed funding and strong organizational backing. With experience distinguishing genetic differences since the 1960s, Seal proposed that the system emphasize quantifiable molecular markers to maximize diversity, an insight that has become the basis for species survival plans ever since. Beginning with red wolves and eventually attracting widespread attention with Florida panthers, the use of genetic markers to reduce the risk of introduction of subspecies hybridization aroused consternation in the community of wildlife conservation (DeBlieu 1993). Questions of how to preserve wild populations that had dwindled to a few dozen individuals – now facing further threats from habitat destruction, disease, and toxins in their environment – demanded scientifically justifiable answers and brought biologists back to fundamental questions about how they would define a species (Tudge 1992). Zoo researchers, with access to populations in captivity across the country, offered the greatest hope of providing answers. The need for study subjects, whether kept in zoos, aquariums, or other research facilities became a component of the modern biological community.

While zoos had always engaged in the study of exotic animals and their requirements, more scientists began to suggest that the primary purpose of these institutions should focus on captive breeding that could even lead to reintroduction into the wild. The California condor, red wolf, and Arabian oryx became examples of early successes (Barrow 2009). Conservationists realized habitat protection for some species may no longer be possible, and so captive breeding became a way of increasing biodiversity in small populations. Such biodiversity provided stability for captive and wild populations alike, when animals were moved from one setting to the other. To succeed with the utilitarian goals of collecting wild animals for breeding in captivity, and returning captive animals to the wild, biological information about the individuals and populations in question became paramount. By the early 1990s, scientists studying animals in both contexts worked to establish a relative priority of determining how essential captive breeding may be to conservation efforts (Tudge 1992). Many of these studies were based in zoos and their affiliated research programs.

Critiques of Science as a Justification for Animal Captivity

Questions of whether and how animals could be appropriately kept in zoos and aquariums, and the extent to which scientific work matched the mission of these institutions, has turned on the widely cited conclusions of Heini Hediger in Switzerland. In many respects, scientists could reframe the public view of animals in captivity. This reframing changed the way exhibits were built, as well as the way scientists participated in the planning and administration of institutions. Zoo design, as a subfield of landscape architecture, came to intersect with the goals of scientific research. Planning exhibits after the middle of the twentieth century involved architects who sought to find inspiration in natural expressions of animal habitat. They also had to contend with the public perception that captive animals were less content in their surroundings than they would be in nature. To resolve these changing impressions of captivity, landscape architects drew directly from emerging scientific understanding of the best conditions for animals. The public also came to expect a shift in what they saw at the zoo. New zoo designs responded to a wide range of conceptual and sociological realities (Hyson 2000). In general, institutions that kept captive animals were expected to show how their exhibits allowed those animals to flourish. The architect and the zoologist shared expertise that could enhance the zoo experience for the public by enshrining animals in more or less naturalistic exhibits, and by enriching the lives of those animals with improved behavioral stimulation and veterinary care. In creating animal exhibits designed to cater primarily to animals' behavioral and physical needs, some new exhibits in the 1960s resembled shimmering white-tiled bathrooms or sterile hospital suites. Although far less naturalistic than Hagenbeck's barless enclosures, designers made the case that animals were kept in better health and provided with vital behavioral stimulation (Hyson 2000). Those designs suggested architectural and scientific advances, and they were embraced by a public who saw them as good for the animals and conducive to easy viewing.

The design efforts of the 1960s opened zoos and aquariums to renewed criticism about their ultimate purpose. Whether in exhibits with stark surroundings that enhanced the health and viewability of animals, or in more naturalistic spaces, the very notion of captivity advanced by Hediger came under fire. The critique of zoos that followed indicated that the reality of captivity had become unpalatable – even inhumane – to enlightened observers. Animal captivity, like human slavery, had run its course, according to those critics, and efforts to end it should be in full swing (Jamieson 1994; Malamud 1998). Meanwhile, as scientists and zoo designers rapidly reintroduced naturalistic features, another generation of viewers could see anew how far these institutions had come from the "bad old days" when caged animals seemed to recognize the futility of their lives in captivity (Hyson 2000). Even in zoos that had maintained natural outdoor enclosures, new exhibits in the 1970s and beyond highlighted ecologically appropriate settings. The zoo became a reflection of nature studied and understood in rational terms. The exhibit, once realistic by virtue of merely containing the live animal, became doubly realistic with rock formations, water features, live vegetation, simulated plant matter, and spaces designed for feeding, sleeping, and activities exhibited as play (Mullen and Marvin 1987).

By the late twentieth century, zoos and aquariums became places where off-site viewers (with online, live, animal-cams) as well as conventional visitors had the opportunity and even an expectation of being "immersed" in nature. The design of exhibits that provided immersive experiences became less about a scientific understanding of nature, earned through research and collaboration within the scientific community, but more about the technology that could reduce the complexity of nature to an acceptable simulation. Nature was not to be merely accurate in these exhibits. Rather, the public anticipating an immersive experience would expect a live version of nature television, where the star attraction was at the center of the frame at all times, both visible and *doing something* (Rothfels 2002). Paying substantial fees for viewing privileges (not to mention parking and a basket of chicken tenders) meant visitors expected to walk through climate-controlled passages to see active animals. As such, the goal of twenty-first-century zoos was to provide entertainment and a *sense* of connection to nature. The sense that there would be scientists working behind the scenes remained essential, but even that work might be simulated, represented by signage and interactive screens depicting the work of zoo scientists.

The scientific community, in addition to providing information and studies that might enhance zoo exhibits, also became more deeply involved in justifications of institutional practices of zoos as attitudes toward the treatment of animals evolved rapidly in the 1970s. Scientists in leading administrative positions advocated for scientific work in zoos and aquariums, pointing to behavioral and physiological studies, conservation efforts, and captive breeding programs. Coordination of this work also helped individual institutions navigate the increasingly complex rules put in place by the Endangered Species Act. That law in the United States and others like it elsewhere made it difficult for zoos to continue practices of exchanging – or even sharing for breeding purposes – species on the list. Transporting listed animals was heavily restricted in most cases, if not prohibited, even for zoos. Institutions coordinated their efforts, often with clearly justified scientific arguments aligned with breeding requirements. By the 1980s, scientists devised Species Survival Plans based on existing stocks of animals and measures of genetic difference that would help to guide viable breeding programs across institutions. As the broader scientific community and the public began to understand and appreciate the importance of biodiversity, zoo scientists were able to capitalize on this awareness by pointing out that Species Survival Plans were a means of maintaining the greatest possible diversity in populations of animals that might already be held in collections across the country or even in other parts of the world. The protection of genetic diversity, rather than the exhibiting of animals, became the central conservation message of the AAZPA and its member institutions, now operating under a shortened name, the American Zoo and Aquarium Association – with its acronym abbreviated to simply AZA (Donahue and Trump 2006).

Zoos and aquariums have long provided an encounter with nature. However much the designers of exhibits had to improvise and simulate natural conditions in order to appeal to the viewer, their primary goal remained always to keep animals alive. They offered a closer experience with real nature than visitors would ever expect to see otherwise. The authenticity of displays in recreating nature suggested a substantial degree of scientific credibility (Hanson 2002). Visitors are reminded of the scientific basis of their experience. The scientific community, since the earliest public exhibits in the United States offered that kind of experience, has operated to support zoos and aquariums filled with the widest possible array of animal life.

Bibliographic Essay

Organizing this chapter to focus on connections between the American scientific community and institutions of zoos and aquariums has proven overwhelming at times, given the abundant literature on the scientific community on the one hand, and zoos on the other. Finding points of connection reveals an important area for further research.

Elizabeth Hanson (2002) has written the most comprehensive account of American zoos as a component of the scientific enterprise. Her work provides detailed institutional histories of key examples of zoos, as well as important scientific contributions of zookeepers. In broad strokes, she characterizes the way nature was put on display. Historians can use this as a vital starting point in examining more carefully how zoos and aquariums have helped scientists to know the natural world.

Nigel Rothfels (2002) explores the modern zoo as an expression of how humans describe the lives of animals through the design of exhibits and the broader experience of visiting animals "in the wild." The *un*natural history of animals in zoos suggests an answer to the scholarly question of what role these institutions play in society. Rothfels offers an impressive historiographic account of zoos, and it remains reasonably current as a starting point.

In a snapshot of American zoo history sponsored by the Smithsonian Institution in 1996, R.J. Hoage and William Deiss (1996) assembled essays that nicely illustrate the growth of zoos as cultural institutions. An essay by Sally Gregory Kohlstedt demonstrates that science has played a role in these institutions throughout their existence, emphasizing examples from the late nineteenth century as important starting points. Other essays focus on European menageries and zoos that served as precursors to the National Zoo, and on American zoos that have benefited from early efforts in Washington, DC. The collection is scholarly and rich, by no means a mere celebration of the importance of the National Zoo on its own. The volume cannot, however, be more than a single line on a plane of the historical context of the work of the scientific community in zoos and aquariums.

Robert Kohler (2002) considers field stations as an important boundary between laboratory research and broader field observations and collection. This work provides a useful model for considering zoos and aquariums similarly as boundary objects. That model has been taken up already for natural history museums, but zoos remain largely unexamined by this approach.

Although not focused on an American context, Lynn Nyhart (2009) offers a thorough analysis of how a scientific community can intersect with the development of zoos as public institutions. She finds it useful to examine zoos in exploring the emergence of what she calls a "biological perspective" in German society in the late nineteenth century. Zoos become a medium for viewing social reform. Nyhart skillfully demonstrates the significance of natural history among scientific networks in Germany.

Natural history collecting and specimen trade has received examination by historians primarily in relation to museums and educational institutions. Mark Barrow (2000) provides a synthesis of some key examples in the second half of the nineteenth century in America. He also offers an engaging account of how concerns about extinction have shaped conservation (Barrow 2009). Paul Farber articulates a broader account of the natural history tradition as it relates to taxonomy, collection, and the increasing specialization of scientific work (2000). His narrative reveals the ongoing importance of natural history and the vibrancy on integrated, observational studies in biology that persist in a variety of institutional settings.

Focusing on institutional histories, Karen Rader and Victoria Cain have studied museum collections in more detail most recently (2014). Their work focuses on how changing configurations of museum exhibits driven – in some cases at least – by scientific understanding created corresponding changes in the institutions themselves. This approach provides a viable model for a more detailed study of zoos and aquariums focused on scientific work done within these institutions. Kohlstedt pioneered these studies with her work on Henry Ward (1980). For zoos, the career and impact of Carl Hagenbeck is examined in depth by Rothfels (2002), and Richard Flint (1996) illuminates an earlier, brief comparison of American and European collectors. Vernon Kisling (1998) provides a more detailed account of how scientific and commercial collection efforts often coincided with governmental and quasi-governmental exploration. Kisling acknowledges the difficulties of transport of exotic animals, highlighting the primary scientific questions of that era that would focus on meeting the basic needs of those animals. The essay provides a particularly useful starting point in this regard, suggesting the importance of the emergence of zoological gardens as a source of demand for those animals, and an endpoint for their successful transport.

George Rabb and Carol Saunders (2005) outline both the historical emergence of conservation of wild animals as a theme for zoos and aquariums as well as the future importance of expanding and emphasizing that theme. They suggest that the twenty-first-century zoo or aquarium must be a "conservation centre," an evolutionary product that has developed out of centuries of menageries and zoological parks.

Eric Baratay and Elisabeth Hardouin-Fugier (2002) have assembled a beautiful and useful survey of the history of zoological parks. Their comprehensive account of zoos in Europe and North America includes discussion of leading scientists who have played a role in developing particular collections within institutions where they became heavily involved. They include examples in nineteenth-century France, Germany, and England, as well as twentieth-century America.

Debates over scientific conservation arising from zoos or aquariums are arranged between the option of natural habitats or in zoological gardens and research stations. Problems that arise out of this limited choice are discussed by Irus Braverman (2014), who posits that conservation options are as conflated as the language over nature and wilderness.

Among critics of animals in captivity, Randy Malamud (1998) can be taken as a starting point: zoos reflect our human culture; holding animals in captivity tells us more about ourselves as captors than it can about the animals we hold (and perhaps study). He reminds readers that zoos are more about zoo goers than they are about animals.

BIBLIOGRAPHY

Abbate, Janet (ed.) (2003) "Women and Gender in the History of Computing." *Annals of the History of Computing* 25, 4.

Abbate, Janet (2012) *Recoding Gender: Women's Changing Participating in Computing*. Cambridge: MIT Press.

Abbott, Andrew (1999) *Department and Discipline: Chicago Sociology at One Hundred*. Chicago: University of Chicago Press.

Abbot, Andrew and Sparrow, James T. (2007) "Hot War, Cold War: The Structures of Sociological Action, 1940–1955," in C. J. Calhoun (ed.), *Sociology in America: A History*. Chicago: University of Chicago Press, 282–313.

Abbott, P. et al. (2011) "Inclusive Fitness Theory and Eusociality." *Nature* 471, 7339: E1–E4.

Abelove, Henry, Michele Barale, and David Halperin (eds.) (1993) *The Lesbian and Gay Studies Reader*. New York: Routledge.

Abir-Am, Pnina (1982) "The Discourse of Physical Power and Biological Knowledge in the 1930s: A Reappraisal of the Rockefeller Foundation's 'Policy' in Molecular Biology." *Social Studies of Science* 12, 3: 341–82.

Abir-Am, Pnina (1992) "The Politics of Macromolecules: Molecular Biologists, Biochemists, and Rhetoric." *Osiris* 7: 164–91.

Abir-Am, Pnina, and Outram, Dorinda (eds.) (1987) *Uneasy Careers and Intimate Lives: Women in Science, 1789–1979*. New Brunswick: Rutgers University Press.

Abraham, Itty (1996) "Science and Power in the Postcolonial State." *Alternatives: Global, Local, Political* 21, 3: 321–39.

Abraham, Itty (1998) *The Making of the Indian Atomic Bomb: Science, Secrecy and the Postcolonial State*. London: Zed Books.

Abraham, Itty (2006) "The Contradictory Spaces of Postcolonial Techno-Science." *Economic and Political Weekly* (21 January): 210–17.

Abraham, Itty (2013) "Clashing Cultures of Science." *Postcolonial Studies* 16, 4: 406–8.

Adam, Alison (1998) *Artificial Knowing: Gender and the Thinking Machine*. New York: Routledge.

Adams, Hazard (1970) "Criticism, Politics, and History: The Matter of Yeats." *The Georgia Review* 24, 2: 158–82.

Adams, J.F.A. (1887) "Is Botany a Suitable Subject for Young Men?" *Science* 209S: 116–17.

Adams, Mark B. (ed.) (1994) *The Evolution of Theodosius Dobzhansky: Essays on His*

A Companion to the History of American Science, First Edition.
Edited by Georgina M. Montgomery and Mark A. Largent.
© 2016 John Wiley & Sons, Ltd. Published 2020 by John Wiley & Sons, Ltd.

Life and Thought in Russia and America. Princeton: Princeton University Press.

Adams, Vincanne (2002) "Randomized Controlled Crime: Postcolonial Sciences in Alternative Medicine Research." *Social Studies of Science* 32: 659–90.

Adas, Michael (1989) *Machines as the Measure of Men: Science, Technology, and Ideologies of Western Dominance.* Ithaca: Cornell University Press.

Adas, Michael (2004) "Contested Hegemo: The Great War and the Afro-Asian Assault on the Civilizing Mission Ideology." *Journal of World History* 15, 1: 31–63.

Agar, John (2003) *The Government Machine: A Revolutionary History of the Computer.* Cambridge: MIT Press.

Agar, John (2006) "What Difference Did Computers Make?" *Social Studies of Science* 36, 6: 869–907.

Agassiz, Elizabeth Cary (ed.) (1885) *Louis Agassiz: His Life and Correspondence.* Boston: Houghton Mifflin.

Agassiz, Louis (1854) "Sketch of the Natural Provinces of the Animal World and Their Relation to the Different Types of Man," in J.C. Nott and George R. Gliddon (eds.), *Types of Mankind.* Philadelphia. Lippincott, Grambo, & Co, LVIII–LXVIII.

Agassiz, Louis (1860) "Prof. Agassiz on The Origin of Species." *American Journal of Science* 30: 142–54.

Agassiz, Louis (1866) *The Structure of Animal Life.* New York: Charles Scribner.

Agassiz, Louis. (1962) *Essay on Classification.* Cambridge: Harvard University Press.

Aiken, Howard (1999) "Proposed Automatic Calculating Machine," in I.B. Cohen and G. Welch (eds.), *Makin' Numbers: Howard Aiken and the Computer.* Cambridge: MIT Press, 9–29.

Akera, Atsushi (2000) "Engineers or Managers? The Systems Analysis of Electronic Dataprocessing in the Federal Bureaucracy," in Agatha Hughes and Thomas Hughes (eds.), *Systems, Experts, and Computers: The Systems Approach in Management and Engineering, World War II and After.* Cambridge: MIT Press, 191–220.

Alagona, Peter S. (2013) *After The Grizzly: Endangered Species and the Politics of Place in California.* Berkeley: University of California Press.

Albanese, Catherine (1990) *Nature Religion in America: From the Algonkian Indians to the New Age.* Chicago: University of Chicago Press.

Alberti, Samuel J.M.M. (2005) "Objects and the Museum." *Isis* 96, 4: 559–71.

Alberti, Samuel J.M.M. (2008) "Constructing Nature Behind Glass." *Museum and Society* 6, 2: 73–97.

Alberts, Bruce (1983) *Molecular Biology of the Cell.* New York: Garland.

Alberts, B.M. et al. (2014) "Rescuing US Biomedical Research from its Systemic Flaws." *Proceedings of the National Academy of Sciences* 111: 5773–7.

Alcock, J. (2001) *The Triumph of Sociobiology.* Oxford: Oxford University Press.

Aldridge, A. Owen. (1992) "Benjamin Franklin: The Fusion of Science and Letters," in Robert J. Scholnick (ed.), in *American Literature and Science.* Lexington: University Press of Kentucky, 39–57.

Alexander, Amir (2002) *Geometrical Landscapes: The Voyages of Discovery and the Transformation of Mathematical Practice.* Palo Alto: Stanford University Press.

Alexander, Edward P. (1996) *Museums in Motion: An Introduction to the History and Functions of Museums.* Walnut Creek: Altamira Press.

Allard, Dean (1978) *Spencer Fullerton Baird and the U.S. Fish Commission.* New York: Arno Press.

Allard, Dean (2000) "Spencer Baird and Support for American Marine Science, 1871–1887." *Earth Sciences History* 19, 1: 44–57.

Allee, W.C. (1931) *Animal Aggregations: A Study in General Sociology.* Chicago: University of Chicago Press.

Allee, W.C. et al. (1949) *Principles of Animal Ecology.* Philadelphia: W.B. Saunders.

Allen, Garland E. (1974) "Opposition to the Mendelian-Chromosome Theory: The Physiological and Developmental Genetics of Richard Goldschmidt." *Journal of the History of Biology* 7: 49–92.

Allen, Garland E. et al. (1975) "Letter to the Editor." *New York Review of Books* 18: 43–4.

Allen, Garland E. (1975a) *Life Science in the Twentieth Century*. Cambridge: Cambridge University Press.

Allen, Garland E. (1978a) *Life Science in the Twentieth Century*. Cambridge: Cambridge University Press.

Allen, Garland E. (1978b) *Thomas Hunt Morgan: The Man and His Life*. Princeton: Princeton University Press.

Allen, Garland E. (1979) "Naturalists and Experimentalists: The Genotype and the Phenotype." *Studies in History of Biology* 3: 179–209.

Allen, Garland E. (1983) "The Misuse of Biological Hierarchies: The American Eugenics Movement, 1900–1940." *History and Philosophy of Life Sciences* 5: 105–28.

Allen, Garland E. (1986) "The Eugenics Record Office at Cold Spring Harbor, 1910–1940: An Essay in Institutional History." *Osiris* 2: 225–64.

Allen, Garland E. (1991) "Old Wine in New Bottles: From Eugenics to Population Control in the Work of Raymond Pearl," in Keith R. Benson, Jane Maienschein, and Ronald Rainger (eds.), *The Expansion of American Biology*, 231–61. New Brunswick: Rutgers University Press.

Allen, John Logan (1991) *Lewis and Clark and the Image of the American Northwest*. New York: Dover.

Alvares, Claude (1992) *Science, Development, and Violence: The Revolt Against Modernity*. New York: Oxford University Press.

Amadae, S.M. (2003) *Rationalizing Capitalist Democracy: The Cold War Origins of Rational Choice Liberalism*. Chicago: University of Chicago Press.

American College of Obstetricians and Gynecologists (2007) "ACOG Practice Bulletin No. 88, December 2007. Invasive Prenatal Testing for Aneuploidy." *Obstetrics and Gynecology* 110: 1459.

American College of Obstetricians and Gynecologists (2009) "ACOG Committee Opinion No. 446: Array Comparative Genomic Hybridization in Prenatal Diagnosis." *Obstetrics and Gynecology* 114: 1161–3.

American College of Obstetricians and Gynecologists (2013) "ACOG Committee Opinion No. 581: The Use of Chromosomal Microarray Analysis in Prenatal Diagnosis." *Obstetrics and Gynecology* 122: 1374–7.

American Meteorological Society (1964) "Our Corporation Members: The Travelers Research Corporation." *Bulletin of the American Meteorological Society* 45: 608–9.

American Museum of Natural History (2014) "About The Exhibition," Pterosaurs: Flight in the Age of Dinosaurs. www.amnh.org/Exhibitions/Current-Exhibitions/Pterosaurs-Flight-in-The-Age-of-Dinosaurs/About-The-Exhibition

Amsel, Abram and Michael Rashotte (1984) *Mechanisms of Adaptive Behavior: Clark L. Hull's Theoretical Papers, with Commentary*. New York: Columbia University Press.

Amundson, Ronald (2005) *The Changing Role of the Embryo in Evolutionary Thought: Roots of Evo-Devo*. New York: Cambridge University Press.

Anderson, Arthur J.O., Frances Berdan, and James Lockhart (eds.) (1976) *Beyond the Codices: The Nahua View of Colonial Mexico*. Berkeley: University of California Press.

Anderson, Garrett and Rachel Ivie (2013) *Demographics Survey of 2013 U.S. AAAS Members Summary Results*. Washington: American Institute of Physics.

Anderson, Katharine (2005) *Predicting the Weather: Victorians and the Science of Meteorology*. Chicago: University of Chicago Press.

Anderson, Katharine (2006) "Mapping Meteorology," in James Fleming, Vladimir Jankovic and Deborah R. Coen (eds.), *Intimate Universality: Local and Global Themes in the History of Weather and Climate*. Sagamore Beach: Science History Publications, 69–92.

Anderson, M. Kat (2005) *Tending the Wild: Native American Knowledge and the Management of California's Natural Resources*. Berkeley: University of California Press.

Anderson, Philip (1972) "More is Different." *Science* 177: 393–6.

Anderson, R.G.W., J. A. Bennett and W.F. Ryan (eds.) (1993) *Making Instruments Count: Essays on Historical Scientific Instruments Presented to Gerard L'Estrange Turner.* Aldershot: Variorum.

Anderson, R.S. and B.M. Morrison (1982) *Science, Politics, and Agricultural Revolution in Asia.* Boulder: Westview Press.

Anderson, Warwick (1995) "Excremental Colonialism: Public Health and the Poetics of Pollution." *Critical Inquiry* 21, 3: 640–69.

Anderson, Warwick (1998) "Where is the Postcolonial History of Medicine?" *Bulletin of the History of Medicine* 79, 3: 522–30.

Anderson, Warwick (2002a) *The Cultivation of Whiteness: Science, Health, and Racial Destiny in Australia.* Durham: Duke University Press.

Anderson, Warwick (2002b) "Introduction: Postcolonial Technoscience." *Social Studies of Science* 32: 643–58.

Anderson, Warwick (2006) *Colonial Pathologies: American Tropical Medicine, Race, and Hygiene in the Philippines.* Durham: Duke University Press.

Anderson, Warwick (2009) "From Subjugated Knowledge to Conjugated Subjects: Science and Globalization or Postcolonial Studies of Science?" *Postcolonial Studies* 12, 4: 389–400.

Anderson, Warwick (2012) "Racial Hybridity, Physical Anthropology, and Human Biology in the Colonial Laboratories of the United States." *Current Anthropology* 53, S5: S95–S107.

Anderson, Warwick and Vincanne Adams (2007) "Pramoedya's Chickens: Postcolonial Studies of Technoscience," in Edward J. Hackett, Olga Amsterdamska, Michael Lynch, and Judy Wajcman (eds.), *Handbook of Science and Technology Studies.* Cambridge: MIT Press, 181–203.

Andrews, Seth (2013) "The Thinking Atheist Radio Podcast #131: The Richard Dawkins Interview." http://www.thethinkingatheist.com/forum/thread-podcast-131-the-richard-dawkins-interview

Angell, M. (2004) *The Truth About the Drug Companies.* New York: Random House.

Ankeny, Rachel A. (2007) "Wormy Logic: Model Organisms as Case-Based Reasoning," in Angela N.H. Creager, Elizabeth Lunbeck, and M. Norton Wise (eds.), *Science Without Laws: Model Systems, Cases, Exemplary Narratives.* Durham: Duke University Press, 46–58.

Anonymous (1850) "The Natural History of Man." *The United States Magazine and Democratic Review* 26: 327–45.

Anonymous (1874) "'Scientist,' The American Word." *The Academy.* 321.

Anonymous (1898) "The Large Refractors of the World." *The Observatory* 21: 239–41, 70–71.

Antonovics, Janis (1987) "The Evolutionary Dys-Synthesis: Which Bottles for Which Wine?" *The American Naturalist* 129: 321–31.

Appadurai, Arjun (1996) *Modernity at Large: Cultural Dimensions of Globalization.* Minneapolis: University of Minnesota Press.

Appel, Toby (1988) "Organizing Biology: The American Society of Naturalists and its 'Affiliated Societies,' 1883–1923," in Ronald Rainger, Keith R. Benson, and Jane Maienschein (eds.), *The American Development of Biology,* 87–120. Philadelphia: University of Pennsylvania Press.

Appel, Toby (1988a) "Jeffries Wyman, Philosophical Anatomy, and the Scientific Reception of Darwinism." *Journal of the History of Biology* 21: 69–94.

Appel, Toby A. (1994) "Physiology in American Women's Colleges: The Rise and Fall of a Female Subculture." *Isis* 85: 26–56.

Appel, Toby (2000) *Shaping Biology: The National Science Foundation and American Biological Research, 1945–1975.* Baltimore: Johns Hopkins University Press.

Apple, Rima D. (1990) *Women, Health, and Medicine in America: A Historical Handbook.* New York: Garland.

Apple, Rima D. (1996) *Vitamania: Vitamins in American Culture.* New Brunswick: Rutgers University Press.

Armitage, Kevin C. (2009) *The Nature Study Movement: The Forgotten Popularizers of*

America's Conservation Ethic. Lawrence: University Press of Kansas.

Arnold, David. (1993) *Colonizing the Body: State Medicine and Epidemic Disease in Nineteenth-Century India.* Berkeley: University of California Press.

Arnold, David (1996) *Warm Climates and Western Medicine: The Emergence of Tropical Medicine, 1500–1900.* Amsterdam: Rodopi.

Arnold, David (2005) "Europe, Technology, and Colonialism in the 20th Century." *History and Technology* 21: 85–106.

Artigas, M., T.F. Martinez, and R.A. Glick (2006) *Negotiating Darwin: The Vatican Confronts Evolution, 1877–1902.* Baltimore: Johns Hopkins University Press.

Arvin, M.R. (2013) "Pacifically Possessed: Scientific Production and Native Hawaiian Critique of the 'Almost White' Polynesian Race." PhD Dissertation: University of California San Diego.

Asad, Talal (1973) *Anthropology and the Colonial Encounter.* London: Ithaca Press.

Aspiz, Harold (1980) *Walt Whitman and the Body Beautiful.* Urbana-Champaign: University of Illinois Press.

Aspray, William (1988) "An Annotated Bibliography of Secondary Sources on the History of Software." *IEEE Annals of the History of Computing* 9, 3/4): 291–343.

Aspray, William (ed) (1990a) *Computing before Computers.* Ames: Iowa State University Press.

Aspray, William (1990b) *John von Neumann and the Origins of Modern Computing.* Cambridge: MIT Press.

Aspray, William (2000a) "Was Early Entry a Competitive Advantage? U.S. Universities that Entered Computing in the 1940s." *IEEE Annals of the History of Computing* 22, 3: 42–87.

Aspray, William (2000b) "The Institute for Advanced Study Computer: A Case Study in the Application of Concepts from the History of Technology," in Raúl Rojas and Ulf Hashagen (eds.), *The First Computers: History and Architectures.* Cambridge: MIT Press, 179–94.

Aspray, William and Bernard Williams (1994) "Arming American Scientists: NSF and the Provision of Scientific Computing Facilities for Universities, 1950–1973." *IEEE Annals of the History of Computing* 16, 4: 60–74.

Assmus, Alexi (1992) "The Americanization of Molecular Physics." *Historical Studies in the Physical and Biological Sciences* 23: 1–34.

Astore, William (1996) "Gentle Skeptics? American Catholic Encounters with Polygenism, Geology, and Evolutionary Theories from 1845 to 1875." *Catholic Historical Review* 82: 40–76.

Astronomy Survey Committee (1972). *Astronomy and Astrophysics for the 1970s.* Washington: National Academy Press.

Astronomy Survey Committee (1982) *Astronomy and Astrophysics for the 1980s, Vol. 1: Report of the Astronomy Survey Committee.* Washington: National Academy Press.

Astronomy Survey Committee (1983) *Astronomy and Astrophysics for the 1980s, Vol. 2: Reports of the Panels.* Washington: National Academy Press.

Astronomy and Astrophysics Survey Committee (1991) *The Decade of Discovery in Astronomy and Astrophysics.* Washington: National Academy Press.

Atkins, P. (2012) "The Charmed Circle," in P. Atkins (ed.), *Animal Cities: Beastly Urban Histories.* Burlington: Ashgate, 53–77.

Atwater, W.O. (1887) "The Chemistry of Food and Nutrition." *Century Magazine* 34: 59–74.

Atwater, W.O. (1893) U.S. Department of Agriculture "Suggestions for the Establishment of Food Laboratories in Connection with the Agricultural Experiment Stations of the United States." Office of Experiment Stations. Bulletin 17. Washington: Government Printing Office.

Atwater, W.O. (1895) U.S. Department of Agriculture, "Food and Diet." *1894 Yearbook of Agriculture.* Washington: Government Printing Office.

Atwater, W.O. (1896) U.S. Department of Agriculture. "The Chemical Composition of American Food Materials." *Office of the Experiment Stations* Bulletin 28. Washington: Government Printing Office.

Atwater, W.O. (1902) U.S. Department of Agriculture "Principles of Nutrition and Nutritive Value of Food." Farmers' Bulletin

No. 142. Washington: Government Printing Office.

Atwater, W.O. and A.P. Bryant (1898) U.S. Department of Agriculture, "Dietary Studies in Chicago in 1895 and 1896." Office of Experiment Stations. Bulletin 55. Washington: Government Printing Office.

Atwater, W.O. and E.B. Rosa (1899) U.S. Department of Agriculture "Description of a New Respiration Calorimeter and Experiments on the Conservation of Energy in the Human Body." Bulletin of the Office of the Experiment Stations No. 63. Washington: Government Printing Office.

Aubin, David, Charlotte Big, and H. Otto Sibum (eds.) (2010) *The Heavens on Earth: Observatories and Astronomy in Nineteenth-Century Science and Culture.* Durham: Duke University Press.

Audubon, John J. (1840–1844) *The Birds of America, from Drawings Made in the United States and Their Territories.* New York: J. J. Audubon.

Auletta, Gennaro (2011) *Biological Evolution: Facts and Theories: A Critical Appraisal 150 Years after "The Origin of Species."* Roma: Gregorian & Biblical Press.

Austing, Richard, Bruce Barnes, and Gerald Engel (1977) "A Survey of the Literature in Computer Science Education since Curriculum '68." *Communications of the ACM* 20, 1: 13–21.

Aveni, Anthony (2008) *Foundations of New World Cultural Astronomy.* Boulder: University Press of Colorado.

Ayala, Francisco J. (1985) "Theodosius Dobzhansky, January 25, 1900–December 18, 1975." *Biographical Memoirs of the National Academy of Sciences* 55: 163–213.

Baber, Zaheer (1996) *The Science of Empire: Scientific Knowledge, Civilization, and Colonial Rule in India.* Albany: SUNY Press.

Bachman, John (1850) *The Doctrine of the Unity of the Human Race.* Charleston: C. Canning.

Backhouse, Roger E. and Bradley W. Bateman (2011) *Capitalist Revolutionary: John Maynard Keynes.* Cambridge: Harvard University Press.

Backhouse, Roger E. and P. Fontaine (2010) *The History of the Social Sciences since 1945.* Cambridge: Cambridge University Press.

Badash, Lawrence (1979) *Radioactivity in America: Growth and Decay of a Science.* Baltimore: Johns Hopkins University Press.

Bailey, Liberty Hyde (1903) *The Nature-Study Idea: Being an Interpretation of the New School Movement to Put the Child in Sympathy with Nature.* New York: Doubleday, Page and Co.

Baird, Davis (2004) *Thing Knowledge: A Philosophy of Scientific Instruments.* Berkeley: University of California Press.

Baird, Spencer (1873) *Report of the Condition of the Sea Fisheries of the South Coast of New England in 1871 and 1872.* Washington: Government Printing Office.

Baldwin, D.L. (2004) "Black Belts and Ivory Towers: The Place of Race in U.S. Social Thought, 1892–1948." *Critical Sociology* 30, 2.

Ballard, Robert D. (2000) *The Eternal Darkness.* Princeton: Princeton University Press.

Balmer, R. (2006) *Thy Kingdom Come: How the Religious Right Distorts the Faith and Threatens America.* New York: Basic Books.

Bannister, R.C. (1979) *Social Darwinism: Science and Myth in Anglo-American Social Thought.* Philadelphia: Temple University Press.

Bannister, R.C. (1987) *Sociology and Scientism: The American Quest for Objectivity, 1880–1940.* Chapel Hill: University of North Carolina Press.

Baratay, Eric and Elisabeth Hardouin-Fugier (2002) *Zoo: A History of Zoological Gardens in the West.* London: Reaktion.

Barber, William J. (1985) *From New Era to New Deal: Herbert Hoover, the Economists, and American Economic Policy, 1921–1933.* Cambridge: Cambridge University Press.

Barber, William J. (ed.) (1988) *Breaking the Academic Mould: Economists and American Higher Learning in the Nineteenth Century.* Middletown: Wesleyan University Press.

Barber, William J. (1996) *Designs within Disorder: Franklin D. Roosevelt, the Economists, and the Shaping of American Economic*

Policy, 1933–1945. Cambridge: Cambridge University Press.

Barber, Willam J. (2003) "American Economics to 1900," in Warren J. Samuels, Jeff E. Biddle, and John B. Davis (eds.), *A Companion to the History of Economic Thought*. Malden: Blackwell, 231–45.

Barbour, Michael (1995) "Ecological Fragmentation in the Fifties," in William Cronon (ed.), *Uncommon Ground: Toward Reinventing Nature*. New York: W.W. Norton, 233–55.

Barbrook, Richard (2007) "New York Prophecies: The Imaginary Future of Artificial Intelligence." *Science as Culture* 16: 151–67.

Barkow, Jerome H., Leda Cosmides, and John Tooby (1992) *The Adapted Mind: Evolutionary Pyschology and the Generation of Culture*. Oxford: Oxford University Press.

Bashe, Charles, Lyle Johnson, John Palmer, and Emerson Pugh (1989) *IBM's Early Computers*. Cambridge: MIT Press.

Barkan, Elazar (1992) *The Retreat of Scientific Racism: Changing Concepts of Race in Britain and the United States Between World Wars*. New York: Cambridge University Press.

Barley, Stephen R. and Becky A. Bechky (1994) "In the Backrooms of Science." *Work and Occupations* 21: 85–126.

Barrow, Mark V. (1998) *A Passion for Birds: American Ornithology after Audubon*. Princeton: Princeton University Press.

Barrow, Mark V. (2000) The Specimen Dealer: Entrepreneurial Natural History in America's Gilded Age. *Journal of the History of Biology* 33: 493–534.

Barrow, Mark V. (2009) *Nature's Ghosts: Confronting Extinction from the Age of Jefferson to the Age of Ecology*. Chicago: University of Chicago Press.

Barth, Kai-Henrik (2003) "The Politics of Seismology Nuclear Testing, Arms Control, and the Transformation of a Discipline." *Social Studies of Science* 33, 5: 743–81.

Bartky, Ian R. (1989) "The Adoption of Standard Time." *Technology and Culture* 30: 25–56.

Bartky, Ian R. (2000) *Selling the True Time: Nineteenth-Century Timekeeping in America*. Stanford: Stanford University Press.

Bartlett, Richard A. (1962) *Great Surveys of the American West*. Norman: University of Oklahoma Press.

Bartram, John (1942) "Diary of a Journey through the Carolinas, Georgia, and Florida, from July 1, 1765, to April 10, 1766." Transactions of the American Philosophical Society XXXIII, Pt. I.

Bartram, William (1794) *Travels through North & South Carolina, Georgia, East & West Florida, the Cherokee Country, the Extensive Territories of the Muscogulges, or Creek Confederacy, and the Country of the Chactaws*. London: J. Johnson.

Bartusiak, Marcia (2000) *Einstein's Unfinished Sympho: Listening to the Sounds of Spacetime*. Washington: Joseph Henry.

Basalla, George (1967) "The Spread of Western Science." *Science* 156: 611–22.

Bateman, Bradley W. (1998) "Clearing the Ground: The Demise of the Social Gospel Movement and the Rise of Neoclassicism in American Economics." *History of Political Economy* 30 (Supplement): 29–52.

Bateman, Bradley W. (2001) "Make A Righteous Number: Social Surveys, the Men and Religion Forward Movement, and Quantification in American Economics." *History of Political Economy* 33 (Supplement): 57–85.

Bateman, Bradley W. and Ethan B. Kapstein (1999) "Between God and the Market: The Religious Roots of the American Economic Association." *Journal of Economic Perspectives* 13, 4: 249–58.

Bates, H.W. (1863) *The Naturalist on the River Amazon*. London: John Murray.

Baum, Richard, and William Sheehan (1997) *In Search of Planet Vulcan: The Ghost in Newton's Clockwork Universe*. New York: Plenum.

Bay, M. (1998) "The World was Thinking Wrong about Race," in M.B. Katz and T.J. Sugrue (eds.), *W.E.B. Dubois, Race, and the City: The Philadelphia Negro and its Legacy*. Philadelphia: University of Pennsylvania Press, 41–60.

Bazerman, Charles (2002) *The Languages of Edison's Light*. Cambridge: MIT Press.

Beam, Lura and Robert Latou Dickinson (1931) *A Thousand Marriages*. Baltimore: Williams & Wilkens.

Beam, Lura and Robert Latou Dickinson (1934) *The Single Woman: A Medical Study in Sex Education*. Baltimore: Williams & Wilkens.

Beardsley, Edward H. (1964) *The Rise of the American Chemistry Profession, 1850–1900*. Gainesville: University of Florida Press.

Beattie, Donald A. (2001) *Taking Science to the Moon: Lunar Experiments and the Apollo Program*. Baltimore: Johns Hopkins University Press.

Beatty, John (1984) "Chance and Natural Selection." *Philosophy of Science* 51: 183–211.

Beatty, John (1987) "Natural Selection and the Null Hypothesis," in John Dupré (ed.), *The Latest on the Best*. Cambridge: MIT Press, 53–75.

Beatty, John (1991) "Genetics in the Atomic Age: The Atomic Bomb Casualty Commission, 1947–1956," in Keith R. Benson, Jane Maienschein, and Ronald Rainger (eds.), *The Expansion of American Biology*, 284–324. New Brunswick: Rutgers University Press.

Bechtel, Stefan (2012) *Mr. Hornaday's War: How a Peculiar Victorian Zookeeper Waged a Lonely Crusade for Wildlife that Changed the World*. Boston: Beacon Press.

Bechtel, William (2006) *Discovering Cell Mechanisms: The Creation of Modern Cell Biology*. Cambridge: Cambridge University Press.

Beck, U. (1992) *Risk Society: Towards a New Modernity*. New Delhi: Sage.

Beckel, Annamarie L. (1987) "Breaking New Waters: A Century of Limnology at the University of Wisconsin," in Carl N. Haywood (ed.), *The Transactions of the Wisconsin Academy of Sciences, Arts and Letters*.

Becker, Gary S. (1975) *Human Capital: A Theoretical and Empirical Analysis, with Special Reference to Education*. New York: Columbia University Press.

Bederman, Gail (2008) *Manliness and Civilization: A Cultural History of Gender and Race in the United States, 1880–1917*. Chicago: University of Chicago Press.

Bedini, Silvio A. (1964) *Early American Scientific Instruments and Their Makers*. Washington: Smithsonian Institution.

Bedini, Silvio (1971) *The Life of Benjamin Banneker*. New York: Scribner.

Bedini, Silvio A. (1975) *Thinkers and Tinkers: Early American Men of Science*. New York: Scribner.

Bedini, Silvio A. (1984) *At the Sign of the Compass and Quadrant: The Life and Times of Anthony Lamb*. Philadelphia: American Philosophical Society.

Bedini, Silvio A. (1990) *Thomas Jefferson: Statesman of Science*. New York: Macmillan.

Bedini, Silvio A. (2001) *With Compass and Chain: Early American Surveyors and Their Instruments*. Frederick: Professional Surveyors Publishing Company.

Beer, Gillian (1983) *Darwin's Plots: Evolutionary Narrative in Darwin, George Eliot and Nineteenth-Century Fiction*. Cambridge: Cambridge University Press.

Behar, Ruth, and Deborah A. Gordon (1995) *Women Writing Culture*. Berkeley: University of California Press.

Behe, Michael (1996) *Darwin's Black Box: The Biochemical Challenge to Evolution*. New York: Free Press.

Bekelman, J., Y. Lee and C. Gross (2003) "Scope and Impact of Financial Conflicts of Interest in Biomedical Research." *Journal of the American Medical Association* 289: 454–65.

Benedict, Ruth (1934) *Patterns of Culture*. New York: Houghton Mifflin Harcourt.

Benedict, Ruth (1946) *The Chrysanthemum and the Sword: Patterns of Japanese Culture*. Boston: Houghton Mifflin.

Benesch, Klaus (2002) *Romantic Cyborgs: Authorship and Technology in the American Renaissance*. Amherst: University of Massachusetts Press.

Benjamin, Ludy T. (2006) *A Brief History of Modern Psychology*. New York: John Wiley & Sons.

Benjamin, Ludy T., Maureen Durkin et al. (1992). "Wundt's American Doctoral Students." *The American Psychologist* 47, 2.

Benjamin, Marina (ed.) (1991) *Science and Sensibility: Gender and Scientific Enquiry, 1780–1945*. Oxford: Blackwell.

Benjamin, Ruha (2009) "A Lab of Their Own: Genomic Sovereignty as Postcolonial Science Policy." *Policy & Society* 28, 4: 341–55.

Bennett, J.A. (2003) "Knowing and Doing in the Sixteenth Century: What were Instruments for?" *British Journal for the History of Science* 36: 129–50.

Bennett, Tony (1995) *The Birth of the Museum: History, Theory, Politics.* New York: Routledge.

Benson, Etienne (2010) *Wired Wilderness: Technologies of Tracking and the Making of Modern Wildlife.* Baltimore: Johns Hopkins University Press.

Benson, Etienne (2012) "Endangered Science: The Regulation of Research by the U.S. Marine Mammal Protection and Endangered Species Acts." *Historical Studies in the Natural Sciences* 42, 1: 30–61.

Benson, Keith (1979) "William Keith Brooks (1848–1908): A Case Study of Morphology and the Development of American Biology." PhD Dissertation: Oregon State University.

Benson, Keith (1988a) "Laboratories on the New England Shore: The "Somewhat Different Direction" of American Marine Biology." *New England Quarterly* 61: 55–78.

Benson, Keith (1988) "From Museum Research to Laboratory Research: The Transformation of Natural History into Academic Biology," in Ronald Rainger, Keith R. Benson, and Jane Maienschein (eds.), *The American Development of Biology.* Baltimore: Johns Hopkins University Press, 49-83.

Benson, Keith (1992) "Experimental Ecology on the Pacific Coast: Victor Shelford and His Search for Appropriate Methods." *History and Philosophy of the Life Sciences* 14: 73–91.

Benson, Keith (2001) "Summer Camp, Seaside Station, and Marine Laboratory: Marine Biology and its Institutional Identity" *Historical Studies in the Physical and Biological Sciences* 32, 1.

Benson, Keith (2002) "Marine Biology or Oceanography: Early American Developments in Marine Science on the West Coast," in Keith R. Benson and Philip F. Rehbock (eds.), *Oceanographic History: The Pacific and Beyond.* Seattle: University of Washington Press.

Benson, Keith, Jane Maienschein, and Ronald Rainger (eds.) (1991) *The Expansion of American Biology.* New Brunswick: Rutgers University Press.

Benson, Keith and Philip Rehbock 9eds.) (2002) *Oceanographic History: The Pacific and Beyond.* Seattle: University of Washington Press.

Benson, Maxine (1986) *Martha Maxwell: Rocky Mountain Naturalist.* Lincoln: University of Nebraska Press.

Bentley, A. (1998) *Eating for Victory: Food Rationing and the Politics of Domesticity.* Urbana: University of Illinois Press.

Bentley, Madison (1945) "Sanity and Hazard in Childhood." *American Journal of Psychology* 58: 212–46.

Bentley, Madison (1950) "Brief Comment Upon Recent Books." *American Journal of Psychology* 63: 310–16.

Berezow, A. and H. Campbell (2012) *Science Left Behind.* New York: Public Affairs.

Berg, Paul et al. (1974) "Potential Biohazards of Recombinant DNA Molecules." *Science* 185: 303.

Bergland, Renée (2008) *Maria Mitchell and the Sexing of Science: An Astronomer among the American Romantics.* Boston: Beacon Press.

Bergman, James (2013) "Climates on the Move: Climatology and the Problem of Economic and Environmental Stability in the Career of C. W. Thornthwaite, 1933–1963." PhD Dissertation: Harvard University.

Berle, Adolf and Gardiner Means (1932) *The Modern Corporation and Private Property.* New York: Macmillan.

Bernasconi, Robert (ed.) (2001) *Concepts of Race in the Eighteenth Century* (8 volumes). Bristol: Thoemmes Press.

Bernasconi, Robert (ed.) (2002) *American Theories of Polygenesis.* Bristol: Thoemmes Press.

Bernasconi, Robert (ed.) (2003) *Race and Anthropology.* Bristol: Thoemmes Press.

Bernasconi, Robert (2008) "A Haitian in Paris: Anténor Firmin as a Philosopher against

Racism." *Patterns of Prejudice* 42, 4-5: 365–84.

Bernasconi, Robert (2009) "'Our Duty to Conserve:' W.E.B. Du Bois's Philosophy of History in Context." *South Atlantic Quarterly* 108: 519–40.

Bernasconi, Robert (2012) "Crossed Lines in the Racialization Process: Race as a Border Concept." *Research in Phenomenology* 42: 206–38.

Bernasconi, Robert and Kristie Dotson (eds.) (2003) Race, Hybridity, and Miscegenation, (3 volumes). Bristol: Thoemmes Continuum

Bernstein, Jay II. (2002) "First Recipients of Anthropological Doctorates in the United States, 1891–1930." *American Anthropologist* 104, 2: 551–64.

Bernstein, Michael A. (2001) *A Perilous Progress: Economists and Public Purpose in Twentieth-Century America.* Princeton: Princeton University Press.

Berry, Dominic (2014) "The Plant Breeding Industry after Pure Line Theory: Lessons from the National Institute of Agricultural Botany." *Studies in History and Philosophy of Biological and Biomedical Sciences.* 46: 25–37.

Berry, Evan (2013) "Religious Environmentalism and Environmental Religion in America." *Religion Compass* 7: 454–66.

Berry, Roberta M. (2007) *The Ethics of Genetic Engineering.* New York: Routledge.

Berry, Wendell (1977) *The Unsettling of America: Culture and Agriculture.* San Francisco: Sierra Club Books.

Bharadwaj, Aditya (2012) "The Other Mother: Supplementary Wombs, Surrogate State and Arts in India," in M. Knecht, M. Klotz, and S. Beck (eds.), *Reproductive Technologies as Global Form: Ethnographies of Knowledge, Practice, and Transnational Encounters.* Chicago: Chicago University Press.

Bharadwaj, Aditya (2013) "Ethic of Consensibility, Subaltern Ethicality: The Clinical Applications of Embryonic Stem Cells in India." *Biosocieties* 8, 1: 25–40.

Biddle, J. (2014) "Intellectual Property in the Biomedical Sciences," in J. Arras, E. Fenton, and R. Kukla (eds.), *Routledge Companion to Bioethics.* London: Routledge.

Bieder, Robert E. (2003) *Science Encounters the Indian, 1820–1880: The Early Years of American Ethnology.* Norman: University of Oklahoma Press.

Biesecker, Barbara Bowles and Theresa M. Marteau (1999) "The Future of Genetic Counseling: An International Perspective." *Nature Genetics* 22: 133–7.

Bilstein, Roger E. (1989) *Orders of Magnitude: A History of the NACA and NASA, 1915–1990.* Washington: NASA SP-4406.

Biltekoff, C. (2013) *Eating Right in America: The Cultural Politics of Food and Health.* Durham: Duke University Press.

Bimber, B. (1996) *Politics of Expertise in Congress: The Rise and Fall of the Office of Technology Assessment.* Albany: SUNY Press.

Binet, A. (1889) *The Psychic Life of Micro-Organisms: A Study in Experimental Psychology.* Chicago: Open Court.

Birdsell, Joseph B. (1953) "Some Environmental and Cultural Factors Influencing the Structuring of Australian Aboriginal Populations." *American Naturalist*: 171–207.

Birke, Lynda (1986) *Women, Feminism and Biology: The Feminist Challenge.* Brighton: Wheatsheaf.

Bittel, Carla (2009). *Mary Putnam Jacobi and the Politics of Medicine in Nineteenth Century America.* Chapel Hill: University of North Carolina Press.

Bix, Amy Sue (1997) "Experiences and Voices of Eugenics Field-Workers: 'Women's Work' in Biology." *Social Studies of Science* 27: 625–68.

Black, Edwin (2003) *War Against the Weak: Eugenics and America's Campaign to Create a Master Race.* New York: Four Walls Eight Windows.

Black, R.D. Collison, A.W. Coats, and Craufurd D.W., Goodwin (eds.) (1973) *The Marginal Revolution in Economics: Interpretation and Evaluation.* Durham: Duke University Press.

Blackwell, Antoinette Brown (1875) *The Sexes Throughout Nature.* New York: G. Putnam's Sons.

Blaug, Mark (1997) *Economic Theory in Retrospect*. New York: Cambridge University Press.

Bleeker, Johan A.M., Johannes Geiss and Martin C.E. Huber (eds.) (2001) *The Century of Space Science*. Dordrecht: Kluwer Academic.

Bleichmar, Daniela, Paula De Vos, Kristin Huffine, and Kevin Sheehan (eds.) (2008) *Science in the Spanish and Portuguese Empires, 1500–1800*. Stanford: Stanford University Press.

Bleichmar, Daniela (2012) *Visible Empire: Botanical Expeditions & Visual Culture in the Hispanic Enlightenment*. Chicago: University of Chicago Press.

Bleier, Ruth (1984) *Science and Gender: A Critique of Biology and its Theories on Women*. New York: Pergamon Press.

Boakes, Robert (1984) *From Darwin to Behaviourism: Psychology and the Minds of Animals*. Cambridge: Cambridge University Press.

Boas, Franz (1904) "The History of Anthropology." *Science* 20, 512: 513–24.

Boas, Franz (1911) *The Mind of Primitive Man*. New York: Macmillan.

Boas, Franz (1912) "Changes in the Bodily Form of Descendants of Immigrants." *American Anthropologist* 14, 3: 530–62.

Boas, Franz (1940) *Race, Language, and Culture*. Chicago: University of Chicago Press.

Boaz, Noel (1981) "History of American Paleoanthropological Research on Early Hominidae, 1925–1980." *American Journal of Physical Anthropology* 56: 397–405.

Boaz, Noel (1982) "American Research on Australopithecines and Early Homo, 1925–1980," in Frank Spencer (ed.), *A History of American Physical Anthropology 1930–1980*. New York: Academic Press, 239–60.

Bocking, Stephen (1995) "Ecosystems, Ecologists, and the Atom: Environmental Research at Oak Ridge National Laboratory." *Journal of the History of Biology* 28: 1–47.

Bocking, Stephen (1997) *Ecologists and Environmental Politics: A History of Contemporary Ecology*. New Haven: Yale University Press.

Bocking, Stephen (2004) *Nature's Experts: Science, Politics, and the Environment*. New Brunswick: Rutgers University Press.

Boffey, Phillip M. (1976) "International Biological Program: Was it Worth the Cost and Effort?" *Science* 193: 866–8.

Bolton, H. (1917) "The Mission as a Frontier Institution in the Spanish-American Colonies." *American Historical Review* 23, 1: 42–61.

Bonner, Thomas N. (1995) *Becoming a Physician: Medical Education in Britain, France, Germany, and the United States, 1750–1945*. Oxford: Oxford University Press.

Bonta, Marcia Myers (1991) *Women in the Field: America's Pioneering Women Naturalists*. College Station: Texas A&M University Press.

Boole, George (1958) *An Investigation of the Laws of Thought on Which are Founded the Mathematical Theories of Logic and Probabilities*. New York: Dover.

Borell, Merrily (1989) *The Biological Sciences in the Twentieth Century*. New York: Charles Scribner's Sons.

Bormann, Herbert F. and Gene E. Likens (1979) *Patterns and Process in a Forested Ecosystem: Disturbance, Development, and the Steady State Based on the Hubbard Brook Ecosystem Study*. New York: Springer-Verlag.

Borrello, M. (2010) *Evolutionary Restraints: The Contentious History of Group Selection*. Chicago: University of Chicago Press.

Boudin, Jean (1857) *Traité De Géographie et de Statistique Médicales et des Maladies*. Paris: J.B. Baillière.

Boudreau, E.B. (2005) " 'Yea, I Have a Goodly Heritage': Health versus Heredity in the Fitter Family Contests, 1920–1928." *Journal of Family History* 30, 4: 366–87.

Bowler, Peter J. (1984) *Evolution: The History of an Idea*. Berkeley: University of California Press.

Bowler, Peter J. (1986) *Theories of Human Evolution: A Century of Debate, 1844–1944*. Baltimore: Johns Hopkins University Press.

Bowler, Peter J. (1989) *The Mendelian Revolution: The Emergence of Hereditarian Concepts in Modern Science and Society*. London: The Athlone Press.

Bowler, Peter J. (1996) *Life's Splendid Drama: Evolutionary Biology and the Reconstruction of Life's Ancestry, 1860–1940*. Chicago: University of Chicago Press.

Bowler, Peter J. (2000) *The Earth Encompassed: A History of the Environmental Sciences*. New York: Norton.

Bowler, Peter J. (2009) *Science for All: The Popularization of Science in Early Twentieth-Century Britain*. Chicago: University of Chicago Press.

Boxer, Marilyn Jacoby (1998) *Women Ask the Questions: Creating Women's Studies in America*. Baltimore: Johns Hopkins University Press.

Boyd, William (2001) "Making Meat: Science, Technology, and American Poultry Production." *Technology and Culture* 42, 4: 631–64.

Bracken, Paul (2012) *The Second Nuclear Age: Strategy, Danger, and the New Power Politics*. New York: Times Books.

Brady, Catherine (2009) *Elizabeth Blackburn and the Story of Telomeres: Deciphering the Ends of DNA*. Cambridge: MIT Press.

Bramwell, Anna (1989) *Ecology in the Twentieth Century: A History*. New Haven: Yale University Press.

Brandl, Bernhard R., Remko Stuik, and J.K. Katgert-Merkelijn (eds.) (2010) *400 Years of Astronomical Telescopes: A Review of History, Science and Technology*. Dordrecht: Springer.

Brandt, Allan (1987) *No Magic Bullet: A Social History of Venereal Disease in the United States Since 1880*. New York: Oxford University Press.

Brantlinger, Patrick (2003) *Dark Vanishings: Discourse on the Extinction of Primitive Races, 1800–1930*. Ithaca: Cornell University Press.

Brantz, D. (2011) "Domestication of Empire: Human–Animal Relations at the Intersection of Civilization and Acclimatization in the Nineteenth Century," in K. Kete (ed.), *A Cultural History of Animals in the Age of Empire*. New York: Bloomsbury.

Brattain, Michelle (2007) "Race, Racism, and Antiracism: UNESCO and the Politics of Presenting Science to the Postwar Public." *The American Historical Review* 112, 5: 1386–413.

Braverman, Irus (2014) "Conservation without Nature: The Trouble with In Situ versus Ex Situ Conservation." *Geoforum* 51: 47–57.

Bray G.A., Nielsen J.N., and Popkin B.M. (2004) "Consumption of High-Fructose Corn Syrup in Beverages May Play a Role in the Epidemic of Obesity." *American Journal of Clinical Nutrition* 79: 537–44.

Brieger, Gert H. (1966) "American Surgery and the Germ Theory of Disease." *Bulletin of the History of Medicine* 40: 135–45.

Briggs, Laura (2002) *Reproducing Empire: Race, Sex, Science, and U.S. Imperialism in Puerto Rico. American Crossroads*. Berkeley: University of California Press.

Brinkman, Paul D. (2010) *The Second Jurassic Dinosaur Rush: Museums & Paleontology in America at the Turn of the Twentieth Century*. Chicago: University of Chicago Press.

Brinkman, Paul D. (2013) "Red Deer River Shakedown: A History of the Captain Marshall Field Paleontological Expedition to Alberta, 1922." *Earth Sciences History* 32, 2: 204–34.

Broad, W. and N. Wade (1982) *Betrayers of the Truth: Fraud and Deceit in the Halls of Science*. New York: Simon & Schuster.

Broca, Paul (1859) "Des Phenomenes d'hybridité dans le Genre Humaine." *Journal de la Physiologie de L'homme et des Animaux* 2: 601–25.

Broca, Paul (1860) "Des Phenomenes d'hybridité dans le Genre Humaine." *Journal de la Physiologie de L'homme et des Animaux* 3: 392–439.

Brock, Emily (2004) "The Challenge of Reforestation: Ecological Experiments in the Douglas Fir Forest, 1920–1940." *Environmental History* 9: 57–79.

Brock, W. H. (1985) *From Protyle to Proton: William Prout and the Nature of Matter, 1785–1985*. Bristol: A. Hilger.

Brody, Howard (1987) *Stories of Sickness*. New Haven: Yale University Press.

Broks, Peter (2006) *Understanding Popular Science*. Maidenhead: Open University Press.

Bromberg, Joan (1991) *The Laser in America*. Cambridge: MIT Press.

Bromberg, Joan (2006) "Device Physics Vis-À-Vis Fundamental Physics in Cold War America: The Case of Quantum Optics." *Isis* 97: 237–59.

Brooke, John Hedley (1991) *Science and Religion: Some Historical Perspectives*. Cambridge: Cambridge University Press.

Brooke, John Hedley and Ronald L. Numbers (eds.) (2011) *Science and Religion Around the World*. Oxford: Oxford University Press.

Brooks, C.E.P. (1970) *Climate Through the Ages: A Study of the Climatic Factors and Their Variations*. New York: Dover.

Brooks, H. (1964) "The Scientific Advisor," in Robert Gilpin and Christopher Wright (eds.), *Scientists and National Policymaking*. New York: Columbia University Press.

Browman, David L. (2002) "The Peabody Museum, Frederic W. Putnam, and the Rise of U.S. Anthropology, 1866–1903." *American Anthropologist* 104, 2: 508–19.

Brown, Charles Brockden (1805) "Terrific Novels." *The Literary Magazine and American Register* 3: 19.

Brown, J.D. and Stone, D.B. (1964) "Tolbutamide-Induced Hypoglycemia." *American Journal of Clinical Nutrition*, 15: 144–8.

Brown, Laurie, Max Dresden, and Lillian Hoddeson (1989) "Pions to Quarks: Particle Physics in the 1950s," in Laurie Brown, Max Dresden, and Lillian Hoddeson (eds.), *Pions to Quarks: Particle Physics in the 1950s*. Cambridge: Cambridge University Press, 3–39.

Browne, Charles A. (ed.) (1926) *A Half-Century of Chemistry in America, 1876–1926: An Historical Review Commemorating the Fiftieth Anniversary of the American Chemical Society*. Easton: American Chemical Society.

Browne, Janet (2001) "Darwin in Caricature: A Study in the Popularization and Dissemination of Evolution." *Proceedings of the American Philosophical Society* 145: 496–509.

Browne, Janet (2003) "Charles Darwin as a Celebrity." *Science in Context* 16: 175–94.

Browne, Janet (2009) "Looking at Darwin: Portraits and the Making of an Icon." *Isis* 100: 542–70.

Brownell, K. and K. Horgen (2004) *Food Fight: The Inside Story of the Food Industry, America's Obesity Crisis, and What We Can Do About It*. New York: McGraw-Hill.

Bruce, Robert V. (1987) *The Launching of Modern American Science, 1846–1876*. Ithaca: Cornell University Press.

Bruinius, Harry (2006) *Better for All the World: The Secret History of Forced Sterilization and America's Quest For Racial Purity*. New York: Alfred A. Knopf.

Brulle, R. (2014) Institutionalizing Delay: Foundation Funding and the Creation of U.S. Climate Change Counter-Movement Organizations. *Climatic Change* 122: 681–94.

Brush, Stephen G. (1979) "Looking Up, the Rise of Astronomy in America." *American Studies* 20: 41–67.

Brush, Stephen G. (1996a) "The Reception of Mendeleev's Periodic Law in America and Britain." *Isis* 87: 595–628.

Brush, Stephen G. (1996b) *A History of Modern Planetary Physics*. New York: Cambridge University Press.

Brush, Stephen G. (2002) "How Theories Became Knowledge: Morgan's Chromosome Theory of Heredity in America and Britain." *Journal of the History of Biology* 35: 471–535.

Brush, Stephen G. and C. Stewart Gillmor (1995) "Geophysics," in Laurie M. Brown (ed.), *Twentieth Century Physics*. Philadelphia: Institute of Physics Publications.

Brush, Stephen G. and Helmut E. Landsberg (1985) *The History of Geophysics and Meteorology: An Annotated Bibliography*. New York: Garland.

Brydon, Diana (ed.) (2000) *Postcolonialism: Critical Concepts in Literary and Cultural Studies, Vols I-V*. New York: Routledge.

Buckley, Kerry W. (1989) *Mechanical Man: John Broadus Watson and the Beginnings of Behaviorism*. New York: Guilford Press.

Bud, Robert (1993) *The Uses of Life: A History of Biotechnology*. Cambridge: Cambridge University Press.

Bud, Robert (1998) "Molecular Biology and the Long-Term History of Biotechnology," in Arnold Thackray (ed.), *Private Science: Biotechnology and the Rise of the Molecular Sciences*. Philadelphia: University of Pennsylvania Press.

Bud, Robert and Deborah Warner (eds.) (1998) *Instruments of Science: A Historical Encyclopedia*. New York: Garland.

Buffetaut, Eric (1987) *A Short History of Vertebrate Palaeontology*. London: Croon Helm.

Bug, Amy (2000) "Gender and Physical Science: A Hard Look at a Hard Science," in Joy Bart (ed.), *Women Succeeding in the Sciences: Theories and Practices Across The Disciplines*. West Lafayette: Purdue University Press, 221–44.

Bugos, Glenn E. and Daniel J. Kevles (1992) "Plants as Intellectual Property: American Practice, Law, and Policy in World Context." *Osiris* 7: 74–104.

Buhs, Joshua Blu (2004) *The Fire Ant Wars: Nature, Science, and Public Policy in Twentieth-Century America*. Chicago: University of Chicago Press.

Burgin, Angus (2012) *The Great Persuasion: Reinventing Free Markets Since the Depression*. Cambridge: Harvard University Press.

Burkhalter, Louis Wood (1965) *Gideon Lincecum*. Austin: University of Texas Press.

Burks, Alice Rowe (2002) *Who Invented the Computer? The Legal Battle that Changed Computing History*. New York: Prometheus.

Burnett, D. Graham (2012) *The Sounding of the Whale: Science and Cetaceans in the Twentieth Century*. Chicago: University of Chicago Press.

Burnham, John C. (1987) *How Superstition Won and Science Lost: Popularizing Science and Health in the United States*. New Brunswick: Rutgers University Press.

Burns, Arthur F. and Wesley C. Mitchell (1946) *Measuring Business Cycles*. New York: National Bureau of Economic Research.

Burns, J. Conor (2008) "Networking Ohio Valley Archaeology in the 1880s: The Social Dynamics of Peabody and Smithsonian Centralization." *Histories of Anthropology Annual* 4: 1–33.

Burns, Matthe W. (2014) "Hundreds of Eugenics Victims Lose Initial Compensation Bids." http://www. wral. com/ hundreds-of-eugenics-victims-lose-initial-compensation-bid/13874521/

Burrell, Herbert (1908) "A New Duty of the Medical Profession: The Education of the Public in Scientific Medicine." *Journal of the American Medical Association* 1, 23: 1873–6.

Burroughs, John (1914) "Science and Literature." *North American Review* 199: 415–24.

Bursey, Maurice (1989) *Francis Preston Venable of the University of North Carolina*. Chapel Hill: The Chapel Hill Historical Society.

Bush, V. (1945) *Science—The Endless Frontier*. Washington: Government Printing Office.

Butler, J. (1736) *The Analogy of Religion Natural and Revealed*. London: John Beecroft and Robert Horsfield.

Butler, Judith (1990) *Gender Trouble: Feminism and the Subversion of Identity*. New York: Routledge.

Bystydzienski, Jill M. and Sharon R. Bird (eds.) (2006) *Removing Barriers: Women in Academic Science, Technology, Engineering, and Mathematics*. Bloomington: Indiana University Press.

Cabell, J.L. (1858) *The Testimony of Modern Science to the Unity of Mankind*. New York: Robert Carter.

Cahan, David and M. Eugene Rudd (2000) *Science at the American Frontier: A Biography of Dewitt Bristol Brace*. Lincoln: University of Nebraska Press.

Cain, Joseph Allen (1993) "Common Problems and Cooperative Solutions: Organizational Activity in Evolutionary Studies, 1936–1947." *Isis* 84: 1–25.

Cain, Joseph Allen (2002) "Epistemic and Community Transition in American Evolutionary Studies: The 'Committee on Common Problems of Genetics, Paleontology, and Systematics' (1942–1949)." *Studies in History and Philosophy of Biological and Biomedical Sciences*. 33: 283–313.

Cain, Joseph Allen and Michael Ruse (eds.) (2009) *Descended from Darwin: Insights into the History of Evolutionary Studies*,

1900–1970. Philadelphia: American Philosophical Society.

Caldwell, Bruce (2003) *Hayek's Challenge: An Intellectual Biography of F. A. Hayek*. Chicago: University of Chicago Press.

Caldwell, Charles (1814) "An Essay on the Causes of the Variety of Complexion and Figure in the Human Species." *The Port Folio*, 4, 8–33, 148–63, 252–71, 362–82, 447–57.

Caldwell, Charles (1852) *Thoughts on the Original Unity of the Human Race*. Cincinnati: J.A. and U.P. James.

Calhoun, C.J. (2007) *Sociology in America: A History*. Chicago: University of Chicago Press.

Calvino, Italo (1986) "Two Interviews on Science and Literature," in *The Uses of Literature: Essays* (Trans. Patrick Creagh). New York: Harcourt Brace, 28–38.

Campbell, Mary B. (1999). *Wonder and Science: Imagining Worlds in Early Modern Europe*. Ithaca: Cornell University Press.

Campbell-Kelly, Martin (2007) "The History of the History of Software." *IEEE Annals of the History of Computing* 29, 4: 40–51.

Campos, Paul (2004) *The Obesity Myth: Why America's Obsession with Weight is Hazardous to Your Health*. New York City: Gotham Books.

Canaday, Margot (2011) *The Straight State: Sexuality and Citizenship in Twentieth-Century America*. Princeton: Princeton University Press.

Canel, Annie, Ruth Oldenziel and Karin Zachmann (eds.) (2000) *Crossing Boundaries, Building Bridges: Comparing the History of Women Engineers, 1870s–1990s*. Amsterdam: Harwood Academic.

Cann, Rebecca, Mark Stoneking, and Allan Wilson (1987) "Mitochondrial DNA and Human Evolution." *Nature* 325: 31–6.

Cannon, Susan Faye (1978) *Science in Culture: The Early Victorian Period*. New York: Science History Publications.

Capshew, James H. (1992) "Psychologists on Site: A Reconnaissance of the Historiography of the Laboratory." *American Psychologist* 47, 2: 132–42.

Capshew, James H. and Karen A. Rader (1992) "Big Science: Price to the Present." *Osiris* 7: 2–25.

Card, David and Alan B. Krueger (1994) "Minimum Wages and Employment: A Case Study of the Fast-Food Industry in New Jersey and Pennsylvania." *American Economic Review* 84, 4: 772–93.

Carey, Daniel (2003) "Anthropology's Inheritance: Renaissance Travel, Romanticism and the Discourse of Identity." *History and Anthropology* 14, 2: 107–26.

Carey, Henry C. (1851) *The Harmony of Interests, Agricultural, Manufacturing, and Commercial*. Philadelphia: J.S. Skinner.

Carlson, John B. (2008) "Romancing the Stone or Moonshine on the Sun Dagger," in Anthony Aveni (ed.), *Foundations of New World Cultural Astronomy*. Boulder: University Press of Colorado, 637–710.

Carpenter, K. (1994) "The Life and Times of W.O. Atwater (1844–1907)" *Journal of Nutrition* 124: 1707S–14S.

Carroll, P. Thomas (1982) "Academic Chemistry in America, 1876–1976: Diversification, Growth, and Change." PhD Dissertation: University of Pennsylvania.

Carson, Elof Axel (2011) "The Hoosier Connection: Complusory Sterilization as Moral Hyigene," in Paul A. Lombardo (ed.), *A Century of Eugenics: From the Indiana Experiment to the Humane Genome Era*. Bloomington: Indiana University Press, 11–25.

Carson, John (2007) *The Measure of Merit: Talents, Intelligence, and Inequality in the French and American Republics, 1750–1940*. Princeton: Princeton University Press.

Carson, Rachel (1962) *Silent Spring*. Boston: Houghton Mifflin

Carter, Edward C. (ed.) (1999) *Surveying the Record: North American Scientific Exploration to 1930*. Philadelphia: American Philosophical Society.

Carter, K. Codell (2003) *The Rise of Causal Concepts of Disease: Case Histories*. Aldershot: Ashgate.

Caspari, Rachel (2003) "From Types to Populations: A Century of Race, Physical

Anthropology, and the American Anthropological Association." *American Anthropologist* 105: 65–76.

Caspari, Rachel and Milford Wolpoff (1996) "Weidenreich, Coon, and Multiregional Evolution." *Human Evolution* 11: 261–8.

Cassedy, James and Deborah J. Warner (1995) "Medical Microscopy in Antebellum America." *Bulletin of the History of Medicine* 69: 367–86.

Cassidy, David (2005) *J. Robert Oppenheimer and the American Century*. New York: Pi Press.

Cassidy, James G. (2000) *Ferdinand V. Hayden: Entrepreneur of Science*. Lincoln: University of Nebraska Press.

Castañeda, Quetzil E. (2003) "Stocking's Historiography of Influence: The 'Story of Boas', Gamio and Redfield at the Cross-'Road to Light'." *Critique of Anthropology* 23, 3: 235–63.

Cat, Jordi (1998) "The Physicists' Debates on Unification in Physics at the End of the 20th Century." *Historical Studies in the Physical and Biological Sciences* 28: 253–99.

Catesby, Mark (1731–1743) *Natural History of Carolina, Florida and The Bahama Islands*. London: Printed At The Expence of the Author, and Sold By W. Innys and R. Manby.

Cenadelli, David (2010) "Solving the Giant Stars Problem: Theories of Stellar Evolution from the 1930s to the 1950s." *Archive for History of Exact Sciences* 64: 203–67.

Ceruzzi, Paul (1983) *Reckoners: The Prehistory of the Digital Computer, from Relays to The Stored Program Concept, 1935–1945*. Santa Barbara: Greenwood.

Ceruzzi, Paul (1991) "When Computers were Human." *Annals of the History of Computing* 13: 242.

Ceruzzi, Paul (1999) "Electronics Technology and Computer Science, 1940–1975: A Coevolution." *IEEE Annals of the History of Computing* 10, 4: 257–75.

Ceruzzi, Paul (2000) "Nothing New since Von Neumann: A Historian Looks at Computer Architecture, 1945–1995," in Raúl Rojas and Ulf Hashagen (eds.), *The First Computers: History and Architectures*. Cambridge: MIT Press, 195–217.

Ceruzzi, Paul (2003) *A History of Modern Computing*. Cambridge: MIT Press.

Chadwick, Edwin (1846) "Evidence." *Great Britain, Parliamentary Papers* 10: 651.

Chakrabarty, Dipesh (1992) "Postcoloniality and the Artifice of History; Who Speaks for Indian Pasts" *Representations* 37, 1.

Chamberlain, Alexander F. (1907) "Thomas Jefferson's Ethnological Opinions and Activities." *American Anthropologist* 9: 199–509.

Chamberlain, Von Del (1982) *When Stars Came Down to Earth: Cosmology of the Skidi Pawnee Indians of North America*. Los Altos: Ballena Press.

Chamberlin, Edward (1933) *Theory of Monopolistic Competition*. Cambridge: Harvard University Press.

Chambers, David Wade and Richard Gillespie (2000) "Loyalty in the History of Science: Colonial Science, Technoscience, and Indigenous Knowledge." *Osiris* 15: 221–40.

Chandler, Alfred D. (2005) *Shaping the Industrial Century: The Remarkable Story of the Modern Chemical and Pharmaceutical Industries*. Cambridge: Harvard University Press.

Chapin, Charles V. (1902) "The End of the Filth Theory of Disease." *Popular Science Monthly* 60: 234–9.

Charles, Asselin (trans.) (2000) *The Equality of the Human Races*. New York: Garland.

Charnley, Berris (2013) "Experiments in Empire-Building: Mendelian Genetics as a National, Imperial, and Global Agricultural Enterprise." *Studies in History and Philosophy of Science* 44: 292–300.

Charnley, Berris and Gregory Radick (2013) "Intellectual Property, Plant Breeding and the Making of Mendelian Genetics." *Studies in History and Philosophy of Science* 44: 222–33.

Chassy B.M. (2007) "The History and Future of GMOs in Food and Agriculture." *Cereal Foods World*, 52, 4: 169–72.

Checkovich, Alex (2004) "Mapping the American Way: Geographical Knowledge and the Development of the United States, 1890–1950." PhD Dissertation: University of Pennsylvania.

Cheyfitz, Eric (2002) "The Post-Colonial Predicament of Native American Studies." *Interventions* 4, 3: 405–27.

Christen, Catherine A. (2002) "At Home in the Field: Smithsonian Tropical Science Field Stations in the U.S. Panama Canal Zone and the Republic of Panama." *The Americas* 58: 537–75.

Christie, J.R.R. (1990) "Feminism and the History of Science," in R.C. Olby, G.N. Cantor, J.R. R. Christie and M.J.S Hodge (eds.), *Companion to the History of Modern Science*. Cambridge: Cambridge University Press, 100–9.

Cipolloni, Marco (2007) *The Anthropology of the Enlightenment*. Stanford: Stanford University Press.

Cittadino, Eugene (1980) "Ecology and the Professionalization of Botany in America, 1890–1905." *Studies in the History of Biology* 4: 171–98.

Cittadino, Eugene (1993) "A 'Marvelous Cosmopolitan Preserve': The Dunes, Chicago, and the Dynamic Ecology of Henry Chandler Cowles." *Perspectives on Science* 1: 520–59.

Clancey, Greg (2006) *Earthquake Nation: The Cultural Politics of Japanese Seismicity, 1868–1930*. Berkeley: University of California Press.

Clark, Constance Areson (2001) "Evolution for John Doe: Pictures, the Public, and the Scopes Trial Debate." *Journal of American History* 87: 1275–303.

Clark, Constance Areson (2008) *God or Gorilla: Images of Evolution in the Jazz Age*. Baltimore: Johns Hopkins University Press.

Clark, Constance Areson (2009) "'You Are Here': Missing Links, Chains of Being, and the Language of Cartoons." *Isis* 100: 571–89.

Clark, Eugenie (1953) *Lady with a Spear*. New York: Harper.

Clark, John Bates (1886) *The Philosophy of Wealth*. Boston: Ginn.

Clark, John Bates (1899) *The Distribution of Wealth: A Theory of Wages, Interest and Profits*. New York: Macmillan.

Clark, John Maurice (1923) *Studies in the Economics of Overhead Costs*. Chicago: University of Chicago Press.

Clark, John Maurice (1926) *Social Control of Business*. Chicago: University of Chicago Press.

Clarke, Adele E. (1998) *Disciplining Reproduction: Modernity, American Life Sciences, and the Problems of Sex*. Berkeley: University of California Press.

Clarke, Adele E., and Joan H. Fujimura (1992) *The Right Tools for the Job: At Work in the Twentieth-Century Life Sciences*. Princeton: Princeton University Press.

Clarke, Adele E., Laura Mamo, Jennifer Ruth Fosket, Jennifer R. Fishman, and Janet K. Shim (eds.) (2010) *Biomedicalization: Technoscience, Health, and Illness in the U.S.* Durham: Duke University Press.

Clarke, Frank Wigglesworth (1882) *A Recalculation of the Atomic Weights*. Washington, DC: Smithsonian Institution.

Clarke, Frank Wigglesworth (1927) "Biographical Memoir, Edward Williams Morley, 1838–1923." *Biographical Memoirs, National Academy of Sciences* 10: 1–8.

Clarke, S. (1704) *A Discourse Concerning the Being and Attributes of God*. Glasgow: Richard Griffin.

Clay, Reginald S. and Thomas H. Court (1932) *The History of the Microscope*. London: Charles Griffin.

Clements, Frederick E. (1916) *Research Methods in Ecology*. Washington: Carnegie Institute.

Clerke, Agnes (1902) *A Popular History of Astronomy during the Nineteenth Century*. London: Adam and Charles Black.

Clifford, James, and George E. Marcus (1986) *Writing Culture: The Poetics and Politics of Ethnography*. Berkley: University of California Press.

Clifford, James (1988) *The Predicament of Culture: Twentieth-Century Ethnography, Literature, and Art*. Cambridge: Harvard University Press.

Cloud, John (2000) "Crossing the Olentangy River: The Figure of the Earth and the Military–Industrial–Academic Complex, 1947–1972." *Studies in History and Philosophy of Modern Physics* 31, 3: 371–404.

Coase, Ronald H. (1960) "The Problem of Social Cost." *Journal of Law and Economics* 3: 1–44.

Coats, A.W. (1960) "The First Two Decades of the American Economic Association." *American Economic Review* 50, 4: 556–74.

Coats, A.W. (1961) "The Political Economy Club: A Neglected Episode in American Economic Thought." *American Economic Review* 51, 4: 624–37.

Coats, A.W. (1963) "The American Economic Association, 1904–29." *American Economic Review* 54, 4: 261–85.

Coats, A.W. (1988) "The Educational Revolution and the Professionalization of American Economics," in William J. Barber (ed.), *Breaking the Academic Mould: Economists and American Higher Learning in the Nineteenth Century.* Middletown: Wesleyan University Press, 340–75.

Cobb, Charles W. and Paul H. Douglas (1928) "A Theory of Production." *American Economic Review* 18, 1 (Supplement): 139–65.

Cochrane, Willard W. (1993) *Development of American Agriculture: A Historical Analysis.* Minneapolis: University of Minnesota Press.

Coen, Deborah R. (2007) *Vienna in the Age of Uncertainty: Science, Liberalism, and Private Life.* Chicago: University of Chicago Press.

Coen, Deborah R. (2010) "Climate and Circulation in Imperial Austria." *e Journal of Modern History* 82: 839–75.

Coen, Deborah R. (2011) "Imperial Climatographies from Tyrol to Turkestan." *Osiris* 26: 45–65.

Coen, Deborah R. (2013) *The Earthquake Observers: Disaster Science from Lisbon to Richter.* Chicago: University of Chicago Press.

Cohen-Cole, Jamie (2007) "Cybernetics and the Machinery of Rationality". *British Journal for the History of Science* 41, 1: 109–14.

Cohen, A.M. (1998) *The Shaping of American Higher Education: Emergence and Growth of the Contemporary System.* San Francisco: Jossey-Bass.

Cohen, Benjamin R. (2004) "The Element of the Table: Visual Discourse and the Preperiodic Representation of Chemical Classification." *Configurations,* 12: 41–75.

Cohen, Benjamin R. (2006) "Surveying Nature: Environmental Dimensions of Virginia's First Scientific Survey, 1835–1842." *Environmental History* 11: 37–69.

Cohen, Benjamin R. (2009) *Notes from the Ground: Science, Soil and Society in the American Countryside.* New Haven: Yale University Press.

Cohen, I. Bernard (1941) *Benjamin Franklin's Experiments: A New Edition of Franklin's Experiments and Observations on Electricity.* Cambridge: Harvard University Press.

Cohen, I. Bernard (1950) *Some Early Tools of American Science: An Account of the Early Scientific Instruments and Mineralogical and Biological Collections in Harvard University.* Cambridge: Harvard University Press.

Cohen, I. Bernard (1956) *Franklin and Newton: An Inquiry into Speculative Newtonian Experimental Science and Franklin's Work in Electricity as an Example Thereof.* Memoirs of the American Philosophical Society. Philadelphia: American Philosophical Society.

Cohen, I. Bernard (1999) *Howard Aiken: Portrait of a Computer Pioneer.* Cambridge: MIT Press.

Cohen, I. Bernard (2000) "Howard Aiken and the Dawn of the Computer Age," in Raúl Rojas and Ulf Hashagen (eds.), *The First Computers: History and Architectures.* Cambridge: MIT Press, 107–120.

Cohen, Lawrence (1994) "Whodunit? – Violence and the Myth of Fingerprints: Comment On Harding." *Configurations* 2: 343–47.

Cohen, Lawrence (2005) "Operability, Bioavailability, and Exception," in Aihwa Ong and Stephen Collier (eds.), *Global Assemblages: Technology, Politics and Ethics as Anthropological Problems.* Oxford: Blackwell, 79–90.

Colander, David C. and Harry Landreth (1996) *The Coming of Keynesianism to America: Conversations with the Founders of Keynesian Economics.* Cheltenham: Edward Elgar.

Coleman, William (1971) *Biology in the Nineteenth Century: Problems of Form, Function and Transformation.* Cambridge: Cambridge University Press.

Collins, Francis S. (2006) *The Language of God: A Scientist Presents Evidence for Belief.* New York: Free Press.

Collins, Francis S. (2008) "Victor A. Mckusick." *Science* 321: 925.

Collins, Harry (2004) *Gravity's Shadow: The Search for Gravitational Waves.* Chicago: University of Chicago Press.

Collins, Harry (2014) *Gravity's Ghost and Big Dog: Scientific Discovery and Social Analysis in the Twenty-First Century.* Chicago: University of Chicago Press.

Collins, Harry and R. Evans (2007) *Rethinking Expertise.* Chicago: University of Chicago Press.

Columella and H.B. Ash (1941) *Columella: On Agriculture, Volume 1, Books 1–IV.* Cambridge: Harvard University Press.

Comfort, Nathaniel (2001) *The Tangled Field: Barbara Mcclintock's Search for the Patterns of Genetic Control.* Cambridge: Harvard University Press.

Comfort, Nathaniel (2006) ""Polyhybrid Heterogeneous Bastards": Promoting Medical Genetics in America in the 1930s and 1940s." *Journal of the History of Medicine and Allied Sciences* 61: 415–55.

Comfort, Nathaniel (2012) *The Science of Human Perfection: How Genes Became the Heart of American Medicine.* New Haven: Yale University Press.

Committee on Astronomy and Astrophysics, Board on Physics and Astronomy and Space Studies Board, Commission on Physical Sciences, Mathematics, and Applications, and National Research Council (2000) *Federal Funding of Astronomical Research.* Washington, DC: National Academy Press.

Committee for a Decadal Survey of Astronomy and Astrophysics (2010) *New Worlds, New Horizons in Astronomy and Astrophysics.* Washington: National Academy Press.

Committee on a Decadal Strategy for Solar and Space Physics (2013) *Solar and Space Physics: A Science for a Technological Society.* Washington: National Academy Press.

Committee on Maximizing the Potential of Women in Academic Science and Engineering (2007) *Beyond Bias and Barriers: Fulfilling the Potential of Women in Academic Science and Engineering.* Washington: National Academy Press.

Committee on Planetary and Lunar Exploration (COMPLEX 1978) *Strategy for Exploration of the Inner Planets: 1977–1987.* Washington: National Academy Press.

Committee on the Planetary Science Decadal Survey (2011) *Visions and Voyages for Planetary Science in the Decade 2013–2022.* Washington: National Academy Press.

Committee on Space Astronomy and Astrophysics (1978) *A Strategy for Space Astronomy and Astrophysics for the 1980s.* Washington: National Academy Press.

Commons, John R. (1924) *Legal Foundations of Capitalism.* New York: Macmillan.

Compton, W. David and Charles D. Benson (1983). *Living and Working in Space: A History of Skylab.* Washington: NASA SP-4208.

Comstock, Anna Botsford (1911) *Handbook of Nature Study.* Ithaca: Cornell University Press.

Condran, G.A. (1987) "Declining Mortality in the United States in the Late Nineteenth and Early Twentieth Centuries." *Annales de Démographie Historique.* Paris: Socieéteé de deémographie historique, 119–41.

Conkin, Paul K. (1998) *When All the Gods Trembled: Darwinism, Scopes, and American Intellectuals.* Maryland: Rowman and Littlefield.

Conn, Steven (1998) *Museums and American Intellectual Life, 1876–1926.* Chicago: University of Chicago Press.

Conn, Steven (2010) *Do Museums Still Need Objects?* Philadelphia: University of Pennsylvania Press.

Connell, R. (1997) Why is Classical Theory Classical? *American Journal of Sociology* 102, 6: 1511–57.

Connell, R. (2007) *Southern Theory: The Global Dynamics of Knowledge in the Social Sciences.* Crows Nest: Allen & Unwin.

Conrad, J. and R. Becker (2012) Enhancing Credibility of Chemical Safety Studies: Emerging Consensus on Key Assessment Criteria. *Environmental Health Perspectives* 119: 757–64.

Conway, Erik M. (2007) "From Rockets to Spacecraft: Making JPL a Place for Planetary Science." *Engineering & Science* 40, 4: 2–10.

Conway, Erik M. (2008) *Atmospheric Science at NASA: A History*. Baltimore: Johns Hopkins University Press.

Conway, Erik M. (2010) "The International Geophysical Year and Planetary Science," in Roger D. Launius (ed.), *Making Polar Science Global*. New York: Palgrave Macmillan.

Conway, Erik M. (2014) "Bringing NASA Back to Earth: A Search for Relevance during the Cold War," in Naomi Oreskes and John Krige (eds.), *Nation and Knowledge: Science and Technology in the Global Cold War*. Cambridge: MIT Press, 251–72.

Conway, Erik M. (2015) *Exploration and Engineering: The Jet Propulsion Laboratory and the Quest for Mars*. Baltimore: Johns Hopkins University Press.

Cook, Robert Edward (1977) "Raymond Lindeman and the Trophic Dynamic Aspect in Ecology." *Science* 198: 22–6.

Cook, Stephen (1971) "The Complexity of Theorem-Proving Procedures." Proceedings of the Third Annual ACM Symposium on Theory of Computing. New York: Association of Computing Machinery, 151–8.

Cook-Deegan, Robert (1994) *Gene Wars: Science, Politics, and the Human Genome*. New York: Norton.

Cook-Deegan, Robert (1995) *The Gene Wars: Science, Politics, and the Human Genome*. New York: W.W. Norton.

Coon, Carleton (1962) *The Origin of Races*. New York: Knopf.

Cooper, Dan (1999) *Enrico Fermi and the Revolutions of Modern Physics*. Oxford: Oxford University Press.

Cooper, Melinda (2008) *Life As Surplus: Biotechnology and Capitalism in the Neoliberal Era*. Washington: University of Washington Press.

Cooper, William Skinner (1926) "The Fundamentals of Vegetational Change." *Ecology* 7, 4: 391–410.

Cooter, Roger and Stephen Pumfrey (1994) "Separate Spheres and Public Places: Reflections on the History of Science Popularization and on Science in Popular Culture." *History of Science* 32: 232–67.

Cope, Edward (1868) "On the Origin of Genera." *Proceedings of the Academy of Natural Sciences of Philadelphia* 20: 242–300.

Cope, Edward (1888) "Archaeology and Anthropology." *American Naturalist* 22: 660–3.

Cope, Edward (1893) "The Genealogy of Man." *American Naturalist* 27: 321–35.

Copeland, Jack (ed.) (2006). *Colossus: The Secrets of Bletchley Park's Codebreaking Computers*. Oxford: Oxford University Press.

Copeland, Jack (2014) *Turing: Pioneer of the Information Age*. Oxford: Oxford University Press.

Cordeschi, Robert (2002) *The Discovery of the Artificial: Behavior, Mind and Machines before and beyond Cybernetics*. Dortrecht: Kluwer.

Corner, George and Sophia Aberle (1953) *Twenty-Five Years of Sex Research: History of the National Research Council Committee for Research in Problems of Sex, 1922–1947*. Philadelphia: Saunders.

Cortada, James. (2000) *Before the Computer: IBM, NCR, Burroughs, & Remington Rand and the Industry they Created, 186–1956*. Princeton: Princeton University Press.

Cosmides, L. (1989) "The Logic of Social Exchange: Has Natural Selection Shaped How Humans Reason? Studies with the Wason Selection Task." *Cognition* 31: 187–276.

Cosmides, L. and J. Toby (1996) "Are Humans Good Intuitive Statisticians After All? Rethinking Some Conclusions from the Literature on Judgment under Uncertainty." *Cognition* 58: 1–73.

Count, Earl W. (ed.) (1950) *This is Race*. New York: Henry Schuman.

Couvet, D., F. Jiguet, R. Julliard, H. Levrel, and A. Teyssedre (2008) "Enhancing Citizen Contributions to Biodiversity and Public Policy." *Interdisciplinary Science Reviews* 33: 95–103.

Cowan, Ruth Schwartz (1994) "Women's Roles in the History of Amniocentesis and Chorionic Villus Sampling," in Karen H. Rothenberg, and Elizabeth Jean Thomson (eds.), *Women and Prenatal Testing: Facing The Challenges of Genetic Technology*. Columbus: Ohio State University Press.

Cowan, Ruth Schwartz (2008) *Heredity and Hope: The Case for Genetic Screening*. Cambridge: Harvard University Press.

Cowles, Henry Chandler (1899) "The Ecological Relations of the Vegetation on the Sand Dunes of Lake Michigan." *Botanical Gazette* 27: 95–117, 167–202, 281–308, 361–91.

Cowles, Henry Chandler (1901) "Physiographic Ecology of Chicago and the Vicinity: A Study of Origin, Development, and Classification of Plant Societies." *Botanical Gazette* 31: 73–108, 145–82.

Cox, Allan (ed.) (1973) *Plate Tectonics and Geomagnetic Reversals*. San Francisco: W.H. Freeman.

Craig, Gerald S. (1940) *A New Science Program for Elementary Schools*. New York: Ginn and Co.

Cravens, Hamilton (1976) "American Science Comes of Age: An Institutional Perspective, 1850–1930." *American Studies* 17, 2: 49–70.

Cravens, Hamilton (1978) *The Triumph of Evolution: American Scientists and the Heredity–Environment Controversy, 1900–1941*. Philadelphia: University of Pennsylvania Press.

Cravens, Hamilton (1993) *Before Head Start: The Iowa Station & America's Children*. Chapel Hill: University of North Carolina Press.

Cravens, Hamilton and A. Marcus (1996) "Introduction: Technical Knowledge in American Culture: An Analysis," in H. Cravens, A. Marcus, and D. Katzman (eds.), *Technical Knowledge in American Culture: Science, Technology and Medicine Since the Early 1800s*. Tuscaloosa: University of Alabama Press, 1–18.

Creager, Angela N.H. (1996) "Wendell Stanley's Dream of a Free-Standing Biochemistry." *Journal of the History of Biology* 29: 331–60.

Creager, Angela N.H. (2002) *The Life of a Virus: Tobacco Mosaic Virus as an Experimental Model, 1930–1965*. Chicago: University of Chicago Press.

Creager, Angela N.H. (2010) "The Paradox of the Phage Group." *Journal of the History of Biology* 43: 183–93.

Creager, Angela N.H. (2013) *Life Atomic: A History of Radioisotopes in Science and Medicine*. Chicago: University of Chicago Press.

Creager, Angela N.H., Elizabeth Lunbeck, and Londa Schiebinger (eds.) (2001) *Feminism in Twentieth-Century Science, Technology, and Medicine*. Chicago: University of Chicago Press.

Crease, Robert P. (1999) *Making Physics: A Biography of Brookhaven National Laboratory, 1946–1972*. Chicago: University of Chicago Press.

Crelinsten, Jeffrey (2006) *Einstein's Jury: The Race to Test Relativity*. Princeton: Princeton University Press.

Cremin, Lawrence (1961) *The Transformation of the Schools: Progressivism in American Education, 1876–1957*. New York: Knopf.

Critser, G. (2003) *Fat Land: How Americans Became the Fattest People in the World*. New York: Houghton-Mifflin.

Croker, Robert A. (1991) *Pioneer Ecologist: The Life and Work of Victor Ernest Shelford, 1877–1968*. Washington: Smithsonian University Press.

Cronon, William (1983) *Changes in the Land: Indians, Colonists, and the Ecology of New England*. New York: Hill and Wang.

Cronon, William (1995) *Uncommon Ground: Toward Reinventing Nature*. New York: W.W. Norton.

Crosby, Alfred (1986) *Ecological Imperialism: The Biological Expansion of Europe, 900–1900*. Cambridge: Cambridge University Press.

Crowe, Nathan (2014) "Cancer, Conflict, and the Development of Nuclear Transplantation Techniques." *Journal of the History of Biology* 47: 63–105.

Crowther-Heyck, Hunter (2005) *Herbert A. Simon: The Bounds of Reason in Modern America*. Baltimore: Johns Hopkins University Press.

Cruickshank, Dale P. and Joseph W. Chamberlain (1999) "The Beginnings of the Division for Planetary Sciences of the American Astronomical Society," in David H. DeVorkin (ed.), *The American Astronomical Society's First Century*. Washington: The American Astronomical Society, 252–68.

Cubberley, Ellwood P. (1909) *Changing Conceptions of Education*. Cambridge: Riverside Press.

Cunningham, Andrew and Perry Williams, (eds.) (1992) *The Laboratory Revolution in Medicine*. New York: Cambridge University Press.

Currell, Susan and Christina Cogdell (eds.) (2006) *Popular Eugenics: National Efficiency and American Mass Culture in the 1930s*. Athens: Ohio University Press.

Dahlberg, Frances (1981) *Woman the Gatherer*. New Haven: Yale University Press.

Dahlberg, K.A. (1988) "Ethical and Value Issues in International Agricultural Research." *Agriculture and Human Values* 5, 1&2: 101–111.

Dalley-Crozier, Ivan (2000) "Havelock Ellis, Eonism, and the Patient's Discourse; Or, Writing a Book about Sex." *History of Psychiatry* 11: 142–54.

Daly, M, and M Wilson (1998) *The Truth about Cinderella: A Darwinian View of Parental Love*. London: Weidenfeld and Nicholson.

Daniels, George H. (1968) *American Science in the Age of Jackson*. Tuscaloosa: University of Alabama Press.

Darnell, Regna (1971) "The Professionalization of American Anthropology: A Case Study in the Sociology of Knowledge." *Social Science Information* 10, 2: 83–103.

Darnell, Regna (1990) *Edward Sapir: Linguist, Anthropologist, Humanist*. Berkeley: University of California Press.

Darnell, Regna (1998) *Along Came Boas: Continuity and Revolution in Americanist Anthropology*. Amsterdam: John Benjamin.

Darnell, Regna (2001) *Invisible Genealogies: A History of Americanist Anthropology*. Lincoln: University of Nebraska Press.

Darwin, Charles (1859) *On the Origin of Species by Means of Natural Selection, or the Preservation of Favoured Races in the Struggle for Life*. London: John Murray.

Darwin, Charles (1871) *The Descent of Man, and Selection in Relation to Sex*. London: John Murray.

Davenport, Charles (1911) *Heredity in Relation to Eugenics*. New York: Henry Holt.

Davenport, Charles and Morris Steggerda (1929) *Race Crossing in Jamaica*. Washington: Carnegie Institute.

Davies, Kevin (2001) *Cracking the Genome: Inside the Race to Unlock Human DNA*. Baltimore: Johns Hopkins University Press.

Davies, Margery W. (1982) *Woman's Place is at the Typewriter: Office Work and Office Workers, 1870–1930*. Philadelphia: Temple University Press.

Davies, Paul Sheldon (2003) *Norms of Nature Naturalism and the Nature of Functions*. Cambridge: MIT Press.

Davis, Edward B. (2008) "Fundamentalist Cartoons, Modernist Pamphlets, and the Religious Image of Science in the Scopes Era," in Charles L. Cohen and Paul S. Boyer (eds.), *Religion and the Culture of Print in Modern America*. Madison: University of Wisconsin Press, 175–98.

Davis, Frederick Rowe (2007) *The Man Who Saved Sea Turtles: Archie Carr and the Origins of Conservation Biology*. Oxford: Oxford University Press.

Davis, Katharine Bement (1929) *Factors in the Sex Life of Twenty Two Hundred Women*. New York: Harper & Row.

Davis, Martin (2000) *Engines of Logic: Mathematicians and the Origin of the Computer*. New York: W.W. Norton.

Davis, Susan G. (1997) *Spectacular Nature: Corporate Culture and the Sea World Experience*. Berkeley: University of California Press.

Dawkins, Richard (1976) *The Selfish Gene*. Oxford: Oxford University Press.

Dawkins, Richard (1996) *The Blind Watchmaker: Why the Evidence of Evolution Reveals a Universe Without Design*. London: W.W. Norton.

Dawkins, Richard (2006) *The God Delusion*. London: Bantam.

Dawson, Virginia P. (1991) *Engines and Innovation: Lewis Laboratory and American Propulsion Technology*. Washington: NASA.

De Buffon, Georges Louis Le Clerc, and Louis Jean Marie Daubenton (1766). *Histoire Naturelle, Générale et Particulière: Avec la Description du Cabinet du Roi*. Paris: Imperial Royale.

De Chadarevian, Soraya (2002) *Designs for Life: Molecular Biology after World War II*. Cambridge: Cambridge University Press.

De Chadarevian, Soraya (2006) "Mice and the Reactor: The 'Genetics Experiment' in 1950s Britain." *Journal of the History of Biology* 39: 707–35.

De Chadarevian, Soraya (2010) "Mutations in the Nuclear Age," in Luis Campos and Alexander Von Schwerin (eds.), *Making Mutations: Objects, Practices, Contexts, Preprint 393*, Berlin: Max Plank Institute for the History of Science.

De Chadarevian, Soraya and Jean-Paul Gaudillière (1996) "The Tools of the Discipline: Biochemists and Molecular Biologists." *Journal of the History of Biology* 29: 327–30.

De Chadarevian, Soraya and Harmke Kamminga (1998) *Molecularizing Biology and Medicine: New Practices and Alliances, 1910s–1970s*. Amsterdam: Harwood.

De Jong, Antina, Wybo J. Dondorp, Merryn V.E. Macville, Christine E.M. De Die-Smulders, Jan M.M. Van Lith, and Guido M.W.R. de Wert (2014) "Microarrays as a Diagnostic Tool in Prenatal Screening Strategies: Ethical Reflection." *Human Genetics* 133: 163–72.

De Kruif, Paul (1926) *The Microbe Hunters*. New York: Harcourt, Brace.

De Lange, D.J. and C.P. Joubert (1964) "Assessment of Nicotinic Acid Status of Population Groups." *American Journal of Clinical Nutrition*. 15: 169–74.

Deacon, Margaret (1997) *Scientists and the Sea 1650–1900: A Study of Marine Science*. Waltham: Academic Press.

Dear, Peter (2009) *Revolutionizing the Sciences: European Knowledge and its Ambitions, 1500–1700*. Princeton: Princeton University Press.

Dear, Peter (2012) "Science is Dead: Long Live Science." *Osiris* 27: 37–55.

Deblieu, Jan (1993) *Meant to be Wild: The Struggle to Save Endangered Species through Captive Breeding*. Golden, Colorado: Fulcrum.

Deboer, G.E. (1991) *A History of Ideas in Science Education: Implications for Practice*. New York: Teachers College Press.

Deegan, M. J. (1988) *Jane Addams and the Men of the Chicago School, 1892–1918*. New Brunswick: Transaction.

Degler, Carl N. (1992) *In Search of Human Nature: The Decline and Revival of Darwinism in American Social Thought*. New York: Oxford University Press.

Delbourgo, James (2006). *A Most Amazing Scene of Wonders: Electricity and Enlightenment in Early America*. Cambridge: Harvard University Press.

Delgado, Richard and Jean Stefancic (2012) *Critical Race Theory: An Introduction*. New York: New York University Press.

Delind, L.B. (2011) "Are Local Food and the Local Food Movement Taking Us Where We Want to Go? Or are We Hitching Our Wagons to the Wrong Stars?" *Agriculture and Human Values* 28: 273–83.

Delisle, Richard (1995) "Human Palaeontology and the Evolutionary Synthesis during the Decade 1950–1960," in R. Corbey and B. Theunissen (eds.), *Ape, Man, Apeman: Changing Views Since 1600*. Department of Prehistory, Leiden University, 217–28.

Delisle, Richard (2001) "Adaptationism versus Cladism in Human Evolution Studies," in Raymond Corbey and Wil Roebroeks (eds.), *Studying Human Origins: Disciplinary History and Epistemology*. Amsterdam: Amsterdam University Press, 107–21.

Delisle, Richard (2007) *Debating Humankind's Place in Nature, 1860–2000: The Nature of PaleoAnthropology*. Upper Saddle River: Pearson Prentice Hall.

Deluzio, Crista (2007) *Female Adolescence in American Scientific Thought, 1830–1930*. Baltimore: Johns Hopkins University Press.

Dembski, William A., and Michael Ruse (2004) *Debating Design: From Darwin to DNA*. New York: Cambridge University Press.

Denby, David (2005) "Herder: Culture, Anthropology, and the Enlightenment." *History of the Human Sciences* 18, 1: 55–76.

Denevan, William M. (1992) "The Pristine Myth: The Landscape of the Americas in 1492." *Annals of the Association of American Geographers* 82, 3: 369–85.

Dennett, Daniel (1995) *Darwin's Dangerous Idea: Evolution and the Meanings of Life.* New York: Simon & Schuster.

Dennis, Michael Aaron (1994) 'Our First Line of Defense': Two University Laboratories in the Postwar American State." *Isis* 85: 427–55.

Des Jardins, Julie (2010). *The Madame Curie Complex: The Hidden History of Women in Science.* New York: Feminist Press at the City University of New York.

Desmond, Adrian (1975). *The Hot-Blooded Dinosaurs.* London: Blond & Briggs.

Dettelbach, Michael (1996) "Humboldtian Science," in N. Jardine, J.A. Secord, and E.C. Spary (eds.), *Cultures of Natural History.* New York: Cambridge University Press, 287–304.

Devonis, David C. (2012) "Bentley, Madison," in Robert W. Rieber (ed.), *Encyclopedia of the History of Psychological Theories.* New York: Springer, 113–15.

DeVorkin, David H. (1982) *The History of Modern Astronomy and Astrophysics: A Selected, Annotated Bibliography.* New York: Garland.

DeVorkin, David H. (1992) *Science with a Vengeance: How the Military Created the US Space Sciences after World War II.* New York: Springer-Verlag.

DeVorkin, David H. (ed.) (1999) *The American Astronomical Society's First Century.* Washington: The American Astronomical Society.

DeVorkin, David H. (2000) "Who Speaks for Astronomy? How Astronomers Responded to Government Funding After World War II." *Historical Studies in the Physical and Biological Sciences* 31, 1: 55–92.

DeVorkin, David H. (2000a) *Henry Norris Russell: Dean of American Astronomers.* Princeton: Princeton University Press.

DeVorkin, David H. (2000b) "Who Speaks for Astronomy? How Astronomers Responded to Government Funding after World War II." *Historical Studies in the Physical and Biological Sciences,* 31: 55–92.

DeVorkin, David H. (2004) "Solar Physics from Space," in John Logsdon et al. (eds.), *Exploring the Unknown: Selected Documents in the History of the U.S. Civil Space Program, Vol. VI, Space and Earth Science.* Washington. NASA SP-2004-4407, 1–36.

Dewbury, Adam (2007) "The American School and Scientific Racism in Early American Anthropology." *Histories of Anthropology Annual* 3, 1: 121–47.

Diamond, Lisa M. (2008) *Sexual Fluidity: Understanding Women's Love and Desire.* Cambridge: Harvard University Press.

Dick, Stephanie (2011). "Aftermath: The Work of Proof in the Age of Human–Machine Collaboration." *Isis* 102, 3: 494–505.

Dick, Stephanie (2015). "Models and Machines: Implementing Bounded Rationality." *Isis* (forthcoming September 2015).

Dick, Steven J. (1991) "John Quincy Adams, the Smithsonian Bequest, and the Founding of the U.S. Naval Observatory." *Journal for the History of Astronomy* 22: 31–44.

Dick, Steven J. (1996) *The Biological Universe: The Twentieth-Century Extraterrestrial Life Debate and the Limits of Science.* Cambridge: Cambridge University Press.

Dick, Steven J. (2003) *Sky and Ocean Joined: The U.S. Naval Observatory, 1830–2000.* Cambridge: Cambridge University Press.

Didier, Emmanuel (2011) "Counting on Relief: Industrializing the Statistical Interviewer During the New Deal." *Science in Context* 24: 281–310.

Dietrich, Michael (1994) "The Origins of the Neutral Theory of Molecular Evolution." *Journal of the History of Biology* 27: 21–59.

Dietrich, Michael (1996) "On the Mutability of Genes and Geneticists: The 'Americanization' of Richard Goldschmidt and Victor Jollos." *Perspectives on Science* 4: 321–45.

Dietrich, Michael (2011) "Reinventing Richard Goldschmidt: Reputation, Memory, and Biography." *Journal of the History of Biology* 44: 693–712.

Dimand, Robert W. and John Geanakoplos (eds.) (2005) *Celebrating Irving Fisher: The Legacy of a Great Economist.* Oxford: Blackwell.

Dinerstein, Joel (2003) *Swinging the Machine: Technology, Modernity, and African American Culture Between the World Wars.* Amherst: University of Massachusetts Press.

Dingus, Lowell and Mark A. Norell (2010) *Barnum Brown: The Man Who Discovered Tyrannosaurus Rex.* Berkeley: University of California Press.

Division for Planetary Sciences of the American Astronomical Society, Meeting Statistics (2014). http://dps.aas.org/meetings/statistics.

Dixon, Thomas (2008) *Science and Religion: A Very Short Introduction.* Oxford: Oxford University Press.

Dixon, Thomas, Geoffrey Cantor, and Stephen Pumfrey (eds.) (2010) *Science and Religion: New Historical Perspectives.* Cambridge: Cambridge University Press.

Dobbs, David (2005) *Reef Madness: Charles Darwin, Alexander Agassiz, and the Meaning of Coral.* New York: Pantheon.

Dobzhansky, Theodosius (1944) "On Species and Races of Living and Fossil Man" *American Journal of Physical Anthropology* 2: 251–65.

Dobzhansky, Theodosius (1937, 1941, 1951) *Genetics and the Origin of Species.* New York: Columbia University Press.

Dobzhansky, Theodosius (1971) *Genetics of the Evolutionary Process.* New York: Columbia University Press.

Doel, Ronald E. (1996) *Solar System Astronomy in America: Communities, Patronage, and Interdisciplinary Science, 1920–1960.* New York: Cambridge University Press.

Doel, Ronald E. (2003a) "The Earth Sciences and Geophysics" in John Krige and Dominique Pestre (eds.), *Companion to Science in the Twentieth Century.* New York: Routledge, 391–416.

Doel, Ronald E. (2003b) "Constituting the Postwar Earth Sciences: The Military's Influence on the Environmental Sciences in the USA after 1945." *Social Studies of Science* 33, 5: 635–66.

Doel, Ronald E., Tanya J. Levin, and Mason K. Marker (2006) "Extending Modern Cartography to the Ocean Depths: Military Patronage, Cold War Priorities, and the Heezen-Tharp Mapping Project, 1952–1959." *Journal of Historical Geography* 32: 605–26.

Donahue, Jesse and Erik Trump (2006) *The Politics of Zoos: Exotic Animals and Their Protectors.* Dekalb: Northern Illinois University Press.

Donald, D.H. (ed.) (1960) *Why the North Won the Civil War.* New York: Simon & Schuster.

Dorfman, Joseph (1946) *The Economic Mind in American Civilization.* New York: Viking Press.

Doris, Ellen Elizabeth (2002) "The Practice of Nature Study: What Reformers Imagined and What Teachers Do." PhD Dissertation: Harvard University.

Dörries, Matthias and Christophe Masutti (eds.) (2006) "Changing Climate – Modeling Climate." *Historical Studies in the Physical and Biological Sciences* 37, 1.

Dorsey, P.A. (1998) "Going to School with Savages: Authorship and Authority among the Jesuits of New France." *The William and Mary Quarterly* 55, 3: 399–420.

Douglass, A.E. (1919) *Climatic Cycles and Tree Growth.* Washington: Carnegie Institution of Washington.

Douglass, Frederick (1886) "The Future of the Colored Race." *North American Review* 142: 437–40.

Douglass, Frederick (1982) "The Claims of the Negro Ethnologically Considered," in John W. Blassingame (ed.), *The Frederick Douglass Papers*, Series 1, Volume 2. New Haven: Yale University Press, 497–525.

Douglas, Heather (2009) *Science, Policy and the Value-Free Ideal.* Pittsburgh: University of Pittsburgh Press.

Doyle, J. (1985) *Altered Harvest.* New York: Viking Penguin.

Doyle, Laura (2014) "Inter-Imperiality: Dialectics in a Postcolonial World History." *Interventions: International Journal of Postcolonial Studies* 16, 3: 159–96.

Draper, John William (1874) *History of the Conflict Between Religion and Science.* London: Forgotten Books.

Dreger, Alice Domurat (1998) *Hermaphrodites and the Medical Invention of Sex.* Cambridge: Harvard University Press.

Dronamraju, K.R. and C.A. Francomano (2012) *Victor Mckusick and the History of Medical Genetics.* New York: Springer.

Droney, Damien (2014) "Ironies of Laboratory Work During Ghana's Second Age of Optimism" *Cultural Anthropology* 29, 2: 363–84.

Drucker, Donna (2014) *The Classification of Sex: Alfred Kinsey and the Organization of Knowledge.* Pittsburgh: University of Pittsburgh Press.

Du Bois, W.E.B. (1897) *The Conservation of Races.* Washington: American Negro Academy.

Du Bois, W.E.B. (1932) "Education and Work." *Journal of Negro Education* 1, 1: 60–74.

Duffy, John (1990) *The Sanitarians: A History of American Public Health.* Urbana: University of Illinois Press.

Dugdale, Robert (1877) *The Jukes: A Study in Crime, Pauperism, Disease, and Heredity.* New York: G.P. Putnam's.

Duncan, Anthony and Michel Janssen (2007) "On the Verge of Umdeutung in Minnesota: Van Vleck and the Correspondence Principle, Part One." *Archive for History of Exact Sciences* 61: 553–624.

Dunkleberg, Pete (2014) "Irreducible Complexity Demystified." http://www.talkdesign.org/faqs/icdmyst/ICDmyst.html.

Dunlap, Thomas R. (1988) *Saving America's Wildlife: Ecology and the American Mind, 1850–1990.* Princeton: Princeton University Press.

Dunlap, Thomas (2011) *In The Field, Among the Feathered: A History of Birders and Their Guides.* New York: Oxford University Press.

Dunn, L.C. (1965) *A Short History of Genetics: The Development of Some of the Main Lines of Thought: 1864–1939.* Ames: Iowa State University Press.

Dupré, Sven (ed.) (2005) "Optics, Instruments and Painting, 1420–1720: Reflections on the Hockney–Falco Thesis." *Special issue of Early Science and Medicine* 10: 125–339.

Dupree, A. Hunter (1957) *Science in the Federal Government: A History of Policies and Activities to 1940.* Cambridge: Belknap Press.

Dupree, A. Hunter (1959) *Asa Gray, 1810–1888.* Cambridge: Harvard University Press.

Dupree, A. Hunter (ed.) (1963) *Science and the Emergence of Modern America, 1865–1916.* Chicago: Rand McNally.

Dupree, A. Hunter (1986) *Science in the Federal Government: A History of Policies and Activities.* Baltimore: Johns Hopkins University Press.

Dupree, A. Hunter (2011) "Freiberg and the Frontier: Louis Janin, German Engineering, and "Civilisation" in the American West." *Annals of Science* 68, 3: 295–323.

Dutton, Dennis (2010) *The Art Instinct: Beauty, Pleasure, and Human Evolution.* New York: Bloomsbury Press.

Dym, Warren Alexander (2011) "Freiberg and the Frontier: Louis Janin, German Engineering, and 'Civilisation' in the American West." *Annals of Science* 68, 3: 295–323.

Dyson, George (2012) *Turing's Cathedral: The Origins of the Digital Universe.* London: Penguin.

Earman, John and Clark Glymour (1980a) "Relativity and Eclipses: The British Eclipse Expeditions of 1919 and their Predecessors." *Historical Studies in the Physical Sciences* 11: 49–85.

Earman, John and Clark Glymour (1980b) "The Gravitational Redshift as a Test of General Relativity: History and Analysis." *Studies in the History and Philosophy of Science* 11: 175–214.

Easlea, Brian (1981) *Science and Sexual Oppression: Patriarchy's Confrontation with Woman and Nature.* London: Weidenfeld and Nicholson.

Easlea, Brian (1983) *Fathering the Unthinkable: Masculinity, Scientists, and the Nuclear Arms Race.* London: Pluto Press.

Ecklund, Elaine Howard (2010) *Science vs. Religion: What Scientists Really Think.* Oxford: Oxford University Press.

Edmonson, James M. (1997) *American Surgical Instruments: The History of Their Manufacture and a Directory of Instrument Makers to 1900.* San Francisco: Norman.

Edmondson, Frank K. (1997) *AURA and its U.S. National Observatories*. Cambridge: Cambridge University Press.

Edwards, Paul (1996) *The Closed World: Computers and the Politics of Discourse in Cold War America*. Cambridge: MIT Press.

Edwards, Paul (2003) "Industrial Genders: Soft/Hard," in N. Lerman, R. Oldenziel, and A. Mohun (eds.), *Gender & Technology: A Reader*. Baltimore: Johns Hopkins University Press, 177–204.

Edwards, Paul (2010) *A Vast Machine: Computer Models, Climate Data, and the Politics of Global Warming*. Cambridge: MIT Press.

Eisenstaedt, Jean (1989a). "The Low Water-Mark of General Relativity, 1925–1955," in J. Stachel and Don Howard (eds.), *Einstein and the History of General Relativity*. Proceedings of the 1986 Osgood Hill Conference. Boston: Birkhäuser, 277–92.

Eisenstaedt, Jean (1989b) "The Early Interpretation of the Schwarzschild Solution," in J. Stachel and Don Howard (eds.) *Einstein and the History of General Relativity*. Proceedings of the 1986 Osgood Hill Conference. Einstein Studies. Boston: Birkhäuser, 213–33.

Eglash, Ron (2007) "Broken Metaphor: The Master–Slave Analogy in Technical Literature." *Technology and Culture* 48, 2: 360–9.

Egorova, Yulia (2013) "The Substance that Empowers? DNA in South Asia." *Contemporary South Asia* 21, 3: 291–303.

Ehrlich, Paul R. (1968) *The Population Bomb*. New York: Ballantine Books.

Eldredge, Niles (1995) *Reinventing Darwin: The Great Debate at the High Table of Evolutionary Theory*. New York: Wiley.

Eldredge, Niles (2004) *Why We Do It: Rethinking Sex and the Selfish Gene*. New York: Norton.

Eldredge, Niles and Stephen Jay Gould (1972) "Punctuated Equilibria: An Alternative to Phyletic Gradualism," in Thomas Schopf (ed.), *Models in Paleobiology*. San Francisco: Freeman, Cooper, 82–115.

Eldredge, Niles and Ian Tattersall (1975). "Evolutionary Models, Phylogenetic Reconstruction, and another Look at Hominid Phylogeny," in F. Szalay (ed.), *Approaches to Primate Paleobiology*. Basel: Karger, 218–42.

Ellingson, Terry Jay (2001) *The Myth of the Noble Savage*. Berkeley: University of California Press.

Elliott, Kevin (2008) Scientific Judgment and the Limits of Conflict-of-Interest Policies. *Accountability in Research* 15: 1–29.

Elliott, Kevin (2009) "Respect for Lay Perceptions of Risk in the Hormesis Case." *Human and Experimental Toxicology* 28: 21–6.

Elliott, Kevin (2011) *Is a Little Pollution Good For You? Incorporating Societal Values in Environmental Research*. New York: Oxford University Press.

Elliott, Kevin (2014) "Financial Conflicts of Interest and Criteria for Research Credibility." *Erkenntnis* 79: 917–37.

Elliott, Kevin (forthcoming a) "Environment," in A.J. Angulo (ed.), *Miseducation: A History of Ignorance-Making in American and Beyond*. Baltimore: Johns Hopkins University Press.

Elliott, Kevin (forthcoming b) "Standardized Study Designs, Value Judgments, and Financial Conflicts of Interest." *Perspectives on Science* 23.

Ellis, Erik (2007) "What is Marine Biology?: Defining a Science in the United States in the Mid-20th Century" *History and Philosophy of the Life Sciences* 29, 4: 469–93.

Elton, Charles (1927) *Animal Ecology*. London: Sidgwick and Jackson.

Ely, Richard T. (1884) "The Past and Present of Political Economy." *Johns Hopkins University Studies in Historical and Political Science* 2: 5–64.

Ely, Richard T. (1886) "Ethics and Economics." *Science* 7 (June 11): 529–33.

Ely, Richard T. (1893) *Outlines of Economics*. Meadville, PA: Flood and Vincent.

Emerson, A.E. (1939) Social Coordination and the Superorganism. *American Midland Naturalist* 21, 1: 182–209.

Emerson, George B. (1846) *A Report on the Trees and Shrubs Growing Naturally in the Forests of Massachusetts*. Boston: Little, Brown.

Emmett, Ross B. (2009) *Frank Knight and the Chicago School in American Economics*. New York: Routledge.

Emmett, Ross B. (ed.) (2010) *The Elgar Companion to the Chicago School of Economics*. Cheltenham: Edward Elgar.

Endersby, Jim (2007) *A Guinea Pig's History of Biology*. Cambridge: Harvard University Press.

Endersby, Jim (2008) *Imperial Nature: Joseph Hooker and the Practices of Victorian Science*. Chicago: University of Chicago Press.

Endfield, Georgina (2011) "Reculturing and Particularizing Climate Discourses: Weather, Identity, and the Work of Gordon Manley." *Osiris* 26: 142–62.

Engerman (2003) *Staging Growth: Modernization, Development, and the Global Cold War. Culture, Politics, and the Cold War*. Amherst: University of Massachusetts Press.

Ensmenger, Nathan (2001) "The Question of Professionalism in the Computer Fields." *IEEE Annals of the History of Computing* 23, 4: 56–74.

Ensmenger, Nathan (2010a) *The Computer Boys Take Over: Computers, Programmers, and the Politics of Technical Expertise*. Cambridge: MIT Press.

Ensmenger, Nathan (2010b) "Making Programming Masculine," in Thomas J. Misa (ed.), *Gender Codes: Why Women are Leaving Computing*. Hoboken: John Wiley & Sons, 115–24.

Ensmenger, Nathan (2012a) "The Digital Construction of Technology: Rethinking the History of Computers in Society." *Technology and Culture* 53: 753–76.

Ensmenger, Nathan (2012b) "Is Chess the Drosophila of Artificial Intelligence? A Social History of an Algorithm." *Social Studies of Science* 42, 1: 5–30.

Epstein, S. (1998) *Impure Science: AIDS, Activism, and the Politics of Knowledge*. Berkeley: University of California Press.

Ericksen, Julia A. and Sally Steffen (2002) *Kiss and Tell: Surveying Sex in the Twentieth Century*. Cambridge: Harvard University Press.

Escobar, Arturo (1995) *Encountering Development: The Making and Unmaking of the Third Word*. Princeton: Princeton University Press.

Etzkowitz, H. (2002) *MIT and the Rise of Entrepreneurial Science*. London: Routledge.

Evans, Howard Ensign (1993) *Pioneer Naturalists: The Discovery and Naming of North American Plants and Animals*. New York: Henry Holt.

Ewald, Paul (2002) *Plague Time: The New Germ Theory of Disease*. New York: Anchor.

Ewan, Joseph (1950) *Rocky Mountain Naturalists*. Denver: University of Denver Press.

Ewan, Joseph (1969) *A Short History of Botany in the United States*. New York: Hafner.

Ewan, Joseph (1970) *Plant Collectors in America: Backgrounds for Linnaeus*. Utrecht: International Association for Plant Taxonomy.

Fabian, Ann (2010) *The Skull Collectors: Race, Science, and America's Unburied Dead*. Chicago: University of Chicago Press.

Fabian, Johannes (1983) *Time and the Other: How Anthropology Makes Its Object*. New York: Columbia University Press.

Fairchild, Amy L. (2003) *Science at the Borders: Immigrant Medical Inspection and the Shaping of the Modern Industrial Labor Force*. Baltimore: Johns Hopkins University Press.

Fairlie, S. (2010) *Meat: A Benign Extravagance*. White River Junction: Chelsea Green.

Falk, D. (2004) *Braindance: New Discoveries about Human Origins and Brain Evolution*. Gainsville: University of Florida Press.

Falk, Raphael (2009) *Genetic Analysis: A History of Genetic Thinking*. New York: Cambridge University Press.

Fama, Eugene F. (1970) "Efficient Capital Markets: A Review of Theory and Empirical Work." *Journal of Finance* 25, 2: 383–417.

Fancher, Raymond E. and Alexandra Rutherford (2012) *Pioneers of Psychology: A History*. New York: W.W. Norton.

Farber, Paul Lawrence (2000) *Finding Order in Nature: The Naturalist Tradition from Linnaeus to E. O. Wilson*. Baltimore: Johns Hopkins University Press.

Farber, Paul Lawrence (2011) *Mixing Races: From Scientific Racism to Modern Evolutionary Ideas*. Baltimore: Johns Hopkins University Press.

Farley, John (2003) *To Cast Out Disease: A History of the International Health Division of the Rockefeller Foundation, 1913–1951*. Oxford: Oxford University Press.

Fausto-Sterling, Anne (1985) *Myths of Gender: Biological Theories about Women and Men*. New York: Basic Books.

Fausto-Sterling, Anne (2012) *Sex/Gender: Biology in A Social World*. New York: Routledge.

Fealing, K., J. Lane, J. Marburger III, and S. Shipp (2011) *The Science of Science Policy: A Handbook*. Stanford: Stanford University Press.

Featherstone, Katie and Paul Atkinson (2012) *Creating Conditions: The Making and Remaking of a Genetic Syndrome*. London: Routledge.

Fedigan, Linda (1986) "The Changing Role of Women in Models of Human Evolution." *Annual Review of Anthropology* 15: 25–66.

Fee, Elizabeth (1973) "The Sexual Politics of Victorian Social Anthropology." *Feminist Studies* 1, 3/4: 23–39.

Fee, Elizabeth (1976) "Science and the Woman Problem: Historical Perspectives," in Michael S. Teitelbaum (ed.), *Sex Differences: Social and Biological Perspectives*. Garden City: Anchor Books, 175–223.

Fee, Elizabeth (1979) "Nineteenth-Century Craniology: The Study of the Female Skull." *Bulletin of the History of Medicine* 53: 415–33.

Fee, Elizabeth (1987) *Disease and Discovery: A History of the Johns Hopkins School of Hygiene and Public Health, 1916–1939*. Baltimore: Johns Hopkins University Press.

Feenstra, G. (2002) "Creating Space for Sustainable Food Systems: Lessons from the Field." *Agriculture and Human Values* 19, 2: 99–106.

Feigenbaum, Edward, and Julian Feldman (eds.) (1963) *Computers and Thought*. New York: McGraw-Hill.

Fritz, Barkley (1996) "The Women of ENIAC." *IEEE Annals of the History of Computing* 18, 3: 13–28.

Ferguson-Smith, Malcolm (2003) "Interviews with Human and Medical Geneticists Series, Special Collections and Archives," Peter Harper (int.). Cardiff University, UK.

Ferngren, Gary (ed.) (2002) *Science and Religion: A Historical Introduction*. Baltimore: Johns Hopkins University Press.

Findlen, Paula (1994) *Possessing Nature: Museums, Collecting, and Scientific Culture in Early Modern Italy*. Berkeley: University of California Press.

Fine, Gary (2007) *Authors of the Storm: Meteorologists and the Culture of Prediction*. Chicago: University of Chicago Press.

Firmin, Anténor (1885) *De L'égalité des Races Humaines*. Paris: F. Pichon.

Fischer, Ernst Peter and Carol Lipson (1988) *Thinking About Science: Max Delbrück and the Origins of Molecular Biology*. New York: Norton.

Fischer, Eugen (1913) *Die Rehobother Bastards und das Bastardierungsproblem beim Menschen*. Jena: Gustav Fischer.

Fisher, R.A. (1930) *Genetical Theory of Natural Selection*. Oxford: Clarendon Press.

Fitelson, Branden, Christopher Stephens, and Elliott Sober (1999) "How Not to Detect Design: The Design Inference of William A. Dembski." *Philosophy of Science* 66, 3: 472.

Flader, Susan (1994) *Thinking Like a Mountain: Aldo Leopold and the Evolution of an Ecological Attitude toward Deer, Wolves, and Forests*. Madison: University of Wisconsin.

Fleagle, John and William Jungers (1982) "Fifty Years of Higher Primate Phylogeny," in Frank Spencer (ed.), *A History of American Physical Anthropology 1930–1980*. New York: Academic Press, 187–230.

Fleagle, John and William Jungers (2000) "The Century of the Past: One Hundred Years in the Study of Primate Evolution." *Evolutionary Anthropology* 9: 87–100.

Fleming, Donald (1964) "The Judgment upon Copernicus in Puritan New England." *Mélanges Alexandre Koyré* 2: 160–75.

Fleming, James Rodger (1990) *Meteorology in America, 1800–1870*. Baltimore: Johns Hopkins University Press.

Fleming, James Rodger (ed.) (1996) *Historical Essays on Meteorology, 1919–1995: The Diamond Anniversary History Volume of the*

American Meteorological Society. Boston: American Meterological Society.

Fleming, James Rodger (1998) *Historical Perspectives on Climate Change.* New York: Oxford University Press.

Fleming, James Rodger (2000) "T.C. Chamberlin, Climate Change, and Cosmogony." *Studies in History and Philosophy of Science Part B: Studies in History and Philosophy of Modern Physics* 31, 3: 293–308.

Fleming, James Rodger (2004) "Sverre Petterssen and the Contentious (and Momentous) Weather Forecasts for D-Day." *Endeavor* 28, 2: 59–63.

Fleming, James Rodger (2007) *The Callendar Effect: The Life and Times of Guy Stewart Callendar (1898–1964), the Scientist Who Established the Carbon Dioxide Theory of Climate Change.* Boston: American Meteorological Society.

Fleming, James Rodger (2010) *Fixing the Sky: The Checkered History of Weather and Climate Control.* New York: Columbia University Press.

Fleming, James Rodger and Vladimir Jankovic (2011) "Introduction: Revisiting Klima." *Osiris* 26: 1–15.

Fleming, James Rodger, Vladimir Jankovic and Deborah R. Coen (eds.) (2006) *Intimate Universality: Local and Global Themes in the History of Weather and Climate.* Sagamore Beach: Science History Publications.

Flint, Richard W. (1996) "American Showmen and European Dealers: Commerce in Wild Animals in Nineteenth-Century America," in R.J. Hoage and William A. Deiss (eds.), *New Worlds, New Animals: From Menagerie to Zoological Park in the Nineteenth Century.* Baltimore: Johns Hopkins University Press, 97–108.

Fogel, Robert, Enid M. Fogel, Mark Guglielmo, and Nathaniel Grotte (2013) *Political Arithmetic: Simon Kuznets and the Empirical Tradition in Economics.* Chicago: University of Chicago Press.

Foley, Robert (2001) "In the Shadow of the Modern Synthesis? Alternative Perspectives on the Last Fifty Years of Paleoanthropology." *Evolutionary Anthropology* 10: 5–14.

Foltz, E.E., C.J. Barbooka, and A.C. Ivy (1944) "The Level of Vitamin B-Complex in the Diet at which Detectable Symptoms of Deficiency Occur in Man." *Gastroenterology* 2: 323–44.

Foner, Eric (1974) "The Causes of the American Civil War: Recent Interpretations and New Directions." *Civil War History* 20, 3: 197–214.

Forel, Auguste (1905) *The Sexual Question: A Scientific, Psychological, Hygienic and Sociological Study.* Munich: E. Reinhardt.

Forman, Paul N. (1987) "Behind Quantum Electronics: National Security as Basis for Physical Research in the United States, 1940–1960." *Historical Studies in the Physical Sciences* 18, 1: 149–229.

Forman, P. (2007) "The Primacy of Science in Modernity of Technology in Postmodernity, and of Ideology in the History of Technology." *History and Technology: An International Journal* 23, 1–2: 1–152.

Fortun, Kim (2001) *Advocacy after Bhopal: Environmentalism, Disaster, New Global Orders.* Chicago: University of Chicago Press.

Foster, Laura A. (2012) "Patents, Biopolitics, and Feminisms: Locating Patent Law Struggles over Breast Cancer Genes and the Hoodia Plant." Special Invited Issue on Patent Law and Property. *International Journal of Cultural Property* 19, 3: 371–400.

Foster, Mike (1994) *Strange Genius: The Life of Ferdinand Vandeveer Hayden.* Niwot: Roberts Rinehart.

Foucault, Michel (1975) *The Birth of the Clinic: An Archaeology of Medical Perception.* A.M. Sheriden Smith trans. New York: Vintage.

Foucault, Michel (1976) *Histoire De La Sexualité, I: La Volonté De Savoir.* Paris: Gallimard.

Foucault, Michel (1980) *Power/Knowledge: Selected Interviews and Other Writings, 1972–1977.* New York: Pantheon.

Foucault, Michel (1990) "Nietzsche, Freud, Marx," in Gayle Ormiston and Alan D. Schrift (eds.), *Transforming the Hermeneutic Context: From Nietzsche to Nancy.* Albany: SUNY Press, 59–67.

Foucault, Michel (2010) "The Policing of Race Mixing: The Place of Biopower within the History of Racisms." *Bioethics Inquiry* 7:, 205–16.

Fought, Leigh (2003) *Southern Womanhood and Slavery. A Biography of Louisa S. McCord*. Columbia: University of Missouri Press.

Fountain, Henry (2014) "A Lost World is Resurrected at Brown." New York Times, 29 July. http://www.nytimes.com/2014/07/29/science/discarded-museum-lives-again-at-brown-university.html

Fourcade, Marion (2009) *Economists and Societies: Discipline and Profession in the United States, Britain, and France, 1890s to 1990s*. Princeton: Princeton University Press.

Frankel, Henry (2009) "Plate Tectonics," in Peter J. Bowler and John V. Pickstone (eds.), *The Cambridge History of Science*. New York: Cambridge University Press, 383–94.

Franklin, Allan (1993) *The Rise and Fall of the Fifth Force*. College Park: American Institute of Physics.

Frayling, Christopher (2005) *Mad, Bad and Dangerous? The Scientist and the Cinema*. London: Routledge.

Freedland, R.A. (1967) "Effect of Progressive Starvation on Rat Liver Enzyme Activities." *Journal of Nutrition* 91: 489–95.

Frehner, Brian (2011) *Finding Oil: The Nature of Petroleum Geology, 1859–1920*. Lincoln: University of Nebraska Press.

Frey, Donald E. (2000) "The Puritan Roots of Daniel Raymond's Economics." *History of Political Economy* 32, 3: 607–29.

Frey, Donald E. (2002) "Francis Wayland's 1830s Textbooks: Evangelical Ethics and Political Economy." *Journal of the History of Economic Thought* 24, 2: 215–31.

Frey, Donald E. (2009) *America's Economic Moralists: A History of Rival Ethics and Economics*. Albany: State University of New York Press.

Fried, Barbara H. (1998) *The Progressive Assault on Laissez Faire: Robert Hale and the First Law and Economics Movement*. Cambridge: Harvard University Press.

Friedberg, Errol (2010) *Sydney Brenner: A Biography*. Cold Spring Harbor: Cold Spring Harbor Laboratory Press.

Friedman, Milton (1953) "The Methodology of Positive Economics," in *Essays in Positive Economics*. Chicago: University of Chicago Press, 3–43.

Friedman, Milton and Simon Kuznets (1954) *Income from Independent Professional Practice*. New York: National Bureau of Economic Research.

Friedman, Robert Marc (1989) *Appropriating the Weather: Vilhelm Bjerknes and the Construction of a Modern Meteorology*. Ithaca: Cornell University Press.

Fruton, Joseph S. (1992) *A Skeptical Biochemist*. Cambridge: Harvard University Press.

Fuerst, John A. (1982) "The Role of Reductionism in the Development of Molecular Biology: Peripheral of Central?" *Social Studies of Science* 12: 241–78.

Fujimura, Joan H. (1996) *Crafting Science: A Sociohistory of the Quest for the Genetics of Cancer*. Cambridge: Harvard University Press.

Fujimura, Joan. H. (2000) "Transnational Genomics: Transgressing the Boundary Between the "Modern/West" and the 'Premodern/East," in Roddy Reid and Sharon Traweek (eds.), *Doing Science + Culture*. New York: Routledge, 71–92.

Fuller, R. and Wayland, F. (1845) *Domestic Slavery Considered as a Scriptural Institution*. New York: Sheldon.

Fuller, Stephen (2008) *Dissent Over Descent: Intelligent Design's Challenge to Darwinism*. London: Icon.

Furner, Mary O. (1975) *Advocacy & Objectivity: A Crisis in the Professionalization of American Social Science, 1865–1905*. Lexington: University Press of Kentucky.

Fyfe, Aileen and Bernard Lightman (eds.) (2007) *Science in the Marketplace: Nineteenth-Century Sites and Experiences*. Chicago: University of Chicago Press.

Galison, Peter (1987) *How Experiments End*. Chicago: University of Chicago Press.

Galison, Peter (1994) "The Ontology of the Enemy: Norbert Wiener and the Cyberneticvision." *Critical Inquiry* 21, 1: 228–66.

Galison, Peter (1997) *Image and Logic: A Material Culture of Microphyiscs*. Chicago: University of Chicago Press.

Galison, Peter (2001) "War Against the Center." *Grey Room* 4: 5–33.

Galison, Peter, and Bruce Hevly (eds.) (1992) *Big Science: The Growth of Large-Scale Research*. Stanford: Stanford University Press.

Galison, Peter and David Stump (eds.) (1996). *The Disunity of Science: Boundaries, Contexts, and Power*. Stanford: Stanford University Press.

Gallagher, Nancy L. (1999) *Breeding Better Vermonters: The Eugenics Project in the Green Mountain State*. Hanover: University Press of New England.

Gallup Poll (2014) "Evolution, Creationism, Intelligent Design." http: //www. gallup. com/poll/21814/evolution-creationism-intelligent-design. aspx

Galton, Francis (1883) *Inquiries into Human Faculty and its Development*. London: Macmillan.

Galton, Francis (1904) "Eugenics: Its Definitions, Scopes, and Aims." *American Journal of Sociology* 10, 1: 1–25.

Gamble, Eliza Burt (1894) *The Evolution of Woman: An Inquiry into the Dogma of Her Inferiority to Man*. New York: G. P. Putnam's Sons.

García-Sancho, M. (2012) *Biology, Computing, and the History of Molecular Sequencing: From Proteins to DNA, 1945–2000*. New York: Palgrave Macmillan.

Garey, Michael and David Johnson (1979) *Computers and Intractability: A Guide to the Theory of NP-Completeness*. New York: Bell Telephone Laboratories

Gariepy, Thomas P. (1994) "The Introduction and Acceptance of Listerian Antisepsis in the United States." *Journal of the History of Medicine and Allied Sciences* 49: 167–206.

Garn, Stanley (1961) *Human Races*. Chicago: Charles C. Thomas.

Garrett, A. E. (1909) *The Periodic Law*. New York: D. Appleton and Co.

Gates, Barbara (2002) In N*ature's Name: An Anthology of Women's Writing and Illustration, 1780–1930*. Chicago: University of Chicago Press.

Gates, William (trans.) (2000) *An Aztec Herbal: The Classic Codex of 1552*. Mineola: Dover.

Gathorne-Hardy, Jonathan (2004) *Kinsey: Sex the Measure of All Things*. Bloomington: Indiana University Press.

Gaudillière, Jean-Paul (2001) The Pharmaceutical Industry in the Biotech Century: Toward a History of Science, Technology, and Business? *Studies in History and Philosophy of Biological and Biomedical Sciences* 32: 191–201.

Gaudillière, Jean-Paul (2005) Better Prepared than Synthesized: Adolf Butenandt, Schering Ag, and the Transformation of Sex Steroids into Drugs (1930–1946). *Studies in History and Philosophy of Biological and Biomedical Sciences* 36: 612–44.

Gaudillière, Jean-Paul (2006) "Globalization and Regulation in the Biotech World: The Transatlantic Debates over Cancer Genes and Genetically Modified Crops." *Osiris* 21: 251–72.

Gaudillière, Jean-Paul (2009) "New Wine in Old Bottles?: The Biotechnology Problem in the History of Molecular Biology." *Studies in History and Philosophy of Biological and Biomedical Sciences*. 40: 20–8.

Gaw, Jerry L. (1999). "*A Time to Heal*": *The Diffusion of Listerism in Victorian Britain*. Philadelphia: American Philosophical Society.

Geertz, Clifford (1973) *The Interpretation of Cultures: Selected Essays*. New York: Basic Books.

Geiger, Roger L. (1986) *To Advance Knowledge: The Growth of American Research Universities, 1900–1940*. Oxford: Oxford University Press.

Geiger, Roger L. (1992) "Science, Universities, and National Defense, 1945–1960." *Osiris* 7: 26–48.

Geison, Gerald (1987) *Physiology in the American Context, 1850–1940*. New York: Lippincot, Williams, and Wilkins.

Geison, Gerald (1995) *The Private Science of Louis Pasteur*. Princeton: Princeton University Press.

Gelernter, Herbert (1995, 1963, 1959) "Realization of a Geometry-Theorem Proving-machine," in Edward Feigenbaum and

Julian Feldman (eds.), *Computers and Thought*. Palo Alto: AAAI Press, 134–52.

George, Henry (1879) *Progress and Poverty*. San Francisco: Henry George.

George, Sarton (1950) "The History of Science in the Carnegie Institution." *Osiris* 9: 624–38.

Gerbi, Antonello (1973) *The New World: The History of a Polemic, 1750–1900*. Trans. Jeremy Moyle. Pittsburg: University of Pittsburg Press.

Gershenhorn, Jerry (2004) *Melville J. Herskovits and the Racial Politics of Knowledge*. Lincoln: University of Nebraska Press.

Gerstner, Patsy A. (1970) "Vertebrate Paleontology, an Early Nineteenth Century Transatlantic Science." *Journal of the History of Biology* 3: 137–48.

Geschwind, Carl-Henry (2003) *California Earthquakes: Science, Risk, and the Politics of Hazard Mitigation*. Baltimore: Johns Hopkins University Press.

Gevitz, N. (1988) *Other Healers: Unorthodox Medicine in America*. Baltimore: Johns Hopkins University Press.

Ghiselin, Michael T. and Christiane Groeben (2000) "A Bioeconomic Perspective on the Organization of the Naples Marine Station," in Michael T. Ghiselin and Alan E. Leviton (eds.), *Cultures and Institutions of Natural History: Essays in the History and Philosophy of Science*. San Francisco: California Academy of Sciences, 273–85.

Gianquitto, Tina (2007) *American Women and the Scientific Study of the Natural World, 1820–1885*. Athens: University of Georgia Press.

Gibbons, Ann (2006) *The First Human: The Race to Discover Our Earliest Ancestors*. New York: Doubleday.

Giddings, F.H. (1896) *The Principles of Sociology*. New York: Macmillan.

Gieryn, Thomas F. (1983) "Boundary Work and the Demarcation of Science from Non-Science: Strains and Interests in Professional Ideologies of Scientists." *American Sociological Review* 48: 781–95.

Gieryn, Thomas F. (2006) "City as Truth-Spot: Laboratories and Field-Sites in Urban Studies." *Social Studies of Science* 36: 5–38.

Gieryn, Thomas F. (2008) "Laboratory Design for Post-Fordist Science." *Isis* 99: 796–802.

Gilbert, Scott F. (1988) "Cellular Politics: Ernest Everett Just, Richard Goldschmidt, and the Attempt to Reconcile Embryology and Genetics," in Ronald Rainger, Keith R. Benson, and Jane Maienschein (eds.), *The American Development of Biology*, 311–45. Philadelphia: University of Pennsylvania Press.

Gillespie, John C. (1994) *The Causes of Molecular Evolution*. Oxford: Oxford University Press.

Gilman, N. (2003) *Mandarins of the Future: Modernization Theory in Cold War America. New Studies in American Intellectual and Cultural History*. Baltimore: Johns Hopkins University Press.

Gingerich, Owen (ed.) (1984) *Astrophysics and Twentieth-Century Astronomy to 1950: Part A*. Cambridge: Cambridge University Press.

Gingras, Yves (2010) "Revisiting the "Quiet Debut" of the Double Helix: A Bibliometric and Methodological Note on the 'Impact' of Scientific Publications." *Journal of the History of Biology* 43: 159–81.

Gintis, H. (2014) "The Distributed Brain." *Nature* 509: 284–5.

Girard, G.T. (1979) "Impact of NSF Science Curriculum Projects." *School Science and Mathematics* 79, 1: 3–6.

Givens, David B., Patsy Evans, and Timothy Jablonski (1997) "Survey of Anthropology PhDs." American Anthropological Association Survey. http://www.aaanet.org/resources/departments/97Survey.cfm.

Glacken, Clarence J. (1967) *Traces on the Rhodian Shore: Nature and Culture in Western Thought from Ancient Times to the End of the Eighteenth Century*. Berkeley: University of California Press.

Go, J. (2011) *Patterns of Empire: The British and American Empires, 1688 to the Present*. New York: Cambridge University Press.

Go, J. (2013a) *Decentering Social Theory*. Bingley: Emerald.

Go, J. (2013b) "For a Postcolonial Sociology." *Theory and Society* 42: 25–55.

Gobineau, Joseph Arthur De (1856) *The Moral and Intellectual Diversity of Races.* H. Hotz (trans.) Philadelphia: J.B. Lippincott.

Gobineau, Joseph Arthur De (1983) *Essai Sur L'inégalité Des Races Humaines, Oeuvres 1.* Paris: Gallimard.

Goddard, Henry H. (1912) *The Kallikak Family: A Study in the Heredity of Feeble-Mindedness.* New York: Macmillan.

Goetzmann, William H. (1959) *Army Exploration in the American West, 1803–1863.* New Haven: Yale University Press.

Goetzmann, William H. (1966) *Exploration and Empire: The Explorer and the Scientist in the Winning of the American West.* New York: Alfred A. Knopf.

Goetzmann, William H. (1987) *New Lands, New Men: America and the Second Great Age of Discovery.* New York: Penguin.

Goldacre, B. (2012) *Bad Pharma: How Drug Companies Mislead Doctors and Harm Patients.* New York: Faber and Faber.

Goldberg, Joshua N. (1992) "US Air Force Support of General Relativity: 1956–1972," in Jean Eisenstaedt and Anne J. Kox (eds.), *Studies in the History of General Relativity.* Berlin: Birkhauser, 89–102.

Goldberg, Stanley and Roger Stuewer (1988) The Michelson Era in American Science 1870–1930: AIP Conference Proceedings. College Park: American Institute of Physics.

Golden, W. (ed.) (1988) *Science and Technology Advice to the President, Congress, and Judiciary.* Oxford: Pergamon.

Goldie, Terry (2014) *The Man Who Invented Gender: Engaging the Ideas of John Money.* Vancouver: University of British Columbia Press.

Goldman, Joanne Abel (2000) "National Science in the Nation's Heartland: The Ames Laboratory and Iowa State University, 1942–1965." *Technology and Culture* 41: 435–59.

Goldsmith, Julian R. (1991) "Some Chicago Georecollections." Annual Review of Earth and Planetary *Sciences* 19: 1–16.

Goldstein, Carolyn M. (2012) *Creating Consumers: Home Economists in Twentieth-Century America.* Chapel Hill: University of North Carolina Press.

Goldstein, Daniel (1994) 'Yours for Science': The Smithsonian Institution's Correspondents and the Shape of Scientific Community in Nineteenth-Century America." *Isis* 85: 573–99.

Golinski, Jan (2007) *British Weather and the Climate of Enlightenment.* Chicago: University of Chicago Press.

Golinski, Jan (2008) "American Climate and the Civilization of Nature," in James Delbourgo and Nicholas Dew (eds.), *Science and Empire in the Atlantic World.* New York: Routledge, : 153–74.

Good, Gregory A. (1985) "Geomagnetics and Scientific Institutions in 19th Century America." *Eos 66,* 27: 521–6.

Good, Gregory A. (ed.) (1990) "Scientific Sovereignty: Canada, the Carnegie Institution and the Earth's Magnetism in the North." Scientia Canadensis: Canadian Journal of the History of Science, *Technology and Medicine* 14, 1–2: 3.

Good, Gregory A. (ed) (1994) *The Earth, the Heavens, and the Carnegie Institution of Washington 5. History of Geophysics.* Washington: American Geophysical Union.

Good, Gregory A. (1998) *Sciences of the Earth: An Encyclopedia of Events, People, and Phenomena.* New York: Garland.

Good, Gregory A. (2000) "The Assembly of Geophysics: Scientific Disciplines as Frameworks of Consensus." *Studies in History and Philosophy of Modern Physics* 31, 3: 259–92.

Good, Gregory A. (2013) "Vision of a Global Physics: The Carnegie Institution and the First World Magnetic Survey" in Gregory A. Good (ed.), *The Earth, the Heavens, and the Carnegie Institution of Washington.* Washington: American Geophysical Union, 29–36.

Gooday, Graeme (2008) "Placing or Replacing the Laboratory in the History of Science?" *Isis* 99: 783–95.

Goode, George Brown and Sally Gregory Kohlstedt (eds.) (1991) *The Origins of Natural Science in America: The Essays of George Brown Goode.* Washington: Smithsonian Institution.

Gooding, David, Trevor Pinch and Simon Schaffer (eds.) (1989) *The Uses of Experiment: Studies in the Natural Sciences.* Cambridge: Cambridge University Press.

Goodrum, Matthew R. (2013) "History," in David Begun (ed.), *A Companion to PaleoAnthropology.* Oxford: Wiley-Blackwell, 17–33.

Goodstein, Judith R. (1984) "Waves in the Earth: Seismology Comes to Southern California." *Historical Studies in the Physical Sciences* 14, 2: 201–30.

Goonatilake Susantha (1984) *Aborted Discovery: Science and Creativity in the Third World.* London: Zed Books.

Goonatilake Susantha (1998) *Toward Global Science: Mining Civilizational Knowledge.* Bloomington: Indian University Press.

Gordin, Michael D. (2012) *The Pseudo-Science Wars: Immanuel Velikovsky and the Birth of the Modern Fringe.* Chicago: University of Chicago Press.

Gordin, Michael (2014) "The Dostoevsky Machine in Georgetown: Scientific Translation in the Cold War." Annals of Science, forthcoming.

Goudie, Andrew (1983) Environmental Change: Contemporary Problems in Geography. New York: Clarendon Press.

Goudie, Andrew (2000) *The Human Impact on the Natural Environment.* Cambridge, MA: MIT Press.

Grier, David Alan (1996) "The ENIAC, the Verb "To Program" and the Emergence of Digital Computers." *IEEE Annals of the History of Computing* 18, 1: 51–5.

Grier, David Alan (2005) *When Computers Were Human.* Princeton: Princeton University Press.

Gordon, Linda (1990) *Woman's Body, Woman's Right: Birth Control in America.* New York: Penguin.

Gormley, Melinda (2005) "It's in the Blood: A Documentary History of Linus Pauling, Hemoglobin, and Sickle Cell Anemia." Ava Helen and Linus Pauling Papers, Oregon State University Special Collections and Archive Research Center.

Gormley, Melinda (2007) *"Geneticist L.C. Dunn: Politics, Activism, and Community."* PhD Dissertation: Oregon State University.

Gormley, Melinda (2009) "Scientific Discrimination and the Activist Scientist: L. C. Dunn and the Professionalization of Genetics and Human Genetics in the United States." *Journal of the History of Biology* 42, 1: 33–72.

Gormley, Melinda (2014) "Genetics and Genetic Engineering," in Hugh Slotten (ed.), *Oxford Encyclopedia of the History of American Science, Medicine, and Technology*, Vol. 1 Oxford: Oxford University Press, 439–44.

Gossett, Thomas (1995) *Race: The History of an Idea in America.* Oxford: Oxford University Press.

Gottlieb, Robert (1993) *Forcing the Spring: The Transformation of the American Environmental Movement.* Washington: Island Press.

Gottweiss, Herbert (1998) *Governing Molecules: The Discursive Politics of Genetic Engineering in Europe and the United States.* Cambridge: MIT Press.

Gould, Carol Grant (2004) *The Remarkable Life of William Beebe: Naturalist and Explorer* New York: Island Press.

Gould, Meredith, and Rochelle Kern-Daniels (1977) "Toward a Sociological Theory of Gender and Sex." *The American Sociologist* 12: 182–9.

Gould, Stephen Jay (1981) *The Mismeasure of Man.* New York: Norton.

Gould, Stephen Jay (1989) *Wonderful Life: The Burgess Shale and the Nature of History.* New York: W.W. Norton.

Gould, Stephen Jay (2006) *The Mismeasure of Man (Revised & Expanded).* New York: W.W. Norton.

Gould, Stephen Jay and R.C. Lewontin (1979) "The Spandrels of San Marco and the Panglossian Paradigm: A Critique of the Adaptationist Programme." *Proceedings of the Royal Society of London, Series B: Biological Sciences* 205: 581–98.

Gradmann, Christoph (2009) *Laboratory Disease: Robert Koch's Medical Bacteriology.* Baltimore: Johns Hopkins University Press.

Gramsci, Antonio (1971) *Selections from the Prison Notebooks of Antonio Gramsci.* New York: International Publishers.

Grant, Edward (1996) *The Foundations of Modern Science in the Middle Ages.* Cambridge: Cambridge University Press.

Granato, Marcus and Marta C. Lourenço (eds.) (2014) Scientific Instruments in the History of Science: Studies in Transfer, Use and Preservation. Rio de Janeiro: Museu de Astronomia e Ciências Afins.

Gray, Asa (1860) "The Origin of Species by Means of Natural Selection," in *Darwiniana: Essays and Reviews Pertaining to Darwinism.* New York: D. Appleton, 1–62.

Gray, Asa (1864) *Manual of the Botany of the Northern United States, from New England to Wisconsin and South to Ohio and Pennsylvania Inclusive.* New York: Ivison

Gray, Asa (1872) "Sequoia and its History." *The American Naturalist* 6: 577–96.

Gray, Dwight (1963) "The New AIP Handbook." *Physics Today* 16: 40–2.

Gray, Edward G. (1999) *New World Babel: Languages and Nations in Early America.* Princeton: Princeton University Press.

Green, James N. and Peter Stallybrass (2006) *"Benjamin Franklin: Writer and Printer." Benjamin Franklin: Writer and Printer.* Philadelphia: The Library Company of Philadelphia.

Green, Judy and Jeanne LaDuke (2009) *Pioneering Women in American Mathematics: The Pre-1940s Ph.D.'s.* Providence: American Mathematical Society.

Green, Ronald M. (2007) *Babies by Design: The Ethics of Genetic Choice.* New Haven: Yale University Press.

Greenberg, D. (1968) *The Politics of Pure Science.* New York: New American Library.

Greene, John C. (1954) "Some Aspects of American Astronomy, 1750–1815." *Isis* 45: 339–58.

Greene, John C. (1984) *American Science in the Age of Jefferson.* Ames: University of Iowa Press.

Greenson, Ralph R. (1964) "On Homosexuality and Gender Identity." *International Journal of Psychoanalysis* 45: 217–19.

Greger, M. (2011) "Transgenesis in Animal Agriculture: Addressing Animal Health and Welfare Concerns." *Journal of Agricultural and Environmental Ethics* 24, 5: 451–72.

Gregory, Jane and Steve Miller (1998) *Science in Public: Communication, Culture, and Credibility.* Cambridge: Perseus Publishing.

Gregory, William King (1934) *Man's Place among the Anthropoids.* Oxford: Clarendon Press.

Greze, V.N. (1971) "Centennial of the Institute of the Southern Seas at Sevastopol." *Internationale Revue Der Gesamten Hydrobiologie* 56, 5: 811–18.

Grier, David Alan (2005) *When Computers Were Human.* Princeton: Princeton University Press

Griffin, Susan (1978) *Woman and Nature: The Roaring Inside Her.* New York: Harper and Row.

Grob, Bart and Hans Hooijmaijers (eds.) (2006) *Who Needs Scientific Instruments.* Leiden: Museum Boerhaave.

Grogan, Jessica (2012) *Encountering America: Humanistic Psychology, Sixties Culture, and the Shaping of the Modern Self.* New York: Harper Collins.

Gronim, Sara S. (1999). "At the Sign of Newton's Head: Astronomy and Cosmology in British Colonial New York." *Pennsylvania History* 66: 55–85.

Gronim, Sara S. (2007) *Everyday Nature: Knowledge of the Natural World in Colonial New York.* New Brunswick: Rutgers University Press.

Gross, Ariela J. (2008) *What Blood Won't Tell.* Cambridge MA: Harvard University Press.

Grove, Richard. (1995) *Green Imperialism: Colonial Expansion, Tropical Island Edens, and the Origins of Environmentalism, 1600–1860.* New York: Cambridge University Press.

Guenebault, J.H. (1837) *Natural History of the Negro Race.* Charleston: D.H. Dowling.

Gundling, Tom (2005) *First in Line: Tracing Our Ape Ancestry.* New Haven: Yale University Press.

Gupta, Gopal (2007) "Computer Science Curriculum Developments in the 1960s." *IEEE Annals of the History of Computing* 29, 2: 40–54.

Guralnick, S.M. (1974) "Sources of Misconception on the Role of Science in the

Nineteenth-Century American College." *Isis* 65, 3: 352–66.

Guston, D. (2000) *Between Politics and Science: Assuring the Integrity and Productivity of Research*. Cambridge: Cambridge University Press.

Habib, Irfan S. and Dhruv Raina (1989) "The Introduction of Scientific Rationality in India: A Study of Master Ramachandra, Urdu Journalist and Mathematician and Educationist." *Annals of Science* 46, 6: 597–610.

Hackett, E., O. Amsterdamska, M. Lynch, J. Wajcman, and W. Bijker (2008) *The Handbook of Science and Technology Studies*. Cambridge: MIT Press.

Hacking, Ian (1999) "Making Up People," in Mario Biagioli (ed.), *The Science Studies Reader*. New York: Routledge, 161–71.

Haeckel, Ernst (1866) *Generelle morphologie der organismen* [General Morphology of the Organisms]. Berlin: G. Reimer. [Translated by author]

Hagen, Joel (1988) "Organism and Environment: Frederic Clements' Vision of a Unified Physiological Ecology," in Ronald Rainger, Keith Benson and Jane Maienschein (eds.), *The American Development of Science*. Philadelphia: University of Pennsylvania Press, 257–80.

Hagen, Joel (1992a) "Clementsian Ecologists: The Internal Dynamics of a Research Group." *Osiris* 8: 178–95.

Hagen, Joel B. (1992b) *An Entangled Bank: The Origins of Ecosystem Ecology*. New Brunswick: Rutgers University Press.

Hagen, Joel B. (1999) "Naturalists, Molecular Biologists, and the Challenges of Molecular Evolution." *Journal of the History of Biology* 32: 321–41.

Hagen, Joel (2009) "Descended from Darwin? George Gaylord Simpson, Morris Goodman, and Primate Systematics," in Joe Cain andMichael Ruse (eds.), *Descended from Darwin: Insights into the History of Evolutionary Studies, 1900–1970*. Philadelphia: American Philosophical Society, 93–109.

Hagen, Joel (2010) "Waiting for Sequences: Morris Goodman, Immunodiffusion Experiments, and the Origins of Molecular Anthropology." *Journal of the History of Biology* 43: 697–725.

Hager, Lori (1997) "Sex and Gender in PaleoAnthropology," in Lori D. Hager (ed.), *Women in Human Evolution*. New York: Routledge, 1–28.

Haig, David (2004) "The Inexorable Rise of Gender and the Decline of Sex: Social Change in Academic Titles, 1945–2001." *Archives of Sexual Behavior* 33, 2: 87–96.

Haigh, Thomas (2001) "Inventing Information Systems: The Systems Men and the Computer, 1950–1968." *The Business History Review* 75, 1: 15–61.

Haigh, Thomas (2002) "Software in the 1960s as Concept, Service, and Product." *IEEE Annals of the History of Computing* 24, 1: 5–13.

Haigh, Thomas (2003) "Technology, Information and Power: Managerial Technicians in Corporate America, 1917–2000." PhD Dissertation: University of Pennsylvania.

Haigh, Thomas (2010a) "Masculinity and the Machine Man: Gender in the History of Data Processing," in Thomas Misa (ed.), *Gender Codes: Why Women are Leaving Computing*. Hoboken: John Wiley & Sons, 51–72.

Haigh, Thomas (2010b) "Computing the American Way: Contextualizing the Early US computer Industry." *IEEE Annals of the History of Computing* 32, 2: 8–20

Haigh, Thomas (2011) "The History of Information Technology." *Annual Review of Information Science and Technology* 45, 1: 431–87.

Haigh, Thomas (2014) "Actually, Turing did not Invent the Computer." *Communications of the ACM* 57, 1: 36–41.

Haigh, Thomas, E. Kaplan, and C. Seib (2007) "Sources for ACM History: What, Where, Why." *Communications of the ACM* 50, 5: 36–41.

Haigh, Thomas, Mark Priestly, and Crispin Rope (2013) "Reconsidering the Stored-Program Concept." *IEEE Annals of the History of Computing* 36, 1: 4–17.

Haigh, Thomas, Mark Priestly, and Crispin Rope (2014) "Los Alamos Bets on ENIAC:

Nuclear Montecarlo Simulations, 1947–1948." *IEEE Annals of the History of Computing* 36, 3: 42–63.

Hale, Nathan (2000) *The Rise and Crisis of Psychoanalysis in the United States: Freud and the Americans, 1917–1985.* New York: Replica Books.

Hall, Cargill (1977) *Lunar Impact: A History of Project Ranger.* Washington: NASA SP-4210.

Hall, David D. (1990) *Worlds of Wonder, Days of Judgment: Popular Religious Beliefs in Early New England.* Cambridge: Harvard University Press.

Hall, Granville Stanley (1927) *Life and Confessions of a Psychologist.* New York: Appleton.

Hall, Steven (1987) Invisible Frontiers: The Race to Synthesize a Human Gene. New York: Atlantic Monthly Press.

Hall, Stuart (1996) "When was the Post-Colonial? Thinking at the Limit," in Iain Chambers and Linda Curtis (eds.), *The Postcolonial Question: Common Skies, Divided Horizons* London: Routledge, 242–60.

Hallowell, A. Irving (1965) "The History of Anthropology as an Anthropological Problem." *Journal of the History of the Behavioral Sciences* 1, 1: 24–38.

Hamblin, Jacob D. (2005) *Oceanographers and the Cold War: Disciples of Marine Science.* Seattle: University of Washington Press.

Hamblin, Jacob D. (2013) *Arming Mother Nature: The Birth of Catastrophic Environmentalism.* Oxford: Oxford University Press.

Hamera Judith (2012) *Parlor Ponds: The Cultural Work of the American Home Aquarium, 1850–1970.* Ann Arbor: University of Michigan Press.

Hamilton, Andrew and Quentin D. Wheeler (2008) "Taxonomy and Why History of Science Matters for Science: A Case Study." *Isis* 99, 2: 331–40.

Hamilton, Gilbert V. (1929) *A Research in Marriage.* New York: Boni.

Hamilton, Gilbert V. and Kenneth Macgowan (1929) *What is Wrong with Marriage?* New York: Boni.

Hamilton, Walton H. (1919) "The Institutional Approach to Economic Theory." *American Economic Review* 9, 1: 309–18.

Hamilton, William D. (1964a) "The Genetical Evolution of Social Behaviour: I." *Journal of Theoretical Biology* 7: 1–16.

Hamilton, William D. (1964b) "The Genetical Evolution of Social Behaviour. II." *Journal of Theoretical Biology,* 7: 17–52.

Hamilton, William D. (1967) "Extraordinary Sex Ratios." *Science* 156: 477–88.

Hamlin, Kimberly A. (2014) *From Eve to Evolution: Darwin, Science, and Women's Rights in Gilded Age America.* Chicago: University of Chicago Press.

Hammonds, Evelynn (1999) "The Logic of Difference: A History of Race in Science and Medicine in the United States." Talk at the Women's Studies Program, UCLA.

Hammonds, Evelynn M. (1999a) *Childhood's Deadly Scourge: The Campaign to Control Diphtheria in New York City, 1880–1930.* Baltimore: Johns Hopkins University Press.

Hammonds, Evelynn (1994) "Black (W)Holes and the Geometry of Black Female Sexuality (More Gender Trouble: Feminism Meets Queer Theory)." *Differences: A Journal of Feminist Cultural Studies* 6, 2–3: 126–46.

Hammonds, Evelynn and Rebecca M. Herzig (2009) *The Nature of Difference: Sciences of Race in the United States from Jefferson to Genomics.* Cambridge: MIT Press.

Hammond, J. Daniel and Claire H. Hammond (eds.) (2006) *Making Chicago Price Theory: Friedman–Stigler Correspondence, 1945–1957.* London: Routledge.

Hancocks, David (2001) *A Different Nature: The Paradoxical World of Zoos and Their Uncertain Future.* Berkeley: University of California Press.

Handler, Richard (ed.) (2004) *Significant Others: Interpersonal and Professional Commitments in Anthropology.* Madison: University of Wisconsin Press.

Handler, Richard (ed.) (2006) *Central Sites, Peripheral Visions: Cultural and Institutional Crossings in the History of Anthropology.* Madison: University of Wisconsin Press.

Haney, D.P. (2008) *The Americanization of Social Science: Intellectuals and Public*

Responsibility in the Postwar United States. Philadelphia: Temple University Press.

Hann, Julius (1903) *Handbook of Climatology.* New York: Macmillan.

Hansen, Alvin H. (1938) "Economic Progress and Declining Population Growth." *American Economic Review* 29, 1: 1–15.

Hansen, Alvin H. (1953) *A Guide to Keynes.* New York: McGraw-Hill.

Hansen, Bert (2002) "Public Careers and Private Sexuality: Some Gay and Lesbian Lives in the History of Medicine and Public Health." *American Journal of Public Health* 92: 36–44.

Hansen, Bert (2009) *Picturing Medical Progress from Pasteur to Polio: A History of Mass Media Images and Popular Attitudes in America.* New Brunswick: Rutgers University Press.

Hansen, James R. (1987). *Engineer in Charge: A History of the Langley Aeronautical Laboratory, 1917–1958.* Washington: NASA SP-4305.

Hansen, James R. (1995) *Spaceflight Revolution: NASA Langley Research Center from Sputnik to Apollo.* Washington: NASA SP-4308.

Hanson, Elizabeth (2002) *Animal Attractions: Nature on Display in American Zoos.* Princeton: Princeton University Press.

Hanson, Victor (1999) *The Other Greeks: The Family Farm and the Agrarian Roots of Western Civilization.* Berkeley: University of California Press.

Haramundanis, Katherine (ed.) (1996) *Cecila Payne-Gaposhkin: An Autobiography and Other Recollections.* Cambridge: Cambridge University Press.

Haraway, Donna (1976, 2004) *Crystals, Fabrics, and Fields: Metaphors That Shape Embryos.* Berkeley: North Atlantic Books.

Haraway, Donna (1984) "Teddy Bear Patriarchy: Taxidermy in the Garden of Eden, New York City, 1908–1936." *Social Text* 11: 20–64.

Haraway, Donna (1988a) "Situated Knowledges: The Science Question in Feminism and the Privilege of Partial Perspective." *Feminist Studies* 14: 575–99.

Haraway, Donna (1988b) "Remodelling the Human Way of Life: Sherwood Washburn

and the New Physical Anthropology," in George Stocking (ed.), *Bones, Bodies, Behavior: Essays on Biological Anthropology.* Madison: University of Wisconsin Press, 206–59.

Haraway, Donna (1989) *Primate Visions: Gender, Race, and Nature in the World of Modern Science.* New York: Routledge.

Haraway, Donna (1991) *Simians, Cyborgs, and Women: The Reinvention of Nature.* New York: Routledge & Kegan Paul.

Haraway, Donna (1997) *Modest_Witness@ Second_Millennium. Femaleman© Meets Oncomouse™: Feminism and Technoscience.* New York: Routledge.

Haraway, Donna (2008) *When Species Meet.* Minneapolis: University of Minnesota Press.

Hardin, Garret (1968) "Tragedy of the Commons." *Science* 162: 1243–8.

Harding, Sandra (1986) *The Science Question in Feminism.* Ithaca: Cornell University Press.

Harding, Sandra (ed.) (1991) *Whose Science? Whose Knowledge? Thinking From Women's Lives.* Ithaca: Cornell University Press.

Harding, Sandra (ed.) (1993) *The "Racial" Economy of Science: Toward a Democratic Future.* Bloomington: Indiana University Press.

Harding, Sandra (1997) "Women's Standpoints on Nature: What Makes Them Possible?" *Osiris* 12: 186–200.

Harding, Sandra (2008) *Sciences from Below: Feminism, Postcolonialities, and Modernities,* Durham: Duke University Press.

Harding, Sandra (2009) "Postcolonial and Feminist Philosophies of Science and Technology: Convergences and Dissonances." *Postcolonial Studies* 12, 4: 401–21.

Harding, Sandra (2012) *The Postcolonial Science and Technology Studies Reader.* Durham: Duke University Press.

Hardy, Quentin (2013) "Why Big Ag Likes Big Data." New York Times, October 2.

Harman, Oren (2010) *The Price of Altruism: George Price and the Search for the Origins of Kindness.* London: Vintage

Harper, Kristine C. (2003) "Research from the Boundary Layer: Civilian Leadership, Military Funding and the Development of Numerical Weather Prediction (1946–55)." *Social Studies of Science* 33: 667–96.

Harper, Kristine C. (2006) "Meteorology's Struggle for Professional Recognition in the USA (1900–1950)." *Annals of Science* 63: 179–99.

Harper, Kristine C. (2008a) *Weather by the Numbers: The Genesis of Modern Meteorology*. Cambridge: MIT Press.

Harper, Peter S. (2008b) *A Short History of Medical Genetics*. Oxford: Oxford University Press.

Harper, R. (1949) "Tables of American Doctorates in Psychology." *American Journal of Psychology* 62: 579–87.

Harris, Anna; Susan E. Kelly and Sally Wyatt (2013) "Counseling Customers: Emerging Roles for Genetic Counselors in the Direct-to-Consumer Genetic Testing Market." *Journal of Genetic Counseling* 22: 277–88.

Harris, Henry (1997) *The Birth of the Cell*. New Haven: Yale University Press.

Harris, Marvin (1968) *The Rise of Anthropological Theory: A History of Theories of Culture*. New York: Thomas Crowell.

Harris, S.J. (2005). "Jesuit Scientific Activity in the Overseas Missions, 1540–1773." *Isis* 96, 1: 71–9.

Harris, William C. (1974) "The Creed of the Carpetbaggers: The Case of Mississippi." *Journal of Southern History* 40, 2: 199–224.

Harrison, Mark (1996) "A Question of Locality: The Identity of Cholera in British India, 1860–1890." *Clio Medica* 35: 133–59.

Hartmann, Betsy (1999) *Reproductive Rights and Wrongs: The Global Politics of Population Control*. Brooklyn: South End Press.

Harvey, Mark W.T. (2000) *A Symbol of Wilderness: Echo Park and the American Conservation Movement*. Seattle: University of Washington Press.

Harvey, Sean P. (2010) "'Must Not Their Languages Be Savage and Barbarous Like Them?' Philology, Indian Removal, and Race Science." *Journal of the Early Republic* 30, 4: 505–32.

Harwood, Jonathon (1993) *Style of Scientific Thought: The German Genetics Community, 1900–1933*. Chicago: University of Chicago Press.

Hasian, Marouf Arif (1996) *The Rhetoric of Eugenics in Anglo-American Thought*. Athens: University of Georgia Press.

Haskell, T.L. (2000) *The Emergence of Professional Social Science: The American Social Science Association and the Nineteenth-Century Crisis of Authority*. Baltimore: Johns Hopkins University Press.

Haskins, R.W. (1839) *History and Progress of Phrenology*. Buffalo: Steele and Peck.

Hausman, Berenice (1995) *Changing Sex: Transsexualism, Technology, and the Idea of Gender*. Chapel Hill: Duke University Press.

Hayden, Cori (2003) *When Nature Goes Public: The Making and Unmaking of Bioprospecting in Mexico*. Princeton: Princeton University Press.

Hayes-Conroy, A. and J. Hayes-Conroy (2013) *Doing Nutrition Differently: Critical Approaches to Diet and Dietary Intervention*. Burlington: Ashgate Press.

Hayes, E.C. (1911) "The 'Social Forces' Error." *American Journal of Sociology* 16: 613–25.

Hayles, N. Katherine (1984) *The Cosmic Web: Scientific Field Models and Literary Strategies in the Twentieth Century*. Ithaca: Cornell University Press.

Hayles, N. Katherine (1990) *Chaos Bound: Orderly Disorder in Contemporary Literature and Science*. Ithaca: Cornell University Press.

Hays, Samuel P. (1959) *Conservation and the Gospel of Efficiency: The Progressive Conservation Movement, 1890–1920*. Cambridge: Harvard University Press.

Hays, Samuel P. (2000) *A History of Environmental Politics Since 1945*. Pittsburgh: University of Pittsburgh Press.

Hays, Samuel P. and Barbara D. Hays (1987) *Beauty, Health, and Permanence: Environmental Politics in the United States, 1955–1985*. New York: Cambridge University Press.

Hayward, Tim (1995) *Ecological Thought: An Introduction*. Cambridge, MA: Blackwell.

Healy, D. and D. Catell (2003) "Interface between Authorship, Industry, and Science in the Domain of Therapeutics. *British Journal of Psychiatry* 183: 22–7.

Hearnshaw, J.B. (1996) *The Measurement of Starlight: Two Centuries of Astronomical Photometry.* Cambridge: Cambridge University Press.

Hearnshaw, J.B. (2014) *The Analysis of Starlight: Two Centuries of Astronomical Spectroscopy.* New York: Cambridge University Press.

Hecht, Gabrielle (2002) "Rupture Talk in the Nuclear Age: Conjugating Colonial Power in Africa." *Social Studies of Science* 32: 691–727.

Hedeen, Stanley (2008) *Big Bone Lick: The Cradle of American Paleontology.* Lexington: University Press of Kentucky.

Hediger, Heini (1942) *Wild Animals in Captivity: An Outline of the Biology of Zoological Gardens.* New York: Dover.

Hediger, Heini (1969) Man and Animal in the Zoo: Zoo Biology. Gwynne Vevers and Winwood Reade (trans.). New York: Delacorte.

Heering, Peter and Roland Wittje (eds.) (2011). *Learning by Doing: Experiments and Instruments in the History of Science Teaching.* Stuttgart: Verlag.

Heilbron, J.L. and Robert W. Seidel (1989) *Lawrence and His Laboratory: A History of the Lawrence Berkeley Laboratory.* Berkeley: University of California Press.

Heimler, Audrey (1997) "An Oral History of the National Society of Genetic Counselors." *Journal of Genetic Counseling* 6: 315–36.

Hein, Hilde S. (1990) *The Exploratorium: The Museum as Laboratory.* Washington: Smithsonian Institution Press.

Helmreich, Stefan (2009) *Alien Ocean: Anthropological Voyages in Microbial Seas.* Berkeley: University of California Press.

Henke, Christopher R. (2000) "Making a Place for Science: The Field Trial." *Social Studies of Science* 30: 483–511.

Henke, Winfried (2007) "Historical Overview of Palaeoanthropological Research," in W. Henke and I. Tattersall (eds.), *Handbook of PalaeoAnthropology.* New York: Springer, 1–45.

Henson, Pamela M. (1993) "Comstock's Research School at Cornell University, 1874–1930." *Osiris* 8: 158–77.

Henson, Pamela (1996) "The Comstocks of Cornell: A Marriage of Interests," in Helena Pycior, Nancy G. Slack, and Pnina Abir-Am (eds.), *Creative Couples in the Sciences.* New Brunswick: Rutgers University Press.

Henson, Pamela M. (1999) "'Objects of Curious Research:' The History of Science and Technology at the Smithsonian." Isis, 90 Supplement: S249–S269.

Henson, Pamela M. (2004) "A National Science and a National Museum," in A.E. Leviton and M.L. Aldrich (eds.), *Museums and Other Institutions of Natural History: Past, Present, and Future.* San Francisco: California Academy of Sciences, 34–57.

Hentschel, Klaus (1993) "The Conversion of St. John: A Case Study on the Interplay of Theory and Experiment." *Science in Context* 6: 137–94.

Herken, G. (2000) *Cardinal Choices: Presidential Science Advising from the Atomic Bomb to SDI.* Stanford: Stanford University Press.

Herman, Ellen (1995) *The Romance of American Psychology.* Berkeley: University of California Press.

Herman, Ellen (1996) *The Romance of American Psychology: Political Culture in the Age of Experts.* Berkeley: University of California Press.

Herre, E.A., C.A. Machado, and S.A. West (2001) "Selective Regime and Fig Wasp Sex Ratios: Toward Sorting Rigor from Pseudo-Rigor in Tests of Adaptation," in S.H. Orzack, and E. Sober (eds.), *Adaptation and Optimality.* Cambridge: Cambridge University Press, 191–218.

Herren, R.V. and M.C. Edwards (2002) "Whence We Came: The Land-Grant Tradition – Origin, Evolution, and Implications for the 21st Century." *Journal of Agricultural Education* 43, 4: 88–98.

Herskovits, Melville J. (1926) "The Cattle Complex in East Africa." *American Anthropologist* 28, 1: 230–72.

Herskovits, Melville J. (1941) *The Myth of the Negro Past.* New York: Harper & Brothers.

Herzig, Rebecca M. (2005a) *Suffering for Science: Reason and Sacrifice in Modern America.* New Brunswick: Rutgers University Press.

Herzig, Rebecca (2005b) "Gender and Technology," in Carroll Pursell (ed.), *A Companion to American Technology*. Malden: Blackwell, 199–211.

Hess, David J. (1995) *Science and Technology in a Multicultural World: The Cultural Politics of Facts and Artifacts*. New York: Columbia University Press.

Hessen, Boris (1931) The Social and Economic Roots of Newton's "Principia." First published as pp. 149–212 in Science at the Cross Roads: Papers Presented to the International Congress of the History of Science and Technology Held in London from June 29th to July 3, 1931, by the Delegates of the U.S.S.R. Edited by N.I. Bukharin. London: Kniga.

Hetherington, Norriss S. (1976) "Cleveland Abbe and a View of Science in Mid-Nineteenth-Century America." *Annals of Science* 33: 31–49.

Hetherington, Norriss S. (1983) "Mid-Nineteenth-Century American Astronomy: Science in a Developing Nation." *Annals of Science* 40: 61–80.

Hewlett, Richard and Jack Holl (1989) *Atoms for Peace and War, 1953–1961: Eisenhower and the Atomic Energy Commission*. Berkeley: University of California Press.

Heyck, Hunter (2008) "Defining the Computer: Herbert Simon and the Bureaucratic Mind, Part 1." *IEEE Annals of the History of Computing* 30, 2: 42–51.

Heyne, Paul (2008) "Clerical Laissez-Faire: A Case Study in Theological Economics," in H.G. Brennan and A.M.C. Waterman (eds.), *"Are Economists Basically Immoral?" And Other Essays on Economics, Ethics, and Religion*. Indianapolis: Liberty Fund, 238–64.

Hicks, Robert D. (ed.) (2007) "Science and Early Jamestown." *Rittenhouse: Journal of the Scientific Instrument Enterprise* 21: 65–144.

High Energy Astrophysics Division (2011) "Newsletter." http://www.aas.org/head/newsletters.html

Hightower, J. (1975) "The Case for the Family Farm," in C. Lerza and M. Jacobson (eds.), *Food for People, Not for Profit*. New York: Ballantine.

Hilbert, David and Wilhelm Ackermann (1928) *Grundzüge Der Theoretischen Logik*. Berlin: Springer.

Hilgard, Ernest Ropiequet (1987) Psychology in America: A Historical Survey. New York: Harcourt Brace Jovanovich.

Hilgartner, Stephen (1990) "The Dominant View of Popularization: Conceptual Problems, Political Uses." *Social Studies of Science* 20: 519–39.

Hillison, John (1996) "The Origins of Agriscience: Or Where Did All That Scientific Agriculture Come From?" *Journal of Agricultural Education* 37(4): 8–13.

Hindle, Brooke (1964) *David Rittenhouse*. Princeton: Princeton University Press.

Hines, Neal O. (1962) *Proving Ground: An Account of the Radiobiological Studies in the Pacific, 1946–1961*. Seattle: University of Washington.

Hinsley, Curtis M. (1981) *Savages and Scientists: The Smithsonian Institution and the Development of American Anthropology, 1846–1910*. Washington, DC: Smithsonian Institution Press.

Hirsch, Richard F. (1983) *Glimpsing an Invisible Universe: The Emergence of X-Ray Astronomy*. Cambridge: Cambridge University Press.

Hitzhusen, Gregory E. and Mary Evelyn Tucker (2013) "The Potential of Religion for Earth Stewardship." *Frontiers in Ecology and the Environment* 11: 368–76.

Hoage, R.J. and Deiss, William A. (eds.) (1996) *New Worlds, New Animals: From Menagerie to Zoological Park in the Nineteenth Century*. Baltimore: Johns Hopkins University Press.

Hockey, Thomas (ed.) (2007) *The Biographical Encyclopedia of Astronomers*. New York: Springer.

Hoddeson, Lillian and Adrienne Kolb (2000) "The Superconducting Super Collider's Frontier Outpost, 1983–1988." *Minerva* 38: 271–310.

Hoddeson, Lillian and Vicki Daitch (2002). *True Genius: The Life and Science of John Bardeen*. Washington: Joseph Henry Press.

Hoddeson, Lillian, Adrienne Kolb, and Catherine Westfall (2008) *Fermilab:*

Physics, the Frontier, and Megascience. Chicago: University of Chicago Press.

Hoddeson, Lillian; Ernst Braun, Jurgen Teichmann, and Spencer Weart (eds) (1992) *Out of the Crystal Maze: Chapters from the History of Solid-State Physics.* New York: Oxford University Press.

Hodge, Clifton F. (1902) *Nature Study and Life.* Boston: Ginn and Co.

Hodgen, Margaret T. (2011) *Early Anthropology in the Sixteenth and Seventeenth Centuries.* Philadelphia: University of Pennsylvania Press.

Hoffman, Darleane C., Albert Ghiorso, and Glenn T. Seaborg (2000) *The Transuranium People: The Inside Story.* London: ICP.

Hoffman, Frederick (1892) "Vital Statistics of the Negro." *The Arena* 29: 529–42.

Hogan, Andrew J. (2013a) "Locating Genetic Disease: The Impact of Clinical Nosology on Biomedical Conceptions of the Human Genome (1966–1990)." *New Genetics and Society* 32: 78–96.

Hogan, Andrew J. (2013b) *"Chromosomes in the Clinic: The Visual Localization and Analysis of Genetic Disease in the Human Genome."* PhD Dissertation: University of Pennsylvania.

Hogan, Andrew J. (2013c) "Set Adrift in the Prenatal Diagnostic Marketplace: Analyzing the Role of Users and Mediators in the History of a Medical Technology." *Technology and Culture* 54: 62–89.

Hogan, Andrew J. (2014) "The 'Morbid Anatomy' of the Human Genome: Tracing the Observational and Representational Approaches of Postwar Genetics and Biomedicine." *Medical History* 58: 315–36.

Holden, Norman E. (2004a) "Atomic Weights and the International Committee: A Brief Historical Review." *Chemistry International* 26: 4–7.

Holden, Norman E. (2004b) "Atomic Weights and the International Committee – A Historical Review." http://www.iupac.org/publications/ci/2004/2601/1_holden.html

Holl, Jack M. (1997) *Argonne National Laboratory, 1946–96.* Urbana: University of Illinois Press.

Holmes, Frederic Lawrence (2006) *Reconceiving the Gene: Seymour Benzer's Adventures in Phage Genetics.* New Haven: Yale University Press.

Holmes, Oliver Wendell (1861) *Elsie Venner: A Romance of Destiny.* Boston: Ticknor & Fields.

Holton, Gerald (1981) "The Formation of the American Physics Community in the 1920s and the Coming of Albert Einstein." *Minerva* 19: 569–81.

Home, Henry (1774) *Sketches of the History of Man.* Edinburgh: W. Creech.

Hooke, Robert (1964) "Description of Cells," in Thomas S. Hall (ed.), *A Source Book in Animal Biology.* New York: Hafner, 431–7.

Hooton, Earnest Albert (1936) "Plain Statements About Race." *Science* 83, 2161: 511–13.

Hooton, Earnest (1937) *Up from the Ape.* New York: Macmillan.

Hooton, Earnest (1942) *Man's Poor Relations.* Garden City: Doubleday.

Hopper, Grace (1999) "Commander Aiken and My Favorite Computer," in I.B. Cohen and G. Welch (eds.), *Makin' Numbers: Howard Aiken and the Computer.* Cambridge: MIT Press, 185–94.

Hornaday, William T. (1913) *Our Vanishing Wild Life: Its Extermination and Preservation.* New York: Charles Scribner's Sons.

Horner John R. and Edwin Dobb (1997) *Dinosaur Lives.* New York: Harcourt, Brace.

Horowitz, Roger (ed.) (2001) *Boys and Their Toys? Masculinity, Technology, and Class in America.* New York: Routledge.

Horsman, Reginald (1981) *Race and Manifest Destiny.* Cambridge: Harvard University Press.

Horsman, Reginald (1987) *Josiah Nott of Mobile.* Baton Rouge: Louisiana University Press.

Hounshell, David (1997) "The Cold War, RAND, and the Generation of Knowledge, 1946–1962." *Historical Studies in the Physical and Biological Sciences* 27: 237–67.

Hounshell, David and John Kenly Smith (1988) *Science and Corporate Strategy: Du Pont R&D, 1902–1980.* Cambridge: Cambridge University Press.

Hovenkamp, Herbert (1978) *Science and Religion in America, 1800–1860.* Philadelphia: University of Pennsylvania Press.

Howard, Ted and Jeremy Rifkin (1977) *Who Should Play God? The Artificial Creation of Life and What it Means for the Future of the Human Race.* New York: Delacorte Press.

Howe, Jas Lewis (1900) "Second Report of the Committee of the German Chemical Society on Atomic Weights." *Science* 12: 246–47.

Howe, Joshua P. (2014) *Behind the Curve: Science and the Politics of Global Warming.* Seattle: University of Washington Press.

Howell, F. Clark (1952) "Pleistocene Glacial Ecology and the Evolution of 'Classic Neandertal' Man." *Southwest Journal of Anthropology* 8: 377–410.

Howell, F. Clark (1958) "Upper Pleistocene Men of the Southwest Asian Mousterian," in Ralf Von Koenigswald (ed.), *Hundert Jahre Neanderthaler. Neanderthal Centenary. 1856–1956.* Köln: Böhlau-Verlag, 185–98.

Howell, F. Clark (1960) "European and Northwest African Middle Pleistocene Hominids." *Current Anthropology* 1: 195–232.

Howells, William White (1973) "Cranial Variation in Man: A Study by Multivariate Analysis of Patterns of Difference among Recent Human Populations." Papers of the Peabody Museum of Archaeology and Ethnology 67.

Howes, Ruth H. and Caroline Herzenberg (1999) *Their Day in the Sun: Women of the Manhattan Project.* Philadelphia: Temple University Press.

Howkins, Adrian (2011) "Melting Empires? Climate Change and Politics in Antarctica Since the International Geophysical Year." *Osiris* 26: 180–97.

Hrdlička, Aleš (1918) "Physical Anthropology: Its Scope and Aims; Its History and Present Status in America." *American Journal of Physical Anthropology* 1, 1: 3–23.

Hrdlička, Aleš (1927) "The Neanderthal Phase of Man." *Journal of the Royal Anthropological Institute* 57: 249–74.

Hrdlička, Aleš (1930) *The Skeletal Remains of Early Man.* Washington: Smithsonian Institution.

Hrdy, S.B. (1978) *The Langurs of Abu: Female and Male Strategies of Reproduction.* Cambridge: Harvard University Press.

Hrdy, S.B. (1999) *Mother Nature: A History of Mothers, Infants, and Natural Selection.* New York: Pantheon.

Hubbard, Jennifer (2006) *A Science on the Scales: The Rise of Canadian Atlantic Fisheries Biology, 1898–1939.* Toronto: University of Toronto Press.

Hubbard, Jennifer (2014) "In the Wake of Politics: The Political and Economic Construction of Fisheries Biology, 1860–1970." *Isis* 105, 2: 364–78.

Hubbard, Ruth (1990) *The Politics of Women's Biology.* New Brunswick: Rutgers University Press.

Hubbell, John G. and Robert W. Smith (1992) "Neptune in America: Negotiating Discovery." *Journal for the History of Astronomy* 23: 261–91.

Hubby, John L. and Richard C. Lewontin (1966) "A Molecular Approach to the Study of Genic Heterozygosity in Natural Populations. I. The Number of Alleles at Different Loci in Drosophila Pseudoobscura." *Genetics* 54: 577–94.

Hufbauer, Karl (1991) *Exploring the Sun: Solar Science since Galileo.* Baltimore: Johns Hopkins University Press.

Hughes, Agatha C. and Thomas P. Hughes, (eds.) (2000). *Systems, Experts, and Computers: The Systems Approach in Management and Engineering, World War II and After.* Cambridge: MIT Press.

Hughes, Sally Smith (2011) *Genentech: The Beginnings of Biotech.* Chicago: University of Chicago Press.

Hughes, Thomas (2004) *Human-Built World: How to Think about Technology and Culture.* Chicago: University of Chicago Press.

Hull, Clark L. (1935) "The Conflicting Psychologies of Learning – A Way Out." *Psychological Review* 42, 6: 491–516.

Hull, Clark L. (1950) "A Primary Social Science Law." *The Scientific Monthly* 71: 221–8.

Hull, David L. (1998) *Science as a Process: An Evolutionary Account of the Social and Conceptual Development of Science.* Chicago: University of Chicago Press.

Hulme, Mike (2008) "Geographical Work at the Boundaries of Climate Change." *Transactions of the Institute of British Geographers* 33: 5–11.

Hulme, Mike (2009) *Why We Disagree about Climate Change: Understanding Controversy, Inaction and Opportunity*. Cambridge: Cambridge University Press.

Hume, Brad D. (2011) "Evolutionisms: Lewis Henry Morgan, Time, and the Question of Sociocultural Evolutionary Theory." *Histories of Anthropology Annual* 7, 1: 91–126.

Humphreys, Margaret (1999) *Yellow Fever and the South*. Baltimore: Johns Hopkins University Press.

Humphreys, Margaret (2013) *Marrow of Tragedy: The Health Crisis of the American Civil War*. Baltimore: Johns Hopkins University Press.

Hunner, Jon (2004) *Inventing Los Alamos: The Growth of an Atomic Community*. Norman: University of Oklahoma Press.

Hunt, C. (1917) U.S. Department of Agriculture, *How to Select Foods*. Bulletin 808. Washington: Government Printing Office.

Hunter, Kathryn Montgomery (1991) *Doctors' Stories: The Narrative Structure of Medical Knowledge*. Princeton: Princeton University Press.

Huntington, Ellsworth (1922) *Civilization and Climate*. New Haven: Yale University Press.

Huxley, Julian (1942) *Evolution: The Modern Synthesis*. London: Allen and Unwin.

Hyatt, Alpheus (1884) "The Evolution of the Cephalopoda." *Science* 3: 145–9.

Hynes, H. Patricia (1982) "Toward a Laboratory of One's Own: Lesbians in Science," in Margaret Cruikshank (ed.), *Lesbian Studies: Present and Future*. New York: The Feminist Press at New York University, 174–8.

Hyson, Jeffrey (2000) "Jungles of Eden: The Design of American Zoos," in Michel Conan (ed.), *Environmentalism in Landscape Architecture*. Washington: Dumbarton Oaks, 23–44.

Igo, Sarah (2008) *The Averaged American: Surveys, Citizens, and the Making of a Mass Public*. Cambridge: Harvard University Press.

Ingram, Carl (2003) "State Issues Apology for Policy of Sterilization." Los Angeles Times, March 12.

Irvine, Janice M. (1990) *Disorders of Desire: Sex and Gender in Modern American Sociology*. Philadelphia: Temple University Press.

Irwin, A. (1995) *Citizen Science: A Study of People, Expertise, and Sustainable Development*. London: Routledge.

Irwin, Alan, and Brian Wynne (1996) *Misunderstanding Science? The Public Reconstruction of Science and Technology*. Cambridge: Cambridge University Press.

Isaac, J. (2012) *Working Knowledge: Making the Human Sciences from Parsons to Kuhn* Cambridge: Harvard University Press.

Isaacson, Walter (2008) *Einstein: His Life and Universe*. New York: Simon & Schuster.

Isager, Signe and Jens Skydsgaard. (1995) *Ancient Greek Agriculture: An Introduction*. New York. Routledge.

Israel, Paul (1992) *From Machine Shop to Industrial Laboratory: Telegraphy and the Changing Context of American Invention, 1830–1920*. Baltimore: Johns Hopkins University Press.

Jackman, Wilbur (1891) *Nature-Study for the Common Schools*. New York: Henry Holt.

Jacknis, Ira (2002) "The First Boasian: Alfred Kroeber and Franz Boas, 1896–1905." *American Anthropologist* 104, 2: 520–32.

Jackson, D.L. and L.L. Jackson (2002) *The Farm as Natural Habitat: Reconnecting Food Systems with Ecosystems*. Washington, DC: Island Press.

Jackson, John P. (2001) "'In Ways Unacademical:' The Reception of Carleton S. Coon's The Origin of Races." *Journal of the History of Biology* 34, 2: 247–85.

Jackson, W. (1980) *New Roots for Agriculture*. San Francisco: Friends of the Earth.

Jackson, W. (1990) *Gunnar Myrdal and America's Conscience: Social Engineering and Racial Liberalism, 1938–1987*. Chapel Hill: University of North Carolina Press.

Jacobi, Abraham (1886) "Inaugural Address." *Transactions of the New York Academy of Medicine* 5: 149–69.

Jacobs, J.A. and S. Frickel (2009) "Interdisciplinarity: A Critical Assessment." *Annual Review of Sociology* 35: 43–65.

Jacobson, M.F. (2004) "High-Fructose Corn Syrup and the Obesity Epidemic." *American Journal of Clinical Nutrition* 80: 108.

Jacoby, Karl (2001) *Crimes Against Nature: Squatters, Poachers, Thieves, and the Hidden History of American Conservation*. Berkeley: University of California Press.

Jaffe, Mark (2000) *The Gilded Dinosaur: The Fossil War Between E.D. Cope and O.C. Marsh and the Rise of American Science*. New York: Crown.

Jager, R. (2004) *The Fate of Family Farming: Variations on an American Idea*. Hanover: University Press of New England.

James, C. and A.F. Krattiger (1996) *Global Review of the Field Testing and Commercialization of Transgenic Plants, 1986 to 1995: The First Decade of Crop Biotechnology*. Ithaca: ISAAA Briefs No. 1.

James, Frank A. J. L. (ed.) (1989) *The Development of the Laboratory: Essays on the Place of Experiment in Industrial Civilization*. New York: American Institute of Physics.

James, Mary Ann (1987) *Elites in Conflict: The Antebellum Clash over the Dudley Observatory*. New Brunswick: Rutgers University Press.

James, W. (1880) "Great Men, Great Thoughts, and the Environment." *Atlantic Monthly* 46, 276: 441–59.

James, William (1879) "Are We Automata?" *Mind* 13: 1–22.

James, William (1918) *The Principles of Psychology*. London: Dover.

Jamestown-Yorktown Foundation (2007) *The World of 1607*. Williamsburg: Jamestown-Yorktown Foundation.

Jamieson, Dale (1994) "Against Zoos," in Ed Gruen and Dale Jamieson (eds.), *Reflecting on Nature: Readings in Environmental Philosophy*. New York: Oxford University, 291–9.

Jankovic, Vladimir (2000) *Reading the Skies: A Cultural History of English Weather, 1650–1820*. Manchester: Manchester University Press.

Jankovic, Vladimir (2010a) "Climates as Commodities: Jean Pierre Purry and the Modelling of the Best Climate on Earth." *Studies in the History and Philosophy of Modern Physics* 41: 201–7.

Jankovic, Vladimir (2010b) *Confronting the Climate: British Airs and the Making of Environmental Medicine*. New York: Palgrave Macmillan.

Jankovic, Vladimir and Michael Hebbert (2012) "Hidden Climate Change – Urban Meteorology and the Scales of Real Weather." *Climatic Change* 113: 23–33.

Jardine, Nicholas, James A. Secord, and E.C. Spary (eds.) (1996) *Cultures of Natural History*. Cambridge: Cambridge University Press.

Jasanoff, Sheila (1990) *The Fifth Branch: Science Advisers as Policy Makers*. Cambridge: Harvard University Press.

Jasanoff, Sheila (2006) "Biotechnology and Empire: The Global Power of Seeds and Science." *Osiris* 21: 273–92.

Jasanoff, Sheila (2013) "Watching the Watchers: Lessons for the Science of Science Advice." http://www.theguardian.com/science/political-science/2013/Apr/08/lessons-science-advice

Jefferson, Thomas (1782) *Notes On Virginia*. Paris: S. N.

Jefferson, Thomas (1787) *Notes on the State of Virginia*. London: John Stockdale.

Jefferson, Thomas (1803) "Transcript: Jefferson's Instructions for Meriwether Lewis," in Gerard W. Gawalt (ed.), *Rivers, Edens, Empires: Lewis & Clark and The Revealing of America*. Manuscript Division, Library of Congress.

Jefferson, Thomas (1984) *Notes on Virginia in Writings*. New York: Library of America.

Jenkins, E.W. (1981) "Science, Sentimentalism, or Social Control? The Nature Study Movement in England and Wales, 1888–1914." *History of Education* 10: 33–43.

Jennings, Herbert Spencer (1906) *Behavior of the Lower Organisms*. New York: Columbia University Press.

Jepsen, Glenn Lowell (ed.) (1949) *Genetics, Paleontology, and Evolution*. Princeton: Princeton University Press.

Jerome, Fred (2003) *The Einstein File: J. Edgar Hoover's Secret War Against the World's Most Famous Scientist*. New York: St. Martin's Press.

Jerome, Fred and Rodger Taylor (2006) *Einstein on Race and Racism*. New Brunswick: Rutgers University Press.

Jesse, Jerry (2013) "Radiation Ecologies: Bombs, Bodies, and Environment during the Atmospheric Nuclear Weapons Testing Period in the United States, 1942–1965." PhD dissertation: Montana State University.

Jiang, Lijing (2012) "History of Apoptosis Research," in *Encyclopedia of Life Sciences*. Chichester: John Wiley.

Jiang, Lijing (2014) "Causes of Aging are Likely to be Many: Robin Holliday and Changing Molecular Approaches to Cell Aging, 1963–1988." *Journal of the History of Biology* 47, 4: 547–84.

Jillson, B.C. (1873) "The Physical Features of the Mississippi Valley." *Pennsylvania School Journal* 22, 3.

Joas, Christian (2011) "Campos Que Interagem: Física Quântica E A Transferência De Conceitos Entre Física De Partículas, Nuclear E Do Estado Sólido," in Olival Freire Jr. et al. (eds.), *Teoria Quântica: Estudos Históricos E Implicações Culturais*. Campina Grande, Brazil: Livraria Da Física, 109–51.

Johns, J. Adam (2008) *The Assault on Progress: Technology and Time in American Literature*. Tuscaloosa: University of Alabama Press.

Johnson, Ann (2008) "What if We Wrote the History of Science from the Perspective of Applied Science?" *Historical Studies in the Natural Sciences* 38: 610–20.

Johnson, G.L. (1984) *Academia Needs a New Covenant for Serving Agriculture*. Mississippi State Agricultural and Forestry Experiment Station Special Publication.

Johnson, Kristin (2009). "The Return of the Phoenix: The 1963 International Congress of Zoology and American Zoologists in the Twentieth Century." *Journal of the History of Biology* 42, 3: 417–56.

Johnston, Lucas F. and Whitney A. Bauman (eds.) (2014) *Science and Religion: One Planet, Many Possibilities*. New York: Routledge.

Jones, Bessie Zaban and Lyle Gifford Boyd (1971) *The Harvard College Observatory: The First Four Directorships, 1839–1919*. Cambridge: Belknap Press.

Jones, Daniel Stedman (2012). *Masters of the Universe: Hayek, Friedman, and the Birth of Neoliberal Politics*. Princeton: Princeton University Press.

Jones, James (2004) *Alfred C. Kinsey: A Life*. W.W. Norton.

Jordan-Young, Rebecca (2010) *Brain Storm: The Flaws in the Science of Sex Difference*. Cambridge: Harvard University Press.

Jordanova, Ludmilla (1989) *Sexual Visions: Images of Gender in Science and Medicine Between the Eighteenth and Twentieth Centuries*. Madison: University of Wisconsin Press.

Jordanova, Ludmilla (1993) "Gender and the Historiography of Science." *British Journal for the History of Science* 16: 469–83.

Jordanova, Ludmilla and Roy Porter (1979) *Images of the Earth: Essays in the History of the Environmental Sciences*. Chalfont St. Giles: British Society for the History of Science.

Judd, Richard William (2009) *The Untilled Garden: Natural History and the Spirit of Conservation in America, 1740–1840*. New York: Cambridge University Press.

Judson, Horace Freeland (1979). *The Eighth Day of Creation: Makers of the Revolution in Biology*. New York: Simon & Schuster.

Kahn, Jonathan (2005) "Bidil: False Promises." *Gene Watch* 18, 6: 6–9.

Kaiser, David (2002) "Cold War Requisitions, Scientific Manpower, and the Production of American Physicists after World War II." *Historical Studies in the Physical and Biological Sciences* 33, 1: 131–59.

Kaiser, David (2004) "The Postwar Suburbanization of American Physics." *American Quarterly* 56: 851–88.

Kaiser, David (2005a) *Drawing Theories Apart: The Dispersion of Feynman Diagrams in Postwar Physics*. Chicago: University of Chicago Press.

Kaiser, David (2005b) "Training and the Generalist's Vision in the History of Science." *Isis* 96: 244–51.

Kaiser, David (2010) "Elephant on the Charles: Postwar Growing Pains," in David Kaiser (ed.), *Becoming MIT: Moments of Decision*. Cambridge: MIT Press, 103–22.

Kaiser, David (2012) "Booms, Busts, and the World of Ideas: Enrollment Pressures and the Challenges of Specialization." *Osiris* 27: 276–302.

Kareiva, Peter and Michelle Marvier (2012) "What is Conservation Science?" *Bioscience* 62, 11: 962–9.

Kass-Simon, G., and Patricia Farnes (eds.) (1990) *Women of Science: Righting the Record.* Bloomington: Indiana University Press.

Kasson, John F. (1976) *Civilizing the Machine: Technology and Republican Values in America, 1776–1900.* New York: Grossman.

Katz, M.B. and Sugrue, T. J. (1998) *W. E. B. Dubois, Race, and the City: The Philadelphia Negro and its Legacy.* Philadelphia: University of Pennsylvania Press.

Kauffman, George B. (1969) "American Forerunners of the Periodic Law." *Journal of Chemical Education* 46: 128–35.

Kauffman, George B. (1970) "The Reception of Mendeleev's Ideas in the United States and Mendeleev's Correspondence with American Scientists." *Archives Internationales d'Histoire Des Sciences* 23: 87–106.

Kay, Lily E. (1993, 1996) *The Molecular Vision of Life: Caltech, the Rockefeller Foundation and the Rise of the New Biology.* New York: Oxford University Press.

Kay, Lily E. (2000) *Who Wrote the Book of Life? A History of Genetic Code.* Stanford: Stanford University Press.

Keating, Peter, and Alberto Cambrosio (2004) "Does Biomedicine Entail the Successful Reduction of Pathology to Biology?" *Perspectives in Biology and Medicine* 47: 357–71.

Keen, M.F. (2004) *Stalking Sociologists: J. Edgar Hoover's FBI Surveillance of American Sociology.* Transaction.

Keeney, Elizabeth B. (1992) *The Botanizers: Amateur Scientists in Nineteenth-Century America.* Chapel Hill: University of North Carolina Press.

Keiner, Christine (2010) *The Oyster Question: Scientists, Watermen, and the Maryland Chesapeake Bay Since 1880.* Athens: University of Georgia Press.

Keith, J. (1995) *Country People in the New South: Tennessee's Upper Cumberland.* Chapel Hill: University of North Carolina Press.

Keller, Evelyn Fox (1974) "Women in Science: A Social Analysis." *Harvard Magazine* 77, 2: 14–19.

Keller, Evelyn Fox (1977) "The Anomaly of a Woman in Physics," in Sara Ruddick and Pamela Daniels (eds.), *Working it Out: Twenty-Three Writers, Scientists and Scholars Talk About Their Lives.* New York: Pantheon, 77–91.

Keller, Evelyn Fox (1978) "Gender and Science." *Psychoanalysis and Contemporary Thought* 1: 409–33.

Keller, Evelyn Fox (1983) *A Feeling for the Organism: The Life and Work of Barbara Mcclintock.* New York: W. H. Freeman.

Keller, Evelyn Fox (1985) *Reflections on Gender and Science.* New Haven: Yale University Press.

Keller, Evelyn Fox (1990) "Physics and the Emergence of Molecular Biology: A History of Cognitive and Political Synergy." *Journal of the History of Biology* 23, 3: 389–409.

Keller, Evelyn Fox (1992) *Secrets of Life, Secrets of Death: Essays on Language, Gender, and Science.* New York: Routledge.

Keller, Evelyn Fox (1995a) "The Origin, History, and Politics of the Subject Called 'Gender and Science'." In Sheila Jasanoff, Gerald E. Markle, James C. Peterson and Trevor Pinch (eds.), *Handbook of Science and Technology Studies.* Thousand Oaks: Sage, 80–94.

Keller, Evelyn Fox (1995b) *Refiguring Life: Metaphors of Twentieth-Century Biology.* New York: Columbia University Press.

Keller, Evelyn Fox (2000) *The Century of the Gene.* Cambridge: Harvard University Press.

Keller, Evelyn Fox (2010) *The Mirage of a Space Between Nature and Nurture.* Durham: Duke University Press.

Keller, Evelyn Fox and Helen Longino (eds.) (1996) *Feminism and Science, Oxford Readings in Feminism.* Oxford: Oxford University Press.

Kennedy, Joseph C.G. (1864) *Population of the United States in 1860*. Washington: Government Printing Office.

Kennefick, Daniel (2007) *Travelling at the Speed of Thought: Einstein and the Quest for Gravitational Waves*. Princeton: Princeton University Press.

Kenney, Martin (1986) *Biotechnology: The University–Industrial Complex*. New Haven: Yale University Press.

Kessler, Suzanne (1998) *Lessons Learned from the Intersexed*. New Brunswick: Rutgers University Press.

Kevles, Daniel (1985) *In the Name of Eugenics: Genetics and the Uses of Human Heredity*. Cambridge: Harvard University Press.

Kevles, Daniel (1987) *The Physicists*. Cambridge: Harvard University Press.

Kevles, Daniel (1990) "Cold War and Hot Physics: Science, Security, and the American State, 1945–56." *Historical Studies in the Physical and Biological Sciences* 20: 239–64.

Kevles, Daniel (1993) "Renato Dulbecco and the New Animal Virology: Medicine, Methods, and Molecules." *Journal of the History of Biology* 26: 409–42.

Kevles, Daniel (1995) *In the Name of Eugenics: Genetics and the Uses of Human Heredity*. Cambridge, MA: Harvard University Press.

Kevles, Daniel (1997) "Big Science and Big Politics in the United States: Reflections on the Death of the SSC and the Life of the Human Genome Project." *Historical Studies of the Physical and Biological Sciences* 27: 269–97.

Kevles, Daniel (1998) *The Baltimore Case: A Trial of Politics, Science, and Character*. New York: W.W. Norton.

Kevles, Daniel (2001) *The Physicists: The History of a Scientific Community in Modern America*. New York: Alfred A. Knopf.

Kevles, Daniel (2002a) "European Group on Ethics in Science and New Technologies to the European Commission, European Commission. A History of Patenting Life in the United States with Comparative Attention to Europe and Canada: A Report to the European Group on Ethics in Science and New Technologies." Office for Official Publications of the European Communities, Lanham: Bernan Associates.

Kevles, Daniel (2002b) "Of Mice & Money: The Story of the World's First Animal Patent." *Daedalus* 131: 78–88.

Kevles, Daniel (2003) "Biotechnology," in John Heilbron (ed.), *The Oxford Companion to the History of Modern Science*. Oxford: Oxford University Press, 96–8.

Kevles, Daniel (2007) "Patents, Protections, and Privileges: The Establishment of Intellectual Property in Animals and Plants." *Isis* 92: 323–31.

Kevles, Daniel and Leroy Hood (eds.) (1992) *The Code of Codes: Scientific and Social Issues in the Human Genome Project*. Cambridge: Harvard University Press.

Key, S. (1996) "Economics or Education: The Establishment of American Land-Grant Universities." *Journal of Higher Education* 67, 2: 196–220.

Keynes, John Maynard (1936) *The General Theory of Employment, Interest and Money*. New York: Harcourt, Brace.

Khrgian, Aleksandr Khristoforovich (1970) *Meteorology: A Historical Survey*. Israel Program for Scientific Translations.

Kidd, Colin (2006) *The Forging of Races. Race and Scripture in the Protestant Atlantic World, 1600–2000*. Cambridge: Cambridge University Press.

Kidwell, Peggy Aldrich, Amy Ackerberg-Hastings, and David Lindsay Roberts (2008) *Tools of American Mathematics Teaching, 1800–2000*. Baltimore: Johns Hopkins University Press.

Kimmelman, Barbara A. (1983) "The American Breeders' Association: Genetics and Eugenics in an Agricultural Context, 1903–13." *Social Studies of Science* 13: 163–204.

Kimmelman, Barbara A. (1987) "*A Progressive Era Discipline: Genetics at American Agricultural Colleges and Experiment Stations, 1900–1920*." PhD Dissertation: University of Pennsylvania.

Kimmelman, Barbara A. (2006) "Mr. Blakeslee Builds His Dream House: Agricultural Institutions, Genetics, and Careers, 1900–1915." *Journal of the History of Biology* 39: 241–80.

Kimura, Motoo (1968) "Evolutionary Rate at the Molecular Level." *Nature* 217: 624–6.

Kimura, Motoo (1983) *The Neutral Theory of Molecular Evolution.* Cambridge: Cambridge University Press.

Kinchy, Abby J. (2006) "On the Borders of Post-War Ecology: Struggles over the Ecological Society of America's Preservation Committee, 1917–1946." *Science as Culture* 15: 23–44.

King, Henry C. (1955) *The History of the Telescope.* Cambridge: Sky Publishing Corporation.

King, Jack L. and Thomas H. Jukes (1969) "Non-Darwinian Evolution." *Science* 164: 788–98.

Kingsbury, N. (2009) *Hybrid: The History and Science of Plant Breeding.* Chicago: University of Chicago Press.

Kingsland, Sharon E. (1991) "Defining Ecology as a Science," in Leslie H. Real and James H. Brown (eds.), *Foundations of Ecology: Classic Papers with Commentary.* Chicago. University of Chicago Press, 1–13.

Kingsland, Sharon E. (1995) *Modeling Nature: Episodes in the History of Population Ecology.* Chicago: University of Chicago Press.

Kingsland, Sharon E. (2005) *The Evolution of American Ecology, 1890–2000.* Baltimore: Johns Hopkins University Press.

Kingsland, Sharon E. (2009) "Frits Went's Atomic Age Greenhouse: The Changing Landscape on the Lab-Field Border." *Journal of the History of Biology* 42: 289–324.

Kingsolver, B., C. Kingsolver, and S. Hopp. (2008) *Animal, Vegetable, Miracle: A Year of Food Life.* New York: Harper.

Kintigh, Keith W. (1992) "I Wasn't Going To Say Anything, But Since You Asked: Archaeoastronomy and Archaeology." *Archaeoastronomy and Ethnoastronomy News* 5, 1: 4.

Kirsch, Scott (2007) "Ecologists and the Experimental Landscape: The Nature of Science at the U.S. Department of Energy's Savannah River Site." *Cultural Geographies* 14: 485–510.

Kisling, Vernon, Jr. (1998). "Colonial Menageries and the Exchange of Exotic Faunas." *Archives of Natural History* 25, 3: 303–20.

Kisling, Vernon, Jr. (2001) *Zoo and Aquarium History: Ancient Animal Collections to Zoological Gardens.* Boca Raton: CRC Press.

Kitcher, P. (1985) *Vaulting Ambition: Sociobiology and the Quest for Human Nature* Cambridge: MIT Press.

Klein, L.R. (2006) "Paul Samuelson as a 'Keynesian' Economist," in Michael Szenberg, Lall Ramrattan, and Aron A. Gottesman (eds.), *Samuelsonian Economics and the Twenty-First Century.* Oxford: Oxford University Press, 165–77.

Klein, Ursula and Wolfgang Lefèvre (2007) *Materials in Eighteenth-Century Science: A Historical Ontology.* Cambridge: MIT Press.

Kleinman, Daniel (1995) *Politics on the Endless Frontier: Postwar Research Policy in the United States.* Durham: Duke University Press.

Kline, Ronald (2011) "Cybernetics, Automata Studies, and the Dartmouth Conference on Artificial Intelligence." *IEEE Annals of the History of Computing* 33, 4: 5–16.

Kline, Wendy (2001) *Building a Better Race: Gender, Sexuality, and Eugenics from the Turn of the Century to the Baby Boom.* Berkeley: University of California Press.

Klingle, Matthew (1998) "Plying Atomic Waters: Lauren Donaldson and the 'Fern Lake Concept' of Fisheries Management." *Journal of the History of Biology* 31: 1–32.

Kloppenburg, Jack Ralph (2004) *First the Seed: The Political Economy of Plant Biotechnology.* Madison: University of Wisconsin Press.

Kluchin, Rebecca M. (2009) *Fit to be Tied: Sterilization and Reproductive Rights in America, 1950–1980.* New Brunswick: Rutgers University Press.

Kneeland, S. (1851) "Introduction," in Charles Hamilton Smith (ed.), *The Natural History of the Human Species.* Boston: Gould & Lincoln, 15–98.

Knight, Frank H. (1921) *Risk, Uncertainty, and Profit.* Boston: Houghton Mifflin.

Knight, Frank H. (1935) *The Ethics of Competition.* New York: Harper & Bros.

Knorr-Cetina, Karin (1981) *The Manufacture of Knowledge: An Essay on the Constructivist and Contextual Nature of Science*. Oxford: Pergamon.

Knox, Robert (1850) *The Races of Men: A Fragment*. Philadelphia: Lea and Blanchard.

Koerner, E.F.K. (1992) "The Sapir–Whorf Hypothesis: A Preliminary History and a Bibliographical Essay." *Journal of Linguistic Anthropology* 2, 2: 173–98.

Koerner, E.F.K. (2004) *Essays in the History of Linguistics*. Amsterdam: John Benjamin.

Kohl, Michael F. and John S. Mcintosh, (eds.) (1997) *Discovering Dinosaurs in the Old West: The Field Journals of Arthur Lakes*. Washington: Smithsonian Institution Press.

Kohler, Robert E. (1976) "The Management of Science: The Experience of Warren Weaver and the Rockefeller Foundation Programme in Molecular Biology." *Minerva* 14: 279–306.

Kohler, Robert E. (1982) *From Medical Chemistry to Biochemistry: The Making of a Biomedical Discipline*. Cambridge: Cambridge University Press.

Kohler, Robert E. (1991) *Partners in Science: Foundations and Natural Scientists 1900–1945*. Chicago: University of Chicago Press.

Kohler, Robert E. (1994) *Lords of the Fly: Drosophila Genetics and the Experimental Life*. Chicago: University of Chicago Press.

Kohler, Robert E. (2002) *Landscapes and Labscapes: Exploring the Lab–Field Border in Biology*. Chicago: University of Chicago Press.

Kohler, Robert E. (2005) "A Generalist's Vision." *Isis* 96: 224–9.

Kohler, Robert E. (2006) *All Creatures: Naturalists, Collectors, and Biodiversity, 1850–1950*. Princeton: Princeton University Press.

Kohler, Robert E. (2008) "Lab History: Reflections." *Isis* 99: 761–8.

Kohlstedt, Sally Gregory (1976) *The Formation of the American Scientific Community: The American Association for the Advancement of Science, 1848–1860*. Urbana: University of Illinois Press.

Kohlstedt, Sally Gregory (1978a) "Maria Mitchell: The Advancement of Women in Science." *The New England Quarterly* 51: 39–63.

Kohlstedt, Sally Gregory. (1978b) "In from the Periphery: American Women in Science, 1830–1880." *Signs* 4: 81–96.

Kohlstedt, Sally Gregory (1980) "Henry A. Ward: The Merchant Naturalist and American Museum Development." *Journal of the Society for the Bibliography of Natural History* 9: 647–61.

Kohlstedt, Sally Gregory (1990) "Parlors, Primers, and Public Schooling: Science Education in Nineteenth-Century America." *Isis* 81: 424–95.

Kohlstedt, Sally Gregory (1992) "Entrepreneurs and Intellectuals: Natural History in Early American Museums," in William T. Alderson (ed.,) *Mermaids, Mummies, and Mastodons: The Emergence of the American Museum*. Washington: American Association of Museums.

Kohlstedt, Sally Gregory (1995) "Women in the History of Science: An Ambiguous Place." *Osiris* 10: 39–58.

Kohlstedt, Sally Gregory (1996) "Reflections on Zoo History," in R.J. Hoage and William A. Deiss (eds.), *New Worlds, New Animals: From Menagerie to Zoological Park in the Nineteenth Century*. Baltimore: Johns Hopkins University Press, 3–8.

Kohlstedt, Sally Gregory (1997) "Nature Study in North America and Australasia, 1890–1945." *Historical Records of Australian Science* 11: 459–64.

Kohlstedt, Sally Gregory (ed.) (1999) *History of Women in the Sciences: Readings from Isis*. Chicago: University of Chicago Press.

Kohlstedt, Sally Gregory (2005a) "Nature not Books: Scientists and the Origins of the Nature Study Movement in the 1890s." *Isis* 96: 324–52.

Kohlstedt, Sally Gregory (2005b) "Thoughts in Things: Modernity, History, and North American Museums." *Isis* 96: 586–601.

Kohlstedt, Sally Gregory (2006) "Sustaining Gains: Reflections on Women in Science and Technology in the Twentieth-Century United States," in Jill M. Bystydzienski and Sharon R. Bird (eds.), *Removing Barriers: Women in Academic Science, Technology, Engineering, and Mathematics*. Bloomington: Indiana University Press, 23–45.

Kohlstedt, Sally Gregory (2010) *Teaching Children Science: Hands-On Nature Study in North America, 1890–1930*. Chicago: University of Chicago Press.

Kohlstedt, Sally Gregory (2013) "Creative Niche Scientists: Women Educators in North American Museums, 1880–1930" *Centaurus* 55: 153–74.

Kohlstedt, Sally Gregory and Helen E. Longino (eds.) (1997) "Women, Gender, and Science: New Directions." *Osiris* 12.

Kohlstedt, Sally Gregory and Margaret W. Rossiter (eds.) (1985) "*Historical Writing on American Science,*" Osiris Second Series, Vol. 1. Philadelphia: History of Science Society.

Kohlstedt, Sally Gregory, Michael M. Sokal, and Bruce V. Lewenstein (1999) *The Establishment of Science in America: 150 Years of the American Association for the Advancement of Science*. New Brunswick: Rutgers University Press.

Kohout, Amy (2013) "From the Aviary: *Haliaeetus Leucocephalus.*" *The Appendix* 1, 2: 64–6.

Koopmans, Tjalling C. (1947) "Measurement Without Theory." *Review of Economics and Statistics* 29, 3: 161–72.

Koppes, Clayton. (1982) *JPL and the American Space Program*. New Haven: Yale University Press.

Kopplin, Z. (2014) "Texas Public Schools are Teaching Creationism." *Slate*. http://www. slate.com/articles/health_and_science/science/2014/01/creationism _in_texas_public_schools_undermining_the _charter_movement.html

Kottler, Malcolm Jay (1974) "From 48 to 46: Cytological Technique, Preconception, and the Counting of Human Chromosomes." *Bulletin of the History of Medicine* 48: 465–502.

Koyré, Alexandre (1943) "Galileo and Plato," *Journal of the History of Ideas* 4: 400–28.

Kragh, Helge (1996) *Cosmology and Controversy: The Historical Development of Two Theories of the Universe*. Princeton: Princeton University Press.

Kragh, Helge (1999) *Quantum Generations: A History of Physics in the Twentieth Century*. Princeton: Princeton University Press.

Kragh, Helge (2009) "Contemporary History of Cosmology and the Controversy over the Multiverse." *Annals of Science*, 66: 529–51.

Kragh, Helge. (2011) *Higher Speculations: Grand Theories and Failed Revolutions in Physics and Cosmology*. New York: Oxford University Press.

Kraut, Alan, (1994) *Silent Travelers: Germs, Genes and the "Immigrant Menace."* New York: Basic Books.

Krementsov, Nikolai (2005) *International Science Between the World Wars: The Case of Genetics*. New York: Routledge.

Krimsky, Sheldon (1991) *Biotechnics and Society: The Rise of Industrial Genetics*. Santa Barbara: Praeger.

Krinksky, Sheldon (2003) *Science in the Public Interest: Has the Lure of Profits Corrupted Biomedical Research?* Lanham: Rowman and Littlefield.

Krings, M. et al. (1997) "Neanderthal DNA Sequences and the Origin of Modern Humans." *Cell* 90: 19–30.

Kroeber, A.L. (1917) "The Superorganic." *American Anthropologist* 19, 2: 163–213.

Kroeber, A.L. (1959) "The History of the Personality of Anthropology." *American Anthropologist* 61, 3: 398–404.

Kroeber, Theodora (2004) *Ishi in Two Worlds: A Biography of the Last Wild Indian in North America*. Berkeley: University of California Press.

Kroll, Gary (2008) *America's Ocean Wilderness: A Cultural History of Twentieth-Century Exploration*. Lawrence: University Press of Kansas.

Krupp, E.C. (1994) *Echoes of the Ancient Skies: The Astronomy of Lost Civilizations*. New York: Oxford University Press.

Kuhn, Thomas S. (1962) *The Structure of Scientific Revolutions*. Chicago: University of Chicago Press.

Kuhn, Thomas S. (1996) *The Structure of Scientific Revolutions*. Chicago: University of Chicago Press.

Kuklick, Henrika (2006) "'Humanity in the Chrysalis Stage': Indigenous Australians in the Anthropological Imagination, 1899–1926." *British Journal for the History of Science* 39, 4: 535–68.

Kuklick, Henrika (ed.) (2008) *A New History of Anthropology*. Malden: Blackwell.

Kuklick, Henrika (2014) "History of Anthropology," in Roger Blackhouse and Philippe Fontaine (eds.), *A Historiography of the Modern Social Sciences*. Cambridge: Cambridge University Press.

Kuklick, Henrika and Robert E. Kohler (eds.) (1996) *Science in the Field, Osiris 11*. Chicago: University of Chicago Press.

Kuletz, Valerie (1998) *The Tainted Desert: Environmental and Social Ruin in the American West*. New York: Routledge.

Kumar, Deepak (1995) *Science and the Raj, 1857–1905*. Delhi: Oxford University Press.

Kupperman, Karen Ordahl (1982) "The Puzzle of the American Climate in the Early Colonial Period." *American Historical Review* 87: 1262–89.

Kutzbach, Gisela (1979) *The Thermal.Theory of Cyclones*. Washington: American Meteorological Society.

Laats, Adam (2010) *Fundamentalism and Education in the Scopes Era: God, Darwin, and the Roots of America's Culture Wars*. New York: Palgrave Macmillan.

Lacqueur, Thomas (1990) *Making Sex: Body and Gender from the Greeks to Freud*. Cambridge: Harvard University Press.

Ladd-Taylor, Molly (2011) "Eugenics and Social Welfare in New Deal Minnesota," in Paul A. Lombardo (ed.), *A Century of Eugenics: From the Indiana Experiment to the Humane Genome Era*. Bloomington: Indiana University Press, 117–41.

Lafollette, Marcel C. (1990) *Making Science Our Own: Public Images of Science 1910–1955*. Chicago: University of Chicago Press.

Lafollette, Marcel C. (2008) *Science on the Air: Popularizers and Personalities on Radio and Early Television*. Chicago: University of Chicago Press.

Lafollette, Marcel C. (2009) "Crafting a Communications Infrastructure: Scientific and Technical Publishing in the United States," in Carl F. Kaestle and Janice A. Radway (eds.), *A History of the Book in America Volume 4: Print in Motion: The Expansion of Publishing and Reading in the United States, 1880–1940*. Chapel Hill: University of North Carolina Press, 235–59.

Lafollette, Marcel C. (2013) *Science on Television: A History*. Chicago: University of Chicago Press.

Laidler, David (1999) *Fabricating the Keynesian Revolution: Studies of the Inter-War Literature on Money, the Cycle, and Unemployment*. Cambridge: Cambridge University Press.

Lamb, H.H. (1982) *Climate, History, and the Modern World*. New York: Methuen.

Landecker, Hannah (2001) "On Beginning and Ending with Apoptosis: Cell Death and Biomedicine," in Sarah Franklin and Margaret Lock (eds.), *Remaking Life and Death: Toward an Anthropology of the Biosciences*. Sante Fe: School of American Research Press, 23–59.

Landecker, Hannah (2007) *Culturing Life: How Cells Became Technologies*. Cambridge: Harvard University Press.

Lane, K. Maria D. (2010) *Geographies of Mars: Seeing and Knowing the Red Planet*. Chicago: University of Chicago Press.

Lanham, Url (1973) *The Bone Hunters: The Heroic Age of Paleontology in the American West*. New York: Columbia University Press.

Lankford, John (1984) "The Impact of Photography on Astronomy," in Owen Gingerich (ed.), *Astrophysics and Twentieth-Century Astronomy to 1950: Part A*. Cambridge: Cambridge University Press, 16–39.

Lankford, John and Ricky L. Slavings (1996) "The Industrialization of American Astronomy, 1880–1940." *Physics Today*, 49, 1: 34–40.

Lankford, John and Ricky L. Slavings (1997) *American Astronomy: Community, Careers, and Power, 1859–1940*. Chicago: Chicago University Press.

Lanouette, William and Silard, Bela (1994) *Genius in the Shadows: A Biography of Leo Szilard, the Man Behind the Bomb*. Chicago: University of Chicago Press.

Lantzer, Jason S. (2011) "The Indiana Way of Eugenics: Sterilization Laws, 1907–74," in A. Lombardo (ed.), *A Century of Eugenics: From the Indiana Experiment to the Humane Genome Era*. Bloomington: Indiana University Press, 26–44.

Laporte, Léo F. 1991. "George G. Simpson, Paleontology, and the Expansion of Biology," in Keith R. Benson, Jane Maienschein, and Ronald Rainger (eds.), *The Expansion of American Biology*, 80–106. New Brunswick: Rutgers University Press.

Laqueur, Thomas (1990) *Making Sex: Body and Gender from the Greeks to Freud*. Cambridge: Harvard University Press.

Larabee, D.F. (1986) "Curriculum, Credentials, and the Middle Class: A Case Study of a Nineteenth-Century High School." *Sociology of Education* 59, 1: 42–57.

Largent, Mark A. (2008) *Breeding Contempt: The History of Coerced Sterilization in the United States*. New Brunswick: Rutgers University Press.

Largent, Mark A. (2012) *Vaccine: The Debate in Modern America*. Baltimore: Johns Hopkins University Press.

Largent, Mark A. (2015) *Keep Out of Reach of Children: Reye's Syndrome, Aspirin, and the Politics of Public Health*. New York: Bellevue Literary Press.

Larson, Edward J. (1989) *Trial and Error: The American Controversy over Creation and Evolution*. New York: Oxford University Press.

Larson, Edward J. (1995) *Sex, Race, and Science: Eugenics in the Deep South*. Baltimore: Johns Hopkins University Press.

Larson, Edward J. (1997) *Summer for the Gods: The Scopes Trial and America's Continuing Debate over Science and Religion*. New York: Harper Collins.

Larson, Edward J. (2004) *Evolution: The Remarkable History of a Scientific Theory*. New York: Modern Library.

Larson, Edward J. and Larry Witham (1999) "Scientists and Religion in America." *Scientific American* 281: 88–93.

Laslett, Barbara, Sally Gregory Kohstedt, Helen Longino, and Evelyn Hammonds (eds.) (1996) *Gender and Scientific Authority*. Chicago: University of Chicago Press.

Latham, M.E. (2000) *Modernization as Ideology: American Social Science and "Nation Building" in Kennedy Era*. Chapel Hill: University of North Carolina Press.

Latham, M.E. (2011) *The Right Kind of Revolution: Modernization, Development, and U.S. Foreign Policy from the Cold War to the Present*. Ithaca: Cornell University Press.

Latimer, Joanna et al. (2006) "Rebirthing the Clinic: The Interaction of Clinical Judgment and Genetic Technology in the Production of Medical Science." *Science, Technology, & Human Values* 31: 599–630.

Latour, Bruno (1983) "Give Me a Lab and I Will Raise the World," in Karin Knorr-Cetina and Michael Mulkay (eds.), *Science Observed: Perspectives on the Social Studies of Science*. Beverly Hills: Sage.

Latour, Bruno (1988) *The Pasteurization of France*. Alan Sheridan and John Law (trans). Cambridge: Harvard University Press.

Latour, Bruno (1993) *We Have Never Been Modern*. Cambridge: Harvard University Press

Latour, Bruno (2004) *Politics of Nature: How to Bring the Sciences into Democracy*. Cambridge, MA: Harvard University Press.

Latour, Bruno and Steve Woolgar (1986) *Laboratory Life: The Construction of Scientific Facts*. Princeton: Princeton University Press.

Laubichler, Manfred D. and Jane Maienschein (eds.) (2007) *From Embryology to Evo-Devo: A History of Developmental Evolution*. Cambridge: MIT Press.

Laudan, Larry (1982) "Commentary: Science at the Bar – Causes for Concern." *Science, Technology, and Human Values* 7, 41:16–19.

Laughlin, Harry H. (1922) *Eugenical Sterilization in the United States*. Chicago: Psychopathic Laboratory of the Municipal Court of Chicago.

Laughlin, Harry H. (1939) *Immigration and Conquest*. New York: The Special Committee on Immigration and Naturalization of the Chamber of Commerce of the State of New York.

Lave, Rebecca; Philip Mirowski and Samuel Randalls (2010) "Introduction: STS and Neoliberal Science." *Social Studies of Science* 40: 659–75.

Lavender, Linda. (1997) "A History of Nature Study in Texas." Master's Thesis: Texas Woman's College.

Lavine, Matthew (2013) *The First Atomic Age: Scientists, Radiations, and the American Public, 1895–1945*. New York: Palgrave Macmillan.

Lawrence, Jane (2000) "The Indian Health Service and the Sterilization of Native American Women." *The American Indian Quarterly* 3, 24: 400–19.

Lawyer, Lee C., Charles Carpenter Bates, and Robert B. Rice (2001) *Geophysics in the Affairs of Mankind: A Personalized History of Exploration Geophysics*. Tulsa: Society of Exploration Geophysicists.

Layton, Robert (1997) *An Introduction to Theory in Anthropology*. Cambridge: Cambridge University Press.

Ladurie, Emmanuel Le Roy (1971) *Times of Feast, Times of Famine: A History of Climate Since the Year 1000*, Barbara Bray (trans.). Garden City: Doubleday.

Leach, William R. (1994) *Land of Desire: Merchants, Power, and the Rise of a New American Culture*. New York: Random House.

Lear, Linda J. (1997) *Rachel Carson: Witness for Nature*. New York: Henry Holt.

Lears, Jackson (1981) *No Place of Grace: Antimoderism and the Transformation of American Culture, 1880–1920*. New York: Pantheon.

Leavitt, Judith Walzer (1996) *Typhoid Mary: Captive to the Public's Health*. Boston: Beacon Press.

Lederman, Muriel and Ingrid Bartsch (eds.) (2001) *The Gender and Science Reader*. New York: Routledge.

Lee, Richard B. and Irven Devore (1968) *Man the Hunter*. Chicago: Aldine.

Lemaine, Gerard, Roy Macleod, Michael Mulkay, and Peter Weingart (eds.) (1976) *Perspectives on the Emergence of Scientific Disciplines*. The Hague: Mouton.

Lemann, Nicholas (2000) *The Big Test: The Secret History of the American Meritocracy*. New York: Macmillan.

Leonard, Robert (2012) *Von Neumann, Morgenstern, and the Creation of Game Theory: From Chess to Social Science, 1900–1960*. Cambridge: Cambridge University Press.

Leopold, Aldo (1933) *Game Management*. Madison: University of Wisconsin Press.

Leopold, Aldo (1949) *A Sand County Almanac*. New York: Oxford University Press.

Lerner, Gerda (1986) *The Creation of Patriarchy*. New York: Oxford University Press.

Leslie, Stuart W. (1993) *The Cold War and American Science: The Military–Industrial-Academic Complex at MIT and Stanford*. New York: Columbia University Press.

LeVay, Simon (1996) *Queer Science: The Use and Abuse of Research into Homosexuality*. Cambridge: MIT Press.

Levenstein, Harvey (1993) *Paradox of Plenty: A Social History of Eating in Modern America*. Berkley: University of California Press.

Levenstein, Harvey (2003) *Revolution at the Table: The Transformation of the American Diet*. Berkley: University of California Press.

Leventhal, Herbert (1976) *In the Shadow of the Enlightenment: Occultism and Renaissance Science in Eighteenth-Century America*. New York: New York University Press.

Levere, Trevor H. (1993) *Science and the Canadian Arctic: A Century of Exploration*. Cambridge: Cambridge University Press.

Levere, Trevor H. (2001) *Transforming Matter: A History of Chemistry from Alchemy to the Buckyball*. Baltimore: Johns Hopkins University Press.

Lévi-Strauss, Claude (1949) *Les Stuctures Élémentaires De Le Parenté*. Paris: Presses Universitaires De France.

Lévi-Strauss, Claude (1963) *Structural Anthropology*. New York: Basic Books.

Levin, David (1998) "Giants in the Earth: Science and the Occult in Cotton Mather's Letters to the Royal Society." *The William and Mary Quarterly* 45: 751–70.

Levine, George (1987) "One Culture: Science and Literature," in George Levine (ed.), *One Culture: Essays in Science and Literature*. Madison: University of Wisconsin Press, 3–32.

Levine, George (1988) *Darwin and the Novelists: Patterns of Science in Victorian Fiction*. Cambridge: Harvard University Press.

Levy, Steven (2010) *Hackers: Heroes of the Computer Revolution*. Sebastopol: O'Reiley Media, Inc.

Lewenstein, Bruce V. (1987) "Was There Really a Popular Science 'Boom'?" *Science, Technology, and Human Values* 12: 29–41.

Lewenstein, Bruce V. (1988) "Why isn't Popular Science More Popular?" *American Scientist* 76: 447–9.

Lewenstein, Bruce V. (1991) *When Science Meets the Public: Proceedings of a Workshop Organized by the American Association for the Advancement of Science Committee on Public Understanding of Science and Technology*. Washington: American Association for the Advancement of Science.

Lewenstein, Bruce V (2001) "What Kind of 'Public Understanding of Science' Programs Best Serve a Democracy?" in Sabine Maasen and Matthias Winterhager (eds.), *Science Studies: Probing the Dynamics of Scientific Knowledge*. Bielefeld: Transcript, 237–55.

Lewenstein, Bruce V. (2007) *Victorian Popularizers of Science: Designing Nature for New Audiences*. Chicago: University of Chicago Press.

Lewenstein, Bruce V. (2009) "Science Books Since 1945," in David Paul Nord, Joan Shelley Rubin, and Michael Schudson (eds.), *A History of the Book, Volume 5: The Enduring Book: Print Culture in Postwar America*. Chapel Hill: University of North Carolina Press, 347–60.

Lewenstein, Bruce V. (2011) "Science and its Public," in Peter Harrison, Ronald L. Numbers, and Michael H. Shank (eds.), *Wrestling with Nature: From Omens to Science*. Chicago: University of Chicago Press, 337–75.

Lewenstein, Bruce V. and Steven Allison-Bunnell (2000) "Creating Knowledge in Science Museums: Serving Both Public and Scientific Communities," in Bernard Schiele and Emlyn H. Koster (eds.), *Science Centers for This Century*. St. Foy: Editions Multimondes.

Lewin, Roger (1987) *Bones of Contention: Controversies in the Search for Human Origins*. New York: Simon & Schuster.

Lewis, Caroline Herbst (2013) *Prescription for Heterosexuality: Sexual Citizenship in the Cold War Era*. Chapel Hill: University of North Carolina Press.

Lewis, Daniel (2012) *The Feathery Tribe: Robert Ridgway and the Modern Study of Birds*. New Haven: Yale University Press.

Lewis, Herbert (2008) "Franz Boas: Boon or Bane?" *Reviews in Anthropology* 37, 2–3): 169–200.

Lewis, Michael L. (2004) *Inventing Global Ecology: Tracking the Biodiversity Ideal in India, 1947–1997*. Athens: Ohio University Press.

Lewis, Sinclair (1925) *Arrowsmith*. New York: Harcourt, Brace.

Lewontin, Richard C. (1978) "Adaptation." *Scientific American* 239(3).

Lewontin, Richard C. (1981) *Dobzhansky's Genetics of Natural Populations I–XLIII*. New York: Columbia University Press.

Lewontin, Richard C. and John L. Hubby (1966) "A Molecular Approach to the Study of Genic Heterozygosity in *Natural Populations. II. Amount of Variation and Degree of Heterozygosity in Natural Populations of Drosophila Pseudoobscura*." *Genetics* 54: 595–609.

Liebersohn, Harry (2008) "Anthropology before Anthropology," in Henrika Kuklick (ed.), *A New History of Anthropology*. Malden: Blackwell, 17–32.

Life Magazine (1943) "Advertisement: Brer Rabbit Molasses." January 11: 86.

Light, Jennifer (1999) "When Computers were Women." *Technology and Culture* 40, 3: 455–83.

Lightman, Alan and Roberta Brawer (1992) *Origins: The Lives and World of Modern Cosmologists*. Cambridge: Harvard University Press.

Lightman, Bernard (ed.) (1997) *Victorian Science in Context*. Chicago: University of Chicago Press.

Lightman, Bernard (2007) *Victorian Popularizers of Science: Designing Nature for New Audiences*. Chicago: University of Chicago Press.

Lightman, Bernard (2011) "Science and the Public," in Peter Harrison, Ronald L. Numbers, and Michael H. Shank (eds.), *Wrestling with Nature: From Omens to Science*. Chicago: University of Chicago Press, 337–75.

Lightman, Bernard (2014) "The Popularization of Evolution and Victorian Culture," in Bernard Lightman and Bennett Zon (eds.), *Evolution and Victorian Culture*. Cambridge: Cambridge University Press, 286–311.

Limon, John (1990) *The Place of Fiction in the Time of Science: A Disciplinary History of American Writing*. Cambridge: Cambridge University Press.

Lindee, M. Susan (1994) *Suffering Made Real: American Science and the Survivors at Hiroshima*. Chicago: University of Chicago Press.

Lindee, Susan (2005) *Moments of Truth in Genetic Medicine*. Baltimore: Johns Hopkins University Press.

Lindeman, Raymond (1942) "The Trophic Dynamic Aspect of Ecology." *Ecology* 23: 399–418.

Linné, Carl von (1735) *Systema Naturae Per Regna Tria Naturæ, Secundum Classes, Ordines, Genera, Species, Cum Characteribus, Differentiis, Synonymis, Locis*. Stockholm: Laurentius Salvius.

Lindqvist, Svante (ed.) (2000) *Museums of Modern Science: Nobel Symposium 112*. Canton: Science History Publications/USA.

Little, Michael A. and Kenneth A.R. Kennedy (2009) *Histories of American Physical Anthropology in the Twentieth Century*. London: Lexington Books.

Little, Michael and Kenneth A.R. Kennedy (eds.) (2010) *Histories of American Physical Anthropology in the Twentieth Century*. Plymouth: Rowman & Littlefield.

Livingstone, David (1992) *The Preadamite Theory and the Marriage of Science and Religion*. Philadelphia: American Philosophical Society.

Livingstone, David (1993) *The Geographical Tradition: Episodes in the History of a Contested Enterprise*. Cambridge, MA: Blackwell.

Livingstone, David (1999) "Tropical Climate and Moral Hygiene: The Anatomy of a Victorian Debate." *British Journal for the History of Science* 32: 93–110.

Livingstone, David (2002) "Race, Space and Moral Climatology: Notes Toward A Genealogy." *Journal of Historical Geography* 28: 159–80.

Livingstone, David. (2008) *Adam's Ancestors: Race, Religion, and the Politics of Human Origins*. Baltimore: Johns Hopkins University Press.

Loe, Meike (2006) *The Rise of Viagra: How the Little Blue Pill Changed Sex in America*. New York: New York University Press.

Logsdon, John M. (2013) "The Survival Crisis of the U.S. Solar System Exploration Program in the 1980s," in Roger Launius (ed.), *Exploring the Solar System: The History and Science of Planetary Exploration*. New York: Palgrave Macmillan.

Lohse, K. and Frantz L.A.F. (2014) "Neandertal Admixture in Eurasia Confirmed by Maximum Likelihood Analysis of Three Genomes." *Genetics* 114: 1241–51.

Lombardo, Paul A. (2010) *Three Generations, No Imbeciles: Eugenics, the Supreme Court, and Buck v. Bell*. Baltimore: Johns Hopkins University Press.

Lombardo, Paul A. (ed.) (2011a) *A Century of Eugenics: From the Indiana Experiment to the Humane Genome Era*. Bloomington: Indiana University Press.

Lombardo, Paul A. (2011b) "Introduction: Looking Back at Eugenics" in *A Century of Eugenics: From the Indiana Experiment to the Humane Genome Era*. Bloomington: Indiana University Press, 1–11.

London, Jack (1903) *The Call of the Wild*. New York: Macmillan.

Longair, Malcolm S. (2006) *The Cosmic Century: A History of Astronomy and Cosmology*. Cambridge: Cambridge University Press.

Longino, Helen (1990) *Science as Social Knowledge: Values and Objectivity in Scientific Inquiry*. Princeton: Princeton University Press.

Lorenz, K. (1966) *On Aggression*. London: Methuen.

Lounsbury, Richard C. (ed.) (1995) *Political and Social Essays*. Charlottesville: University Press of Virginia.

Louv, Richard (2005) *Last Child in the Woods: Saving Our Children from Nature-Deficit Disorder*. Chapel Hill: Algonquin.

Lowenthal, David (2000) *George Perkins Marsh, Prophet of Conservation.* Seattle: University of Washington Press.

Löwy, Ilana (2014) "How Genetics Came to the Unborn: 1960–2000." *Studies in History and Philosophy of Science Part C: Studies in History and Philosophy of Biological and Biomedical Sciences* 47: 290–9.

Lucier, Paul (2008) *Scientists and Swindlers: Consulting on Coal and Oil in America, 1820–1890.* Baltimore: Johns Hopkins University Press.

Lucier, Paul (2009) "The Professional and the Scientist in Nineteenth-Century America." *Isis* 100, 4: 699–732.

Ludmerer, Kenneth M. (1972) *Genetics and American Society: A Historical Appraisal.* Baltimore: Johns Hopkins University Press.

Lunbeck, Elizabeth (1995) *The Psychiatric Persuasion: Knowledge, Gender, and Power in Modern America.* Princeton: Princeton University Press.

Lundberg, D. and H.F. May (1976) "The Enlightened Reader in America." *American Quarterly* 28, 2: 262–93.

Lupfer, Eric (2002) "Reading Nature Writing: Houghton Mifflin Company, The Ohio Teachers' Reading Circle, and In American Fields and Forests (1909)." *Harvard Library Bulletin* 13: 37–58.

Lurie, Edward (1988) *Louis Agassiz: A Life in Science.* Baltimore: Johns Hopkins University Press.

Lutkehaus, Nancy (2008) *Margaret Mead: The Making of an American Icon.* Princeton: Princeton University Press.

Lutts, Ralph H. (1990) *Nature Fakers: Wildlife, Science, and Sentiment.* Golden: Fulcrum.

Lykknes, Annette, Donald L. Opitz, and Brigitte Van Tiggelen (eds.) (2012) *For Better or for Worse? Collaborative Couples in the Sciences.* Basel: Birkhäuser.

Lynas, M. and C. Tudge (2014) "Gmos: A Solution or a Problem?" *Journal of International Affairs* 67, 2: 131–8.

Lynch, Michael (1985) *Art and Artifact in Laboratory Science: A Study of Shop Work and Shop Talk in a Research Laboratory.* London: Routledge & Kegan Paul.

Lynch, Peter (2006) *The Emergence of Numerical Weather Prediction: Richardson's Dream.* Cambridge: Cambridge University Press.

Lyons, Andrew P. and Harriet Lyons (2004) *Irregular Connections: A History of Anthropology and Sexuality.* Lincoln: University of Nebraska Press.

Lyson, T. (2004) *Civic Agriculture: Reconnecting Farm, Food, and Community.* Medford: Tufts University Press.

MacArthur, Robert H. and Edward O. Wilson (1963) "An Equilibrium Theory of Insular Zoogeography." *Evolution* 17: 373–87.

Macleod, Roy (2000) "Nature and Empire: Science and the Colonial Enterprise." *Osiris* 15: 1–13.

Macdonald, Sharon (ed.) (1998a) *A Companion to Museum Studies.* Chichester: Wiley.

Macdonald, Sharon (ed.) (1998b) *The Politics of Display: Museums, Science, Culture.* New York: Routledge.

Mack, Pamela Etter (1990) *Viewing the Earth: The Social Construction of the Landsat Satellite System.* Cambridge: MIT Press.

MacKenzie, David (1991) "Agroethics and Agricultural Research," in P.B. Thompson and B.A. Stout (eds.), *Beyond the Large Farm: Ethics and Research Goals for Agriculture.* Boulder: West View Press, 3–33.

MacKenzie, Donald (1995) "The Automation of Proof: A Historical and Sociological Exploration." *IEEE Annals of the History of Computing* 17, 3: 7–29.

MacKenzie, Donald (2001) *Mechanizing Proof: Computing, Risk, and Trust.* Cambridge: MIT Press.

MacKenzie, Donald and Garrel Pottinger (1997) "Mathematics, Techology, and Trust: Formal Verification, Computer Security, and the U.S. Military." *IEEE Annals of the History of Computing* 19, 3: 41–59.

Macleod, Roy (ed.) (2000) "Nature and Empire: Science and the Colonial Enterprise." *Osiris* 15.

Madden, E.H. (1962) "Francis Wayland and the Limits of Moral Responsibility." *Proceedings of the American Philosophical Society* 106, 4: 348–59.

Maddox, Brenda (2002) *Rosalind Franklin: The Dark Lady of DNA*. New York: Harper-Collins.

Maher, Neil (2008) *Nature's New Deal: The Civilian Conservation Corps and the Roots of the American Environmental Movement*. New York: Oxford University Press.

Mahoney, Michael (1990) "The Roots of Software Engineering." *CWI Quarterly* 3, 4: 325–34.

Mahoney, Michael (2005) "The Histories of Computing(s)." *Interdisciplinary Science Reviews* 30: 119–35.

Mahoney, Michael (2011) *Histories of Computing*. Cambridge: Harvard University Press.

Maienschein, Jane (1991a) *Transforming Traditions in American Biology, 1880–1915*. Baltimore: Johns Hopkins University Press.

Maienschein, Jane (1991b) "Cytology in 1924: Expansion and Collaboration," in Keith R. Benson, Jane Maienschein, and Ronald Rainger (eds.), *The Expansion of American Biology*. New Brunswick: Rutgers University Press, 23–51.

Maienschein, Jane (2003) *Whose View of Life? Embryos, Cloning, and Stem Cells*. Cambridge: Harvard University Press.

Maienschein, Jane, Ronald Rainger, and Keith R. Benson (1981) "Were American Morphologists in Revolt? Introduction of the Special Section on American Morphology at the Turn of the Century." *Journal of the History of Biology* 14, 1: 83–7.

Malamud, Randy (1998) *Reading Zoos: Representations of Animals and Captivity*. New York: New York University Press.

Mallis, Arnold B. (1971) *American Entomologists*. New Brunswick: Rutgers University Press.

Malphrus, Benjamin K. (1996) *The History of Radio Astronomy and the National Radio Astronomy Observatory: Evolution Toward Big Science*. Malabar: Krieger.

Manganaro, Christine L. (2012) "Assimilating Hawai'i: Racial Science in a Colonial "Laboratory," 1919–1939." PhD Dissertation: University of Minnesota.

Mann, Walter, M. (1930) *Wild Animals In and Out of the Zoo, Volume 6*. New York: Smithsonian Institution Series.

Manning Kenneth R. (1983) *Black Apollo of Science: The Life of Earnest Everett Just*. Oxford: Oxford University Press,.

Manning, Melanie and Louanne Hudgins (2007) "Use of Array-Based Technology in the Practice of Medical Genetics." *Genetics in Medicine* 9: 650–3.

Manning, Thomas G. (1967) *Government in Science: The U.S. Geological Survey, 1867–1894*. Lexington: University of Kentucky Press.

Manson, Neil A. (ed.) (2003) *God and Design: The Teleological Argument and Modern Science*. New York: Routledge.

Marché, Jordan D., II (2005) *Theaters of Time and Space: American Planetaria, 1930–1970*. New Brunswick: Rutgers University Press.

Marchese, Francis T. (2013) "Periodicity, Visualization, Design." *Foundations of Chemistry* 15: 31–55.

Marcus, Alan I. (1985) *Agricultural Science and the Quest for Legitimacy: Farmers, Agricultural Colleges, and Experiment Stations, 1870–1890*. Ames: Iowa State University Press.

Mark, Joan (1988) *A Stranger in Her Native Land: Alice Fletcher and the American Indians*. Lincoln: University of Nebraska Press.

Markell, Howard (2004) *When Germs Travel: Six Major Epidemics That Have Invaded America Since 1900 and the Fears They Have Unleashed*. New York: Pantheon.

Markowitz, G. and D. Rosner (2002) *Deceit and Denial: The Deadly Politics of Industrial Pollution*. Berkeley: University of California Press.

Marks, Harry (1997) *The Progress of Experiment: Science and Therapeutic Reform in the United States, 1900–1990*. New York: Cambridge University Press.

Marks, Jonathan (1994) "Blood Will Tell (Won't It?): A Century of Molecular Discourse in Anthropological Systematics." *American Journal of Physical Anthropology* 94: 59–79.

Marks, Jonathan (1995) *Human Biodiversity: Genes, Race, and History*. New York: Walter De Gruyter.

Marks, Jonathan (1996) "The Legacy of Serological Studies in American Physical

Anthropology." *History and Philosophy of the Life Sciences*: 345–62.

Marks, Jonathan (2012) "The Origins of Anthropological Genetics." *Current Anthropology* 53: S161–S172.

Marks, Lara (2010) *Sexual Chemistry: A History of the Contraceptive Pill*. New Haven: Yale University Press.

Marsden, George M. (2006) *Fundamentalism and American Culture*. New York: Oxford University Press.

Marsh, George Perkins (1965) *Man and Nature: Or, Physical Geography as Modified by Human Action*. Boston: Harvard University Press.

Marshall, David (ed.) (2012) *Science and Religion: Christian and Muslim Perspectives*. Washington: Georgetown University Press.

Marshall, Gloria A. (1993) "Racial Classifications," in Sandra Harding (ed.), *The 'Racial' Economy of Science*. Bloomington: Indiana University Press, 116–27.

Martin, Aryn (2004) "Can't Any Body Count? Counting as an Epistemic Theme in the History of Human Chromosomes." *Social Studies of Science* 34: 923–48.

Martin, Dianne C. (1995/1996) "ENIAC: Press Conference That Shook the World." *IEEE Technology and Society Magazine*: 3–10.

Martin, Emily (1987) *The Woman in the Body: A Cultural Analysis of Reproduction*. Boston: Beacon Press.

Martin, Joseph D. (2015) "What's in a Name Change? Solid State Physics, Condensed Matter Physics, and Materials Science." *Physics in Perspective* 17: 3–32.

Marx, Leo (1964) *The Machine in the Garden: Technology and the Pastoral Ideal in America*. New York: Oxford University Press.

Mauchly, John (1942) "The Use of High Speed Vacuum Tube Devices for Calculating," in Brian Randall (ed.), *The Origins of Digital Computers*. New York: Springer-Verlag, 355–8.

May, Elaine Tyler (2010) *America and the Pill: A History of Promise, Peril, and Liberation*. New York: Basic Books.

May, Henry F. (1949) *Protestant Churches and Industrial America*. New York: Harper.

Mayall, R. Newton and Margaret W. Mayall (1938) *Sundials: How to Know, Use, and Make Them*. Boston: Hale, Cushman & Flint.

Mayberry, Maralee, Banu Subramaniam, and Lisa Weasel (eds.) (2001) *Feminist Science Studies: A New Generation*. New York: Routledge.

Maynard Smith, J. (1964) "Group Selection and Kin Selection." *Nature* 201, 4924: 1145–7.

Maynard Smith, J. (1982) *Evolution and the Theory of Games*. Cambridge: Cambridge University Press.

Mayr, Ernst (1942) *Systematics and the Origin of Species from the Viewpoint of a Zoologist*. New York: Columbia University Press.

Mayr, Ernst (1950) "Taxonomic Categories in Fossil Hominids." *Cold Spring Harbor Symposia on Quantitative Biology* 15: 109–18.

Mayr, Ernst (1995) "Darwin's Impact on Modern Thought." *Proceedings of the American Philosophical Society* 1394: 317–25.

Mayr, Ernst (1999) *Systematics and The Origin of Species from the Point of View of a Zoologist*. Cambridge: Harvard University Press.

Mayr, Ernst and William B. Provine (eds.) (1980) *The Evolutionary Synthesis: Perspectives on the Unification of Biology*. Cambridge: Harvard University Press.

Mazurs, Edward G. (1974) *Graphic Representations of the Periodic System during One Hundred Years*. University: University of Alabama Press.

McCord, Louisa S. (1851) "Diversity of the Races: Its Bearing Upon Negro Slavery." *The Southern Quarterly Review* 3: 392–419.

McCorduck, Pamela (1975) John Clifford Shaw (int.). Carnegie Mellon Archives, Pamela McCorduck Collection, Series III, Transcripts.

McCorduck, Pamela (2004) *Machines Who Think: A Personal Inquiry into the History and Prospects of Artificial Intelligence* New York: A.K. Peters.

McCray, W. Patrick (2004) *Giant Telescopes: Astronomical Ambition and the Promise of Technology*. Cambridge: Harvard University Press.

McCray, W. Patrick (2008) *Keep Watching The Skies! The Story of Operation Moonwatch and the Dawn of the Space Age*. Princeton: Princeton University Press.

McCright, Aaron and Riley Dunlap (2010) "Anti-Reflexivity: The American Conservative Movement's Success in Undermining Climate Science and Policy." *Theory, Culture, and Society* 27: 100–33.

McCurdy, Howard E. (2001) *Faster Better Cheaper: Low-Cost Innovation in the U.S. Space Program*. Baltimore: Johns Hopkins University Press.

McEvoy, Arthur F. (1990) *The Fisherman's Problem: Ecology and Law in California's Fisheries, 1850–1980*. New York: Cambridge University Press.

McGarity, T. and W. Wagner (2008) *Bending Science: How Special Interests Corrupt Public Health Research*. Cambridge: Harvard University Press.

McGerr, Michael E. (2003) *A Fierce Discontent: The Rise and Fall of the Progressive Movement in America, 1870–1920*. New York: Free Press.

McGillivray, G., J.A. Rosenfeld, R.J. Mckinlay Gardner, and L.H. Gillam (2012) "Genetic Counseling and Ethical Issues with Chromosome Microarray Analysis in Prenatal Testing." *Prenatal Diagnosis* 32: 389–95.

McIntosh, Elaine (1995) *American Food Habits in Historical Perspective*. New York: Praeger.

McIntosh, Robert P. (1985) *The Background of Ecology: Concept and Theory*. New York: Cambridge University Press.

McKeown, Thomas (1976) *The Modern Rise of Population*. New York: Academic Press.

McKibben, Bill (1990) *The End of Nature*. New York: Anchor Books.

McKusick, V.A. (1970) "Birth Defects: Prospects for Progress," in F. Clarke Fraser and V.A. McKusick (eds.), *Congenital Malformations: Proceedings of the Third International Conference*. Amsterdam: Excerpta Medica.

McKusick, V.A. (1975) *Mendelian Inheritance in Man: Catalogues of Autosomal Dominant, Autosomal Recessive, and X-Linked Phenotypes*. Baltimore: Johns Hopkins University Press.

McKusick, V.A. (1982) "The Human Genome through the Eyes of a Clinical Geneticist." *Cytogenetic and Genome Research* 32: 7–23.

McKusick, V.A. (1984) "Diseases of the Genome." *Journal of the American Medical Association* 252: 1041–8.

McKusick, V.A. (1986) "The Morbid Anatomy of the Human Genome: A Review of Gene Mapping in Clinical Medicine (First of Four Parts)." *Medicine* 65: 1–33.

McKusick, V.A. (2001) "The Anatomy of the Human Genome." *Journal of the American Medical Association* 286: 2289–95.

McKusick, V.A. (2006) "A 60-Year Tale of Spots, Maps, and Genes." *Annual Reviews of Genomics and Human Genetics* 7: 1–27.

McLester J.S. (1939) Borderline States of Nutritive Failure. *Journal of the American Medical Association* 112: 2110–14.

McNeil, Maureen (2005) "Introduction: Postcolonial Technoscience." *Science as Culture* 14, 2: 105–12.

Mead, Margaret (1928) *Coming of Age in Samoa: A Psychological Study of Primitive Youth for Western Civilization*. New York: William Morrow.

Mead, Margaret (1949) *Male and Female: A Study of the Sexes in a Changing World*. New York: William Morrow.

Medema, Steven G. (2009) *The Hesitant Hand: Taming Self-Interest in the History of Economic Ideas*. Princeton: Princeton University Press.

Meine, Curt (1988) *Aldo Leopold: His Life and Work*. Madison: University of Wisconsin.

Meine, Curt, Michael Soulé, and Reed Noss (2006) "'A Mission-Driven Discipline': The Growth of Conservation Biology." *Conservation Biology* 20, 3: 631–51.

Melia, Fulvia (2009) *Cracking the Einstein Code: Relativity and the Birth of Black Hole Physics*. Chicago: University of Chicago Press.

Melosi, Martin V. (2000) *The Sanitary City: Urban Infrastructure in America from Colonial Times to the Present*. Baltimore: Johns Hopkins University Press.

Melosi, Martin V. (2005) *Garbage in the Cities: Refuse, Reform, and the Environment*. Pittsburgh: University of Pittsburgh Press.

Menand, Louis (2002) *The Metaphysical Club: A Story of Ideas in America*. New York: Macmillan.

Mencken, H.L. (1925) "Mencken Finds Daytonians Full of Sickening Doubts About Value of Publicity." Baltimore Evening Sun, July 9.

Mendelsohn, J. Andrew (2003) "Lives of the Cell." *Journal of the History of Biology* 36: 1–37.

Mendenhall, Chase D., Daniel S. Karp, Christoph F. J. Meyer, Elizabeth A. Hadly, and Gretchen C. Daily (2014) "Predicting Biodiversity Change and Averting Collapse in Agricultural Landscapes." *Nature* 509, 7499: 213–17.

Merchant, Carolyn (1990) *The Death of Nature: Women, Ecology, and the Scientific Revolution*. San Francisco: Harper & Row.

Merchant, Carolyn (2006) "The Scientific Revolution and the Death of Nature." *Isis* 97: 513–33.

Mergen, Bernard (2008) *Weather Matters: An American Cultural History Since 1900*. Lawrence: University Press of Kansas.

Merrill, George P. (1924) *The First One Hundred Years of American Geology*. New Haven: Yale University Press.

Merson, John, (2001) "Intellectual Property Rights and Biodiversity in a Colonial and Postcolonial Context." *Osiris* 15.

Merton, Robert K. (1938) Science, Technology and Society in Seventeenth Century England. Osiris 4: 360–632.

Metropolis, N. and E.C. Nelson (1982) "Early Computing at Los Alamos." *IEEE Annals of the History of Computing* 4, 4: 348–57.

Meyer, William B. (2000) *Americans and Their Weather*. New York: Oxford University Press.

Meyerowitz, Joanne (2002) *How Sex Changed: A History of Transsexuality in the United States*. Cambridge: Harvard University Press.

Michaels, David (2008) *Doubt is their Product*. New York: Oxford University Press.

Midwinter, Charles and Michel Janssen (2013) "Kuhn Losses Regained: Van Vleck from

Spectra to Susceptibilities," in Massimiliano Badino and Jaume Navarro (eds.), *Research and Pedagogy: A History of Early Quantum Physics through its Textbooks*. Berlin: Edition Open Access, 137–205.

Mikels-Carrasco, Jessica (2012) "Sherwood Washburn's New Physical Anthropology: Rejecting the 'Religion of Taxonomy'." *History and Philosophy of the Life Sciences* 34, 1–2: 79–101.

Milam, Erika Lorraine and Robert A. Nye (eds.) (2015) "Scientific Masculinities." *Osiris* 15.

Miller, Clark A. (2001) "Scientific Internationalism in American Foreign Policy: The Case of the International Geophysical Year," in Clark A. Miller and Paul Edwards (eds.), *Changing the Atmosphere: Expert Knowledge and Environmental Governance*. Cambridge: MIT Press, 167–217.

Miller, Clark A. and Paul Edwards (eds.), *Changing the Atmosphere: Expert Knowledge and Environmental Governance*. Cambridge: MIT Press.

Miller, David H. (1988) "In Memoriam: John B. Leighly, 1895–1986." *Annals of the Association of American Geographers* 78.

Miller, Howard S. (1970) *Dollars for Research: Science and its Patrons in Nineteenth-Century America*. Seattle: University of Washington Press.

Miller, Kenneth R. (2008) *Only a Theory: Evolution and the Battle for America's Soul*. New York: Viking Penguin.

Mills, Eric (1989) *Biological Oceanography: An Early History, 1870–1960*. Toronto: University of Toronto Press.

Mills, John A. (2000) *Control: A History of Behavioral Psychology*. New York: New York University Press.

Milunsky, Aubrey (1993) "Commercialization of Clinical Genetic Laboratory Services: In Whose Best Interest?" *Obstetrics & Gynecology* 81: 627–9.

Mindell, David (2004) *Between Human and Machine: Feedback, Control, and Computing before Cybernetics*. Baltimore: Johns Hopkins University Press.

Minker, Jack (2007) "Developing a Computer Science Department at the University of

Maryland." *IEEE Annals of the History of Computing* 29, 4: 64–75.

Minton, Henry (2001) *Departing from Deviance*. Chicago: University of Chicago Press.

Mintz, Sidney W. (1985) "*Sweetness and Power: The Place of Sugar in Modern History*." New York: Penguin.

Mirowski, Philip (2002) *Machine Dreams: How Economics became a Cyborg Science*. Cambridge: Cambridge University Press.

Mirowski, P. (2011) *Science-Mart: Privatizing American Science*. Cambridge: Harvard University Press.

Mirowski, Philip, and Dieter Plehwe (eds) (2009) *The Road from Mont Pelerin: The Making of the Neoliberal Thought Collective*. Cambridge: Harvard University Press.

Mishkin, Andrew (2003) *Sojourner: An Insider's View of the Mars Pathfinder Mission*. New York: Berkeley.

Mitchell, John (1744) "An Essay Upon the Causes of the Different Colours of People in Different Climates." *Philosophical Transactions* 43: 102–50.

Mitman, Gregg (1992) *The State of Nature: Ecology, Community, and American Social Thought, 1900–1950*. Chicago: University of Chicago Press.

Mitman, Gregg (1993) "Cinematic Nature: Hollywood Technology, Popular Culture, and the American Museum of Natural History." *Isis* 84: 637–61.

Mitman, Gregg (1996) "When Nature is the Zoo: Vision and Power in the Art and Science of Natural History." *Osiris* 11: 117–43.

Mitman, Gregg (1999) *Reel Nature: America's Romance with Wildlife on Film*. Cambridge: Harvard University Press.

Mizelle, B. (2011) *Pig*. London: Reaktion Books.

Mody, Cyrus C.M. (2011) *Instrumental Community: Probe Microscopy and the Path to Nanotechnology*. Cambridge: MIT Press.

Money, John (1955) "Hermaphroditism, Gender and Precocity in Hyperadrenocorticism: Psychologic Findings." *Bulletin of the Johns Hopkins Hospital* 96: 253–64.

Monmonier, Mark S. (1999) *Air Apparent: How Meteorologists Learned to Map, Predict, and Dramatize Weather*. Chicago: University of Chicago Press.

Monosson, Emily (ed.) (2008) *Motherhood, the Elephant in the Laboratory: Women Scientists Speak Out*. Ithaca: Cornell University Press.

Montagu, Ashley (1942) *Man's Most Dangerous Myth. The Fallacy of Race*. New York: Columbia University Press.

Montgomery, Georgina (2005) "Place, Practice and Primatology: Clarence Ray Carpenter, Primate Communication and the Development of Field Methodology, 1931–1945." *Journal of the History of Biology* 38: 495–533.

Mooney, Chris (2005) *The Republican War on Science*. New York: Basic Books.

Moore, Randy (2001) "The Lingering Impact of the Scopes Trial on High School Biology Textbooks." *Bioscience* 51: 790–6.

Morange, Michel (1993) "The Discovery of Cellular Oncogenes." *History and Philosophy of the Life Sciences* 15, 1: 45–58.

Morange, Michel (1998) *A History of Molecular Biology*. Matthew Cobb (trans). Cambridge: Harvard University Press.

Morange, Michel (2001) *The Misunderstood Gene*. Matthew Cobb (trans). Cambridge: Harvard University Press.

Morange, Michel (2008) "The Death of Molecular Biology?" *History and Philosophy of the Life Sciences* 30, 1: 31–42.

Morantz, Regina Markell (1977) "The Scientist as Sex Crusader: Alfred Kinsey and American Culture." *American Quarterly* 29: 563–89.

Morantz-Sanchez, Regina Markell (1985) *Sympathy and Science: Women Physicians in American Medicine*. New York: Oxford University Press.

Morbid Anatomy Museum (2014) "About Us." http://morbidanatomymuseum. org/visitor-info/about-us/

More, Ellen Singer (1999) *Women Physicians and the Profession of Medicine, 1850–1995*. Cambridge: Harvard University Press.

Morgan, A.F. and E.O. Madsen (1933) "A Comparison of Apricots and their Carotene Equivalent as Sources of Vitamin A." *Journal of Nutrition* 6: 83–93.

Morgan, Gregory (1998) "Emile Zuckerkandl, Linus Pauling, and the Molecular Evolutionary Clock, 1959–1965." *Journal of the History of Biology* 3: 155–78.

Morgan, Lewis (1851) *League of the Ho-De-No-Sau-Nee Or Iroquois*. Rochester: Sage and Brother.

Morgan, Mary S. (1991) *The History of Econometric Ideas*. Cambridge: Cambridge University Press.

Morgan, Mary S. (2012) *The World in the Model: How Economists Work and Think*. Cambridge: Cambridge University Press.

Morgan, Mary S. and Malcolm Rutherford (eds.) (1998) *From Interwar Pluralism to Postwar Neoclassicism*. Durham: Duke University Press.

Morgan, Mary S. and Malcom Rutherford (1998a) "American Economics: The Character of the Transformation." *History of Political Economy* 30 (Supplement): 1–26.

Morgan, Ruth A. (2011) "Diagnosing the Dry: Historical Case Notes from Southwest Western Australia, 1945–2007." *Osiris* 26: 89–108.

Morris, Peter J.T. and Klaus Staubermann (eds.) (2010) *Illuminating Instruments in Artefacts: Studies in the History of Science and Technology*. Washington: Smithsonian Institution.

Morrison-Low, A.D. (2007) *Making Scientific Instruments in the Industrial Revolution*. Aldershot: Ashgate.

Morrison-Low, A. D. et al. (2012). *From Earth-Bound to Satellite: Telescopes, Skills, and Networks. Vol. 2 of Scientific Instruments and Collections*. Leiden: Brill.

Morton, Samuel George (1839) *Crania Americana; Or, a Comparative View of the Skulls of Various Aboriginal Nations of North and South America*. Philadelphia: J. Dobson.

Morton, Samuel George (1847) "Hybridity in Animals, Considered in Reference to the Question of the Unity of the Human Species." *American Journal of Science and Arts* 3: 39–50, 203–12.

Morus, Iwan Rhys (2005) *When Physics Became King*. Chicago: University of Chicago Press.

Motulsky, A.G. (1987) "Presidential Address. Human and Medical Genetics: Part, Present, and Future," in *Human Genetics: Proceedings of the 7th International Conference, Berlin 1986*. Berlin: Springer-Verlag.

Mudry, J. (2009) *Measured Meals: Nutrition in America*. Albany: SUNY Press.

Muka, Samantha (2014a) *Working at Water's Edge: Life Sciences at American Marine Stations, 1880–1930*. PhD Dissertation: University of Pennsylvania.

Muka, Samantha (2014b) "Portrait of an Outsider: Class, Gender, and the Scientific Career of Ida M. Mellen." *Journal of the History of Biology* 47, 1: 29–61.

Mukerji, Chandra (1989) *A Fragile Power: Scientists and the State*. Princeton: Princeton University Press.

Mullen, Bob and Garry Marvin (1987) *Zoo Culture*. London: George Weidenfeld.

Müller-Wille, Staffan and Hans-Jörg Rheinberger (eds.) (2007) *Heredity Produced: At the Crossroads of Biology, Politics, and Culture, 1500–1870*. Cambridge: MIT Press.

Müller-Wille, Staffan and Hans-Jörg Rheinberger (2012) *The Cultural History of Heredity*. Chicago: University of Chicago Press.

Multhauf, Robert P. et al. (1962) *Holcomb, Fitz, and Peate: Three 19th Century American Telescope Makers*. Bulletin 228, United States National Museum. Washington: Smithsonian Institution.

Munns, David D.P. (2003) "If We Build It, Who Will Come? Radio Astronomy and the Limitations of 'National' Laboratories in Cold War America." *Historical Studies in the Physical and Biological Sciences* 34: 95–113.

Munns, David D.P. (2013) *A Single Sky: How an International Community Forged the Science of Radio Astronomy*. Cambridge: MIT Press.

Murray, Margaret A.M. (2000) *Women Becoming Mathematicians: Creating a Professional Identity in Post-World War II America*. Cambridge: MIT Press.

Musto, David F. (1967) "A Survey of the American Observatory Movement, 1800–1850." *Vistas in Astronomy* 9: 87–92.

Mutua, Kagendo, and Beth Blue Swadener (2004) *Decolonizing Research in Cross-Cultural Contexts: Critical Personal Narratives*. New York: SUNY Press.

Nadis, Fred (2005) *Wonder Shows: Performing Science, Magic, and Religion in America*. New Brunswick: Rutgers University Press.

Nanda, Meera (1998) "Reclaiming Modern Science for Third World Progressive Social Movements." *Economic and Political Weekly* 33, 16: 915–22.

Nanda, Meera (2001) "We are All Hybrids Now: The Dangerous Epistemology of Post-Colonial Populism." *Journal of Peasant Studies* 28, 2: 162–86.

Nandy, Ashis (1980) *Alternative Sciences: Creativity and Authenticity in Two Indian Scientists*. Delhi: Allied.

Nandy, Ashis (1988) *Science, Hegemony and Violence: A Requiem for Modernity*. Delhi: Oxford University Press.

NASA (2008) *Science Mission Directorate Management Handbook*. Washington: National Aeronautics and Space Administration.

Nasar, Sylvia (1998) *A Beautiful Mind*. New York: Simon & Schuster.

Nash, Catherine (2012) "Genetics, Race, and Relatedness: Human Mobility and Human Diversity in the Genographic Project." *Annals of the Association of American Geographers* 103, 2: 667–84.

Nash, Linda Lorraine (2006) *Inescapable Ecologies: A History of Environment, Disease, and Knowledge*. Berkeley: University of California Press.

Nash, Roderick Frazier (1967) *Wilderness and the American Mind*. New Haven: Yale University Press.

Nash, Roderick Frazier (1989) *The Rights of Nature: A History of Environmental Ethics*. Madison University of Wisconsin Press.

National Cyclopaedia of American Biography (1984) Current Series (N-63). New York: J.T. White.

National Research Council (1996) *Understanding Risk: Informing Decisions in a Democratic Society*. Washington: National Academy Press.

National Science Board (1971) *Environmental Science: Challenge for the Seventies*. Washington: Government Printing Office.

Naugle, John E. (1991) *First Among Equals: The Selection of NASA Space Science Experiments*. Washington: NASA SP-4215.

Nebeker, Frederic (1995) *Calculating the Weather: Meteorology in the 20th Century*. New York: Academic Press.

Needell, Allan A. (2000) *Science, Cold War and the American State: Lloyd V. Berkner and the Balance of Professional Ideals*. Washington: Smithsonian Institution.

Needham, Joseph (1969) *The Grand Titration: Science and Society in East and West*, London: Allen and Unwin.

Neel, James V. (1987) "Curt Stern: August, 30, 1902–October 23, 1981." *Biographical Memoirs of the National Academy of Sciences* 56: 442–73.

Neel, James V. (1994) *Physician to the Gene Pool: Genetic Lessons and Other Stories*. New York: John Wiley.

Nelkin, Dorothy (1995) *Selling Science: How the Press Covers Science and Technology*. New York: W.H. Freeman.

Nelkin, Dorothy and M. Susan Lindee (2004) *The DNA Mystique: The Gene as a Cultural Icon*. Ann Arbor: University of Michigan Press.

Nelson, P. et al. (1930) "Meat in Nutrition: I. Preliminary Report on Beef Muscle" *Journal of Nutrition*. 3: 303–11.

Nestle, M. (2002) *Food Politics: How the Food Industry Influences Nutrition and Health*. Berkeley: University of California Press.

Neufeld, Michael J. (2014) "Transforming Solar System Exploration: The Origins of the Discovery Program, 1989–1993." *Space Policy* 30: 5–12.

Neushul, Peter (1993) "Science, Government, and the Mass Production of Penicillin." *Journal of the History of Medicine and Allied Sciences* 48: 371–95.

Newcomb, Simon (1884) "The Two Schools of Political Economy." *Princton Review* 60 (November): 291–301.

Newcomb, Simon (1886) "Aspects of the Economic Discussion." *Science* 7 (June 18): 538–42.

Newell, Allen, John Clifford Shaw and Herbert Simon (1958) "Chess-Playing Programs and the Problem of Complexity." *IBM Journal of Research and Development* 2, 4: 320–35.

Newell, Homer E. (1980) *Beyond the Atmosphere: Early Years of Space Science.* Washington: NASA SP-4211.

Newman, Greg, Jim Graham, Alycia Crall, and Melinda Laituri (2011) "The Art and Science of Multi-Scale Citizen Science Support." *Ecological Informatics* 6: 217–27.

Ngai, Mae (2004) *Impossible Subjects: Illegal Aliens and the Making of Modern America.* Princeton: Princeton University Press.

NGSS Lead States (2013) *Next Generation Science Standards: For States, by States.* Washington: National Academies Press.

Nichols, Herbert (1893) "The Psychological Laboratory at Harvard." *Mcclure's Magazine* 1: 399–409.

Nicks, Oran (1985) *Far Travelers: The Exploring Machines.* Washington: NASA SP-480.

Nier, Alfred and John Van Vleck (1975) "John Torrence Tate: 1889–1950," in *Biographical Memoirs of the National Academy of Sciences,* Volume 47. Washington: National Academies Press, 461–84.

Nies, Betsy L. (2006) "Defending Jeeter: Conservative Arguments Against Eugenics in the Depression Era South," in Susan Currell and Christina Cogdell (eds.), *Popular Eugenics: National Efficiency and American Mass Culture in the 1930s.* Athens: Ohio University Press, 120–39.

Nisbett, Catherine E. (2007) "Business Practice: The Rise of American Astrophysics, 1859–1919." PhD Dissertation: Princeton University.

Noble, David F. (1992) *A World without Women: The Christian Clerical Culture of Western Science.* New York: Oxford University Press.

Nofre, David (2010) "Unraveling Algol: US, Europe, and the Creation of a Programming Language." *IEEE Annals of the History of Computing* 32, 2: 58–68.

Noll, Stephanie (2013) "Broiler Chickens and a Critique of the Epistemic Foundations of Animal Modification." *Journal of Agricultural Environmental Ethics* 26, 1: 273–80.

Noll, Steven (1995) *Feeble-Minded in Our Midst: Institutions for the Mentally Retarded in the South, 1900–1940.* Chapel Hill: University of North Carolina Press.

Norberg, Arthur (1990) "High Technology Calculation in the Early Twentieth Century: Punched Card Machinery in Business and Government." *Technology and Culture* 31: 753–79.

Nordlund, Christer (2011) *Hormones of Life: Endocrinology, the Pharmaceutical Industry, and the Dream of a Remedy for Sterility, 1930–1970.* Sagamore Beach: Science History Publications.

Norfre, David (2010) "Unraveling Algol: US, Europe, and the Creation of a Programming Language." *IEEE Annals of the History of Computing* 32, 2: 58–68.

North, John D. (2008) *Cosmos: An Illustrated History of Astronomy and Cosmology.* Chicago: University of Chicago Press.

Nott, J.C. (1843) "The Mulatto a Hybrid – Probable Extermination of the Two Races if the Whites and Blacks are Allowed to Intermarry." *American Journal of Medical Sciences* 6: 252–6.

Nott, J.C. (1844) *Two Lectures on the Natural History of the Caucasian and Negro Races.* Mobile: Dade and Thompson.

Nott, J.C. and George Gliddon (eds.) (1854) *Types of Mankind.* Philadelphia: Lippincott, Grambo, & Co.

Nowotny, H., P. Scott and M. Gibbons (2001) *Re-Thinking Science: Knowledge and the Public in an Age of Uncertainty.* Cambridge: Polity Press.

Numbers, Ronald L. (1977) *Creation by Natural Law: Laplace's Nebular Hypothesis in American Thought.* Seattle: University of Washington Press.

Numbers, Ronald (2006) *The Creationists: From Scientific Creationism to Intelligent Design.* Cambridge: Harvard University Press.

Nye, Robert A. (1993) *Masculinity and Male Codes of Honor in Modern France.* New York: Oxford University Press.

Nye, Robert A. (1997) "Medicine and Science as Masculine 'Fields of Honor'." *Osiris* 12: 60–79.

Nye, Bill and Ken Ham (2014) "Bill Nye Debates Ken Ham." https://www.youtube.com/watch?v=z6kgvhG3AkI.

Nyhart, Lynn K. (2009) *Modern Nature: The Rise of the Biological Perspective in Germany.* Chicago: University of Chicago Press.

O'Neill, Jean and Elizabeth P. Mclean (2008) *Peter Collinson and the Eighteenth-Century Natural History Exchange.* Philadelphia: American Philosophical Society.

O'Connor, A. (2001) *Poverty Knowledge: Social Science, Social Policy, and the Poor in Twentieth-Century U.S. History. Politics and Society in Twentieth-Century America.* Princeton: Princeton University Press.

O'Donnell, John M. (1985) *The Origins of Behaviorism: American Psychology, 1870–1920.* New York: New York University Press.

O'Malley, Henry (1927) *Report of the U.S. Commissioner of Fishes.* Washington: Government Printing Office.

Oakley, Ann (1972) *Sex, Gender and Society.* New York: Harper and Row.

Odum, Eugene (1953) *The Fundamentals of Ecology.* Philadelphia: Saunders.

Odum, Eugene (1964) "The New Ecology." *Bioscience* 4: 14–16.

Offen, Karen (2006) "Le Gender Est-Il Une Invention Américaine?" *Clio: Femmes, Genre, Histoire* 24: 291–304.

Office of Technology Assessment (1984) *Commercial Biotechnology: An International Analysis* (1984). Washington: U.S. Congress, OTA-BA-218.

Ogilvie, Marilyn Bailey (1986) *Women in Science: Antiquity Through the Nineteenth Century: A Biographical Dictionary with Annotated Bibliography.* Cambridge: MIT Press.

Ogilvie, Marilyn Bailey and Joy Harvey (eds.) (2000) *The Biographical Dictionary of Women in Science: Pioneering Lives from Ancient Times to the Mid-20th Century.* New York: Routledge.

Ogren, Christine (2005) *The American Normal School: "An Instrument of Great Good."* New York: Palgrave Macmillan.

Ohta, Tomoko (2000) "Mechanisms of Molecular Evolution." *Philosophical Transactions of the Royal Society of London (Part B)* 355: 1623–6.

Olby, Robert (1974) *The Path to the Double Helix: The Discovery of DNA.* Seattle: University of Washington Press.

Olby, Robert (1985) *Origins of Mendelism.* Chicago: University of Chicago Press.

Olby, Robert (2003) "Quiet Debut for the Double Helix." *Nature* 421, 6921: 402–5.

Oldenziel, Ruth (1999) *Making Technology Masculine: Men, Women, and Modern Machines in America, 1870–1945.* Thousand Oaks: Sage.

Oldroyd, David (2009) "Geophysics and Geochemistry," in Peter J. Bowler and John V. Pickstone (eds.), *Cambridge History of Science*, 6. New York: Cambridge University Press, 395–415.

Olesko, K.M. (2006) "Science Pedagogy as a Category of Historical Analysis: Past, Present, and Future." *Science & Education* 15, 7–8: 863–80.

Oleson, Alexandra, and Sanborn C. Brown (1976) *The Pursuit of Knowledge in the Early Republic: American Scientific and Learned Societies from Colonial Times to the Civil War.* Baltimore: Johns Hopkins University Press.

Olmstead, A.L. and P.W. Rhode (2008) *Creating Abundance: Biological Innovation and American Agricultural Development.* Cambridge: Cambridge University Press.

Oosterhuis, Harry (2000) *Stepchildren of Nature: Krafft-Ebing, Psychiatry, and the Making of Sexual Identity.* Chicago: University of Chicago Press.

Opitz, Donald L. (2012) "Co-operative Comradeships versus Same-Sex Partnerships: Historicizing Collaboration among Homosexual Couples in the Sciences," in Annette Lykknes, Donald L. Opitz, and Brigitte Van Tiggelen (eds.), *For Better or for Worse? Collaborative Couples in the Sciences.* New York: Birkhäuser, 245–69.

Opitz, Donald L., Staffan Bergwik, and Brigitte Van Tiggelen (eds.) (2015) *Domesticity in the Making of Modern Science.* Basingstoke: Palgrave Macmillan.

Opitz, J.M. (1977) "The American Journal of Medical Genetics – Forward." *American Journal of Medical Genetics* 1: 1–2.

Oppenheimer, Jane M. (1980). "Some Historical Backgrounds for the Establishment of the Stazione Zoologica at Naples," in Mary Sears and Daniel Merrimen (eds.), *Oceanography: The Past*. New York: Springer, 179–87.

Oreskes, Naomi (1996) "Objectivity or Heroism? On the Invisibility of Women in Science." *Osiris* 11: 87–113.

Oreskes, Naomi (1999) *The Rejection of Continental Drift: Theory and Method in American Earth Science*. New York: Oxford University Press.

Oreskes, Naomi (ed.) (2001) *Plate Tectonics: An Insider's History of the Modern Theory of the Earth*. Boulder: Westview Press.

Oreskes, Naomi (2015) *Science on a Mission: American Oceanography in the Cold War and Beyond*. Chicago: University of Chicago Press.

Oreskes, Naomi and Erik Conway (2010) *Merchants of Doubt: How a Handful of Scientists Obscured the Truth on Issues from Tobacco Smoke to Global Warming*. New York: Bloomsbury.

Oreskes, Naomi and Ronald E. Doel (2003) "Physics and Chemistry of the Earth," in Mary Jo Nye (ed.), *Cambridge History of Science, Volume 5: The Modern Physical and Mathematical Sciences*. New York: Cambridge University Press, 538–52.

Oreskes, Naomi and James R. Fleming (2000) "Why Geophysics?" *Studies in History and Philosophy of Science Part B: Studies in History and Philosophy of Modern Physics* 31, 3: 253–7.

Oreskes, Naomi and H.E. Legrand (2001) *Plate Tectonics: An Insider's History of the Modern Theory of the Earth*. Boulder: Westview Press.

Orosz, Joel J. (1990) *Curators and Culture: The Museum Movement in America, 1740–1870*. Tuscaloosa: University of Alabama Press.

Orsi, Jared (2014) *Citizen Explorer: The Life of Zebulon Pike*. Oxford: Oxford University Press.

Ortner, Sherry B. (1972) "Is Female to Male as Nature is to Culture?" *Feminist Studies* 1, 2: 5–31.

Osborn, Henry Fairfield (1916) *Men of the Old Stone Age*. New York: Charles Scribner's Sons.

Osborn, Henry Fairfield (1931) *Cope: Master Naturalist*. Princeton: Princeton University Press.

Osborne, Michael A. (1994) *Nature, the Exotic, and the Science of French Colonialism*. Bloomington: Indiana University Press.

Oshinsky, David (2006) *Polio: An American Story*. Oxford: Oxford University Press.

Osterbrock, Donald (1984) *James E. Keeler, Pioneer American Astrophysicist, and the Early Development of American Astrophysics*. New York: Cambridge University Press.

Osterbrock, Donald (1993) *Pauper & Prince: Ritchey, Hale, and Big American Telescopes*. Tucson: University of Arizona Press.

Osterbrock, Donald (1995) "Founded in 1895 by George E. Hale and James E. Keeler: The Astrophysical Journal Centennial." *Astrophysical Journal* 438: 1–7.

Osterbrock, Donald (1997) *Yerkes Observatory, 1892–1950: The Birth, Near Death, and Resurrection of a Scientific Research Institution*. Chicago: University of Chicago Press.

Osterbrock, Donald, John R. Gustafson, and W. J. Shiloh Unruh (1998) *Eye on the Sky: Lick Observatory's First Century*. Berkeley: University of California Press.

Ostrom, John H. and John S. Mcintosh (1966) *Marsh's Dinosaurs: The Collections from Como Bluff*. New Haven: Yale University Press.

Otis, Laura (2002) *Literature and Science in the Nineteenth Century: An Anthology*. New York: Oxford University Press.

Ott, Katherine (1999) *Fevered Lives: Tuberculosis in American Culture Since 1870*. Cambridge: Harvard University Press.

Oudshoorn, Nelly (1994) *Beyond the Natural Body: An Archaeology of Sex Hormones*. New York: Routledge.

Outram, Dorinda (2013) *The Enlightenment*. Cambridge: Cambridge University Press.

Owens, Larry (1985) "Pure and Sound Government: Laboratories, Gymnasia, and Playing Fields in Nineteenth-Century America." *Isis* 76: 182–94.

Packard, Alpheus (1888) "On Certain Factors of Evolution." *The American Naturalist* 22: 808–21.

Page, E. and L. Phipard (1956) *U.S. Department of Agriculture. Essentials of an Adequate Diet.* Bulletin No. 160. Washington: Government Printing Office.

Page, L. and L. Fincher (1960) *Food and Your Weight.* Washington: Government Printing Office.

Paine, Thomas (1795) *The Age of Reason.* London: H.D. Symonds.

Pais, Abraham (1988) *Inward Bound: Of Matter and Forces in the Physical World.* New York: Oxford University Press.

Pais, Abraham (1997) *A Tale of Two Continents.* Princeton: Princeton University Press.

Pais, Abraham (2005) *Subtle is the Lord: The Science and the Life of Albert Einstein.* New York: Oxford University Press.

Paleontology Research Institution (2003) "Hyde Park Mastodon Fast Facts." http://www. priweb.org/mastodon/HP_Mast/HP_Mast_Facts.html.

Paley, William (1785) *Moral and Political Philosophy.* London: R. Faulder.

Paley, William (1794) *View of the Evidences of Christianity.* London: R. Faulder.

Paley, William (1802) *Natural Theology, Or, Evidences of the Existence and Attributes of the Deity collected from the Appearances of Nature.* London: R. Faulder.

Paley, William (1831) *Natural Theology: Or Evidences of the Existence and Attributes of the Deity with Additional Notes Original and Selected for this Edition.* Boston: Lincoln and Edmands.

Palladino, Paolo (1996) "People, Institutions, and Ideas: American and British Geneticists at Cold Spring Harbor Symposium on Quantitative Biology, June 1955." *History of Science* 34: 411–50.

Pallen, M.J. and U. Gophna (2007) "Bacterial Flagella and Type III Secretion: Case Studies in the Evolution of Complexity," in Jean-Nicolas Volff (ed.), *Gene and Protein Evolution*, Volume 3. Basel: Karger, 30–47.

Palmer, Steven (2010) *Launching Global Health: The Caribbean Odyssey of the Rockefeller Foundation.* Ann Arbor: University of Michigan Press.

Palmer, W.P. (2007) "Dissent at the University of Iowa: Gustavus Detlef Hinrichs – Chemist and Polymath." *Khymia* 16: 534–53.

Pandora, Katherine (2009a) "The Children's Republic of Science in the Antebellum Literature of Samuel Griswold Goodrich and Jacob Abbott." *Osiris* 24: 75–98.

Pandora, Katherine (2009b) "Popular Science in National and Transnational Perspective: Suggestions from the American Context." *Isis* 100: 346–58.

Pandora, Katherine and Karen Rader (2008) "Science in the Everyday World: Why Perspectives from the History of Science Matter." *Isis* 99: 350–64.

Panel of Astronomical Facilities (Whitford 1964). *Ground-Based Astronomy: A Ten-Year Program.* Washington: National Academy Press.

Pang, Alex Soojung-Kim (1996) "Gender, Culture, and Astrophysical Fieldwork: Elizabeth Campbell and the Lick Observatory-Crocker Eclipse Expeditions." *Osiris* 11: 17–43.

Pantalony, David, Richard L. Kremer, and Francis J. Manasek (2005) *Study, Measure, Experiment: Stories of Scientific Instruments at Dartmouth College.* Norwich: Terra Nova Press.

Parezo, Nancy J. and Don D. Fowler (2007) *Anthropology Goes to the Fair: The 1904 Louisiana Purchase Exposition.* Lincoln: University of Nebraska Press.

Park, Hyung Wook (2008) "Edmund Vincent Cowdry and the Making of Gerontology as a Multidisciplinary Scientific Field in the United States." *Journal of the History of Biology* 41: 529–72.

Parthasarathy, Shobita (2007) *Building Genetic Medicine: Breast Cancer, Technology, and the Comparative Politics of Health Care.* Cambridge: MIT Press.

Patterson, Thomas Carl (2001) *A Social History of Anthropology in the United States.* Oxford: Berg.

Paul, Diane B. (1991) "The Rockefeller Foundation and the Origins of Behavior Genetics," in Keith R. Benson, Jane

Maienschein, and Ronald Rainger (eds.), *Expansion of American Biology*, 262–83. New Brunswick: Rutgers University Press.

Paul, Diane B. (1995) *Controlling Human Heredity, 1865 to the Present*. New York: Humanities Press.

Paul, Diane B. (1998) *The Politics of Heredity: Essays on Eugenics, Biomedicine, and the Nature–Nurture Debate*. Alba: State University of New York Press.

Paul, Diane B. and Jeffrey P. Brosco (2013) *The PKU Paradox: A Short History of a Genetic Disease*. Baltimore: Johns Hopkins University Press.

Paul, Diane B. and Barbara A. Kimmelman (1988) "Mendel in America: Theory and Practice, 1900–1919," in Ronald Rainger, Keith R. Benson, and Jane Maienschein (eds.), *American Development of Biology*. Philadelphia: University of Pennsylvania Press.

Pauly, Philip J. (1984) "The Appearance of Academic Biology in Late Nineteenth-Century America." *Journal of the History of Biology* 17, 3: 369–97.

Pauly, Philip J. (1987) *Controlling Life: Jacques Loeb & the Engineering Ideal in Biology*. New York: Oxford University Press.

Pauly, Philip J. (1988) "Summer Resort and Scientific Discipline: Woods Hole and the Structure of American Biology," in Ronald Rainger, Keith R. Benson, and Jane Maienschein (eds.), *American Development of Biology*. Philadelphia: University of Pennsylvania Press, 121–50.

Pauly, Philip J. (1991) "The Development of High School Biology: New York City, 1900–1925." *Isis* 82, 4: 662–88.

Pauly, Philip J. (2000) *Biologists and the Promise of American Life: From Meriwether Lewis to Alfred Kinsey*. Princeton: Princeton University Press.

Pawley, Emily (2010) "Accounting with the Fields: Chemistry and Value in Nutriment in American Agricultural Improvement, 1835–1860." *Science as Culture* 19: 461–82.

Peale, C. Willson, R. Peale and J. J. Hawkins (1800) *Discourse Introductory to a Course of Lectures on the Science of Nature: With Original Music, Composed for, and Sung on, the Occasion: Delivered in the Hall of the Universiy [Sic] of Pennsylvania, Nov. 8, 1800*. Philadelphia: Zachariah Poulson, Jr.

Pegram, G.B. (1925) "Chandler and the Columbia School of Mines." *Science*. 62, 1614: 501–3.

Perkins, Charlotte Gilman (1898) *Women and Economics: A Study of the Economic Relation between Men and Women as a Factor in Social Evolution*. Boston: Small and Maynard.

James Petiver (1705) "An Account of Animals and Shells Sent from Carolina to Mr James Petiver, F. R. S." *Philosophical Transactions of the Royal Society* 24: 1952–60.

Petryna, Adriana (2009) *When Experiments Travel: Clinical Trials and the Global Search for Human Subjects*. Princeton: Princeton University Press.

Petrayna, Adriana (2013) *Life Exposed: Biological Citizens after Chernobyl*. Princeton: Princeton University Press.

Petrella, Frank (1987) "Daniel Raymond, Adam Smith, and Classical Growth Theory: An Inquiry into the Nature and Causes of the Wealth of America." *History of Political Economy* 19, 2: 239–59.

Petzhold, Charles (2008) *The Annotated Turing: A Guided Tour through Alan Turing's Historic Paper on Computability and the Turing Machine*. Hoboken: Wiley.

Pfafflin, J.R., Paul Baham, and F.S. Gill (2008) *Dictionary of Environmental Science and Engineering*. Amsterdam: Gordon and Breach.

Pfiefer, Edward J. (1965) "The Genesis of American neo-Lamarckism." *Isis* 56: 157–67.

Pfiefer, Edward J. (1972) "United States," in T.F. Glick (ed.), *The Comparative Reception of Darwinism*. Austin: University of Texas Press, 169–205.

Phalkey, Jahnavi (2013) "How May We Study Science and the State in Postcolonial India?" in B. Lightman, G. Mcouat, and L. Stewart (eds.), *Circulation of Knowledge Between Britain, India, and China: The Early-Modern World to the Twentieth Century*. Boston: Brill, 263–84.

Phalkey, Jahnavi (2013) "Introduction: Science, History, and Modern India." *Isis* 104: 330–6.

Phelps, Amira Hart Lincoln (1829) *Familiar Lectures on Botany*. Hartford: H. and F.J. Huntington.

Philbrick, M. and J. Barandiaran (2009) "The National Citizens' Technology Forum: Lessons for the Future." *Science and Public Policy* 36: 335–47.

Philbrick, Nathaniel (2003) *Sea of Glory: America's Voyages of Discovery, the U.S. Exploring Expedition, 1838–1842*. New York: Viking.

Philip, Kavita (2004) *Civilizing Natures: Race, Resources, and Modernity in Colonial South India*. New Brunswick: Rutgers University Press.

Philip, Kavita, L. Irani, and P. Dourish (2010) "Postcolonial Computing: A Tactical Survey." *Science, Technology and Human Values* 37: 3–29

Philippon, Daniel J. (2005) *Conserving Words: How American Writers Shaped the Environmental Movement*. Athens: University of Georgia Press.

Phillips, C.J. (2014) "In Accordance with a 'More Majestic Order' the New Math and the Nature of Mathematics at Midcentury." *Isis* 105, 3: 540–63.

Phillips, Rodney (1994) *The Chicago Plan and New Deal Banking Reform*. New York: M. E. Sharpe.

Pickering, Andrew (ed.) (1992) *Science as Practice and Culture*. Chicago: University of Chicago Press.

Pielke, R. Jr. (2005) "Science Policy," in Carl Mitcham (ed.), *Encyclopedia of Science, Technology, and Ethics*, Vol. 4. Detroit: Macmillan Reference, 1699–705.

Pielke, R. Jr. (2007) *The Honest Broker: Making Sense of Science in Policy and Politics*. New York: Cambridge University Press.

Pielke, R. and M. Betsill (1997) "Policy for Science for Policy: Ozone Depletion and Acid Rain Revisited." *Research Policy* 26: 157–68.

Pietruska, Jamie (2009) "Propheteering: A Cultural History of Prediction in the Gilded Age." PhD Dissertation: Massachusetts Institute of Technology.

Pietruska, Jamie (2011) "U.S. Weather Bureau Chief Willis Moore and the Reimagination of Uncertainty in Long-Range Forecasting. *Environment and History* 17: 79–105.

Pigliucci, Massimo and Gerd Mäller (2010) *The Extended Synthesis*. Cambridge: MIT Press.

Pimbert, M. (2008) *Towards Food Sovereignty: Reclaiming Autonomous Food Systems*. London: International Institute for Environment and Development.

Pinkoski, Marc (2011) "Back To Boas." *Histories of Anthropology Annual* 7, 1: 127–69.

Plantinga, Alvin. (2011) *Where the Conflict Really Lies: Science, Religion, and Naturalism*. New York: Oxford University Press.

Plotkin, Howard (1978a) "Edward C. Pickering, the Henry Draper Memorial, and the Beginnings of Astrophysics in America." *Annals of Science* 35: 365–77.

Plotkin, Howard (1978b) "Edward C. Pickering and the Endowment of Scientific Research in America." *Isis* 69: 44–57.

Plotkin, Howard (1980) "Henry Tappan, Franz Brünnow, and the Founding of the Ann Arbor School of Astronomers, 1892–1863." *Annals of Science* 37: 287–302.

Poe, Edgar Allan (2000) *The Complete Poems of Edgar Allan Poe*. Thomas Ollive Mabbott (ed.). Urbana: University of Illinois Press.

Polachek, Harry (1997) "Before the ENIAC." *IEEE Annals of the History of Computing* 19, 2: 25–30.

Polanyi, M. (1962) "The Republic of Science." *Minerva* 1: 54–73.

Poliquin, Rachel (2012) *The Breathless Zoo: Taxidermy and the Cultures of Longing*. University Park: Pennsylvania State University Press.

Pollan, Michael (2009) *A Defense of Food: An Eater's Manifesto*. New York: Penguin.

Pollan, Michael and M. Kalman (2011) *Food Rules: An Eater's Manual*. New York: Penguin.

Pollock, Anne. "Places of Pharmaceutical Knowledge-Making: Global Health, Postcolonial Science, and Hope in South African Drug Discovery." *Social Studies of Science* 44, 6: 848–73.

Porter, Charlotte M. (1986) *The Eagle's Nest: Natural History and American Ideas, 1812–1842*. Tuscaloosa: University of Alabama Press.

Portolano, Marlana (2000) "John Quincy Adams's Rhetorical Crusade for Astronomy." *Isis* 91: 480–503.

Potter, Elizabeth (2001) *Gender and Boyle's Law of Gases*. Bloomington: Indiana University Press.

Pouchet, Georges (1858) *De la Pluralité des Races Humaines*. Paris: J.B. Baillière.

Pouchet, Georges (1864) *De la Pluralité des Races Humaines*. Paris: Victor Masson.

Powell, John Wesley (1875) *Exploration of the Colorado River of the West and its Tributaries*. Washington, DC: Government Printing Office.

Powell, John Wesley (1877) *Introduction to the Study of Indian Languages, with Words, Phrases, and Sentences to be Collected*. Washington, DC: Government Printing Office.

Powell, John Wesley (1888) "From Barbarism to Civilization." *American Anthropologist* A1, 2: 97–124.

Prakash, Gyan (1999) *Another Reason: Science and the Imagination of Modern India*. Princeton: Princeton University Press.

Prasad, Amit (2008) "Science in Motion: What Postcolonial Science Studies Can Offer." *Electronic Journal of Communication Information & Innovation in Health* 2, 2: 35–47.

Prasad, Amit (2014) *Imperial Technoscience: Transnational Histories of MRI in the United States, Britain, and India*. Cambridge: MIT Press.

Pratt, Mary Louise (1992) *Imperial Eyes: Travel Writing and Transculturation*. New York: Routledge.

Price, David H. (2008) *Anthropological Intelligence: The Deployment and Neglect of American Anthropology in the Second World War*. Durham: Duke University Press.

Prichard, James Cowles (1826) *Researches into the Physical History of Mankind*. London: John and Arthur Arch.

Prichard, James Cowles (1836) *Researches into the Physical History of Mankind*. London: John and Arthur Arch.

Priestly, Mark (2011) *A Science of Operations: Machines, Logic and the Invention of Programming*. Berlin: Springer.

Pritchard, James A. (1999) *Preserving Yellowstone's Natural Conditions: Science and the Perception of Nature*. Lincoln: University of Nebraska.

Proctor, Robert (2003) "Three Roots of Human Recency: Molecular Anthropology, the Refigured Acheulean, and the UNESCO Response to Auschwitz." *Current Anthropology* 44: 213–40.

Proctor, Robert (2008) "Agnotology: A Missing Term to Describe the Cultural Production of Ignorance (and its Study)," in Robert Proctor and Londa Schiebinger (eds.), *Agnotology: The Making & Unmaking of Ignorance*. Stanford: Stanford University Press.

Proctor, Robert (2012) *Golden Holocaust: Origins of the Cigarette Catastrophe and the Case for Abolition*. Berkeley: University of California Press.

Proctor, Robert and L. Schiebinger (2008) *Agnotology: The Making and Unmaking of Ignorance*. Redwood City: Stanford University Press.

Prosser, Jodicus W. (2009) "*Bigger Eyes in a Wider Universe: The American Understanding of Earth in Outer Space, 1893–1941*." PhD Dissertation: Texas A&M University.

Provine, William B. (1971) *The Origins of Theoretical Population Genetics*. Chicago: University of Chicago Press.

Provine, William B. (1986) *Sewall Wright and Evolutionary Biology*. Chicago: University of Chicago Press.

Provine, William B. (1986a) *Evolution*. Chicago: University of Chicago Press.

Pycior, Helena M., Nancy G. Slack, and Pnina Abir-Am (1996) *Creative Couples in the Sciences*. New Brunswicksssss: Rutgers University Press.

Pyeritz, Reed E. (2011) "The Family History: The First Genetic Test, and Still Useful after All Those Years?" *Genetics in Medicine* 14: 3–9.

Pyne, Stephen J. (2007) *Grove Karl Gilbert: A Great Engine of Research*. Iowa City: University of Iowa Press.

Quick, J.A. and E.W. Murphy (1982) *The Fortification of Foods: A Review*. Washington: U.S. Department of Agriculture, Food Safety and Inspection Service.

Quintero, Camilo (2011) "Trading in Birds: Imperial Power, National Pride, and the Place of Nature in U.S.—Colombia Relations." *Isis* 102: 421–45.

Rabb, G. and C. Saunders (2005) "The Future of Zoos and Aquariums: Conservation and Caring." *International Zoo Yearbook* 39: 1–26.

Rabbitt, Mary C. and Survey Geologica (1979) *A Brief History of the U.S. Geological Survey*. Reston: Department of the Interior, Geological Survey.

Rabeharisoa, Vololona and Pascale Bourret (2009) "Staging and Weighting Evidence in Biomedicine." *Social Studies of Science* 39: 691–715.

Rabin, Michael and Dana Scott (1959) "Finite Automata and their Decision Problems." *IBM Journal of Research and Development* 2, 2: 114–25.

Rabinow, Paul (1996) *Making PCR: A Story of Biotechnology*. Chicago: University of Chicago Press.

Raby, Megan (2012) "*Making Biology Tropical: American Science in the Caribbean, 1898–1963*." PhD Dissertation: University of Wisconsin.

Rachman, Stephen (2010) "Poe and Origins of Detective Fiction," in Catherine Ross Nickerson (ed.), *Cambridge Companion to American Crime Fiction*. New York: Cambridge University Press, 17–28.

Rader, Karen (2004) *Making Mice: Standardizing Animals for American Biomedical Research, 1900–1955*. Princeton: Princeton University Press.

Rader, Karen A. and Victoria E.M. Cain (2008) "From Natural History to Science: Display and the Transformation of American Museums of Science and Nature." *Museum and Society* 6, 2: 152–71.

Rader, Karen A. and Victoria E.M. Cain (2014) *Life on Display: Revolutionizing U.S. Museums of Science and Natural History in the United States in the Twentieth Century*. Chicago: University of Chicago Press.

Radin, Paul (1927) *Primitive Man as Philosopher*. New York: D. Appleton.

Raina, Dhruv (1996) "Reconfiguring the Center: The Structure of Scientific Exchanges Between Colonial India and Europe." *Minerva* 34: 161–76.

Raina, Dhruv (1999) "From West to Non-West? Basalla's Three-Stage Model Revisited." *Science as Culture* 8: 497–516.

Raina, Dhruv (2003) *Images and Contexts: The Historiography of Science and Modernity in India*. Delhi: Oxford University Press.

Rainger, Ronald (1991) *An Agenda for Antiquity: Henry Fairfield Osborn and Vertebrate Paleontology at the American Museum of Natural History, 1890–1935*. Tuscaloosa: University of Alabama Press.

Rainger, Ronald (1992) "The Rise and Decline of a Science: Vertebrate Paleontology at Philadelphia's Academy of Natural Sciences, 1820–1900." *Proceedings of the American Philosophical Society* 136, 1: 1–32.

Rainger, Ronald (1997) "Everett C. Olsen and the Development of Vertebrate Paleoecology and Taphonomy." *Archives of Natural History* 24, 3: 373–96.

Rainger, Ronald, Keith Rodney Benson, and Jane Maienschein (1988) *The American Development of Biology*. Philadelphia: University of Pennsylvania Press.

Raj, Kapil (2001) "Colonial Encounters and the Forging of New Knowledge and National Identities: Great Britain and India, 1760–1850." *Osiris* 15: 119–34.

Raj, Kapil (2006) *Relocating Modern Science: Circulation and the Construction of Scientific Knowledge in South Asia and Europe*. Delhi: Permanent Black.

Raj, Kapil (2013) "Beyond Postcolonialism … and Postpostivism: Circulation and the Global History of Science." *Isis* 104: 337–47.

Ralson, Anthony (2004) "Four Editions and Eight Publishers: A History of the Encyclopedia of Computer Science." *IEEE Annals of the History of Computing* 26, 1: 42–52.

Ralston, Anthony et al. (1976) *Encyclopedia of Computer Science*. Jacksonville: Mason/Charter.

Ramsey, Paul (1970) *Fabricated Man: The Ethics of Genetic Control.* New Haven: Yale University Press.

Randalls, Samuel (2010) "Weather Profits: Weather Derivatives and the Commercialization of Meteorology." *Social Studies of Science* 40: 706–30.

Rapp, Rayna (1999) *Testing Women, Testing the Fetus: The Social Impact of Amniocentesis in America.* New York: Routledge.

Rasmussen, Nicolas (1997) *Picture Control: The Electron Microscope and the Transformation of Biology in America, 1940–1960.* Stanford: Stanford University Press.

Rasmussen, Nicolas (1999) *Picture Control: The Electron Microscope and the Transformation of Biology in America, 1940–1960.* Palo Alto: Stanford University Press.

Rasmussen, Nicolas (2001) "Biotechnology before the 'Biotech Revolution:' Life Scientists, Chemists, and Product Development in 1930s–1940s America," in C. Reinhardt (ed.), *Chemical Sciences in the 20th Century.* Weinheim: Wiley-VCH Verlag, 201–27.

Rasmussen, Nicolas (2002) "Steroids in Arms: Science, Government, Industry, and the Hormones of the Adrenal Cortex in the United States, 1930–1950." *Medical History* 46: 299–324.

Rasmussen, Nicolas (2014) *Gene Jockeys: Life Science and the Rise of Biotech Enterprise.* Baltimore: Johns Hopkins University Press.

Rau, Charles (1876) *Early Man in Europe.* New York: Harper.

Ray, John (1691) *Wisdom of God Manifested in the Works of the Creation.* London: John Ray Society.

Raymond, Daniel (1820) *Thoughts on Political Economy.* Baltimore: Fielding Lucas, Jrs.

Rea, Tom (2001) *Bone Wars: The Excavation and Celebrity of Andrew Carnegie's Dinosaur.* Pittsburgh: University of Pittsburgh Press.

Reader, John (2011) *Missing Links: In Search of Human Origins.* Oxford: Oxford University Press.

Reaka-Kudla, Marjorie L., Don E. Wilson, and Edward O. Wilson (1997) *Biodiversity II.* Washington: Joseph Henry.

Reardon, Jenny (2004) "Decoding Race and Human Difference in a Genomic Age." *Differences: A Journal of Feminist Cultural Studies.* 15: 38–65.

Reardon, Jenny (2005) *Race to the Finish: Identity and Governance in an Age of Genomics.* Princeton: Princeton University Press.

Reardon, Jenny (2012) "The Human Genome Diversity Project: What Went Wrong?" in Sandra Harding (ed.), *The Postcolonial Science and Technology Studies Reader.* Durham: Duke University Press.

Reardon, Jenny (2013) "Indigenous Body Parts, Mutating Temporalities, and the Half-Lives of Postcolonial Technoscience." *Social Studies of Science* 43, 4: 465–83.

Reardon, Jenny and Kim Tallbear (2012) "Your DNA is Our History: Genomics, Anthropology and the Construction of Whiteness as Property." *Current Anthropology* 53: S5.

Reder, Melvin W. (1982) "Chicago Economics: Permanence and Change." *Journal of Economic Literature* 20, 1: 1–38.

Redfield, Robert (1930) *Tepoztlan, a Mexican Village: A Study in Folk Life.* Chicago: University of Chicago Press.

Redfield, Robert (1956) *Peasant Society and Culture: An Anthropological Approach to Civilization.* Chicago: University of Chicago Press.

Reed, James (1978) *From Private Vice to Public Virtue: The Birth Control Movement and American Society since 1830.* New York: Basic Books.

Rees, Amanda (2009) *The Infanticide Controversy: Primatology and the Art of Field Science.* Chicago: University of Chicago Press.

Reese, Kenneth (ed.) (2002) *The American Chemical Society at 125: A Recent History, 1976–2001.* Washington, DC: American Chemical Society.

Regal, Brian (2002) *Henry Fairfield Osborn: Race and the Search for the Origins of Man.* Burlington: Ashgate.

Reich, Leonard (1985) *The Making of American Industrial Research: Science and Business at GE and Bell, 1876–1926.* Cambridge: Cambridge University Press.

Reichenbach, Count Oscar (1864) "On the Vitality of the Black Race, Or the Coloured People in the United States, According to the Census" *Journal of the Anthropological Society of London* 2: LXV–LXXIII.

Reider, Rebecca (2009) *Dreaming the Biosphere: The Theater of All Possibilities.* Albuquerque: University of New Mexico Press.

Reidy, Michael S., Gary Kroll, and Erik M. Conway (2007) *Exploration and Science: Social Impact and Interaction.* Santa Barbara: ABC-CLIO Press.

Reilly, Phillip (1991) *The Surgical Solution: A History of Involuntary Sterilization in the United States.* Baltimore: Johns Hopkins University Press.

Reingold, Nathan (1976) "Definitions and Speculations: The Professionalization of Science in America in the Nineteenth Century," in Alexandra Oleson and Sanborn C. Brown (eds.), *The Pursuit of Knowledge in the Early American Republic: American Scientific Learned Societies from Colonial Times to the Civil War.* Baltimore: Johns Hopkins University Press.

Reingold, Nathan (1979) *The Sciences in the American Context: New Perspectives.* Washington: Smithsonian Institution Press.

Reingold, Nathan (ed.) (1985) *Science in Nineteenth-Century America: A Documentary History.* Chicago: University of Chicago Press.

Reingold, Nathan (1991) *Science, American Style.* New Brunswick: Rutgers University Press.

Reinhardt, Carsten (2006) *Shifting and Rearranging: Physical Methods and the Transformation of Modern Chemistry.* Sagamore Beach: Science History Publications.

Reis, Elizabeth (2009) *Bodies in Doubt: An American History of Intersex.* Baltimore: Johns Hopkins University Press.

Reiss, Michael J. (2009) "The Relationship Between Evolutionary Biology and Religion." *Evolution* 63: 1934–41.

Reumann, Miriam (2005) *American Sexual Character: Sex, Gender, and National Identity in the Kinsey Reports.* Berkeley: University of California Press.

Reverby, Susan (2009) *Examining Tuskegee: The Infamous Syphilis Study and its Legacy.* Chapel Hill: University of North Carolina Press.

Reynolds, Andrew (2007) "The Cell's Journey: From Metaphorical to Literal Factory." *Endeavour* 31, 2: 65–70.

Reynolds, D.R. (2002) *There Goes the Neighborhood: Rural School Consolidation at the Grass Roots in Early Twentieth-Century Iowa.* Iowa City: University of Iowa Press.

Rezneck, Samuel (1970) "The European Education of an American Chemist and its Influence in 19th-Century America: Eben Norton Horsford." *Technology and Culture* 11: 366–88.

Rheinberger, Hans-Jörg (1996) "Comparing Experimental Systems: Protein Synthesis in Microbes and in Animal Tissue at Cambridge (Ernest F. Gale) and at the Massachusetts General Hospital (Paul C. Zamecnik), 1945–1960." *Journal of the History of Biology* 29: 387–417.

Rheinberger, Hans-Jörg (1997) *Toward a History of Epistemic Things: Synthesizing Proteins in the Test Tube.* Stanford: Stanford University Press.

Rheinberger, Hans-Jörg (2010) *An Epistemology of the Concrete: Twentieth-Century Histories of Life.* Durham: Duke University Press.

Rhodes, Richard (1986) *The Making of the Atomic Bomb.* New York: Simon & Schuster.

Rhodes, Richard (1995) *Dark Sun: The Making of the Hydrogen Bomb.* New York: Simon & Schuster.

Rice, John and Saul Rosen (2004) "Computer Sciences at Purdue University – 1962 to 2000." *IEEE Annals of the History of Computing* 26, 2: 48–61.

Richards, Robert J. (1987) *Darwin and the Emergence of Evolutionary Theories of Mind and Behavior.* Chicago: University of Chicago Press.

Richards, Robert J. and Michael Ruse (2015) *Debating Darwin: Mechanist or Romantic?* Chicago: University of Chicago Press.

Richards, Theodore W. (1999) "Atomic Weights," in *Nobel Lectures, Chemistry, 1901–1921.* Singapore: World Scientific, 280–92.

Richardson, Joseph G. (1878) *The Germ Theory of Disease, and Its Present Bearing upon*

Public and Personal Hygiene. Philadelphia: Philadelphia Social Science Association.

Richardson, Robert D., Jr. (1992) "Thoreau and Science," in Robert J. Scholnick (ed.), *American Literature and Science.* Lexington: University Press of Kentucky, 110–27.

Richardson, Sarah S. (2013) *Sex Itself: The Search for Male and Female in the Human Genome.* Chicago: University of Chicago Press.

Richerson, P. and R. Boyd (2005) *Not By Genes Alone: How Culture Transformed Human Evolution.* Chicago: University of Chicago Press.

Richmond, Phyllis A. (1954) "American Attitudes toward the Germ Theory of Disease, 1860–1880." *Journal of the History of Medicine and Allied Sciences* 9: 428–54.

Rickles, Dean (2015) *Covered in Deep Mist: The Development of Quantum Gravity, 1916–1956.* Oxford: Oxford University Press.

Rieppel, Lukas Benjamin (2012) *"Dinosaurs: Assembling an Icon of Science."* PhD Dissertation: Harvard University.

Rifkin, Jeremy (1984) *Algeny: A New Word – A New World.* New York: Penguin.

Rifkin, Jeremy (1998) *The Biotech Century: Harnessing the Gene and Remaking the World.* New York: Gollancz.

Rimoin, D.L. (2008) "Victor A. McKusick 1921–2008." *Nature Genetics* 40, 9: 1037.

Riordan, Michael (2000)"The Demise of the Superconducting Super Collider." *Physics in Perspective* 2: 411–25.

Riordan, Michael (2001) "A Tale of Two Cultures: Building the Superconducting Super Collider, 1988–1993." *Historical Studies in the Physical and Biological Sciences* 32: 125–44.

Riordan, Michael; Lillian Hoddeson and Conyers Herring (1999) "The Invention of the Transistor." *Reviews of Modern Physics* 71: S336–S345.

Risse, Guenter B. (2012) *Plague, Fever, and Politics in San Francisco's Chinatown.* Baltimore: Johns Hopkins University Press.

Roberts, Adam (2006) *Science Fiction: The New Critical Idiom.* New York: Routledge.

Roberts, Adam (2007) *The History of Science Fiction.* New York: Palgrave.

Roberts, Cecilia (2007) *Messengers of Sex: Hormones, Biomedicine and Feminism.* Cambridge: Cambridge University Press.

Roberts, Dorothy (2012) *Fatal Invention: How Science, Politics, and Big Busines Recreate Race in the Twenty-First Century.* New York: The New Press.

Roberts, J.H. (1988) *Darwinism and the Divine in America: Protestant Intellectuals and Organic Evolution, 1859–1900.* Madison: University of Wisconsin Press.

Robinson, Daniel N. (1995) *An Intellectual History of Psychology.* Madison: University of Wisconsin Press.

Robinson, Gloria (1979) *A Prelude to Genetics: Theories of a Material Substance of Heredity, Darwin to Weismann.* Lawrence: Coronado Press.

Robinson, Michael F. (2006) *The Coldest Crucible: Arctic Exploration and American Culture.* Chicago: University of Chicago Press.

Rodgers, Daniel T. (1982) "In Search of Progressivism." *Reviews in American History* 10, 4: 113–32.

Rodgers, Daniel T. (1998) *Atlantic Crossings: Social Politics in a Progressive Age.* Cambridge: Harvard University Press.

Rogers, Naomi (1989) "Germs with Legs: Flies, Disease, and the New Public Health." *Bulletin of the History of Medicine* 63: 599–617.

Rogers, William Barton and Louis Agassiz (1860) "William Barton Rogers and Louis Agassiz Debate Evolution." *Proceedings of the Boston Society of Natural History* 7: 231–76.

Rojas, Raúl and Ulf Hashage (eds.) (2000) *The First Computers: History and Architectures.* Cambridge: MIT Press.

Roland, Alex (1985) *Model Research: The National Advisory Committee for Aeronautics, 1915–1958, Volume 1.* Washington: NASA SP-4103.

Roll-Hansen, Nils (1978) "Drosophila Genetics: A Reductionist Research Program." *Journal of the History of Biology* 11: 159–210.

Rome, Adam Ward (2001) *The Bulldozer in the Countryside: Suburban Sprawl and the*

Rise of American Environmentalism. Cambridge: Cambridge University Press.

Rome, Adam Ward (2013) *The Genius of Earth Day: How a 1970 Teach-in Unexpectedly Made the First Green Generation.* New York: Hill and Wang.

Romo, A.A. (2010) *Brazil's Living Museum: Race, Reform, and Tradition in Bahia.* Chapel Hill: University of North Carolina Press.

Ronda, J.P. (1972) "The European Indian: Jesuit Civilization Planning in New France." *Church History* 41, 3: 385–95.

Root, Waverly and Richard de Rochemont (1976) *Eating in America: A History.* New York: Harper Collins.

Rosario, Vernon (ed.) (1997) *Science and Homosexualities.* New York: Routledge.

Rosario, Vernon (2002) *Homosexuality and Science: A Guide to the Debates.* Santa Barbara: ABC-CLIO.

Rose Polytechnic Institute: Memorial Volume (1909) Terre Haute: Rose Polytechnic Institute.

Rose, Hillary (1994) *Love, Power, and Knowledge: Towards a Feminist Transformation of the Sciences.* Bloomington: Indiana University Press.

Rosen, Christine (2004) *Preaching Eugenics: Religious Leaders and the American Eugenics Movement.* New York: Oxford University Press.

Rosen, George (1993) *A History of Public Health.* Baltimore: Johns Hopkins University Press.

Rosenberg, Charles (1962) *The Cholera Years: The United States in 1832, 1849, and 1866.* Chicago: Chicago University Press.

Rosenberg, Charles (1977) "Rationalization and Reality in the Shaping of American Agricultural Research." *Social Studies of Science* 7: 401–22.

Rosenberg, Charles (1997) *No Other Gods: On Science and American Social Thought.* Baltimore: Johns Hopkins University Press.

Rosenberg, Charles and Janet Golden (1992) *Framing Disease: Studies in Cultural History* New Brunswick: Rutgers University Press.

Rosenberg, Rosalind (1982) *Beyond Separate Spheres: Intellectual Roots of American Feminism.* New Haven: Yale University Press.

Rosenof, Theodore (1997) *Economics in the Long Run: New Deal Theorists and their Legacies, 1933–1993.* Chapel Hill: University of North Carolina Press.

Ross, Dorothy 1991. *The Origins of American Social Science.* Cambridge: Cambridge University Press.

Rosser, Sue V. (1990) *Female-Friendly Science: Applying Women's Studies Methods and Theories to Attract Students.* New York: Pergamon.

Rossi, Michael (2010) "Fabricating Authenticity: Modeling a Whale at the American Museum of Natural History, 1906–1974." *Isis* 101, 2: 338–61.

Rossiter, Margaret W. (1971) "Benjamin Silliman and the Lowell Institute: The Popularization of Science in Nineteenth-Century America." *New England Quarterly* 44: 602–26.

Rossiter, Margaret W. (1974) "Women Scientists in America Before 1920." *American Scientist* 62: 312–23.

Rossiter, Margaret W. (1975). *The Emergence of Agricultural Science: Justus Liebig and the Americans, 1840–1880.* New Haven: Yale University Press.

Rossiter, Margaret W. (1980) "'Women's Work' in Science, 1880–1910. *Isis* 71: 381–98.

Rossiter, Margaret W. (1982) *Women Scientists in America: Struggles and Strategies to 1940.* Baltimore: Johns Hopkins University Press.

Rossiter, Margaret W. (1993) "The Matthew Matilda Effect in Science." *Social Studies of Science* 23: 325–41.

Rossiter, Margaret W. (1995) *Women Scientists in America: Before Affirmative Action, 1940–1972.* Baltimore: Johns Hopkins University Press.

Rossiter, Margaret W. (2002) "Writing Women into Science" in Jonathan Monroe (ed.), *Writing and Revising the Disciplines.* Ithaca: Cornell University Press, 54–72.

Rossiter, Margaret W. (2003) "A Twisted Tale: Women in the Physical Sciences in the Nineteenth and Twentieth Centuries," in Mary Jo Nye (ed.), *Cambridge History of Science.*

Cambridge: Cambridge University Press, 54–71.

Rossiter, Margaret W. (2012) *Women Scientists in America: Forging a New World Since 1972*. Baltimore: Johns Hopkins University Press.

Rothenberg, Marc (1985) "History of Astronomy." *Osiris* 1: 117–31.

Rothfels, Nigel (2002) *Savages and Beasts: The Birth of the Modern Zoo*. Baltimore: Johns Hopkins University Press.

Rothman, Hal K. (2000) *Saving the Planet: The American Response to the Environment in the Twentieth Century*. Chicago: Ivan R. Dee.

Rothman, Hal K., Gerald D. Nash, and Richard W. Etulain (1998) *The Greening of a Nation? Environmentalism in the United States since 1945*. Fort Worth: Harcourt Brace.

Roughgarden, Joan (2004) *Evolution's Rainbow: Diversity, Gender, and Sexuality in Nature and People*. Berkley: University of California Press.

Rowe, John Howland (1965) "The Renaissance Foundations of Anthropology." *American Anthropologist* 67, 1: 1–20.

Rowland, Henry (1883) "A Plea for Pure Science." *Science* 2: 242–50.

Roy, Srirupa (2007) *Beyond Belief: India and the Politics of Postcolonial Nationalism*. Durham: Duke University Press.

Royal Society (1985) *The Public Understanding of Science*. London: Royal Society.

Rozwadowski, Helen M. (2002) *The Sea Knows No Boundaries: A Century of Marine Sciences under ICES*. Seattle: University of Washington Press.

Rozwadowski, Helen (2005) *Fathoming the Ocean: Human Enterprise and the Opening of the Deep Sea*. Cambridge: Harvard University Press.

Rubin, Gayle (1975) "The Traffic in Women: Notes on the 'Political Economy' of Sex," in Rayna R. Reiter (ed.), *Toward and Anthropology of Women*. New York: Monthly Press Review, 157–210.

Rubner, M. (1913) "Nutrition for the People." *Journal of Home Economics* 1: 1–25.

Ruckmich, C. (1912) "The History and Status of Psychology in the United States." *American Journal of Psychology* 23: 517–31.

Ruckmich, C. (1926) "Development of Laboratory Equipment in Psychology in the United States." *American Journal of Psychology* 23: 582–92.

Ruddick, Sara and Pamela Daniels (eds.) (1977) *Working it Out: Twenty-Three Writers, Scientists and Scholars Talk about Their Lives*. New York: Pantheon.

Rudolph, J.L. (2002) *Scientists in the Classroom: The Cold War Reconstruction of American Science Education*. New York: Palgrave.

Rudolph, J.L. (2005a). "Epistemology for the Masses: The Origins of the 'Scientific Method' in American Schools." *History of Education Quarterly* 45, 3: 341–76.

Rudolph, J.L. (2005b) "Turning Science to Account: Chicago and the General Science Movement in Secondary Education, 1905–1920." *Isis* 96, 3: 353–89.

Rudorf, George (1900) *The Periodic Classification and the Problem of Chemical Evolution*. London: Wittaker & Co.

Ruffin, J. Rixey (1997) "'Urania's Dusky Veils': Heliocentrism in Colonial Almanacs, 1700–1735." *New England Quarterly* 70: 306–13.

Rufus, W.C. (1924) "Astronomical Observatories in the United States prior to 1848." *Scientific Monthly* 19: 120–39.

Ruggles, Clive L.N. and Nicholas J. Saunders (eds.) (1993) *Astronomies and Cultures: Papers Derived from the Third Oxford International Symposium on Archaeoastronomy, St. Andrews, UK, September 1990*. Niwot: University Press of Colorado.

Ruley, John D. (2013) "Homer Newell and the Origins of Planetary Science in the United States" in Roger Launius (ed.), *Exploring the Solar System: The History and Science of Planetary Exploration*. New York: Palgrave Macmillan, 25–44.

Rumore, Gina (2009) "A Natural Laboratory, a National Monument: Carving Out a Place for Science in Glacier Bay, Alaska, 1879–1859." PhD Dissertation: University of Minnesota.

Rumore, Gina (2012) "Preservation for Science: The Ecological Society of America and the Campaign for Glacier Bay National Monument." *Journal of the History of Biology* 45: 613–50.

Ruse, Michael (1979) Sociobiology: Sense or Nonsense? Dordrecht: Reidel.

Ruse, Michael (1996) *Monad to Man: The Concept of Progress in Evolutionary Biology.* Cambridge: Harvard University Press.

Ruse, Michael (1999) *The Darwinian Revolution: Science Red in Tooth and Claw.* Chicago: University of Chicago Press.

Ruse, Michael (2000) The Evolution Wars: A Guide to the Debates. Santa Barbara: ABC-CLIO.

Ruse, Michael (2005) *The Evolution–Creation Struggle.* Cambridge: Harvard University Press.

Ruse, Michael (2006) *Darwinism and its Discontents.* Cambridge: Cambridge University Press.

Ruse, Michael (ed.) (2009) *Philosophy After Darwin: Classic and Contemporary Readings.* Princeton: Princeton University Press.

Ruse, Michael (2012) *The Philosophy of Human Evolution.* Cambridge: Cambridge University Press.

Ruse, Michael (ed.) (2013a) *The Cambridge Companion of Darwin and Evolutionary Thought.* Cambridge: Cambridge University Press.

Ruse, Michael (2013b) *The Gaia Hypothesis: Science on a Pagan Planet.* Chicago: University of Chicago Press.

Ruse, Michael and Robert J. Richards (eds.) (forthcoming) *The Cambridge Handbook of Evolutionary Ethics.* Cambridge: Cambridge University Press.

Russell, Richard Joel (1934) "Climatic Years." *Geographical Review* 24: 92–103.

Russett, Cynthia Eagle (1976) *Darwin in America: The Intellectual Response, 1865–1912.* San Francisco: W.H. Freeman.

Russett, Cynthia Eagle (1989) *Sexual Science: The Victorian Construction of Womanhood.* Cambridge: Harvard University Press.

Rutherford, Malcolm (2011) *The Institutionalist Movement in American Economics, 1918–1947.* Cambridge: Cambridge University Press.

Rydell, Robert W. (1984) *All the World's a Fair: Visions of Empire at American International Expositions, 1876–1916.* Chicago: University of Chicago Press.

Sachs, Aaron (2006) *The Humboldt Current: Nineteenth-Century Exploration and the Roots of American Environmentalism.* New York: Viking.

Sachs, C. (1996) *Gendered Fields: Rural Women, Agriculture, and Environment.* Boulder: Westview Press.

Sack, D. (2001) *Whitebread Protestants: Food and Religion in American Culture.* New York: Palgrave-Macmillan.

Sackman, Douglas Cazaux (2010) *Wild Men: Ishi and Kroeber in the Wilderness of Modern America.* Oxford: Oxford University Press.

Sahlins, Marshall (1999) "What is Anthropological Enlightenment? Some Lessons of the Twentieth Century." *Annual Review of Anthropology* 28: i–xxii.

Saint-Arnaud P. and P. Feldstein (2009) *African American Pioneers of Sociology: A Critical History.* Toronto: University of Toronto Press.

Sale, Kirkpatrick and Eric Foner (1993) *The Green Revolution: The American Environmental Movement, 1962–1992.* New York: Hill and Wang.

Samuelson, Paul A. (1948) *Economics: An Introductory Analysis.* New York: McGraw-Hill.

Sandage, Allan (2004) *The Mount Wilson Observatory: Breaking the Code of Cosmic Evolution. Centennial History of the Carnegie Institution of Washington.* Cambridge: Cambridge University Press.

Santesmases, Maria Jesus (2010) "Size and the Centromere: Translocations and Visual Cultures in Early Human Genetics," in Luis Campos and Alexander Von Schwerin (eds.), *Making Mutations: Objects, Practices, Contexts.* Berlin: Max Plank Institute for the History of Science.

Sapp, Jan (1987) *Beyond the Gene: Cytoplasmic Inheritance and the Struggle for Authority in Genetics.* New York: Oxford University Press.

Sarewitz, Daniel (1996) *Frontiers of Illusion: Science, Technology, and the Politics of Progress.* Philadelphia: Temple University Press.

Saunders, F.H. and G. Stanley Hall (1900) "Pity." *American Journal of Psychology* 11: 534–91.

Sbicca, J. (2012) "Growing Food Justice by Planting an Anti-Oppression Foundation: Opportunities and Obstacles for a Budding Social Movement." *Agriculture and Human Values* 29: 455–66.

Scerri, Eric R. (2007) *The Periodic Table: Its Story and its Significance.* New York: Oxford University Press.

Scerri, Eric R. (2011) *The Periodic Table: A Very Short Introduction.* New York: Oxford University Press.

Schechner, Sara J (1982). "John Prince and Early American Scientific Instrument Making," in Frederick S. Allis, Jr. and Philip C.F. Smith (eds.), Sibley's Heir: A Volume in Memory of Clifford Kenyon Shipton. Vol. 59 of Publications of the Colonial Society of Massachusetts. Boston: Colonial Society of Massachusetts, 431–503.

Schechner, Sara J. (1992) "From Heaven's Alarm to Public Appeal: Comets and the Rise of Astronomy at Harvard," in Clark A. Elliott and Margaret W. Rossiter (eds.), *Science at Harvard University: Historical Perspectives.* Bethlehem: Lehigh University Press, 28–54.

Schechner, Sara J. (1996) "Tools for Teaching and Research: John Prince, the Deerfield Academy, and Educational Reform in the Early Republic." *Rittenhouse* 10: 97–120.

Schechner, Sara J. (1997) *Comets, Popular Culture, and the Birth of Modern Cosmology.* Princeton: Princeton University Press.

Schechner, Sara J. (2001) "The Material Culture of Astronomy in Daily Life: Sundials, Science, and Social Change." *Journal for the History of Astronomy* 32: 189–222.

Schechner, Sara J (2004) "Sundials of Newfoundland." Paper Presented at the Annual Meeting of the North American Sundial Society, Tenafly, New Jersey, August 19–22.

Schechner, Sara J (2005) "Between Knowing and Doing: Mirrors and their Imperfections in the Renaissance," in Optics, Instruments and Painting, 1420–1720: Reflections on the Hockney–Falco Thesis. Vol. 10 of Early Science and Medicine, 137–62.

Schechner, Sara J (2006). "Benjamin Franklin and a Tale of Two Electrical Machines," in Benjamin Franklin: A How-To Guide. Vol. 17, nos. 1-2 of Harvard Library Bulletin, 33–40.

Schechner, Sara J (2007) "The Adventures of Captain John Smith among the Mathematical Practitioners: Cosmology, Mathematics, and Power at the Time of Jamestown," in *Science and Early Jamestown. Rittenhouse: Journal of the Scientific Instrument Enterprise* 21: 126–44.

Schechner, Sara J. (2009) "Telescopes in Colonial and Federal America, 1620–1820." Paper presented at the Annual Meeting of the Historical Astronomy Division of the American Astronomical Society, Long Beach, California, January 4–6 and at the Stellafane Convention, Hartness House History of Astronomy Workshop, Springfield, Vermont, August 13–16.

Schechner, Sara J. (2012) "David P. Wheatland (1898–1993): Scholar, Author, Avid Collector, Sine qua non for the Collection of Historical Scientific Instruments." Collection of Historical Scientific Instruments, Harvard University. http://chsi.harvard.edu/chsi_wheatland.html

Schechner, Sara J. (2014) "How Telescopes Came to New England, 1620–1740," in Marcus Granato and Marta C. Lourenço (eds.), *Scientific Instruments in the History of Science: Studies in Transfer, Use and Preservation.* Rio de Janeiro: Museu de Astronomia e Ciências Afins, 69–78.

Scheper-Hughes, Nancy (2000) "The Global Traffic in Human Organs." *Current Anthropology* 41: 191–224.

Schiebinger, Londa (1989) *The Mind Has No Sex? Women in the Origins of Modern Science.* Cambridge: Harvard University Press.

Schiebinger, Londa (1999) *Has Feminism Changed Science?* Cambridge: Harvard University Press.

Schiebinger, Londa (2004) *Plants and Empire: Colonial Bioprospecting in the Atlantic World.* Cambridge: Harvard University Press.

Schiebinger, Londa (2005) "Forum Introduction: The European Colonial Science Complex." *Isis* 96: 52–5.

Schiebinger, Londa (ed.) (2008) *Gendered Innovations in Science and Engineering*. Stanford: Stanford University Press.

Schiebinger, Londa (ed.) (2014) *Women and Gender in Science and Technology*. New York: Routledge.

Schiebinger, Londa, and Claudia Swan (eds.) (2005) *Colonial Bota: Science, Commerce, and Politics in the Early Modern World*. Philadelphia: University of Philadelphia Press.

Schneider, Thomas D. (2000) "Evolution of Biological Information." *Nucleic Acids Research* 28, 14: 2794–9.

Schneider, Thomas D. (2014) "Refuting Michael Behe's 'Irreducible Complexity' with Roman Arches." http://schneider. ncifcrf. gov/paper/ev/behe/

Schloegel, J. and H. Schmidgen (2002) "General Physiology, Experimental Psychology, and Evolutionism: Unicellular Organisms as Objects of Psychophysiological Research, 1877–1918." *Isis* 93: 614–45.

Schmitt, Peter J. (1990) *Back to Nature: The Arcadian Myth in Urban America*. Baltimore: Johns Hopkins University Press.

Schneider, Daniel W. (2000) "Local Knowledge, Environmental Politics, and the Founding of Ecology in the United States: 'Stephen Forbes and the Lake as a Microcosm (1887)'." *Isis* 91, 4: 681–705.

Schoen, Johanna (2005) *Choice and Coercion: Birth Control, Sterilization, and Abortion in Public Health and Welfare*. Chapel Hill: University of North Carolina Press.

Scholnick, Robert J. (1992) "Permeable Boundaries: Literature and Science in America," in Robert J. Scholnick (ed.), *American Literature and Science*. Lexington: University Press of Kentucky, 1–17.

Schrödinger, Erwin (1944) *What is Life?* Cambridge: Cambridge University Press.

Schroeder, H.A., J.J. Balassa, and W.H. Vinton, Jr. (1965) "Chromium, Cadmium and Lead in Rats: Effects on Life Span, Tumors and Tissue Levels." *Journal of Nutrition* 86: 51–66.

Schuchert, Charles and Clara M. Levene (1940) *O.C. Marsh: Pioneer in Paleontology*. New Haven: Yale University Press.

Schucking, Englebert (1989) "The First Texas Symposium on Relativistic Astrophysics." *Physics Today*: 46–52.

Schulten, Susan (2001) *Geographical Imagination in America, 1880–1950*. Chicago: University of Chicago Press.

Schultz, T.W. (1961) "Investment in Human Capital." *American Economic Review* 51, 1: 1–17.

Schuurbiers, D. and E. Fisher (2009) "Lab-Scale Intervention." *EMBO Reports* 10: 424–7.

Schwartz, Neena B. (2010) *A Lab of My Own*. Amsterdam: Rodophi.

Schweber, Silvan (1986) "The Empiricist Temper Regnant: Theoretical Physics in the United States, 1920–1950." *Historical Studies in the Physical and Biological Sciences* 17: 55–98.

Scott, D. (1995) "Colonial Governmentality." *Social Text* 43: 191–220.

Scott, Joan W. (1986) "Gender: A Useful Category of Historical Analysis." *American Historical Review* 91: 1053–75.

Scrinis, G. (2013) *Nutritionism: The Science and Politics of Dietary Advice*. New York: Columbia University Press.

Scroggs, Jack B. (1961) "Carpetbagger Constitutional Reform in the South Atlantic States, 1867–1868." *Journal of Southern History* 27, 4: 475–93.

Seaborg, Glenn T. (1945) "The Chemical and Radioactive Properties of the Heavy Elements." *Chemical & Engineering News* 23: 2190–3.

Seaborg, Glenn T. (1998) *A Chemist in the White House: From the Manhattan Project to the End of the Cold War*. Washington, DC: American Chemical Society.

Seaborg, Glenn T. (2001) *Adventures in the Atomic Age: From Watts to Washington*. New York: Farrar, Straus and Giroux.

Sears, Paul B. (1935) *Deserts on the March*. Norman: University of Oklahoma Press.

Sears, Paul B. (1964) "Ecology – A Subversive Subject." *Bioscience* 14: 11–13.

Secord, Anne (1994) "Science in the Pub: Artisan Botanists in Early Nineteenth-Century Lancashire." *History of Science* 32: 269–315.

Secord, James A. (2000) *Victorian Sensation: The Extraordinary Publication, Reception and Secret Authorship of Vestiges of the Natural History of Creation*. Chicago: University of Chicago Press.

Secord, James A. (2004) "Knowledge in Transit." *Isis* 95: 654–72.

Segerstrale, U. (2000) *Defenders of the Truth: The Battle for Science in the Sociobiology Debate and Beyond*. Oxford: Oxford University Press.

Selin, Helaine (ed.) (2000) *Astronomy across Cultures: The History of Non-Western Astronomy*. Dordrecht: Kluwer Academic.

Sell, S. (2003) *Private Property, Public Law: The Globalization of Intellectual Property Rights*. New York: Cambridge University Press.

Sellers, Charles Coleman (1980) *Mr. Peale's Museum: Charles Wilson Peale and the First Popular Museum of Natural Science and Art*. New York: W.W. Norton.

Semonin, Paul (2000) *American Monster. How the Nation's First Prehistoric Creature Became a Symbol of National Identity*. New York: New York University Press.

Semple, Ellen Churchill (1911) *Influences of Geographic Environment, on the Basis of Ratzel's System of Anthropogeography*. New York: H. Holt.

Sepkoski, David (2012) *Rereading the Fossil Record: The Growth of Paleobiology as an Evolutionary Discipline*. Chicago: University of Chicago Press.

Sepkoski, David (2014) "Paleontology at the 'High Table'? Popularization and Disciplinary Status in Recent Paleontology." *Studies in History and Philosophy of Biological and Biometrical Sciences* 45: 133–8.

Sepkoski, David and Michael Ruse (eds.) (2009) *The Paleobiological Revolution: Essays on the Growth of Modern Paleontology*. Chicago: University of Chicago Press.

Servos, John W. (1983) "To Explore the Borderland: The Foundation of the Geophysical Laboratory of the Carnegie Institution of Washington." *Historical Studies in the Physical Sciences* 14: 147–85.

Servos, John W. (1986) "Mathematics and the Physical Sciences in America, 1880–1930." *Isis* 77: 611–29.

Servos, John W. (1990) *Physical Chemistry from Ostwald to Pauling: The Making of a Science in America*. Princeton: Princeton University Press.

Seth, Sanjay, Leela Gandhi, and Michael Dutton (1998) "Postcolonial Studies: A Beginning." *Postcolonial Studies* 1: 7–11.

Seth, Suman (2009) "Putting Knowledge in its Place: Science, Colonialism, and the Postcolonial." *Postcolonial Studies* 12, 4: 373–88.

Shackley, Simon and Brian Wynne (1996) "Representing Uncertainty in Global Climate Change Science and Policy: Boundary-Ordering Devices and Authority." *Science, Technology, & Human Values* 21: 275–302.

Shapin, Steven (1990) "Science and the Public," in R.C. Olby, G.N. Cantor, J.R.R. Christie, and M.J.S. Hodge (eds.), *Companion to the History of Modern Science*. London: Routledge, 990–1007.

Shapin, Steven and Simon Schaffer (1989) *Leviathan and the Air Pump: Hobbes, Boyle, and the Experimental Life*. New York: University Press.

Shapiro, Adam R. (2008) "Civic Biology and the Origin of the School Antievolution Movement." *Journal of the History of Biology* 41, 3: 409–33.

Shapiro, Adam R. (2010) "State Regulation of the Textbook Industry," in A.R. Nelson and J.L. Rudolph (eds.), *Education and the Culture of Print in Modern America*. Madison: University of Wisconsin Press.

Shapiro, Adam R. (2013). *Trying Biology: The Scopes Trial, Textbooks, and the Antievolution Movement in American Schools*. Chicago: University of Chicago Press.

Shapiro, Harry L. (1929) *Descendants of the Mutineers of the Bounty*. Honolulu: The Bishop Museum.

Shapiro, Harry L. (1952) "Revised Version of UNESCO Statement on Race." *American Journal of Physical Anthropology* 10, 3: 363–8.

Shapiro, L. (2001) *Perfection Salad: Women and Cooking at the Turn of the Century*. New York: Modern Library.

Shapley, Harlow (1918) "The Age of the Earth." *Publications of the Astronomical Society of the Pacific* 30, 177: 283–98.

Sharp, J.P. (2009) *Geographies of Postcolonialism: Spaces of Power and Representation.* London: Sage.

Sheets-Pyenson, Susan (1988) *Cathedrals of Science: The Development of Colonial Natural History Museums during the Late Nineteenth Century.* Kingston: McGill-Queen's University Press.

Sheffield, Suzanne Le-May (2004) Women and Science: Social Impact and Interaction. Santa Barbara: ABC-CLIO.

Sheldon, Myrna Perez (2014) "Claiming Darwin: Stephen Jay Gould in Contests over Evolutionary Orthodoxy and Public Perception." *Studies in History and Philosophy of Biological and Biometrical Sciences* 45: 139–47.

Shelford, Victor E. (1926) *The Naturalist's Guide to the Americas.* Baltimore: Williams and Wilkins.

Shelford, Victor E. (1929) *Laboratory and Field Ecology: The Responses of Animals as Indicators of Correct Working Methods.* Baltimore: Williams and Wilkins.

Shelford, Victor E. (1963) *The Ecology of North America.* Champaign-Urbana: University of Illinois Press.

Sherkat, Darren E. (2011) "Religion and Scientific Literacy in the United States." *Social Science Quarterly* 92: 1134–50.

Shindell, Matthew B. (2010) "Instruments and Practices in the Development of Planetary Geology." *Spontaneous Generations: A Journal for the History and Philosophy of Science* 4, 1: 191–230.

Shindell, Matthew B. (2015) "From the End of the World to the Age of the Earth: The Cold War Development of Isotope Geochemistry at the University of Chicago and Caltech," in Naomi Oreskes and Erik Conway (eds.), *Nation and Knowledge: Science and Technology in the Global Cold War.* Cambridge: MIT Press.

Shipman, Pat (1994) *The Evolution of Racism: The Human Differences and the Use and Abuse of Science.* Cambridge: Harvard University Press.

Shirk, J., H. Ballard, C. Wilderman, T. Phillips, A. Wiggins, R. Jordan, E. Mccallie et al. (2012) "Public Participation in Scientific Research: A Framework for Deliberate Design." *Ecology and Society* 17: 29.

Shiva, Vandana (1988) "Reductionst Science as Epistemological Violence," in Ashis Nandy (ed.), *Science, Hegemony and Violence: A Requiem for Modernity.* New York: Oxford University Press.

Shiva, Vandana (1989) *Staying Alive: Women, Ecology, and Development.* London: Zed Books.

Shiva, Vandana (1992) *The Violence of the Green Revolution.* London: Zed Books.

Shiva, Vandana (2000) *Stolen Harvest: The Highjacking of the Global Food Supply.* Cambridge: South End Press.

Shoemaker, Philip Stanley (1991) "Stellar Impact: Ormsby Macknight Mitchel and Astronomy in Antebellum America." PhD Dissertation: University of Wisconsin-Madison.

Shortland, Michael (1993) Science and Nature: Essays in the History of the Environmental Sciences. Stanford in the Vale: British Society for the History of Science.

Shrader-Frechette, K. (2007) *Taking Action, Saving Lives: Our Duties to Protect Environmental and Public Health.* New York: Oxford University Press.

Shryock, Richard (1972) "Germ Theories in Medicine Prior to 1870: Further Comments on Continuity in Science." *Clio Medica* 7: 81–109.

Shteir, Ann B. (1996) *Cultivating Women, Cultivating Science: Flora's Daughters and Botany in England, 1760–1860.* Baltimore: Johns Hopkins University Press.

Shteir, Ann B. and Bernard Lightman (2006) *Figuring it Out: Science, Gender, and Visual Culture.* Lebanon: Dartmouth College Press

Shukers, C.F. et al. (1931) "Food Intake in Pregnancy, Lactation, and Reproductive Rest in the Human Mother." *Journal of Nutrition* 4: 399–410.

Shuster, Evelyne (2007) "Microarray Genetic Screening: A Prenatal Roadblock for Life?" *Lancet* 369, 9560: 526–9.

Sidgwick, H. (1876) The Theory of Evolution in its Application to Practice. *Mind* 1: 52–67.

Silverman, Sydel (1981) *Totems and Teachers: Key Figures in the History of Anthropology.* New York: Columbia University Press.

Sime, Ruth Lewin (1996) *Lise Meitner: A Life in Physics.* Berkeley: University of California Press.

Simpson, George Gaylord (1942) "The Beginnings of Vertebrate Paleontology in North America." *Proceedings of the American Philosophical Society* 86, 1: 130–88.

Simpson, George Gaylord (1944) *Tempo and Mode in Evolution.* New York: Columbia University Press.

Simpson, George Gaylord (1949) *The Meaning of Evolution: A Study of the History of Life and of its Significance for Man.* New Haven: Yale University Press.

Sismondo, S. (2004) *An Introduction to Science and Technology Studies.* Malden: Blackwell.

Sismondo, S. (2008) "Pharmaceutical Company Funding and its Consequences: A Qualitative Systematic Review." *Contemporary Clinical Trials* 29: 109–13.

Sismondo, S. (2009) "Ghosts in the Machine: Publication Planning in the Medical Sciences." *Social Studies of Science* 39: 171–98.

Sklar, K.K. (1991) "Hull House Maps and Papers: Social Science as Women's Work in the 1890s," in M. Bulmer, K. Bales and K.K. Sklar (eds.), *The Social Survey in Historical Perspective,* 1880–1940. Cambridge: Cambridge University Press.

Skolnik, Herman (ed.) (1976) *A Century of Chemistry: The Role of Chemists and the American Chemical Society.* Washington, DC: American Chemical Society.

Skopek, Jeffrey M. (2011) "Principles, Exemplars, and Uses of History in Early 20th-Century Genetics." *Studies in History and Philosophy of Biological and Biomedical Sciences* 42: 210–25.

Slaughter, Aimee (2013) "Harnessing the Modern Miracle: Physicists, Physicians, and the Making of American Radium Therapy." PhD Dissertation: University of Minnesota.

Slaughter, S. and G. Rhoades (2004) *Academic Capitalism and the New Economy.* Baltimore: Johns Hopkins University Press.

Slayton, Rebecca (2012) "From a 'Dead Albatross' to Lincoln Labs: Applied Research and the Making of a Normal Cold War University." *Historical Studies in the Natural Sciences* 42: 255–82.

Slayton, Rebecca (2013) *Arguments that Count: Physics, Computing, and Missile Defense, 1949–2012.* Cambridge: MIT Press.

Sleigh, C. (2002) "Brave New Worlds: Trophallaxis and the Origin of Society in the Early Twentieth Century." *Journal of the History of the Behavioral Sciences* 38: 133–56.

Sleigh, C. (2007) *Six Legs Better: A Cultural History of Myrmecology.* Baltimore: Johns Hopkins University.

Sloan, Phillip R. and Brandon Fogel (2011) *Creating Physical Biology: The Three-Man Paper and Early Molecular Biology.* Chicago: University of Chicago Press.

Slotten, Hugh R. (1994) *Patronage, Practice, and the Culture of American Science: Alexander Dallas Bache and the U. S. Coast Survey.* New York: Cambridge University Press.

Small, A. (1905) *General Sociology.* Chicago: University of Chicago Press.

Smart, Charles E. (1962–1967) *The Makers of Surveying Instruments in America since 1700.* Troy: Regal Art Press.

Smedley, Audrey (1999) *Race in North America: Origin and Evolution of a Worldview.* Boulder: Westview Press.

Smeenk, Christopher (2003) "*Approaching the Absolute Zero of Time: Theory Development in Early Universe Cosmology.*" PhD Dissertation: University of Pittsburgh.

Smith, Adam (1976) *An Inquiry into the Nature and Causes of the Wealth of Nations.* R.H. Campbell and A.S. Skinner (eds). Oxford: Clarendon Press.

Smith, B. (1990) *American Science Policy Since World War II.* Washington: Brookings Institution Press.

Smith, Hugh (1918) *Report of the U.S. Commissioner of Fishes.* Washington: Government Printing Office.

Smith, Michael L. (1987) *Pacific Visions: California Scientists and the Environment, 1850–1915*. New Haven: Yale University Press.

Smith, Robert W. (1982) *The Expanding Universe: Astronomy's "Great Debate," 1900–1931*. Cambridge: Cambridge University Press.

Smith, Robert W. (1989) *The Space Telescope: A Study of NASA, Science, Technology, and Politics*. Cambridge: Cambridge University Press.

Smith, Robert W. (1997) "Engines of Discovery: Scientific Instruments and the History of Astronomy and Planetary Science in the United States in the Twentieth Century." *Journal for the History of Astronomy* 28: 49–77.

Smith, Robert W. (2006) "Beyond the Big Galaxy: The Structure of the Stellar System 1900–1952." *Journal for the History of Astronomy*, 37: 307–42.

Smith, Robert W. (2008) "Beyond the Galaxy: The Development of Extragalactic Astronomy 1885–1965, Part 1." *Journal for the History of Astronomy* 39: 91–119.

Smith, Samuel Stanhope (1787) *An Essay on the Causes of the Variety of Complexion and Figure in the Human Species to which are Added Strictures on Lord Kaim's Discourse of the Original Diversity of Mankind*. Philadelphia: Aitken.

Smith, Samuel Stanhope (1810) *An Essay on the Causes of the Variety of Complexion and Figure in the Human Species*. New Brunswick: *J. Simpson*.

Smith, Tim D. (2007) *Scaling Fisheries: The Science of Measuring the Effects of Fishing, 1855–1955*. Cambridge: Cambridge University Press.

Smocovitis, Vassiliki Betty (1996) *Unifying Biology: The Evolutionary Synthesis and Evolutionary Biology*. Princeton: Princeton University Press.

Smocovitis, Vassiliki Betty (2009) "The 'Plant Drosophila:' E.B. Babcock, the Genus Crepis, and the Evolution of a Genetics Research Program at Berkeley, 1915–1947." *Historical Studies in the Natural Science* 39: 300–55.

Smocovitis, Vassiliki Betty (2012) "Humanizing Anthropology, the Evolutionary Synthesis, and the Prehistory of Biological Anthropology, 1927–1962." *Current Anthropology* 53: S108–S125.

Smocovitis, Vassiliki Betty (2014) "Disciplining and Popularizing: Evolution and its Publics from the Modern Synthesis to the Present." *Studies in History and Philosophy of Biological and Biometrical Sciences* 45: 111–13.

Smyth, Thomas (1910) *The Unity of the Human Races*. New York: R.L. Bryan.

Snell, K.D.M. (1985) *Annals of the Labouring Poor, Social Change and Agrarian England 1660–1900*. Cambridge: Cambridge University Press.

Snow, Charles Percy (1959) *The Two Cultures*. Cambridge: Cambridge University Press.

Sober, Elliott (2007) "What is Wrong with Intelligent Design?" *Quarterly Review of Biology* 82, 1: 3–8.

Sober, Elliott (2008a) *Evidence and Evolution: The Logic Behind the Science*. Cambridge: Cambridge University Press.

Sober, Elliott (2008b) "Fodor's Bubbe Meise against Darwinism." *Mind & Language* 23, 1: 42–9.

Sober, Elliott and D.S. Wilson (1997) *Unto Others: The Evolution and Psychology of Unselfish Behavior*. Cambridge: Harvard University Press.

Solar and Space Physics Survey Committee (2003). *The Sun to the Earth – and Beyond: A Decadal Research Strategy in Solar and Space Physics*. Washington: National Academy Press.

Solar Physics Division (2014) "Membership Statistics." http://spd.aas.org/navbar_members.html.

Solar System Exploration Committee (1983) *Planetary Exploration Through Year 2000: A Core Program*. Washington: NASA.

Solar System Exploration Committee (1986) *Planetary Exploration Through the Year 2000: An Augmented Program*. Washington: NASA.

Solecki, Ralph (1971) *Shanidar, the First Flower People*. New York: Knopf.

Solomon, Barbara Miller (1973) "Historical Determinants in Individual Life

Experiences of Successful Professional Women." *Annals of the New York Academy of Sciences* 208: 170–8.

Solovey, Mark (2001) "Project Camelot and the 1960s Epistemological Revolution: Rethinking the Politics-Patronage-Social Science Nexus." *Social Studies of Science* 31, 2: 171–206.

Solovey, Mark and Hamilton Cravens (2012) *Cold War Social Science: Knowledge Production, Liberal Democracy, and Human Nature.* New York: Palgrave Macmillan.

Somerville, Siobhan (2002) *Queering the Color Line: Race and the Invention of Homosexuality in American Culture.* Durham: Duke University Press.

Sommer, Marianne (2008) "History in the Gene: Negotiations between Molecular and Organismal Anthropology." *Journal of the History of Biology* 41: 473–528.

Sommer, Marianne (2012) "Human Evolution Across the Disciplines: Spotlights on American Anthropology and Genetics." *History and Philosophy of the Life Sciences* 34: 211–36.

Sontag, Susan (1990) *Illness as Metaphor and AIDS and its Metaphors.* New York: Anchor.

Sörlin, Sverker (2011) "The Anxieties of a Science Diplomat: Field Coproduction of Climate Knowledge and the Rise and Fall of Hans Ahlmann's 'Polar Warming.'" *Osiris* 26: 66–88.

Soulé, Michael (1987) "History of the Society for Conservation Biology: How and Why We Got Here." *Conservation Biology* 1, 1: 4–5.

Soulé, Michael and Gary Lease (eds.) (1995) *Reinventing Nature? Responses to Postmodern Deconstruction.* Washington: Island Press.

Soulé, Michael, and Bruce A. Wilcox (eds.) (1980) *Conservation Biology: An Evolutionary-Ecological Perspective.* Sunderland: Sinauer Associates.

Southall, James (1875) *The Recent Origin of Man, as Illustrated by Geology and the Modern Science of Pre-Historic Archæology.* Philadelphia: Lippincott.

Southern, D.W. (1987) *Gunnar Myrdal and Black-White Relations: The Use and Abuse of an American Dilemma, 1944–1969.* Baton Rouge: Louisiana State University Press.

Space Science Board (Hess 1966) *Space Research: Directions for the Future.* Washington: National Academy Press.

Space Science Board (McDonald 1968) *Planetary Exploration, 1968–1975: Report of a Study by the Space Science Board.* Washington: National Academy Press.

Space Science Board (Goody 1976) *Report on Space Science.* Washington: National Academy Press.

Spahn, Hannah (2011) *Thomas Jefferson, Time and History.* Charlottesville: University of Virginia Press.

Spanagel, David I. (2014) *Dewitt Clinton and Amos Eaton: Geology and Power in Early New York.* Baltimore: Johns Hopkins University Press.

Spanier, Bonnie (1995) *Im/Partial Science: Gender Ideology in Molecular Biology.* Bloomington: Indiana University Press.

Spence, Mark David (1999) *Dispossessing the Wilderness: Indian Removal and the Making of the National Parks.* New York: Oxford University Press.

Spencer, Frank (1979) "*Aleš Hrdlička, MD, 1869–1943: A Chronicle of the Life and Work of an American Physical Anthropologist.*" Ph.D. Dissertation: University of Michigan.

Spencer, Frank (1981) "The Rise of Academic Physical Anthropology in the United States, 1880–1980." *American Journal of Physical Anthropology* 56: 353–64.

Spencer, Frank (ed.) (1982) *A History of American Physical Anthropology, 1930–1980.* New York: Academic Press.

Spencer, Frank (1984) "The Neanderthals and their Evolutionary Significance: A Brief Historical Survey," in Fred H. Smith and Frank Spencer (eds.), *Origins of Modern Humans: A World Survey of the Fossil Evidence.* New York: Alan Liss, 1–49.

Spencer, Frank and Fred Smith (1981) "The Significance of Aleš Hrdlicka's 'Neanderthal Phase of Man': A Historical and Current Assessment." *American Journal of Physical Anthropology* 56: 435–59.

Spencer, Herbert (1879) *The Data of Ethics.* London: Williams and Norgate.

Spencer, Herbert (1884) *Man Versus the State*. London: Williams and Norgate.

Sponsel, Alistair (2009) *Coral Reef Formation and the Sciences of Earth, Life, and Sea, c. 1770–1952* PhD Dissertation: Princeton University.

Stage, Sarah and Virginia B. Vincenti (eds.) (1997) *Rethinking Home Economics: Women and the History of a Profession*. Ithaca: Cornell University Press.

Staley, Richard (2010) "Michelson and the Observatory: Physics and the Astronomical Community in Late Nineteenth-Century America," in David Aubin, Charlotte Bigg and H. Otto Sibum (eds.), *The Heavens on Earth*. Durham: Duke University Press, 225–52.

Stanley Goldberg (1987) "Putting New Wine in Old Bottles: The Reception of Relativity in America" in T.F. Glick (ed.), *The Comparative Reception of Relativity*. Dordrecht: Reidel, 1–26.

Stanley, Autumn (1995) *Mothers and Daughters of Invention: Notes for a Revised History of Technology*. New Brunswick: Rutgers University Press.

Stanton, William (1960) *The Leopard's Spots. Scientific Attitudes toward Race in America 1815–1819*. Chicago: University of Chicago Press.

Stanton, Willliam R. (1975) *The Great United States Exploring Expedition of* 1838–1843. Berkeley: University of California Press.

Star, Susan Leigh (1992) "Craft vs. Commodity, Mess vs. Transcendence: How the Right Tool Became the Wrong One in the Case of Taxidermy and Natural History," in Adele E. Clarke and Joan H. Fujimura (eds.), *The Right Tools for the Job: At Work in Twentieth-Century Life Sciences*. Princeton: Princeton University Press.

Stathern, Marilyn (1999) *Property, Substance and Effect: Anthropological Essays on Persons and Things*. London: Athlone Press.

Stearns, Raymond Phineas (1948) "Colonial Fellows of the Royal Society of London, 1661–1788." Notes and Records of the Royal Society of London, 178–246.

Stearns, Raymond Phineas (1970) *Science in the British Colonies of America*. Urbana: University of Illinois Press.

Stebbins, G. Ledyard (1950) *Variation and Evolution in Plants*. New York: Columbia University Press.

Steffen, Will; Paul J. Crutzen and J.R. McNeill (2007) "The Anthropocene: Are Humans Now Overwhelming the Great Forces of Nature?" *Ambio* 36, 8: 614–21.

Stegner, Wallace (1992) *Beyond the Hundredth Meridian: John Wesley Powell and the Second Opening of the West*. New York: Penguin.

Stehlin, D. (1993) "A Little 'Lite' Reading." FDA Consumer. Rockville: U.S. Department of Health and Human Service. DHHS Publication No. (FDA) 93-2262.

Steil, Benn (2013) *The Battle of Bretton Woods: John Maynard Keynes, Harry Dexter White, and the Making of a New World Order*. Princeton: Princeton University Press.

Stein, Barbara R. (2001) *On Her Own Terms: Annie Montague Alexander and the Rise of Science in the American West*. Berkeley: University of California Press.

Steinmetz, G. (2013) *Sociology & Empire: The Imperial Entanglements of a Discipline. Politics, History, and Culture*. Durham: Duke University Press.

Stepan, Nancy (1982) *The Idea of Race in Science: Great Britain, 1800–1960*. Hamden: Archon Books.

Stephens, Carlene E. (2002) *On Time: How America Has Learned to Live by the Clock*. Washington: Smithsonian Institution.

Stephens, Lester D. (2000) *Science, Race, and Religion in the American South. John Bachman and the Charleston Circle of Naturalists 1815–1895*. Chapel Hill: University of North Carolina Press.

Stephens, Lester D. and Dale R. Calder (2006) *Seafaring Scientist: Alfred Goldsborough Mayer, Pioneer in Marine Biology*. Columbia: University of South Carolina Press.

Sterckx, S. (2010) "Knowledge Transfer from Academia to Industry through Patenting and Licensing: Rhetoric and Reality," in H. Radder (ed.), *The Commodification of Academic Research: Science and the Modern University*. Pittsburgh: University of Pittsburgh Press, 44–64.

Sterling, Keir B. (1989) "Builders of the U.S. Biological Survey, 1885–1930." *Journal of Forest History* 33: 180–7.

Sterling, Keir B., Richard P. Harmond, George A. Cevasco, and Lorne F. Hammond (eds.) (1997) *Biographical Dictionary of American and Canadian Naturalists and Environmentalists*. Westport: Greenwood Press.

Stern, Alexandra Minna (2002) "Making Better Babies: Public Health and Race Betterment in Indiana, 1920–1935." *American Journal of Public Health* 92, 5: 742–52.

Stern, Alexandra Minna (2005) *Eugenic Nation: Faults and Frontiers of Better Breeding in Modern America*. Berkeley: University of California Press.

Stern, Alexandra Minna (2007) "We Cannot Make a Silk Purse Out of a Sow's Ear: Eugenics in the Hoosier Heartland." *Indiana Magazine of History* 103: 3–38.

Stern, Alexandra Minna (2012) *Telling Genes: The Story of Genetic Counseling in America*. Baltimore: Johns Hopkins University Press.

Stern, Curt (1949, 1960, 1973) *Principles of Human Genetics*. New York: W.H. Freedman.

Sternberg, Charles H. (1990) *The Life of a Fossil Hunter*. Bloomington: Indiana University Press.

Stevens, Hallam (2003) "Fundamental Physics and its Justifications, 1945–1993." *Historical Studies in the Physical and Biological Sciences* 34: 151–97.

Stevens, Hallam (2013) *Life Out of Sequence: A Data-Driven History of Bioinformatics*. Chicago: University of Chicago Press.

Stevenson, Lloyd (1955) "Science Down the Drain: On the Hostility of Certain Sanitarians to Animal Experimentation, Bacteriology, and Immunology." *Bulletin of the History of Medicine* 29: 1–26.

Steward, Julian Haynes (1955) *Theory of Culture Change: The Methodology of Multilinear Evolution*. Illinois: University of Illinois Press.

Stiebeling, H. (1933) U.S. Department of Agriculture. *Diets at Four Levels of Nutritive Content and Cost*. Circular No. 296. Washington: Government Printing Office.

Stigler, George J. (1966) *The Theory of Price*. London: Macmillan.

Stigler, George J. and Gary S. Becker (1977) "De Gustibus Non Est Disputandum." *American Economic Review* 67, 2: 76–90.

Stillwell, Devon (2013) "Interpreting the Genetic Revolution: A History of Genetic Counseling in the United States, 1930–2000." PhD Dissertation: McMaster University,

Stimpson, Catharine R. and Joan N. Burstyn (eds.) (1978) "Women, Science, and Society." *Signs* 4, 1.

Stocking, George W. (1965) "On the Limits of 'Presentism' and 'Historicism' in the Historiography of the Behavioral Sciences." *Journal of the History of the Behavioral Sciences* 1, 3: 211–18.

Stocking, George W. (1966) "Franz Boas and the Culture Concept in Historical Perspective." *American Anthropologist* 68, 4: 867–82.

Stocking, George W. (1968) *Race, Culture, and Evolution: Essays in the History of Anthropology*. Chicago: University of Chicago Press.

Stocking, George W. (1974) *The Shaping of American Anthropology 1883–1911*. New York: Basic Books.

Stocking, George W. (ed.) (1984) *Observers Observed: Essays on Ethnographic Fieldwork*. Madison: University of Wisconsin Press.

Stocking, George W. (1982) "The Persistence of Polygenist Thought in Post-Darwinian Anthropology," in *Races, Culture, and Evolution. Essays in the History of Anthropology*, Chicago: University of Chicago Press, 42–68.

Stocking, George W. (ed.) (1987) *Malinowski, Rivers, Benedict and Others: Essays on Culture and Personality*. Madison: University of Wisconsin Press.

Stocking, George W. (ed.) (1988) *Objects and Others: Essays on Museums and Material Culture*. Madison: University of Wisconsin Press.

Stocking, George W. (ed.) (1990) *Bones, Bodies and Behavior: Essays in Behavioral Anthropology*. Madison: University of Wisconsin Press.

Stocking, George W. (ed.) (1992) *The Ethnographer's Magic and Other Essays in the*

History of Anthropology. Madison: University of Wisconsin Press.

Stocking, George W. (ed.) (1996) *Volksgeist as Method and Ethic: Essays on Boasian Ethnography and the German Anthropological Tradition*. Madison: University of Wisconsin Press.

Stocking, George W. (2010) *Glimpses into My Own Black Box: An Exercise in Self-Deconstruction*. Madison: University of Wisconsin Press.

Stokes, D. (1997) *Pasteur's Quadrant: Basic Science and Technological Innovation*. Washington: Brookings Institution Press.

Stoler, Ann Laura (2002) *Carnal Knowledge and Imperial Power: Race and the Intimate in Colonial Rule*. Berkeley: University of California Press.

Stoll, Steven (2002) *Larding the Lean Earth: Soil and Society in Nineteenth-Century America*. New York: Hill and Wang.

Stoller, Robert J. (1964) "A Contribution to the Study of Gender Identity." *International Journal of Psychoanalysis* 45: 220–6.

Stoller, Robert J. (1968) *Sex and Gender: On the Development of Masculinity and Femininity*. New York: Science House.

Stouffer, Samuel (1949) *The American Soldier*. Princeton: Princeton University Press

Strang, Cameron B. (2014) "Violence, Ethnicity, and Human Remains during the Second Seminole War." *Journal of American History* 100, 4: 973–94.

Strasser, Bruno J. (1999) "Sickle Cell Anemia, a Molecular Disease." *Science* 286: 1488–90.

Strasser, Bruno J. (2006) *La Fabrique D'une Nouvelle Science: La Biologie Moléculaire À L'âge Atomique (1945–1964)*. Florence: Leo O. Olschki Editore.

Strasser, Bruno J. (2010) "Laboratories, Museums, and the Comparative Perspective: Alan A. Boyden's Quest for Objectivity in Serological Taxonomy, 1924–1962." *Historical Studies in the Natural Sciences* 40: 149–82.

Strasser, Bruno J. (2011) "The Experimenter's Museum: Genbank, Natural History, and the Moral Economies of Biomedicine" *Isis* 102, 1: 60–96.

Strasser, Bruno J. (2012) "Collecting Nature: Practices, Styles, and Narratives." *Osiris* 27, 1: 303–40.

Strauss, Leo (1970) *Xenophon's Socratic Discourse: An Interpretation of the "Oeconomicus."* Ithaca: Cornell University Press.

Strauss, Sarah and Ben Orlove (eds.) (2003) *Weather, Climate, Culture*. Oxford: Berg.

Stroud, Patricia (1992) *Thomas Say: New World Naturalist*. Philadelphia: University of Pennsylvania Press.

Stubbe, Hans (1965) *History of Genetics: From Prehistoric Times to the Rediscovery of Mendel's Laws*. T.R.W. Waters (trans.). Cambridge: MIT Press.

Stuewer, Roger (1975) *The Compton Effect: Turning Point in Physics*. New York: Science History Publications.

Stuewer, Roger (2010) "Act of Creation: The Meitner-Frisch Interpretation of Nuclear Fission," in Shaul Katzir, Christoph Lehner, and Jürgen Renn (eds.), *Traditions and Transformations in the History of Quantum Physics*. Berlin: Edition Open Access, 231–45.

Sturtevant, A.H. (1965) *A History of Genetics*. New York: Harper & Row.

Subramaniam, Banu (2000) "Archaic Modernities: Science, Secularism and Religion in Modern India." *Social Text* 18, 3: 67–86.

Subramaniam, Banu (2014) *Ghost Stories for Darwin: The Science of Variation and the Politics of Diversity*. Champaign-Urbana: Illinois University Press.

Suchman, Lucy (2009). *Human–Machine Reconfigurations: Plans and Situated Actions*. Cambridge: Cambridge University Press.

Sullivan, James (1795) *History of the District of Maine*. Boston: I. Thomas and E.T. Andrews.

Sullivan, Woodruff T. (2009) *Cosmic Noise: A History of Early Radio Astronomy*. Cambridge: Cambridge University Press.

Sulloway, Frank J. (1992) *Freud, Biologist of the Mind: Beyond the Psychoanalytic Legend*. Cambridge: Harvard University Press.

Sunder Rajan, Kaushik (2006) *Biocapital: The Constitution of Postgenomic Life*. Durham: Duke University Press.

Sunderland, Mary E. (2012) "Collections-Based Research at Berkeley's Museum of

Vertebrate Zoology." *Historical Studies in the Natural Sciences* 42: 83–113.

Sur, Abha (2011) *Dispersed Radiance: Caste, Gender, and Modern Science in India*. New Delhi: Navayana Press.

Susalla, Peter J. (2013) "The Last Dim Horizon: Scientific Cosmology in Twentieth-Century America." PhD Dissertation: University of Wisconsin-Madison.

Susalla, Peter J. and James Lattis (2009) *Wisconsin at the Frontiers of Astronomy: A History of Innovation and Exploration*. Madison: Wisconsin Legislative Reference Bureau.

Sussman, Herbert (2009) *Victorian Technology: Invention, Innovation, and the Rise of the Machine*. New York: Praeger.

Sutter, Paul S. (2002) *Driven Wild: How the Fight Against Automobiles Launched the Modern Wilderness Movement*. Seattle: University of Washington Press.

Swanner, Leandra (2015) "Contested Spiritual Landscapes in Modern American Astronomy." *Journal of Religion and Society* 17.

Sweet, John Wood (2007) *Bodies Politic: Negotiating Race in the American North, 1730–1830*. Philadelphia: University of Pennsylvania Press.

Sweetnam, George (2000) *The Command of Light: Rowland's School of Physics and the Spectrum*. Philadelphia: American Philosophical Society.

Sylvester, Christine (2006) "Bare Life as a Development/Postcolonial Problematic." *The Geographical Journal* 172, 1: 66–77.

Sypher, F.J. (2002) *Frederick L. Hoffman. His Life and Works*. Philadelphia: Xlibris.

Takeshita, Chikako (2011) *The Global Politics of the IUD: How Science Constructs Contraceptive Users and Women's Bodies*. Cambridge: MIT Press.

Tallbear, Kim (2013) *Native American DNA*. Minneapolis: University of Minnesota Press.

Tansley, Arthur G. (1935) "The Use and Abuse of Vegetational Concepts and Terms." *Ecology* 16: 284–307.

Tarr, Joel A. (1996) *The Search for the Ultimate Sink: Urban Pollution in Historical Perspective*. Akron: University of Akron Press.

Tatarchenko, Ksenia (2013) "'A House with the Window to the West': The Akademgorodok Computer Center (1958–1993)." PhD Dissertation: Princeton University.

Tatarewicz, Joseph N. (1990) *Space Technology and Planetary Astronomy*. Bloomington: Indiana University Press.

Tattersall, Ian (2000) "Paleoanthropology: The Last Half-Century." *Evolutionary Anthropology* 9: 2–16.

Tattersall, Ian (2009) *The Fossil Trail: How We Know What We Think We Know about Human Evolution*. Oxford: Oxford University Press.

Taub, Liba (ed.) (2009) "On Scientific Instruments." *Studies in History and Philosophy of Science* 40: 337–438.

Taub, Liba (ed.) (2011) "Focus: The History of Scientific Instruments." *Isis* 102: 689–729.

Tauger, Mark (2010) *Agriculture in World History*. New York: Routledge.

Tax, Sol (1955) "The Integration of Anthroplogy." *Yearbook of Anthropology*. Chicago: Chicago University Press.

Taylor, Bron (2001a) "Earth and Nature-Based Spirituality: From Deep Ecology to Radical Environmentalism. Part I." *Religion* 31, 2: 175–93.

Taylor, Bron (2001b) "Earth and Nature-Based Spirituality: From Deep Ecology to Radical Environmentalism. Part II." *Religion* 31, 3: 225–45.

Taylor, Joseph E. (1999) *Making Salmon: An Environmental History of the Northwest Fisheries Crisis*. Seattle: University of Washington Press.

Tedre, Matti (2014) *The Science of Computing: Shaping a Discipline*. Boca Raton: CRC Press.

Teitelbaum, Michael S. (ed.) (1976) *Sex Differences: Social and Biological Perspectives*. Garden City: Anchor Books.

Teresi, Dick (2001) *Lost Discoveries: Ancient Rotts of Modern Science – From the Babylonians to the Mayans*. New York: Simon & Schuster.

Terry, Jennifer (1999) *An American Obsession: Science, Medicine, and Homosexuality in Modern Society*. Chicago: University of Chicago Press.

Terzian, Sevan G. (2013) *Science Education and Citizenship: Fairs, Clubs, and Talent Searches for American Youth, 1918–1958.* New York: Palgrave Macmillan.

Teslow, Tracy (2014) *Constructing Race: The Science of Bodies and Cultures in American Anthropology.* Cambridge: Cambridge University Press.

Thackray, Arnold (ed.) (1998) *Private Science: Biotechnology and the Rise of the Molecular Sciences.* Philadelphia: University of Pennsylvania Press.

Thackray, Arnold, Jeffrey L. Sturchio, P. Thomas Carroll, and Robert Bud (1985). *Chemistry in America, 1876–1976: Historical Indicators.* Dordrecht: D. Reidel.

Theunissen, Bert (2014) "Practical Animal Breeding as the Key to an Integrated View of Genetics, Eugenics and Evolutionary Theory: Arend L. Hagedoorn." *Studies in History and Philosophy of Biological and Biomedical Sciences* 46: 55–64.

Thode, Simon (2013) "The Practices of Observational Science and the Development of the American Nation in the Trans-Appalachian West, 1763–1814." PhD Dissertation: Johns Hopkins University.

Thomas, John H. (1999) "The Solar Physics Division," in David H. DeVorkin (ed.), *The American Astronomical Society's First Century.* Washington: The American Astronomical Society, 238–51.

Thomas, William L., Jr. (ed.) (1956) *Man's Role in Changing the Face of the Earth.* Chicago: University of Chicago Press.

Thompson, Charis (2008) "Stem Cells, Women, and the New Gender and Science," in David H. DeVorkin (ed.), *Gendered Innovations in Science and Engineering.* Stanford: Stanford University Press, 109–30.

Thompson, P.B., G.L. Ellis, and B.A. Stout (1991) "Introduction: Values in the Agricultural Laboratory," in P.B. Thompson and B.A. Stout (eds.), *Beyond the Large Farm: Ethics and Research Goals for Agriculture.* Boulder: West View Press, 3–33.

Thompson, P.B. and S. Noll (2014) "Agricultural Ethics," in J. Britt Holbrook (ed.), *Ethics, Science, Technology, and Engineering:*

A Global Resource. New York: Macmillan, 35–42.

Thomson, Keith S. (2008) *The Legacy of the Mastodon: The Golden Age of Fossils in America.* New Haven: Yale University Press.

Thomson, Keith S. (2012) *Jefferson's Shadow: The Story of His Science.* New Haven: Yale University Press.

Thorndike, E.L. (1898) Animal Intelligence: An Experimental Study of the Associative Processes in Animals. *Psychological Review* 2(4).

Thorne, Kip (1995) *Black Holes and Time Warps: Einstein's Outrageous Legacy.* New York: Norton.

Thorne, Kip (2003) "Warping Spacetime," in G.W. Gibbons, E.P.S. Shellard and S.J. Rankin (eds.), *The Future of Theoretical Physics and Cosmology: Celebrating Stephen Hawking's Contributions to Physics.* Cambridge: Cambridge University Press.

Thurs, Daniel Patrick (2007) *Science Talk: Changing Notions of Science in American Culture* New Brunswick: Rutgers University Press.

Thurs, Daniel Patrick (2011) "Scientific Methods," in Peter Harrison, Ronald L. Numbers, and Michael H. Shank (eds.), *Wrestling with Nature: From Omens to Science.* Chicago: University of Chicago Press, 307–35.

Thurs, Daniel Patrick and Ronald L. Numbers (2011) "Science, Pseudoscience and Science Falsely So-Called," in Peter Harrison, Ronald L. Numbers, and Michael H. Shank (eds.), *Wrestling with Nature: From Omens to Science.* Chicago: University of Chicago Press, 281–306.

Tichi, Cecilia (1987) *Shifting Gears: Technology, Literature, Culture in Modernist America.* Chapel Hill: University of North Carolina Press.

Titchener, Edward Bradford (1980) A Textbook of Psychology (1910). History of Psychology Series, V. 354. Delmar: Scholars' Facsimiles & Reprints.

Tjossem, Sarah Fairbank (1994) "Preservation of Nature and Academic Responsibility: Tensions in the Ecological Society of America, 1915–1979." PhD Dissertation: Cornell University.

Tobey, Ronald C. (1971) *The American Ideology of National Science, 1919–1930*. Pittsburgh: University of Pittsburgh Press.

Tobey, Ronald C. (1981) *Saving the Prairies: The Life Cycle of the Founding School of American Plant Ecology, 1895–1955*. Berkeley: University of California Press.

Tolley, K. (2003) *The Science Education of American Girls: A Historical Perspective*. New York: Routledge.

Tomes, Nancy J. (1999) *Gospel of Germs: Men, Women, and the Microbe in American Life*. Harvard: Harvard University Press.

Tomes, Nancy J. and John H. Warner (1997) "Introduction to Special Issue on Rethinking the Reception of the Germ Theory of Disease: Comparative Perspectives." *Journal of the History of Medicine and Allied Sciences* 52.

Tone, Andrea (2002) *Devices and Desires: A History of Contraceptives in America*. New York: Hill & Wang.

Tonn, Jenna (2015) "Biology Building: Making Space for the Life Sciences at Harvard University, 1870–1930." PhD Dissertation: Harvard University.

Topham, Jonathan R. (1992) "Science and Popular Education in the 1830s: The Role of The 'Bridgewater Treatises'." *The British Journal for the History of Science* 25: 397–430.

Topham, Jonathan R. (2009a) "Introduction." (Focus: Historicizing "Popular Science"). *Isis* 100: 310–18.

Topham, Jonathan R. (2009b) "Rethinking the History of Science Popularization/Popular Science," in Faidra Papanelopoulou, Agusti Nieto-Galan, and Enrique Perdiguero (eds.), *Popularizing Science and Technology in the European Periphery, 1800–2000*. Aldershot: Ashgate, 1–20.

Toumey, C. (2006) "Science and Democracy." *Nature Nanotechnology* 1: 6–7.

Towghi, Fouzieyha and Kalindi Vora (2014) "Bodies, Markets, and the Experimental in South Asia." *Ethnos: Journal of Anthropology* 79: 1–18.

Traweek, Sharon (1988) *Beamtimes and Lifetimes: The World of High Energy Physicists*. Cambridge: Harvard University Press.

Trencher, Susan R. (2000) *Mirrored Images: American Anthropology and American Culture, 1960–1980*. Westport: Bergin & Garvey.

Trimble, Virginia (1999) "The Origin of the Divisions of the American Astronomical Society and the History of the High Energy Astrophysics Division," in David H. DeVorkin (ed.), *The American Astronomical Society's First Century*. Washington: The American Astronomical Society, 223–73.

Trinkaus, Erik (1982) "A History of Homo Erectus and Homo Sapiens Paleontology in America," in Frank Spencer (ed.), *A History of American Physical Anthropology 1930–1980*. New York: Academic Press, 261–80.

Trinkaus, Erik and Pat Shipman (1993) *The Neandertals: Changing the Image of Mankind*. New York: Knopf.

Trivers, R.L. (1971) "The Evolution of Reciprocal Altruism." *Quarterly Review of Biology* 46: 35–57.

Trivers, R.L. (1972) "Parental Investment and Sexual Selection," in B. Campbell (ed.), *Sexual Selection and the Descent of Man*. Chicago: Aldine-Atherton, 136–79.

Trivers, R.L. (1974) "Parent–Offspring Conflict." *American Zoologist* 14: 249–64.

Tsing, Anna (2005) *Friction: An Ethnography of Global Connection*. Princeton: Princeton University Press.

Tsing Anna (2009) "A New Form of Collaboration in Cultural Anthropology: Matsutake Worlds." *American Ethnologist* 36, 2: 380–403.

Tuana, Nancy (ed.) (1989) *Feminism and Science*. Bloomington: Indiana University Press.

Tudge, Colin (1992) *Last Animals at the Zoo: How Mass Extinction Can Be Stopped*. Washington: Island Press.

Turing, Alan (1937) "On Computable Numbers with an Application to the Entscheidungsprobem." *Proceedings of the London Mathematical Society* 42, 1: S2–42.

Turkle, Sherry (2005) *The Second Self: Computers and the Human Spirit*. Cambridge: MIT Press.

Turner, Fred (2006) *From Counterculture to Cyberbulture: Stewart Brand, the Whole*

Earth Network, and the Rise of Digital Utopianism. Chicago: University of Chicago Press.

Turner, Gillian M. (2011) *North Pole, South Pole: The Epic Quest to Solve the Great Mystery of Earth's Magnetism.* New York: The Experiment.

Turner, Roger (2010) "Weathering Heights: The Emergence of Aeronautical Meteorology as an Infrastructural Science." PhD Dissertation: University of Pennsylvania.

Turner, S.P. and J.H. Turner (1990) *The Impossible Science: An Institutional Analysis of American Sociology.* Newbury Park: Sage.

Turrini, Mauro (2014) "The Controversial Molecular Turn in Prenatal Diagnosis. CGH-Array Clinical Approaches and Biomedical Platforms." *Technoscienza* 5, 1: 115–39.

Uchida, Hisao (1993) "Building a Science in Japan: The Formative Decades of Molecular Biology." *Journal of the History of Biology* 26, 3: 499–517.

Ullman, Sharon R. (1998) *Sex Seen: The Emergence of Modern Sexuality in America.* Berkeley: University of California Press.

Ulrich, Laurel Thatcher, Ivan Gaskell, Sara J. Schechner, and Sarah Anne Carter (2015) *Tangible Things: Making History through Objects.* New York: Oxford University Press.

Underhill, Orra Ervin (1941) *The Origin and Development of Elementary School Science.* Chicago: Scott, Freeman.

UNESCO (1950) *UNESCO and its Programme: The Race Question.* Paris: UNESCO.

United States Army, Ordnance Department (1921) *The Manufacture of Optical Glass and of Optical Systems: A War-Time Problem.* Washington: Government Printing Office.

United States Bureau of Education (1892) *Report of the Committee on Secondary School Studies.* Washington: Government Printing Office.

United States Department of Agriculture (1977) *Dietary Goals for the United States.* Washington: Government Printing Office.

United States Department of Health, Education and Welfare (1980) *Nutrition and Your Health: Dietary Guidelines for Americans.* Washington: Government Printing Office.

United States Food and Drug Administration (1991) "Food Labeling: Nutrient Content Claims, General Principles; Health Claims, General Requirements and Other Specific Requirements for Individual Health Claims." Docket 94P-0390: 66208.

United States Food and Drug Administration (1999) "The Food Label." http://www. fda. gov/opacom/back grounders/foodlabel/newlabel.html

Valencius, Conevery Bolton (2002) *The Health of the Country: How American Settlers Understood Themselves and Their Land.* New York: Basic Books.

Van Amringe, William Frederick (1848) *An Investigation of the Theories of the Natural History of Man.* New York: Baker and Scribner.

Van Den Ende, Jan and René Kemp (1999) "Technological Transformations in History: How the Computer Regime Grew Out of Existing Computing Regimes." *Research Policy* 28: 833–51

Van Den Wijngaard, Marianne (1997) *Reinventing the Sexes: The Biomedical Construction of Femininity and Masculinity.* Bloomington: Indiana University Press.

Van der Spiegel, Jan, James Tau, Titiimaea Ala'ilima, and Lin Ping Ang (2002) "The ENIAC: History, Operation, and Reconstruction in VLSI," in Raúl Rojas and Ulf Hashagen (eds.), *The First Computers: History and Architectures.* Cambridge: MIT Press, 121–78.

Van Der Valk, Arnold G. (2014) "From Formation to Ecosystem: Tansley's Response to Clements' Climax." *Journal of the History of Biology* 47: 293–321.

van Dongen, Jeroen (2011) *Einstein's Unification.* Cambridge: Cambridge University Press.

Van Helden, Albert and Thomas L. Hankins (eds.) (1994) "Instruments." *Osiris* 9: 1–250.

Van Horn, Robert; Philip Mirowski, and Thomas A. Stapleford (eds.) (2013) *Building Chicago Economics: New Perspectives on the History of America's Most Powerful*

Economics Program. Cambridge: Cambridge University Press.

Van Overtveldt, Johan (2007) *The Chicago School.* Chicago: Agate.

Van Spronsen, J.W. (1969) *The Periodic System of Chemical Elements: A History of the First Hundred Years.* Amsterdam: Elsevier.

Varga, Donna (2009) "Babes in the Woods: Wilderness Aesthetics in Children's Stories and Toys, 1830–1915." *Society and Animals,* 17: 187–205.

Vaughan, Alden T. (ed.) (1995) *Roots of American Racism: Essays on the Colonial Experience.* Oxford: Oxford University Press.

Veblen, Thorstein (1900) "The Preconceptions of Economic Science – III." *Quarterly Journal of Economics* 14, 2: 240–69.

Venable, F. P. (1889) "Recalculations of the Atomic Weights." *Journal of Analytical Chemistry* 3: 48–61.

Venable, F. P. (1896) *The Development of the Periodic Law.* Easton: Chemical Publishing Co.

Verbrugge, Martha II. (1988) *Able-Bodied Womanhood: Personal Health and Social Change in Nineteenth-Century Boston.* New York: Oxford University Press.

Verbrugge, Martha H. (2012) *Active Bodies: A History of Women's Physical Education in Twentieth-Century America.* New York: Oxford University Press.

Verdon, Michel (2007) "Franz Boas: Cultural History for the Present, or Obsolete Natural History?" *Journal of the Royal Anthropological Institute* 13, 2: 433–51.

Vernon, Keith (2001) "A Truly Taxonomic Revolution? Numerical Taxonomy 1957–1970." *Studies in History and Philosophy of Science Part C: Studies in History and Philosophy of Biological and Biomedical Sciences* 32, 2: 315–41.

Verran, Helen (2002) "A Postcolonial Moment in Science Studies: Alternative Firing Regimes of Environmental Scientists and Aboriginal Landowners." *Social Studies of Science* 32.

Verworn, M. (1889) *Psycho-Physiologische Protistenstudien: Experimentelle Untersuchungen.* Jena: Fischer.

Vetter, Jeremy (2004) "Science Along the Railroad: Expanding Field Work in the U.S. Central West." *Annals of Science* 61, 2: 187–211.

Vetter, Jeremy (2008) "Cowboys, Scientists, and Fossils: The Field Site and Local Collaboration in the American West." *Isis* 99: 273–303.

Vetter, Jeremy (ed.) (2010) *Knowing Global Environments: New Historical Perspectives on the Field Sciences.* New Brunswick: Rutgers University Press.

Vetter, Jeremy (2011) "Lay Observers, Telegraph Lines, and Kansas Weather: The Field Network as a Mode of Knowledge Production." *Science in Context* 24: 259–80.

Vetter, Jeremy (2012) "Labs in the Field? Rocky Mountain Biological Stations in the Early Twentieth Century." *Journal of the History of Biology* 45: 587–611.

Veysey, Lawrence (1975) "Who's a Professional? Who Cares?" *Reviews in American History* 3: 419–23.

Virey, J.J. (1824) *Histoire Naturelle du Genre Humaine.* Paris: Crochard.

Viswanathan, Shiv (1997) *A Carnival for Science.* New Delhi: Oxford University Press.

Vogel, Brant (2009) "Bibliography of Recent Literature in the History of Meteorology: Twenty Six Years, 1983–2008." *History of Meteorology* 5: 23–125.

Vogel, Shawna (1995) *Naked Earth: The New Geophysics.* New York: Dutton.

Vogeler, Alfred G. (1900) "Hydrogen or Oxygen as Basis of Atomic Weights – Which?" *Western Druggist* 22: 414–16.

Volmar, Daniel (2015) The Power of the Atom: US Nuclear Command, Control, and Communications at the Dawn of the Nuclear Age (Forthcoming).

Von Humboldt, Alexander (1850) *Cosmos: A Sketch of a Physical Description of the Universe.* New York: Harper.

Von Neumann, John (1993) "First Draft of a Report on the EDVAC." *IEEE Annals of the History of Computing* 15, 4: 27–75.

Waddell, Craig (ed.) (2000) *And No Birds Sing: Rhetorical Analysis of Rachel Carson's Silent Spring.* Carbondale: Southern Illinois University Press.

Wadewitz, Lissa (2012) *The Nature of Borders: Salmon, Boundaries, and Bandits on the*

Salish Sea. Seattle: University of Washington Press.

Waitz, Theodor (1859) *Anthropologie Der Naturvölker, Erster Theil*. Leipzig: Friedrich Fleischer.

Walker, Alan and Richard E. Leakey (1993) *The Nariokotome Homo erectus Skeleton*. Cambridge: Harvard University Press.

Wallace, B. Alan (ed.) (2013) *Buddhism and Science: Breaking New Ground*. New York: Columbia University Press.

Wallace, David Rains (1999) *The Bonehunters' Revenge: Dinosaurs, Greed, and the Greatest Scientific Feud of the Gilded Age*. Boston: Houghton Mifflin.

Waller, John (2002) *Fabulous Science: Fact and Fiction in the History of Scientific Discovery*. New York: Oxford University Press.

Walls, Laura Dassow (2009) *The Passage to Cosmos: Alexander von Humboldt and the Shaping of America*. Chicago: University of Chicago Press.

Wang, J. (1995) "Liberals, the Progressive Left, and the Political Economy of Postwar American Science: The National Science Foundation Debate Revisited." *Historical Studies in the Physical and Biological Sciences* 26, 1: 139–66.

Wang, Jessica (1999) *American Science in an Age of Anxiety: Scientists, Anticommunism, and the Cold War*. Chapel Hill: University of North Carolina Press.

Wang, Zuoyue (1995) "The Politics of Big Science in the Cold War: PSAC and the Founding of SLAC." *Historical Studies in the Physical and Biological Sciences* 25: 329–56.

Wansink, B. (2007) *Marketing Nutrition: Soy, Functional Foods, Biotechnology and Obesity*. Chicago: University of Illinois Press.

Ward, Arthur (2012) "Trouble for Natural Design," in R.L. Gordon, Swan Stillwaggon and J. Seckbach (eds.), *Origins of Design in Nature*. Berlin: Springer.

Ward, L.F. (1894) "Contributions to Social Philosophy: Social Genesis." *American Journal of Sociology* 2: 532–46.

Warming, Eugenius (1909) *Oecology of Plants: An Introduction to the Study of Plant-Communities*. London: Oxford.

Warner, Deborah Jean (1968) *Alvan Clark & Sons: Artists in Optics*. Washington, DC: Smithsonian Institution Press.

Warner, Deborah Jean (1978) "Science Education for Women in Antebellum America." *Isis* 69: 58–67.

Warner, Deborah Jean (1979) "Astronomy in Antebellum America," in Nathan Reingold (ed.), *The Sciences in the American Context: New Perspectives*. Washington, DC: Smithsonian Institution Press, 55–75.

Warner, Deborah Jean (1985) "Optics in Philadelphia during the Nineteenth Century." *Proceedings of the American Philosophical Society* 129: 291–9.

Warner, Deborah Jean (1990) "What is a Scientific Instrument, When Did it Become One, and Why?" *British Journal for the History of Science* 23: 83–93.

Warner, Deborah Jean and Robert B. Ariail (1995) *Alvan Clark & Sons, Artists in Optics*. Richmond: Willmann-Bell.

Warner, Michael (1991) "Fear of a Queer Planet." *Social Text* 29: 3–17.

Warren, Leonard (1998) *Joseph Leidy: The Last Man Who Knew Everything*. New Haven: Yale University Press.

Warren, Louis S. (1997) *The Hunter's Game: Poachers and Conservationists in Twentieth-Century America*. New Haven: Yale University Press.

Washburn, J. (2005) *University, Inc.* New York: Basic Books.

Washburn, Sherwood (1951a) "The New Physical Anthropology." *Transactions of the New York Academy of Science* 13: 298–304.

Washburn, Sherwood (1951b) "Section of Anthropology: The New Physical Anthropology." *Transactions of the New York Academy of Sciences* 13 : 298–304.

Watkins, Elizabeth S. (1998) *On the Pill: Social History of Oral Contraceptives, 1950–1970*. Baltimore: Johns Hopkins University Press.

Watson, John B. (1903) *Animal Education*. Chicago: Chicago University Press.

Watson, John B (1913) "Psychology as the Behaviorist Views it." *Psychological Review* 20: 158–77.

Watson, John B. (1930) *Behaviorism*. Chicago: University of Chicago Press.

Way, Albert G. (2011) *Conserving Southern Longleaf: Herbert Stoddard and the Rise of Ecological Land Management.* Athens: University of Georgia Press.

Way, Michael and Diedre Hunter (2013) Origins of the Expanding Universe. ASP Conference Series, 471. San Francisco: Astronomical Society of the Pacific.

Wayland, Francis (1837) *Elements of Political Economy.* New York: Leavitt, Lord.

Wazeck, Milena (2014) *Einstein's Opponents.* New York: Cambridge University Press.

Weart, Spencer R. (1992) "The Solid Community," in Lillian Hoddeson, Earnest Braun, Jürgen Teichmann and Spencer Weart (eds.), *Out of the Crystal Maze: Chapters from the History of Solid-State Physics.* New York: Oxford University Press, 617–68.

Weart, Spencer R. (2008) *The Discovery of Global Warming.* Cambridge: Harvard University Press.

Weart, Spencer R. (2012) *The Rise of Nuclear Fear.* Cambridge: Harvard University Press.

Weatherall, D.J. (2012) "Memories of the Moore Clinic, 1960–1965," in Krishna R., Dronamraju and Clair A. Francomano (eds.), *Victor McKusick and the History of Medical Genetics.* New York: Springer, 41–52.

Webley, K. (2013) "College Costs: Would Tuition Discounts Get More Students to Major in Science?" *Time Magazine.*

Webster, Sandra and S.R. Coleman (1992) "Contributions to the History of Psychology: LXXXVI. Hull and His Critics: The Reception of Clark L. Hull's Behavior Theory, 1943–1960." *Psychological Reports* 70, 3c: 1063–71.

"The Wedge Document." http://ncse.com/creationism/general/wedge-document

Weeks, Mary Elvira (1968) "The Discovery of the Elements." *Journal of Chemical Education.*

Wege, Klaus and Peter Winkler (2005) "The Societas Meteorologica Palatina (1780–1795) and the Very First Beginnings of Hohenpeissenberg Observatory." *Algorismus* 52: 45–54.

Weidman, Nadine (2011) "Popularizing the Ancestry of Man: Robert Ardrey and the Killer Instinct." *Isis* 102: 269–99.

Weintraub, E. Roy (1991) *Stabilizing Dynamics: Constructing Economic Knowledge.* Cambridge: Cambridge University Press.

Weintraub, E. Roy (2002) *How Economics Became a Mathematical Science.* Durham: Duke University Press.

Weis, Judith S. (1990) "The Status of Undergraduate Programs in Environmental Science." *Environmental Science & Technology* 24, 8: 1116–21.

Welch, Margaret (1998) *Book of Nature: Natural History in the United States, 1825–1875.* Boston: Northeastern University Press.

Wellerstein, Alex (2008) "Patenting the Bomb: Nuclear Weapons, Intellectual Property, and Technological Control." *Isis* 99: 57–87.

Wells, William Charles (1818) "An Account of the Female of the White Race of Mankind Part of Whose Skin Resembles that of a Negro," in Two Essays. Edinburgh: Archibald Constable, 423–9.

Werkheiser, I. and S. Noll. (2014) "From Food Justice to a Tool of the Status Quo: Three Sub-Movements within Local Food." *Journal of Agricultural and Environmental Ethics* 27, 2: 201–10.

Weschler, Lawrence (1995) *Mr. Wilson's Cabinet of Wonder: Pronged Ants, Horned Humans, Mice on Toast, and Other Marvels of Jurassic Technology.* New York: Vintage.

Westfall, Catherine (2003a) "Civilian Nuclear Power on the Drawing Board: The Development of Experimental Breeder Reactor-II." Argonne National Laboratory Report ANL/HIST-1.

Westfall, Catherine (2003b) "Rethinking Big Science: Modest, Mezzo, Grand Science and the Development of the Bevalac, 1971–1993." *Isis* 94: 30–56.

Westfall, Catherine and John Krige (1998) "The Path of Post-War Physics," in Gordon Fraser (ed.), *The Particle Century.* London: Institute of Physics, 1–11.

Westwick, Peter J. (2003) *The National Labs: Science in an American System, 1947–1974.* Cambridge: Harvard University Press.

Westwick, Peter J. (2007) *Into the Black: JPL and the American Space Program, 1976–2004*. New Haven: Yale University Press.

"What is Life?" (1848) *Scientific American*: 1.

Wheatland, David. P. (1968) *The Apparatus of Science at Harvard, 1765–1800*. Cambridge: Collection of Historical Scientific Instruments, Harvard University Press.

Wheatland, David P. and I. Bernard Cohen (1949) *A Catalogue of Some Early Scientific Instruments at Harvard University: Placed on Exhibition in the Edward Mallinckrodt Chemical Laboratory, February 12, 1949*. Cambridge: Harvard University Press.

Wheeler, Alwyne C. (ed.) (1983) *Contributions to the History of North American Natural History: Papers from the First North American Conference of the Society for the Bibliography of Natural History, 21–23 October 1981*. London: Society for the Bibliography of Natural History.

Wheeler, W.M. (1911) "The Ant Colony as an Organism." *Journal of Morphology* 22: 307–26.

White, Edward A. (1968) *Science and Religion in American Thought: The Impact of Naturalism*. New York: AMS Press.

White, Andrew Dickson (1896) *History of the Warfare of Science with Theology in Christendom*. Amhurst: Prometheus.

White, Leslie (1959) "The Concept of Culture." *American Anthropologist* 61, 2: 227–51.

White, Leslie (1966) "The Social Organization of Ethnological Theory." *Rice University Studies* 52, 4.

Whitesell, Patricia S. (1998) *A Creation of His Own: Tappan's Detroit Observatory*. Ann Arbor: Bentley Historical Library.

Whitley, Richard (1985) "Knowledge Producers and Knowledge Acquirers: Popularization as a Relation Between Scientific Fields and Their Publics," in Terry Shinn and Richard Whitley (eds.), *Expository Science: Forms and Functions of Popularization*. Dordrecht: Reidel, 3–28.

Whitman, Walt (1996) *Poetry and Prose*. New York: Library of America.

Whitnah, Donald Robert (1961) *A History of the United States Weather Bureau*. Urbana: University of Illinois Press.

Whorf, Benjamin Lee (1956) *Language, Thought, and Reality: Selected Writings of Benjamin Lee Whorf*. Cambridge: MIT Press.

Whorton, J. (1982) *Crusaders for Fitness: The History of American Health Reformers*. Princeton: Princeton University Press.

Whorton, J. (2000) *Inner Hygiene: Constipation and the Pursuit of Health in Modern Society*. New York: Oxford University Press.

Whorton, J. (2002) *Natural Cures: The History of Alternative Medicine in America*. New York: Oxford University Press.

Wiber, Melanie (1997). *Erect Men, Undulating Women: The Visual Imagery of Gender, "Race" and Progress in Reconstructive Illustrations of Human Evolution*. Waterloo: Wilfrid Laurier University Press.

Wiebe, R.H. (1967) *The Search for Order, 1877–1920*. New York: Hill and Wang.

Wilcox, Clifford (2006) *Robert Redfield and the Development of American Anthropology*. Lanham: Lexington Books.

Wilensky, H.L. (1964) "The Professionalization of Everyone?" *American Journal of Sociology* 70, 2: 137–58.

Will, Clifford (1993a) *Theory and Experiment in Gravitational Physics*. Cambridge: Cambridge University Press.

Will, Clifford (1993b) *Was Einstein Right? Putting General Relativity to the Test*. New York: Basic Books.

Willard, Elizabeth O.G. (1867) *Sexology as the Philosophy of Life: Including Social Organization and Government*. Chicago: Private Printing.

Willard, Emma (1826) *Geography for Beginners*. Hartford: O.D. Cooke.

Williams, Geroge C. (1966) *Adaptation and Natural Selection: A Critique of Some Current Evolutionary Thought*. Princeton: Princeton University Press.

Williams, Kathleen Broome (2004) *Grace Hopper: Admiral of the Cyber Sea*. Annapolis: Naval Institute Press.

Williams, Raymond (1980) "Ideas of Nature," in Raymond Williams (ed.), *Problems in Materialism and Culture*. London: Verso.

Williams, Raymond (1983) *Keywords: A Vocabulary of Culture and Society*. New York: Oxford University Press.

Williams, R.D., H.L. Mason, and R.M. Wilder (1943) "The Minimum Daily Requirement of Thiamin in Man." *Journal of Nutrition* 25: 71–97.

Williams, Thomas R. and Michael Saladyga (2011) *Advancing Variable Star Astronomy: The Centennial History of the American Association of Variable Star Observers.* Cambridge: Cambridge University Press.

Wilson, Alexander (1808–1814) *American Ornithology.* Philadelphia: Bradford and Inskeep.

Wilson, Allan and Vincent Sarich (1967) "Immunological Time Scale for Hominid Evolution." *Science* 158: 1200–3.

Wilson, David Sloan and E. Sober (1989) "Reviving the Superorganism." *Journal of Theoretical Biology* 136: 337–56.

Wilson, David Sloan and Edward O. Wilson (2007) "Rethinking the Theoretical Foundation of Sociobiology." *Quarterly Review of Biology* 82: 327–48.

Wilson, Edmund B. (1897) *The Cell in Development and Inheritance.* Ann Arbor: University of Michigan Library.

Wilson, Edward O. (1975) *Sociobiology: The New Synthesis.* Cambridge: Harvard University Press.

Wilson, Edward O. (1978) *On Human Nature.* Cambridge: Harvard University Press.

Wilson, Edward O. (1993) *The Diversity of Life* New York: W.W. Norton.

Wilson, Edward O. (1994) *Naturalist.* Washington: Island Press.

Wilson, Edward O. (2006) *The Creation: An Appeal to Save the Earth.* New York: W.W. Norton.

Wilson, Edward O. (2012) *The Social Conquest of Earth.* New York: W.W. Norton.

Wilson, Edward O. (2014) *The Meaning of Human Existence.* New York: Liveright.

Wilson, Edward O. and Frances M. Peter (eds.) (1988) *BioDiversity.* Washington: National Academy Press.

Wilson, Edward O. and M.A. Nowak (2014) "Natural Selection Drives the Evolution of Ant Life Cycles." *Proceedings of the National Academy of Sciences* 111: 12585–90.

Wilson, Elizabeth (2009) "'Would I Had Him With Me Always': Affects of Longing in Early Artificial Intelligence." *Isis* 100: 839–47.

Wilson, J. Tuzo (1961) *I.G.Y.: The Year of the New Moons.* New York: Knopf.

Wilson, Joan Hoff (1973) "Dancing Dogs of the Colonial Period: Women Scientists." *Early American Literature* 7: 225–35.

Wilson, Thomas (1891) "The International Congress of Anthropology and Prehistoric Archeology of Paris, 1889." *The American Naturalist* 25: 764–8.

Winchell, Alexander (1880) *Preadamites: Or, a Demonstration of the Existence of Men before Adam.* Chicago: S.C. Griggs.

Winslow, Charles-Edward Amory (1943) *The Conquest of Epidemic Disease: A Chapter in the History of Ideas.* Madison: University of Wisconsin Press.

Winsor, Mary P. (1991) *Reading the Shape of Nature: Comparative Zoology at the Agassiz Museum.* Chicago: University of Chicago Press.

Winsor, Mary P. (2009) "Museums," in Peter J. Bowler and John V. Pickstone (eds.), *The Cambridge History of Science.* Cambridge: Cambridge University Press.

Winston, Johnnie (2011) "Science, Practice, and Policy: The Committee on Rare and Endangered Wildlife Species and the Development of U.S. Federal Endangered Species Policy, 1956–1973." PhD Dissertation: Arizona State University.

Wise, George (1985) *Willis R. Whitney, General Electric, and the Origins of U.S. Industrial Research.* New York: Columbia University Press.

Wolf, Eric (1982) *Europe and the People without History.* Berkeley: University of California Press.

Wolfe, Audra (2012) *Competing with the Soviets.* Baltimore: Johns Hopkins University Press.

Wolfe, Douglas A. (2001) *A History of the Federal Laboratory at Beaufort, North Carolina 1899–1999.* Washington: U.S. National Oceanic and Atmospheric Administration.

Wolfe, Patrick (1999) *Settler Colonialism and the Transformation of Anthropology: The Politics and Poetics of an Ethnographic Event.* London: Cassell.

Wolpoff, Milford and Rachel Caspari (1997) *Race and Human Evolution*. New York: Simon & Schuster.

Wonders, Karen (1993) *Habitat Dioramas: Illustrations of Wilderness in Museums of Natural History*. Stockholm: Almquist and Wikseil.

Wood, Helen (1995) "Computer Society Celebrates 50 Years." *IEEE Annals of the History of Computing* 17, 4: 6.

Yates, Joanne (1993) *Control through Communication: The Rise of System in American Management*. Baltimore: Johns Hopkins University Press.

Woodson, C.G. (1919) *The Education of the Negro Prior to 1861*. Washington: Associated Publishers.

Woody, T. (1954). "The Country Schoolmaster of Long Ago." *History of Education Journal* 5, 2: 41–53.

Woolf, Harry (1972) "Science for the People: Copernicanism and Newtonianism in the Almanacs of Early America." *Studia Copernica* 5: 293–309.

Worboys, Michael (2000) *Spreading Germs: Disease Theories and Medical Practice in Britain, 1865–1900*. Cambridge: Cambridge University Press.

Worster, Donald (1990) "Transformations of the Earth: Toward an Agro-ecological Perspective in History." *Journal of American History* 76, 4: 1087–106.

Worster, Donald (1994) *Nature's Economy: A History of Ecological Ideas*. New York: Cambridge University Press.

Worster, Donald (1996) "The Two Cultures Revisited: Environmental History and the Environmental Sciences." *Environment and History* 2, 1: 3–14.

Worster, Donald (2001) *A River Running West: The Life of John Wesley Powell*. Oxford: Oxford University Press.

Worster, Donald (2004) *Dust Bowl: The Southern Plains in the 1930s*. New York: Oxford University Press.

Wright, Gilbert (1813) *Natural History and Antiquities of Selborne*. London: H.C. Bohn.

Wright, Helen (1966) *Explorer of the Universe: A Biography of George Ellery Hale*. New York: Dutton.

Wright, Robert (1994) *The Moral Animal. Why We Are, the Way We Are: The New Science of Evolutionary Psychology*. London: Vintage.

Wright, Sewall (1931) "Evolution in Mendelian Populations." *Genetics* 16: 97–159.

Wright, Sewall (1932) "The Roles of Mutation, Inbreeding, Crossbreeding and Selection in Evolution." *Proceedings of the Sixth Annual Congress of Genetics*, 1: 356–66.

Wright, Susan (1994) *Molecular Politics: Developing American and British Regulatory Policy for Genetic Engineering, 1972–1982*. Philadelphia: University of Pennsylvania Press.

Wright, W.W. (1860) "Amalgamation" *De Bow's Review* 29: 1–20.

Wyer, Mary, Mary Barbercheck, Donna Cookmeyer, Hatice Ozturk, and Marta Wayne (eds.) (2001) *Women, Science, and Technology: A Reader in Feminist Science Studies*. New York: Routledge.

Wylie, Caitlin Donahue (2012) "Teaching Nature Study on the Blackboard in Late Nineteenth and Early Twentieth-Century England." *Archives of Natural History* 39: 59–76.

Wyman, Bruce C. and L. Harold Stevenson (2007) *The Facts on File Dictionary of Environmental Science*. New York: Facts on File.

Wynne, B. (1992) "Public Understanding of Science Research: New Horizons or Hall of Mirrors?" *Public Understanding of Science* 1: 37–43.

Wynne, B. (2005) "Risk as Globalizing 'Democratic' Discourse? Framing Subjects and Citizens," in M. Leach, I. Scoones, and B. Wynne (eds.), *Science and Citizens: Globalization and the Challenge of Engagement*. London: Zed Books, 66–82.

Yanni, Carla (1999) *Nature's Museums: Victorian Science and the Architecture of Display*. Baltimore: Johns Hopkins University Press.

Yeomans, Donald K. (1977) "The Origin of North American Astronomy – Seventeenth Century." *Isis* 68: 414–25.

Yerkes, R. (1907) *The Dancing Mouse: A Study in Animal Behavior*. New York: Macmillan.

Yerkes, R. (1916) *The Dancing Mouse: A Study in Animal Behavior*. New York: Macmillan.

Yerkes, R. (1925) *Almost Human.* New York: Century.

Yerkes R. and A.W. Yerkes (1929) *The Great Apes: A Study of Anthropoid Life.* New Haven: Yale University Press.

Yi, Doogab (2008) "Cancer, Viruses, and Mass Migration: Paul Berg's Venture into Eukaryotic Biology and the Advent of Recombinant DNA Research and Technology, 1967–1980." *Journal of the History of Biology* 41: 589–636.

Yochelson, Ellis (1985) *The National Museum of Natural History: 75 Years in the Natural History Building.* Washington: Smithsonian Institution Press.

Yoder, Hatten S. (1994) "Development and Promotion of the Initial Scientific Program for the Geophysical Laboratory," in Gregory Good (ed.), *The Earth, the Heavens, and the Carnegie Institution of Washington, DC.* Washington, DC: American Geophysical Union, 21–8.

Yoder, Hatten S. (2005) *Centennial History of the Carnegie Institution of Washington, DC: Volume 3, the Geophysical Laboratory.* New York: Cambridge University Press.

Young, Allyn A. (1928) "Increasing Returns and Economic Progress." *Economic Journal* 38, 152: 527–42.

Young, Christian C. (1998) "Defining the Range: The Development of Carrying Capacity in Management Practice." *Journal of the History of Biology* 31: 61–83.

Young, Christian C. (2002) *In the Absence of Predators: Conservation and Controversy on the Kaibab Plateau.* Lincoln: University of Nebraska Press.

Young, Christian C. (2005) *The Environment and Science: Social Impact and Interaction.* Santa Barbara: ABC-CLIO Press.

Young, Christian C. and Mark A. Largent (2007) *Evolution and Creationism: A Documentary and Reference Guide.* Westportss: Greenwood Press.

Young, Robert J.C. (1995) *Colonial Desire. Hybridity in Theory, Culture and Race* London: Routledge.

Young, Robert Maxwell (1970) *Mind, Brain, and Adaptation in the Nineteenth Century: Cerebral Localization and its Biological Context from Gall to Ferrier.* New York: Oxford University Press.

Yoxen, Edward J. (1979) "Where does Schrödinger's 'What is Life?' Belong in the History of Molecular Biology?" *History of Science* 17: 17–52.

Yu, H. (2001) *Thinking Orientals: Migration, Contact, and Exoticism in Modern America.* Oxford: Oxford University Press.

Zalasiewicz, Jan, Colin N. Waters, and Mark Williams (2014) "Human Bioturbation, and the Subterranean Landscape of the Anthropocene." *Anthropocene* 6: 3–9.

Zammito, John H. (2002) *Kant, Herder, and the Birth of Anthropology.* Chicago: University of Chicago Press.

Zapffe, Carl A. (1969) "Gustavus Hinrichs, Precursor of Mendeleev." *Isis* 60: 461–76.

Zeller, Benjamin E. (2010) *Prophets and Protons: New Religious Movements and Science in Late Twentieth-Century America.* New York: New York University Press.

Zeller, Susan (2000) "The Colonial World as Geological Metaphor: Strata(Gems) of Empire in Victorian Canada." *Osiris* 15: 85–107.

Zenderland, Leila (2001) *Measuring Minds: Henry Herbert Goddard and the Origins of American Intelligence Testing.* New York: Cambridge University Press.

Zimmerman, A. (2010) *Alabama in Africa: Booker T. Washington, the German Empire, and the Globalization of the New South.* Princeton: Princeton University Press.

Zunz, Olivier (1990) *Making America Corporate, 1870–1920.* Chicago: University of Chicago Press.

INDEX

A Companion to the History of American Science, First Edition.
Edited by Georgina M. Montgomery and Mark A. Largent.
© 2016 John Wiley & Sons Ltd. Published 2020 by John Wiley & Sons Ltd.